Biological Wastewater Treatment

Examples and Exercises

Biological Wastewater Treatment

Examples and Exercises

Carlos M. Lopez-Vazquez
Damir Brdjanovic
Eveline I.P. Volcke
Mark C.M. van Loosdrecht
Di Wu
Guanghao Chen

PUBLISHING

Published by

IWA Publishing
Unit 104–105, Export Building
1 Clove Crescent
London E14 2BA, UK
Telephone: +44 (0)20 7654 5500
Fax: +44 (0)20 7654 5555
Email: publications@iwap.co.uk
Web: www.iwapublishing.com

First published 2023
© 2023 IWA Publishing

British Library Cataloguing in Publication Data
A CIP catalogue record for this book is available from the British Library

Cover design: Damir Brdjanovic
Cover image: Francisco Rubio Rincon
Graphic design: Synopsis d.o.o.
English language editor: Claire Taylor

ISBN: 9781789062298 (Paperback)
ISBN: 9781789062304 (eBook)

Editors: Carlos M. Lopez-Vazquez   Mark C.M. van Loosdrecht
            Damir Brdjanovic   Di Wu
            Eveline I.P. Volcke   Guanghao Chen

Authors: Adrian Oehmen   Edward J.H. van Dijk   Hui Lu   Michael K. Stenstrom
            Bengt Carlsson   Ernest R. Blatchley III   Jiao Zhang   Michele Laureni
            Carlos M. Lopez-Vazquez   Eveline I.P. Volcke   Jules B. van Lier   Nidal Mahmoud
            Christine M. Hooijmans   Fangang Meng   Kang Xiao   Pernille Ingildsen
            Coenraad Pretorius   Ferenc Házi   Kim H. Sørensen   Shuai Liang
            David G. Weissbrodt   Francisco Rubio Rincon   Kimberly Solon   Tianwei Hao
            Damir Brdjanovic   Guanghao Chen   Laurence Strubbe   Victor S. Garcia Rea
            Diego Rosso   George A. Ekama   Manel Garrido-Baserba   Xia Huang
            Di Wu   Gustaf Olsson   Mark C.M. van Loosdrecht Yves Comeau
            Eberhard Morgenroth   Ho Kwong Chui   Merle de Kreuk

This eBook was made Open Access in April 2023.

Doi: 10.2166/9781789062304

In memoriam
George A. Ekama

This book is dedicated to our late colleague, co-author and friend, Emeritus Professor George A. Ekama. For more than 40 years George was at the forefront of developments in biological nitrogen removal activated sludge systems modelling, filamentous bulking, secondary settling tank design, and modelling and anaerobic systems. He was instrumental in the process of making the first edition of the book and the online course Biological Wastewater Treatment: Principles, Modelling and Design in 2008, and his contribution remained the backbone of the book's second edition in 2020. Many examples and exercises from George's extensive archive have also been used in the present book, which unfortunately he was unable to see published. His legacy will continue to live on in our thoughts, books and courses, and the many students and young professionals worldwide who have benefitted from his considerable knowledge and experience.

The Authors

Authors

1. Wastewater treatment development
Carlos M. Lopez-Vazquez
Mark C.M. van Loosdrecht

2. Basic microbiology and metabolism
David G. Weissbrodt
Michele Laureni
Mark C.M. van Loosdrecht
Yves Comeau

3. Wastewater characteristics
Kimberly Solon
Carlos M. Lopez-Vazquez
Eveline I.P. Volcke

4. Organic matter removal
Carlos M. Lopez-Vazquez

5. Nitrogen removal
Carlos M. Lopez-Vazquez
Eveline I.P. Volcke
Mark C.M. van Loosdrecht

6. Enhanced biological phosphorus removal
Carlos M. Lopez-Vazquez
Francisco Rubio Rincon
Adrian Oehmen

7. Innovative sulphur-based wastewater treatment
Di Wu
Hui Lu
Tianwei Hao
Ho Kwong Chui
George A. Ekama[†]
Mark C.M. van Loosdrecht
Guanghao Chen

8. Wastewater disinfection
Ernest R. Blatchley III

9. Aeration and mixing
Diego Rosso
Eveline I.P. Volcke
Manel Garrido-Baserba
Coenraad Pretorius
Michael K. Stenstrom

10. Bulking sludge
Eveline I.P. Volcke
Laurence Strubbe
Carlos M. Lopez-Vazquez
Mark C.M. van Loosdrecht

11. Aerobic granular sludge
Laurence Strubbe
Merle de Kreuk
Edward J.H. van Dijk
Mark C.M. van Loosdrecht
Eveline I.P. Volcke

12. Final settling
Ferenc Házi
Eveline I.P. Volcke

13. Membrane bioreactors
Xia Huang
Fangang Meng
Kang Xiao
Shuai Liang
Jiao Zhang

14. Modelling activated sludge processes
Damir Brdjanovic
Carlos M. Lopez-Vazquez
Christine M. Hooijmans

15. Process control
Gustaf Olsson
Pernille Ingildsen
Bengt Carlsson

16. Anaerobic wastewater treatment
Jules B. van Lier
Nidal Mahmoud
Victor S. Garcia Rea

17. Biofilm modelling and biofilm reactors
Eberhard Morgenroth
Kim Helleshøj Sørensen

Editors

Carlos M. Lopez-Vazquez, PhD

Prof. Damir Brdjanovic, PhD

Carlos M. Lopez-Vazquez is Associate Professor of Sanitary Engineering at IHE Delft Institute for Water Education. In 2009 he received his doctoral degree on Environmental Biotechnology (*cum laude*) from Delft University of Technology and UNESCO-IHE Institute for Water Education. During his professional career, he has taken part in different advisory and consultancy projects for both public and private sectors concerning municipal and industrial wastewater treatment systems. After working for a couple of years in the Water R&D Department of Nalco Europe on industrial water and wastewater treatment applications, he re-joined IHE Delft in 2009. Since then, he has been involved in education, capacity building and research projects guiding several MSc and PhD students. He has edited and authored over 100 scientific publications and book chapters, including five books in the field of wastewater treatment (*e.g. Experimental methods in wastewater treatment, Applications of activated sludge models*, and the Spanish version of *Biological wastewater treatment: principles, modelling and design*). By applying mathematical modelling as an essential tool, he has a special focus on the development and transfer of innovative and cost-effective wastewater treatment technologies to developing countries, countries in transition and industrial applications.

Damir Brdjanovic is Professor of City-wide Inclusive Sanitation at IHE Delft Institute for Water Education and Endowed Professor at Delft University of Technology. Areas of his expertise include pro-poor and emergency sanitation, faecal sludge management, urban drainage, and wastewater treatment. He is a pioneer in the practical application of models in wastewater treatment practice in developing countries. He is co-inventor of the Shit Killer® device for excreta management in emergencies, the award-winning eSOS® Smart Toilet and the medical toilet MEDiLOO®, with funding by the Bill & Melinda Gates Foundation. He has initiated the development and implementation of innovative didactic approaches and novel educational products (including e-learning) at IHE Delft. Brdjanovic is co-founder and director of the Global Sanitation Graduate School, the world's largest network for postgraduate education on sanitation. In addition to 20 PhD fellows, in excess of 200 MSc students have graduated under his supervision so far. Prof. Brdjanovic has a substantial publication record, is co-initiator of the IWA Journal of Water, Sanitation and Hygiene for Development, and is the initiator, author and editor of ten books in the wastewater treatment and sanitation field. He received the IWA Publishing Award in 2018, the IWA Water and Development Award in 2021, and the IWA Publishing Best Scientific Book Prize (shared award) in 2022.

Prof. Eveline I.P. Volcke, PhD

Prof. Mark C.M. van Loosdrecht, PhD

Eveline Volcke is a full professor at Ghent University. Her interdisciplinary research expertise can best be described as process engineering for biological wastewater treatment and resource recovery processes. Her aim is process optimization through physical-based modelling and simulation, data treatment techniques and experimental studies. With her BioCo research group and through international collaborations, she has provided major insights concerning greenhouse gas emissions from wastewater treatment, innovative nitrogen removal techniques and granular sludge reactors, among others. She has co-authored over 120 publications in international peer-reviewed journals, besides several IWA book chapters. She has been the promotor of over 15 PhD students and over 50 MSc students, and has served on an honorary basis on various committees, including dozens of (international) PhD examination committees. Eveline Volcke is an IWA Distinguished Fellow and has held active and leading roles in various IWA task groups, specialist groups, as co-chair of the IWA World Water Congress Program Committee and in the scientific (review) committee of over 30 more IWA conferences. Her influential and consistent volunteer work contributing to IWA for over 20 years led her to receive the IWA Outstanding Service Award in 2022.

Mark C.M. van Loosdrecht is a scientist recognised for his significant contributions to the study of reducing energy consumption and the footprint of wastewater treatment plants through the patented and award-winning technologies Sharon®, Anammox® and Nereda®. His main work focuses on the use of microbial communities within the environmental process-engineering field, with a special emphasis on nutrient removal, biofilm and biofouling. His research interests include granular sludge systems, microbial storage polymers, wastewater treatment, gas treatment, soil treatment, microbial conversion of inorganic compounds, production of chemicals from waste, and modelling. He is a full professor and Group Leader of Environmental Biotechnology at TU Delft in the Netherlands. He is also a Fellow of the Royal Dutch Academy of Arts and Sciences (KNAW), as well as the Netherlands (AcTI), USA (NAE) and Chinese (CAE) academies of engineering. Professor van Loosdrecht has won the Stockholm and Singapore water prizes and was awarded an honorary doctorate by ETH in Zurich and Ghent University. Apart from his other achievements, he has published over 800 papers, supervised 85 PhD students so far and is an honorary professor at the University of Queensland. He was the Editor-in-Chief of Water Research and is Advisor to IWA Publishing.

Di Wu, PhD

Di Wu is an Associate Professor in Environmental Technology at the Department of Green Chemistry and Technology and Centre for Environmental and Energy Research (CEER), Ghent University Global Campus. In 2014 he received his doctoral degree in Civil Engineering from the Hong Kong University of Science and Technology (HKUST). His main research work focuses on water science, environmental biotechnology, and resource recovery processes. He is active in combining green and environmental organic chemistry with wastewater engineering for education and research. So far, he has published in excess of 70 peer-reviewed papers and trained dozens of MSc and PhD students. He is also an adjunct Associate Professor of HKUST and associate member of the Centre for Advanced Process Technology for Urban Resource Recovery, Belgium.

Prof. Guanghao Chen, PhD

Guanghao Chen, a PhD graduate of Kyoto University, is currently a Chair Professor of Civil and Environmental Engineering, a director of the Water Technology Centre, and Head of the Hong Kong Branch of the Chinese National Engineering Research Centre in the water engineering field at the Hong Kong University of Science and Technology. His research interests include sulphur-cycle-based biological treatment process development, seawater for toilet flushing, a total water system with seawater for toilet flushing and cooling, greywater reuse, seawater-catalysed urine phosphorus recovery, and modelling of in-sewer sulphide control. He pioneered the invention and development of the SANI® process that helps to close the water loop between fresh and brackish/saline water supplies. In 2015 he was elected an IWA Distinguished Fellow and between 2010-2018 he received three IWA awards as well as the Hong Kong Green Innovation Gold Award 2018. Prof. Chen has published 250 journal papers and supervised 40 PhD and more than 27 MPhil students so far.

Preface

I am very pleased to see publication of this compliment to the well-known textbook *Biological Wastewater Treatment: Principles, Modelling and Design*. This companion text, *Biological Wastewater Treatment: Examples and Exercises* is a necessary compliment to the textbook. The worked examples in the 'Examples and Exercises' book supplement the fundamental information presented in the textbook and help users of these two books to better understand both the required fundamental knowledge and how to translate this knowledge into practice.

As described in Mogen's preface to the second edition of the textbook, knowledge and practice in water management is advancing quite rapidly, driven both by advances in the underlying science and by the global needs for improved wastewater management. I am quite certain that both the textbook and the examples and exercises book will be of inestimable value, not only to students but also to more experienced practitioners as they continue to learn and advance their practice. These two books will clearly serve a broad audience.

My congratulations to the authors of this new complement to the second edition of the textbook, as well as the authors of the textbook itself. We are solving our water problems and depend on a solid understanding of the fundamentals of the technologies we use, along with an understanding of how to apply this knowledge.

Glen T. Daigger, Ph.D., P.E., BCEE, NAE, CAE
Professor of Engineering Practice
University of Michigan

About the book

The first edition of the textbook *Biological Wastewater Treatment: Principles, Modelling and Design* was published in 2008 and it went on to become IWA Publishing's bestseller to date. In 2020, the 2nd updated and extended edition of the textbook was published because, since 2008, the knowledge and understanding of wastewater treatment had advanced extensively and moved away from empirically-based approaches to a fundamental first-principles approach based on chemistry, microbiology, physical and bioprocess engineering, mathematics and modelling. Like the first edition, for a whole new generation of young scientists and engineers entering the wastewater treatment profession, the 2nd edition of the textbook assembled and integrated the postgraduate course material of a dozen or so professors from research groups around the world who have made significant contributions to the advances in wastewater treatment. While all the chapters of the first edition have been updated to accommodate the latest advances and developments, some, such as granular sludge, membrane bioreactors, sulphur conversion-based bioprocesses and biofilm reactors which were new in 2008, have matured into new approaches in the industry and were also included in the 2020 2nd edition. The updated edition has already been available for more than two years and the feedback from readers has been overwhelming – the textbook won the IWA Publishing Best Scientific Book Prize in 2022. This inspired the authors to embark on a new challenge – to prepare this complementary book *Biological Wastewater Treatment: Examples and Exercises*. This new book is an extension of the 2nd edition textbook; each chapter corresponds to a chapter in the textbook and is structured similarly around five sections, namely, Introduction, Learning objectives, Examples, and Exercises, with solutions provided in an annex. The overall objective of the book is to deepen, expand and test the knowledge of the reader through a set of worked out examples, followed up by exercises and questions with provided answers. Where applicable, the book is supplemented with MS Office Excel files. The book is open access and can be downloaded (together with supplements) at the publisher's website[1]. The target readership of the book remains young water professionals, who will still be active in the field of protecting our precious water resources long after the aging professors who are leading some of these advances have retired. The authors are aware that cleaning dirty water has become more complex but also that it is even more urgent now than before, and offer this new book to help young water professionals engage with the scientific and bioprocess engineering principles of wastewater treatment science and technology with deeper insight, advanced knowledge and greater confidence built on stronger competence.

The Editors

---

[1] https://www.iwapublishing.com/books/9781789062298/biological-wastewater-treatment-principles-modelling-and-design-examples

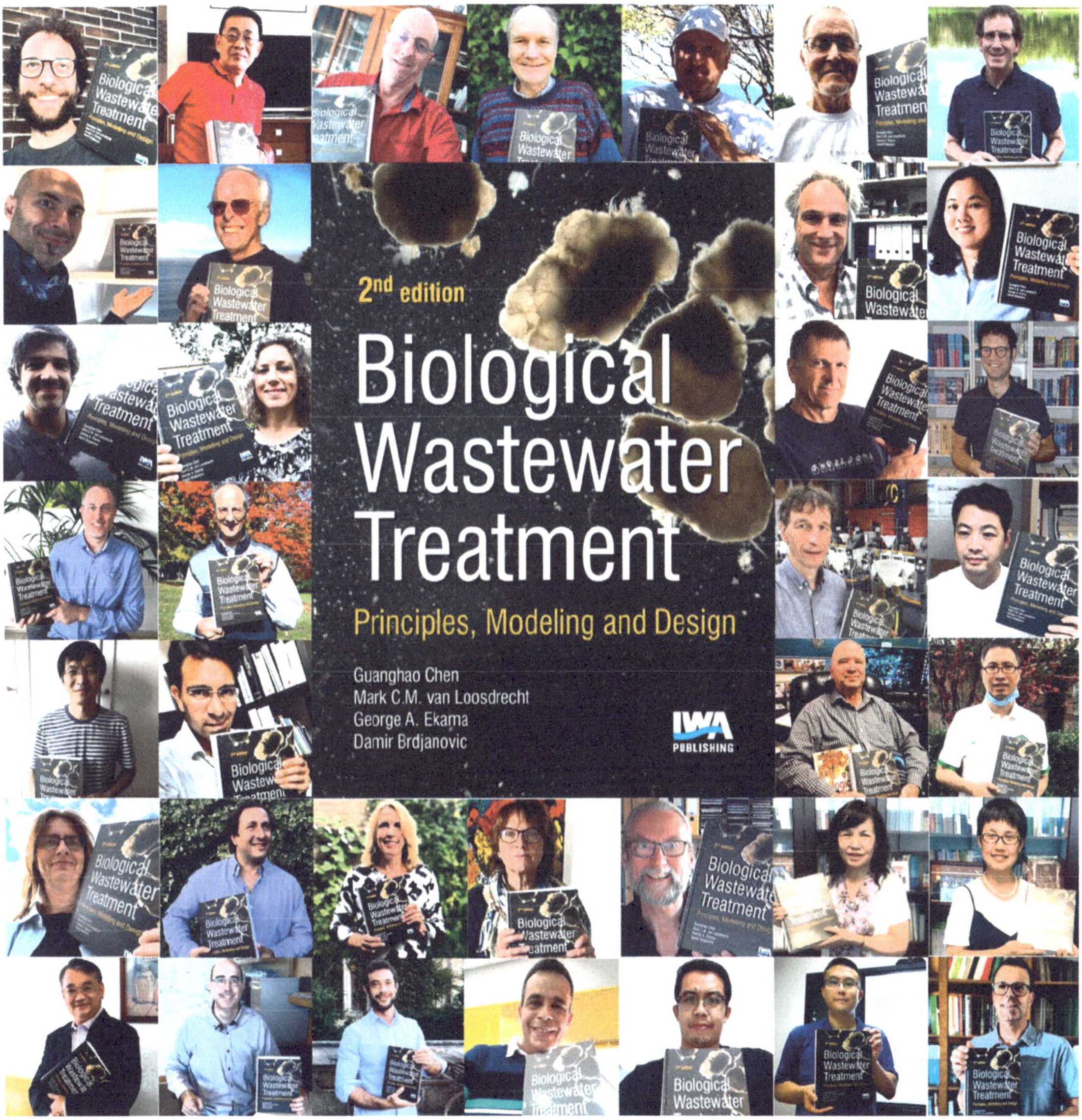

The first edition of the textbook Biological Wastewater Treatment: Principles, Modelling and Design (Henze *et al.*, 2008) was translated into Spanish, Chinese, Arabic, Korean, and Russian. The 2nd edition has already been published in Chinese, Portuguese and Japanese. On behalf of 39 editors and authors, Prof. Mark van Loosdrecht and Prof. Damir Brdjanovic received the Best Scientific Book Prize for the textbook Biological Wastewater Treatment: Principles, Modelling and Design 2nd edition (Chen *et al.*, 2020) at IWA President Dinner during the IWA World Water Congress and Exhibition, 10-15 September 2022, in Copenhagen.

Table of contents

doi: 10.2166/9781789062304_0001

1

Wastewater treatment development

Carlos M. Lopez-Vazquez and Mark C.M. van Loosdrecht

1.1 INTRODUCTION

Chapter 1 on Wastewater Treatment Development in the book *Biological Wastewater Treatment: Principles, Modelling and Design* (Chen *et al.*, 2020) (referred to hereafter as the textbook) explains how current wastewater treatment technologies have evolved over time. It describes the main drivers for sanitation on the historical journey from ancient cultures, passing through the Middle Ages and into the 20th century, thus providing deeper insight into the wastewater treatment technologies that have been developed. This chapter aims to guide readers through the contents of Chapter 1 in the main textbook in order to emphasize the factors that have driven and supported the development of the wastewater treatment technologies available up to now, and also to increase their understanding of how and why new technologies and applications will be developed in the near future.

1.2 LEARNING OBJECTIVES

After the successful completion of this chapter, the reader will be able to:

1. Describe the main purposes and drivers of sanitation and wastewater treatment technologies.
2. Discuss the development of sewage collection and wastewater treatment systems.
3. Define the main characteristics and advantages and disadvantages of existing wastewater treatment technologies.
4. Explain the main factors that have supported and led to the development of nutrient removal systems, instrumentation, control and automation, disinfection and micropollutant removal.
5. Distinguish different resources that have been or could be recovered from wastewater.

1.3 EXERCISES

Exercise 1.3.1
What are the main global drivers for sanitation?

Exercise 1.3.2
Describe ancient practices or applications used to recover resources from wastewater.

Exercise 1.3.3
What were the sanitation conditions in the Sanitary Dark Ages?

Exercise 1.3.4
In the Modern Era, when were the first sanitary collection systems installed or put into practice?

Exercise 1.3.5
What was the first water-free vacuum collection system implemented?

Exercise 1.3.6
What were the first biological wastewater treatment systems?

Exercise 1.3.7
Why were activated sludge systems named this way?

Exercise 1.3.8
Why is the Biochemical Oxygen Demand (BOD) determined over a duration of 5 days?

Exercise 1.3.9
What was one of the first applications of mathematical modelling in the wastewater treatment field?

Exercise 1.3.10
What led to the introduction of nitrification in wastewater treatment systems?

Exercise 1.3.11
What were the first treatment systems used to perform nitrification?

Exercise 1.3.12
Describe the main disadvantage of the first wastewater treatment systems applied to perform nitrification.

Exercise 1.3.13
What is eutrophication and why is it an undesirable process in receiving waters?

Exercise 1.3.14
What was and continues to be one of the main applications of the Monod kinetic expression?

Exercise 1.3.15
What supported the development of the nitrification-denitrification processes?

Exercise 1.3.16
What led to the development of the pre-denitrification systems?

Exercise 1.3.17
What was the first activated sludge configuration to combine the pre-denitrification and post-denitrification processes?

Exercise 1.3.18
Describe the main characteristics of the Pasveer ditch system developed in 1959.

Exercise 1.3.19
Why is it important to also remove phosphorus in order to prevent eutrophication?

Exercise 1.3.20
Explain why it is assumed that the Enhanced Biological Phosphorus Removal (EBPR) process was discovered by accident.

Exercise 1.3.21
What is the main characteristic of the Phoredox system developed by Barnard in 1976?

Exercise 1.3.22
What led to the development of anaerobic wastewater treatment technologies?

Exercise 1.3.23
Describe the factors leading to the development of the latest biofilm-based treatment systems.

Exercise 1.3.24
How is the solid-liquid separation process carried out in a membrane bioreactor (MBR) system?

Exercise 1.3.25
What is the Nereda® process and what removal processes does it perform?

Exercise 1.3.26
What was the reason to upgrade existing wastewater treatment plants by carrying out the removal of nutrients in the sludge treatment line?

Exercise 1.3.27
List at least five resources that can be recovered or generated from wastewater.

Exercise 1.3.28
What is the main advantage of the development and implementation of instrumentation, control and automation (ICA)?

Exercise 1.3.29
Why have disinfection and the removal of micropollutants of emerging concern received an increasing amount of interest in recent decades?

Exercise 1.3.30
Describe the main driver behind the use of seawater for toilet flushing and the advantages of this practice.

ANNEX 1: SOLUTIONS

Solution 1.3.1

Mainly, good public health by minimising waterborne diseases. In addition, together with access to safe water, it is essential to eradicate poverty, building liveable and prosperous societies.

Solution 1.3.2

In China from around 200 BC and up to the 1970s, due to recognition of its fertilizing value, the vast majority of agricultural land was fertilized by human faeces from latrines. In the Indus valley, in the Euphrates region and Greece, sewage and stormwater were being collected in basins outside the cities and used for irrigation purposes and to fertilise crops and orchards from before 2000 BC.

Solution 1.3.3

In the Sanitary Dark Ages, sanitation conditions were rather precarious: waste was simply disposed of in the streets, often by emptying buckets from second-storey windows.

Solution 1.3.4

In the Modern Era, the first sanitary collection systems were put into practice in several cities, driven by the city dwellers who no longer wanted to put up with the stench. Carts drove through the streets to empty buckets that were full of waste. Farmers located around the cities made use of this practice because they used the 'humanure' to fertilise. However, spillages during transportation and emptying did not help to reduce the smell.

Solution 1.3.5

The Liernur pneumatic sewer system was developed by Mr. Liernur in around 1900. The system collected toilet water using a vacuum sewer.

Solution 1.3.6

Biological filters were the first biological treatment systems to treat the sewage from towns and cities, mostly in the United States and United Kingdom, and they were introduced between 1893 and 1901.

Solution 1.3.7

Based on fill-and-draw wastewater treatment experiments by Ardern and Locket (1914), a highly treated wastewater effluent was produced resembling a sludge. Believing that the working principle was similar to activated carbon, the sludge was therefore called activated sludge.

Solution 1.3.8

In the first half of the 20th century, the river into which the (treated) wastewater was discharged was considered an integral part of the treatment process. Since the longest time that water spent in the rivers of the UK before it reached the sea is 5 days, this was chosen as the duration of the BOD test.

Solution 1.3.9

A mathematical model presented by Phelps (1944) in the book *Stream Sanitation*. It was applied to calculate the maximum organic load to a river from the oxygen sag curve. This was to prevent the dissolved oxygen (DO) concentration falling below a minimum value at a defined point downstream the wastewater discharge point.

Solution 1.3.10

To decrease the oxygen demand in rivers and the toxic effect that ammonia has on aquatic species.

Solution 1.3.11

Low-loaded trickling filters plants in the USA, Europe and South Africa.

Solution 1.3.12

The low-loaded trickling filters failed to nitrify consistently throughout the year, in particular due to the lower temperatures experienced in winter.

Solution 1.3.13

Eutrophication is the excessive growth of algae and other plants in surface water bodies due to the fertilizing effect of nitrogen (N) and phosphorus (P). It is an undesirable process: during the day there is a large photosynthetic production, and during the night oxygen depletion occurs and therefore plants and fish die off. The decaying biomass contributes even more to oxygen shortage. In addition, cyanobacteria (which also proliferate during the eutrophication processes) generate toxins that have a major deleterious effect on aquatic organisms and this affects the use of the water body as a source of (potable) water. Consequently, eutrophication may decrease water availability, affecting key sectors and activities (such as food production, industry and even tourism and recreation).

Solution 1.3.14

Its application to describe the growth rate of bacteria as a function of the substrate concentration. In particular, in 1964, Downing *et al.* used it to show that the nitrification process depends on the maximum specific growth rate of autotrophic organisms (Downing *et al.*, 1964). This demonstrated that their growth is slower than that of ordinary heterotrophic organisms and that biological wastewater treatment systems had to be designed and operated at sludge ages long enough to enable the growth of autotrophic organisms to achieve consistently low effluent ammonia concentrations.

Solution 1.3.15

The advanced studies on bioenergetics carried out by McCarthy (McCarthy, 1964). He showed that nitrate generated by the nitrification process could be used as an alternative for oxygen by certain heterotrophic organisms and it is thereby converted to dinitrogen gas. Unaerated sections were included in activated sludge systems to induce denitrification, thus saving aeration energy and removing nitrogen.

Solution 1.3.16

Ludzack and Ettinger (1962) proposed an unaerated stage prior to the aerated stage in order to increase the denitrification rate by utilizing the organics present in wastewater (see Figure 5.13B in Chen *et al.,* 2020). This configuration was preferred to the post-denitrification system proposed by Wuhrmann (1964) which had an unaerated section after the aerobic nitrification stage and used methanol as its external carbon source to increase the denitrification rate. Due to the methanol addition, the post-denitrification configuration had higher operational costs and it was contradictory to add organics to the unaerated stage after the ones present in the influent wastewater had been removed in the aerated section. Thus, pre-denitrification configurations became more popular than post-denitrification ones. However, systems with post-denitrification stages are able to achieve lower effluent total nitrogen concentrations (e.g. lower than 5 mgN/l) if required (Chen *et al.,* 2020).

Solution 1.3.17

The 4-stage Bardenpho system (Barnard, 1973). This system, developed in South Africa by James Barnard in 1972, combined the pre- and post-denitrification reactors and introduced recycle flows to control the nitrate entering the pre-denitrification unit (see Figure 5.13C in Chen *et al.,* 2020).

Solution 1.3.18

This was a simple and economical system solely composed of one treatment tank with no primary settler or secondary settling tank. It followed the fill-and-draw principle developed by Ardern and Locket in the UK in 1914 (Pasveer, 1959). Moreover, if operated with continuous feeding, it was able to achieve simultaneous nitrification and denitrification. These continuously operated oxidation ditch and carrousel systems evolved from the Pasveer system but included a secondary settling tank.

Solution 1.3.19

Because phosphorus has been identified as the main enabling element for eutrophication in several ecosystems, removing only nitrogen is therefore insufficient to prevent it. Microorganisms (especially blue-green algae) can use nitrogen gas as a nitrogen source, and therefore phosphorus is the main growth-limiting compound in surface water.

Solution 1.3.20

Because the first indication of the occurrence of EBPR in activated sludge systems was observed by Srinath *et al.* in India (Srinath *et al.,* 1959) in a treatment plant where the aeration in the first stage of the activated sludge plant was compromised. They merely noticed that the sludge showed an excessive uptake of phosphorus (beyond that required for biomass synthesis) when it was aerated. It was also shown that it was a biological process since it was oxygen-dependent and it was inhibited by toxic substances.

Solution 1.3.21

The Phoredox system developed by James Barnard (Barnard, 1976) consists of one anaerobic stage (that receives the influent wastewater) followed by one aerobic reactor. Thus, mixed liquor activated sludge is cycled through the anaerobic-aerobic configuration of the system (see Figure 6.20D in Chen *et al.,* 2020). This development built on the pioneering research carried out by Levin and Shapiro (1965) who coined the term 'luxury uptake' to describe the induced biological phosphorus removal in excess of the metabolic needs of activated sludge when alternating anaerobic and aerobic conditions.

Solution 1.3.22

The development of anaerobic wastewater treatment technologies was motivated by the energy crisis experienced in the 1970s together with an increased demand for industrial wastewater treatment. Furthermore, the invention of upflow anaerobic sludge blanket reactors (UASB) by Lettinga and colleagues (Lettinga *et al.,* 1980) led to a breakthrough in anaerobic treatment. Not only was this technology feasible for industrial wastewater treatment but anaerobic treatment of low-strength municipal wastewater could also be efficiently introduced in tropical regions of South America, Africa and Asia.

Solution 1.3.23

The main cause was the need to develop more compact wastewater treatment plants since rapid urbanization has led to a lower availability of land. Also industries, often with land limitations, started to treat their own wastewater. This caused the development of a whole range of new biofilm-based processes such as biological aerated filters, fluid-bed reactors, moving-bed bioreactors and granular sludge processes, among others.

Solution 1.3.24

The solid-liquid separation process is carried out by a membrane (either submerged in the main aerobic reactor of the MBR system or located externally). The membrane enables the solids to be retained and produces a clarified treated effluent (Yamamoto *et al.*, 1989).

Solution 1.3.25

The Nereda® process is an aerobic granular sludge technology that allows a more efficient and compact removal of nutrients. To minimize costs, it is a sequencing technology based on the fill-and-draw principles of Ardern and Locket (1914) and Pasveer (1959) so all the biological conversion and settling processes occur in one single reactor.

Solution 1.3.26

One of the main factors was the need to upgrade existing plants in order to comply with the new stricter effluent discharge standards instead of building new treatment systems. Thus, it was observed that considerable nitrogen and phosphorus concentrations were released in the sludge handling facilities, which returned to the main-stream wastewater treatment line through internal recycle flows. The development and implementation of different side-stream processes, which take advantage of the particular characteristics of the side-stream streams (e.g. highly concentrated, higher temperatures and lower flow rates), promoted and facilitated the cost-effective removal of nitrogen and phosphorus. Processes such as the high activity ammonium removal over nitrite (SHARON®), the anaerobic ammonia oxidation (ANAMMOX®) and Crystalactor® (for improved nitrogen removal and mineral crystallization for phosphorus precipitation, recovery and reuse), are some of the technologies that have contributed to these developments.

Solution 1.3.27

In view of an increasing interest in the last decade, in addition to the recovery of water and biogas, cellulose, hydrogen, polyhydroxyalkanoates, nitrogen, phosphates, proteins, extracellular polymers, and even heat, have been identified as recoverable resources from wastewater (Van Loosdrecht and Brdjanovic, 2014).

Solution 1.3.28

The main advantage is to facilitate the operation of existing wastewater treatment plants increasing their reliability to meet stricter effluent standards. It also contributes to a reduction in the operational costs, savings, and recovery of resources.

Solution 1.3.29

Because water reclamation and reuse has been seen as an alternative in order to alleviate water scarcity. Thus, for instance, UV and ozonation technologies (among others) have become increasingly interesting as disinfection processes as well as for the removal of pollutants of emerging concern.

Answer 1.3.30

The main driver is coping with water scarcity and the need to save fresh water through the search and implementation of alternative water sources for non-potable water related activities. This practice relies on the fact that, in current sanitation systems, on average at least one third of the water consumed in a household is used for toilet flushing and does not require water of drinking quality (Chen *et al.*, 2012; Van Loosdrecht *et al.*, 2012).

REFERENCES

Ardern E. and Lockett W.T. (1914). Experiments on the oxidation of sewage without the aid of filters. *Journal of the Society of Chemical Industry*, 33, 523.

Barnard J.L. (1973). Biological denitrification. *Water Pollut. Control*, 72, 705-720.

Barnard J.L. (1976). A review of biological phosphorus removal in the activated sludge process. *Water SA*, 2(3), 136-144.

Chen G. (2020) Wastewater treatment development. In: *Biological wastewater treatment: principles, modelling and design*. 2nd edition. Edited by G. Chen, M.C.M. van Loosdrecht, G.A. Ekama and D. Brdjanovic. ISBN: 9781789060355. Published by IWA Publishing, London, UK.

Chen G.H., Chui H.K., Wong C.L., Tang D.T.W., Lu H., Jiang F. and Van Loosdrecht M.C.M. (2012). An Innovative Triple Water Supply System and a Novel SANI® Process to Alleviate Water Shortage and Pollution Problem for Water-scarce Coastal Areas in China. *Journal of Water Sustainability*, 2(2), 121–129.

Downing A.L., Painter H.A. and Knowles G. (1964). Nitrification in the activated sludge process. *J. Proc. Inst. Sewage Purif.*, 64(2), 130-158.

Lettinga G., Van Velsen A.F.M., Hobma S.W., De Zeeuw, W. and Klapwijk A. (1980). Use of the upflow sludge blanket (USB) reactor concept for biological wastewater treatment, especially for anaerobic treatment. *Biotechnology and Bioengineering*, 22, 699-734.

Levin G.V. and Shapiro J. (1965). Metabolic uptake of phosphorus by wastewater organisms. *Journal of the Water Pollution Control Federation*, 37, 800-821.

Ludzack F.J. and Ettinger M.B. (1962). Controlling operation to minimize activated sludge effluent nitrogen. *Water Pollution Control Federation Journal*, 34, 920-931.

McCarthy P.L. (1964). Thermodynamics of biological synthesis and growth. Procs. 2nd International Conference on Water Pollution Control, 2, 169-199.

Pasveer A. (1959). A contribution to the development in activated sludge treatment. *J. Proc. Inst. Sew. Purif.*, 4, 436.

Phelps E.B. (1944). *Stream Sanitation*. John Wiley and Sons Inc., New York, USA.

Srinath E.G., Sastry C.A. and Pillai S.C. (1959). Rapid removal of phosphorus from sewage by activated sludge. *Experientia*, 15(9), 339-340.

Van Loosdrecht M.C.M. and Brdjanovic D. (2014). Anticipating the next century of wastewater treatment. *Science*, 344(6191):1452-1453.

Van Loosdrecht M.C.M., Brdjanovic D., Chui H.K. and Chen. G.H. (2012). A source for toilet flushing and for cooling, sewage treatment benefits, and phosphorus recovery: direct use of seawater in an age of rapid urbanisation. *Water 21*, 14(5), 17-20.

Wuhrmann K. (1964). Hauptwirkugen und Wechselwirkugen einiger Betriebsparameter Belebtschlammsystem: Ergibnisse mehrjäriger grossversuche. *Schweigerische Zeitschrift für Hydrologie*, XXVI(2) 218.

Yamamoto K., Hiasa M., Mahmood T. and Matsuo T. (1989). Direct solid-liquid separation using hollow fiber membrane in an activated sludge aeration tank. *Water Science and Technology*, 21, 43-54.

doi: 10.2166/9781789062304_0009

2

Basic microbiology and metabolism

David G. Weissbrodt, Michele Laureni, Mark C.M. van Loosdrecht and Yves Comeau

2.1 INTRODUCTION

In Chapter 2 on basic microbiology and metabolism of the 2nd edition of the textbook *Biological Wastewater Treatment: Principles, Modelling and Design* (Chen *et al.,* 2020), we learned about microorganisms involved in biological nutrient removal (BNR), their trophic groups, metabolisms, and growth systems. You should now be able to answer:

- What is a microbial cell and what are microbial populations?
- How do they grow and metabolically function?
- How can a growth system be defined by a set of substrates, nutrients, and products?
- How can microbial growth and conversions be mathematically modelled and predicted?
- How can microorganisms and their metabolisms be tracked in microbial communities?

In the textbook, we introduced the fundamentals of microbiology and metabolism, as well as the mathematical formulations of stoichiometry, thermodynamics, and kinetics in order to characterise and model microbial growth. In addition, we covered the microbial ecology and ecophysiology methods necessary to track microorganisms and their functionalities in microbial communities of activated sludge, biofilms, and granular sludge.

Here, you will practise the principles of microbial growth, selection, and interactions in WWTP biomasses. This chapter focuses mainly on BNR microorganisms involved in carbon and nitrogen conversions.

2.2 LEARNING OBJECTIVES

We will implement simple conceptual and mathematical models to open the microbial and metabolic black box of BNR biomass. After completing this chapter, you will be able to:

1. Describe the microbiology and metabolisms of BNR microorganisms.
2. Conceptualise a BNR microbial ecosystem by linking their metabolisms.
3. Define their growth systems with appropriate C source, N source, e-donor couple, e-acceptor couple, and partner compounds.
4. Formulate their anabolic, catabolic, and metabolic reactions, by deriving stoichiometries by elemental, charge, electron, and/or Gibbs free energy balances.
5. Implement thermodynamic relations to estimate stoichiometric and kinetic parameters.
6. Formulate a volumetric growth process rate using saturation/inhibition switching functions.
7. Use Herbert-Pirt kinetic relations to predict material allocations to maintenance and growth.
8. Simulate simple mathematical models to predict their growth, selection, and interactions in ideal (dis)continuous bioreactors such as batch, sequencing batch, and chemostat.
9. Propose methods to measure microbial selection and conversions.

Together with the other chapters of this book, this will enhance your understanding of the microbial processes in WWTP biomasses necessary for process design, operation, monitoring and control. The examples and exercises are solved on paper, with Excel spreadsheets, and using the Aquasim software.

2.3 EXAMPLES

Conceptual and mathematical models to describe the growth of BNR microorganisms

In a scaffolded approach, we will formulate conceptual and mathematical models to describe the growth, selection, and interactions between ordinary heterotrophs, nitrifiers, denitrifiers, and anammox bacteria, involved in organic matter and nitrogen removal from wastewater.

Example 2.3.1

Describe BNR organisms, their trophic groups, primary metabolisms, and growth systems

Ordinary heterotrophic organisms (OHOs), ammonium-oxidising organisms (AOOs), nitrite-oxidising organisms (NOOs), complete ammonium-oxidising organisms (CMOs), denitrifying heterotrophic organisms (DHOs), and anaerobic ammonium-oxidising organisms (AMOs) are used to remove organic matter and nitrogen from wastewater. Systematics in microbial naming in wastewater engineering have been proposed (Corominas *et al.,* 2010). Engineering and microbiological terms are often intermixed.

a) Characterise these guilds by accurately describing their trophic group in microbiological terms by highlighting the main dissimilation pathway (fermentative/anaerobic respiring/aerobic respiring), energy source (chemo-/photo-), electron donor (litho-/organo-), and carbon source (auto-/hetero-) involved in their respective metabolisms.
b) Establish their growth systems by listing the e-donor couple, e-acceptor couple, carbon source, nitrogen source, and accompanying compounds involved in their metabolism.

Solution

In microbiology, accurate formulations of trophic groups provide hints on growth characteristics, related to the energy-electron-carbon triangle of microbial life (Figure 2.1).

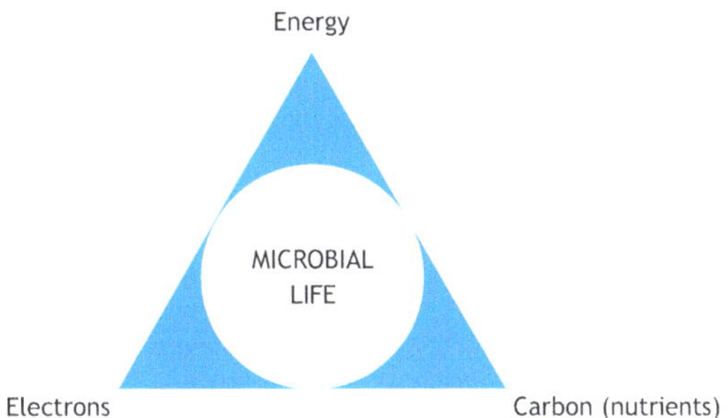

Figure 2.1 Energy-electron-carbon triangle of microbial life: growth results from energy, electrons, and carbon (and nutrients such as N and P).

Table 2.1 provides the terms used to accurately name trophic groups.

Table 2.1 Microbiological systematics in naming of trophic groups to highlight the main components of the basic growth system of microorganisms along the energy-electron-carbon triangle of microbial life.

Dissimilation pathway	Energy generation	Electron donor	Carbon source	
Aerobic respiring Anaerobic respiring Fermentative	Chemo- Photo-	Organo- Litho-	Hetero- Auto-	-troph

The trophic descriptions of the BNR guilds are compiled in Table 2.2. The components of their growth systems are given, based on reflections on BNR process configurations and recirculation loops. Additional hints are provided:

- e-donors (eD) and e-acceptors (eA) are indicated with their respective redox couple:
  - Use the biochemical reference conditions: at pH 7.0, bicarbonate (HCO_3^-), ammonium (NH_4^+) and acetate ($C_2H_3O_2^-$) are dominant chemical species according to their pK_a values:
    H_2CO_3 *vs.* $HCO_3^- = 6.4$, HCO_3^- *vs.* $CO_3^{2-} = 10.2$; NH_4^+ *vs.* $NH_3 = 9.25$; $C_2H_4O_2$ *vs.* $C_2H_3O_2^- = 4.75$.
- Anammox bacteria are respiring ammonium with nitrite. Nitric oxide (NO) is however the central intermediate which could also be integrated as the terminal e-acceptor.
- Denitrifiers can respire organics using different nitrogen oxides from nitrate (NO_3^-) to nitrite (NO_2^-), nitric oxide (NO), nitrous oxide (N_2O), eventually releasing molecular nitrogen (N_2). Different e-acceptors can be considered depending on the conditions and physiologies.

Table 2.2 Trophic groups and growth systems of microbial guilds removing organic matter and nitrogen.

Microbial guild in environmental engineering (and its acronym)	Trophic group (in accurate microbiology terms)	Energy generation	Electron donor couple	Electron acceptor couple	Carbon source	Nitrogen source
Ordinary heterotrophic organisms (OHOs)	Aerobic respiring chemoorganoheterotrophs	Chemical redox reaction	Organic matter $(C_{11}H_{18}O_5)$[a] / Bicarbonate (HCO_3^-)	Molecular oxygen (O_2) / Water (H_2O)	Organic matter $(C_{11}H_{18}O_5)$[a]	Ammonium (NH_4^+)
Complete ammonium-oxidising organisms (CMOs)	Aerobic respiring chemolithoautotrophs	Chemical redox reaction	Ammonium (NH_4^+) / Nitrate (NO_3^-)	Molecular oxygen (O_2) / Water (H_2O)	Bicarbonate (HCO_3^-)	Ammonium (NH_4^+)
Ammonium-oxidising organisms (AOOs)	Aerobic respiring chemolithoautotrophs	Chemical redox reaction	Ammonium (NH_4^+) / Nitrite (NO_2^-)	Molecular oxygen (O_2) / Water (H_2O)	Bicarbonate (HCO_3^-)	Ammonium (NH_4^+)
Nitrite-oxidising organisms (NOOs)	Aerobic respiring chemolithoautotrophs	Chemical redox reaction	Nitrite (NO_2^-) / Nitrate (NO_3^-)	Molecular oxygen (O_2) / Water (H_2O)	Bicarbonate (HCO_3^-)	Ammonium (NH_4^+) if remaining or nitrite (NO_2^-)
Anaerobic ammonium-oxidising organisms (AMOs)	Anaerobic respiring chemolithoautotrophs	Chemical redox reaction	Ammonium (NH_4^+) / Dinitrogen (N_2)	Nitrite (NO_2^-) / Dinitrogen (N_2)	Bicarbonate (HCO_3^-)	Ammonium (NH_4^+)
Denitrifying heterotrophic organisms (DHOs)	Anaerobic respiring chemoorganoheterotrophs	Chemical redox reaction	Organic matter $(C_{11}H_{18}O_5)$ / Bicarbonate (HCO_3^-)	Nitrogen oxides $(NOx^-$ or $N_yO)$ / Dinitrogen (N_2) or other nitrogen oxides $(NOx^-$ or $N_yO)$	Organic matter $(C_{11}H_{18}O_5)$	Ammonium (NH_4^+)

[a] The generic elemental formula of organic matter ($C_{11}H_{18}O_5$ or $C_1H_{1.63}O_{0.46}$) is given here as e-donor and C source for OHOs and DHOs. An alternative is to indicate acetate ($C_2H_3O_2^-$), often used as a model compound in synthetic wastewater.

Example 2.3.2

Outline the growth systems of BNR guilds and conceptualise their ecosystem in a WWTP

Microbial growth and interactions are efficiently understood by outlining the growth systems of microorganisms. This can help conceptualise the ecosystem that they can form.

a) Outline the growth system of each guild and highlight possible symbiotic and competitive interactions between them.

b) Outline a conceptual model of an ecosystem formed by these guilds.

c) Describe technical ecosystems involving these guilds in, *e.g.*, flocculent activated sludge, biofilm carriers, granules, and/or hybrid systems of your choice(s).

Solution

a) Outlines of growth systems of microbial guilds

Growth systems for each guild are outlined in Figure 2.2 from the compounds listed in Table 2.2.

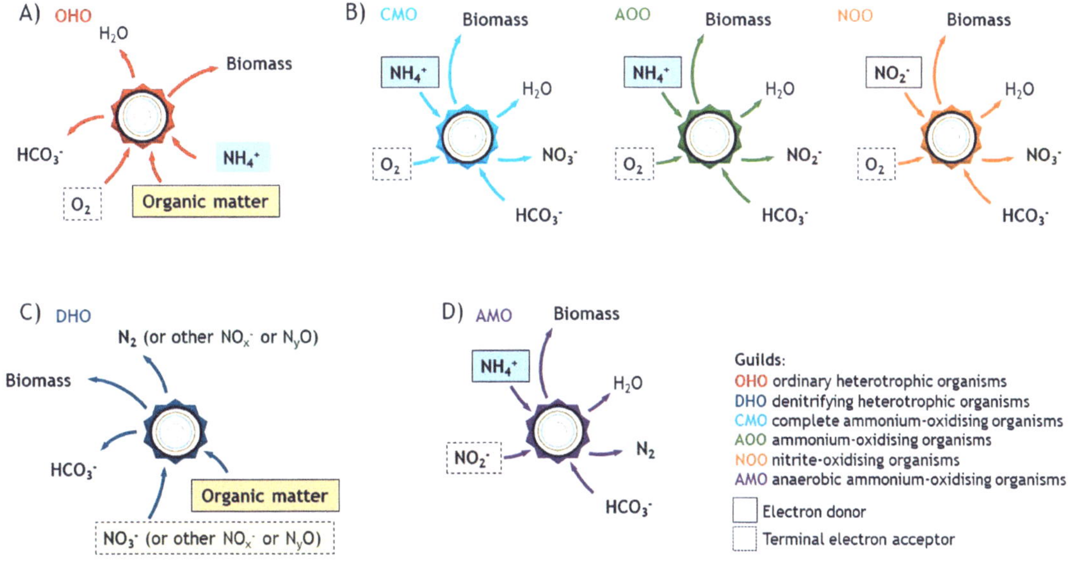

Figure 2.2 Outlines of growth systems of microbial guilds involved in organic matter and nitrogen removal: A) ordinary heterotrophic organisms (OHOs), B) complete ammonium-oxidising organisms (CMOs), ammonium-oxidising organisms (AOOs) and nitrite-oxidising organisms (NOOs), C) denitrifying heterotrophic organisms (DHOs), and D) anaerobic ammonium-oxidising organisms (AMOs). Growth systems entail the e-donor couple, e-acceptor couple, carbon source, nitrogen source, and biomass formed.

b) Conceptual model of an ecosystem of OHOs, CMOs, DHOs and AMOs

Growth systems of each guild are connected in Figure 2.3 to represent their symbiotic and competitive interactions, and the ecosystem that they form.

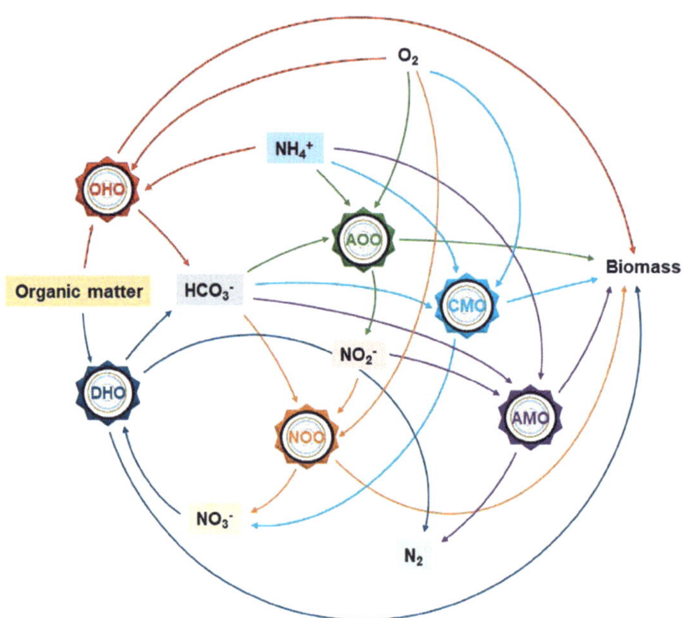

Figure 2.3 Conceptual ecosystem model of symbiotic and competitive metabolic interactions between BNR guilds. The wheels represent microorganisms and their genomes. The arrows represent flows of materials converted.

c) *Conceptual ecosystems of technical BNR processes*

Design and operation rely on the selection of BNR guilds. Flow schemes of biological secondary treatments match with chemical inputs and outputs related to their growth systems (Figure 2.4).

In an anoxic-aerobic (A/O) system operated for pre-denitrification and nitrification (see Chapter 5 in the textbook, Figure 2.4A), DHOs anaerobically respire the organic matter supplied by the influent as e-donor and C source using nitrogen oxides as terminal e-acceptors such as nitrate that is recirculated from the nitrification tank where ammonium is respired aerobically by nitrifiers to the denitrification tank.

In partial nitritation and anammox (PN/A) (see Chapter 5 in the textbook) at the main line (Figure 2.4B), organic matter is first removed by OHOs in a high-rate aeration tank prior to aerobic and anaerobic respirations of ammonium by AOOs and AMOs, respectively.

In biofilms or granules (see chapters 11, 17 and 18 in the textbook, Figure 2.4C), microbial niches establish along gradients of dissolved materials generated across the biofilm depth by diffusional resistances. This is conceived in a multilayer 'onion' model. Heterotrophs and nitrifiers respire organics and ammonium, respectively, in the outer aerobic biovolume. Nitrogen oxides produced by nitrifiers are used by denitrifiers to respire organics (if not limiting) anaerobically in an internal anoxic biovolume. When e-acceptors are depleted, residual organics can be fermented in the core by fermentative organisms and methanogens. When substrates are depleted, biomass inactivates and decays.

In a hybrid biofilm-floc system (Figure 2.4D), microbial guilds occupy different niches on biofilms and flocs depending on their growth physiologies. In single-sludge PN/A, faster-growing aerobic AOOs occupy flocs where dioxygen and ammonium are rapidly accessed. Slower-growing and oxygen-sensitive AMOs occupy the inner O_2-depleted biovolumes of biofilms.

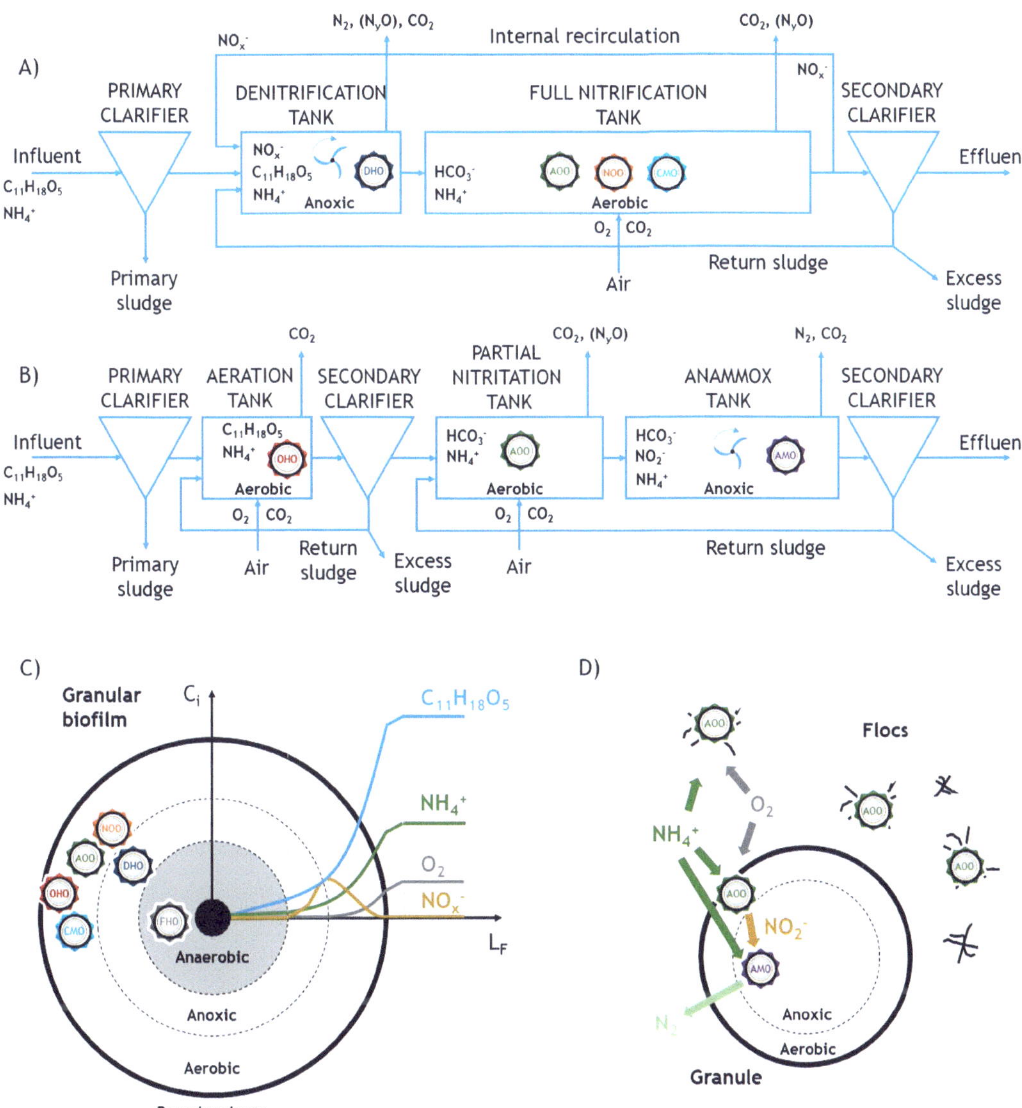

Figure 2.4 Conceptual models of technical ecosystems involving microbial guilds removing organic matter and nitrogen: flocculent sludge systems for A) pre-denitrification and nitrification and B) high-rate removal of organic matter followed by partial nitritation and anammox (PN/A); C) granules for C/N removal, and D) hybrid floc-granule biomass for PN/A. Guilds are displayed by wheels at locations where they are primarily active. In activated sludge flow-through systems, the biomass moves with the water flow across all the tanks; all the microbial guilds are present (but not active) in all tanks.

Example 2.3.3

Develop the mathematical expressions necessary to model the growth of BNR microorganisms

Predicting microbial growth, selection, and interactions is facilitated by simple mathematical models. Formulate the growth stoichiometries and kinetics of the guilds.

a) Write in literal (non-mathematical) forms their overall growth, anabolic, and independent catabolic reactions. Reflect on the nature of each compound participating in the metabolism.
b) Calculate their independent anabolic and catabolic reactions.
c) Formulate their overall growth reaction by calculating the elemental, charge and/or degree of reduction balances. If the number of unknowns is higher than the number of conservation balances, retrieve a measured biomass yield on substrate ($Y_{X/S}^{obs}$) and/or any other useful observed yield ($Y_{x/y}^{obs}$) for the targeted organisms from literature.
d) Calculate and compare their yield of Gibbs free energy dissipated during the growth (ΔG_{Diss}^{01}) under biochemical reference conditions (25 °C, 1 atm, pH 7.0).
e) Formulate their overall growth reactions using empirical formulas based on thermodynamics. Compare the reactions obtained via thermodynamics with those obtained in step c. using elemental balances and the observed yield.
f) Calculate their biomass-specific maintenance rate on the e-donating substrate (m_S) and their maximum growth rate (μ_{max}) using thermodynamic relations.
g) Formulate the Herbert-Pirt relations for each compound involved in their metabolism.
h) Formulate in literal terms the volumetric rates of their growth process using saturation (*i.e.*, Monod terms) and inhibition (*e.g.*, inverse Monod terms) switching functions.

Compile their reactions and rates in a stoichiometric-biokinetic matrix (known as the Petersen matrix or Gujer matrix) for an efficient comparison of metabolisms.

Critically address your results by reflecting on stoichiometries and kinetics to predict microbial selection and interactions, and their integration for process design and operation.

Hint: in the calculations, organic matter is considered as acetate ($C_2H_3O_2^-$), often present in wastewater and used as a model compound in synthetic wastewater for lab experiments.

Option 1: test the effects of temperature and pH on growth stoichiometries and kinetics.

Option 2: perform the same calculations using the elemental formula of organic matter ($C_{11}H_{18}O_5$ or $C_1H_{1.63}O_{0.46}$) representative for real municipal wastewater.

Solution

a) Literal expressions of overall growth, anabolic and catabolic reactions

Based on the growth system defined in Table 2.2, the overall growth reaction (Table 2.3), anabolic reaction (Table 2.4), and independent catabolic reaction (Table 2.5) can be formulated literally for each guild.

Reactions are formulated mathematically with stoichiometric coefficients ($v_{i,j}$) for each material i and microbial process j, that take negative ($v_{i,j} = -Y_{i/n,j} < 0$, consumption) or positive ($v_{i,j} = Y_{i/n,j} > 0$, production) yield values per unit of reference compound n (biomass in overall growth and anabolism, e-donor in catabolism). This aids model implementation in simulation software.

Overall growth reactions

Overall growth reactions (Table 2.3) include materials involved in anabolisms and catabolisms.

Partner compounds (HCO_3^-, H_2O, H^+) are often present either as reactants or products depending on the metabolisms and used to balance elements and charges.

Some compounds act with multiple functions. Chemoorganoheterotrophs (OHOs, DHOs) involve organics as the e-donor and C source. With nitrifiers (AOOs, NOOs, CMOs), the N source is often the e-donor. In pre-denitrification, ammonium is the N source, and nitrogen oxides (*e.g.*, nitrate, nitrite) are e-acceptors recirculated from the nitrification tank. In post-denitrification, nitrogen oxides are both e-acceptors and N sources; an organic e-donor and C source is supplied.

The overall growth reaction formulated using stoichiometric coefficients ($v_{i,j}$) is:

$$
\begin{aligned}
&v_{C,j}\ \text{carbon source} + v_{N,j}\ \text{nitrogen source} + v_{eD,j}\ \text{e-donor} + v_{eA,j}\ \text{e-acceptor} \\
&+ v_{X,j}\ \text{biomass} + v_{ox\text{-}eD,j}\ \text{oxidised e-donor} + v_{red\text{-}eA,j}\ \text{reduced e-acceptor} \\
&+ v_{i,j}\ \text{partner compounds } (e.g.,\ HCO_3^-,\ H_2O,\ H^+)
\end{aligned}
$$

and by replacing them by yield values ($\pm Y_{i/X,j}$) per unit of biomass becomes:

$$
\begin{aligned}
&-Y_{C/X,j}\ \text{carbon source} - Y_{N/X,j}\ \text{nitrogen source} - Y_{eD/X,j}\ \text{e-donor} - Y_{eA/X,j}\ \text{e-acceptor} \\
&+ 1\ \text{biomass} + Y_{ox\text{-}eD/X,j}\ \text{oxidised e-donor} + Y_{red\text{-}eA/X,j}\ \text{reduced e-acceptor} \\
&\pm Y_{i/X,j}\ \text{partner compounds } (e.g.,\ HCO_3^-,\ H_2O,\ H^+)
\end{aligned}
$$

Table 2.3 Overall growth reactions in literal forms. Acetate ($C_2H_3O_2^-$) is used as the model organic compound. Several expressions can be found in the examples in the textbook (Chen *et al.*, 2020).

a – Metabolism of aerobic respiring chemoorganoheterotrophic OHOs		
C source	$C_2H_3O_2^-$	
N source	NH_4^+	
eD couple	$C_2H_3O_2^-$ / HCO_3^-	$- Y_{S/OHO}\ C_2H_3O_2^- - Y_{NH4/OHO}\ NH_4^+ - Y_{O2/OHO}\ O_2$
eA couple	O_2 / H_2O	$+ 1\ C_1H_{1.8}O_{0.5}N_{0.2} + Y_{CO2/OHO}\ HCO_3^- \pm Y_{H2O/OHO}\ H_2O \pm Y_{H+/OHO}\ H^+$
Biomass	$C_1H_{1.8}O_{0.5}N_{0.2}$	
Partner compounds	H^+	

b – Metabolism of aerobic respiring chemolithoautotrophic CMOs		
C source	HCO_3^-	
N source	NH_4^+	
eD couple	NH_4^+ / NO_3^-	$- Y_{NH4/ANO}\ NH_4^+ - Y_{O2/CMO}\ O_2 - Y_{CO2/CMO}\ HCO_3^-$
eA couple	O_2 / H_2O	$+ 1\ C_1H_{1.8}O_{0.5}N_{0.2} + Y_{NO3/CMO}\ NO_3^- \pm Y_{H2O/CMO}\ H_2O \pm Y_{H+/CMO}\ H^+$
Biomass	$C_1H_{1.8}O_{0.5}N_{0.2}$	
Partner compounds	H^+	

c – Metabolism of aerobic respiring chemolithoautotrophic AOOs

C source	HCO_3^-
N source	NH_4^+
eD couple	NH_4^+ / NO_2^-
eA couple	O_2 / H_2O
Biomass	$C_1H_{1.8}O_{0.5}N_{0.2}$
Partner compounds	H^+

$$- Y_{NH4/AOO}\,NH_4^+ - Y_{O2/AOO}\,O_2 - Y_{CO2/AOO}\,HCO_3^- + 1\,C_1H_{1.8}O_{0.5}N_{0.2} + Y_{NO2/AOO}\,NO_2^- \pm Y_{H2O/AOO}\,H_2O \pm Y_{H+/AOO}\,H^+$$

d – Metabolism of aerobic respiring chemolithoautotrophic NOOs

C source	HCO_3^-
N source	NO_2^-
eD couple	NO_2^- / NO_3^-
eA couple	O_2 / H_2O
Biomass	$C_1H_{1.8}O_{0.5}N_{0.2}$
Partner compounds	H^+

$$- Y_{NO2/NOO}\,NO_2^- - Y_{O2/NOO}\,O_2 - Y_{CO2/NOO}\,HCO_3^- + 1\,C_1H_{1.8}O_{0.5}N_{0.2} + Y_{NO3/NOO}\,NO_3^- \pm Y_{H2O/NOO}\,H_2O \pm Y_{H+/NOO}\,H^+$$

e – Metabolism of anaerobic respiring chemoorganoheterotrophic DHOs

C source	HCO_3^-
N source	NO_3^-
eD couple	$C_2H_3O_2^-$ / HCO_3^-
eA couple	NO_3^- / N_2
Biomass	$C_1H_{1.8}O_{0.5}N_{0.2}$
Partner compounds	H_2O, H^+

$$- Y_{S/DHO}\,C_2H_3O_2^- - Y_{NO3/DHO}\,NO_3^- + 1\,C_1H_{1.8}O_{0.5}N_{0.2} + Y_{N2/DHO}\,N_2 + Y_{CO2/DHO}\,HCO_3^- \pm Y_{H2O/DHO}\,H_2O \pm Y_{H+/DHO}\,H^+$$

f – Metabolism of anaerobic respiring chemolithoautotrophic AMOs

C source	HCO_3^-
N source	NH_4^+
eD couple	NH_4^+ / N_2
eA couple	NO_2^- / N_2
Biomass	$C_1H_{1.8}O_{0.5}N_{0.2}$
Partner compounds	H_2O, H^+
Extra compounds	NO_3^- [a)]

$$- Y_{NH4/AMO}\,NH_4^+ - Y_{CO2/AMO}\,HCO_3^- - Y_{NO2/AMO}\,NO_2^- + 1\,C_1H_{1.8}O_{0.5}N_{0.2} + Y_{N2/AMO}\,N_2 + Y_{NO3/AMO}\,NO_3^- \pm Y_{H2O/AMO}\,H_2O \pm Y_{H+/AMO}\,H^+$$

[a)] From wet-lab experiments with anammox enrichment cultures (Strous *et al.*, 1999; Lotti *et al.*, 2015), nitrite (NO_2^-) is used as an extra e-donor to reduce bicarbonate into biomass in the anabolism of AMOs, besides its main role in catabolism. The extra oxidation of nitrite in the anabolism produces nitrate (described in the textbook).

Anabolic reactions

Anabolisms (Table 2.4) include materials involved in biomass synthesis from nutrients. However, anabolic reactions remain redox reactions, where electrons are transferred from the substrates into the biomass. The e-balance should be closed.

A close look at the C source and its degree of reduction (γ) compared to biomass is needed. Autotrophs need an extra e-donor to reduce the fully oxidised carbon dioxide into biomass. For heterotrophs, acetate (4.0 mol e- C-mol^{-1}) is in a reduction state close to biomass (4.2 mol e- C-mol^{-1}); here no extra e-acceptor is needed in the anabolisms of OHOs and DHOs. If their C source (*e.g.*, organic matter $C_1H_{1.63}O_{0.46}$, 4.71 mol e- C-mol^{-1}) is more reduced than biomass, the electron excess is dissipated on an extra e-acceptor involved in their anabolism.

The anabolic reaction formulated with stoichiometric coefficients ($v_{i,j}^{ana}$) is:

$$v_{C,j}^{ana}\ \text{carbon source} + v_{N,j}^{ana}\ \text{nitrogen source} + (v_{eD,j}^{ana}\ \text{extra e-donor}) + (v_{eA,j}^{ana}\ \text{extra e-acceptor})$$
$$+\ v_{X,j}^{ana}\ \text{biomass} + (v_{ox\text{-}eD,j}^{ana}\ \text{oxidised extra e-donor}) + (v_{red\text{-}eA,j}^{ana}\ \text{reduced extra e-acceptor})$$
$$+\ v_{i,j}^{ana}\ \text{partner compounds } (e.g., HCO_3^-, H_2O, H^+)$$

and with yield values ($\pm Y_{i/X,j}^{ana}$) normalised per unit of biomass formed becomes:

$$-Y_{C/X,j}^{ana}\ \text{carbon source} - Y_{N/X,j}^{ana}\ \text{nitrogen source} - (Y_{eD/X,j}^{ana}\ \text{extra e-donor})$$
$$-(Y_{eA/X,j}^{ana}\ \text{extra e-acceptor}) + 1\ \text{biomass} + (Y_{ox\text{-}eD/X,j}^{ana}\ \text{oxidised extra e-donor})$$
$$+(Y_{red\text{-}eA/X,j}^{ana}\ \text{reduced extra e-acceptor}) \pm Y_{i/X,j}^{ana}\ \text{partner compounds } (e.g., HCO_3^-, H_2O, H^+)$$

Table 2.4 Anabolic reactions in literal forms. Acetate ($C_2H_3O_2^-$) is used as a model organic compound. The information given in red highlights the differences in degrees of reduction of the carbon source and of the biomass. It indicates the need to include (or not) an extra e-donor couple or e-acceptor couple in the anabolism besides the C and N sources and the partner compounds, depending on the trophic groups.

a – Heterotrophic anabolism of OHOs

C source	$C_2H_3O_2^-$ (8 mol e- mol^{-1} = 4.0 mol e- C-mol^{-1}) a)	$-\ Y_{S/OHO}^{ana}\ C_2H_3O_2^- - Y_{NH4/OHO}^{ana}\ NH_4^+$
N source	NH_4^+	$+\ 1\ C_1H_{1.8}O_{0.5}N_{0.2} + Y_{CO2/OHO}^{ana}\ HCO_3^- \pm Y_{H2O/OHO}^{ana}\ H_2O$
Biomass	$C_1H_{1.8}O_{0.5}N_{0.2}$ (4.2 mol e- C-mol^{-1})	$\pm\ Y_{H+/OHO}^{ana}\ H^+$
Partner compounds	HCO_3^-, H_2O, H^+	
Extra eD couple	-	
Extra eA couple	-	

a) The C source used here by OHOs is acetate. It contains 4.0 mol e- C-mol^{-1} which is lower than 4.2 mol e- C-mol^{-1} in biomass. No extra e-acceptor is needed in the anabolic reaction.

b – Autotrophic anabolism of CMOs

C source	HCO_3^- (0 mol e- mol^{-1}) b)	$-\ Y_{CO2/CMO}^{ana}\ HCO_3^- - Y_{NH4/CMO}^{ana}\ NH_4^+$
N source	NH_4^+ c)	$+\ 1\ C_1H_{1.8}O_{0.5}N_{0.2} + Y_{NO3/CMO}^{ana}\ NO_3^- \pm Y_{H2O/CMO}\ H_2O$
Biomass	$C_1H_{1.8}O_{0.5}N_{0.2}$ (4.2 mol e- mol^{-1})	$\pm\ Y_{H+/CMO}\ H^+$
Partner compounds	H_2O, H^+	
Extra eD couple	$NH_4^+\ /\ NO_3^-$ c)	
Extra eA couple	-	

a) The C source used by nitrifiers is bicarbonate (HCO_3^-) ($\gamma = 0$ mol e- mol^{-1}) and is fully oxidised (no electrons to donate).
b) Ammonium is the nitrogen source for CMOs.
c) An additional e-donor is needed to fix and condense bicarbonate into biomass (4.2 mol e- mol^{-1}). Ammonium (NH_4^+) is assumed to be the anabolic e-donor in addition to the N source. Although the degree of reduction of ammonium is 0 mol e- mol^{-1} (ammonium is used as the reference in the degree of reduction calculations), ammonium can still donate electrons by being oxidised (contrary to bicarbonate which cannot be oxidised further). NO_3^- is produced from this extra oxidation of ammonium.

c – Autotrophic anabolism of AOOs		
C source	HCO_3^-	
	(0 mol e- mol^{-1})	
N source	NH_4^+ [b]	$- Y_{CO2/AOO}^{ana} HCO_3^- - Y_{NH4/AOO}^{ana} NH_4^+$
Biomass	$C_1H_{1.8}O_{0.5}N_{0.2}$	$+ 1\ C_1H_{1.8}O_{0.5}N_{0.2} + Y_{NO3/AOO}^{ana} NO_2^- \pm Y_{H2O/AOO} H_2O$
	(4.2 mol e- mol^{-1})	$\pm Y_{H+/AOO} H^+$
Partner compounds	H_2O, H^+	
Extra eD couple	NH_4^+ / NO_2^- [a]	
Extra eA couple	-	

[a] Nitrite (NO_2^-) is assumed to be the product of the extra oxidation of NH_4^+ in the anabolism of AOOs.
[b] Ammonium is the nitrogen source for AOOs.

d – Autotrophic anabolism of NOOs		
C source	HCO_3^-	
	(0 mol e- mol^{-1})	
N source	NO_2^- [b]	$- Y_{CO2/NOO}^{ana} HCO_3^- - Y_{NO2/NOO}^{ana} NO_2^-$
Biomass	$C_1H_{1.8}O_{0.5}N_{0.2}$	$+ 1\ C_1H_{1.8}O_{0.5}N_{0.2} + Y_{NO3/NOO}^{ana} NO_3^- \pm Y_{H2O/NOO}^{ana} H_2O$
	(4.2 mol e- mol^{-1})	$\pm Y_{H+/NOO}^{ana} H^+$
Partner compounds	H_2O, H^+	
Extra eD couple	NO_2^- / NO_3^- [a]	
Extra eA couple	-	

[a] Nitrite is the extra anabolic e-donor and is oxidised into nitrate (NO_3^-) to supply electrons to assimilate bicarbonate in the anabolism of NOOs.
[b] Nitrite is considered here to be the N source for NOOs, assuming an initial full oxidation of the ammonium by the AOOs.

e – Heterotrophic anabolism of DHOs		
C source	$C_2H_3O_2^-$	
	(8 mol e- mol^{-1} =	
	4.0 mol e- C-mol^{-1}) [a]	$- Y_{S/DHO}^{ana} C_2H_3O_2^- - Y_{NO3/DHO}^{ana} NO_3^-$
N source	NO_3^- [b]	$+ 1\ C_1H_{1.8}O_{0.5}N_{0.2} + Y_{CO2/DHO}^{ana} HCO_3^- \pm Y_{H2O/DHO}^{ana} H_2O$
Biomass	$C_1H_{1.8}O_{0.5}N_{0.2}$	$\pm Y_{H+/DHO}^{ana} H^+$
	(4.2 mol e- mol^{-1})	
Partner compounds	HCO_3^-, H_2O, H^+	
Extra eD couple	-	
Extra eA couple	-	

[a] Same reasoning as for OHOs.
[b] Nitrate is considered as the nitrogen source, assuming full oxidations of ammonium by AOOs and nitrite by NOOs.

f – Autotrophic anabolism of AMOs		
C source	HCO_3^-	
	(0 mol e- mol^{-1})	
N source	NH_4^+ [a]	$- Y_{CO2/AMO}^{ana} HCO_3^- - Y_{NH4/AMO}^{ana} NH_4^+ - Y_{NO2/AMO}^{ana} NO_2^-$
Biomass	$C_1H_{1.8}O_{0.5}N_{0.2}$	$+ 1\ C_1H_{1.8}O_{0.5}N_{0.2} + Y_{NO3/AMO}^{ana} NO_3^- \pm Y_{H2O/AMO}^{ana} H_2O$
	(4.2 mol e- mol^{-1})	$\pm Y_{H+/AMO}^{ana} H^+$
Partner compounds	H_2O, H^+	
Extra eD couple	NO_2^- / NO_3^- [b]	
Extra eA couple	-	

[a] AMOs are fed with ammonium and nitrite. Ammonium is the preferential N source since it is already in a reduction state of proteins.
[b] Similar reasoning as for nitrifiers: in anammox enrichment cultures, nitrate is formed, resulting from the extra oxidation of nitrite as the anabolic e-donor to assimilate bicarbonate into the AMO biomass.

Independent catabolic reactions

Independent catabolic reactions (Table 2.5) involve materials of the e-donor couple and e-acceptor couple of the redox reaction, and partner compounds for elemental/charge balances.

The catabolic reaction formulated with stoichiometric coefficients ($v_{i,j}^{cat}$) is:

$$v_{eD,j}^{cat} \text{ e-donor} + v_{eA,j}^{cat} \text{ e-acceptor}$$
$$+ v_{ox\text{-}eD,j}^{cat} \text{ oxidised e-donor} + v_{red\text{-}eA,j}^{cat} \text{ reduced e-acceptor}$$
$$+ v_{i,j}^{cat} \text{ partner compounds } (e.g., HCO_3^-, H_2O, H^+)$$

and with yield values ($\pm Y_{i/eD,j}^{cat}$) normalised per unit of e-donor consumed becomes:

$$- 1 \text{ e-donor} - Y_{eA/eD,j}^{cat} \text{ e-acceptor}$$
$$+ Y_{ox\text{-}eD/eD,j}^{cat} \text{ oxidised e-donor} + Y_{red\text{-}eA/eD,j}^{cat} \text{ reduced e-acceptor}$$
$$\pm Y_{i/eD,j}^{cat} \text{ partner compounds } (e.g., HCO_3^-, H_2O, H^+)$$

Table 2.5 Independent catabolic reactions in literal forms.

a – Catabolism of aerobic respiring chemoorganotrophic OHOs		
eD couple	$C_2H_3O_2^- / HCO_3^-$	$- 1\ C_2H_3O_2^- - Y_{O2/OHO}^{cat}\ O_2$
eA couple	O_2 / H_2O	$+ Y_{CO2/OHO}^{cat}\ HCO_3^- + Y_{H2O/OHO}^{cat}\ H_2O \pm Y_{H+/OHO}^{cat}\ H^+$
Partner compounds	H^+	

b – Catabolism of aerobic respiring chemolithotrophic CMOs		
eD couple	NH_4^+ / NO_3^-	$- 1\ NH_4^+ - Y_{O2/CMO}^{cat}\ O_2$
eA couple	O_2 / H_2O	$+ Y_{NO3/CMO}^{cat}\ NO_3^- + Y_{H2O/CMO}^{cat}\ H_2O \pm Y_{H+/CMO}^{cat}\ H^+$
Partner compounds	H^+	

c – Catabolism of aerobic respiring chemolithotrophic AOOs		
eD couple	NH_4^+ / NO_2^-	$- 1\ NH_4^+ - Y_{O2/AOO}^{cat}\ O_2$
eA couple	O_2 / H_2O	$+ Y_{NO2/AOO}^{cat}\ NO_2^- + Y_{H2O/AOO}^{cat}\ H_2O \pm Y_{H+/AOO}^{cat}\ H^+$
Partner compounds	H^+	

d – Catabolism of aerobic respiring chemolithotrophic NOOs		
eD couple	NO_2^- / NO_3^-	$- 1\ NO_2^- - Y_{O2/NOO}^{cat}\ O_2$
eA couple	O_2 / H_2O	$+ Y_{NO3/NOO}^{cat}\ NO_3^- + Y_{H2O/NOO}^{cat}\ H_2O \pm Y_{H+/NOO}^{cat}\ H^+$
Partner compounds	H^+	

e – Catabolism of anaerobic respiring chemoorganotrophic DHOs		
eD couple	$C_2H_3O_2^- / HCO_3^-$	$- 1\ C_2H_3O_2^- - Y_{NO3/DHO}^{cat}\ NO_3^-$
eA couple	NO_3^- / N_2	$+ Y_{CO2/DHO}^{cat}\ HCO_3^- + Y_{N2/DHO}^{cat}\ N_2 \pm Y_{H2O/DHO}\ H_2O \pm Y_{H+/DHO}\ H^+$
Partner compounds	H_2O, H^+	

f – Catabolism of anaerobic respiring chemolithotrophic AMOs		
eD couple	NH_4^+ / N_2	$- 1\ NH_4^+ - Y_{NO2/AMO}^{cat}\ NO_2^-$
eA couple	NO_2^- / N_2	$+ Y_{N2/AMO}^{cat}\ N_2 \pm Y_{H2O/AMO}^{cat}\ H_2O \pm Y_{H+/AMO}^{cat}\ H^+$
Partner compounds	H_2O, H^+	

Overview of metabolic, anabolic, and catabolic reactions

Table 2.6 compiles all the literal metabolic stoichiometries for each guild in a compact matrix form, analogous to the Petersen or Gujer matrix used in activated sludge models (Henze *et al.,* 2000). It enables the growth and interaction processes of microorganisms to be quickly read. Here the matrix lists the materials involved in reactions; their elemental, charge, electron, Gibbs free energies of formation, and mass conservatives necessary for balances; and the microbial processes that convert them, as described by their stoichiometry.

Table 2.6 Stoichiometric matrix compiling the metabolisms of OHOs, nitrifiers, DHOs and AMOs in literal form and their anabolic and independent catabolic reactions. The nature of the compounds involved is addressed. The stoichiometric coefficients ($v_{i,j}$) are provided with their yield values ($\pm Y_{i,j}$): $v_{i,j} = -Y_{i,j} <0$ for substrates that are consumed, $v_{i,j} = Y_{i,j} >0$ for products that are produced (see page 23 and page 24).

Phase [a]	Solutes				
Materials i	Acetate	Ammonium	Dioxygen	Bicarbonate	Nitrite
Elemental formula	$C_2H_3O_2^-$	NH_4^{+} [b]	O_2	HCO_3^- [b]	NO_2^-
Units [c]	mol	mol	mol	mol	mol
Conservatives k	Conservation coefficients ($\iota_{k,i}$)				
C (mol C mol^{-1} i)	2	0	0	1	0
H (mol H mol^{-1} i)	3	4	0	1	0
O (mol O mol^{-1} i)	2	0	2	3	2
N (mol N mol^{-1} i)	0	1	0	0	1
Charge + (mol + mol^{-1} i)	-1	+1	0	-1	-1
γ (mol e- mol^{-1} i)	8	0	-4	0	-6
G^{f0} (kJ mol^{-1} i)	-369.4	-79.4	0	-586.9	-32.2
Molecular mass (g i mol^{-1} i)	59.04	18	32	61	46

Phase [a]	Solutes				Solids
Materials i	Nitrate	Dinitrogen	Water	Protons	Biomass
Elemental formula	NO_3^-	N_2	H_2O	H^+	$C_1H_{1.8}O_{0.5}N_{0.2}$
Units [c]	mol	mol	mol	mol	(C-)mol
Conservatives k	Conservation coefficients ($\iota_{k,i}$)				
C (mol C mol^{-1} i)	0	0	0	0	1
H (mol H mol^{-1} i)	0	0	2	1	1.8
O (mol O mol^{-1} i)	3	0	1	0	0.5
N (mol N mol^{-1} i)	1	2	0	0	0.2
Charge + (mol + mol^{-1} i)	-1	0	0	+1	0
γ (mol e- mol^{-1} i)	-8	-6	0	0	4.2
G^{f0} (kJ mol^{-1} i)	-111.3	0	-237.2	0	-67.0
Molecular mass (g i mol^{-1} i)	62	28	18	1	24.6

[a] Dissolved materials are metabolised by microorganisms. Biomass is a solid phase suspended in the mixed liquor.

[b] Calculations are primarily made under biochemical reference conditions, *i.e.,* at 25 °C, 1 atm, and pH 7.0. Ammonium (NH_4^+) and bicarbonate (HCO_3^-) are primarily involved rather than ammonia (NH_3) and carbon dioxide (CO_2).

[c] All units are provided here in moles. Stoichiometric/kinetic/thermodynamic formulations are processed on a mole basis. In environmental engineering, mass balances are preferred. Stoichiometric coefficients can be converted using molecular masses.

[d] Yield values are normalised per unit of biomass in metabolism and anabolism, and per unit of e-donor in catabolism.

Table 2.6 Continued from page 22.

Guilds j	Stoichiometric coefficients ($v_{i,j} = \pm Y_{i/n,j}$ in mol i mol^{-1} n) [d]							

a – Aerobic respiring chemoorganoheterotrophic OHOs

Materials involved	Organics	Ammonium	Dioxygen	Bicarbonate		Water	Protons [e]	Biomass
Metabolism	$-Y_{S/OHO}$ <0 e-donor C source	$-Y_{NH4/OHO}$ <0 N source [f]	$-Y_{O2/OHO}$ <0 e-acceptor	$Y_{CO2/OHO}$ >0 Oxidised e-donor		$\pm Y_{H2O/OHO}$ $<>0$ Reduced e-acceptor and partner	$\pm Y_{H-/OHO}$ $<>0$ Partner	$Y_{OHO/OHO}$ $+1$ [d] Biomass
Anabolism	$-Y_{S/OHO}^{ana}$ <0 C source	$-Y_{NH4/OHO}^{ana}$ <0 N source		$Y_{CO2/OHO}^{ana}$ >0 Partner		$\pm Y_{H2O/OHO}^{ana}$ $<>0$ Partner	$\pm Y_{H-/OHO}^{ana}$ $<>0$ Partner	$Y_{OHO/OHO}^{ana}$ $+1$ [d] Biomass
Independent catabolism	$-Y_{S/OHO}^{cat}$ -1 [d] e-donor		$-Y_{O2/OHO}^{cat}$ <0 e-acceptor	$Y_{CO2/OHO}^{cat}$ >0 Oxidised e-donor		$Y_{H2O/OHO}^{cat}$ >0 Reduced e-acceptor	$\pm Y_{H-/OHO}^{cat}$ $<>0$ Partner	

b – Aerobic respiring chemolithoautotrophic CMOs

Materials involved	Ammonium	Dioxygen	Bicarbonate	Nitrate	Water	Protons	Biomass
Metabolism	$-Y_{NH4/CMO}$ <0 e-donor N source	$-Y_{O2/CMO}$ <0 e-acceptor	$-Y_{CO2/CMO}$ <0 C source	$Y_{NO3/CMO}$ >0 Oxidised e-donor	$\pm Y_{H2O/CMO}$ $<>0$ Reduced e-acceptor and partner	$\pm Y_{H-/CMO}$ $<>0$ Partner	$Y_{CMO/CMO}$ $+1$ Biomass
Anabolism	$-Y_{NH4/CMO}^{ana}$ <0 N source and additional anabolic e-donor		$-Y_{CO2/CMO}^{ana}$ <0 C source	$Y_{NO3/CMO}^{ana}$ >0 Additional anabolic oxidised e-donor	$\pm Y_{H2O/CMO}^{ana}$ $<>0$ Partner	$\pm Y_{H-/CMO}^{ana}$ $<>0$ Partner	$Y_{CMO/CMO}^{ana}$ $+1$ Biomass
Independent catabolism	$-Y_{NH4/CMO}^{cat}$ -1 e-donor	$-Y_{O2/CMO}^{cat}$ <0 e-acceptor		$Y_{NO3/CMO}^{cat}$ >0 Oxidised e-donor	$Y_{H2O/CMO}^{cat}$ >0 Reduced e-acceptor	$\pm Y_{H-/CMO}^{cat}$ $<>0$ Partner	

c – Aerobic respiring chemolithoautotrophic AOOs

Materials involved	Ammonium	Dioxygen	Bicarbonate	Nitrite	Water	Protons	Biomass
Metabolism	$-Y_{NH4/AOO}$ <0 e-donor N source	$-Y_{O2/AOO}$ <0 e-acceptor	$-Y_{CO2/AOO}$ <0 C source	$Y_{NO2/AOO}$ >0 Oxidised e-donor	$\pm Y_{H2O/AOO}$ $<>0$ Reduced e-acceptor and partner	$\pm Y_{H-/AOO}$ $<>0$ Partner	$Y_{AAO/AOO}$ $+1$ Biomass
Anabolism	$-Y_{NH4/AOO}^{ana}$ <0 N source and additional anabolic e-donor		$-Y_{CO2/AOO}^{ana}$ <0 C source	$Y_{NO2/AOO}^{ana}$ >0 Additional anabolic oxidised e-donor	$Y_{H2O/AOO}^{ana}$ >0 Partner	$\pm Y_{H-/AOO}^{ana}$ $<>0$ Partner	$Y_{AOO/AOO}^{ana}$ $+1$ Biomass
Independent catabolism	$-Y_{NH4/AOO}^{cat}$ -1 e-donor	$-Y_{O2/AOO}^{cat}$ <0 e-acceptor		$Y_{NO2/AOO}^{cat}$ >0 Oxidised e-donor	$Y_{H2O/AOO}^{cat}$ >0 Reduced e-acceptor	$\pm Y_{H-/AOO}^{cat}$ $<>0$ Partner	

d – Aerobic respiring chemolithoautotrophic NOOs

Materials involved	Dioxygen	Bicarbonate	Nitrite	Nitrate	Water	Protons	Biomass
Metabolism	$-Y_{O2/NOO}$ <0 e-acceptor	$-Y_{CO2/NOO}$ <0 C source	$-Y_{NO2/NOO}$ <0 e-donor N source [f]	$Y_{NO3/NOO}$ >0 Oxidised e-donor	$\pm Y_{H2O/NOO}$ >0 Reduced e-acceptor and partner	$\pm Y_{H+/NOO}$ $<>0$ Partner	$Y_{NOO/NOO}$ $+1$ Biomass
Anabolism		$-Y_{CO2/NOO}^{ana}$ <0 C source	$-Y_{NO2/NOO}^{ana}$ <0 N source and additional anabolic e-donor	$Y_{NO3/NOO}^{ana}$ >0 Additional anabolic oxidised e-donor	$\pm Y_{H2O/NOO}^{ana}$ $<>0$ Partner	$\pm Y_{H+/NOO}^{ana}$ $<>0$ Partner	$Y_{NOO/NOO}^{ana}$ $+1$ Biomass
Independent catabolism	$-Y_{O2/NOO}^{cat}$ <0 e-acceptor		$-Y_{NO2/NOO}^{cat}$ -1 e-donor	$Y_{NO3/NOO}^{cat}$ >0 Oxidised e-donor	$Y_{H2O/NOO}^{cat}$ >0 Reduced e-acceptor	$\pm Y_{H+/NOO}^{cat}$ $<>0$ Partner	

e – Anaerobic respiring chemoorganoheterotrophic DHOs

Materials involved	Organics	Bicarbonate	Nitrate	Dinitrogen	Water	Protons	Biomass
Metabolism	$-Y_{S/DHO}$ <0 e-donor C source	$Y_{CO2/DHO}$ >0 Oxidised e-donor	$-Y_{NO3/DHO}$ <0 e-acceptor N source [f]	$Y_{N2/DHO}$ >0 Reduced e-acceptor	$\pm Y_{H2O/DHO}$ $<>0$ Partner	$\pm Y_{H+/DHO}$ $<>0$ Partner	$Y_{DHO/DHO}$ $+1$ Biomass
Anabolism	$-Y_{S/DHO}^{ana}$ <0 C source	$Y_{CO2/DHO}^{ana}$ >0 Partner	$-Y_{NO3/DHO}^{ana}$ <0 N source		$\pm Y_{H2O/DHO}^{ana}$ $<>0$ Partner	$\pm Y_{H+/DHO}^{ana}$ $<>0$ Partner	$Y_{DHO/DHO}^{ana}$ $+1$ Biomass
Independent catabolism	$-Y_{S/DHO}^{cat}$ -1 e-donor	$Y_{CO2/DHO}^{cat}$ >0 Oxidised e-donor	$-Y_{NO3/DHO}^{cat}$ <0 e-acceptor	$Y_{N2/DHO}^{cat}$ >0 Reduced e-acceptor	$\pm Y_{H2O/DHO}^{cat}$ $<>0$ Partner	$\pm Y_{H+/DHO}^{cat}$ $<>0$ Partner	

f – Anaerobic respiring chemolithoautotrophic AMOs

Materials involved	Ammonium	Dioxygen	Bicarbonate	Nitrite	Nitrate	Dinitrogen	Water	Protons	Biomass
Metabolism	$-Y_{NH4/AMO}$ <0 e-donor N source		$-Y_{CO2/AMO}$ <0 C source	$-Y_{NO2/AMO}$ <0 e-acceptor and additional anabolic e-donor	$Y_{NO3/AMO}$ >0 Additional anabolic oxidised e-donor	$Y_{N2/AMO}$ >0 Oxidised e-donor Reduced e- acceptor	$\pm Y_{H2O/AMO}$ $<>0$ Partner	$\pm Y_{H+/AMO}$ $<>0$ Partner	$Y_{AMO/AMO}$ $+1$ Biomass
Anabolism	$-Y_{NH4/AMO}^{ana}$ <0 N source		$-Y_{CO2/AMO}^{ana}$ <0 C source	$-Y_{NO2/AMO}^{ana}$ <0 Additional anabolic e-donor	$Y_{NO3/AMO}^{ana}$ >0 Additional anabolic oxidised e-donor		$\pm Y_{H2O/AMO}^{ana}$ $<>0$ Partner	$\pm Y_{H+/AMO}^{ana}$ $<>0$ Partner	$Y_{AMO/AMO}^{ana}$ $+1$ Biomass
Independent catabolism	$-Y_{NH4/AMO}^{cat}$ -1 e-donor			$-Y_{NO2/AMO}^{cat}$ <0 e-acceptor		$Y_{N2/AMO}^{cat}$ >0 Oxidised e-donor Reduced e- acceptor	$\pm Y_{H2O/AMO}^{cat}$ $<>0$ Partner	$\pm Y_{H+/AMO}^{cat}$ $<>0$ Partner	

[d] Yield values are normalised per unit of biomass in metabolism and anabolism, and per unit of e-donor in catabolism.

[e] Protons (H$^+$) are 'partner compounds' either produced or consumed in most metabolisms. They are needed for charge balance. Knowledge from literature can help predict whether protons will be produced (*e.g.*, nitrification) or consumed (*e.g.*, denitrification). When acetate (as a dissociate form) is consumed, a proton is also consumed.

[f] Nitrogen sources depend on conditions and process configurations. Ammonium is the preferred N source for microorganisms since it is bound in this form in the protein fraction of the biomass. In municipal wastewater nitrogen is mainly present as ammonium from the ammonification of urea released with urine. Anaerobic digesters are rich in ammonium or ammonia (depending on the pH) formed from the digestion of proteins. The N source for denitrifiers depends on pre-/simultaneous/post-denitrification configurations.

b) Calculation of anabolic and independent catabolic reactions

Elemental, charge and/or degree of reduction balances to solve stoichiometries

Stoichiometric coefficients of anabolisms and independent catabolisms (tables 2.4 and 2.5) are obtained by elemental balances (*e.g.*, C, H, O, N), charge balance, and/or degree of reduction (γ) balances.

The γ balance is not an extra independent equation since γ is calculated from elements and charges. When using it, one other conservation equation (an element or charge) is removed from the system of equations. It often simplifies calculations since several compounds (NH_4^+, CO_2, HCO_3^-, H_2O, H^+, OH^-, SO_4^{2-}, PO_4^{3-}, *etc.*) are references with a $\gamma = 0$ mol e- mol^{-1}. This helps verify electron flows and balances.

For all conservatives involved in a reaction: $\Sigma\, (\iota_{k,i} \cdot \nu_{i/j}) = 0$ $\hfill$ (2.1)

with $\iota_{k,i}$ the conservation coefficients of element k in compound i, and $\nu_{i,j}$ the stoichiometric coefficient of compound i in the microbial process j.

Stoichiometric coefficients are obtained by solving the system of conservation equations, either by hand or using a solver that multiplies the matrix of conservatives by the vector of stoichiometric coefficients (*e.g.*, an online solver, Excel, MathCAD, Matlab, or Python). The mass or mol balance over the reaction is useful to check the stoichiometric calculations. The resolution is given for the anabolism and independent catabolism of OHOs (Table 2.7).

Table 2.7 Elemental, charge and/or degree of reduction balances to solve the anabolism and independent catabolism of OHOs.

OHO anabolism	5 unknown stoichiometric coefficients to solve with 5 conservation equations
Anabolic reaction in literal form:	
$\nu_{S/OHO}^{ana}\,C_2H_3O_2^- + \nu_{NH4/OHO}^{ana}\,NH_4^+ + 1\,C_1H_{1.8}O_{0.5}N_{0.2} + \nu_{CO2/OHO}^{ana}\,HCO_3^- + \nu_{H2O/OHO}^{ana}\,H_2O + \nu_{H+/OHO}^{ana}\,H^+$	
$-\,Y_{S/OHO}^{ana}\,C_2H_3O_2^- -\, Y_{NH4/OHO}^{ana}\,NH_4^+ + 1\,C_1H_{1.8}O_{0.5}N_{0.2} + Y_{CO2/OHO}^{ana}\,HCO_3^- \pm Y_{H2O/OHO}^{ana}\,H_2O \pm Y_{H+/OHO}^{ana}\,H^+$	
Conservatives k	
C	$2\cdot\nu_{S/OHO}^{ana} + 1\cdot\nu_{CO2/OHO}^{ana} + 1\cdot(+1) = 0$
H	$3\cdot\nu_{S/OHO}^{ana} + 4\cdot\nu_{NH4/OHO}^{ana} + 1\cdot\nu_{CO2/OHO}^{ana} + 2\cdot\nu_{H2O/OHO}^{ana} + 1\cdot\nu_{H+/OHO}^{ana} + 1.8\cdot(+1) = 0$
O	$2\cdot\nu_{S/OHO}^{ana} + 3\cdot\nu_{CO2/OHO}^{ana} + 1\cdot\nu_{H2O/OHO}^{ana} + 0.5\cdot(+1) = 0$
N	$1\cdot\nu_{NH4/OHO}^{ana} + 0.2\cdot(+1) = 0$
Charge +	$(-1)\cdot\nu_{S/OHO}^{ana} + 1\cdot\nu_{NH4/OHO}^{ana} + (-1)\cdot\nu_{CO2/OHO}^{ana} + 1\cdot\nu_{H+/OHO}^{ana} = 0$
Degree of reduction γ [a)]	$8.0\cdot\nu_{S/OHO}^{ana} + 4.2\cdot(+1) = 0$
Calculated stoichiometric coefficients (mol i mol^{-1} X):	
$\nu_{X,OHO}^{ana} = Y_{X/X\_OHO}^{ana} = +1$ (fixed)	
$\nu_{S,OHO}^{ana} = -Y_{S/X\_OHO}^{ana} = -0.525$	
$\nu_{NH4,OHO}^{ana} = -Y_{NH4/X\_OHO}^{ana} = -0.200$	
$\nu_{CO2,OHO}^{ana} = Y_{CO2/X\_OHO}^{ana} = +0.050$	
$\nu_{H2O,OHO}^{ana} = Y_{H2O/X\_OHO}^{ana} = +0.400$	
$\nu_{H+,OHO}^{ana} = Y_{H+/X\_OHO}^{ana} = -0.275$	
Anabolic reaction of OHOs:	
$-0.525\,C_2H_3O_2^- - 0.200\,Y_{NH4/OHO}^{ana}\,NH_4^+ - 0.275\,H^+ + 1\,C_1H_{1.8}O_{0.5}N_{0.2} + 0.050\,HCO_3^- + 0.400\,H_2O$	
Conservation checks on anabolic reaction:	
Charge +: $\quad -0.525\cdot(-1) - 0.200\cdot(+1) - 0.275\cdot(+1) + 0.050\cdot(-1) = 0$ $\hfill$ OK!	
Electrons (γ): $\quad -0.525\cdot 8 - 0.200\cdot 0 - 0.275\cdot 0 + 1\cdot 4.2 + 0.050\cdot 0 + 0.400\cdot 0 = 0$ $\hfill$ OK!	
Mass: $\quad -0.525\cdot 59.04 - 0.200\cdot 18.04 - 0.275\cdot 1.01 + 1\cdot 24.63 + 0.050\cdot 61.02 + 0.400\cdot 18.02 \approx 0$ OK!	

OHO catabolism	4 unknown stoichiometric coefficients to solve with 4 conservation equations

Independent catabolic reaction in literal form:

$- 1\ C_2H_3O_2^- + v_{O2/OHO}^{cat}\ O_2 + v_{CO2/OHO}^{cat}\ HCO_3^- + v_{H2O/OHO}^{cat}\ H_2O + v_{H+/OHO}^{cat}\ H^+$

$- 1\ C_2H_3O_2^- - Y_{O2/OHO}^{cat}\ O_2 + Y_{CO2/OHO}^{cat}\ HCO_3^- + Y_{H2O/OHO}^{cat}\ H_2O \pm Y_{H+/OHO}^{cat}\ H^+$

Conservatives k	
C	$2\cdot(-1) + 1\cdot v_{CO2/OHO}^{cat} = 0$
H	$3\cdot(-1) + 1\cdot v_{CO2/OHO}^{cat} + 2\cdot v_{H2O/OHO}^{cat} + 1\cdot v_{H+/OHO}^{cat} = 0$
O	$2\cdot(-1) + 2\cdot v_{O2/OHO}^{cat} + 3\cdot v_{CO2/OHO}^{cat} + 1\cdot v_{H2O/OHO}^{cat} = 0$
Charge +	$(-1)\cdot(-1) + (-1)\cdot v_{CO2/OHO}^{cat} + 1\cdot v_{H+/OHO}^{cat} = 0$
Degree of reduction γ [a]	$4.0\cdot(-1) + (-4)\cdot v_{O2/OHO}^{cat} = 0$

Calculated stoichiometric coefficients (mol i mol^{-1} S):

$v_{S,OHO}^{cat}\quad = -Y_{S/S}^{cat}\quad = -1 \text{ (fixed)}$

$v_{O2,OHO}^{cat}\quad = Y_{O2/S}^{cat}\quad = -2$

$v_{CO2,OHO}^{cat}\quad = Y_{CO2/S}^{cat}\quad = +2$

$v_{H2O,OHO}^{cat}\quad = Y_{H2O/S}^{cat}\quad = 0$

$v_{H+,OHO}^{cat}\quad = Y_{H+/S}^{cat}\quad = +1$

Independent catabolic reaction of OHOs:

$- 1\ C_2H_3O_2^- - 2\ O_2 + 2\ HCO_3^- + 1\ H^+$

Conservation checks on catabolic reaction:

Charge +:	$- 1 \cdot (-1) + 2 \cdot (-1) + 1 \cdot (+1) = 0$	OK!
Electrons (γ):	$-1 \cdot 8 - 2 \cdot (-4) + 2 \cdot 0 + 1 \cdot 0 = 0$	OK!
Mass:	$-1 \cdot 59.04 - 2 \cdot 32.00 + 2 \cdot 61.02 + 1 \cdot 1.01 \approx 0$	OK!

[a] The γ balance is not independent from the elemental and charge balances. It does not provide a 6th independent equation but can be used instead of one of the other conservatives. It is not only useful to verify electron balances, but also often helpful to obtain yields of biomass production, oxygen consumption, and carbon dioxide production on substrate.

Anabolic and independent catabolic stoichiometries for all guilds are compiled in Table 2.8.

Table 2.8 Anabolisms and independent catabolisms of the BNR guilds. All units are mols (for biomass (C-)mol).

Materials Elemental formula	Acetate $C_2H_3O_2^-$	Ammonium NH_4^+	Dioxygen O_2	Bicarbonate HCO_3^-	Nitrite NO_2^-	Nitrate NO_3^-	Dinitrogen N_2	Water H_2O	Protons H^+	Biomass $C_1H_{1.8}O_{0.5}N_{0.2}$
OHOs Anabolism	−0.525	−0.200		+0.050				+0.400	−0.275	+1
OHOs Independent catabolism	−1		−2	+2					+1	
CMOs Anabolism		−0.725		−1		+0.525		+0.925	+0.250	+1
CMOs Independent catabolism		−1	−2			+1		+1	+2	
AOOs Anabolism		−0.900		−1	+0.700			+1.100	+0.600	+1
AOOs Independent catabolism		−1	−1.5		+1			+1	+2	
NOOs Anabolism				−1	−2.9	+2.7		+0.2	−1.2	+1
NOOs Independent catabolism			−0.5		−1	+1				
DHOs Anabolism	−0.725			+0.450		−0.200		+0.200	−0.475	+1
DHOs Independent catabolism	−1			+2		−1.6	+0.8	+0.8	−0.6	
AMOs Anabolism		−0.2		−1	−2.1	+2.1		+0.4	−0.8	+1
AMOs Independent catabolism		−1			−1		+1	+2		

c) Calculation of overall growth reactions using a measured growth yield

For calculating the overall growth reactions (Table 2.3) using elemental, charge and/or degree of reduction balances, 6 (for OHOs, nitrifiers, DHOs) and 7 (for AMOs) unknown stoichiometric coefficients must be computed while only 5 independent conservatives can be used. Additional information is needed from observed yield(s) measured or retrieved from literature (Table 2.9). When retrieving measured yields reported in literature, bear in mind the experimental conditions and do not use values obtained under different conditions.

Table 2.9 Observed yields retrieved from literature.

Guild	Missing information	Observed growth yield (mol X mol^{-1} i)		Reference value	Literature report
OHOs	1 yield	$Y_{X/S,OHO}^{obs}$	1.276	0.67 g COD_X g^{-1} COD_S	Muller *et al.*, 2004
DHOs	1 yield	$Y_{X/S,DHO}^{obs}$	1.029	0.54 g COD_X g^{-1} COD_S	
CMOs	1 yield	$Y_{X/NO3,CMO}^{obs}$	0.100	0.24 g COD_X g^{-1} N oxidised	Henze *et al.*, 1987; Gujer *et al.*, 1999; Henze *et al.*, 2000
AOOs	1 yield	$Y_{X/NH4,AOO}^{obs}$	0.080	0.14 ± 0.02 g VSS g^{-1} N-NH$_4^+$ (*Nitrosomonas*)	Blackburne *et al.*, 2007
NOOs	1 yield	$Y_{X/NO2,NOO}^{obs}$	0.041	0.072 ± 0.01 g VSS g^{-1} N-NO$_2^-$ (*Nitrobacter*)	
AMOs	2 yields	$Y_{X/NH4,AMO}^{obs}$	0.071	0.066 to 0.071 ± 0.010 C-mol X mol^{-1} N-NH$_4^+$	Lotti *et al.*, 2014; Lotti *et al.*, 2015
		$Y_{NO2/NH4,AMO}^{obs}$	1.146	1.146 mol N-NO$_2^-$ consumed mol^{-1} N-NH$_4^+$	
		$Y_{NO3/NH4,AMO}^{obs}$	0.161	0.161 mol N-NO$_3^-$ produced mol^{-1} N-NH$_4^+$	
		$Y_{NO2/X,AMO}^{obs}$	-16.140	$= -\ Y_{NO2/NH4,AMO}^{obs} / Y_{NO3/NH4,AMO}^{obs}$ $= -\ 1.146 / 0.071$ mol N-NO$_2^-$ consumed C-mol^{-1} X	

Using measured yields of Table 2.9, overall growth stoichiometries were obtained (Table 2.10).

Table 2.10 Overall growth stoichiometries calculated by elemental, charge and/or degree of reduction balances and using the observed yields given in Table 2.8 and highlighted here (shaded). All units are mols (for biomass (C-)mol).

Materials Elemental formula	Organics $C_2H_3O_2^-$	Ammonium NH_4^+	Dioxygen O_2	Bicarbonate HCO_3^-	Nitrite NO_2^-	Nitrate NO_3^-	Dinitrogen N_2	Water H_2O	Protons H^+	Biomass $C_1H_{1.8}O_{0.5}N_{0.2}$
OHOs Overall growth	$-1/Y_{X/S,OHO}^{obs}$ $= -0.784$	− 0.2	−0.518	+0.568				+0.4	−0.016	+1
CMOs Overall growth		−10.2	−18.95	−1.0		$+1/Y_{X/NO3,CMO}^{obs}$ $= +10$		+10.4	+19.2	+1
AOOs Overall growth		$-1/Y_{X/NH4,AOO}^{obs}$ $= -12.5$	−17.4	−1.0	+12.3			+12.7	+23.8	+1
NOOs Overall growth			−10.75	−1.0	$-1/Y_{X/NO2,AOO}^{obs}$ $= -24.39$	+24.19		+0.2	−1.2	+1
DHOs Overall growth	$-1/Y_{X/S,DHO}^{obs}$ $= -0.972$			+0.944		−0.5952	+0.1976	+0.3976	−0.6232	+1
AMOs Overall growth		$-1/Y_{X/NH4,AMO}^{obs}$ $= -14.09$	−1.0		$Y_{NO2/X,AMO}^{obs}$ $= -16.14$	+2.19	+13.92	+28.21	−0.86	+1

d) Thermodynamic calculations of Gibbs free energy dissipated during growths

Gibbs free energies dissipated during growths were calculated under biochemical reference conditions (ΔG_{Diss}^{01}) using an empirical formula per trophic group (Table 2.11).

Table 2.11 Empirical formula of Gibbs free energy dissipations during growth (ΔG_{Diss}^{01}) for trophic groups.

Trophic group and formula	Equation
Heterotrophs: *e.g.*, OHOs and DHOs $\Delta G_{Diss}^{01} \quad = 200 + 18 \cdot (6 - NoC_{Cs})^{1.8} + \exp\{[(3.8 - \gamma_{Cs})^2]^{0.16} \cdot (3.6 + 0.4 \cdot NoC_{Cs})\}$ $\quad\quad\quad = $ comprised in general between 200-1,000 kJ mol^{-1} X with NoC_{Cs} the number of carbon atoms in the carbon source (mol C per mol S), and γ_{Cs} the degree of reduction of the carbon source (mol e- per C-mol S).	(textbook 2.23)
Chemolithoautotrophs with strong inorganic e-donor: $\Delta G_{Diss}^{01} \quad \approx 1,000$ kJ mol^{-1} X	(2.2)
Chemolithoautotrophs with weak inorganic e-donor (reverse electron transfer RET is needed): *e.g.*, CMOs, AOOs, NOOs, and AMOs $\Delta G_{Diss}^{01} \quad \approx 3,500$ kJ mol^{-1} X	(2.3)

From ΔG_{Diss}^{01} (Table 2.12), chemoorganoheterotrophs (OHOs, DHOs) grow more efficiently with 8 times less Gibbs free energy dissipated than chemolithoautotrophs (nitrifiers, AMOs).

Table 2.12 Comparison of the Gibbs free energy dissipated (ΔG_{Diss}^{01}) by OHOs, nitrifiers, DHOs and AMOs.

Trophic group	Guild	e-donor	C source	RET needed?	NoC_{Cs} (mol C mol^{-1} S)	γ_{Cs} (mol e- C-mol^{-1} S)	ΔG_{Diss}^{01} (kJ mol^{-1} X)
Chemoorganoheterotrophs	OHOs	$C_2H_3O_2^-$ (strong)	$C_2H_3O_2^-$		2	8.0 / 2 = 4.0	432
	DHOs	$C_2H_3O_2^-$ (strong)	$C_2H_3O_2^-$		2	8.0 / 2 = 4.0	432
Chemolithoautotrophs	CMOs	NH_4^+ (weak)	HCO_3^-	Yes			3,500
	AOOs	NH_4^+ (weak)	HCO_3^-	Yes			3,500
	NOOs	NO_2^- (weak)	HCO_3^-	Yes			3,500
	AMOs	NH_4^+ (weak)	HCO_3^-	Yes			3,500

e) Calculation of overall growth stoichiometries using thermodynamics

Instead of using a measured yield, overall growth stoichiometries can be solved using ΔG_{Diss}^{01} as the 6th independent conservative (Table 2.13).

Table 2.13 Overview of conservatives: elements (C, H, O, N), charge, Gibbs free energy, electrons, and mass. Degree of reduction and mass do not provide extra independent conservatives since calculated based on elements and charge. However, they can be used in place of another conservative and they also make it possible to verify the electron and mass balances over the reactions.

Materials i		Acetate	Ammonium	Dioxygen	Bicarbonate	Nitrite	Nitrate	Dinitrogen	Water	Protons	Biomass	ΔG_{Diss}^{01}
Elemental formula		$C_2H_3O_2^-$	NH_4^+	O_2	HCO_3^-	NO_2^-	NO_3^-	N_2	H_2O	H^+	$C_1H_{1.8}O_{0.5}N_{0.2}$	
Units		(mol)	(mol)	(mol)	(mol)	(mol)	(mol)	(mol)	(mol)	(mol)	((C-)mol)	(kJ mol^{-1} X)
Conservatives k					Conservation coefficients ($\iota_{k,i}$ in mol k mol^{-1} i)							
C	mol C mol^{-1} i	2	0	0	1	0	0	0	0	0	1	
H	mol H mol^{-1} i	3	4	0	1	0	0	0	2	1	1.8	
O	mol O mol^{-1} i	2	0	2	3	2	3	0	1	0	0.5	
N	mol N mol^{-1} i	0	1	0	0	1	1	2	0	0	0.2	
Charge +	mol + mol^{-1} i	-1	+1	0	-1	-1	-1	0	0	+1	0	
Gibbs free energy of formation under standard conditions G_f^0	kJ mol^{-1} i	-369.4	-79.4	0	-586.9	-32.2	-111.3	0	-237.2	0	-67.0	
Degree of reduction γ	mol e- mol^{-1} i	8.0	0	-4	0	-6	-8	-6	0	0	4.2	
Molecular mass	g i mol^{-1} i	59.04	18.04	32.00	61.02	46.01	62.00	28.01	18.02	1.01	24.63	
Guild j					Stoichiometric coefficients ($v_{i,j} = Y_{i,Xj}$ in mol i mol^{-1} X)							
OHO growth		$v_{S,OHO}$	$v_{NH4,OHO}$	$v_{O2,OHO}$	$v_{CO2,OHO}$				$v_{H2O,OHO}$	$v_{H+,OHO}$	+1	432
CMO growth			$v_{NH4,CMO}$	$v_{O2,CMO}$	$v_{CO2,CMO}$		$v_{NO3,CMO}$		$v_{H2O,CMO}$	$v_{H+,ANO}$	+1	3,500
AOO growth			$v_{NH4,AOO}$	$v_{O2,AOO}$	$v_{CO2,AOO}$	$v_{NO2,AOO}$			$v_{H2O,AOO}$	$v_{H+,AOO}$	+1	3,500
NOO growth				$v_{O2,NOO}$	$v_{CO2,NOO}$	$v_{NO2,NOO}$	$v_{NO3,NOO}$		$v_{H2O,NOO}$	$v_{H+,NOO}$	+1	3,500
DHO growth		$v_{S,DHO}$			$v_{CO2,DHO}$		$v_{NO3,DHO}$	$v_{N2,DHO}$	$v_{H2O,DHO}$	$v_{H+,DHO}$	+1	432
AMO growth			$v_{NH4,AMO}$		$v_{CO2,AMO}$	$v_{NO2,AMO}$	$v_{NO3,AMO}$	$v_{N2,AMO}$	$v_{H2O,AMO}$	$v_{H+,AMO}$	+1	3,500

Calculation of Gibbs free energy of formation (G_f^0) of non-tabulated compounds

The standard Gibbs free energy of formation (G_f^0) of certain organic compounds are not provided in reference tables (*e.g.*, Table 2.5 in the textbook). This challenges the calculation of Gibbs free energy changes. An alternative can be used to (roughly) estimate the G_f^0 of a compound by making an enthalpy balance over its full combustion reaction and assuming that the compound is primarily enthalpic, *i.e.*, $\Delta G_R = \Delta H_R - T \cdot \Delta S_R \approx \Delta H_R$. In a combustion reaction, a standard enthalpy (ΔH^0) of 444 kJ mol^{-1} O_2 is dissipated. Two examples are given for acetate ($C_2H_3O_2^-$, for which G_f^0 is provided in reference tables: -369.4 kJ mol^{-1} acetate) and for organic matter ($C_1H_{1.63}O_{0.46}$, G_f^0 is absent from reference tables).

- Verification of G_f^0 of acetate ($C_2H_3O_2^-$):
  Full combustion reaction: $-1\ C_2H_3O_2^- - 1\ H^+ - 2\ O_2 + 2\ CO_2 + 2\ H_2O$

$$\Delta G_R^0 = -2\ mol\ O_2\ mol^{-1}\ acetate \cdot 444\ kJ\ mol^{-1}\ O_2 = -888\ kJ\ mol^{-1}\ acetate$$

Gibbs free energy balance over the combustion reaction:

$$\Sigma\ (v_{i,j} \cdot G_{f,i}^0) = (-1) \cdot G_{f,Ace}^0 + (-1) \cdot G_{f,H^+}^0 + (-2) \cdot G_{f,O2}^0 + 2 \cdot G_{f,CO2}^0 + 2 \cdot G_{f,H2O}^0 = \Delta G_R^0 = -888\ kJ\ mol^{-1}$$

$$(-1) \cdot G_{f,Ace}^0 + (-1) \cdot 0 + (-2) \cdot 0 + 2 \cdot (-394.4) + 2 \cdot (-237.2) = (-1) \cdot G_{f,Ace}^0 - 1263.2 = -888\ kJ\ mol^{-1}$$

$$G_{f,Ace}^0 = 888 - 1263.2 \approx -375\ kJ\ mol^{-1}\ C_2H_3O_2^-$$

This estimate is not far from the tabulated reference value of -369.4 kJ mol^{-1} $C_2H_3O_2^-$.

- Estimation of G_f^0 of organic matter ($C_1H_{1.63}O_{0.46}$), hereafter referred to as OM:
  Full combustion reaction: $-1\ C_1H_{1.63}O_{0.46} - 1.1775\ O_2 + 1\ CO_2 + 0.815\ H_2O$

$$\Delta G_R^0 = -1.1775\ mol\ O_2\ C\text{-}mol^{-1}\ organic\ matter \cdot 444\ kJ\ mol^{-1}\ O_2 = -522.81\ kJ\ C\text{-}mol^{-1}\ organic\ matter$$

Gibbs free energy balance over the combustion reaction:

$$\Sigma\ (v_{i,j} \cdot G_{f,i}^0) = (-1) \cdot G_{f,OM}^0 + (-1.1775) \cdot G_{f,O2}^0 + 1 \cdot G_{f,CO2}^0 + 0.815 \cdot G_{f,H2O}^0 = \Delta G_R^0 = -522.81\ kJ\ C\text{-}mol^{-1}$$

$$(-1) \cdot G_{f,OM}^0 + (-1.1775) \cdot 0 + 1 \cdot (-394.4) + 0.815 \cdot (-237.2) = (-1) \cdot G_{f,OM}^0 - 587.718 = -522.81\ kJ\ C\text{-}mol^{-1}$$

$$G_{f,OM}^0 = 522.81 - 587.718 \approx -65\ kJ\ C\text{-}mol^{-1}\ C_1H_{1.63}O_{0.46}$$

This procedure can be used for any organic for which a reference value of G_f^0 is not available.

Resolution of growth stoichiometries using thermodynamics

For OHOs, nitrifiers and DHOs, the 6 unknown stoichiometric coefficients in the metabolism can be obtained by solving the system of 6 independent conservation equations with C/H/O/N/charge (or γ) and G_f^{01} (corrected for biochemical reference conditions from the standard G_f^0), and using ΔG_{Diss}^{01} values. For AMOs, 7 unknowns can still not be solved with only 6 equations, but an electron balance on anabolism can help (see Example 2.6 in the textbook).

Metabolic reactions are calculated by combining anabolic and independent catabolic reactions (Table 2.7) using the thermodynamics-derived multiplication factor of catabolism (λ_{cat}^*):

$$\lambda_{cat}^* = (\Delta G_{An}^{01} + \Delta G_{Diss}^{01}) / - \Delta G_{Cat}^{01} \qquad \text{(textbook Eq. 2.22)}$$

ΔG_{Diss}^{01} values were calculated in Table 2.12. ΔG_{An}^{01} and ΔG_{Cat}^{01} are calculated using standard $G_{f,i}^0$ of compounds involved in anabolisms and catabolisms, respectively, and by correcting for biochemical reference conditions at 25 °C and pH 7.0 (ΔG_{An}^{01}, ΔG_{Cat}^{01}):

$$\Delta G_{An}^0 = \Sigma \, (G_{f,i}^0 \cdot v_{i,j}^{ana})$$

$$\Delta G_{An}^{01} = \Delta G_{An}^0 + R \cdot T \cdot v_{H+} \cdot \ln \, (10^{-7} / 1)$$

$$\Delta G_{Cat}^0 = \Sigma \, (G_{f,i}^0 \cdot v_{i,j}^{cat})$$

$$\Delta G_{Cat}^{01} = \Delta G_{Cat}^0 + R \cdot T \cdot v_{H+} \cdot \ln \, (10^{-7} / 1) \qquad \text{(adapted from the textbook Eq. 2.9 and Eq. 2.10)}$$

with R the universal ideal gas constant ($8.3145 \cdot 10^{-3}$ kJ mol^{-1} K^{-1}), T the temperature (298 K), v_{H+} the stoichiometric coefficient of protons H$^+$ in the considered reaction ($v_{H+} = \pm \, Y_{H+}$; if H+ consumed <0 or produced >0). If protons are not involved in the reaction $\Delta G^{01} = \Delta G^0$.

Table 2.14 provides an overview of the ΔG_{An}^{01}, ΔG_{Cat}^{01}, ΔG_{Diss}^{01} and resulting λ_{cat}^* values of BNR guilds.

Chemoorganoheterotrophs (OHOs, DHOs) efficiently grow by only running their catabolic reaction 0.5 times. Chemolithoautotrophs (nitrifiers, AMOs) need to run their catabolism 10 to 48 times to grow; they dissimilate a large amount of substrate to produce the ATP for biomass synthesis and only a few building blocks remain for anabolism, resulting in low growth yields. Less (waste) biomass is produced, which is a benefit for the wastewater treatment process.

Table 2.14 Calculations of Gibbs free energy changes and λ_{cat}^* for OHOs, DHOs, nitrifiers and AMOs.

Trophic group	Guild	ΔG_{An}^0 (kJ mol^{-1} X)	ΔG_{Cat}^0 (kJ mol^{-1} eD)	ΔG_{An}^{01} (kJ mol^{-1} X)	ΔG_{Cat}^{01} (kJ mol^{-1} eD)	ΔG_{Diss}^{01} (kJ mol^{-1} X)	λ_{cat}^* (-)
Chemoorganoheterotrophs	OHOs	18.6	−804.4	29.6	−844.3	432	0.547
	DHOs	−88.5	−816.1	−69.5	−792.1	432	0.458
Chemolithoautotrophs	CMOs	299.6	−269.1	289.6	−349.0	3500	10.859
	AOOs	307.9	−190.0	283.9	−269.9	3500	14.021
	NOOs	265.3	−79.1	313.3	−79.1	3500	48.208
	AMOs	274.8	−362.8	306.7	−362.8	3500	10.493

With the λ_{cat}^* values, the anabolic and independent catabolic reactions (tables 2.4 and 2.5) were combined to calculate the metabolic reactions (Table 2.15). The metabolisms obtained via thermodynamics are compared to those previously obtained with an observed yield (Step c).

Table 2.15 Thermodynamic calculations of growth stoichiometries under biochemical reference conditions using λ_{cat}* (shaded raws), and a comparison to stoichiometries obtained using an observed yield from literature (*italic*).

Materials i	$C_2H_3O_2^-$	NH_4^-	O_2	HCO_3^-	NO_2^-	NO_3^-	N_2	H_2O	H^-	$C_1H_{1.8}O_{0.5}N_{0.2}$	ΔG_R^0	ΔG_R^{01}	λ_{cat}*
Units	(mol)	(mol)	(mol)	(mol)	(mol)	(mol)	(mol)	(mol)	(mol)	((C-)mol)	(kJ mol^{-1} fixed i)		-
γ (mol e- mol^{-1} i)	8.0	0	-4	0	-6	-8	-6	0	0	4.2			
G_f^0 (kJ mol^{-1} i)	-369.4	-79.4	0	-586.9	-32.2	-111.3	0	-237.2	0	-67.0			
OHOs													
Anabolism	-0.525	-0.2		-0.05				+0.4	-0.275	+1	+19	+30	
Independent catabolism	-1		-2	+2					+1		-804	-844	0.547
Metabolism (thermo.)	-1.072	-0.2	-1.094	+1.144				+0.4	+0.272	+1		-432	
Metabolism (observed)	*-0.784*	*-0.2*	*-0.518*	*+0.568*				*+0.4*	*-0.016*	*+1*			
CMOs													
Anabolism		-0.725		-1		+0.525		+0.925	+0.25	+1	+300	+290	
Independent catabolism		-1	-2			+1		+1	+2		-269	-349	10.86
Metabolism (thermo.)		-11.6	-21.7	-1		+11.4		11.8	22.0	+1		-3,500	
Metabolism (observed)		*-10.2*	*-18.95*	*-1*		*+10.0*		*+10.4*	*+19.2*	*+1*			
AOOs													
Anabolism		-0.9		-1	+0.7			+1.1	+0.6	+1	+308	+284	
Independent catabolism		-1	-1.5		+1			+1	+2		-190	-270	14.02
Metabolism (thermo.)		-14.9	-21.0	-1	+14.7			+15.1	+28.6	+1		-3,500	
Metabolism (observed)		*-12.5*	*-17.4*	*-1*	*+12.3*			*+12.7*	*+23.8*	*+1*			
NOOs													
Anabolism				-1	-2.9	+2.7		+0.2	-1.2	+1	+265	+313	
Independent catabolism			-0.5		-1	1					-79	-79	48.21
Metabolism (thermo.)			-24.1	-1	-51.1	+50.9		+0.2	-1.2	+1		-3,500	
Metabolism (observed)			*-10.8*	*-1*	*-24.4*	*+24.2*		*+0.2*	*-1.2*	*+1*			
DHOs													
Anabolism	-0.725			+0.45		-0.2		+0.2	-0.475	+1	-89	-70	
Independent catabolism	-1			+2		-1.6	+0.8	+0.8	-0.6		-816	-792	0.458
Metabolism (thermo.)	-1.183			+1.366		-0.932	+0.366	+0.566	-0.750	+1		-432	
Metabolism (observed)	*-0.972*			*+0.944*		*-0.5952*	*+0.1976*	*+0.3976*	*-0.6232*	*+1*			
AMOs													
Anabolism		-0.2		-1	-2.1	+2.1		0.4	-0.8	+1	+275	+307	
Independent catabolism		-1			-1		1	2			-363	-363	10.49
Metabolism (thermo.)		-10.7		-1	-12.6	+2.1	+10.5	+21.4	-0.8	+1		-3,500	
Metabolism (observed)		*-14.1*		*-1*	*-16.1*	*+2.2*	*+13.9*	*+28.2*	*-0.9*	*+1*			

f) Stoichiometric and kinetic parameters of growth calculated by thermodynamics

Thermodynamic relations provide key stoichiometric and kinetic parameters (Table 2.16).

Table 2.16 Formulas used to derive important stoichiometric and kinetic parameters using thermodynamics. The equations are compiled from the textbook. Before using the equations, please read the explanations given in the textbook carefully to understand the equations.

Term	Symbol	Units	Formula	Equation
Biomass-specific rate of Gibbs free energy dissipation for maintenance	m_G	kJ h^{-1} mol^{-1} X	$= 4.5 \cdot \exp(-69/R \cdot (1/T - 1/298))$	(textbook Eq. 2.28)
Biomass-specific rate of substrate consumption for maintenance	m_S	mol S h^{-1} mol^{-1} X	$= -Y_{S/S}^{Cat} \cdot m_G / \Delta G_{cat}^{01}$	(textbook Eq. 2.29)
Maximum biomass-specific rate of electron transfer in transport chain	q_e^{max}	mol e- h^{-1} mol^{-1} X	$= 3 \cdot \exp(-69/R \cdot (1/T - 1/298))$	(textbook Eq. 2.31)
Number of electrons transferred per mol substrate in catabolism	γ_S^*	mol e- transferred mol^{-1} S	Reflection on electron transferred in catabolism.	(explained in textbook p. 68)
Maximum biomass-specific rate of Gibbs free energy dissipation	q_G^{max}	kJ h^{-1} mol^{-1} X	$= q_e^{max} \cdot \Delta G_{cat}^{01} / \gamma_S^*$	(textbook Eq. 2.32)
Maximum growth rate	μ_{max}	mol X h^{-1} mol^{-1} X $= h^{-1}$	$= (q_G^{max} + m_G) / -\Delta G_{Diss}^{01}$	(textbook Eq. 2.35)
Maximum yield of biomass formation on substrate	$Y_{X/S}^{max}$	mol X mol^{-1} S	$= 1 / Y_{S/X}^{max}$ Given by calculated metabolic reaction.	(2.4)
Maximum biomass-specific rate of decay	k_d	h^{-1}	$= Y_{X/S}^{max} \cdot m_S$	(textbook Eq. 2.37)
Maximum biomass-specific rate of substrate consumption	q_S^{max}	mol S h^{-1} mol^{-1} X	$= 1/Y_{X/S}^{max} \cdot \mu_{max} + Y_{S/S}^{Cat} \cdot m_S$	(adapted from textbook Eq. 2.25)
Maximum biomass-specific rate of any material consumption or production	q_i^{max}	mol i h^{-1} mol^{-1} X	$= 1/Y_{X/i}^{max} \cdot \mu_{max} + Y_{i/S}^{Cat} \cdot m_S$ A m_i term is only present if the material i participates in the catabolic reaction.	(textbook Eq. 2.33)

Table 2.17 provides an overview of the parameter values calculated for BNR guilds, comparing metabolic efficiencies. Chemolithoautotrophs (nitrifiers, AMOs) catabolise most of their substrate (as high as 94-98 %) to produce sufficient ATP for biomass synthesis. They synthesise their own endogenous organic carbon source by fixing/reducing CO_2, at an electron and energy price. For chemoorganoheterotrophs (OHOs, DHOs), the exogenous acetate is at a reduction state (4.0 mol e- C-mol^{-1}) close to biomass (4.2 mol e- C-mol^{-1}), making

biosynthesis straightforward. They catabolise only half the acetate; the other half is anabolised. They yield 50× higher biomass production on substrate (which is less favourable for WWTP operation), grow 20× faster, but also decay 5× faster, than chemolithoautotrophs.

Table 2.17 Comparison of stoichiometric and kinetic parameters of microbial growth of BNR guilds.

Growth parameter derived from thermodynamics	Trophic groups and microbial guilds					
	Chemoorganoheterotrophs		Chemolithoautotrophs			
	OHOs	DHOs	CMOs	AOOs	NOOs	AMOs
Resource allocation						
Fraction of main substrate anabolised (%)[a]	49 %	61 %	6 %	6 %	6 %	2 %
Fraction of main substrate catabolised (%)[a]	51 %	39 %	94 %	94 %	94 %	98 %
Gibbs free energy changes						
$-\Delta G_{Diss}^{01}$ (kJ mol^{-1} X)	-432	-432	-3,500	-3,500	-3,500	-3,500
ΔG_{cat}^{01} (kJ mol^{-1} X)	-844	-792	-349	-270	-79	-363
ΔG_{ana}^{01} (kJ mol^{-1} X)	30	-70	290	284	313	307
q_G^{max} (kJ h^{-1} mol^{-1} X)	-158	-297	-131	-135	-119	-181
Maintenance rates						
m_G (kJ h^{-1} mol^{-1} X)	4.5	4.5	4.5	4.5	4.5	4.5
m_S (mol S h^{-1} mol^{-1} X)	0.005	0.006	0.013	0.017	0.057	0.012
Growth, substrate uptake, and decay						
μ_{max} (mol X h^{-1} mol^{-1} X, *i.e.*, h^{-1})	0.722	0.677	0.036	0.037	0.033	0.051
$Y_{X/S}^{max}$ (mol X mol^{-1} S)	0.933	0.845	0.086	0.067	0.020	0.094
$q_{S,max}$ (mol S h^{-1} mol^{-1} X)	-0.780	-0.806	-0.431	-0.573	-1.724	-0.553
k_d (mol X h^{-1} mol^{-1} X, *i.e.*, h^{-1})	0.0050	0.0048	0.0011	0.0011	0.0011	0.0012

[a] The fraction of substrate anabolised was obtained by dividing the stoichiometric coefficients of the substrate in the anabolic and metabolic reactions: $v_{S,j}^{ana}$ / $v_{S,j}^{met}$. The fraction of substrate catabolised is the residual: $v_{S,j}^{cat}$ / $v_{S,j}^{met} = 1 - v_{S,j}^{ana}$ / $v_{S,j}^{met}$. The main substrates are different for the different guilds: acetate as e-donor and C source for OHOs and DHOs; ammonium as e-donor and N source for CMOs, AOOs and AMOs; nitrite as e-donor and N source for NOOs.

g) *Herbert-Pirt relations*

Stoichiometric and kinetic parameters derived by thermodynamics are used to formulate q rates following Herbert-Pirt ($q_s = m_s + 1/Y_{X/S}^{max} \cdot \mu$). It describes substrate allocation for cellular maintenance (m_s; a purely catabolic process) and for growth ($1/Y_{X/S}^{max} \cdot \mu$; combines anabolism and catabolism). The relation is generalised for any compound involved in the metabolism ($q_i = m_i + 1/Y_{X/i}^{max} \cdot \mu$). Only catabolic compounds have a maintenance term (m_i). Figure 2.5 summarises the maintenance rates, maximum material yields on biomass, and maximum q rates of BNR guilds under biochemical reference conditions (298 K, pH 7.0, 1 atm).

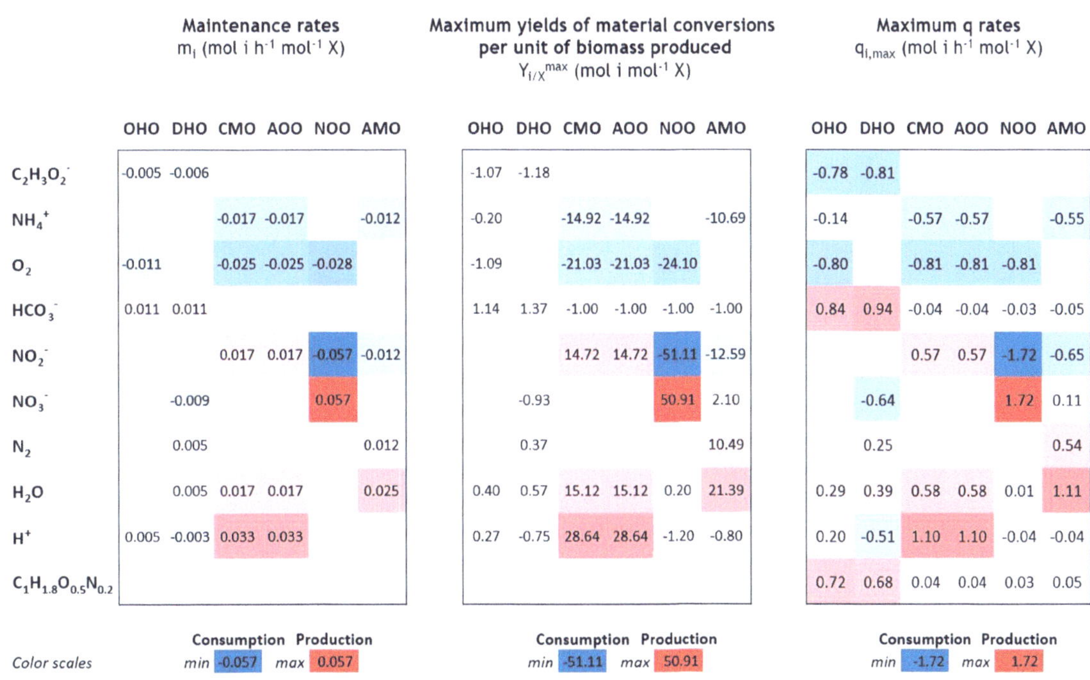

Figure 2.5 Overview of maximum rates and yields of biomass growth of BNR guilds computed using thermodynamics and Herbert-Pirt relations ($q_i = m_i + 1/Y_{X/i} \cdot \mu = m_i + Y_{i/X} \cdot \mu$). Chemoorganoheterotrophs (OHOs, DHOs) grow at a 20× faster rate ($\mu^{max} > 0.5$ h^{-1}) than chemolithoautotrophs (nitrifiers, AMOs). OHOs and DHOs minimise substrate expenditures per unit of biomass produced. Nitrifiers and AMOs catabolise most of their resources but there is little biomass production; they grow slowly but process rapidly the catabolic resources because of their high material consumption yield on biomass. Because of higher catabolic requirements, nitrifiers and AMOs exhibit higher maintenance rates than OHOs and DHOs.

h) *Literal formulations of volumetric rates of growth processes*

The volumetric rate of a microbial growth process (ρ_j in mmol X_j h^{-1} l^{-1}, described in ASM models as the 'process rate') is formulated as the product of the maximum growth rate of the microorganisms (μ_j^{max}, h^{-1}), substrate saturation terms (S_i ($K_{i,j} + S_i)^{-1}$), possible inhibition terms (*e.g.*, I_i ($I_{i,j} + S_i)^{-1}$), and biomass concentration of the targeted microorganism (X_j, mmol X_j l^{-1}). Different saturation and inhibition switching functions can be included depending on the physiological knowledge of the organism. For the saturation term, often one term is given for the limiting compound. If concentration conditions are not yet known, it is possible to develop the process rates with a saturation term for each compound consumed (Table 2.18).

Values of maximum growth rates can be found in literature or calculated using thermodynamics (Table 2.17). When comparing organisms, do not intermix data obtained from literature and from thermodynamics calculations. Affinity and inhibition constants may be found in literature. If not, rough estimates can be used with a very low affinity constant value and a relatively high inhibition value as first guesses.

Table 2.18 Literal expressions of volumetric rates of growth processes ('process rates') defined using saturation (Monod terms) and inhibition (inverse Monod terms) switching functions.

Volumetric rate of growth process	Maximum growth rate · Saturation terms · Inhibition terms · Biomass concentration
ρ_{OHO} (mol X_{OHO} h^{-1} l^{-1})	$\mu_{OHO}^{max} \cdot S_S (K_{S,OHO} + S_S)^{-1} \cdot S_{NH4} (K_{NH4,OHO} + S_{NH4})^{-1} \cdot S_{O2} (K_{O2,OHO} + S_{O2})^{-1} \cdot X_{OHO}$
ρ_{CMO} (mol X_{CMO} h^{-1} l^{-1})	$\mu_{CMO}^{max} \cdot S_{NH4} (K_{NH4,CMO} + S_{NH4})^{-1} \cdot S_{CO2} (K_{CO2,CMO} + S_{CO2})^{-1} \cdot S_{O2} (K_{O2,CMO} + S_{O2})^{-1} \cdot K_{H+} (K_{H+,CMO} + S_{H+})^{-1} \cdot X_{CMO}$
ρ_{AOO} (mol X_{AOO} h^{-1} l^{-1})	$\mu_{AOO}^{max} \cdot S_{NH4} (K_{NH4,AOO} + S_{NH4})^{-1} \cdot S_{CO2} (K_{CO2,AOO} + S_{CO2})^{-1} \cdot S_{O2} (K_{O2,AOO} + S_{O2})^{-1} \cdot K_{H+} (K_{H+,AOO} + S_{H+})^{-1} \cdot X_{AOO}$
ρ_{NOO} (mol X_{NOO} h^{-1} l^{-1})	$\mu_{NOO}^{max} \cdot S_{NO2} (K_{NO2,NOO} + S_{NH4})^{-1} \cdot S_{CO2} (K_{CO2,NOO} + S_{CO2})^{-1} \cdot S_{O2} (K_{O2,NOO} + S_{O2})^{-1} \cdot X_{NOO}$
ρ_{DHO} (mol X_{DHO} h^{-1} l^{-1})	$\mu_{DHO}^{max} \cdot S_S (K_{S,DHO} + S_S)^{-1} \cdot S_{NO3} (K_{NO3,DHO} + S_{NO3})^{-1} \cdot X_{DHO}$
ρ_{AMO} (mol X_{AMO} h^{-1} l^{-1})	$\mu_{AMO}^{max} \cdot S_{NH4} (K_{NH4,AMO} + S_{NH4})^{-1} \cdot S_{CO2} (K_{CO2,AMO} + S_{CO2})^{-1} \cdot S_{NO2} (K_{NO2,AMO} + S_{NO2})^{-1} \cdot K_{O2} (K_{O2,AMO} + S_{O2})^{-1} \cdot X_{AMO}$

EXAMPLE 2.3.4

Simulate the growth models, and analyse the selections and conversions in the mixed culture
Implement the growth models of BNR organisms in software to simulate and analyse their selection/interactions in ideal discontinuous (batch) and continuous-flow (chemostat) reactors.

a) Review the main principles of microbial selection and mass balances in a bioreactor.
b) Address the selection and competition in the batch. Identify the main stoichiometric and/or kinetic parameters that impact microbial selection.
c) Address the selection and competition in the chemostat. Identify the main stoichiometric and/or kinetic parameters that impact the microbial selection.

Option 1: test the impact of stoichiometric and kinetic parameter values on microbial selection.

Option 2: test the effects of temperature and pH using the adapted stoichiometries and kinetics.

Solution

Using the calculated growth stoichiometries and kinetics, we can simulate the growth, selection, conversions, and interactions of BNR guilds in bioreactors from pure cultures to mixed cultures.

a) Principles of mass balances and microbial selection in an ideal bioreactor
Two ideal, discontinuous (*i.e.*, batch) and continuous-flow (*i.e.*, chemostat) stirred-tank reactors are used to simulate and analyse growths and interactions of BNR guilds.

<u>Mass balances in batch and chemostat</u>
Conversions are followed by mass or mol balances:

$$\text{ACCUMULATION} = \text{TRANSPORT} \pm \text{REACTION} = \text{IN} - \text{OUT} \pm \text{REACTION} \qquad (2.5)$$

$$\begin{pmatrix} \text{Mass accumulated} \\ \text{per unit of time} \end{pmatrix} = \begin{pmatrix} \text{Mass transported} \\ \text{per unit of time} \end{pmatrix} \pm \begin{pmatrix} \text{Mass converted} \\ \text{per unit of time} \end{pmatrix}$$

Across reactor boundaries, masses enter and leave, and react over time. Material consumptions/productions relate to negative/positive accumulation, with a negative/positive reaction term value. The generalised mass balance is given as a differential equation over time:

Mass balance:
$$dm_i / dt \qquad = \dot{m}_{i,in} - \dot{m}_{i,out} + R_i \qquad\qquad (2.6)$$
$$d(V \cdot C_i) / dt \qquad = Q_{in} \cdot C_{i,in} - Q_{out} \cdot C_{i,out} + r_i \cdot V$$
$$= Q_{in} \cdot C_{i,in} - Q_{out} \cdot C_{i,out} + q_i \cdot C_X \cdot V$$

If $Q_{in} = Q_{out} = Q$:
$$d(V \cdot C_i) / dt \qquad = (C_{i,in} - C_{i,out}) \cdot Q + r_i \cdot V$$
$$= (C_{i,in} - C_{i,out}) \cdot Q + q_i \cdot C_X \cdot V$$

If V is constant[1]:
$$dC_i / dt \qquad = (C_{i,in} - C_{i,out}) \cdot Q / V + r_i$$
$$= (C_{i,in} - C_{i,out}) \cdot Q / V + q_i \cdot C_X$$
$$= (C_{i,in} - C_{i,out}) \cdot D + q_i \cdot C_X$$

with mass flow rate of material i ($\dot{m}_i$ with dimensions[2] as $M_i\ T^{-1}$), reactor working volume (V as L^3), volumetric flow rate (Q as $L^3\ T^{-1}$), dilution rate (D as T^{-1}, *i.e.*, inverse of hydraulic retention time HRT as T), concentration of material i (C_i as $M_i\ L^{-3}$), biomass concentration (C_X as $M_X\ L^{-3}$), total conversion rate (R_i as $M_i\ T^{-1}$), volumetric conversion rate (r_i as $M_i\ T^{-1}\ L^{-3}$) and biomass-specific conversion rate (q_i as $M_i\ T^{-1}\ M_X^{-1}$) of material i.

Mass balances in batch (no inflow, no outflow) and chemostat (inflow and outflow) are given in Table 2.19. These differential equations express the dynamics of materials over reactor operation. A batch is a non-stationary and homogenous reactor: concentrations evolve over time ($dC/dt \neq 0$) but at one time point are equal at any geographical point. No steady state is achieved, other than when conversions are stopped (*e.g.*, when limiting substrate is depleted, while neglecting decay). A chemostat is stationary and homogenous: at steady state, concentrations are constant ($dC/dt = 0$) at any point in reactor. Influent and effluent volumetric flow rates are equal ($Q_{in} = Q_{out} = Q$). Concentrations in effluent are equal to concentrations in the tank.

---

[1] Often the volume of a reactor is considered constant. However, in laboratory and engineering practices, volumes vary with evaporation and sampling. Always consider mass (or mol) balances (instead of 'concentration balances').
[2] Managing dimensions and units across all calculations is crucial. Dimensions are given as mass (M), length (L), and time (T). The related SI units use kg (or mol), m, and s, respectively. In practice different units are used.

Table 2.19 Mass (or mol) balance equations for ideal, discontinuous (batch), semi-continuous (fed-batch), and continuous-flow (chemostat) stirred-tank reactors. The mass balances are developed for biomass (X) and the limiting substrate (S). The equations can be applied to any material transported/converted in the system.

Reactor scheme	Reactor regime and state variable	Mass balance equations
Batch (discontinuous)	Batch Biomass (X) Substrate (S)	ACCUMULATION = REACTION $dm_X / dt = r_X \cdot V = \mu \cdot C_X \cdot V$ $dm_S / dt = r_S \cdot V = q_S \cdot C_X \cdot V$ with m: mass
Chemostat (continuous)	Chemostat Biomass (X) Substrate (S)	ACCUMULATION = IN − OUT + REACTION $dm_X / dt = Q_{in} \cdot C_{X,in} (\sim 0) - Q_{out} \cdot C_{X,out} + r_X \cdot V = C_X \cdot (\mu \cdot V - Q)$ $dm_S / dt = Q_{in} \cdot C_{S,in} - Q_{out} \cdot C_{S,out} + r_S \cdot V = Q \cdot (C_{S,in} - C_S) + q_S \cdot C_X \cdot V$ At steady state: $\quad 0 = dm_X / dt = C_X \cdot (\mu \cdot V - Q)$ $\qquad$ (1) $\qquad\qquad 0 = dm_S / dt = Q \cdot (C_{S,in} - C_S) + q_S \cdot C_X \cdot V = 0$ $\qquad$ (2) From (1): $\mu = Q / V = D$ From (2): the analytical solutions of C_X and C_S are obtained at steady state.
		For all reactor regimes, q rates relate to Monod and Herbert-Pirt relations: $q_X = \mu$ (historical term defined by Monod) $= \mu_{max} \cdot C_s / (K_s + C_s) \cdot I_i / (I_i + C_i)$ $q_S = q_{s,max} \cdot C_s / (K_s + C_s) \cdot I_i / (I_i + C_i)$ ('Monod') $= - 1/Y_{X/s} \cdot \mu + m_s$ ('Herbert-Pirt')

Selection principles in mixed cultures in a batch and in a chemostat

Basic principles (Kuenen, 2019; Rombouts, *et al.,* 2019a and 2019b) to predict microbial growth, selection, and conversions in a mixed-culture bioreactor are briefly restated here:

- Batch: Substrate is supplied as a pulse. The substrate concentration remains much higher than the K_s value during most of the batch reaction period. Substrate becomes limiting only close to the end of the batch reaction time. Microorganisms deploy their maximum growth rate. The organism with fastest μ_{max} scavenges the substrate and dominates the community if the substrate is directly coupled to growth. In one single batch, one can detect selection when starting with a low biomass concentration. In a sequencing batch reactor (SBR), the sequence of batches drives the effective selection in the long run (see Exercise 2.4.1h).
- Chemostat: μ is set by D. Only microorganisms withstanding D remain in the system (D < μ_{max}). The low residual concentration of substrate (C_S) is a function of D and affinity properties (μ_{max}/K_s) of microorganisms. The affinity for substrate is governed by both the affinity constant (*i.e.*, highest affinity at lowest K_S value) and the maximum biomass-specific growth rate (μ_{max}). This dictates the selection. The organism with the highest affinity for substrate and which makes the lowest substrate concentration should take the lead.

Implementation of model simulations in Aquasim

Growth models were simulated using Aquasim (Reichert, 1994) (Figure 2.6) to analyse the selection, conversions, and interactions of BNR guilds in batch and chemostat mixed cultures.

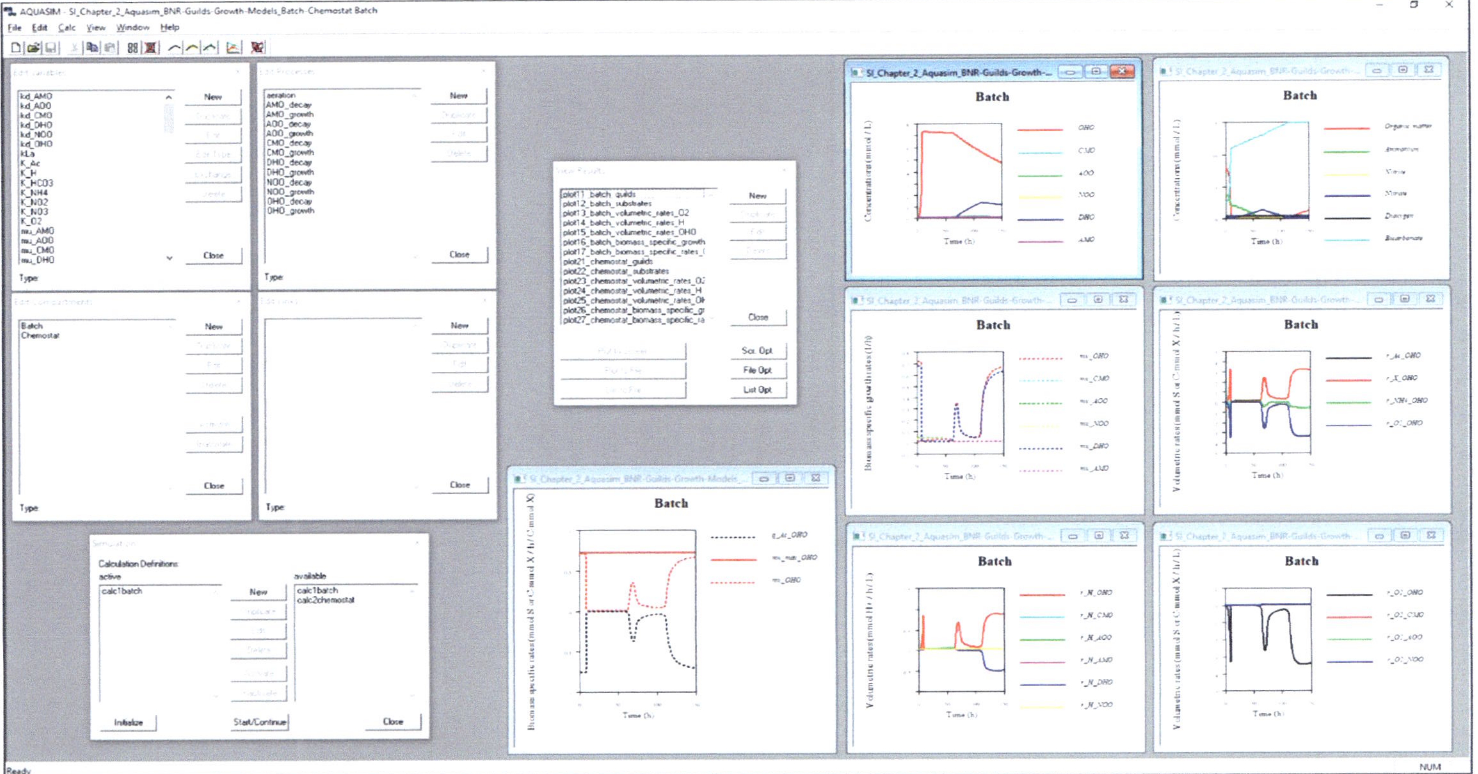

Figure 2.6 Overview of the Aquasim software interface for the implementation and simulation of the mathematical models of microbial growth of the OHO, CMO, AOO, NOO, DHO and AMO guilds. The 4 squares on the left help edit (*i*) variables and constants, (*ii*) microbial processes (*e.g.*, growth and decay), (*iii*) reactor compartments (*e.g.*, batch and chemostat), and when applicable (*iv*) possible transport links between different reactor tanks (not used here). After implementation, the model simulations are initialised and launched, and the results are visualised (boxes in the middle). Simulation outputs are displayed in graphs. The plots highlighting here are the growth of the different microbial guilds and the conversions of the main substrates and products in the batch, in terms of concentrations and rates.

Note: Aquasim can be downloaded here: https://www.eawag.ch/en/department/siam/software/. After downloading, verify that the folder contains the necessary .dll files; if not these can be downloaded elsewhere (*e.g.*, www.dlldownloader.com). Aquasim is launched from the folder as an executable. While developed for Windows, Mac users can run Aquasim in different ways: (*i*) use a Windows parallel desktop or dual boot; (*ii*) use a Windows weblogin to your university virtual PC; or (*iii*) download a Windows virtual machine that fits your Mac version.

b) Batch simulations of growth, selection, and interactions of BNR guilds

<u>Simulation conditions in batch</u>

The simulation conditions for the batch are given in Table 2.20a.

Table 2.20a Constants used for batch and chemostat simulations: operation.

Operational conditions	Symbol	Units	Batch simulations	Chemostat simulations
Simulation time	Δt	h	150	5,000-25,000
Reactor working volume	V	l	100	100
Volumetric flow rate	Q	$l\ h^{-1}$	0	1
Dilution rate	$D = Q / V$	h^{-1}	0	0.010
Volumetric mass transfer coefficient of O_2	k_La	h^{-1}	100	100
Saturation concentration of dissolved oxygen	C_{O2}^{sat}	mmol $O_2\ l^{-1}$	0.28125	
Initial concentration of biomass at inoculation	$C_{X,0}$	C-mmol X l^{-1}	0.01[a]	
Initial concentration of acetate	$C_{S,0}$	mmol l^{-1}	7.8125 [b]	
Initial concentration of ammonium	$C_{NH4,0}$	mmol N l^{-1}	3.5714 [b]	
Initial concentration of bicarbonate	$C_{HCO,0}$	mmol l^{-1}	2.5	
Initial concentrations of all other materials	$C_{i,0}$	mmol i l^{-1}	0	

[a] The reactor is inoculated with a very low amount of cells. An initial concentration of 0.01 C-mmol X l^{-1} corresponds to 0.2 mg VSS l^{-1} which is about 10,000× lower than typical concentrations of activated sludge of 2-3 g VSS l^{-1}.

[b] Concentrations of 500 mg COD l^{-1} of acetate and 50 mg N-NH_4^+ l^{-1} typically used in laboratory studies with synthetic wastewater relates to 7.8125 mmol acetate l^{-1} and 3.57 mmol N l^{-1}, respectively.

Conversions of acetate, bicarbonate, nitrogen compounds (ammonium, nitrite, nitrate) and dissolved oxygen by the BNR guilds were simulated over 150 h. Other materials were not limiting, and their state variable was not simulated. An aeration process was included to deliver dissolved oxygen in the bulk liquid phase (eq. 2.7). Inhibitions, pH, acid-base speciation, and liquid/gas phase equilibria (*e.g.*, CO_2) were not implemented.

$$\text{Oxygen transfer rate:} \qquad OTR = k_La \cdot (C_{O2}^{sat} - C_{O2}) \tag{2.7}$$

with volumetric mass transfer coefficient (k_La, h^{-1}), saturation concentration of dissolved oxygen (C_{O2}^{sat}, mmol $O_2\ l^{-1}$), actual concentration of dissolved oxygen (C_{O2}, mmol $O_2\ l^{-1}$).

BNR growth processes were implemented using stoichiometric and kinetic constants (yields, maximum growth rate, decay rate) calculated via thermodynamics (tables 2.15 and 2.17) and summarised in Table 2.20b. Decay was assumed to generate soluble substrate (simplified here as additional acetate). Affinity constants (K_S values) were assumed by default at 0.1 mmol l^{-1}.

Table 2.20b Constants used for batch and chemostat simulations: stoichiometry and kinetics.

Stoichiometric and kinetic constants	Symbol	Units	OHOs	DHOs	CMOs	AOOs	NOOs	AMOs
Limiting substrate [a]			Acetate	Acetate	Ammonium	Ammonium	Nitrite	Ammonium
Maximum yield of biomass growth on acetate	$Y_{X/Acc}^{max}$	(C-mmol X mmol^{-1} Ace)	0.933	0.845				
Maximum yield of biomass growth on ammonium	$Y_{X/NH4}^{max}$	(C-mmol X mmol^{-1} N)	5		0.086	0.067		0.094
Maximum yield of biomass growth on nitrite	$Y_{X/NO2}^{max}$	(C-mmol X mmol^{-1} N)					0.020	0.079
Maximum growth rate	μ_{max}	(h^{-1})	0.722	0.677	0.036	0.037	0.033	0.051
Decay rate	k_d	(h^{-1})	0.0050	0.0048	0.0011	0.0011	0.0011	0.0012
Half saturation constant for limiting substrate [b]	K_{Si}	(mmol S$_i$ l^{-1})	0.1	0.1	0.1	0.1	0.1	0.1

[a] Based on the conditions of the simulations, the limiting substrate was set here as the e-donor of each guild.

[b] A low value *ca.* 100× lower than the initial concentration was considered as the K_S value. Some K_S values can be found in literature, but their measurement is affected by experimental conditions and diffusional resistances in bioaggregates.

Simulation results in batch

Simulation outputs in batch are given in Figure 2.7. Chemoorganoheterotrophs (OHOs and to some extent DHOs) lead in the batch. OHOs are selected by rapidly consuming acetate over the first 10 h. They consume ammonium up to half of its initial concentration. As soon as the acetate is depleted, OHOs stop growing. Over the next 50 h, OHOs maintain themselves to some extent by growing on additional soluble organic matter generated by their decay, when ammonium is still available as the N source. When ammonium is depleted after 65 h, OHOs decay. During OHO growth, the O_2 transfer rate is not sufficient to compensate for the O_2 consumption rate as indicated by the dip in O_2 profile. The aerator programmed is still relatively efficient ($k_L a$ of 100 h^{-1}), corresponding to a lab stirred-tank reactor with agitation to break air bubbles and maximise their surface area. You can play around with the $k_L a$ value in the provided Aquasim file to test less efficient aerations such as in shake flasks: O_2 concentration will rapidly drop to microaerophilic conditions during OHO growth.

Bicarbonate is produced by OHOs and used as the C source by chemolithoautotrophs (nitrifiers, AMOs). AOO activity is displayed by the second shoulder in the ammonium profile. AOOs grow on ammonium and produce nitrite. Nitrite is used by NOOs, producing nitrate. CMOs consume ammonium and produce nitrate. Nitrifiers are aerobic and use O_2, but are less active than OHOs, thus not significantly impacting the O_2 profile. As soon as nitrite is produced by AOOs and ammonium is still available, AMOs can potentially grow by anaerobically respiring ammonium with nitrite. In reality, AMOs are inhibited by O_2 and their growth does not happen when facing the O_2 level present at saturation in the batch simulation. As soon as ammonium is depleted, nitrifiers and anammox bacteria decay.

DHOs establish mainly thanks to the additional organic matter supplied by the decay of OHOs which rapidly scavenge acetate, and by using nitrate produced by nitrifiers. When ammonium is depleted no more nitrate is produced. When the residual nitrate is depleted, denitrifiers decay.

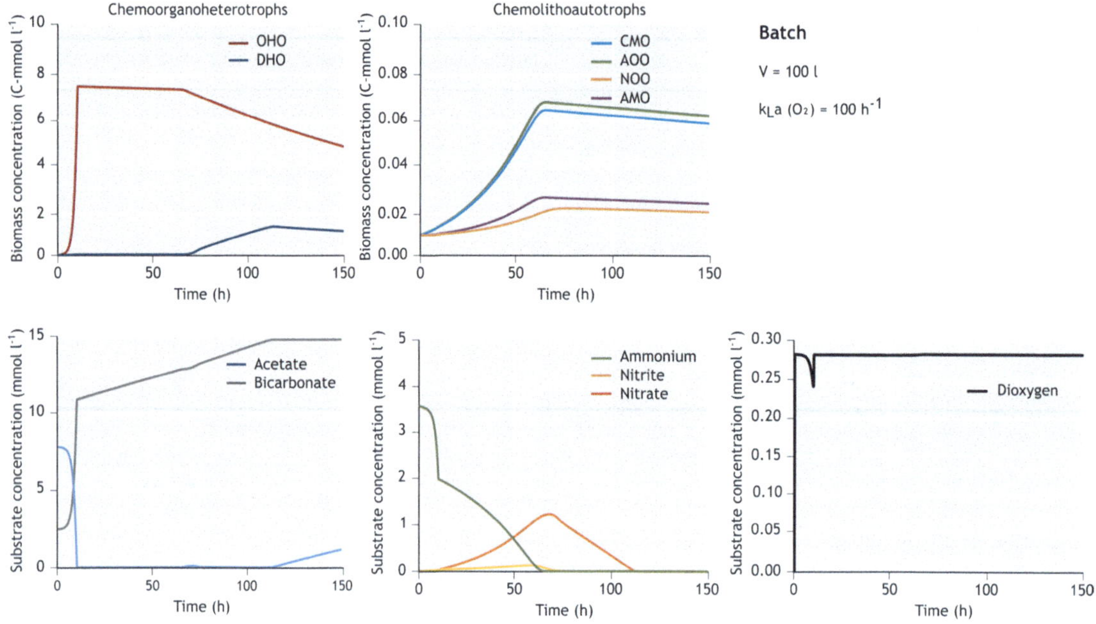

Figure 2.7 Simulation of microbial growth, selection, conversions, and interactions of the BNR guilds in the batch over 150 h. The mathematical model was implemented with growth and decay processes. An aeration process was included to transfer O₂ in the liquid phase.

c) *Chemostat simulations of growth, selection, and interaction of BNR guilds*

<u>Simulation conditions in chemostat</u>

The same 100-litre stirred-tank reactor used for the batch was simulated as a chemostat over 3 years by including an influent and effluent flow rate (Q) set initially at 1.0 l h^{-1} (Table 2.20). The low dilution rate (D = Q/V) of 0.01 h^{-1} (or long HRT of 100 h) should allow all guilds to grow, with respect to their theoretical μ_{max} from 0.033 (NOOs) to 0.722 (OHOs) h^{-1}.

<u>Simulation results in chemostat</u>

The simulation results in chemostat are given in Figure 2.8. Chemoorganoheterotrophs (OHOs, DHOs) rapidly establish and predominate in the chemostat, reaching their steady-state biomass concentration after *ca.* 500 h (*i.e.*, 20 d). They consume their limiting acetate substrate to a residual concentration of 0.007 mmol l^{-1}. OHOs use O₂ which is continuously supplied and reaches a level close to saturation at steady state. DHOs grow by respiring acetate with nitrate produced by nitrifiers. Since acetate is continuously supplied with the influent, DHOs grow as soon as nitrifiers are active. Bicarbonate is produced by OHOs and DHOs, reaching 12 mmol HCO$_3^-$ l^{-1} at steady state. This inorganic C source supports chemolithoautotrophs (nitrifiers and AMOs). In this multi-species culture, organisms face simultaneously multiple limitations. In simulations, all K$_S$ values were set at 0.1 mmol l^{-1} for all guilds. The cumulated limitations hamper chemolithoautotrophs to achieve an effective μ (0.0056-0.0085 h^{-1}) at the level of the imposed dilution rate (0.01 h^{-1}) although initially set low versus the respective μ_{max} (Table 2.21). While AOOs and NOOs are selected, CMOs and AMOs display the lowest μ and cannot persist and compete for nitrogen. AMOs and CMOs are most impacted by the multiple limitations and are outcompeted by AOOs and NOOs.

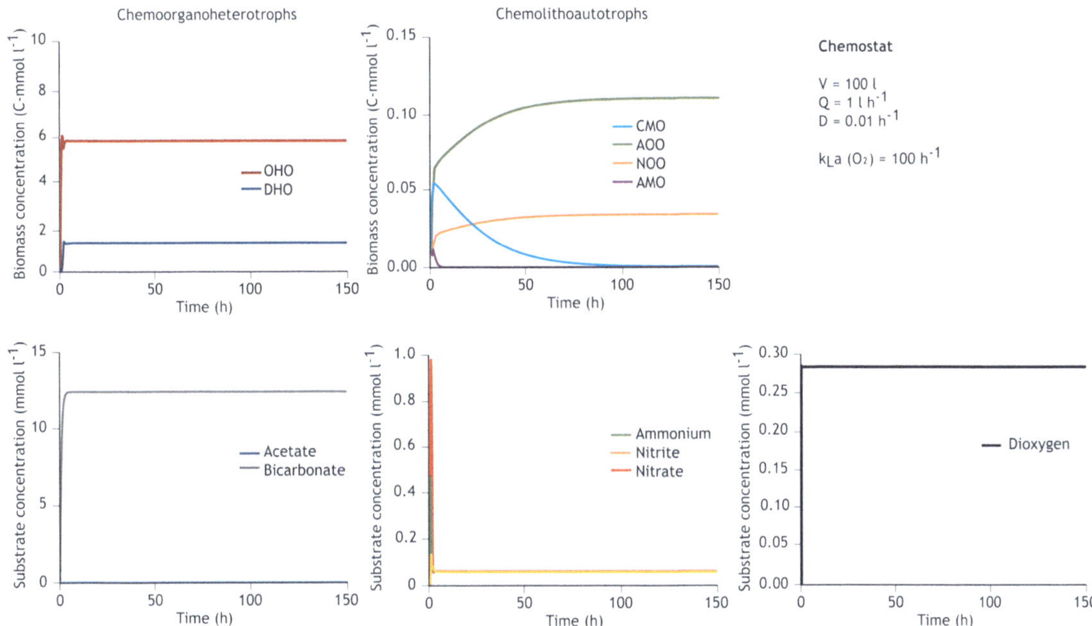

Figure 2.8 Simulation of microbial growth, selection, conversions, and interactions of the BNR guilds in the chemostat operated at a low dilution rate of 0.01 h^{-1} that should allow all guilds to grow (μ_{max} = 0.033-0.722 h^{-1}). Simulations were performed over 3 years (25,000 h) to reach steady states for all microbial guilds. The mathematical model was implemented with growth and decay processes. An aeration process was included to transfer O_2 in the liquid phase.

Table 2.21 Cumulated effects of multiple limitations on growth rates of guilds studied in the chemostat.

	μ_{max} (h^{-1})	Cumulated Monod terms effect [a]	Resulting μ (h^{-1}) = $\mu_{max} \cdot$ Monod terms	Dilution rate D (h^{-1})
Chemoorganoheterotrophs				0.01
OHOs	0.722	0.016	0.0114	
DHOs	0.677	0.022	0.0148	
Chemolithoautotrophs				
CMOs	0.036	0.229	0.0082	
AOOs	0.037	0.229	0.0085	
NOOs	0.033	0.257	0.0085	
AMOs	0.051	0.109	0.0056	

[a] Calculated as the multiplication of Monod terms ($M_i = S_i (K_i + S_i)^{-1}$) for an organism:
OHOs: $M_{Ace} \cdot M_{O2} \cdot M_{NH4}$; DHOs: $M_{Ace} \cdot M_{NO3}$; CMOs and AOOs: $M_{NH4} \cdot M_{O2} \cdot M_{CO2}$; NOOs: $M_{NO2} \cdot M_{O2} \cdot M_{CO2}$;
AMOs: $M_{NH4} \cdot M_{NO2} \cdot M_{CO2}$

Variations of Q (and D) were tested on microbial selections (Table 2.22). OHOs are selected on most of the D values (<0.7 h^{-1}) until D becomes too close to their μ_{max}. At 0.03 h^{-1}, OHOs, AOOs and AMOs are selected. NOOs and CMOs are outcompeted: nitrate is not produced and DHOs cannot grow. By neglecting O_2-inhibition, AMOs use nitrite produced by AOOs. This is the only condition in which AMOs grow.

At 0.01-0.02 h^{-1}, OHOs, AOOs, NOOs and DHOs are selected. DHOs thrive on nitrate supplied by NOOs. At 0.005 h^{-1}, CMOs also establish. At 0.001 h^{-1}, only OHOs are selected with the limited supply of substrates.

Table 2.22 Effect of the dilution rate (D) on the selection of BNR guilds in the chemostat. Legend: selected guilds ('+') and washed-out guilds ('−').

V (l)	Q (l h^{-1})	D (h^{-1})	OHOs	DHOs	AOOs	NOOs	CMOs	AMOs
100	70	0.700	−	−	−	−	−	−
	60	0.600	+	−	−	−	−	−
	50	0.500	+	−	−	−	−	−
	5.0	0.050	+	−	−	−	−	−
	4.0	0.040	+	−	−	−	−	−
	3.0	0.030	+	−	+	−	−	+
	2.0	0.020	+	+	+	+	−	−
	1.0	0.010	+	+	+	+	−	−
	0.5	0.005	+	+	+	+	+	−
	0.1	0.001	+	−	−	−	−	−
	0	0	−	−	−	−	−	−

Outlook from growth simulations
Simulations highlight the power of simple growth models to analyse microbial selection and interactions in a BNR mixed culture and predict microbial processes at higher scale in WWTPs.

The implemented models can be used to address questions on microbial selection that you can address critically, creatively, and collectively by exchanging with peers. You might test effects of, *e.g.*, reactor regimes (different D in a chemostat), physiological parameters values (μ_{max}, K_S, k_d), environmental conditions (T, pH), and number and type of microbial guilds present (from pure to co- and mixed cultures), among many other possibilities.

Model development and testing requires a structured investigation approach. Write down all the implementations and modifications in an (electronic) lab journal. Just as wet-lab experiments, dry-lab computations require an 'experimental' design and traceability.

As for any mathematical model, calibration of model outputs with experimental data is important: see chapters 14 and 17 in the textbook on modelling of activated sludge processes. Building the model at the start of the investigations can provide a preliminary understanding of microbial phenomena. This helps to reflect on the key variables, parameters, and factors to elucidate, and supports the design of experiments for measuring these to improve the model.

Example 2.3.5

Propose measurements for the microbial conversions and selections in the BNR mixed culture
Propose analytical methods to measure the conversions of materials, the selection of microorganisms, and their ecophysiologies in the BNR mixed culture. Highlight the pros/cons of the methods identified.

Option: outline the bioreactor system and your analytical plan.

Solution

Measurements of microbial conversions and selections

Conversions of materials can be measured from the liquid, solid, and gas phases of the bioreactor using physicochemical analyses either in-line (*e.g.*, pH and dissolved oxygen sensors), on-line (*e.g.*, O_2, CO_2, N_2 and N_yO_x from the off-gas by FT-IR and MS gas analysers) or off-line (*e.g.*, dissolved substrates and products by colorimetric kits and liquid or ion chromatographs; biomass by measurements of total/inorganic/volatile suspended solids) (Table 2.23).

Table 2.23 Analytical methods to measure targeted materials involved in the metabolisms of BNR organisms.

Phase	Material	Formula	In-line sensors	On-line analysers	Off-line measurements	
Gas	Dioxygen [a]	O_2 (g)		FT-IR spectroscopy		
	Carbon dioxide [a]	CO_2 (g)		Micro-GC		
	Dinitrogen	N_2 (g)		Mass spectrometer		
	N_yO_x intermediates	NO, N_2O (g)				
Liquid	Temperature (T) [b]	-	Temperature sensor			
	Redox (rH) [b]	-	Redox sensor			
	Electrical conductivity (σ) [b]	-	Electrical conductivity sensor			
	Protons	H^+ (aq)	pH sensor			
	Dioxygen [a]	O_2 (aq)	Dissolved oxygen sensor			
	Bicarbonate [a]	HCO_3^- (aq)			Alkalinity measurement	
	Organic matter	$C_1H_{1.63}O_{0.45}$			COD	
	Acetate	$C_2H_3O_2^-$ (aq)			colourimetric kits, TOC	HPLC, Ion chromatograph
	Ammonium	NH_4^+ (aq)	Ion selective electrode [c]			
	Nitrite	NO_2^- (aq)	Ion selective electrode? [c]			
	Nitrate	NO_3^- (aq)	Ion selective electrode [c]		NH_4^+, NO_2^-, NO_3^-, PO_4^{3-} colourimetric kits	
	Phosphate	$H2PO_4^-$, HPO_4^{2-} (aq)	Phosphate-selective polymer membrane electrode? [d] Electrical conductivity sensor? [d]			Discrete analyser, Ion chromatograph
Solid	Biomass	$C_1H_{1.8}O_{0.5}N_{0.2}$ (s)			TSS, ISS (ash), VSS measurements	

[a] Dioxygen and carbon dioxide can be measured from both the gas and liquid phase.

[b] Temperature, redox, and electrical conductivity measurements can provide important additional information on the bioreactor environment of microorganisms.

[c] Ion selective electrodes (ISEs) are handy for in-line measurements of dissolved ammonium and nitrogen oxides. However, since ISEs tend to progressively drift on a continuous process, these measurements mostly provide trends and should be verified by punctual off-line measurements. ISEs are frequently used in WWTPs for process control based on ammonium and nitrate measurements. Nitrite ISEs have lately been developed for use in laboratories.

[d] Phosphate is an important additional parameter to measure. Although not modelled here, phosphate is an important component of wastewater as: (*i*) a nutrient pollutant that needs to be removed, (*ii*) an important medium component since assimilated in biomass, and (*iii*) a buffer capacity agent. ISEs are not available for phosphate measurements. Recently, phosphate-selective polymer membrane electrodes have been developed for implementation for wastewater analyses. An alternative is to measure phosphate indirectly by electrical conductivity: this works well for EBPR systems where phosphate evolutions mostly compose the conductivity profile; but this is more challenging in full BNR since conductivity is impacted by the sum of ions present.

For further information, refer to standard methods for the examination of water and wastewater (APHA-AWWA-WEF, 2012) and experimental methods for wastewater treatment (Van Loosdrecht *et al.,* 2016).

Analytical targets can be identified from growth stoichiometries (Table 2.15) and simulations (figures 2.7 and 2.8). Measuring the limiting substrate and biomass helps to derive rates and verify yields. Measuring organics helps to track chemoorganotrophic activities (OHOs, DHOs). Ammonium, nitrite, and nitrate help to follow conversions by nitrifiers, denitrifiers, and anammox (either directly from the reactor or in separate batch tests), and check nitrogen removals and balances. Accurate on-line measurements of gas compounds help to verify stoichiometries.

Molecular biology analyses help track microbial populations in the mixed culture. A first set of 3-4 simple methods (microscopy, amplicon sequencing, qPCR and/or FISH) is suggested to measure microbial selection on top of chemical conversions.

Microscopy. Phase-contrast microscopy provides a first overview of the aspect of the microbial community. It delivers information on the predominance of prokaryotic or eukaryotic cells, and on the aspect of microorganisms (*e.g.*, if unfavourable filamentous bacteria are present).

Amplicon sequencing. 16S (archaea, bacteria) and 18S (eukaryotes) rRNA gene amplicon sequencings provide fingerprints of microbial communities. The DNA pool is extracted from mixed liquor samples. Primers for PCR amplification of hypervariable regions of rRNA genes are tested upfront *in silico* for their taxonomic coverage. Main populations forming the BNR guilds are identified among other flanking lineages, together with their relative abundance.

qPCR and FISH. Quantitative PCR (qPCR) and rRNA-targeted fluorescence *in-situ* hybridisation (FISH) track and localise specific populations over time and in bioaggregates, using specific qPCR primer pairs and FISH probes. qPCR provides high sensitivity to detect microbial selection early. Although called 'quantitative', qPCR is prone to variations depending on operators, labs, and protocols (Agrawal *et al.,*2021). FISH provides a visual, semi-quantitative information on relative abundances of guilds/populations in the biomass and their localisation in bioaggregates (Nielsen *et al.,*2009). Advanced meta-omics can then provide a deep scientific understanding of metabolic processes.

Meta-omics. Metagenomics provides the overview of microbial populations and functional genes present in the biomass. Metagenome-assembled genomes can be binned and annotated to uncover the functional genetic potential of single lineages. Analysing genetic expressions into RNA (RT-qPCR, metatranscriptomics) and proteins (metaproteomics) gives information on metabolic activation and regulation of populations (Cerruti *et al.,* 2021). RNA and protein sequences are associated with functions (genes, enzymes) and populations. Combining metagenomics and metaproteomics can be used, *e.g.*, to identify denitrifiers active on the denitrification pathways. When imposing reactor perturbation (*e.g.*, temperature), metagenomics and time-resolved metatranscriptomics can provide information on the up/down-regulation of catabolic pathways of populations of the nitrogen cycle. Intermediate intra/extra-cellular metabolites can be tracked by mass spectrometry (metabolomics). However, relating metabolites to specific lineages remains challenging.

All molecular methods are susceptible to biases, notably related to the extraction of the highly ordered informational macromolecules (*e.g.*, DNA, RNA, proteins) from biomass.

<u>Ecophysiology.</u> Metabolic traits of microbes are tracked by ecophysiology. Biomass samples are incubated with a labelled substrate. Stable isotope probing (SIP) uses non-radioactive labels (*e.g.*, ^{13}C, ^{15}N). Microautoradiography uses radioactive heavy isotopes (*e.g.*, ^{14}C, ^{3}H, $^{32/33}P$, ^{35}S). It is possible to identify populations that assimilate (anabolism) the (in)organic carbon and nitrogen resources. SIP followed by sequencing (DNA-SIP, RNA-SIP) or mass spectrometry (protein-SIP) help identify these active populations. Microautoradiography and FISH help localise them.

Continuous and batch experimentations in the lab serve to validate hypotheses derived from the theoretical approach proposed in this chapter, or to estimate unknown parameters that can be used in modelling, and to calibrate mathematical models (photo: IHE Delft).

2.4 EXERCISES

From individual metabolisms to ecosystem interactions

Exercise 2.4.1

Microbial competition

In this exercise, you study the competition for a single, growth-limiting substrate between chemolithoautotrophic anammox bacteria (AMOs; hereafter referred to as AMX) which reduce nitrite with ammonium as e-donor, and chemoorganoheterotrophic denitrifiers (DHOs; hereafter DEN) which reduce nitrite with acetate as e-donor.

First, all the required stoichiometric and kinetic parameters to describe the metabolism and growth of the two guilds are derived. Subsequently, the obtained parameters are implemented into simple mass balances for stirred-tank reactors. Finally, the competition outcomes for continuous-flow stirred-tank reactors (CSTR, or chemostat) and sequencing batch reactors (SBR) are compared. For the chemostat, the results of both the dynamic and the steady-state analyses are discussed.

All solutions are provided in Annex 1. The calculations are implemented in three interactive spreadsheets provided as online supplementary material (SI_Chapter_2_AMX_DEN_[...].xlsx). An overview of the spreadsheets structure is provided in Figure 2.9, where the stoichiometric and kinetic parts build on the approach proposed by Kleerebezem and van Loosdrecht (2010).

a) Stoichiometry and growth-limiting compound

Derive the individual stoichiometry of the catabolic and anabolic reactions of AMX and DEN. Next, derive the stoichiometry of their overall metabolism using the thermodynamic-based method to estimate the maximum biomass yield. Consider that the two guilds differ in their anabolic carbon source: AMX are autotrophs (inorganic C source) and DEN are heterotrophs (organic C source). Assume that both use ammonium as the anabolic nitrogen source. The operational temperature is 30 °C. Based on the derived stoichiometries, identify the growth-limiting compound for both guilds in the case of an influent containing 1 mmol l^{-1} of nitrite (*i.e.*, 14.0 mg NO_2^--N l^{-1}), 1.05 mmol l^{-1} of acetate (*i.e.*, 70.4 g COD l^{-1}), and 1.2 mmol l^{-1} of ammonium (*i.e.*, 16.8 mg NH_4^+-N l^{-1}).

Note: the proposed concentrations are in the range of those encountered in aerobically pre-treated municipal wastewaters used as influents to mainstream anammox applications (Laureni *et al.*, 2015). Nevertheless, these concentrations were primarily chosen in this exercise because they give the best visual display of the conversions in the output graphs.

b) Kinetic parameters estimation

Derive the biomass-specific maximum uptake rate for the identified growth-limiting compound j (q_j^{max}), and the biomass-specific substrate consumption rate for maintenance (m_S) for both guilds. For the affinity constant for the identified growth-limiting compound j (K_j), assume a value of 0.4 mmol l^{-1} for AMX and 4 mmol l^{-1} for DEN.

c) General mass balance

Establish the general mass balance equation for any soluble compound (C_i) or suspended material (X_i), involved in the active biological reactions, in a stirred-tank reactor with coinciding influent and effluent flows.

d) Mass balances for the biomass and the growth-limiting compound
Establish the two mass balances that define the behaviour of the entire system, *i.e.*, for the biomass and the growth-limiting compound, and subsequently derive the mass balances for all the other non-limiting compounds for each guild.

e) Chemostat dynamics: individual guild
Consider an influent containing 1 mmol l^{-1} of nitrite, 1.05 mmol l^{-1} of acetate, and 1.2 mmol l^{-1} of ammonium, and an operational temperature of 30 °C. Resolve how the concentrations of the biomass and the growth-limiting compound evolve over time in a chemostat run at a dilution rate D of 0.03 and 0.10 h^{-1}. Assume that AMX and DEN are individually inoculated at 0.2 $mmol_X$ l^{-1}, and that the initial concentrations of ammonium, nitrite and acetate in the reactor (C_i at t=0) equal the ones in the influent (C_i^0). Hint: if you approximate the derivatives with finite differences, you can use a simple spreadsheet to visualise the results.

f) Chemostat dynamics: multiple guild competition
Building on the previous task 1e, resolve the competition dynamics in the same chemostat when AMX and DEN are inoculated simultaneously at an initial concentration of 0.2 $mmol_X$ l^{-1} each.

g) Chemostat steady state: single and multiple guilds
Individually resolve for AMX and DEN the steady-state concentrations of the different compounds (*i.e.*, ammonium, nitrite, acetate, and biomass) for D ranging between 0-0.3 h^{-1}. Based on the results, identify the D value marking the transition from an AMX-dominated to a DEN-dominated enrichment.

h) SBR dynamics: multiple guild competition
Resolve the competition outcome between AMX and DEN in a SBR operated with the same initial conditions (*i.e.*, 1 mmol l^{-1} of nitrite, 1.05 mmol l^{-1} of acetate, and 1.2 mmol l^{-1} of ammonium, and 0.2 $mmol_X$ l^{-1} for each guild) and fed with the same influent at a D of 0.03 h^{-1}. Consider a SBR cycle length of 4 h and compare the results with the chemostat case. Hint: for the SBR, in order to impose a desired D similarly to the chemostat case, you can make the following assumptions:

- the feeding and discharge steps are instantaneous;
- the SBR cycle does not comprise a settling phase, thus the bulk is always homogeneous;
- the hydraulic and solids retention times coincide (HRT = SRT), and are a direct function of the fraction of homogeneously mixed bulk volume removed at the end of each cycle;
- the fraction of biomass and any other dissolved compound instantaneously removed at the end of each cycle is equal to the product of the selected cycle length (h) and the imposed D (h^{-1}). For example, at a cycle length of 4 h and a D of 0.03 h^{-1}, 12 % of each compound is removed at the end of a cycle;
- the feeding results in a concentration increase for each compound at the beginning of each cycle equivalent to its influent concentration, e.g., $\Delta C_{NO_2^-} = C_{NO_2^-}^0 = 1.0 \ mmol_{NO_2^-} l^{-1}$.

Exercise 2.4.2
Microbial commensalism
In this exercise, you explore the metabolic interaction between ammonium-oxidising bacteria (AOO; hereafter AOB) and nitrite-oxidising bacteria (NOO; hereafter NOB). Both guilds use dissolved oxygen as the terminal e-acceptor in their catabolism, yet their e-donors differ. AOB use the ammonium present in the influent, while NOB rely on the nitrite resulting from ammonium oxidation by AOB. Assume that the provided oxygen is not

limiting (*i.e.*, AOB and NOB do not compete for it): this is a typical example of microbial commensalism, where one guild relies on the product of another guild that remains unaffected.

Just as in Exercise 2.4.1, all the stoichiometric and kinetic parameters are derived first, and subsequently implemented into a simple mass balance for a chemostat. After solving the dynamic behaviour of the system, use the steady-state analysis to reflect on the different (co-)existence regions defined by the chosen dilution rate D.

The solutions are provided in Annex 1. Calculations are implemented in an interactive spreadsheet with the same structure as in Figure 2.9, provided as online supplementary material (SI_Chapter_2_AOB_NOB _Chemostat_dynamics.xlsx).

a) Stoichiometry and kinetics
Any microbial growth system can be fully described by one stoichiometric and three kinetic parameters. Enumerate and estimate, for each guild, the three parameters for which a thermodynamic-based estimation framework is available. Assume an operational temperature of 25 °C and that the system is oxygen replete: the dissolved oxygen concentration at any given time exceeds by far the oxygen affinity constant of both AOB and NOB. Under these conditions, ammonium and nitrite become the growth-limiting compound for AOB and NOB, respectively. Assume a value of 0.5 $\mathrm{mmol}_{NH_4^+}\,l^{-1}$ and 0.5 $\mathrm{mmol}_{NO_2^-}\,l^{-1}$ for their respective affinity constants. For the benefit of graph readability, the suggested values are significantly higher than the ones commonly reported in literature (Vannecke and Volcke, 2015).

You are encouraged to repeat the calculations with different values.

b) Chemostat dynamics and steady state
Consider a fully aerated chemostat operated at 25 °C and receiving an influent with 1.0 mmol l^{-1} of ammonium. AOB and NOB are inoculated at 0.2 $\mathrm{mmol}_X\,l^{-1}$ each. The initial ammonium concentration in the bulk equals the influent one.

Just as in Exercise 2.4.1, (*i*) resolve how the concentrations of nitrite and the two AOB and NOB biomasses evolve over time at a D of 0.012 h^{-1}, and (*ii*) discuss the steady-state concentrations of nitrite and the two biomasses for D in the range 0-0.3 h^{-1}.

Subsequently, consider the case of an influent containing 1.0 mmol l^{-1} of both ammonium and nitrite, and discuss the implications on NOB abundance in the absence or presence of an accompanying AOB population. Resolve the dynamics of nitrite and biomass concentrations over time for two D of 0.017 and 0.022 h^{-1}, prior to discussing their steady-state concentrations for D in the range 0-0.3 h^{-1}.

ANNEX 1: SOLUTIONS TO EXERCISES

From individual metabolisms to ecosystem interactions

Solution 2.4.1

Microbial competition

a) Stoichiometry and growth-limiting compound

The values of all the stoichiometric coefficients and the sign preceding them in each reaction are derived from elemental and charge balances (see the online supplementary material: SI_Chapter_2_AMX_DEN _Chemostat_dynamics.xlsx and Figure 2.9):

- first, the elemental balance of the central atom of the e-donor and e-acceptor couples is closed by manually defining the corresponding stoichiometric coefficients (*e.g.*, $-1/+0.5$ for NO_2^-/N_2; values in blue font);
- next, the elemental balances for O, H, and the charge are balanced by calculating the required stoichiometric coefficients for H_2O, H^+ and e^-, in this given order;
- finally, the balanced half reactions are used to derive the full stoichiometry of each reaction and its thermodynamic properties.

The anammox stoichiometry is derived and discussed in examples 2.12 and 2.13 in the textbook. The overall metabolism derived from thermodynamics is reiterated here:

$$-10.7\ NH_4^+ - 12.6\ NO_2^- - HCO_3^- - 0.8\ H^+ + 10.5\ N_2 + 2.1\ NO_3^- + CH_{1.8}O_{0.5}N_{0.2} + 21.4\ H_2O = 0$$

with a $\Delta G^{01}(30\ °C) = -327.9\ kJ\ mol^{-1}_{NH_4^+}$.

The metabolism of the chemoorganoheterotrophic denitrifiers oxidising acetate (*i.e.*, the catabolic e-donor and anabolic C source) with nitrite as e-acceptor, and growing on ammonium as the N source is calculated like this:

Cat: $\ -C_2H_3O_2^{-1} - 2.67\ NO_2^- - 1.67\ H^+ + 1.33\ N_2 + 2\ HCO_3^- + 1.33\ H_2O = 0$

An: $\ -0.53\ C_2H_3O_2^{-1} - 0.2\ NH_4^+ - 0.28\ H^+ + CH_{1.8}O_{0.5}N_{0.2} + 0.05\ HCO_3^- + 0.4\ H_2O = 0$

Based on the derived stoichiometries, one can calculate $\Delta G_{Cat}^{01}(30\ °C) = -968.9\ kJ\ mol^{-1}_{C_2H_3O_2^-}$ and $\Delta G_{An}^{01}(30\ °C) = 29.2\ kJ\ mol^{-1}_X$, and $\Delta G_{Diss}^{01} = 432.1\ kJ\ mol^{-1}_X$. Importantly, ΔG_{Cat}^{01} and ΔG_{An}^{01} always need to be corrected for the actual temperature, while ΔG_{Diss}^{01} can be considered temperature-independent (Tijhuis *et al.*, 1993). Accordingly, λ_{Cat}^* equals $0.48\ mol_{C_2H_3O_2^-}\ mol^{-1}_X$ and the overall denitrifiers metabolism at 30 °C becomes:

$$-C_2H_3O_2^{-1} - 1.27\ NO_2^- - 0.2\ NH_4^+ - 1.07\ H^+ + CH_{1.8}O_{0.5}N_{0.2} + 0.63\ N_2 + HCO_3^- + 1.03\ H_2O = 0$$

with an associated $\Delta G^{01}(30\ °C) = -432.1\ kJ\ mol^{-1}_{C_2H_3O_2^-}$ expressed per mole of acetate substrate, and a $Y_{X/C_2H_3O_2^-}^{max} = 0.5\ mol_X\ C\text{-}mol^{-1}_{C_2H_3O_2^-}$ consistent with reported experimental values.

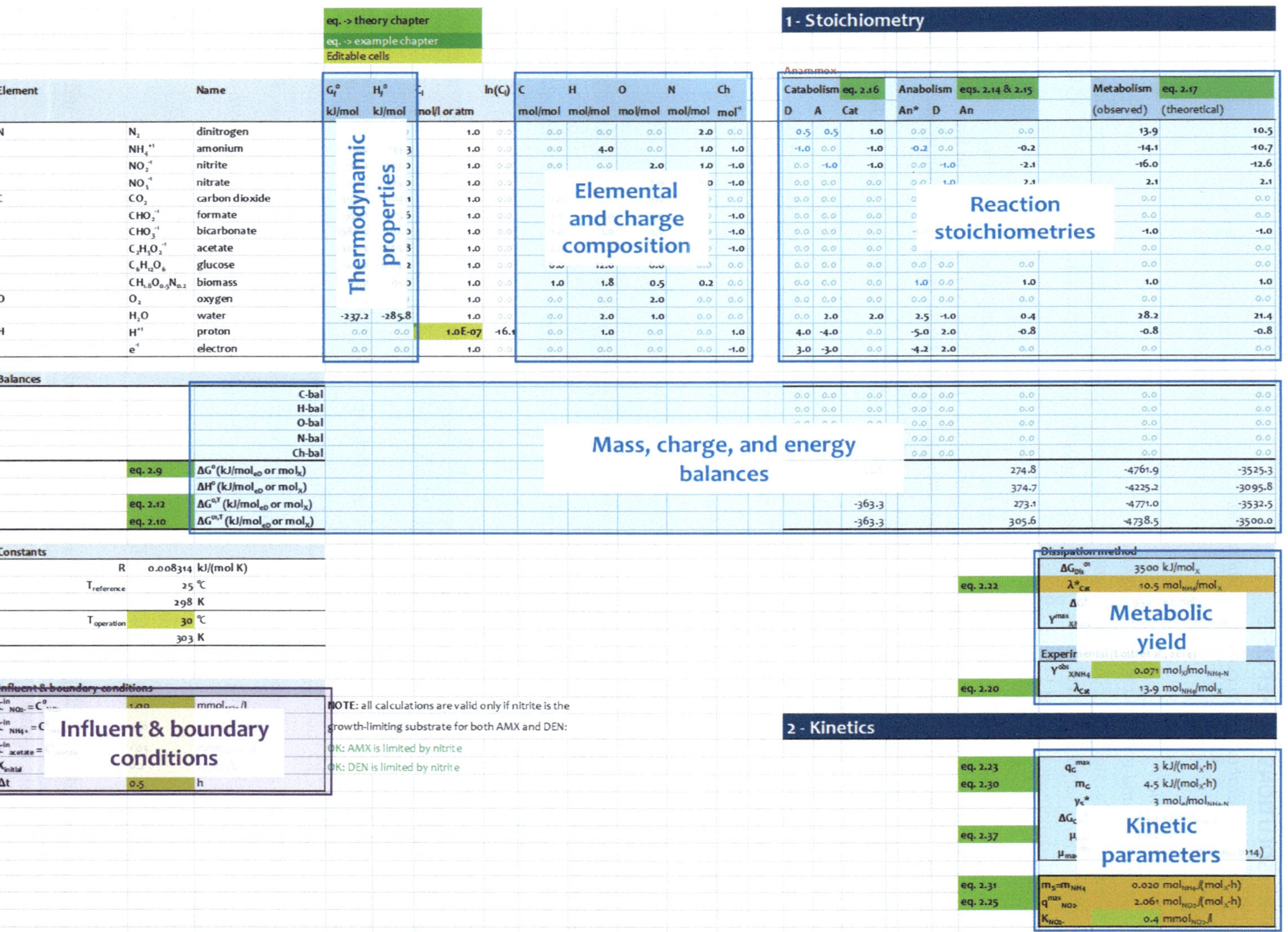

eq. -> theory chapter
eq. -> example chapter
Editable cells

1 - Stoichiometry

Anammox

Thermodynamic properties — Elemental and charge composition — Reaction stoichiometries

Element		Name	G_f° kJ/mol	H_f° kJ/mol	C_i mol/l or atm	$\ln(C_i)$	C mol/mol	H mol/mol	O mol/mol	N mol/mol	Ch mol$^+$	Catabolism eq. 2.16 D	A	Cat	Anabolism eqs. 2.14 & 2.15 An*	D	An	Metabolism eq. 2.17 (observed)	(theoretical)
N	N_2	dinitrogen			1.0	0.0	0.0	0.0	0.0	2.0	0.0	0.5	0.5	1.0	0.0	0.0	0.0	13.9	10.5
	NH_4^{+1}	amonium			1.0	0.0	0.0	4.0	0.0	1.0	1.0	-1.0	0.0	-1.0	-0.2	0.0	-0.2	-14.1	-10.7
	NO_2^{-1}	nitrite			1.0	0.0	0.0	0.0	2.0	1.0	-1.0	0.0	-1.0	-1.0	0.0	-1.0	-2.1	-16.0	-12.6
	NO_3^{-1}	nitrate			1.0						-1.0	0.0	0.0	0.0	0.0	1.0	2.1	2.1	2.1
C	CO_2	carbon dioxide			1.0						0.0	0.0	0.0	0.0				0.0	0.0
	CHO_2^{-1}	formate			1.0						-1.0	0.0	0.0	0.0				0.0	0.0
	CHO_3^{-1}	bicarbonate			1.0						-1.0	0.0	0.0	0.0				-1.0	-1.0
	$C_2H_3O_2^{-1}$	acetate			1.0						-1.0	0.0	0.0	0.0				0.0	0.0
	$C_6H_{12}O_6$	glucose			1.0						0.0	0.0	0.0	0.0	0.0	0.0	0.0	0.0	0.0
	$CH_{1.8}O_{0.5}N_{0.2}$	biomass			1.0		1.0	1.8	0.5	0.2	0.0	0.0	0.0	0.0	1.0	0.0	1.0	1.0	1.0
O	O_2	oxygen			1.0		0.0	0.0	2.0	0.0	0.0	0.0	0.0	0.0	0.0	0.0	0.0	0.0	0.0
	H_2O	water	-237.2	-285.8	1.0		0.0	2.0	1.0	0.0	0.0	0.0	2.0	2.0	2.5	-1.0	0.4	28.2	21.4
H	H^{+1}	proton	0.0	0.0	1.0E-07	-16.1	0.0	1.0	0.0	0.0	1.0	4.0	-4.0	0.0	-5.0	2.0	-0.8	-0.8	-0.8
	e^{-1}	electron	0.0	0.0	1.0	0.0	0.0	0.0	0.0	0.0	-1.0	3.0	-3.0	0.0	-4.2	2.0	0.0	0.0	0.0

Balances

Mass, charge, and energy balances

	quantity				
	C-bal	0.0	0.0	0.0	0.0
	H-bal	0.0	0.0	0.0	0.0
	O-bal	0.0	0.0	0.0	0.0
	N-bal	0.0	0.0	0.0	0.0
	Ch-bal	0.0	0.0	0.0	0.0
eq. 2.9	ΔG° (kJ/mol$_{eD}$ or mol$_x$)		274.8	-4761.9	-3525.3
	ΔH° (kJ/mol$_{eD}$ or mol$_x$)		374.7	-4225.2	-3095.8
eq. 2.12	$\Delta G^{0,T}$ (kJ/mol$_{eD}$ or mol$_x$)	-363.3	273.1	-4771.0	-3532.5
eq. 2.10	$\Delta G^{0x,T}$ (kJ/mol$_{eD}$ or mol$_x$)	-363.3	305.6	-4738.5	-3500.0

Constants

R	0.008314	kJ/(mol K)
$T_{reference}$	25	°C
	298	K
$T_{operation}$	30	°C
	303	K

Influent & boundary conditions

$C^{in}_{NO2-} = C^\circ_{\dots}$	1.00	mmol$_{\dots}$/l
$C^{in}_{NH4+} = C_{\dots}$		
$C^{in}_{acetate} = $		
$X_{initial}$		
Δt	0.5	h

NOTE: all calculations are valid only if nitrite is the growth-limiting substrate for both AMX and DEN:
OK: AMX is limited by nitrite
OK: DEN is limited by nitrite

Dissipation method — Metabolic yield

	ΔG_{Dis}^{01}	3500	kJ/mol$_x$
eq. 2.22	λ^*_{Cat}	10.5	mol$_{NH4}$/mol$_x$
	Δ		
	$Y^{max}_{X/\dots}$		

Experimental (Lotti et al., 2014)

	$Y^{obs}_{X/NH4}$	0.071	mol$_x$/mol$_{NH4-N}$
eq. 2.20	λ_{Cat}	13.9	mol$_{NH4}$/mol$_x$

2 - Kinetics

Kinetic parameters

eq. 2.23	q_C^{max}	3	kJ/(mol$_x$·h)
eq. 2.30	m_C	4.5	kJ/(mol$_x$·h)
	y_c^*	3	mol$_a$/mol$_{NH4-N}$
	ΔG_C		
eq. 2.37	μ		
	μ_{max}		(…14)
eq. 2.31	$m_S = m_{NH4}$	0.020	mol$_{NH4}$/(mol$_x$·h)
eq. 2.25	q^{max}_{NO2-}	2.061	mol$_{NO2}$/(mol$_x$·h)
	K_{NO2-}	0.4	mmol$_{NO2-}$/l

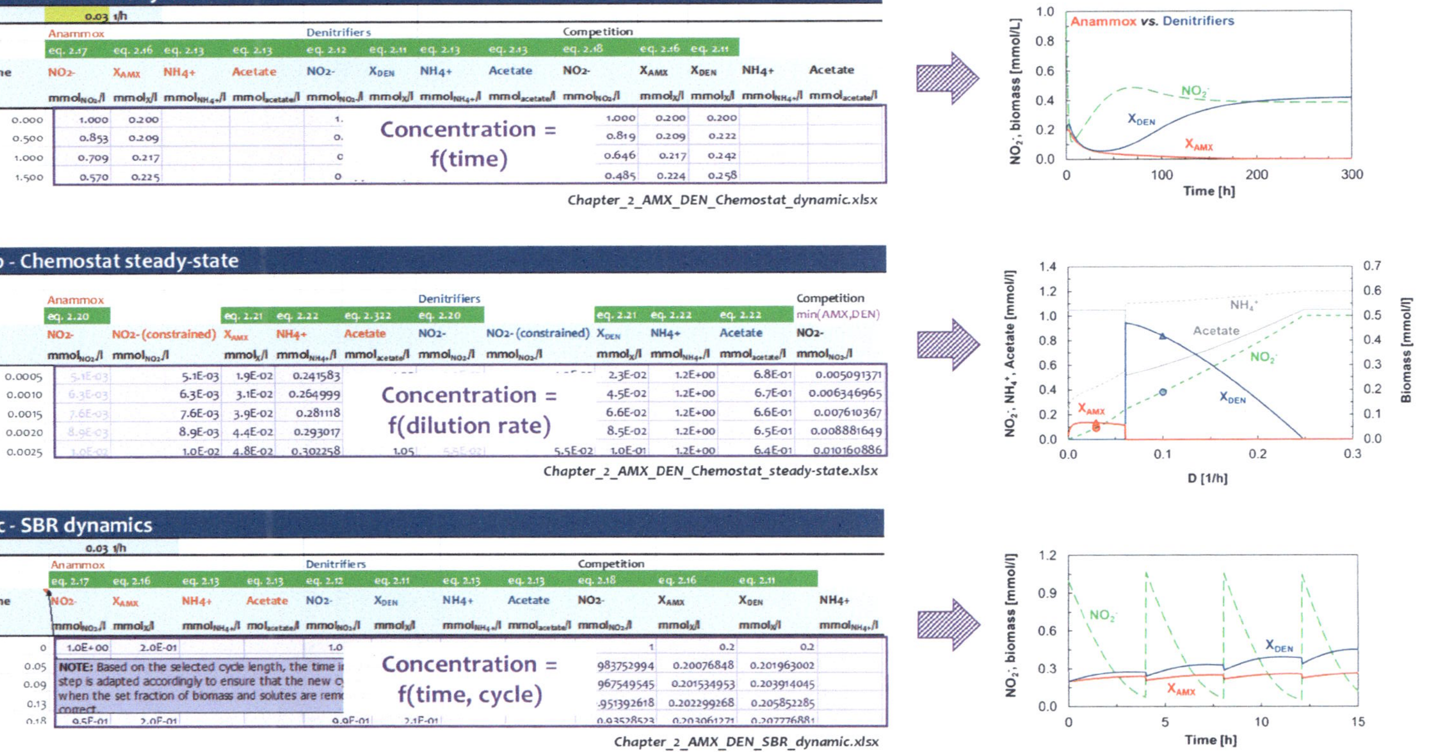

Figure 2.9 Overview of the spreadsheet structure for the calculation of reaction stoichiometries, their thermodynamic properties, and the estimation of the associated kinetic parameters (light blue shaded areas in the previous page; based on Kleerebezem and van Loosdrecht, 2010) and for the calculation of the dynamic and steady-state values for substrates, products and biomass concentrations for the chemostat and sequencing batch reactor (SBR) cases (purple shaded areas). Yellow cells highlight the four parameters required to fully describe a microbial growth system, namely the maximum biomass-specific growth yield, the maximum biomass specific substrate uptake rate, the biomass-specific substrate consumption rate for maintenance, and the affinity constant for the growth-limiting compound. Light and dark green cells refer to the number of the corresponding equations in the chapter in the textbook and in this example & exercise chapter, respectively. Chartreuse coloured cells can be edited. The password to make the whole spreadsheet editable is BWWT_EE. The name of the corresponding spreadsheet in the online supplementary materials is also provided.

Based on the two derived growth stoichiometries of the anammox and denitrifiers, and on the defined influent characteristics, the limiting influent compound can be identified. Anammox and denitrifiers need $12.6 / 10.7 = 1.18$ $mol_{NO_2^-}$ $mol_{NH_4^+}^{-1}$ and 1.27 $mol_{NO_2^-}$ $mol_{C_2H_3O_2^-}^{-1}$, respectively. The influent contains less than one mole of nitrite per mole of ammonium or acetate. Therefore, nitrite is the limiting compound for both guilds.

b) Kinetic parameters estimation

A maximum biomass-specific growth rate (μ_{AMX}^{max}) of 0.162 h^{-1} (30 °C) has been estimated for anammox in Example 2.17 in the textbook. It will be used for the calculations in the following sections. Be aware that the estimated value is one order of magnitude higher than the one experimentally observed under optimal conditions (*e.g.*, 0.0142 h^{-1} at 30 °C) (Lotti *et al.*,2015; Zhang *et al.*,2017).

Similarly, the maximum biomass-specific growth rate of denitrifiers (μ_{DEN}^{max}) at 30 °C can be calculated based on eq. 2.35 in the textbook, considering a catabolic Gibbs free energy change ($\Delta G_{Cat}^{01}(30\ °C)$) of -967.8 kJ $mol_{C_2H_3O_2^-}^{-1}$, a Gibbs energy dissipation (ΔG_{Diss}^{01}) of 432.1 kJ mol_X^{-1}, and the eight electrons that are transferred in the oxidation of acetate to CO_2 ($\gamma_S^* = 8$ mol_e $mol_{C_2H_3O_2^-}^{-1}$):

$$\mu^{max} = \frac{q_e^{max} \cdot \dfrac{\Delta G_{Cat}^{01}}{\gamma_S^*} + m_G}{-\Delta G_{Diss}^{01}} - \frac{3\ mol_e\ mol_{X_{DEN}}^{-1}\ h^{-1} \cdot \dfrac{-967.8\ kJ\ mol_{C_2H_3O_2^-}^{-1}}{8\ mol_e\ mol_{C_2H_3O_2^-}^{-1}} + 4.5\ kJ\ mol_{X_{DEN}}^{-1}\ h^{-1}}{-432.1\ kJ\ mol_{X_{DEN}}^{-1}}$$

$$\cdot \exp\left\{-\frac{69\ kJ\ mol^{-1}}{0.008314\ kJ\ mol^{-1}\ K^{-1}} \cdot \left(\frac{1}{303K} - \frac{1}{298K}\right)\right\} = 1.313\ h^{-1}$$

In this case as well, the value is higher than the ones commonly reported in literature (*e.g.*, 0.086 h^{-1} at 20 °C) (van den Berg *et al.*, 2016). Yet, μ_{DEN}^{max} remains in all cases consistently higher than μ_{AMX}^{max}. As we will see later, this provides DEN with a strong competitive advantage in environments where the growth rate plays a prominent role.

As an additional exercise, you could explore the competition outcome with the experimentally determined values. Pay attention to the temperature at which the parameters were determined.

Next, the biomass-specific substrate consumption rate for maintenance can be calculated based on eqs. 2.28 and 2.29 in the textbook. At 30 °C, for anammox it becomes:

$$m_{AMX,NH_4^+} = -Y_{NH_4^+/NH_4^+}^{Cat} \cdot \frac{m_G}{\Delta G_{Cat}^{01}} =$$

$$= -\left(1\ \frac{mol_{NH_4^+}}{mol_{NH_4^+}}\right) \cdot \frac{4.5\ kJ\ mol_{X_{AMX}}^{-1}\ h^{-1} \cdot \exp\left\{-\dfrac{69\ kJ\ mol^{-1}}{0.008314\ kJ\ mol^{-1}\ K^{-1}} \cdot \left(\dfrac{1}{303K} - \dfrac{1}{298K}\right)\right\}}{-363.3\ kJ\ mol_{NH_4^+}^{-1}}$$

$$= 0.02\ mol_{NH_4^+}\ mol_{X_{AMX}}^{-1}\ h^{-1}$$

Similarly, the acetate consumption rate for maintenance for the denitrifiers ($m_{DEN,C_2H_3O_2^-}$) at 30 °C is $0.007\ mol_{C_2H_3O_2^-}\ mol_{X_{DEN}}^{-1}\ h^{-1}$.

The biomass-specific consumption rate for the growth-limiting compound (*i.e.*, nitrite; $q_{NO_2^-}^{max}$) can now be calculated for both guilds based on eq. 2.25 in the textbook and the respective μ^{max} and m_S calculated above. The resulting rates at 30 °C equal $2.06\ mol_{NO_2^-}\ mol_{X_{AMX}}^{-1}\ h^{-1}$ for AMX and $1.68\ mol_{NO_2^-}\ mol_{X_{DEN}}^{-1}\ h^{-1}$ for DEN.

In these calculations, you should pay attention to the units of all the individual terms in the equations and to the involved stoichiometric yields.

Lastly, to fully characterise the growth systems, the affinity constant for the growth-limiting substrate ($K_{NO_2^-}$) is needed. However, there is no bioenergetic framework for its estimation. Thus, $K_{NO_2^-}$ needs to be determined experimentally or taken from literature. For this exercise, and for the benefit of graphical readability, we assume $K_{NO_2^-}^{AMX} = 0.4\ mmol_{NO_2^-}\ l^{-1}$ and $K_{NO_2^-}^{DEN} = 4\ mmol_{NO_2^-}\ l^{-1}$. Note that one order of magnitude difference in affinity constant values is often reported in literature.

You are invited to further explore the impact of different values of the affinity constants on the competition outcome, e.g., 0.002-$0.3\ mmol_{NO_2^-}\ l^{-1}$ for anammox (Oshiki *et al.*, 2016) and $0.015\ mmol_{NO_2^-}\ l^{-1}$ for denitrifiers (Hiatt and Grady, 2008).

All the calculations can be found in supporting information (SI_Chapter_2_AMX_DEN _Chemostat_dynamics.xlsx).

c) General mass balance
The mass balance for any compound *i* in a biologically active stirred-tank reactor with equal incoming and outgoing liquid flows is defined as follows:

$$V \cdot \frac{dC_i}{dt} = Q \cdot \left(C_i^0 - C_i\right) + \left(\pm q_i\right) \cdot C_X \cdot V$$

where, V is the working volume of the reactor (l), Q is the influent and effluent volumetric flow rate (l h^{-1}), C_i^0 and C_i are the concentration of compound *i* in the influent and bulk, respectively (mol$_i$ l^{-1}), q_i is the biomass-specific consumption or production rate of compound i (mol$_i$ mol$_X^{-1}$ h^{-1}), and C_X is the biomass concentration of the guild catalysing the reaction (mol$_X$ l^{-1}; hereafter referred to as X).

The q_i rates are the ones derived in task b. By definition, the rate values are positive. A preceding positive or negative sign is used if compound *i* is produced or consumed, respectively.

If the volume is constant, the same mass balance equation can be rearranged as follows:

56

$$\frac{dC_i}{dt} = D \cdot \left(C_i^0 - C_i\right) + \left(\pm q_i\right) \cdot X \tag{2.8}$$

where D is the dilution rate (Q/V in h^{-1}). D makes it possible to discuss the system without the need to define a volumetric flow or reactor volume.

d) Mass balances for the biomass and the growth-limiting compound
We assume here a single growth-limiting compound j (not necessarily coinciding with the catabolic e-donor) for each guild. In this framework, the overall behaviour of any growth system is defined by the mass balances of the growth-limiting compound and the biomass (catalyst), whose growth is limited by it. The growth-limiting compound is identified based on the given influent concentrations and the derived growth process stoichiometry. The behaviour of all the other compounds is then directly derived based on the overall growth stoichiometry.

Considering that $q_X = \mu$ ($mol_X\ mol_X^{-1}\ h^{-1}$ or h^{-1}), the biomass mass balance is defined as:

$$\frac{dX}{dt} = D \cdot \left(X^0 - X\right) + \mu \cdot X$$

According to the Herbert-Pirt equation (eq. 2.25 in the textbook), the biomass-specific consumption rate of any compound *i* can be defined as follows:

$$q_i = Y_{i/X}^{max} \cdot \mu + Y_{i/S}^{Cat} \cdot m_S \tag{textbook Eq. 2.33}$$

where $Y_{i/X}^{max}$ ($mol_X\ mol_X^{-1}$) and $Y_{i/S}^{Cat}$ ($mol_i\ mol_S^{-1}$) are the stoichiometric coefficients of compound i in the overall metabolic and catabolic reactions defined per unit biomass and substrate, respectively; m_S ($mol_S\ mol_X^{-1}\ h^{-1}$) is the biomass-specific substrate (catabolic e-donor) consumption rate for maintenance. $Y_{i/S}^{Cat}$, often equal to unity (when i = S), is here explicitly included for the general case where the limiting compound does not coincide with the e-donor of the catabolism ($i = j \neq S$). The maintenance-related term equals zero when compound *i* is solely involved in the anabolism.

Also, we know that the biomass-specific consumption rate of the growth-limiting compound (q_j) is a function of the actual concentration of the growth-limiting compound (C_j):

$$q_j = q_j^{max} \cdot \frac{C_j}{C_j + K_j} \tag{2.9; adapted from the textbook Eq. 2.30}$$

where q_j^{max} is the maximum biomass-specific consumption rate of the growth-limiting compound j ($mol_j\ mol_X^{-1}\ h^{-1}$), and K_j is the affinity constant for the growth-limiting compound j ($mol_j\ l^{-1}$). By combining eq. 2.33 in the textbook with eq. 2.9 here, the biomass-specific growth rate can be expressed as:

$$\mu = \frac{1}{Y^{max}_{j/X}} \cdot \left(q^{max}_j \cdot \frac{C_j}{C_j + K_j} - Y^{Cat}_{j/S} \cdot m_S \right) \tag{2.10}$$

and the mass balance for the biomass becomes:

$$\frac{dX}{dt} = D \cdot \left(X^0 - X \right) + \frac{1}{Y^{max}_{j/X}} \cdot \left(q^{max}_j \cdot \frac{C_j}{C_j + K_j} - Y^{Cat}_{j/S} \cdot m_S \right) \cdot X \tag{2.11}$$

where commonly $X^0 = 0$ (*i.e.*, influent is devoid of biomass). Similarly, by combining eqs. 2.8 and 2.9, the mass balance for the limiting compound j becomes:

$$\frac{dC_j}{dt} = D \cdot \left(C_j^0 - C_j \right) - q_j \cdot X = D \cdot \left(C_j^0 - C_j \right) + \left(- q^{max}_j \cdot \frac{C_j}{C_j + K_j} \right) \cdot X \tag{2.12}$$

where a negative sign precedes the reaction term as compound j is consumed. Differently, the sign is positive in eq. 2.11 as the biomass is produced. Moreover, as discussed in Exercise 2.4.2, when more guilds are involved, the growth-limiting compound j consumed by one guild might be produced by a co-existing guild.

Similarly, the mass balance of any other non-limiting compound i, as previously mentioned, is now simply derived by stoichiometry, combining eq. 2.8 here with eq. 2.33 from the textbook:

$$\frac{dC_i}{dt} = D \cdot \left(C_i^0 - C_i \right) \pm \left(Y^{max}_{i/X} \cdot \mu + Y^{Cat}_{i/S} \cdot m_S \right) \cdot X \tag{2.13}$$

Importantly, the above derived equations highlight how any microbial growth system is fully described by one parameter defining the overall growth stoichiometry, such as the maximum yield of consumption of the growth-limiting compound j per biomass produced ($Y^{max}_{j/X}$), and three kinetic parameters related to the growth-limiting compound j, namely the maximum biomass-specific consumption rate (q^{max}_j), the biomass-specific consumption rate for maintenance ($m_j = Y^{Cat}_{j/S} \cdot m_S$), and the affinity constant (K_j).

Specifically, considering the derived stoichiometries and the given influent concentrations, nitrite (NO_2^-) was already identified as the limiting compound for both AMX and DEN. Using AMX as an example along with the stoichiometric and kinetic values previously calculated, the mass balances for the AMX biomass and nitrite become:

58

$$\frac{dX_{AMX}}{dt} = D \cdot \left(X_{AMX}^0 - X_{AMX}\right) + \frac{1}{Y_{NO_2^-/X}^{max}} \cdot \left(q_{AMX,NO_2^-}^{max} \cdot \frac{C_{NO_2^-}}{C_{NO_2^-} + K_{NO_2^-}^{AMX}} - Y_{NO_2^-/NH_4^+}^{Cat} \cdot m_{AMX,NH_4^+}\right) \cdot X_{AMX}$$

$$= D \cdot \left(X_{AMX}^0 - X_{AMX}\right) + \frac{1}{12.6\ mol_{NO_2^-}\ mol_{X_{AMX}}^{-1}} \tag{2.14}$$

$$\cdot \left(2.06\ mol_{NO_2^-}\ mol_{X_{AMX}}^{-1}\ h^{-1} \cdot \frac{C_{NO_2^-}}{C_{NO_2^-} + 0.4\ mol_{NO_2^-}\ l^{-1}} - 1\ mol_{NO_2^-}\ mol_{NH_4^+}^{-1} \cdot 0.02\ mol_{NH_4^+}\ mol_{X_{AMX}}^{-1}\ h^{-1}\right) \cdot X_{AMX}$$

and

$$\frac{dC_{NO_2^-}}{dt} = D \cdot \left(C_{NO_2^-}^0 - C_{NO_2^-}\right) + \left(- q_{AMX,NO_2^-}^{max} \cdot \frac{C_{NO_2^-}}{C_{NO_2^-} + K_{NO_2^-}^{AMX}}\right) \cdot X_{AMX}$$

$$= D \cdot \left(C_{NO_2^-}^0 - C_{NO_2^-}\right) + \left(- 2.06\ mol_{NO_2^-}\ mol_{X_{AMX}}^{-1}\ h^{-1} \cdot \frac{C_{NO_2^-}}{C_{NO_2^-} + 0.4\ mol_{NO_2^-}\ l^{-1}}\right) \cdot X_{AMX} \tag{2.15}$$

For AMX, the catabolic substrate is ammonium ($S = NH_4^+$). Analogous equations can be derived for DEN with acetate instead being the catabolic substrate.

e) Chemostat dynamics: individual guild
The dynamics of the biomass and growth-limiting compound can be easily followed directly in a spreadsheet by solving the differential eqs. 2.11 and 2.12 by approximating the derivatives with finite differences (*i.e.*, $dC / dt \approx \Delta C / \Delta t$) (see supporting information, SI_Chapter_2_AMX_DEN_Chemostat_dynamics.xlsx).

For AMX (eq. 2.14 and eq. 2.15), this approach results in:

$$\frac{dX_{AMX}}{dt} \approx \frac{X_{AMX}(t + \Delta t) - X_{AMX}(t)}{\Delta t}$$

$$= D \cdot \left(-X_{AMX}(t)\right) + \frac{1}{Y_{NO_2^-/X_{AMX}}^{max}} \cdot \left(q_{AMX,NO_2^-}^{max} \cdot \frac{C_{NO_2^-}(t)}{C_{NO_2^-}(t) + K_{NO_2^-}^{AMX}} - Y_{NO_2^-/NH_4^+}^{Cat} \cdot m_{AMX,NH_4^+}\right) \cdot X_{AMX}(t) \tag{2.16}$$

and

$$\frac{dC_{NO_2^-}}{dt} \approx \frac{C_{NO_2^-}(t + \Delta t) - C_{NO_2^-}(t)}{\Delta t}$$

$$= D \cdot \left(C_{NO_2^-}^0 - C_{NO_2^-}(t)\right) + \left(- q_{AMX,NO_2^-}^{max} \cdot \frac{C_{NO_2^-}(t)}{C_{NO_2^-}(t) + K_{NO_2^-}^{AMX}}\right) \cdot X_{AMX}(t) \tag{2.17}$$

where no biomass is assumed to enter with the influent ($X_{AMX}^0 = 0$), and the two boundary conditions for the initial concentrations of biomass (0.2 $mmol_X\ l^{-1}$) and nitrite (1 mmol l^{-1}; equal to influent) are considered.

The same approach is used for DEN.

Figure 2.10 is obtained by explicitly calculating (and plotting) the concentrations at time $t + \Delta t$ as a function of the ones at time t.

You are invited to explore the impact of the chosen Δt in the finite differences approximation by increasing the used value of $\Delta t = 0.5$ h to, *e.g.*, 2 h.

Figure 2.10 shows how higher steady-state concentrations of nitrite (C_{NO_2}) are required to sustain the growth of both AMX and DEN at higher D, while resulting in lower biomass concentrations (X). In the next sections, you will see if this apparent monotonic relationship holds true for both the biomass and the growth-limiting compound over the entire range of D enabling growth.

At a dilution of 0.03 h^{-1}, the residual concentration of nitrite (0.096 $mmol_{NO_2}\,l^{-1}$; Figure 2.10A) resulting from AMX growth is lower than the one resulting from DEN growth (0.143 $mmol_{NO_2}\,l^{-1}$; Figure 2.10B). The opposite is true for a D of 0.1 h^{-1}, with 0.654 $mmol_{NO_2}\,l^{-1}$ for AMX and 0.383 $mmol_{NO_2}\,l^{-1}$ for DEN (Figure 2.10D,E). As a result, one may expect AMX to win the competition at 0.03 h^{-1} and DEN at 0.1 h^{-1}. This will be further discussed in the next section.

You are invited to study the effect of D close to 0.06 h^{-1} on the time needed to reach the steady state. At this D, the conditions favouring the two guilds are very similar and thus the washout of the least-fit guild takes much longer.

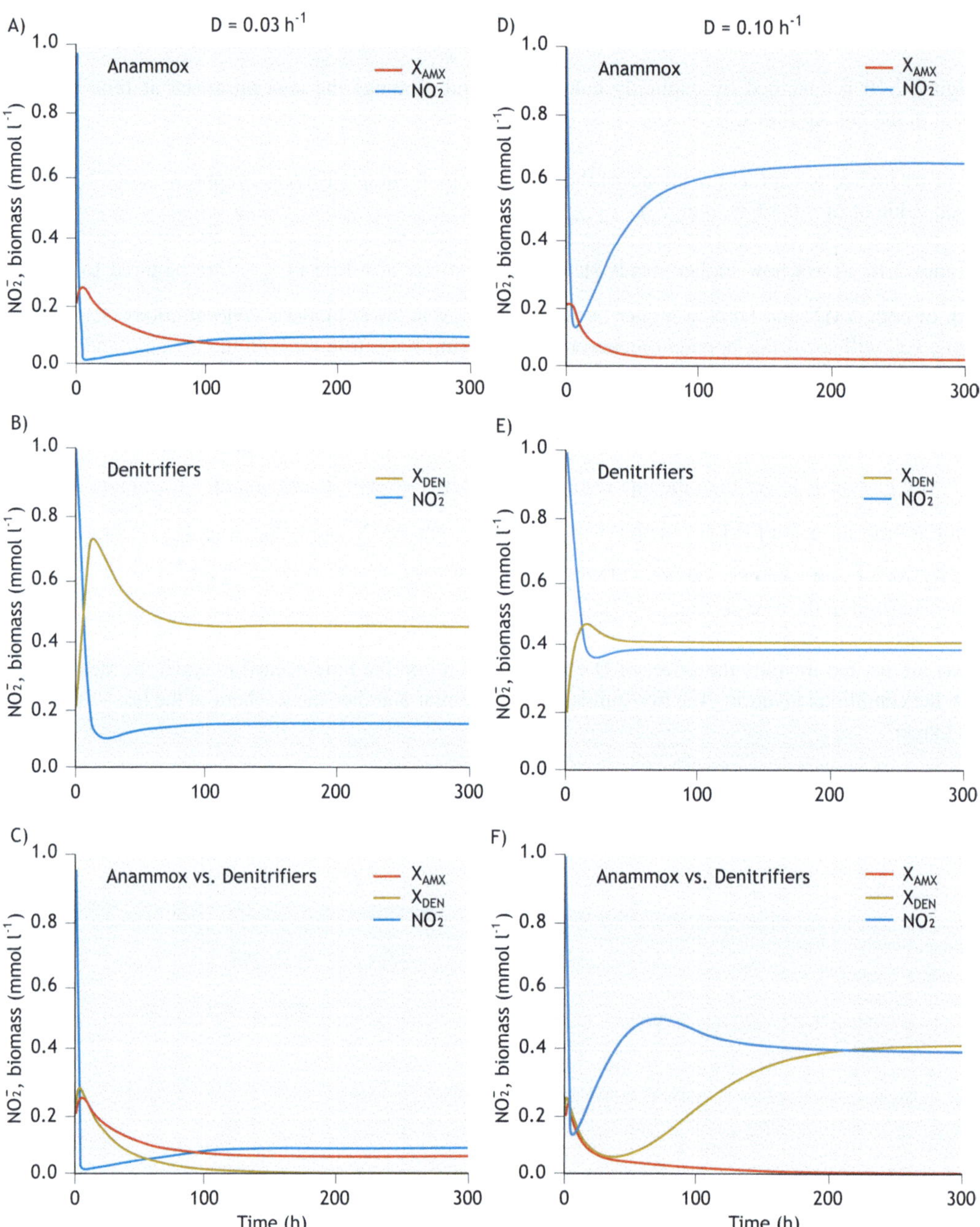

Figure 2.10. Dynamics of nitrite and biomass concentrations for the cases where AMX [A) and D)] and DEN [B) and E)] are individually present, or co-exist [C) and F)], at the two distinct dilution rates of 0.03 and 0.1 h⁻¹. For convenience, the concentrations of all soluble compounds at time zero (boundary conditions) are assumed equivalent to the ones in the influent (SI_Chapter_2_AMX_DEN_Chemostat_dynamics.xlsx).

f) Chemostat dynamics: multiple guilds competition

The approach presented in task 1e can directly be applied to study the competition between AMX and DEN in a chemostat (see SI_Chapter_2_AMX_DEN _Chemostat_dynamics.xlsx). The mass balance equations for the two biomasses (X_{AMX} and X_{DEN}), now initially co-existing in the reactor, remain the same and depend on the actual nitrite concentration. The only difference lies in the mass balance equation of nitrite, which is now consumed simultaneously as the growth-limiting compound by both AMX and DEN:

$$\frac{dC_{NO_2}}{dt} \approx \frac{C_{NO_2}(t+\Delta t) - C_{NO_2}(t)}{\Delta t} = D \cdot \left(C_{NO_2}^0 - C_{NO_2}(t)\right)$$

$$+ \left(-q_{AMX,NO_2}^{max} \cdot \frac{C_{NO_2}(t)}{C_{NO_2}(t) + K_{NO_2}^{AMX}}\right) \cdot X_{AMX}(t) + \left(-q_{DEN,NO_2}^{max} \cdot \frac{C_{NO_2}(t)}{C_{NO_2}(t) + K_{NO_2}^{DEN}}\right) \cdot X_{DEN}(t) \qquad (2.18)$$

The flexibility of this approach allows any additional nitrite-consuming or nitrite-producing guild k to be directly included in the equation by simply adding the corresponding q_{k,NO_2} term preceded by a negative or positive sign, respectively, while leaving the accumulation and transport term unchanged. In general, for any compound *i* the mass balance equation becomes:

$$\frac{dC_i}{dt} = D \cdot \left(C_i^0 - C_i\right) + \sum_k \left(\pm q_{k,i} \cdot X_k\right) \qquad (2.19)$$

where $q_{k,i}$ is calculated according to eq. 2.33 in the textbook if compound *i* is not limiting the growth of guild k, or eq. 2.9 if *i* is the growth-limiting compound for guild k.

This is equivalent to the stoichiometric-biokinetic matrix approach previously discussed in Example 2.3.3a (Table 2.6) and further detailed in Chapter 14 in the textbook on modelling of activated sludge processes: the overall consumption rate of any given compound results from the sum of the individual consumption or production rates across all the microbial processes involving it.

According to Figure 2.10C and F, and as hypothesised in the previous task 1e, AMX and DEN prevailed at a D of 0.03 and 0.1 h^{-1}, respectively. Both were present at the start and the nitrite concentration was significantly above their corresponding steady-state nitrite concentration. Interestingly, under these conditions, both guilds grew 'enthusiastically' at the beginning of the operation and considerably reduced the nitrite to below the minimum concentration required to sustain their respective growth at the given D. Progressively, each guild then reached its steady state or was washed out.

You are invited to calculate and plot the actual biomass-specific growth rates of AMX and DEN as a function of the concentration of nitrite in the bulk (eq. 2.10), and to compare the dynamics of this kinetic parameter over time with the imposed (constant) D.

g) Chemostat steady state: single and multiple guilds

The steady-state concentration of each compound can be analytically calculated by setting the accumulation term to zero in eqs. 2.11, 2.12 and 2.13, and by considering that $\mu = D$ is obtained from the biomass balance at steady state in a chemostat. All the calculations can be found in the supporting information (SI_Chapter_2_AMX_DEN _Chemostat_steady-state.xlsx).

If a single guild is involved, the effluent concentration of the growth-limiting compound is obtained from the mass balance equation of the biomass (eq. 2.11):

$$C_j = K_j \cdot \dfrac{D + \dfrac{1}{Y^{max}_{j/X}} \cdot \left(Y^{Cat}_{j/S} \cdot m_S\right)}{\dfrac{1}{Y^{max}_{j/X}} \cdot \left(q^{max}_j - Y^{Cat}_{j/S} \cdot m_S\right) - D} \tag{2.20}$$

Importantly, the residual concentration of the growth-limiting compound (C_j) is independent from its concentration in the influent (C_j^0) and from the actual biomass concentration in the reactor (X).

Similarly, the steady-state concentration of biomass can be derived from the mass balance of the growth-limiting compound (eq. 2.12):

$$X = D \cdot \dfrac{\left(C_j^0 - C_j\right)}{q_j} = D \cdot \dfrac{\left(C_j^0 - C_j\right) \cdot \left(C_j + K_j\right)}{q_j^{max} \cdot C_j} = D \cdot \dfrac{\left(C_j^0 - C_j\right)}{\left(Y^{max}_{j/X} \cdot D + Y^{Cat}_{j/S} \cdot m_S\right)} \tag{2.21}$$

Here, instead, the biomass concentration directly depends on the concentration of the growth-limiting compound in the influent (C_j^0). An increase in the substrate concentration in the influent leads to an increase in the biomass produced: the higher the influent concentration of substrate, the higher the biomass concentration in the reactor for a given D.

For all other non-limiting compounds i, their steady-state concentration is directly derived from eq. 2.13:

$$C_i = \dfrac{D \cdot C_i^0 \pm X \cdot \left(Y^{max}_{i/X} \cdot D + Y^{Cat}_{i/S} \cdot m_S\right)}{D} \tag{2.22}$$

For AMX, the derived equations become:

$$C_{NO_2^-} = K^{AMX}_{NO_2^-} \cdot \dfrac{D + \dfrac{1}{Y^{max}_{NO_2^-/X_{AMX}}} \cdot \left(Y^{Cat}_{NO_2^-/NH_4^+} \cdot m_{AMX,NH_4^+}\right)}{\dfrac{1}{Y^{max}_{NO_2^-/X_{AMX}}} \cdot \left(q^{max}_{AMX,NO_2} - Y^{Cat}_{NO_2^-/NH_4^+} \cdot m_{AMX,NH_4^+}\right) - D},$$

$$X_{AMX} = D \cdot \frac{\left(C^0_{NO_2} - C_{NO_2}\right)}{q_{AMX,NO_2}} = D \cdot \frac{\left(C^0_{NO_2} - C_{NO_2}\right) \cdot \left(C_{NO_2} + K^{AMX}_{NO_2}\right)}{q^{max}_{AMX,NO_2} \cdot C_{NO_2}}$$

and, considering ammonium to exemplify a non-limiting compound:

$$C_{NH_4^+} = \frac{D \cdot C^0_{NH_4^+} \pm X_{AMX} \cdot \left(Y^{max}_{NH_4^+ / X_{AMX}} \cdot D + Y^{Cat}_{NH_4^+ / NH_4^+} \cdot m_{AMX,NH_4^+} \right)}{D}$$

For AMX, the effluent concentration of acetate will inevitably be equal to the influent one, as this compound is not consumed in the described AMX metabolism.

Similarly, all steady-state concentrations can be derived for DEN.

The resulting steady-state concentrations as a function of D are presented in Figure 2.11A-B. Consistent with what was previously hypothesised, the residual NO_2^- concentration increases monotonically with D; higher steady-state nitrite concentrations are required to sustain growth, in other words, to guarantee that the actual growth rate is at least equivalent to the imposed dilution rate ($\mu = D$). Conversely, the biomass does not follow a monotonic trend. Initially, the biomass increases for increasing D, *i.e.*, there is more substrate provided per time unit to sustain growth. For higher D, the increasing wash-out outweighs the benefits of higher substrate inflows, and the biomass starts to decrease until the critical D is reached. Above this D, conversion and growth no longer occur (Figure 2.11A-B), and all the effluent concentrations equal the influent ones.

The value of D demarcating the shift between AMX and DEN can be identified by comparing the minimum residual nitrite concentration required by each guild to grow at the different D. According to Figure 2.11B, the minimum residual nitrite required by AMX is lower than the one of DEN up to a D of 0.061 h⁻¹, and for higher D values DEN can grow at lower nitrite concentrations. Thus, as a function of D, the metabolic system initially comprising AMX and DEN will behave according to Figure 2.11C. In the same figure, the two competition outcomes of Figure 2.10C and F are highlighted with red and blue marks, respectively. Importantly, for D higher than 0.249 h⁻¹, DEN are also washed out; their actual growth rate is lower than the imposed D even for nitrite concentrations equivalent to the influent one. This is evident from Figure 2.11D where the actual growth rate of both AMX and DEN are calculated according to eq. 2.10 as a function of the residual nitrite in Figure 2.11C.

You are invited to perform the same calculations with the experimentally determined μ^{max} and K_{NO_2} for both AMX and DEN as provided in the kinetic section.

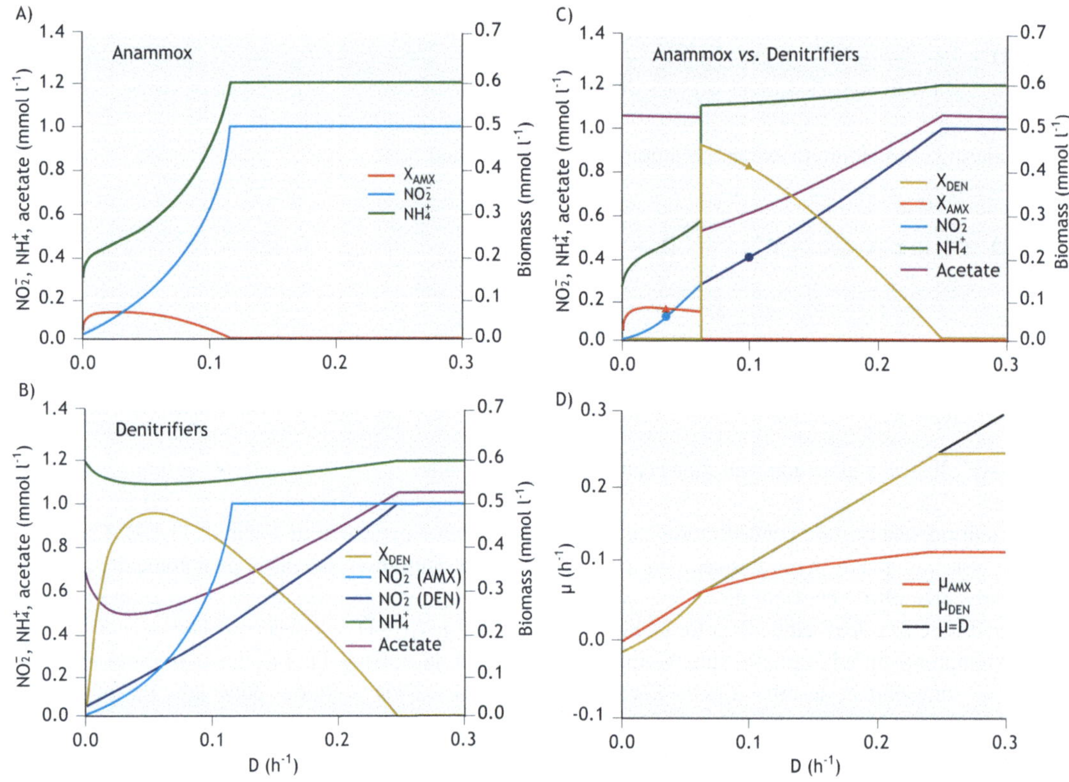

Figure 2.11. Steady-state concentrations of ammonium, nitrite, acetate, and biomass as a function of the imposed dilution rate D for the cases of A) AMX and B) DEN alone, and C) when they compete. In panel C), the steady-state concentrations for the two cases discussed in Figure 2.10C and F are highlighted. In panel D), the actual growth rate of each guild is calculated based on eq. 2.10 and the resulting nitrite concentration calculated in panel C) (SI_Chapter_2_AMX_DEN_Chemostat_steady-state.xlsx).

i)   SBR dynamics: multiple guilds competition

Based on eqs. 2.11, 2.12 and 2.13, a batch system can be simulated by setting the dilution rate D to zero, *i.e.*, there is no influent and no effluent flow. Considering an initial AMX and DEN biomass concentration of 0.2 $mmol_X$ l^{-1}, and initial concentrations of 1 mmol l^{-1} of nitrite, 1.05 mmol l^{-1} of acetate, and 1.2 mmol l^{-1} of ammonium, a single batch results in the concentrations profiles displayed in Figure 2.12.

The operation of a SBR consists of subsequent batches as in Figure 2.12. Part of the volume is removed at the end of a batch and is replaced by new feed at the start of the following one. The length of the batches (*i.e.*, the frequency of discharge and feeding), and the fraction of reactor volume that is exchanged determine the hydraulic retention time (HRT in h). In the absence of a settling phase, the sludge retention time (SRT in h) can be approximated reasonably well by the HRT for most engineering applications. In real settings, the concentration change of any compound after feeding is directly proportional to the volume exchanged and the concentrations in the influent. For the sake of simplicity, we assume here that the concentration increase at the beginning of a cycle is equal to the concentration in the influent.

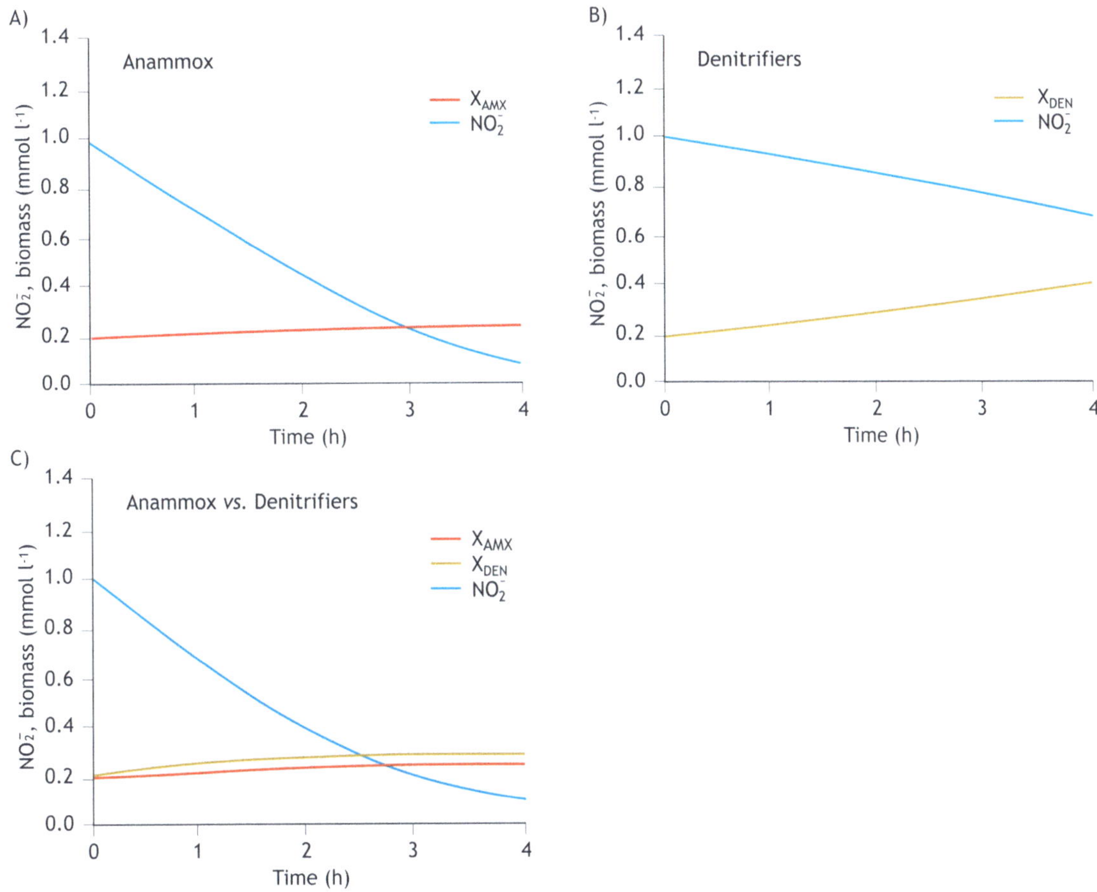

Figure 2.12. Biomass and nitrite concentration profiles over a single batch for A) AMX and B) DEN individually, and C) for the co-inoculation case (SI_Chapter_2_AMX_DEN_SBR_dynamic.xlsx).

According to Figure 2.13, DEN are enriched under the imposed conditions, namely a D of 0.03 h^{-1} and a SBR cycle length of 4 h. This contrasts with the chemostat case where AMX prevailed at a D of 0.03 h^{-1}. This is explained by the fact that the maximum biomass-specific growth rate (μ^{max}) plays a much more prominent role in batch operation where all substrates are present at non-limiting concentrations for most of the time. Conversely, the affinity for the growth-limiting compound (K_j) is the main selection driver in a chemostat.

You are invited to identify the highest D that allows for the retention of AMX in the SBR with all other conditions unchanged (answer: 0.0105 h^{-1}).

Following the cycle definition adopted here, for a given D, an increase in the cycle length results in a proportional increase in the fraction of biomass and dissolved compounds instantaneously removed at the end of each cycle. Also, longer cycles result in less frequent feeding events and thus a lower substrate load. Consequently, the pseudo steady-state concentration of biomass decreases in response to the reduced substrate availability.

You can use the provided spreadsheet (SI_Chapter_2_AMX_DEN _SBR_dynamics.xlsx) to assess the impact of longer SBR cycle lengths and discuss how different cycle durations impact the pseudo steady-state concentrations in the effluent.

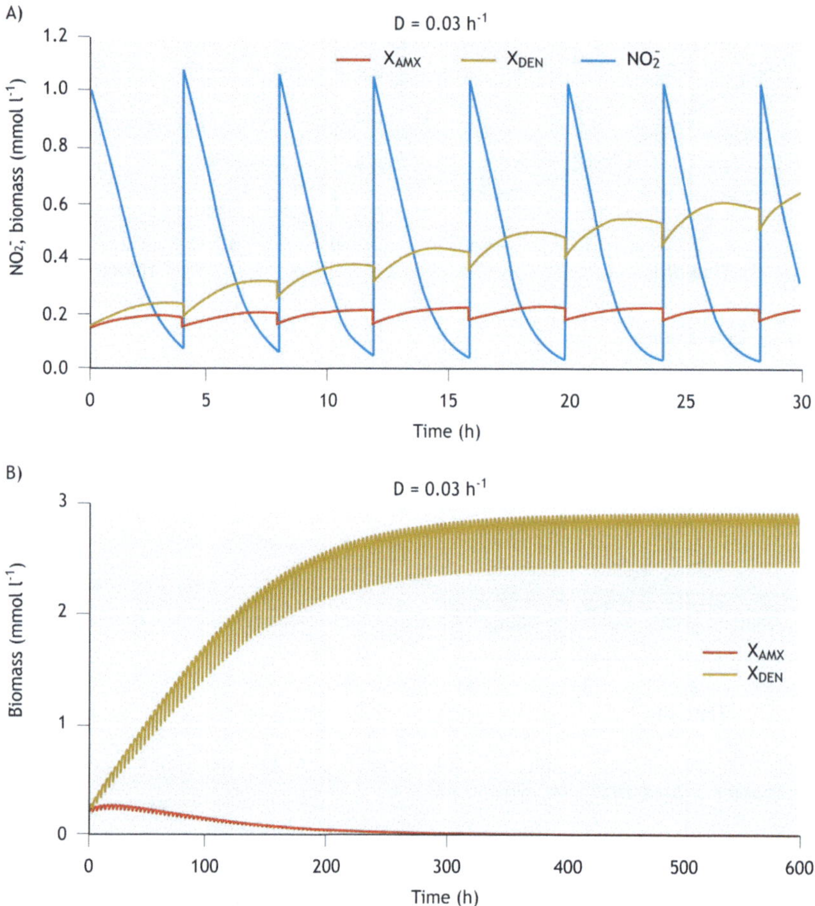

Figure 2.13. Concentration profiles of biomass and nitrite during A) the first cycles of the SBR, and B) their long-term dynamics for a D of 0.03 h⁻¹ (SI_Chapter_2_AMX_DEN_SBR_dynamic.xlsx).

Solution 2.4.2
Microbial commensalism
a) Stoichiometry and kinetics
Among the 4 parameters required to fully describe a microbial growth system, only 3 can be estimated on thermodynamic grounds:

i) maximum biomass yield ($Y_{X/i}^{max}$) on any substrate or product i involved in both the anabolism and catabolism;

ii) maximum biomass-specific uptake rate (q_i^{max}) of compound i, usually but not necessarily coinciding with the catabolic e-donor or the growth-limiting substrate;

iii) biomass-specific consumption rate for maintenance of compound i (m_i), also usually corresponding to the catabolic e-donor.

Subsequently, all other parameters can be expressed as a function of these three based on the defined process stoichiometries.

Conversely, the value of the affinity constant for the growth-limiting substrate (K_S) needs to be estimated experimentally or assumed based on literature.

The metabolism of AOB can be derived based on examples 2.10 and 2.6 in the textbook, describing their catabolism and the general anabolism of chemolithoautotrophs, respectively.

Next, the Gibbs free energy dissipation method yields a λ_{Cat}^{*} of 14.0 $mol_{NH_4^+}$ $mol_{X_{AOB}}^{-1}$, resulting in a $Y_{X/NH_4^+}^{max,AOB}$ = 0.067 mol_X $mol_{NH_4^+}^{-1}$ and the following overall AOB metabolism:

$$-14.9\ NH_4^+ - 21.0\ O_2 - 1.0\ CHO_3^- + CH_{1.8}O_{0.5}N_{0.2} + 14.7\ NO_2^- + 28.6\ H^+ + 15.1\ H_2O = 0$$

A maximum biomass-specific growth rate (μ_{AOB}^{max}) of 0.037 h^{-1} is calculated in Example 2.16 in the textbook. A biomass-specific ammonium consumption rate for maintenance (m_{AOB,NH_4^+}) of 0.017 $mol_{NH_4^+} mol_{X_{AOB}}^{-1} h^{-1}$ can be calculated. By combining these two values, the resulting biomass-specific consumption rate of the growth-limiting compound (*i.e.*, ammonium; $q_{AOB,NH_4^+}^{max}$) equals 0.573 $mol_{NH_4^+} mol_{X_{AOB}}^{-1} h^{-1}$.

Similarly, with an estimated $Y_{X/NO_2}^{max,NOB}$ = 0.020 mol_X $mol_{NO_2}^{-1}$, the overall metabolism of NOB becomes:

$$-50.2\ NO_2^- - 24.1\ O_2 - 0.2\ NH_4^+ - 0.8\ H^+ - 1.0\ CHO_3^- + CH_{1.8}O_{0.5}N_{0.2} + 50.2\ NO_2^- + 0.4\ H_2O = 0$$

along with a maximum biomass-specific growth rate (μ_{NOB}^{max}) of 0.033 h^{-1}, a biomass-specific nitrite consumption rate for maintenance (m_{NOB,NO_2^-}) of 0.057 $mol_{NO_2^-}$ $mol_{X_{NOB}}^{-1} h^{-1}$, and a resulting biomass-specific consumption rate of the growth-limiting compound ($q_{NOB,NO_2^-}^{max}$) of 1.695 $mol_{NO_2^-}$ $mol_{X_{NOB}}^{-1} h^{-1}$.

b) Chemostat dynamics and steady state

The dynamics of AOB, NOB and nitrite over time for the suggested D of 0.012 h^{-1}, and their steady-state concentrations for D in the range 0-0.3 h^{-1} are presented in Figure 2.14. For the detailed development of the underlying equations, you are referred to tasks c-g in the previous exercise 2.4.1.

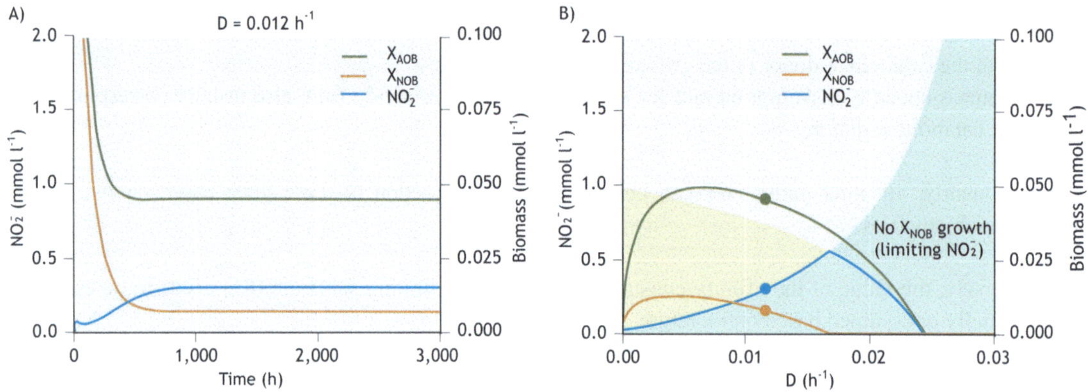

Figure 2.14 A) Dynamics in concentrations of nitrite, AOB, and NOB in the oxygen-replete chemostat operated at a D of 0.012 h⁻¹ and receiving an influent with 1 mmol l⁻¹ of ammonium. B) Steady-state nitrite concentrations available from AOB production (light green) and minimum required for NOB growth (dark green area represents where NOB cannot grow), and steady-state AOB and NOB concentrations. Marked points represent the steady state concentration reached in A). Note: the light green area (difference between NO₂⁻ produced by AOB and minimum NO₂⁻ required by NOB for a given D) is the actual amount of NO₂⁻ available for growth and maintenance (SI_Chapter_2_AOB_NOB_Chemostat_dynamics.xlsx).

Here, nitrite is produced by AOB, thus it is not their growth-limiting compound (eq. 2.13). At the same time, nitrite is the growth-limiting substrate for NOB (eq. 2.12). Accordingly, the equation describing the dynamics of the nitrite concentration over time becomes:

$$\frac{dC_{NO_2^-}}{dt} = D \cdot \left(C_{NO_2^-}^0 - C_{NO_2^-}\right)$$

$$+ \left(Y_{NO_2^-/X_{AOB}}^{max} \cdot D + Y_{NO_2^-/NH_4^+}^{Cat} \cdot m_{AOB,NH_4^+}\right) \cdot X_{AOB} + \left(- q_{NOB,NO_2^-}^{max} \cdot \frac{C_{NO_2^-}}{C_{NO_2^-} + K_{NO_2^-}^{NOB}}\right) \cdot X_{NOB} \qquad (2.23)$$

When calculating the steady-state NOB concentration (eq. 2.21), the 'influent' nitrite concentration for NOB (C_j^0) corresponds here to the steady-state nitrite concentration produced by AOB. The latter is calculated directly based on AOB stoichiometry (eq. 2.22) because ammonium and not nitrite is the growth-limiting compound for AOB.

Under the imposed conditions of influent ammonium oxidation at non-limiting O_2 availability, AOB and NOB co-exist at a D of 0.012 h⁻¹ (Figure 2.14A). The corresponding steady-state nitrite and biomass concentrations are marked in Figure 2.14B. In the latter panel, the border of the dark green-shaded region represents the minimum nitrite concentration required for NOB growth at the given D. It coincides with the expected steady-state effluent nitrite (eq. 2.20). Conversely, the border of the light green-shaded area represents the steady-state nitrite produced by AOB at the given D, and is a function of the influent growth-limiting ammonium concentration (eq. 2.22). The difference between the two areas (*i.e.*, the visible, light green-shaded region in Figure 2.14B) is the nitrite available for NOB growth (eq. 2.21). Consistently, at D above 0.017 h⁻¹, NOB are completely washed out, *i.e.*, the nitrite produced by AOB is not enough to sustain NOB growth. Thus, the resulting steady-state effluent nitrite concentration (blue line) coincides with the minimum required by NOB, until NOB are present, and with the nitrite produced by AOB afterwards. At D above 0.024 h^{-1}, the influent ammonium concentration is below the minimum required by AOB (not plotted) and thus AOB are washed out.

Nitrite can also be present in the influent, *e.g.*, produced in a partial nitritation reactor targeting equal ammonium and nitrite concentrations for subsequent anammox (Laureni *et al.*, 2015). Following the same approach developed in the previous sections, particular attention should be given to how the influent nitrite concentration is accounted for in eq. 2.23 and in the calculation of the steady-state NOB concentration (eq. 2.21). In the latter, C_j^0 actually corresponds to the sum of the constant nitrite in the influent and, if present, the steady-state nitrite concentration produced by AOB at the given dilution rate.

If nitrite is provided with the influent, NOB are independent from AOB for their growth. However, if AOB are also present, the additional nitrite source broadens the range of D allowing for NOB growth and increases the steady-state NOB concentration at any given D. This is shown in Figure 2.15A-B. At a D of 0.017 h^{-1} (Figure 2.15A), the presence of AOB increases the concentration of NOB at steady state from 0.008 (dashed yellow line) to 0.018 (full yellow line) mmol l^{-1}, while at a D of 0.022 h^{-1} (Figure 2.15B) NOB growth is only possible if AOB are present. Provided that NOB are present but irrespective of their concentration, the effluent nitrite remains the same (*i.e.*, no biomass term is present in eq. 2.20). Only when NOB are washed out, the residual concentration of nitrite in the effluent coincides with the influent one.

As in the previous example, the steady-state concentrations of biomasses and nitrite are presented in Figure 2.15C for the D range 0-0.3 h^{-1}. The steady-state concentrations corresponding to the cases in Figure 2.15A-B are also marked.

In the sole presence of influent nitrite, NOB are washed out for D above 0.021 h^{-1}: this occurs when the difference between the violet and dark green-shaded areas, representing the nitrite available for growth, becomes zero. If AOB are also present, NOB growth is possible up to a D of 0.023 h^{-1} owing to the increased nitrite availability.

The dashed and full blue lines in Figure 2.15C represent the expected effluent concentration of nitrite when it is only present in the influent or also produced by AOB, respectively. The steady-state concentration of AOB remains unaffected in all scenarios (Figure 2.14B and Figure 2.15C), and only depends on the influent ammonium.

You are encouraged to use the provided spreadsheet (SI_Chapter_2_AOB_NOB _Chemostat_dynamics.xlsx) to explore the impact of different influent ammonium and/or nitrite concentrations.

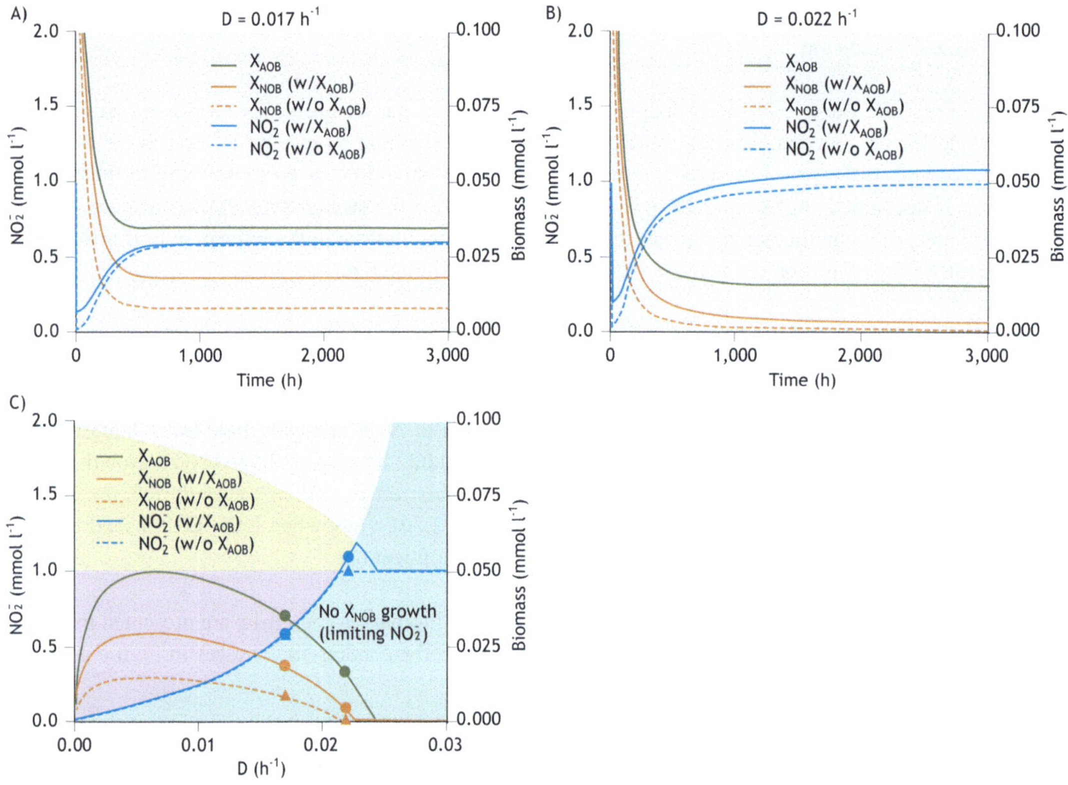

Figure 2.15. Dynamic concentrations of nitrite, AOB, and NOB in the oxygen-replete chemostat operated at a D of 0.017 h[-1] [A)] and 0.022 h[-1] [B)] and receiving an influent with 1 mmol l[-1] of both ammonium and nitrite. The scenarios where AOB are present (w/) or absent (w/o) are presented. (C) Steady-state nitrite concentrations available from the influent (violet) and AOB production (light green) and minimum required for NOB growth (dark green), and steady-state AOB and NOB concentrations. Marked points represent the steady-state concentration reached in A) and B). Note: the violet and light green areas (difference between the nitrite from the influent and produced by AOB, and the minimum nitrite required by NOB for a given D) is the actual amount of nitrite available for growth and maintenance. Note that irrespective of the presence or absence of nitrite in the influent, the steady-state AOB concentration remains unchanged (SI_Chapter_2_AOB_NOB_Chemostat_dynamics.xlsx).

REFERENCES

Agrawal S., Weissbrodt D.G., Annavajhala M., Jensen M.M., Arroyo J.M.C., Wells G., Chandran K., Vlaeminck S.E., Terada A., Smets B.F. and Lackner S. (2021). Time to act – assessing variations in qPCR analyses in biological nitrogen removal with examples from partial nitritation/anammox systems. In: *Water Research* 190, 116604.

APHA-AWWA-WEF (2012). *Standard methods for the examination of water and wastewater*. American Public Health Association, American Water Works Association, Water Environment Federation, Washington DC.

Blackburne R., Vadivelu V.M., Yuan Z. and Keller J. (2007). Determination of growth rate and yield of nitrifying bacteria by measuring carbon dioxide uptake rate. In: *Water Environment Research* 79(12), 2437-45.

Cerruti M., Guo B., Delatolla R., De Jonge N., Hommes-de Vos van Steenwijk A., Kadota P., Lawson C.E., Mao T., Oosterkamp M.J., Sabba F., Stokholm-Bjerregaard M., Watson I., Frigon D. and Weissbrodt D.G. (2021). Plant-wide systems microbiology for the wastewater industry. In: *Environmental Science: Water Research & Technology* 7(10), 1687-706.

Chen G.H., Van Loosdrecht M.C.M., Ekama G.A. and Brdjanovic D. (eds.) (2020). Biological Wastewater Treatment: Principles, Modelling and Design. ISBN: 9781789060355. IWA publishing, London, UK.

Corominas L., Rieger L., Takacs I., Ekama G., Hauduc H., Vanrolleghem P.A., Oehmen A., Gernaey K.V., Van Loosdrecht M.C.M. and Comeau Y. (2010). New framework for standardized notation in wastewater treatment modelling. In: *Water Science and Technology* 61, 841-57.

Gujer W., Henze M., Mino T. and Van Loosdrecht M.C.M. (1999). Activated Sludge Model No. 3. In: *Water Science and Technology*, pp. 183-93.

Henze M., Grady C.P.L., Gujer W., Marais Gv R. and Matsuo T. (1987). *Activated Sludge Model No. 1. IAWPRC Scientific and Technical Reports No. 1*, IWA Publishing, London, UK.

Henze M., Gujer W., Mino T. and Van Loosdrecht M.C.M. (2000). *Activated sludge models ASM1, ASM2, ASM2d and ASM3*. IWA Publishing, London, UK.

Hiatt W.C. and Grady C.P.L. (2008). An Updated Process Model for Carbon Oxidation, Nitrification, and Denitrification. In: *Water Environment Research* 80(11), 2145-56.

Kleerebezem R. and Van Loosdrecht M.C.M. (2010). A Generalized Method for Thermodynamic State Analysis of Environmental Systems. In: *Critical Reviews in Environmental Science and Technology* 40(1), 1-54.

Kuenen J.G. (2019). Continuous Cultures (Chemostats). In: *Encyclopedia of Microbiology (Fourth Edition)* Schmidt T.M. (ed.), Academic Press, Oxford, p.p. 743-61.

Laureni M., Weissbrodt D.G., Szivak I., Robin O., Nielsen J.L., Morgenroth E. and Joss A. (2015). Activity and growth of anammox biomass on aerobically pre-treated municipal wastewater. In: *Water Research* 80, 325-36.

Lotti T., Kleerebezem R., Abelleira-Pereira J.M., Abbas B. and Van Loosdrecht M.C.M. (2015). Faster through training: The anammox case. In: *Water Research* 81, 261-8.

Lotti T., Kleerebezem R., Lubello C. and Van Loosdrecht M.C.M. (2014). Physiological and kinetic characterization of a suspended cell anammox culture. In: *Water Research* 60, 1-14.

Muller A.W., Wentzel M.C. and Ekama G.A. (2004). Experimental determination of the heterotroph anoxic yield in anoxic-aerobic activated sludge systems treating municipal wastewater. In: *Water SA* 30(5), 7-12.

Nielsen P.H., Daims H. and Lemmer H. (2009). *FISH Handbook for Biological Wastewater Treatment - Identification and quantification of microorganisms in activated sludge and biofilms by FISH*. IWA Publishing, London, UK.

Oshiki M., Satoh H. and Okabe S. (2016). Ecology and physiology of anaerobic ammonium oxidising bacteria. In: *Environmental Microbiology* 18(9), 2784-96.

Reichert P. (1994). Aquasim - A tool for simulation and data analysis of aquatic systems. In: *Water Science and Technology* 30(2 pt 2), 21-30.

Rombouts J.L., Mos G., Weissbrodt D.G., Kleerebezem R. and Van Loosdrecht M.C.M. (2019a). Diversity and metabolism of xylose and glucose fermenting microbial communities in sequencing batch or continuous culturing. In: *FEMS Microbiology Ecology* 95(2), fiy233.

Rombouts J.L., Mos G., Weissbrodt D.G., Kleerebezem R. and Van Loosdrecht M.C.M. (2019b). The impact of mixtures of xylose and glucose on the microbial diversity and fermentative metabolism of sequencing-batch or continuous enrichment cultures. In: *FEMS Microbiology Ecology* 95(8), fiz112.

Strous M., Kuenen J.G. and Jetten M.S.M. (1999). Key physiology of anaerobic ammonium oxidation. In: *Applied and Environmental Microbiology* 65(7), 3248-50.

Tijhuis L., Van Loosdrecht M.C.M. and Heijnen J.J. (1993). A thermodynamically based correlation for maintenance Gibbs energy requirements in aerobic and anaerobic chemotrophic growth. In: *Biotechnology and Bioengineering* 42(4), 509-19.

Van den Berg E.M., Boleij M., Kuenen J.G., Kleerebezem R. and Van Loosdrecht M.C.M. (2016). DNRA and Denitrification Coexist over a Broad Range of Acetate/N-NO$_3^-$ Ratios, in a Chemostat Enrichment Culture. In: *Frontiers in Microbiology* 7, 1842.

Van Loosdrecht M.C.M., Nielsen P.H., Lopez Vazquez C.M. and Brdjanovic D. (2016). In: *Experimental Methods in Wastewater Treatment*. IWA Publishing, London, UK.

Vannecke T.P. and Volcke E.I. (2015). Modelling microbial competition in nitrifying biofilm reactors. In: *Biotechnology and Bioengineering* 112(12), 2550-61.

Zhang L., Narita Y., Gao L., Ali M., Oshiki M. and Okabe S. (2017). Maximum specific growth rate of anammox bacteria revisited. In: *Water Research* 116, 296-303.

NOMENCLATURE

Acronym	Description
AMO	Anaerobic ammonium-oxidising (anammox) organism (also referred to as AMX)
A/O	Anoxic-aerobic(oxic)
A2/O	Anaerobic-anoxic-aerobic(oxic)
AOB	Ammonium-oxidising bacteria
AOO	Ammonium-oxidising organism
BNR	Biological nutrient removal
CMO	Complete ammonium-oxidising organism (metabolism similar as early described in Activated Sludge Models for the lumped microbial group of autotrophic nitrifying organisms, ANO)
COD	Chemical oxygen demand
CSTR	Continuous-flow stirred-tank reactor (same as chemostat)
DHO	Denitrifying heterotrophic organism (also referred to as DEN)
eA	Electron acceptor
eD	Electron donor
FISH	Fluorescence *in-situ* hybridisation
HPLC	High-performance liquid chromatography
NOB	Nitrite-oxidising bacteria
NOO	Nitrite-oxidising organism
OHO	Ordinary heterotrophic organism
OM	Organic matter
PN/A	Partial nitritation and anammox
qPCR	Quantitative polymerase chain reaction
SBR	Sequencing batch reactor
SIP	Stable isotope probing
TOC	Total organic carbon
VSS	Volatile suspended solids
WWTP	Wastewater treatment plant

Symbol	Description	Unit	Unit
C_i	Concentration of material i	mol i l^{-1}	mol i/l
D	Dilution rate	h^{-1}	1/h
Δ	Difference		
ΔG^{01}	Gibbs free energy change under biochemical reference conditions	kJ mol^{-1}	kJ/mol
ΔG_{An}^{01}	Gibbs free energy change in anabolism	kJ mol^{-1} X	kJ/mol X
ΔG_{Cat}^{01}	Gibbs free energy change in catabolism	kJ mol^{-1} eD	kJ/mol eD
ΔG_{Diss}^{01}	Gibbs free energy released during growth, under biochemical reference conditions (also referred to as a yield $Y_{G/X}^{01}$)	kJ mol^{-1} X	kJ /mol X
ΔG_R	Gibbs free energy change in reaction	kJ mol^{-1} i	kJ/mol i
ΔH_R	Enthalpy change in reaction	kJ mol^{-1} i	kJ/mol i
Δt	Time step	h	H
ΔS_R	Entropy change in reaction	kJ mol^{-1} i K^{-1}	kJ/mol i.K
$\iota_{k,i}$	Conservation coefficient of element k in material i	mol k mol^{-1} i	mol k/mol i
$G_{f,i}^{0}$	Gibbs free energy of formation of compound i, under standard conditions	kJ mol^{-1} i	kJ/mol i
γ_{Cs}	Degree of reduction of the carbon source	mol e- C-mol^{-1}	mol e-/C-mol
γ_i	Degree of reduction of compound i	mol e- mol^{-1} i	mol e-/mol i
γ_S^*	Number of electrons transferred per mol electron donor in catabolism	mol e- mol^{-1} S	mol e-/mol S
HRT	Hydraulic retention time	h	H
k_d	Biomass-specific decay rate	h^{-1}	1/h
$k_L a$	Volumetric mass transfer coefficient of gas into liquid phase	h^{-1}	1/h
K_i	Affinity constant (half-saturation constant) for material i	mol i l^{-1}	mol i/l
K_S	Affinity constant (half-saturation constant) for the substrate	mol S l^{-1}	mol S/l
λ_{cat}^*	Thermodynamics-derived multiplication factor of catabolism to run the anabolism		
m	Mass	kg	kg
m_i	Biomass-specific rate of maintenance on material i	mol i h^{-1} mol^{-1} X	mol i/h.mol X
$\dot{m}_i$	Mass flow rate of material i	kg h^{-1}	kg/h
m_G	Biomass-specific rate of Gibbs free energy dissipation for maintenance	kJ h^{-1} mol^{-1} X	kJ/h.mol X
m_S	Biomass-specific rate of maintenance on substrate	mol S h^{-1} mol^{-1} X	mol S/h.mol X
μ	Biomass-specific growth rate	h^{-1}	1/h
μ^{max}	Maximum biomass-specific growth rate	h^{-1}	1/h
n	Number of moles	mol	mol
$\dot{n}$	Moles flow rate	mol h^{-1}	mol/h
NoC_{Cs}	Number of carbon atoms in the carbon source	-	-
$\nu_{i,j}$	Stoichiometric coefficient of material i in conversion process of material j	mol i mol^{-1} j	mol i/mol j

pH	Hydrogen potential	-	-
pK_a	Acid dissociation constant	-	-
q_i	Biomass-specific rate of conversion of material i	mol i h^{-1} mol^{-1} X	mol i/ h.mol X
q_e^{max}	Maximum biomass-specific rate of electron transfer in transport chain	mol e- h^{-1} mol^{-1} X	mol i/h.mol X
q_i^{max}	Maximum biomass-specific rate of any material consumption or production	mol i h^{-1} mol^{-1} X	mol i/h.mol X
q_G^{max}	Maximum biomass-specific rate of Gibbs free energy dissipation	kJ h^{-1} mol^{-1} X	kJ/h.mol X
q_S^{max}	Maximum biomass-specific rate of substrate consumption	mol S h^{-1} mol^{-1} X	mol S /h.mol X
Q	Volumetric flow rate	l h^{-1}	l /h
r_i	Volumetric rate of conversion of material i	mol i h^{-1} l^{-1}	mol i /h.l
R	Universal ideal gas constant	kJ mol^{-1} K^{-1}	kJ/mol.K
R_i	Total rate of conversion of material i	mol i h^{-1}	mol i/h
ρ_j	Volumetric rate of growth process j ('process rate')	mol X_j h^{-1} l^{-1}	mol X_j/h.l
SRT	Sludge retention time	h (or d)	h (or d)
t	Time	h	H
T	Temperature	°C or K	°C or K
V	Volume	l	l
VER	Volume exchange ratio	- (or %)	- (or %)
X	Concentration of biomass (or C_X)	mol X l^{-1}	mol X/l
$Y_{X/i}$	Yield of biomass production per conversion of material i	mol X mol^{-1} i	mol X/mol i
$Y_{X/S}$	Yield of biomass production per substrate consumed	mol X mol^{-1} S	mol X/mol S
$Y_{X/S}^{max}$	Maximum yield of biomass production per substrate consumed	mol X mol^{-1} S	mol X/mol S
$Y_{X/S}^{obs}$	Observed yield of biomass production per substrate consumed	mol X mol^{-1} S	mol X/mol S
$Y_{x/y}^{obs}$	Any observed yield of conversion of material x per material y	mol x mol^{-1} y	mol x/mol y

ONLINE SUPPLEMENTARY MATERIALS

Examples

- Aquasim implementation file of the growth models of the examined BNR guilds
- Aquasim output files (.lis, .xlsx) of the simulations

Exercises

Excel spreadsheets (password for editing the excel spreadsheets is BWWT_EE):

- SI_Chapter_2_AMX_DEN_Chemostat_dynamics.xlsx
- SI_Chapter_2_AMX_DEN_Chemostat_steady-state.xlsx
- SI_Chapter_2_AMX_DEN_SBR_dynamic.xlsx
- SI_Chapter_2_AOB_NOB_Chemostat_dynamics.xlsx

doi: 10.2166/9781789062304_0075

3

Wastewater characteristics

Kimberly Solon, Carlos M. Lopez-Vazquez and Eveline I.P. Volcke

3.1 INTRODUCTION

The production of waste from human activities is unavoidable and a significant part of this waste will end up as wastewater. The quantity and quality of wastewater is determined by many factors. Not all humans or industries produce the same amount of waste. The amount and type of waste produced in households is influenced by the behaviour, lifestyle and standard of living of the inhabitants as well as the technical and juridical framework by which people are surrounded. Chapter 3 on wastewater characteristics in the textbook (*Biological Wastewater Treatment: Principles, Modelling and Design* by Chen *et al.*, 2020) gives an overview of the various types of wastewater, and their sources and characteristics.

This chapter guides the reader in familiarizing themselves with wastewater constituents such as microorganisms, organic matter, nutrients, cellulose, micropollutants, metals and toxic organic, as well as the typical concentrations found in wastewater. The reader will also learn how to determine COD, nitrogen and phosphorus fractionations, which are relevant for biological conversions and solid-liquid separation. They will investigate wastewater characteristics using the concepts of population equivalents, the ratios between various wastewater components, and the dynamics of wastewater characteristics, all of which are important for treatment plant design, monitoring and control.

3.2 LEARNING OBJECTIVES

After the successful completion of this chapter, the reader will be able to:

- Analyse overall wastewater characteristics and describe them in terms of population equivalent and person load.

- Calculate the ratios of compounds in the influent based on standard measurements and analyse whether they are within typically reported wastewater ranges.
- Determine the COD, nitrogen and phosphorus fractionation (concentrations and mass flow rates) of wastewater based on standard measurements in terms of their organic and inorganic constituents, as well as on their biodegradable and unbiodegradable, soluble and particulate fractions.
- Identify the strength of relevant sub-streams and the presence of other wastewater constituents, analysing whether they are within typical wastewater ranges and evaluate their importance.
- Describe the sampling procedures and analyse the dynamics of wastewater flows and composition.

3.3 EXAMPLE

A wastewater treatment plant (WWTP) treats the sewage of a town of 250,000 residents. The flow-weighted average concentrations measured in the raw and settled wastewater (WW) over the last year are summarized in Table 3.1. In the same period, the average daily flow rate was 27,000 m^3/d and the average BOD to COD ratio amounted to 0.65.

Table 3.1 Flow-weighted average concentrations of typical constituents for the wastewater being studied.

Tests on wastewater	COD (mgCOD/l)	VFA (mgCOD/l)	TOC (mgC/l)	TKN (mgN/l)	FSA (mgN/l)	TP (mgP/l)	PO_4^{3-}-P (mgP/l)	TSS (mgTSS/l)	ISS (mgISS/l)
Unfiltered (raw WW)	1,250	-	415	87.1	-	26.0	-	755	136
Supernatant (settled WW)	816	-	270	67.8	-	20.8	-	386	46.4
0.45 µm membrane-filtered	313	0	104	62.2	57.0	19.2	18.0	-	-

COD: chemical oxygen demand; VFA: volatile fatty acids; TOC: total organic carbon; TKN: total Kjeldahl nitrogen; FSA: free and saline ammonia; TP: total phosphorus; PO_4^{3-}-P: orthophosphate; TSS: total suspended solids; ISS: inorganic suspended solids; -: not measured.

The raw and settled wastewaters described above were each treated in an activated sludge (AS) system with a long sludge age. The measured filtered effluent concentrations are summarized in Table 3.2. Note that these effluent concentrations are the same for raw and settled wastewater, since filtered concentrations concern soluble compounds, *i.e.* which are not affected by primary settling.

Table 3.2 Filtered effluent concentrations of typical constituents in raw and settled wastewater, after treatment in a long sludge age activated sludge system.

Tests on wastewater	COD (mgCOD/l)	TOC (mgC/l)	TKN (mgN/l)	FSA (mgN/l)	TP (mgP/l)	PO_4^{3-}-P (mgP/l)	TSS (mgTSS/l)	ISS (mgISS/l)
Raw WW AS effluent	63	20.9	1.6	0.5	-	0.0	0.0	0.0
Settled WW AS effluent	63	20.9	1.6	0.5	-	0.0	0.0	0.0

- : not measured.

Additional information:

1. The primary settling tank (PST) underflow is 0.6 % of the raw wastewater influent flow.
2. The particulate unbiodegradable COD fraction ($f_{XU,CODi}$) of the raw and settled wastewater amounts to 0.15 and 0.04, respectively.
3. Inorganic nitrogen in the form of nitrate and nitrite is negligible in the influent wastewater.
4. The COD/VSS ratio of the sludge ($f_{cv,OHO}$) is 1.48 mgCOD/mgVSS, the N/VSS ratio (f_n) is 0.10 mgN/mgVSS, and the P/VSS ratio (f_P) is 0.025 mgP/mgVSS.

In addition to the typical wastewater constituents from Table 3.1, the concentrations of other constituents in the influent wastewater were also measured and overall WWTP removal efficiencies were determined and presented in Table 3.3.

Table 3.3 Flow-weighted average concentrations of specific constituents for the wastewater being studied. Metal concentrations are measured at least once per month and microorganism concentrations at least once per week.

Compound	Raw WW influent concentration	Unit	WWTP removal efficiency (%)	Unit
Sulphur compounds				
Sulphate	60.0	mg/l	-	
Sulphide	0.6	mg/l	-	
Cellulose	289.0	mg/l	-	
Toxic compounds				
PAHs	0.7	µg/l	-	
LAS	5,400	µg/l	-	
DEHP	0.1	µg/l	-	
NPE	0.003	µg/l	-	
Metals				
Cd	0.005	mg/l	41	%
Cr	0.04	mg/l	18	%
Cu	0.08	mg/l	55	%
Pb	0.1	mg/l	58	%
Zn	0.2	mg/l	58	%
Microorganisms			WWTP + ozonation removal efficiency	
Intestinal enterococci	$1.41 \cdot 10^6$	cfu/100 ml	-	
Escherichia coli	$7.18 \cdot 10^6$	cfu/100 ml	3.99	LRV
Legionella	20	cfu/100 ml	1.5	LRV
Helminth eggs	42	egg/l	98.0	%

PAH: polycyclic aromatic hydrocarbons; LAS: linear alkylbenzene sulfonate; DEHP: di(2-ethylhexyl)phthalate; NPE: nonylphenol ethoxylates; - : not measured.

$$LRV = \text{log removal value} = \log_{10}\left(\frac{\text{influent pathogen concentration}}{\text{effluent pathogen concentration}}\right)$$

Using information from the textbook and the specifications above, determine the following:

1. The population equivalent (PE) and person load (PL) in terms of the daily flow rate and in terms of the daily influent BOD observed in the last year.
2. The daily and annual COD loads.
3. If the maximum BOD and COD discharge limits are 20 and 75 mg/l, respectively, what must the BOD and COD removal efficiencies of the plant be to meet these standards?
4. For this wastewater stream, calculate the wastewater concentration ratios given in Table 3.19 (page 94 of the textbook, Chen *et al.*, 2020) and compare them with the typical values shown in that table. Discuss your findings.
5. Set up the mass balances over the primary settler in terms of total mass and the mass of the individual components COD, TOC, TN (TKN), TP and TSS. Complete the table shown below (Table 3.4).

Table 3.4 Mass balances over the primary settler.

	Raw WW		Settled WW		Primary sludge	
Flow rate (m³/d)	27,000					
	Concentration (mg/l)	Mass flow rate (kg/d)	Concentration (mg/l)	Mass flow rate (kg/d)	Concentration (mg/l)	Mass flow rate (kg/d)
COD	1,250					
TOC	415					
TKN	87.1					
TP	26					
TSS	755					
ISS	136					
VSS	619					

Note: 'mass flow rate' was termed 'flux' in the textbook (Chen *et al.*, 2020).

6. Characterize the raw wastewater, settled wastewater and primary sludge COD, N and P concentrations into their biodegradable and unbiodegradable, soluble and particulate concentrations.
7. Based on the typical raw municipal wastewater composition, are the concentrations of the following wastewater constituents typical?

   a. Sulphate and sulphides
   b. Cellulose
   c. Toxic organic compounds
   d. Metals

Concerning the concentrations of metals, assume that the WWTP is located in the EU, for which the best available techniques (BAT) emission levels for metals are summarized in Table 3.5. Is an additional treatment necessary before discharging the WWTP effluent into a surface water?

Table 3.5 BAT-associated emission levels for direct discharges to receiving water body[1]

Metal	Emission level (mg/l)
Arsenic	0.01 - 0.1
Cadmium	0.01 - 0.1
Chromium	0.01 - 0.3
Hexavalent chromium	0.01 - 0.1
Copper	0.05 - 0.5
Lead	0.05 - 0.3
Nickel	0.05 - 1.0
Mercury	0.001 - 0.01
Zinc	0.10 - 2.0

e. Microorganisms. Is the effluent wastewater in this study allowed to be reused for agricultural irrigation? The EU effluent standards relating to agricultural irrigation are summarized in Table 3.6. You can assume that the effluent criteria regarding BOD_5, TSS and turbidity are fulfilled. Therefore, you only need to check the wastewater quality in terms of microorganisms.

Table 3.6 Reclaimed water quality criteria for agricultural irrigation (Alcalde-Sanz and Gawlik, 2017).

Reclaimed water quality class	*Escherichia coli* (cfu/100 ml)	BOD_5 (mg/l)	TSS (mg/l)	Turbidity (NTU)	Additional criteria
Class A	≤ 10 or below detection limit	≤ 10	≤ 10	≤ 5	Legionella: ≤ 1,000 cfu/l when there is risk of aerosolization
Class B	≤ 100				
Class C	≤ 1,000		According to Directive 91/271/EEC		Helminth eggs: ≤ 1 egg/l when irrigation of pastures or fodder for livestock
Class D	≤ 10,000				

Table 3.7 Classes of reclaimed water quality, and the associated agricultural use and irrigation method considered (Alcalde-Sanz and Gawlik, 2017).

Crop category	Minimum reclaimed water quality class	Irrigation method
All food crops, including root crops consumed raw and food crops where the edible portion is in direct contact with reclaimed water	Class A	All irrigation methods allowed
Food crops consumed raw where the edible portion is produced above ground and is not in direct contact with reclaimed water	Class B	All irrigation methods allowed
	Class C	Drip irrigation only
Processed food crops	Class B	All irrigation methods allowed
	Class C	Drip irrigation only
Non-food crops including crops to feed milk- or meat-producing animals	Class B	All irrigation methods allowed
	Class C	Drip irrigation only
Industrial, energy, and seeded crops	Class D	All irrigation methods allowed

---

[1] Official Journal of the European Union, L 208, 17.08.2018
https://eur-lex.europa.eu/legalcontent/EN/TXT/?uri=uriserv:OJ.L_.2018.208.01.0038.01.ENG&toc=OJ:L:2018:208:FULL

8. A factory located in the same town as the WWTP plans to discharge 50 m³ of concentrated wastewater which contains 1,000 kgBOD into the sewer. This wastewater does not contain any toxic or inhibiting compounds. The WWTP has a maximum capacity to handle 50 % more than its current BOD load.
   a. Check if the BOD concentration of the wastewater to be treated is realistic.
   b. What must the maximum discharge flow rate of the concentrated wastewater from the factory into the sewer be in order to reduce the risk of potential process disruptions at the local WWTP?
   c. Based on the maximum suitable flow rate calculated in question a, what is the minimum time it will take to discharge the 50 m³ of concentrated wastewater?
   d. What is the corresponding dilution factor of the concentrated wastewater considering the daily flow rate of the plant (27,000 m³/d)?
9. At the WWTP described in question 8 above, most of the wastewater flow is received between 7:30 h and 18:30 h.
   a. Calculate the maximum hourly flow rate ($Q_{h,max}$) for the given conditions.
   b. What is the maximum hourly constant ($f_{h,max}$)?

Solution

1. Population equivalent and person load

According to Eq. 3.3 in the textbook, 1 PE = 0.2 m³/d.

Thus, the population equivalent (PE), a standard unit referring to the typical contribution of an individual to the wastewater load, based on the daily flow rate is:

Number of PE $\quad$ = Daily flow rate · 1 PE/0.2 m³.d
$\qquad\qquad\qquad$ = 27,000 m³/d · 1 PE/0.2 m³.d
$\qquad\qquad\qquad$ = 135,000 PE

which is lower than the size of the town (250,000 residents) mentioned in the assignment. This suggests that this town generates less wastewater than a typical town.

The person load (PL), *i.e.* the actual contribution from a person living in the sewer catchment to the wastewater production, in terms of the daily flow rate is:

PL $\qquad\qquad\qquad$ = Daily flow rate / Number of persons
$\qquad\qquad\qquad$ = (27,000 m³/d) / 250,000 persons
$\qquad\qquad\qquad$ = 0.108 m³/person.d

Note that the actual contribution, expressed as person load (PL), is different from the population equivalent (PE), which is a fixed value.

In order to calculate the PE and the PL in terms of the influent BOD load, the latter should be determined first. Its value is obtained from the influent COD concentration and influent BOD/COD ratio, which are given.

BOD load $\qquad\qquad$ = Daily influent BOD concentration · Daily flow rate
$\qquad\qquad\qquad$ = 1,250 mgCOD/l · 0.65 mgBOD/mgCOD · 27·10⁶ l/d
$\qquad\qquad\qquad$ = 21,937,500 gBOD/d = 21,937.5 kgBOD/d

According to Eq. 3.4 in the textbook, 1 PE = 60 gBOD/d. Thus, the PE for the WWTP under study, based on the influent BOD, is:

Number of PE   = Daily influent BOD load · 1 PE/60 gBOD.d
                       = 21,937,500 gBOD/d · 1 PE/60 gBOD.d
                       = 365,625 PE

And the PL in terms of the influent BOD is:

PL   = Daily influent BOD load / Number of persons
                       = (21,937,500 gBOD/d) / 250,000 persons
                       = 87.75 gBOD/person.d

Based on the values calculated above, 1 person actually contributes (*i.e.* referring to PL) 0.108 m^3/d and 87.75 gBOD/d. While the wastewater flow rate contribution corresponds to typical variations in load per person (namely 0.05-0.40 m^3/person.d, Table 3.15 in the textbook), the BOD contribution seems on the high side (*i.e.* the typical range is 15-80 gBOD/person.d, Table 3.15 in the textbook).

2. Daily and annual COD loads
Given the average daily influent COD concentration, the daily COD load can be calculated by multiplying by the daily flow rate.

Daily COD load   = Daily influent COD concentration · Daily flow rate
                       = 1,250 mgCOD/l · 27·10^6 l/d
                       = 33,750 kgCOD/d

The annual COD load is simply calculated by multiplying the daily COD load by the number of days in a year.

Annual COD load = Daily COD load · 365 d/yr
                       = 33,750 kgCOD/d · 365 d/yr
                       = 12,318,750 kgCOD/yr

3. Required removal efficiencies
The removal efficiency (E(%)) is defined as the difference between WWTP influent and effluent concentrations, relative to the influent concentration:

E (%)   = (Concentration$_i$ – Concentration$_e$) · (100) / Concentration$_i$

The influent BOD to COD concentration ratio is specified as 0.65; the influent BOD concentration is thus calculated as:

BOD$_i$ = COD$_i$ · 0.65 = 1,250 · 0.65 = 813 mgBOD/l

The minimum COD and BOD removal efficiencies required to meet the discharge limits can be estimated considering the WWTP effluent concentrations needed to meet the discharge limits for COD and BOD, respectively, as follows:

$$E(\%)_{COD} = (COD_i - COD_e) / COD_i$$
$$= (COD_i - COD_{e,limit}) / COD_i$$
$$= (1{,}250 - 75) / 1{,}250 = 0.94 = 94\ \%$$

$$E(\%)_{BOD} = (BOD_i - BOD_e) / BOD_i$$
$$= (BOD_i - BOD_{e,limit}) / BOD_i$$
$$= (813 - 20\ mgBOD/l) / (813) = 0.975 = 97.5\ \%$$

4. Typical wastewater concentration ratios

In order to calculate typical wastewater concentration ratios (Table 3.19 in the textbook), the BOD concentration for settled wastewater remains to be determined (the BOD concentration for raw wastewater was calculated in Section 3), as well as the total nitrogen (TN) and volatile suspended solids (VSS) concentrations for both raw and settled wastewater.

This is calculated based on the following considerations:

- The BOD concentration is 65 % of the COD concentration.
- Inorganic nitrogen in the form of nitrate and nitrite is typically negligible for influent municipal wastewater. Thus, the influent total nitrogen concentration (TN) equals the influent total Kjeldahl nitrogen concentration (TKN, Table 3.1).
- The total suspended solids (TSS) concentration is the sum of volatile suspended solids (VSS) and inorganic suspended solids (ISS) concentrations. Thus, the VSS concentrations can be calculated from the TSS and ISS values given in Table 3.1 as VSS = TSS – ISS.

The resulting concentrations of the raw and settled influent municipal wastewater components in this example are summarized in Table 3.8.

Table 3.8 Summary of concentrations of typical constituents for the wastewater under study (the grey values were given or calculated previously).

	COD (mg/l)	BOD (mg/l)	TOC (mg/l)	TN (mg/l)	TP (mg/l)	TSS (mg/l)	ISS (mg/l)	VSS (mg/l)
Raw WW influent	1,250	813	415	87.1	26.0	755	136	619
Settled WW influent	816	530	270	67.8	20.8	386	46.4	339.6

The corresponding concentration ratios for the wastewater in this example are calculated and compared to the typical wastewater ratios in a municipal wastewater (Table 3.19 in the textbook), as summarized in Table 3.9.

Table 3.9 Wastewater concentration ratios for the given case study and their comparison against typical values.

	Raw WW (this example)	Observations (compared to Table 3.19 in the textbook)	Settled WW (this example)	Observations (compared to Table 3.19 in the textbook)
COD/BOD	1.54	Low	1.54	Low
COD/TN	14.4	High	12.0	Medium-High
COD/TP	48.1	High	39.2	Medium
BOD/TN	9.33	Very high	7.82	High
BOD/TP	31.3	Very high	25.5	High
COD/VSS	2.02	High	2.40	Very high
VSS/TSS	0.820	High	0.880	High
COD/TOC	3.01	Medium-High	3.02	Medium-High

Note: since the VFA concentration was not determined, the influent VFA/COD ratio could not be calculated.

Most of the calculated ratios fall within the typical ranges for municipal wastewater. For the raw wastewater in this example, the BOD/TN and BOD/TP ratios are very high, while the COD/BOD ratio is low. This may result from a relatively high BOD concentration, which is confirmed by the typical BOD concentrations listed in Table 3.17.

5. *Mass balances over the primary settler*
The primary settler is schematically represented in Figure 3.1

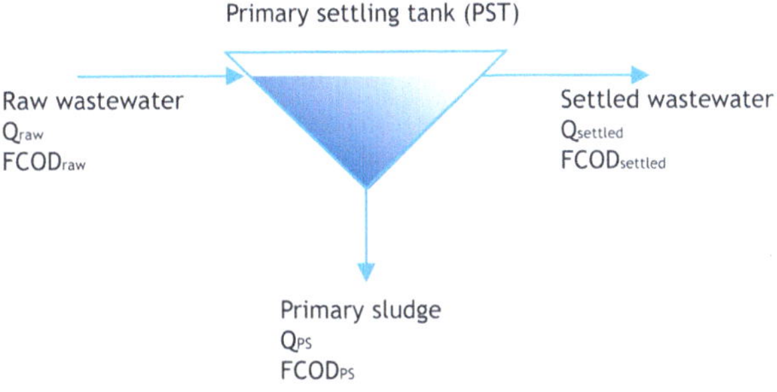

Figure 3.1 Schematic representation of mass balances over the primary settler.

a) The primary settling tank (PST) underflow is specified to be 0.6 % of the raw WW influent flow. Thus, the flow rate of the primary sludge (PS) is calculated as:

$$Q_{PS} = Q_{raw} \cdot 0.6\,\%$$
$$= 27,000\ \mathrm{m^3/d} \cdot 0.006$$
$$= 162\ \mathrm{m^3/d}$$

The flow rate of the settled WW is calculated from the total mass balance over the primary settler. Note that the mass balance is made in terms of volumetric flow rates rather than mass flow rates, which implies that the density is realistically assumed to be constant and the same for all streams.

$$Q_{settled} = Q_{raw} - Q_{PS}$$
$$= 27{,}000 \ m^3/d - 162 \ m^3/d$$
$$= 26{,}838 \ m^3/d$$

b) The individual mass flow rates for all the components (COD, TOC, TKN, TP, TSS, ISS and VSS) in raw and settled WW are calculated as the product of the (volumetric) flow rate (for raw and settled wastewater, respectively) with the concentration of the component under concern.

Individual mass flow rate (for a component) = (Volumetric) flow rate · Concentration (of that component)

For example, the COD mass flow rate is calculated as follows:

$$FCOD_{raw} = Q_{raw} \cdot COD_{raw}$$
$$= 27 \cdot 10^6 \ l/d \cdot 1{,}250 \ mg/l \cdot 10^{-6} \ kg/mg$$
$$= 33{,}750 \ kg/d$$

The remaining individual mass flow rates in the raw and settled wastewater are calculated analogously and summarized in Table 3.10.

c) The individual mass flow rates in the primary sludge can be calculated as the difference between their mass flow rates in the raw and settled wastewater, at least if their masses are conserved over the PST. This is definitely the case for COD, TN and TP, which are inherently conserved quantities. It also holds for TOC, TSS, ISS and VSS, since they do not undergo any transformations in the PST. Also note that the TKN concentration equals the TN concentration in the three PST-related streams, realistically assuming that the influent wastewater does not to contain nitrate or nitrite.

For example, the COD mass flow in the primary sludge is calculated as:

$$FCOD_{PS} = FCOD_{raw} - FCOD_{settled}$$
$$= 33{,}750 \ kg/d - 21{,}900 \ kg/d$$
$$= 11{,}850 \ kg/d$$

The mass flow rates of the remaining components in the primary sludge are calculated analogously and summarized in Table 3.10.

d) The concentrations of the individual components (COD, TOC, TKN, TP, TSS, ISS and VSS) in the primary sludge are simply calculated as the component mass flow rate divided by the volumetric flow rate of the primary sludge stream. For example, the COD concentration in the primary sludge is calculated as:

$$COD_{PS} = FCOD_{PS} / Q_{PS}$$
$$= (11{,}850 \ kg/d) / (0.162 \cdot 10^6 \ l/d \cdot 10^6 \ mg/kg)$$
$$= 73.1 \ mg/l$$

Note that the number of significant digits in the multiplication is 3, corresponding with the lowest amount of significant digits of the constituting factors.

e) Summarizing table. The COD, TOC, TKN, TP, TSS, ISS and VSS concentrations and mass flow rates in the raw wastewater, settled wastewater and primary sludge are summarized in Table 3.10. The mass flow rates were calculated assuming conservation of mass of these components over the PST.

Table 3.10 Results of the mass balances over the primary settler. Values in grey have been given or calculated before. The superscripts [a), b), c), d)] refer to the part of this section where the calculation of this value has been addressed.

	Raw WW		Settled WW		Primary sludge	
Flow rate (m³/d)	27,000		26,838[a)]		162[a)]	
	Concentration	Mass flow rate	Concentration	Mass flow rate[b)]	Concentration[d)]	Mass flow rate[c)]
	(mg/l)	(kg/d)	(mg/l)	(kg/d)	(mg/l)	(kg/d)
COD	1,250	33,750	816	21,900	73,100	11,850
TOC	415	11,205	270	7,246	24,400	3,959
TKN	87.1	2,352	67.8	1,820	3,280	532
TP	26	702	20.8	558	887.0	144
TSS	755	20,385	386	10,359	62,000	10,026
ISS	136	3,672	46.4	1,245	15,000	2,427
VSS	619	16,713	339.6	9,114	46,900	7,599

6. Wastewater fractionation in terms of COD, N and P

COD fractionation

The sequence of the characterization in terms of biodegradable and unbiodegradable, soluble and particulate COD concentrations for the for raw wastewater, settled wastewater and primary sludge are summarized in Figure 3.2 and described below.

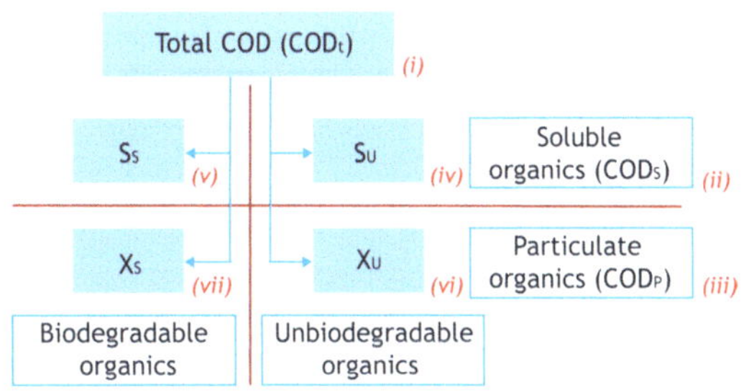

Figure 3.2 COD fractions in wastewater and the sequence in which they are calculated (steps (i)-(vii)).

(i) The total COD concentrations for raw and settled wastewater are given (Table 3.1); the total COD concentration in the primary sludge was calculated in d) in Section 5 from the mass balance over the primary settler.

(ii) The soluble COD concentration ($COD_S = S_S + S_U$) is equivalent to the 0.45 µm membrane-filtered COD and is given as 313 mg/l (Table 3.1). For soluble components, the concentrations in the raw and settled wastewater and primary sludge are the same.

(iii) The particulate COD ($COD_P = X_S + X_U$) is calculated as the difference between the total COD and soluble COD. Thus, $COD_P = COD_{total} - COD_S$, for each of the three streams.

Note that the particulate concentrations are different in the raw wastewater, settled wastewater, and primary sludge. However, they are related through the mass balance for particulate COD over the PST:

$$Q_{raw} \cdot COD_{P,raw} = Q_{settled} \cdot COD_{P,settled} + Q_{PS} \cdot COD_{P,PS}$$

(iv) The soluble unbiodegradable COD concentration (S_U) in the wastewater equals the measured filtered (= soluble) COD concentration of the activated sludge effluent, since it is assumed that all the biodegradable COD was degraded. This is the same for raw wastewater, settled wastewater and primary sludge since it concerns a soluble compound.

(v) The soluble biodegradable COD concentration (S_S) is then calculated as the difference between soluble COD and the soluble unbiodegradable COD: $S_S = COD_S - S_U$.

(vi) The particulate unbiodegradable COD fractions ($f_{XU,CODi}$) of the raw and settled wastewaters are given (as 0.15 and 0.04, respectively). As a result, the particulate unbiodegradable COD concentrations for raw and settled wastewater can be directly calculated as $X_U = COD_{total} \cdot f_{XU,CODi}$. The particulate unbiodegradable COD fraction in the primary sludge is then calculated from the mass balance for particulate unbiodegradable COD over the PST:

$$
\begin{aligned}
X_{U,PS} &= (Q_{raw} \cdot X_{U,raw} - Q_{settled} \cdot X_{U,settled}) / Q_{PS} \\
&= (27,000 \cdot 188 - 26,838 \cdot 33) / 162 \text{ mgCOD/l} \\
&= 25,866 \text{ mgCOD/l}
\end{aligned}
$$

(vii) The particulate biodegradable COD concentrations (X_S) are calculated by subtracting the particulate unbiodegradable fraction from the particulate COD: $X_S = COD_P - X_U$, for each of the three streams.

Table 3.11 COD fractionation for the raw wastewater, settled wastewater and primary sludge in the example.

		Raw WW	Settled WW	Primary sludge
	Flow rate Q (m³/d)	27,000	26,838	162
		Concentration (mgCOD/l)	Concentration (mgCOD/l)	Concentration (mgCOD/l)
i	COD_{total}	1,250	816	73,100
ii	$COD_S = S_S + S_U$	313	313	313.0
iii	$COD_P = X_S + X_U$	937	503	72,800
iv	S_U	63	63	63.0
v	S_S	250	250	250.0
vi	X_U	188	33	25,900
vii	X_S	749	470	47,000

<u>Nitrogen fractionation</u>

The sequence of the nitrogen fractionation for the raw wastewater, settled wastewater and primary sludge is summarized in Figure 3.3 and described lines below.

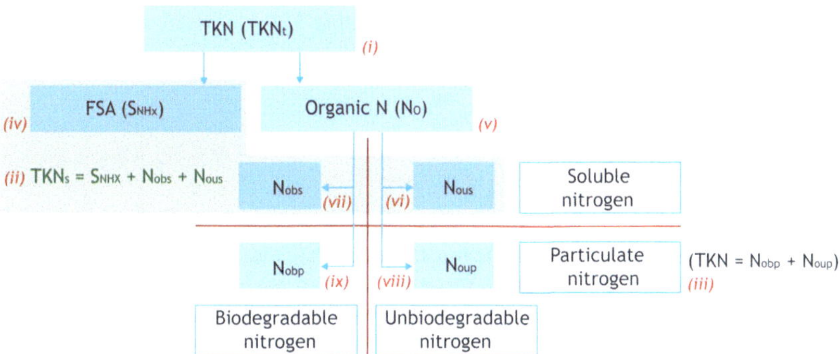

Figure 3.3 Nitrogen fractions in wastewater and indication of the order in which they are calculated (steps *(i)-(ix)*).

(i)   The total Kjeldahl nitrogen (TKN) is the sum of free and saline ammonia (FSA = S_{NHX}) and organically bound nitrogen (N_o), whereas total nitrogen (TN) is the sum of TKN, nitrate and nitrite. In municipal wastewater, influent nitrate and nitrite are typically negligible, thus TKN = TN. The TKN concentrations for raw and settled wastewater are given in Table 3.1; the TKN concentration in the primary sludge was calculated in d)-e) in Section 5.

(ii)   The soluble TKN concentration (TKN$_s$ = S_{NHX} + N_{obs}+ N_{ous}) is equivalent to the 0.45 μm membrane-filtered TKN determined on the influent wastewater and was given as 62.2 mgN/l (Table 3.1). As for all soluble components, the raw and settled wastewater and primary sludge concentrations are the same.

(iii) The particulate TKN concentration (TKN$_p$ = N_{obp}+ N_{oup}) for each of the three streams is calculated as the difference between the total TKN concentration and TKN$_s$: TKN$_p$= TKN – TKN$_s$.

Note that the particulate concentrations are different in the raw wastewater, settled wastewater and primary sludge. However, they are related through the mass balance for particulate nitrogen over the PST, which can be used to double-check the results:

$$Q_{raw} \cdot TKN_{P,raw} = Q_{settled} \cdot TKN_{P,settled} + Q_{PS} \cdot TKN_{P,PS}$$

(iv)   The influent total (*i.e.* free and saline) ammonium concentration (S_{NHX}) is equivalent to the 0.45 μm membrane-filtered FSA concentration, which was given as 57 mg/l (Table 3.1). Since this concerns a soluble component, the ammonium concentration is the same in raw wastewater, settled wastewater and primary sludge.

(v)   The organic nitrogen concentration (N_o) for each of the three streams is the difference between the TKN and ammonium concentration. N_o = TKN – S_{NHX}.

Note: as for all other components, the organic nitrogen concentrations in the three streams are related through the corresponding mass balance over the PST, which can be used to double-check the results:

$$Q_{raw} \cdot N_{o,raw} \quad = Q_{settled} \cdot N_{o,settled} + Q_{PS} \cdot N_{o,PS}$$

(vi) The soluble unbiodegradable organic nitrogen concentration (N_{ous}) is the same for all three streams and, since it is soluble and unbiodegradable, is equal to the soluble unbiodegradable organic nitrogen concentration in the effluent of the whole treatment system ($N_{ous,raw} = N_{ous,settled} = N_{ous,PS} = N_{ous,e}$). The latter is calculated as the difference between the effluent soluble TKN and effluent FSA ($TKN_{s,e}$ and $S_{NHX,e}$, respectively), using the values given in Table 3.2:

$$\begin{aligned} N_{ous,e} \quad &= TKN_{s,e} - S_{NHX,e} \\ &= (1.6 - 0.5) \text{ mgN/l} \\ &= 1.1 \text{ mgN/l} \end{aligned}$$

(vii) The soluble biodegradable organic nitrogen concentration (N_{obs}) is obtained as the difference between the soluble TKN concentration on the one hand and the ammonia and soluble unbiodegradable organic nitrogen concentrations on the other hand:

$$N_{obs} \quad = TKN_s - S_{NHX} - N_{ous}$$

Given that these compounds are soluble, their concentrations are the same for the three streams.

(viii) The particulate unbiodegradable organic nitrogen concentration (N_{oup}) is the nitrogen concentration associated with the particulate unbiodegradable organics (X_U). The nitrogen content of the influent particulate unbiodegradable organics (f_N) is 0.1 mgN/mgVSS, and the COD/VSS ratio (f_{CV}) is 1.48 mgCOD/mgVSS, for raw sewage, settled sewage as well as primary sludge. Thus, $N_{oup} = f_N \cdot X_U / f_{CV}$ can be applied to all three streams.

Note: again, the mass balance over the PST can be used to double-check the results.

(ix) The particulate biodegradable organic nitrogen concentration (N_{obp}) is the difference between the particulate total Kjeldahl nitrogen and the particulate unbiodegradable organic nitrogen concentrations: $N_{obp} = TKN_p - N_{oup}$. Here as well, results can be double-checked through the mass balance over the PST.

Table 3.12 Nitrogen fractionation for the raw wastewater, settled wastewater and primary sludge in the example.

		Raw WW	Settled WW	Primary sludge
Flow rate Q (m³/d)		27,000	26,838	162
		Concentration (mgN/l)	Concentration (mgN/l)	Concentration (mgN/l)
i	TN = TKN	87.1	67.8	3,280.0
ii	TKN_s	62.2	62.2	62.2
iii	TKN_p	24.9	5.6	3,222.0
iv	S_{NHX}	57.0	57.0	57.0
v	N_o	30.1	10.8	3,227.0
vi	N_{ous}	1.1	1.1	1.1
vii	N_{obs}	4.1	4.1	4.1
viii	N_{oup}	12.7	2.2	1,750.0
ix	N_{obp}	12.2	3.4	1,470.0

<u>Phosphorus fractionation</u>
Like for the COD and N fractionations, the sequence of the phosphorus fractionation for the three different flows of study (raw wastewater, settled wastewater and primary) is summarized in Figure 3.4 and described below.

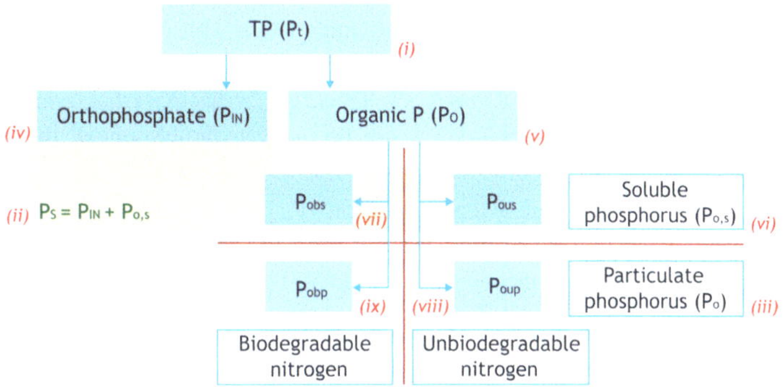

Figure 3.4 Phosphorus fractions in wastewater and indication of the sequence in which they are calculated (steps (i)-(ix)).

(i)   The value of the total phosphorus concentration (P_t) is given in Table 3.1 for raw and settled wastewater and was calculated in Section 5 for the primary sludge. It is the sum of the inorganic phosphorus (P_{IN}) and the organic phosphorus concentration (P_O).

(ii) The soluble phosphorus concentration (P_s) including orthophosphate is equivalent to the 0.45 μm membrane-filtered TP concentration measurement determined for the influent wastewater (Table 3.1). As for all soluble components, its concentrations in the raw and settled wastewater and primary sludge are the same.

(iii) The particulate phosphorus concentration (P_p) is calculated for each of the three streams as the difference between the total phosphorus and the soluble phosphorus: $P_p = P_t - P_s$. The particulate concentrations in the raw wastewater, settled wastewater and primary sludge are related through the mass balance for particulate phosphorus over the PST, which can be used to double-check the results:

$$Q_{raw} \cdot P_{p,raw} = Q_{settled} \cdot P_{p,settled} + Q_{PS} \cdot P_{p,PS}$$

(iv) The inorganic P fraction (P_{IN}) in the influent is typically present as orthophosphate (PO_4^{3-}-P). Reasonably assuming that no other inorganic P fractions are present, thus P_{IN} equals the PO_4^{3-}-P concentration (S_{PO4}) determined on the influent wastewater (Table 3.1) and therefore: $P_{IN} = S_{PO4.}$

(v)   The organic P concentration (P_o) is the difference between total phosphorus and inorganic phosphorus: $P_o = P_t - P_{IN}$.

(vi) The soluble organic phosphorus concentration ($P_{o,s}$) is calculated as the influent total phosphate concentration minus the particulate phosphorus concentration and the orthophosphate concentration: $P_{o,s} = P_t - P_p - P_{IN}$. This also equals the difference between the organic P concentration and the particulate (organic) P concentration: $P_{o,s} = P_o - P_p$. As for all soluble components, the concentration of this fraction is the same for the three streams.

(vii) The soluble organic phosphorus concentration ($P_{o,s}$) can be further subdivided into soluble biodegradable phosphorus and soluble unbiodegradable phosphorus concentrations ($P_{o,s} = P_{ous} + P_{obs}$). The soluble unbiodegradable organic phosphorus concentration in the influent is equal to the one in the effluent. However, no information on the latter is given, nor any other information which could be used to identify these fractions.

(viii) The particulate unbiodegradable organic phosphorus concentration (P_{oup}) is the phosphorus concentration associated with the particulate unbiodegradable organics (X_U). The phosphorus content of the influent unbiodegradable particulate organics (f_P) in the raw and settled sewage are both 0.025 mgP/mgVSS, and the COD/VSS ratio (f_{CV}) is 1.48 mgCOD/mgVSS. Thus, $P_{oup} = f_P \cdot X_U / f_{CV}$. can be applied to all three streams. The mass balance over the PST can again be used to double-check the results (note that the number of significant digits = 2).

(ix) The particulate biodegradable organic P concentration (P_{obp}) is the difference between the particulate (organic) phosphorus and the particulate unbiodegradable organic phosphorus fractions: $P_{obp} = P_p - P_{oup}$. Here as well, results can be double-checked through the mass balance over the PST (note that the number of significant digits = 2).

Table 3.13 Phosphorus fractionation for the raw wastewater, settled wastewater and primary sludge in the example.

		Raw WW	Settled WW	Primary sludge
Flow rate Q (m³/d)		27,000	26,838	162
		Concentration (mgP/l)	Concentration (mgP/l)	Concentration (mgP/l)
i	Total phosphorus (P_t)	26.0	20.8	887.0
ii	Soluble phosphorus including orthophosphate (P_s)	19.2	19.2	19.2
iii	Particulate phosphorus (P_p)	6.8	1.6	868.0
iv	$P_{IN} (= S_{PO4})$	18.0	18.0	18.0
v	P_o	8.0	2.8	869.0
vi	$P_{o,s} (=P_{ous} + P_{obs})$	1.2	1.2	1.2
viii	P_{oup}	3.2	0.56	438.0
ix	P_{obp}	3.6	1.0	430.0

7. Evaluation of other compounds' concentrations

<u>Sulphate and sulphides</u>

The concentrations of sulphate and sulphide reported for this case (Table 3.3) are within typical municipal wastewater ranges as shown in tables 3.13 and 3.17 in the textbook.

	Raw WW (mg/l)	Observation
Sulphate	60.0	Medium-High (Table 3.17)
Sulphide	0.6	Medium (Table 3.13)

<u>Cellulose</u>

The influent cellulose concentration in this case is reported to be 289 mg/l (Table 3.3). This corresponds to 23 % (for 1,250 mgCOD/l, Table 3.1 in this chapter) of the influent COD load and to 38 % (for 755 mgTSS/l, Table 3.1) of the influent total suspended solids load. This corresponds closely with the values given in

Section 3.8 in the textbook, which gives a typical cellulose content for domestic wastewater of 20-30 % of the COD load and 35 % of the suspended solids load.

Toxic organic compounds
The concentrations of toxic organic compounds reported for this case (Table 3.3) are within the typical municipal wastewater ranges in Table 3.14 in the textbook:

	Raw WW (µg/l) (Table 3.3, this chapter)	Observation (Table 3.14 in the textbook)
PAHs	0.7	Low
LAS	5,400	Medium
DEHP	0.1	Low
NPE	0.03	Low-Medium

Metals
The concentrations of metals reported for this case (Table 3.3) are within the typical municipal wastewater ranges given in Table 3.12 in the textbook.

	Raw WW (mg/m^3) (Table 3.3, this chapter)	Observation (Table 3.12 in the textbook)
Cd (Cadmium)	5	High
Cr (Chromium)	40	High
Cu (Copper)	80	Medium
Pb (Lead)	100	High
Zn (Zinc)	200	Medium

To determine if additional treatment regarding metals is necessary before discharging the WWTP effluent into a surface water, the effluent concentrations of the metals are calculated using the given WWTP treatment removal efficiencies:

$$\text{Effluent WW concentration} = \text{Raw WW concentration} \cdot \left(1 - \frac{\text{Removal Efficiency}}{100\ \%} \right)$$

and subsequently compared to the EU BAT values (Table 3.5):

	Removal efficiency (%) (Table 3.3 of this chapter)	Effluent WW (mg/l)	Observation (compared with Table 3.5 in this chapter)
Cd	41	0.003	$< 0.1 \Rightarrow$ direct discharge allowed
Cr	18	0.033	$< 0.3 \Rightarrow$ direct discharge allowed
Cu	55	0.036	$< 0.5 \Rightarrow$ direct discharge allowed
Pb	58	0.042	$< 0.3 \Rightarrow$ direct discharge allowed
Zn	58	0.084	$< 0.3 \Rightarrow$ direct discharge allowed

It is concluded that the given performance of the WWTP regarding metal removal is sufficient and no additional treatment is needed.

<u>Microorganisms</u>

The influent concentrations of intestinal enterococci and *Escherichia coli* in this study are within the typical municipal wastewater ranges given in Table 3.5 in Chapter 3 of the textbook:

	Raw WW (cfu/100 ml) (Table 3.3)	Observation
Escherichia coli	$7.18 \cdot 10^6$	Low (compare with Table 3.5 in the textbook)
Intestinal enterococci	$1.41 \cdot 10^6$	Low (compare with Table 3.5 in the textbook)[a]

[a]In practice, the terms intestinal enterococci, faecal streptococci, enterococci, and enterococcus group may refer to the same bacteria (see https://www.who.int/water_sanitation_health/bathing/srwe1-chap4.pdf).

To determine if the effluent wastewater is allowed to be reused for agricultural irrigation, the concentration of the microorganisms in the effluent should be first calculated based on their log removal value (*LRV*), given for Escherichia coli and Legionella) or removal efficiency (given for helminth eggs).

$$LRV = \log_{10} \left(\frac{\text{Influent pathogen concentration}}{\text{Effluent pathogen concentration}} \right)$$

$$10^{LRV} = \frac{\text{Influent pathogen concentration}}{\text{Effluent pathogen concentration}}$$

$$\text{Effluent pathogen concentration} = \frac{\text{Influent pathogen concentration}}{10^{LRV}}$$

$$\text{Effluent E. coli concentration} = \frac{7.18 \cdot 10^6}{10^{3.99}} = 735 \text{ cfu/100 ml}$$

$$\text{Effluent Legionella concentration} = \frac{20}{10^{1.5}} = 0.6 \text{ cfu/100 ml}$$

$$\text{Effluent concentration} = \text{Raw WW concentration} \cdot \left(1 - \frac{\text{Removal efficiency}}{100\%} \right)$$

$$\text{Effluent concentration}_{\text{helminth eggs}} = 42 \cdot \left(1 - \frac{98}{100} \right) = 0.84 \text{ egg/l}$$

The results are summarized as follows:

	Raw WW (Table 3.3, this chapter)	Removal efficiency (Table 3.3, this chapter)	Effluent WW (calculated)	Observation (Table 3.6, this chapter)
Escherichia coli	$7.18 \cdot 10^6$ cfu/100 ml	LRV = 3.99	735 cfu/100 ml	Reclaimed water quality class C
Legionella	20 cfu/100 ml	LRV = 1.5	0.6 cfu/100 ml	$\leq$ 100 cfu/100 ml $\Rightarrow$ can be used
Helminth eggs	42 egg/l	98 %	0.84 egg/l	$\leq$ 1 egg/l $\Rightarrow$ can be used

Based on the calculated effluent concentrations of *Escherichia coli*, the effluent water is considered class C. This means that the reclaimed water can be used only for the following drip-irrigated crops: (1) food crops consumed raw where the edible portion is produced above ground and is not in direct contact with reclaimed water, (2) processed food crops and (3) non-food crops including crops to feed milk- or meat-producing animals.

8. Additional loads

a) The BOD concentration of the concentrated factory wastewater to be treated amounts to 20,000 gBOD/m^3 (= 20 kgBOD/m^3 =1,000 kgBOD/50 m^3). Comparison with Table 3.21 in the textbook shows that this is a very high value which however falls within realistic ranges for *e.g.* winery or dairy wastewater. Note that municipal wastewater typically has much lower BOD concentrations, *e.g.* up to 560 gBOD/m^3 (see Table 3.17 in the textbook).

b) In order to reduce the risk of potential process upsets at the local WWTP, the BOD load of the WWTP should not exceed its maximum capacity, which is 50 % higher than its current BOD load.

Current BOD load   = 21,937,500 gBOD/d (calculated in Section 1)
                                = 21,937,500 gBOD/d $\cdot$ 1 d/24 h $\cdot$ 1 kg/1,000 g
                                = 914 kgBOD/h

The maximum capacity of the WWTP is 50 % higher than its current BOD load:

Max BOD load capacity = 1.5 $\cdot$ Current BOD load
                                = 1.5 $\cdot$ 914 kgBOD/h
                                = 1,371 kgBOD/h

The maximum additional load allowed is: 1,371 – 914 = 457 kgBOD/h

The maximum discharge flow rate of the concentrated wastewater from the factory is thus calculated as:

$Q_{max,discharge}$ = Maximum additional load / BOD concentration of factory WW allowed
                                = (457 kgBOD/h) / (20 kgBOD/m^3)
                                = 22.85 m^3/h

c) The minimum duration of discharge of the factory wastewater is:

Duration of discharge $= \text{Volume of WW} / Q_{max,discharge}$
$= 50\ \text{m}^3 / (22.85\ \text{m}^3/\text{h})$
$= 2.18\ \text{h}$

d) The corresponding dilution factor of the factory wastewater is:

Dilution factor $= Q_i / Q_{max,discharge}$
$= (27{,}000\ \text{m}^3/\text{d}) / (22.85\ \text{m}^3/\text{h} \cdot 24\ \text{h/d})$
$= 49$

9. *Maximum hourly flow rate, hourly factor, and maximum hourly constant*

a) To determine the maximum hourly flow rate ($Q_{h,max}$), the hourly factor should first be determined. The hourly factor is based on a time window over one day during which wastewater is received in the WWTP, *i.e.* between 7:30 h and 18:30 h, or 11 hours. Thus, the hourly factor is 11 h/d.

The maximum hourly flow rate is calculated as:

$Q_{h,max}$ $= (27{,}000\ \text{m}^3/\text{d}) / (11\ \text{h/d})$
$= 2.455\ \text{m}^3/\text{h}$

b) The maximum hourly constant ($f_{h,max}$) is the ratio between the maximum and average hourly flow rate:

$f_{h,max}$ $= Q_{h,max} / Q_{h,avg}$
$= (2{,}455\ \text{m}^3/\text{h}) / (27{,}000\ \text{m}^3/\text{d} \cdot 1\ \text{d}/24\ \text{h})$
$= (2{,}455\ \text{m}^3/\text{h}) / (1{,}125\ \text{m}^3/\text{h})$
$= 2.18$

Note that the maximum hourly constant ($f_{h,max}$) could also be calculated as the inverse of the fraction of the day during which the WWTP receives most of the wastewater:

$f_{h,max}$ $= 24\text{h} / 11\text{h}$
$= 2.18$

3.4   EXERCISES

Person load and population equivalent (exercises 3.4.1-3.4.3)

Exercise 3.4.1

What is the difference between the concepts of person load (PL) and population equivalent (PE)? What do they have in common?

Exercise 3.4.2

All the wastewater on an island with 240 inhabitants is collected. The daily wastewater flow rate amounts to 60 m³/d and the daily wastewater BOD load is 18 kgBOD/d. What is the person load in terms of the daily flow rate and in terms of the influent BOD?

Exercise 3.4.3

A food industry generates 2,000 m³ of wastewater daily with an average BOD concentration of 450 ppm. What is the population equivalent of this wastewater? If you are asked to design a WWTP to treat this food industry waste, on which value will you base your design?

Ratios of compounds in wastewater (exercises 3.4.4-3.4.5)

Exercise 3.4.4

Consider a WWTP influent with the composition given in Table 3.14. Calculate the following ratios: COD/BOD, COD/TN, BOD/TN, COD/VSS and VSS/TSS. What can you infer from these ratios?

Table 3.14 Composition of an example wastewater stream

Parameter	Symbol	Unit	Value
Biochemical oxygen demand	BOD	g/m³	400
Chemical oxygen demand	COD	g/m³	300
Total suspended solids	TSS	g/m³	350
Volatile suspended solids	VSS	g/m³	280
Total nitrogen	TN	g/m³	65

Exercise 3.4.5

Why are the ratios of various wastewater components important? Give specific examples of insights you can gain from these ratios.

Wastewater fractionation (exercises 3.4.6-3.4.7)

The characteristics of the raw and settled wastewater considered in the textbook are summarized in Table 3.15 below.

Table 3.15 Raw and settled wastewater characteristics considered in the textbook (summarized from tables 4.2, 5.2 and 6.3 in the textbook). The subscript 'i' refers to 'influent'.

Parameter	Symbol	Unit	Raw	Settled
Flow rate	Q_i	m³/d	15,000	14,930
COD concentration	COD_i	mgCOD/l	750	450
Particulate unbiodegradable COD fraction	$f_{XU,CODi}$	-	0.15	0.04
Soluble unbiodegradable COD fraction	$f_{SU,CODi}$	-	0.07	0.12
Soluble unbiodegradable organic nitrogen fraction	$f_{SU,TKNi}$	-	0.03	0.035
Nitrogen content of volatile suspended solids	f_n	mgN/mgVSS	0.1	0.1
Soluble biodegradable COD fraction	$f_{SS,CODi}$	-	0.195	-
TKN concentration	TKN_i	mgN/l	60.0	51.0
FSA fraction	f_{aN}	-	0.75	0.88
Total P concentration	$P_{t,i}$	mgP/l	15.0	12.75
TKN/COD ratio	$f_{TKNi/CODi}$	mgN/mgCOD	0.08	0.117
P/COD ratio	$f_{Pi/CODi}$	mgP/mgCOD	0.02	0.028
Inorganic suspended solids concentration	$X_{FSS,i}$	mgISS/l	47.8	9.5
VSS/TSS ratio of activated sludge	f_{VT}	mgVSS/mgTSS	0.75	0.83

-: value not measured.

Exercise 3.4.6
Determine the COD, TKN, TP, ISS and TSS concentrations and mass flow rates for the raw wastewater, settled wastewater and primary sludge.

Exercise 3.4.7
Characterize the raw wastewater, settled wastewater, and primary sludge from Table 3.15 of this chapter based on COD and N according to the soluble, particulate, biodegradable and unbiodegradable concentrations.

Other constituents (exercises 3.4.8-3.4.9)
Exercise 3.4.8
Metals

The effluent wastewater described in Example 3.3 is being considered for agricultural irrigation. Particular attention to the metal content is required, as addressed in Section 7 in this chapter. According to the Food and Agriculture Organization guidelines summarized in Table 3.16 of this book, is the effluent wastewater allowed to be reused in agricultural irrigation? Assume that the concentrations of Al, As, Be, Co, F, Fe, Li, Mn, Mo, Ni, Pb, Se and V in the effluent wastewater are all below 1 µg/l.

What are the maximum concentrations allowed for Cd, Cr, Cu, Pb and Zn in the influent wastewater such that the effluent can be used for agricultural irrigation?

Table 3.16 Threshold levels of trace elements for crop production according to FAO guidelines (Pescod, 1992). http://www.fao.org/3/T0551E/t0551e04.htm and http://www.fao.org/3/t0551e/t0551e00.htm

Metal	Recommended maximum concentration (mg/l)	Comments
Al	5.00	Can cause non-productivity in acid soils (pH < 5.5), but more alkaline soils at pH > 7.0 will precipitate the ion and eliminate any toxicity.
As	0.10	Toxicity to plants varies widely, ranging from 12 mg/l for Sudan grass to less than 0.05 mg/l for rice.
Be	0.10	Toxicity to plants varies widely, ranging from 5 mg/l for kale to 0.5 mg/l for bush beans.
Cd	0.01	Toxic to beans, beets and turnips at concentrations as low as 0.1 mg/l in nutrient solutions. Conservative limits recommended due to its potential for accumulation in plants and soils in concentrations that may be harmful to humans.
Co	0.05	Toxic to tomato plants at 0.1 mg/l in nutrient solution. Tends to be inactivated by neutral and alkaline soils.
Cr	0.10	Not generally recognized as an essential growth element. Conservative limits recommended due to lack of knowledge on its toxicity to plants.
Cu	0.20	Toxic to a number of plants at 0.1 to 1.0 mg/l in nutrient solutions.
F	1.00	Inactivated by neutral and alkaline soils.
Fe	5.00	Not toxic to plants in aerated soils but can contribute to soil acidification and loss of availability of essential phosphorus and molybdenum. Overhead sprinkling may result in unsightly deposits on plants, equipment and buildings.
Li	2.50	Tolerated by most crops up to 5 mg/l; mobile in soil. Toxic to citrus at low concentrations (<0.075 mg/l). Acts similarly to boron.
Mn	0.20	Toxic to a number of crops from a few tenths to a few mg/l, but usually only in acid soils.
Mo	0.01	Not toxic to plants at normal concentrations in soil and water. Can be toxic to livestock if foraged fodder is grown in soils with high concentrations of available molybdenum.
Ni	0.20	Toxic to a number of plants at 0.5 mg/l to 1.0 mg/l; reduced toxicity at neutral or alkaline pH.
Pb	5.00	Can inhibit plant cell growth at very high concentrations.
Se	0.02	Toxic to plants at concentrations as low as 0.025 mg/l and toxic to livestock if foraged fodder is grown in soils with relatively high levels of added selenium. An essential element for animals but in very low concentrations.
V	0.10	Toxic to many plants at relatively low concentrations.
Zn	2.00	Toxic to many plants at widely varying concentrations; reduced toxicity at pH > 6.0 and in fine textured or organic soils.

Exercise 3.4.9

Microorganisms

The bathing water quality limit is 1,000 *E. coli* per 100 ml of water. Calculate the required dilution in the receiving water to reach the bathing water criteria for a wastewater of high microbial strength.

Internal sub-streams (exercises 3.4.10-3.4.11)

Exercise 3.4.10

Internal loads

What are the sources of internal loads in a wastewater treatment plant? Describe their main characteristics.

Exercise 3.4.11

Non-sewered (onsite) sanitation flows

A septic tank truck carries 20 m^3 septic sludge containing 20 kgBOD/m^3. It takes 1 hour to discharge its load directly into a wastewater treatment plant for 50,000 population equivalents of wastewater. The hourly wastewater flow in the plant is 1,000 m^3 with a BOD concentration of 300 gBOD/m^3. What is the increase in BOD load during that hour?

Wastewater dynamics and sampling (exercises 3.4.12-3.4.14)

Exercise 3.4.12

A wastewater treatment plant is designed to treat an incoming flow of 70,000 m^3/d. The hourly factor is 16. What is the population equivalent based on flow? What is the annual wastewater flow? What is the maximum wastewater flow?

Exercise 3.4.13

9,000 m^3 of sewage is produced by a small town during a 24-hour period. During flow monitoring, it was observed that the peak usage occurs around 19:30 in the evening, with the peak flow measured at 0.25 m^3/s What is the peak factor (f_{max})?

Exercise 3.4.14

List and describe the different wastewater sampling procedures.

ANNEX1: SOLUTION TO EXERCISES

Person load and population equivalent (solutions 3.4.1-3.4.3)

Solution 3.4.1

PE is a standard unit referring to the typical contribution of an individual to the wastewater load (in terms of flow rate or BOD load). PE can also be used to quantify the load of industrial wastewater streams.

PL refers to the actual contribution of persons living in the sewer catchment connected to the WWTP of concern. The PL therefore varies between regions whereas PE (a standard unit) does not.

Both PE and PL are related to the load of a WWTP and based on average contributions, which are obtained from long-term data.

Solution 3.4.2

The person load (PL) in terms of the daily flow rate is:

$$PL = \text{Daily flow rate / Number of persons}$$
$$= (60 \text{ m}^3/\text{d}) / 240 \text{ persons}$$
$$= 0.35 \text{ m}^3/\text{person.d}$$

Note that the PL expressed in wastewater flow rate in this case is higher than the PE, which is 0.2 m^3/person.d.

The PL in terms of the influent BOD is:

$$PL = \text{Daily influent BOD load / Number of persons}$$
$$= (18,000 \text{ gBOD/d}) / 240 \text{ persons}$$
$$= 75 \text{ gBOD/person.d}$$

Note that the PL expressed in BOD load in this case is also higher than the PE, which is 60 gBOD/person.d.

Solution 3.4.3

The population equivalent (PE) based on the daily flow rate is:

$$PE = \text{Daily flow rate} \cdot 1 \text{ PE}/0.2 \text{ m}^3.\text{d}$$
$$= 2,000 \text{ m}^3/\text{d} \cdot 1 \text{ PE}/0.2 \text{ m}^3.\text{d}$$
$$= 10,000 \text{ PE}$$

The PE based on the influent BOD is:

$$PE = \text{Daily influent BOD load} \cdot 1 \text{ PE}/60 \text{ gBOD.d}$$
$$= 450 \text{ gBOD/m}^3 \cdot 4,000 \text{ m}^3/\text{d} \cdot 1 \text{ PE}/60 \text{ gBOD.d}$$
$$= 30,000 \text{ PE}$$

The WWTP design should be based on the highest PE number, in this case the one corresponding to the influent BOD load. Industrial wastewater is typically characterized by a low volumetric flow rate and a high concentration of pollutants (*e.g.* BOD, COD, SS) compared to a municipal wastewater, which means that the BOD load rather than the flow rate is the limiting factor for the design.

Ratios of compounds in wastewater (solutions 3.4.4-3.4.5)
Solution 3.4.4
The calculated ratios are:

	Raw WW	Observation (Table 3.19 in the textbook)
COD/BOD	0.75	Unusually low
COD/TN	4.62	Unusually low
BOD/TN	6.15	High
COD/VSS	1.07	Extremely low
VSS/TSS	0.80	Medium

All the ratios involving COD fall outside the typical range. For example, a range of 1.5-3.5 is reasonable for COD/BOD. Since BOD measures a fraction of the organic matter, COD must be higher than BOD. Therefore, one of the two analytical results must be wrong. Since the BOD/TN ratio is within the expected range, the BOD value is most probably correct. Thus, the COD value is most likely due to an analytical error.

Solution 3.4.5
Ratios in wastewater typically remain relatively constant. Thus, they can be used to detect anomalies that are due to analytical errors or due to special discharges into the sewer system, often from industry. The ratios can also give insights into the functioning of the wastewater treatment processes.

For example:

(i)   Treatment of a wastewater with a low biodegradable organic carbon to nitrogen ratio may require external carbon source addition for biological denitrification.
(ii) A variation in the ratio between readily and slowly biodegradable organic carbon (S_S / X_S) for the same total amount of biodegradable organic carbon ($COD_{b,i} = S_S + X_S$) can significantly affect the nitrate concentration in the effluent.
(iii) A wastewater with relatively high nitrate concentration or low concentration of volatile fatty acids (VFAs) may not be suitable for biological phosphorus removal.
(iv) A high COD to BOD ratio in wastewater indicates that a substantial part of the organic matter will be difficult to degrade biologically.
(v)   A sludge stream with a high VSS to TSS ratio means that it can be successfully treated through anaerobic digestion.

Wastewater fractionation (solutions 3.4.6-3.4.7)

Solution 3.4.6

	Raw WW		Settled WW		Primary sludge	
Flow rate (m³/d)	15,000		14,930		70	
	Concentration (mg/l)	Mass flow rate (kg/d)	Concentration (mg/l)	Mass flow rate (kg/d)	Concentration (mg/l)	Mass flow rate (kg/d)
COD	750	11,250	450	6,719	65,000	4,531
TKN	60	900	51	761	2,000	139
TP	15	225	12.75	190	500	35
ISS	47.8	717	9.5	142	8,200	575
TSS	191.2	2,868	55.88	834	29,000	2,034

Solution 3.4.7

	Raw WW	Settled WW	Primary sludge
Flow rate Q (m³/d)	15,000	14,930	70
	Concentration (mg/l)	Concentration (mg/l)	Concentration (mg/l)
COD_{total}	750	450	65,000
S_U	53	53	53
X_U	113	18	20,000
S_S	146	146	146
X_S	438	233	44,000

<u>Calculation method</u>

(i)   The total COD concentrations for raw and settled wastewater are given (Table 3.15). The total COD concentration in the primary sludge is calculated from the mass balance over the primary settler; note the number of significant digits (= 2).

(ii)   The soluble unbiodegradable COD fraction is calculated as $S_U = f_{SU,CODi} \cdot COD_i$ for raw and/or settled wastewater; its value is the same for all three streams.

(iii)   The particulate unbiodegradable COD fraction is calculated as $X_U = f_{XU,CODi} \cdot COD_i$, for both raw and settled wastewater; the value for primary sludge is calculated from the mass balance over the primary settler

(iv)   The soluble biodegradable COD fraction for raw wastewater is calculated as $S_S = f_{SS,CODi} \cdot COD_i$, realistically assuming that the readily biodegradable COD fraction corresponds with the soluble biodegradable COD fraction. Given that this is a soluble component, its concentration in the settled wastewater and the primary sludge are the same.

(v)   The particulate biodegradable COD fraction is calculated as the difference $X_S = COD_i - S_U - X_U - S_S$ (for all three streams); the resulting values can be double-checked through a mass balance over the primary clarifier.

	Raw WW	Settled WW	Primary sludge
Flow rate Q (m^3/d)	15,000	14,930	70
	Concentration	Concentration	Concentration
	(mg/l)	(mg/l)	(mg/l)
TKN	60.0	51.0	2,000
S_{NHX}	45.0	45.0	45.0
N_{ous}	1.8	1.8	1.8
N_{oup}	7.6	1.2	1,400
$N_{obs} + N_{obp}$	5.6	3.0	550

<u>Calculation method</u>

(i)   The total Kjeldahl nitrogen (TKN) concentrations for raw and settled wastewater are given in Table 3.15 in this book. The TKN concentration in the primary sludge is calculated from the mass balance over the primary settler; note the number of significant digits (= 2). The total (*i.e.* free and saline) ammonium concentration is calculated as $S_{NHX} = f_{aN} \cdot$ TKN for raw and/or settled wastewater. Since this concerns a soluble component, its value is the same for all three streams.

(ii)   The soluble unbiodegradable organic nitrogen concentration is calculated as $N_{ous} = f_{SU,TKNi} \cdot$ TKN for raw and/or settled wastewater. Since this concerns a soluble component, its value is the same for all three streams.

(iii)   The particulate unbiodegradable organic nitrogen concentration is calculated as the nitrogen fraction associated with particulate unbiodegradable organic COD: $N_{oup} = f_n / f_{cv} X_U$, in which $f_{cv} = f_{cv,OHO}$ is 1.48 mgCOD/mgVSS. Note: it is reasonably assumed that all VSS fractions have the same COD/VSS ratio and the same nitrogen content. N_{oup} is calculated as indicated for all three streams; the result can be double-checked through the mass balance over the primary clarifier.

(iv)   Since no information is available on the soluble TKN concentration, it is not possible in this case to calculate the soluble biodegradable organic nitrogen concentration (N_{obs}) and the particulate organic nitrogen (N_{obp}) concentrations separately. However, their sum for all three streams can be calculated as $N_{obs} + N_{obp}.$ = TKN $- S_{NHX} - N_{ous} - N_{oup}.$; the result can be double-checked through the mass balance over the primary clarifier.

<u>Add-on</u>

If the soluble TKN concentration is measured, *e.g.* TKN$_s$ = 48.5 mgN/l, the soluble biodegradable organic nitrogen concentration can be calculated as $N_{obs} = $ TKN$_s - S_{NHX} - N_{ous}$. The particulate biodegradable organic nitrogen concentration results from $N_{obp} = $ TKN $-$ TKN$_s - N_{oup} = $ TKN $- S_{NHX} - N_{ous} - N_{oup} - N_{obs}$.

The resulting values are summarized below:

	Raw WW	Settled WW	Primary sludge
N_{obs}	1.7	1.7	1.7
N_{obp}	3.9	1.3	550

Other constituents (solutions 3.4.8-3.4.9)
Solution 3.4.8
<u>Metals</u>
Given that the concentrations of Al, As, Be, Co, F, Fe, Li, Mn, Mo, Ni, Pb, Se and V in the effluent wastewater are all below 1 µg/l, it is clear from comparison with Table 3.16 that the threshold levels for these components are not exceeded. The metals which remain to be considered are Cd, Cr, Cu, Pb and Zn.

	Removal efficiency (%) (Table 3.3)	Effluent WW (mg/l) (Section 7)	Limit (mg/l) (Table 3.16)	Observation
Cd	41	0.003	0.01	Allowed for agricultural irrigation
Cr	18	0.033	0.10	Allowed for agricultural irrigation
Cu	55	0.036	0.20	Allowed for agricultural irrigation
Pb	58	0.042	5.00	Allowed for agricultural irrigation
Zn	58	0.084	2.00	Allowed for agricultural irrigation

The maximum concentrations allowed for Cd, Cr, Cu, Pb and Zn in the influent wastewater, given the removal efficiencies from Table 3.3, such that the effluent can still be used for agricultural irrigation are:

	Removal efficiency (%)	Limit (mg/l)	Maximum influent concentration (mg/l)
Cd	41	0.01	0.017
Cr	18	0.10	0.12
Cu	55	0.20	0.44
Pb	58	5.000	11.9
Zn	58	2.0	4.76

Solution 3.4.9
<u>Microorganisms</u>
According to Table 3.5 in the textbook, a high-strength wastewater contains around $5{\cdot}10^8$ *E. coli* per 100 ml. Thus, the required dilution is calculated as:

Dilution
$$= 5{\cdot}10^8 /1,000$$
$$= 500,000 \text{ or } 5{\cdot}10^5$$

Wastewater sub-streams (solutions 3.4.10-3.4.11)
Solution 3.4.10
<u>Internal loads</u>
- Supernatant from sludge thickener (before digestion) and digester – high ammonia concentrations.
- Reject water from dewatering unit (following sludge digestion) – high concentration of organic soluble components and soluble nitrogen.
- Filter washing water – high hydraulic load and also high suspended solids concentration.

Solution 3.4.11

<u>Non-sewered (onsite) sanitation flows</u>

The normal BOD load in the wastewater treatment plant is:

$$\text{Normal BOD load in WWTP} = 1{,}000 \text{ m}^3/\text{h} \cdot 300 \text{ gBOD/m}^3$$
$$= 300 \text{ kgBOD/h}$$

The additional BOD load due to the septic sludge is:

$$\text{BOD load of septic sludge} = 20 \text{ kgBOD/m}^3 \cdot 20 \text{ m}^3/\text{h}$$
$$= 400 \text{ kgBOD/h}$$

The increase in BOD load is calculated as:

$$\text{Percentage increase in BOD load} = \left(\frac{\text{Additional BOD load}}{\text{Normal BOD load}} \right) \cdot 100\%$$
$$= \left(\frac{400 \text{ kgBOD/h}}{300 \text{ kgBOD/h}} \right) \cdot 100\%$$
$$= 133\%$$

Wastewater dynamics and sampling (solutions 3.4.12-3.4.14)

Solution 3.4.12

The population equivalent (PE) is:

$$\text{PE} = (70{,}000 \text{ m}^3/\text{d}) / (0.2 \text{ m}^3/\text{d})$$
$$= 350{,}000 \text{ population equivalents}$$

The annual wastewater flow is:

$$Q_{year} = 70{,}000 \text{ m}^3/\text{d} \cdot 365 \text{ d/yr}$$
$$= 25.55 \text{ million m}^3/\text{yr}$$

The maximum wastewater flow is:

$$Q_{max} = (70{,}000 \text{ m}^3/\text{d}) / (16 \text{ h/d})$$
$$= 4{,}375 \text{ m}^3/\text{h}$$

Solution 3.4.13

The average flow rate (Q_{avg}) is:

$$Q_{avg} = 9{,}000 \text{ m}^3/\text{d}$$

The maximum flow rate is (Q_{max}):

$$Q_{max} = 0.25 \text{ m}^3/\text{s} \cdot 86{,}400 \text{ s/d}$$
$$= 21{,}600 \text{ m}^3/\text{d}$$

The peak factor (f_{max}) is calculated as:

$$f_{max} = Q_{max} / Q_{avg}$$
$$= (21,600 \text{ m}^3/\text{d}) / (9,000 \text{ m}^3/\text{d})$$
$$= 2.4$$

Solution 3.4.14

- Grab sampling is the collection of a single sample or individual samples in a recipient at a specific time instant.
- Time proportional sampling is a type of composite sampling, wherein a number of samples are taken at regular time intervals and are then combined in one final sample.
- Flow proportional sampling is another type of composite sampling, wherein a number of samples are taken for each specific volume of wastewater flow. Another method of flow proportional sampling is to take samples at a constant time interval with volumes of the samples proportional to the flow.

REFERENCES

Alcalde-Sanz L. and Gawlik B.M. (2017). *Minimum quality requirements for water reuse in agricultural irrigation and aquifer recharge - Towards a water reuse regulatory instrument at EU level.* EUR 28962 EN, Publications Office of the European Union, Luxembourg. ISBN 978-92-79-77176-7, doi 10.2760/887727, PUBSY No.109291.

European Union (2018). Commission Implementing Decision (EU) 2018/1147 of 10 August 2018 establishing best available techniques (BAT) conclusions for waste treatment. (2018). *Official Journal of the European Union*, L 208, 17.08.2018.

Pescod M.B. (1992). *Wastewater Treatment and Use in Agriculture*. FAO Irrigation and Drainage Paper 47, Food and Agriculture Organization of the United Nations, Rome. https://www.fao.org/3/T0551E/t0551e04.htm

Chen G.H., van Loosdrecht M.C.M., Ekama G.A. and Brdjanovic D. (eds.) (2020). Biological Wastewater Treatment: Principles, Modelling and Design. ISBN: 9781789060355. IWA Publishing, London, UK.

NOMENCLATURE

Symbol	Description	Unit
BOD	BOD concentration	mgBOD/l
$BOD_{e,limit}$	Maximum permissible BOD concentration in the effluent of a treatment system	mgBOD/l
BOD_i	BOD concentration in the influent of a treatment system	mgBOD/l
$COD_{b,i}$	Biodegradable COD concentration in the influent	mgCOD/l
COD	COD concentration	mgCOD/l
$COD_{e,limit}$	Maximum permissible COD concentration in the effluent	mgCOD/l
COD_p	Particulate COD concentration	mgCOD/l
COD_s	Soluble COD concentration	mgCOD/l
COD_t	Total COD concentration	mgCOD/l
E(%)	Removal efficiency	%
$E(\%)_{BOD}$	BOD removal efficiency	%
$E(\%)_{COD}$	COD removal efficiency	%
FCOD	COD mass flow	kgCOD/m^3
f_{aN}	Free and saline ammonia fraction	mgN/mgN
$f_{cv,OHO}$	COD/VSS ratio of the ordinary heterotrophic biomass	kgCOD/kgVSS
f_{cv}	COD/VSS ratio of the sludge	kgCOD/kgVSS

$f_{h,max}$	Maximum hourly constant	-
f_{max}	Peak factor	-
f_n	Nitrogen content of volatile suspended solids, i.e., N/VSS ratio of the sludge	mgN/mgVSS
$f_{Pi/CODi}$	P/COD ratio	mgP/mgCOD
f_p	P/VSS ratio of the sludge	mgP/mgVSS
$f_{SS,CODi}$	Soluble biodegradable COD fraction in the influent	mgCOD/mgCOD
$f_{SU,CODi}$	Soluble unbiodegradable COD fraction in the influent	mgCOD/mgCOD
$f_{SU,TKNi}$	Soluble unbiodegradable organic nitrogen fraction in the influent	mgN/mgN
$f_{TKNi/CODi}$	TKN/COD ratio	mgN/mgCOD
f_{VT}	VSS/TSS ratio of activated sludge	mgVSS/mgTSS
$f_{XU,CODi}$	Particulate unbiodegradable COD fraction in the influent	mgCOD/mgCOD
N_o	Organic nitrogen concentration	mgN/l
N_{obp}	Biodegradable particulate organic nitrogen concentration	mgN/l
N_{obs}	Biodegradable soluble organic nitrogen concentration	mgN/l
N_{oup}	Unbiodegradable particulate organic nitrogen concentration	mgN/l
N_{ous}	Unbiodegradable soluble organic nitrogen concentration	mgN/l
P_{IN}	Inorganic phosphorus concentration	mgP/l
P_o	Organic phosphorus concentration	mgP/l
P_{obp}	Biodegradable particulate organic phosphorus concentration	mgP/l
P_{obs}	Biodegradable soluble organic phosphorus concentration	mgP/l
$P_{o,s}$	Soluble organic phosphorus concentration	mgP/l
P_{ous}	Unbiodegradable soluble organic phosphorus concentration	mgP/l
P_p	Particulate phosphorus concentration	mgP/l
P_{oup}	Unbiodegradable particulate organic phosphorus concentration	mgP/l
P_s	Soluble phosphorus concentration	mgP/l
P_t	Total phosphorus concentration	mgP/l
Q	Flow rate	m^3/d
$Q_{h,avg}$	Average hourly influent flow rate	m^3/h
$Q_{h,max}$	Maximum hourly flow rate	m^3/h
$Q_{max,discharge}$	Maximum wastewater flow rate discharge allowed	m^3/h
Q_{year}	Annual wastewater flow rate	$m^3/year$
S_{NHx}	Free and saline ammonia (FSA) concentration	mgN/l
S_{PO4}	Orthophosphate concentration	mgP/l
S_s	Biodegradable soluble organic concentration	mgCOD/l
S_u	Unbiodegradable soluble organic concentration	mgCOD/l
TKN	Total Kjeldahl nitrogen concentration	mgN/l
TKN_p	Particulate Kjeldahl nitrogen concentration	mgN/l
TKN_s	Soluble Kjeldahl nitrogen concentration	mgN/l
X_{FSS}	Inorganic (=fixed) suspended solids concentration	mgISS/l
X_s	Biodegradable particulate organic concentration	mgCOD/l
X_u	Unbiodegradable particulate organic concentration	mgCOD/l

Subscripts	Description
e	Effluent
i	Influent
PS	Primary settling tank underflow
raw	Raw influent wastewater
settled	Settled influent wastewater

Abbreviation	Description
AS	Activated sludge
BAT	Best available techniques
BOD	Biochemical oxygen demand
COD	Chemical oxygen demand
CAS	Conventional activated sludge
DEHP	Di(2-ethylhexyl)phthalate
FSA	Free and saline ammonia
ISS	Inorganic suspended solids
LAS	Linear alkylbenzene sulfonate
LRV	Log removal value
MLD	Millions of litres per day
NPE	Nonylphenol ethoxylates
PE	Population equivalent
PL	Person load
PAH	Polycyclic aromatic hydrocarbons
PST	Primary settling tank
PS	Primary sludge
TKN	Total Kjeldahl nitrogen
TOC	Total organic carbon
TP	Total phosphorus
TSS	Total suspended solids
VFA	Volatile fatty acids
VSS	Volatile suspended solids
WW	Wastewater
WWTP	Wastewater treatment plant

Proper sampling is an essential step in obtaining representative samples for wastewater and sludge characterization (photo: D. Brdjanovic).

doi: 10.2166/9781789062304_0109

4

Organic matter removal

Carlos M. Lopez-Vazquez

4.1 INTRODUCTION

Chapter 4 on Organic Matter Removal in the textbook *Biological Wastewater Treatment: Principles, Modelling and Design* (Chen *et al.,* 2020) introduces the principles behind the physical, chemical and microbial conversions and key factors affecting the removal of organic matter in activated sludge systems. It provides the basis to understand the mechanisms involved in the removal of organic matter and presents a stoichiometric-based steady-state model for the design of activated sludge systems that carry out most of the aerobic removal of organic matter. The present chapter aims to guide the reader through the principles, microbial mechanisms and the steady-state stoichiometric model to design, assess and evaluate conventional activated sludge (CAS) wastewater treatment systems. All the equations presented here belong to the textbook.

4.2 LEARNING OBJECTIVES

After the successful completion of this chapter, the reader will be able to:

- Describe the basic transformations of organic and inorganic wastewater compounds that take place in CAS wastewater treatment plants.
- Discuss the fundamental principles for the selection, design and control of the sludge age of activated sludge wastewater treatment systems.
- Describe and assess the influence of key system constraints, environmental factors and operational parameters on the performance of CAS wastewater treatment plants.
- Apply a stoichiometric-based steady-state model for the process design and evaluation of activated sludge wastewater treatment plants performing the aerobic removal of organic matter.
- Explain the meaning of the food-to-microorganism ratio and its applicability within the context of the COD stoichiometric-based steady-state model for the process design of CAS wastewater treatment plants.

4.3 EXAMPLES

Example 4.3.1

By applying the COD stoichiometric-based steady-state design model, design an activated sludge system (Figure 4.1) to treat an influent wastewater (Q_i) with a flow rate of 15 MLD (15,000 m³/d). The plant needs to comply with an effluent total COD concentration (COD_e) of 125 mg COD/l. The plant does not have a primary settling tank. The raw wastewater contains a total influent COD concentration (COD_i) of 585 mg/l that contains an influent total Kjeldahl nitrogen concentration (TKN_i) and total phosphorus concentration (P_i) of 44.5 mgN/l and 14.5 mgP/l, respectively. The influent COD fractionation is shown in Figure 4.2. In addition, the influent flow rate contains 35 mg/l of inorganic or fixed suspended solids.

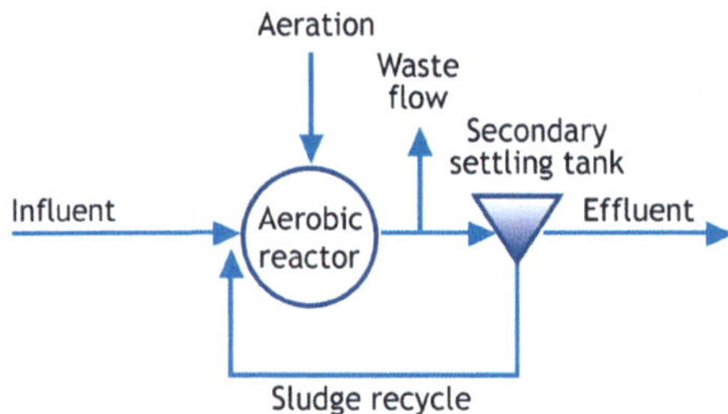

Figure 4.1 Activated sludge system with a single reactor completely-mixing regime.

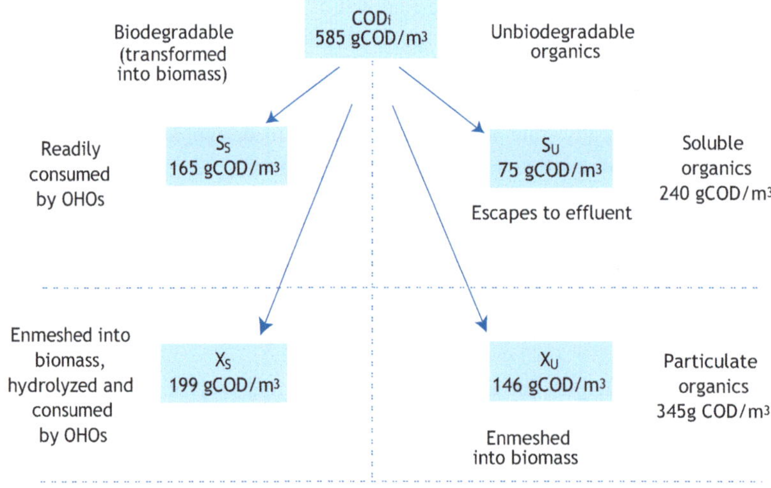

Figure 4.2 Influent COD fractionation for the activated sludge design example 4.3.1.

To carry out the design, assume:

a. A minimum yearly wastewater temperature of 14 °C.
b. An SRT of 6 days.
c. An average MLSS concentration of 4,000 mg/l to size the aerobic reactor volume (V_R).
d. A TSS concentration in the effluent of the secondary settling tank of 15 mgTSS/l.
e. A TSS concentration in the sludge recycle flow of 10,000 mgTSS/l.

Based on the previous information, determine:

a. The s-recycle ratio required to keep an average MLSS concentration of 4,000 mgTSS/l in the system ($s = Q_s/Q_i$; where Q_s is the sludge recycle flow rate and Q_i is the influent flow rate).
b. Total COD flux consumed by OHO.
c. MX_{VSS} and MX_{TSS} masses.
d. Total volume of the system (V_R)
e. Nutrient requirement of the system (N_s and P_s).
f. O_2 requirements of the system.
g. Effluent total COD, total Kjeldahl nitrogen and total phosphorus concentrations (COD_e, TKN_e and P_e, respectively).
h. The COD mass balance in order to confirm the reliability of your design.

Solution

Table 4.1 presents a summary of the main design data and Table 4.2 displays the main stoichiometric and kinetic parameters necessary to carry out the process design.

Table 4.1 Summary of the data to carry out the process design of the activated sludge plant.

Description	Symbol	Value	Unit
Flow rate	Q_i	15	MLD
Total COD	COD_i	585	$gCOD/m^3$
COD concentrations			
- readily biodegradable COD	$S_{S,i}$	165	$gCOD/m^3$
- slowly biodegradable COD	$X_{S,i}$	199	$gCOD/m^3$
- unbiodegradable soluble COD	$S_{U,i}$	75	$gCOD/m^3$
- unbiodegradable particulate COD	$X_{U,i}$	146	$gCOD/m^3$
Influent inorganic (fixed) suspended solids	$X_{FSS,i}$	35	$gFSS/m^3$
Influent TKN concentration	TKN_i	44.5	gN/m^3
Influent total P concentration	P_i	14.5	gP/m^3
Temperature	T	14	°C
Sludge retention time	SRT	6	d
Total suspended solids in the effluent	$X_{TSS,e}$	15	$gTSS/m^3$
Design TSS concentration	X_{TSS}	4,000	$gTSS/m^3$

Table 4.2 Kinetic and stoichiometric parameters for the EBPR design example 4.3.1.

	Parameter	Symbol	Value	Unit
Kinetic	*OHO*			
	Specific endogenous mass loss rate of the OHOs at 20 °C	$b_{OHO,20}$	0.24	gVSS /gVSS.d
	Temperature coefficient for $b_{OHO,T}$	$\theta_{b,OHO}$	1.029	
	Specific endogenous mass loss rate of the OHOs at temperature T	$b_{OHO,T}$	0.202	gVSS /gVSS.d
Stoichiometric	*OHO*			
	Biomass yield of the OHOs	Y_{OHOv}	0.45	gVSS/gCOD
	Fraction of endogenous residue of the OHOs	$f_{XE,OHO}$	0.20	gVSS/gVSS
	Fraction of fixed (inorganic) suspended solids of the OHOs	$f_{FSS,OHO}$	0.15	gFSS/gVSS
	General			
	COD/VSS ratio of the sludge	f_{cv}	1.48	gCOD/gVSS
	Nitrogen content in the volatile suspended solids	f_n	0.10	gN/gVSS
	Phosphorus content in the volatile suspended solids	f_p	0.03	gP/gVSS

Using the data provided, Table 4.3 shows the detailed calculations to carry out the process design to comply with COD_e of less than 125 mg/l.

Table 4.3 Activated sludge system design procedure.

1. System configuration

Aerobic completely-stirred tank reactor configuration operated at 14 °C

2. Influent and sludge recycle composition (based on table 4.1 and figure 4.2)

Q_i	15	MLD	(15,000 m³/d) influent flow rate

2.1 Influent concentrations

Influent and bioreactor data

COD_i	585	gCOD/m³	influent concentration of total COD
$S_{S,i}$	165	gCOD/m³	influent concentration of readily biodegradable COD
$X_{S,i}$	199	gCOD/m³	influent concentration of slowly biodegradable COD
$COD_{b,i}$	364	gCOD/m³	influent concentration of biodegradable COD $(S_{S,i} + X_{S,i})$
$S_{U,i}$	75	gCOD/m³	influent concentration of soluble unbiodegradable COD
$X_{U,i}$	146	gCOD/m³	influent concentration of particulate unbiodegradable COD
$X_{FSS,i}$	35	gFSS/m³	influent concentration of fixed (inorganic) suspended solids
TKN_i	44.5	gN/m³	influent concentration of total Kjeldahl N
P_i	14.5	gP/m³	influent concentration of total P

2.2 Influent flux used for calculations (= Qi · influent concentration of component)

$FCOD_i$	8,775	kgCOD/d	influent daily flux of total COD
$FS_{S,i}$	2,475	kgCOD/d	influent daily flux of RBCOD
$FCOD_{b,i}$	5,460	kgCOD/d	influent daily flux of biodegradable COD $(S_{S,i} + X_{S,i})$
$FX_{U,i}$	2,190	kgCOD/d	influent daily flux of particulate unbiodegradable COD
$FS_{U,i}$	1,125	kgCOD/d	influent daily flux of soluble unbiodegradable COD
$FX_{FSS,i}$	525	kgFSS/d	influent daily flux of fixed (inorganic) suspended solids

2.3 Sludge recycle characteristics

s		$m^3.d/\ m^3.d$	$s = X_{TSS} / (X_{TSS,s} - X_{TSS})$
		$m^3.d/\ m^3.d$	$s = 4,000 / (10,000 - 4,000)$
s	0.67	$m^3.d/\ m^3.d$	sludge recycle ratio (s) with regard to influent flow rate

3. Biomass equations

Corresponds to the biological mass present in the system as synthesized from the influent COD (in g/d), taking into account the cumulative effect of SRT [(g/d) · d = g in the system]

3.1 OHOs

Active mass

Y_{OHOv}	0.45 gVSS/gCOD
$Y_{OHOv,obs}$	$= Y_{OHOv} / (1 + b_{OHO,T} \cdot SRT)$
	$= 0.45 / (1 + 0.202 \cdot 6)$
$Y_{OHOv,obs}$	0.203 gVSS/gCOD

$$MX_{OHOv} = Y_{OHOv,obs} \cdot F_{CODb,i} \cdot SRT \qquad (4.9)$$
$$= 0.203 \cdot 5,460 \cdot 6$$

MX_{OHOv}	6,661 kgVSS

Endogenous mass

$$MX_{E,OHOv} = f_{XE,OHO} \cdot b_{OHO,T} \cdot MX_{OHOv} \cdot SRT \qquad (4.10)$$
$$= 0.20 \cdot 0.202 \cdot 6,661 \cdot 6$$

$MX_{E,OHOv}$	1,616 kgVSS

3.2 Unbiodegradable particulate organics

$$MX_{Uv} = FX_{U,i} \cdot SRT / f_{cv} \qquad (4.11)$$
$$= 2,190 \cdot 6 / 1.48$$

MX_{Uv}	8,878 kgVSS

4. VSS and TSS

4.1 VSS and active fraction

MX_B	$= MX_{OHOv}$
MX_B	6,661 kgVSS

$$MX_{VSS} = MX_{OHOv} + MX_{E,OHOv} + MX_{Uv} \tag{4.12}$$
$$= 6{,}661 + 1{,}616 + 8{,}878$$

MX_{VSS} 17,156 kgVSS

$$f_{av} = MX_B / MX_{VSS}$$
$$= 6{,}661 / 17{,}156$$

f_{av} 39 %

4.2 FSS

$$MX_{FSS} = f_{FSS,OHO} \cdot MX_{OHOv} + FX_{FSS,i} \cdot SRT \tag{4.14a}$$
$$= (0.15 \cdot 6{,}661) + (525 \cdot 6)$$

MX_{FSS} 4,149 kgFSS

4.3 TSS

$$MX_{TSS} = MX_{VSS} + MX_{FSS} \tag{4.15}$$
$$= 17{,}156 + 4{,}149$$

MX_{TSS} 21,305 kgTSS

4.4 f_{VT}

$$f_{VT} = MX_{VSS} / MX_{TSS} \tag{4.16}$$
$$= 17{,}156 / 21{,}305$$

f_{VT} 0.81 gVSS/gTSS

5. Process volume (based on TSS)

X_{TSS} 4 kgTSS / m^3

$$V_R = MX_{TSS} / X_{TSS} \tag{4.20}$$
$$= 21{,}305/4$$

V_R 5,326 m^3

6. Nutrient requirements

6.1 Nitrogen requirements for biomass synthesis

$$N_s = f_n \cdot MX_{VSS} / (Q_i \, SRT) \tag{5.25}$$
$$= (0.10 \cdot 17{,}156) / (15 \cdot 6)$$

N_s 19.1 gN/m^3

6.2 Phosphorus requirements for biomass synthesis

$$P_s = f_p \cdot MX_{VSS} / (Q_i \, SRT)$$
$$= (0.03 \cdot 17{,}156) / (15 \cdot 6)$$

P_s 5.7 gP/ m^3 $\tag{5.31}$

7. Oxygen demand (OD)

OD for synthesis and endogenous respiration

$$FO_{OHO} = FO_{OHO,s} + FO_{OHO,e} \tag{6.30a}$$
$$FO_{OHO,s} = FCOD_{b,OHO} \cdot (1 - f_{cv} \cdot Y_{OHOv})$$

$$= 5{,}460 \cdot (1 - 1.48 \cdot 0.45)$$

$FO_{OHO,s}$ 1,824

$FO_{OHO,e}$ $= FCOD_{b,OHO} \cdot f_{cv} \cdot (1 - f_{XE,OHO}) \cdot b_{OHO,T} \cdot Y_{OHOv,obs} \cdot SRT$

$$= 5{,}460 \cdot 1.48 \cdot (1 - 0.20) \cdot 0.202 \cdot 0.203 \cdot 6$$

$FO_{OHO,e}$ 1,595

FO_{OHO} 3,418 kgO$_2$/d

8. Effluent quality

8.1 COD$_e$

COD_e $= S_{U,i} + f_{cv} \cdot f_{VT} \cdot X_{TSS,e}$ (non-filtered)

$$= 75 + (1.48) \cdot (0.81) \cdot (15)$$

93 gCOD/m^3

8.2 TKN$_e$

TKN_e $= TKN_i - N_s + f_n \cdot f_{VT} \cdot X_{TSS,e}$ (non-filtered)

$$= 44.5 - 19.1 + (0.10) \cdot (0.81) \cdot (15)$$

26.6 gN/m^3

8.3 P$_e$

P_e $= P_i - P_s + f_p \cdot f_{VT} \cdot X_{TSS,e}$ (non-filtered)

$$= 14.5 - 5.7 + (0.03) \cdot (0.81) \cdot (15)$$

9.2 gP/m^3

Overall, the system is able to comply with the effluent total COD discharge limit of 125 mgCOD/l. It also produces a treated effluent that contains a total Kjeldahl concentration of 26.6 mgN/l and total phosphorus concentration of 9.2 mgP/l. This indicates that despite the system not being designed for nutrient removal, it is able to remove approximately 40 % and 3 7% of the nitrogen and phosphorus concentrations present in the influent wastewater, respectively. It is important to underline that these effluent concentrations have been estimated based on the effluent TSS concentration being 15 mgTSS/l. If the effluent TSS is different, then other concentrations can be expected.

9. Waste of activated sludge

Q_W $= V_R/SRT$ (4.1)

$$= 5{,}326/6$$

Q_W 888 m^3/d

10. COD mass balance verification

Input

FCOD$_i$		8,775	kgCOD/d	100 %		IN

Output

O$_2$ demand for synthesis and endogenous respiration of OHO (FO$_{OHO}$ = FO$_c$)

FO$_c$	3,418	kgCOD/d	39.0 %

Soluble unbiodegradable organics leaving via the effluent

FS$_{U,i}$	1,125	kgCOD/d	12.8 %

Sludge	gVSS	kgCOD/d			
		(= gVSS · f_{cv} / SRT = gVSS · 1.48 / 8 = gVSS · 0.185)			
MX_{OHOv}(=MX_B)	6,661	1,643	kgCOD/d	18.7 %	
$MX_{E,OHOv}$	1,616	399	kgCOD/d	4.5 %	
MX_{Uv}	8,878	2,190	kgCOD/d	25.0 %	
$MX_{E,OHO}$ + MX_{Uv}	10,494	2,589		29.5 %	
MX_{VSS}	17,156	4,232	kgCOD/d	48.2 %	
	Sum:	8,775	kgCOD/d	100 %	OUT
	Delta (OUT-IN):	0	kgCOD/d	0 %	

The 100 % mass balance for COD indicates that all the influent COD is accounted for in the calculated values of oxygen demand and sludge production. From the COD mass balance, and for the conditions of the design example, the fate of the influent COD is as follows: 39 % is oxidized with oxygen, 12.8 % escapes in the effluent as soluble unbiodegradable organics and 48.2 % becomes activated sludge. This activated sludge is composed of 38.8 % (6,661 / 17,156) active biomass and 61.1% (10,494 / 17,156) inactive particulate matter of which 84.6 % (8,878 / 10,494) are influent unbiodegradable particulate organics and 15.4 % (1,616 / 10,494) is endogenous residue). A summary of the process design results is presented in Table 4.4.

Table 4.4 Summary of the process design results.

Description	Parameter	Unit	Value
1. Influent and bioreactor			
Type of wastewater	Raw		raw
Temperature	T	°C	14
Influent flow rate	Q_i	MLD	15
Influent total COD	COD_i	$gCOD/m^3$	585
Influent biodegradable COD	$COD_{b,i}$	$gCOD/m^3$	364
Influent TKN	TKN_i	gN/m^3	44.5
Influent total P	P_i	gP/m^3	14.5
Sludge retention time	SRT	d	6
Sludge recycle ratio (s)	s	m^3.d /m^3.d	0.67
2. $COD_{b,i}$ for OHOs			
Flux of $COD_{b,i}$ for OHOs	$FCOD_{b,OHO}$	kgCOD/d	5,460
3. System biomass (VSS) equations			
Mass of OHOs	MX_{OHOv}	kgVSS	6,661
Mass of endogenous residue from OHOs	$MX_{E,OHOv}$	kgVSS	1,616
Mass of unbiodegradable organics from influent	MX_{Uv}	kgVSS	8,878
4. Volatile and total suspended solids (VSS and TSS) in system			
Mass of active biomass (=MX_{OHOv})	MX_B	kgVSS	6,661
Mass of VSS	MX_{VSS}	kgVSS	17,156
MX_B/VSS	f_{av}	gVSS/gVSS	0.39

Mass of fixed or inorganic SS	MX_{FSS}	kgFSS	4,149
Mass of TSS	MX_{TSS}	kgTSS	21,305
Ratio of VSS/TSS	f_{VT}	gVSS/gTSS	0.81

5. Reactor volume

Bioreactor volume	V_R	m^3	5,326

6. Waste of activated sludge

Activated sludge waste flow rate	Q_W	m^3/d	888

7. Nutrient requirement

N requirement concentration for biomass synthesis	N_s	gN/m^3	19.1
P requirement concentration for biomass synthesis	P_s	gP/m^3	5.7

8. Oxygen demand

Flux of carbonaceous O_2 demand	FO_C	kgO_2/d	3,418

9. Treated effluent quality [a]

Effluent total COD	COD_e	$gCOD/m^3$	93.0
Effluent total Kjeldahl Nitrogen	TKN_e	gN/m^3	26.6
Effluent total phosphorus	P_e	gP/m^3	9.2

10. COD mass balance

COD output/COD input	COD mass balance	gCOD/gCOD	100 %

Flow rate is in m^3/d and mass fluxes in kg/d

For a flow rate of 1,000 or greater, mass fluxes can be read in kg/d

[a] Considering an effluent total solids concentration of 15 mgTSS/l.

Example 4.3.2

Design the activated sludge plant designed in Example 4.3.1 following the food-to-microorganism (F/M) ratio.

Table 4.5 Activated sludge system design procedure following the F/M ratio approach (using the same data from Example 4.3.1).

1. System configuration

Aerobic completely-stirred tank reactor configuration process configuration operated at 14 °C

2. Influent composition (from previous example 4.3.1)

Q_i	15	MLD	(15,000 m^3/d) influent flow rate
COD_i	585	$gCOD/m^3$	influent concentration of total COD
$S_{U,i}$	75	$gCOD/m^3$	influent concentration of soluble unbiodegradable COD
$f_{SU,CODi}$	$= S_{U,i}/COD_i$	gCOD/gCOD	influent fraction of soluble unbiodegradable COD
	$= 75/585$		
$f_{SU,CODi}$	0.13	gCOD/gCOD	influent fraction of soluble unbiodegradable COD
$X_{U,i}$	146	$gCOD/m^3$	influent concentration of particulate unbiodegradable COD
$f_{XU,CODi}$	$= X_{U,i}/COD_i$	gCOD/gCOD	influent fraction of particulate unbiodegradable COD
$f_{XU,CODi}$	$= 146/585$		

$f_{XU,CODi}$	0.25	gCOD/gCOD	influent fraction of particulate unbiodegradable COD
$X_{FSS,i}$	35	gFSS/m^3	influent concentration of fixed (inorganic) suspended solids

3. Determination of $A_{ND,TSS}$ ratio

A_{ND}

$$= (1 - f_{SU,CODi} - f_{XU,CODi}) \cdot [Y_{OHOv} \cdot SRT/(1 + b_{OHO,T} \cdot SRT)] \cdot [1 + f_{XE,OHO} \cdot b_{OHO,T} \cdot SRT + f_{FSS,OHO}]$$
$$+ [(f_{XU,CODi} / f_{cv}) + (X_{FSS,i} / COD_i)] \cdot SRT \qquad (4.35)$$
$$= (1 - 0.13 - 0.25) \cdot [(0.45 \cdot 6) / (1 + (0.202 \cdot 6)] \cdot [1 + (0.20 \cdot 0.202 \cdot 6) + 0.15] +$$
$$+ [(0.25 / 1.48) + (35 / 585)] \cdot 6$$

A_{ND} 2.43 kgTSS/kgCOD.d

Since:

A_{ND} $= 1 / (F/M)_{COD,TSS}$

$(F/M)_{COD,TSS}$ $= 1 / A_{ND}$

  $= 1 / 2.43$

$(F/M)_{COD,TSS}$ 0.41 kgCOD.d/kgTSS

4. Reactor volume

$(F/M)_{COD,TSS}$ $= Q_i \cdot COD_i / V_R \cdot X_{TSS}$ (derived from 4.36a)

Then:

V_R $= Q_i \cdot COD_i / (X_{TSS} \cdot (F/M)_{COD,TSS})$

Thus:

V_R $= Q_i \cdot COD_i / (X_{TSS} \cdot (F/M)_{COD,TSS})$

Since $X_{TSS} =$ 4 kgTSS/m^3

  $= (15 \cdot 585) / (4 \cdot 0.41)$

V_R 5,321 m^3

The volume of the reactor is the same as that obtained using the stoichiometric-based design model. This indicates that the two procedures can lead to similar results. Furthermore, the F/M ratio of 0.41 kgCOD.d/kgVSS indicates that it is a high-rate system and, with the SRT of 6d, it only performs the removal of COD (see Figure 4.8 in *Chen et al.*, 2020).

Example 4.3.3

The activated sludge plant designed in Example 4.3.1 only carries out the COD removal process. However, based on the MX_B / MX_{TSS} and f_{VT} ratios (of 0.39 and 0.81, respectively), the system generates an activated sludge waste flow rate of 888 m^3/d that contains a considerable fraction of active biomass (see Table 4.4 in Example 4.3.1). This sludge needs to be properly handled (digested either aerobically or preferably anaerobically) prior to dewatering and disposal. As such, the purpose is to redesign the plant with a longer SRT that should be long enough to decrease the fraction of the active biomass (e.g., to design an extended aeration activated sludge system).

Based on the previous requirements, the system will be designed with an SRT of 25 days. Using the data provided, Table 4.6 shows the detailed calculations to carry out the process design to decrease the active mass fraction and generate stable activated sludge waste.

Table 4.6 Procedure for the design of an activated sludge system designed with a long SRT (extended aeration system).

---

1. System configuration

Aerobic completely-stirred tank reactor configuration process configuration operated at 14 °C

2. Influent and sludge recycle composition (from previous tables)

| Q_i | 15 | MLD | (15,000 m³/d) influent flow rate |

2.1 Influent concentrations

Influent and bioreactor data

COD_i	585	gCOD/m³	influent concentration of total COD
$S_{S,i}$	165	gCOD/m³	influent concentration of readily biodegradable COD
$X_{S,i}$	199	gCOD/m³	influent concentration of slowly biodegradable COD
$COD_{b,i}$	364	gCOD/m³	influent concentration of biodegradable COD ($S_{S,i} + X_{S,i}$)
$S_{U,i}$	75	gCOD/m³	influent concentration of soluble unbiodegradable COD
$X_{U,i}$	146	gCOD/m³	influent concentration of particulate unbiodegradable COD
$X_{FSS,i}$	35	gFSS/m³	influent concentration of fixed (inorganic) suspended solids
TKN_i	44.5	gN/m³	influent concentration of total Kjeldahl N
P_i	14.5	gP/m³	influent concentration of total P

2.2 Influent fluxes used for calculations (= Q_i · influent concentration of component)

$FCOD_i$	8,775	kgCOD/d	influent daily flux of total COD
$FS_{S,i}$	2,475	kgCOD/d	influent daily flux of RBCOD
$FCOD_{b,i}$	5,460	kgCOD/d	influent daily flux of biodegradable COD ($S_{S,i} + X_{S,i}$)
$FX_{U,i}$	2,190	kgCOD/d	influent daily flux of particulate unbiodegradable COD
$FS_{U,i}$	1,125	kgCOD/d	influent daily flux of soluble unbiodegradable COD
$FX_{FSS,i}$	525	kgFSS/d	influent daily flux of fixed (inorganic) suspended solids

2.3 Sludge recycle characteristics

s		m³.d/ m³.d	$s = X_{TSS} / (X_{TSS,s} - X_{TSS})$
		m³.d/ m³.d	$s = 4,000 / (10,000 - 4,000)$
s	0.67	m³.d/ m³.d	sludge recycle ratio (s) with regard to influent flow rate

3. Biomass equations

Corresponds to the biological mass present in the system as synthesized from the influent COD (in g/d), taking into account the cumulative effect of SRT [(g/d) · d = g in the system]

3.1 OHOs

Active mass

Y_{OHOv}	0.45 gVSS/gCOD
$Y_{OHOv,obs}$	$= Y_{OHOv} / (1 + b_{OHO,T} \cdot SRT)$
	$= 0.45 / (1 + 0.202 \cdot 25)$
$Y_{OHOv,obs}$	0.074 gVSS/gCOD

MX_{OHOv}	$= Y_{OHOv,obs} \cdot F_{CODb,i} \cdot SRT$
	$= 0.074 \cdot 5,460 \cdot 25$
MX_{OHOv}	$10,156$ kgVSS

Endogenous mass

$MX_{E,OHOv}$	$= f_{XE,OHO} \cdot b_{OHO,T} \cdot MX_{OHOv} \cdot SRT$	(4.10)
	$= 0.20 \cdot 0.202 \cdot 10,156 \cdot 25$	
$MX_{E,OHOv}$	$10,256$ kgVSS	

3.2 Unbiodegradable particulate organics

MX_{Uv}	$= FX_{U,i} \cdot SRT / f_{cv}$	(4.11)
	$= 2,190 \cdot 25 / 1.48$	
MX_{Uv}	$36,993$ kgVSS	

4. VSS and TSS

4.1 VSS and active fraction

MX_B	$= MX_{OHOv}$	
MX_B	$10,156$ kgVSS	
MX_{VSS}	$= MX_{OHOv} + MX_{E,OHOv} + MX_{Uv}$	(4.12)
	$= 10,156 + 10,256 + 36,993$	
MX_{VSS}	$57,405$ kgVSS	
f_{av}	$= MX_B / MX_{VSS}$	
	$= 10,156 / 57,405$	
f_{av}	$18\ \%$	

With the SRT of 25 days, the active mass fraction comprises 18 % of MX_{VSS}, indicating that the sludge can be considered to be stabilized (assumed to happen when $f_{bio,VSS} < 20\ \%$).

4.2 FSS

MX_{FSS}	$= f_{FSS,OHO} \cdot MX_{OHOv} + FX_{FSS,i} \cdot SRT$	(4.14a)
	$= (0.15 \cdot 10,156) + (525 \cdot 25)$	
MX_{FSS}	$14,648$ kgFSS	

4.3 TSS

MX_{TSS}	$= MX_{VSS} + MX_{FSS}$	(4.15)
	$= 57,405 + 14,648$	
MX_{TSS}	$72,053$ kgTSS	

4.4 f_{VT}

f_{VT}	$= MX_{VSS} / MX_{TSS}$	(4.16)
	$= 57,405 / 72,053$	
f_{VT}	0.80 gVSS/gTSS	

In spite of the long SRT of 25 days, the f_{VT} ratio remains similar to that of the design carried out with SRT of 6 days. The main reason is that, at the extended SRT of 25 d, less active biomass is produced per COD consumed (the $Y_{OHOv,obs}$ decreases from 0.203 to 0.074 kgVSS/kgCOD as the SRT increases from 6 to 25 d). However, simultaneously, the longer SRT also causes a proportionally higher generation and accumulation of the endogenous residue mass ($MX_{E,OHOv}$). This results in a relatively constant f_{VT} ratio regardless of the SRT applied (because $MX_{E,OHOv}$ contributes to the VSS).

5. Process volume (based on TSS)

X_{TSS} $4 \ kgTSS / m^3$

V_R $= MX_{TSS} / X_{TSS}$ (4.20)

             $= 72,053 / 4$

V_R $18,010 \ m^3$

Due to the longer SRT, a larger tank volume is needed.

6. Nutrient requirements

6.1 Nitrogen requirements for biomass synthesis

N_s $= f_n \cdot MX_{VSS} / (Q_i \cdot SRT)$ (5.25)

             $= (0.10 \cdot 57,405) / (15 \cdot 25)$

N_s $15.3 \ gN/m^3$

6.2 Phosphorus requirements for biomass synthesis

P_s $= f_p \cdot MX_{VSS} / (Q_i \cdot SRT)$ (5.31)

             $= (0.03 \cdot 57,405) / (15 \cdot 25)$

P_s $4.6 \ gP/ m^3$

Less nutrients are required as the SRT increases to 25 d.

7. Effluent quality

7.1 COD_e

COD_e $= S_{U,i} + f_{cv} \cdot f_{VT} \cdot X_{TSS,e}$ (non-filtered)

             $= 75 + (1.48 \cdot 0.80 \cdot 15)$

             $96 \ gCOD/m^3$

7.2 Effluent nitrogen concentrations

Since the SRT has been extended, the system may be able to nitrify. This can be evaluated by calculating the minimum required SRT for nitrification (SRT_{MIN}) against the SRT of 25 d.

SRT_{min} $= 1 / [(\mu_{ANOmax,T} / S_f) - b_{ANO,T}]$ (5.16)

If:

$\mu_{ANOmax,20}$ $= 0.45 \ 1/d$

θ_{NIT} $= 1.123$

$\mu_{ANOmax,T}$ $= \mu_{ANOmax,20} \cdot \theta_{NIT}^{(14-20)}$

             $= 0.45 \cdot 1.123^{(14-20)}$

$\mu_{ANOmax,T}$ $0.22 \ 1/d$

$b_{ANO,20}$ $= 0.04 \ 1/d$

$\theta_{b,ANO}$ $= 1.029$

$b_{ANO,T}$ $= b_{ANO,20} \cdot \theta_{b,ANO}^{(T-20)}$

             $= 0.04 \cdot 1.029^{(14-20)}$

$b_{ANO,T}$ $= 0.034 \ 1/d$

And,

S_f $= 1.25$ (suggested)

SRT_{min} $= 1 / [(0.224 / 1.25) - 0.034)]$

             $6.6 \ d$

The applied SRT of 25 days is longer than SRT_{MIN}. Thus, nitrification will take place and the TKN_e will need to be calculated as:

$TKN_e \qquad = S_{NHx,e} + N_{ous,i}$

And, since the system can nitrify, the expected concentration of nitrate in the effluent can be calculated as:

$S_{NO3,e} \qquad = NIT_c = N_i - N_s - TKN_e$

Overall, because the system can nitrify, the influence of the nitrification process needs to be added and taken into account in the design even though the system was initially designed for only carbon removal. For that purpose, the influent nitrogen fractions also need to be known. This is covered in Chapter 5 of this book (Example 5.3.1).

In any case (assuming that there is no nitrate present in the influent), the expected effluent total nitrogen concentration (composed of TKN_e, $S_{NO3,e}$ and organic nitrogen) can be computed as:

$$N_e \qquad = TKN_e + S_{NO3,e} + f_n \cdot f_{VT} \cdot X_{TSS,e} = TKN_i - N_s + f_n \cdot f_{VT} \cdot X_{TSS,e}$$

$$= TKN_i - N_s + f_n \cdot f_{VT} \cdot X_{TSS,e}$$

$$= 44.5 - 15.3 + (0.10 \cdot 0.80 \cdot 15)$$

$N_e \qquad 30.4 \ gN/m^3$

7.3. P_e

$$P_e \qquad = P_i - P_s + f_p \cdot f_{VT} \cdot X_{TSS,e} \quad \text{(non-filtered)}$$

$$= 14.5 - 4.6 + (0.03 \cdot 0.80 \cdot 15)$$

$$10.3 \ gP/m^3$$

Overall, the system is still able to comply with the effluent total COD discharge limit of 125 mgCOD/l. It also produces a treated effluent that contains a total nitrogen and total phosphorus concentration of 30.4 mgN/l and 10.3 mgP/l, respectively.

This indicates that although the system is not designed for nutrient removal, it is able to remove approximately 34 % and 32 % of the nitrogen and phosphorus concentrations present in the influent wastewater, respectively. It is important to underline that these effluent concentrations have been estimated considering that the effluent TSS is 15 mgTSS/l. If the effluent TSS is different, then other concentrations can be expected.

8. Oxygen demand (OD)

OD for synthesis and endogenous respiration

$FO_{OHO} \qquad = FO_{OHO,s} + FO_{OHO,e} \hfill (6.30a)$

$FO_{OHO,s} \qquad = FCOD_{b,OHO} \cdot (1 - f_{cv} \cdot Y_{OHOv})$

$$= 5,460 \cdot (1 - 1.48 \cdot 0.45)$$

$FO_{OHO,s} \qquad 1,824 \ kgO_2/d$

$FO_{OHO,e} \qquad = FCOD_{b,OHO} \cdot f_{cv} \cdot (1 - f_{XE,OHO}) \cdot b_{OHO,T} \cdot Y_{OHOv,obs} \cdot SRT$

$$= 5,460 \cdot 1.48 \cdot (1 - 0.20) \cdot 0.202 \cdot 0.074 \cdot 25$$

$FO_{OHO,e} \qquad 2,429$

$FO_{OHO} \qquad 4,252 \ kgO_2/d$

Although the system was originally designed to only perform the COD removal process, the extended SRT can allow the growth of autotrophic nitrifying organisms (ANO) (see Chapter 5 in the text book). However, while their mass ($MX_{ANO,v}$) can be neglected because they have a low growth yield and the influent TKN concentration is lower than the influent COD, ANOs' oxygen consumption can be extremely high and therefore it needs to be estimated as they will increase the oxygen consumption of the plant. The oxygen consumption of ANO can be calculated as:

$FO_{NIT} \qquad = 4.57 \cdot NIT_c \cdot Q_i \hfill \text{(based on 5.35, 5.42b and 5.43b)}$

To estimate FO_{NIT}, the influent nitrogen fractions need to be known. This is covered in Chapter 5 in this book (Example 5.3.1). Yet, for the purpose of this example, assuming that all TKN_i can be nitrified (which means that $TKN_e = S_{NHx,e} + N_{ous,i} = 0$, which is not correct since $S_{NHx,e}$ cannot be zero and usually $N_{ous,i}$ is present in the influent wastewater), a rough estimation of the maximum oxygen consumed by the nitrification process would be:

NIT_c	$= TKN_i - N_s - TKN_c$
	$= 44.5 - 15.3 - 0$
NIT_c	$29.2 \ gN/m^3$
FO_{NIT}	$= 4.57 \cdot 15{,}000 \cdot 0{,}0292$
FO_{NIT}	$2{,}002 \ kgO_2/d$

The FO_{NIT} of 2,002 kgO_2/d is a rough estimation of the maximum additional oxygen consumption that the system will have when increasing the SRT to 25 d.

9. Waste of activated sludge

Q_W	$= V_R/SRT$	(4.1)
	$= 18{,}010 \ / \ 25$	
Q_W	$720 \ m^3/d$	

10. COD mass balance verification

Input

$FCOD_i$	8,775	kgCOD/d	100 % IN

Output

O_2 demand for synthesis and endogenous respiration

FOc	4,252	kgCOD/d	48.5 %

Soluble unbiodegradable organics leaving via the effluent

$FS_{U,i}$	1,125	kgCOD/d	12.8 %

Sludge	gVSS	gCOD/d	
		$(= gVSS \cdot f_{cv} / SRT = gVSS \cdot 1.48 / 25 = gVSS \cdot 0.0592)$	
$MX_{OHOv}(=MX_B)$	10,156	601	kgCOD/d	6.8 %
$MX_{E,OHOv}$	10,256	607	kgCOD/d	6.9 %
MX_{Uv}	36,993	2,190	kgCOD/d	25.0 %
$MX_{E,OHO} + MX_{Uv}$	47,249	2,797		31.9 %
MX_{VSS}	57,405	3,398	kgCOD/d	38.7 %
	Sum:	8,775	kgCOD/d	100 % OUT
	Delta (OUT-IN):	0	kgCOD/d	0 %

The 100 % mass balance for COD indicates that all the influent COD is accounted for in the calculated values of oxygen demand and sludge production. From the COD mass balance, and for the conditions of the design example, the fate of the influent COD is as follows: 48.5 % is oxidized with oxygen, 12.8 % escapes in the effluent as soluble unbiodegradable organics and 38.7 % becomes activated sludge. This activated sludge is composed of 17.7 % (10,156 / 57,405) active biomass and 82.3 % (47,249 / 57,405) inactive particulate matter. This indicates that the activated sludge waste can be considered stabilized or inactive as a consequence of having increased the SRT to 25 d. A summary of the process design results and a comparison with the design in Example 4.3.1 is presented in Table 4.7.

Table 4.7 Summary and comparison between the design carried out with SRT of 6 and 25 days.

Description	Parameter	Unit	SRT 6 d	SRT 25 d
1. Influent and bioreactor				
Type of wastewater	raw		raw	raw
Temperature	T	ºC	14	14
Influent flow rate	Q_i	MLD	15	15
Influent total COD	COD_i	gCOD/m^3	585	585
Influent biodegradable COD	$COD_{b,i}$	gCOD/m^3	364	364
Influent TKN	TKN_i	gN/m^3	44.5	44.5
Influent total P	P_i	gP/m^3	14.5	14.5
Sludge recycle ratio (s)	s	m^3.d /m^3.d	0.67	0.67
2. $COD_{b,i}$ for OHOs				
Flux of $COD_{b,i}$ for OHOs	$FCOD_{b,OHO}$	kgCOD/d	5,460	5,460
3. System biomass (VSS) equations				
Mass of OHOs	MX_{OHOv}	kgVSS	6,661	10,156
Mass of endogenous residue from OHOs	$MX_{E,OHOv}$	kgVSS	1,616	10,256
Mass of unbiodegradable organics from influent	MX_{Uv}	kgVSS	8,878	36,993
4. Volatile and total suspended solids (VSS and TSS) in system				
Mass of active biomass (=MX_{OHOv})	MX_B	kgVSS	6,661	10,156
Mass of VSS	MX_{VSS}	kgVSS	17,156	57,405
MX_B/VSS	f_{av}	gVSS/gVSS	0.39	0.18
Mass of fixed or inorganic SS	MX_{FSS}	kgFSS	4,149	14,647
Mass of TSS	MX_{TSS}	kgTSS	21,305	72,053
Ratio of VSS/TSS	f_{VT}	gVSS/gTSS	0.81	0.80
5. Reactor volume				
Bioreactor volume	V_R	m^3	5,326	18,010
6. Waste of activated sludge				
Activated sludge waste flow rate	Q_W	m^3/d	888	720
7. Nutrient requirement				
N requirement concentration for biomass synthesis	N_s	gN/m^3	19.1	15.3
P requirement concentration for biomass synthesis	P_s	gP/m^3	5.7	4.6
8. Oxygen demand				
Flux of carbonaceous O_2 demand	FO_C	kgO$_2$/d	3,418	4,252 (+2,002)[a]
9. Treated effluent quality[a]				
Effluent total COD	COD_e	gCOD/m^3	93	93

Effluent total nitrogen	N_c	gN/m^3	26.6	30.4
Effluent total Kjeldahl Nitrogen	TKN_c	gN/m^3	26.6	$\ll 26.6$[a]
				$(S_{NHx,c} + N_{ous,i})$
Effluent nitrate concentration	$S_{NO3,c}$	gN/m^3	0	
Effluent total phosphorus	P_c	gP/m^3	9.2	10.3
10. COD mass balance				
COD output/COD input	COD mass balance	gCOD/gCOD	100 %	100 %

Flow rate is in m^3/d and mass fluxes in kg/d.

For a flow rate of 1,000 or greater, mass fluxes can be read in kg/d.

[a] Since the SRT of 25 days > SRT_{min} (6.6 days), the system will nitrify and the contribution of the ANO to the oxygen consumption and effluent nitrogen concentrations need to be calculated.

When the system is increased from 6 days to 25 days SRT, the system generates approximately 19 % less sludge waste (888 versus 720 m^3/d, correspondingly) that, based on the active biomass ratio (f_{av}), is also less active (0.39 compared to 0.18). At the SRT of 25 days, the generation of less stabilized sludge waste makes it easier to handle and it could possibly be merely dehydrated prior to disposal (whereas at the SRT of 6 days the sludge needs to be digested 'to make it stable'). However, the longer SRT has implications on (*i*) the tank volume (that increased more than 3 times), (*ii*) oxygen requirements (by at least 25 % but that could double if the oxygen requirements due to nitrification are included) and (*iii*) the effluent quality (where the nutrient concentrations increased slightly at the SRT of 25 days, a high concentration of nitrate can be expected). These advantages and disadvantages need to be carefully taken into account when deciding to design an activated sludge plant with short or long SRT (e.g., 6 and 25 d, as in this example).

Example 4.3.4

The plant from Example 4.3.1 was designed with an average total solids concentration in the reactor (X_{TSS}) of 4 $kgTSS/m^3$, resulting in a reactor tank volume (V_R) of 5,326 m^3 to treat the influent wastewater flow rate of 15 MLD (15,000 m^3/d) that has a peak factor (f_q) of 3.5. However, a secondary settling tank has not been designed. Furthermore, the unit price to construct the aeration tanks are (as a function of the volume range):

Volume range (m^3)	Unit price[a] ($€/m^3$)
< 1,250	360
1,251-2,500	270
2,501-5,000	220
5,001-10,000	180
10,001-25,000	160

[a] Unit prices are estimated in The Netherlands in the year 2020 for prefabricated concrete elements.

To design the settling tanks, a circular configuration will be considered, taking into account 2 main treatment lines (N_{AS}), each one with 2 secondary settling tanks (N_{SST}). For this purpose, the following settling properties of the sludge are assumed: a maximum settling velocity (v_o) of 8.6 m/h and a hindrance coefficient (r_{hin}) of 0.486 m^3/kg. Thus, the area of the secondary settling tank can be found as follows:

$$A_{SST} = 1,000 \cdot f_q \cdot (Q_{i,\,ADWF} / 24) / [0.80 \cdot v_o \cdot \exp(-r_{hin} \cdot X_{TSS}) \cdot N_{AS} \cdot N_{SST}] \tag{4.31}$$

The estimated costs of the circular settling tanks, as a function of their diameter, are as follows:

Diameter (m)	Unit price [b] ($€/m^2$)
< 20	790
21-25	710
26-30	650
31-35	610
36-40	580
41-50	560
> 50	540

[b] Estimated unit price in The Netherlands in the year 2020 for prefabricated concrete elements.

The main question is: what would be the optimal X_{TSS} concentration that could minimize the construction costs to build these two treatment units?

Solution

Using the data provided, Table 4.8 shows the detailed calculations to determine the X_{TSS} that would lead to the lowest construction costs of the aerobic reactor and the secondary settling tank.

Table 4.8 Determination of the optimal X_{TSS} concentration to minimize the construction costs of the aerobic reactor and secondary settling tank

1. Aerobic reactor volume costs

$$V_R = MX_{TSS} / X_{TSS} \qquad (4.20)$$
$$= 21,305 / X_{TSS}$$

For different X_{TSS}:

X_{TSS} (kgTSS/m^3)	V_R (m^3)	Unit price ($€/m^3$)	Cost ($€$)
1.0	21,305	160	3,408,800
1.5	14,203	160	2,272,533
2.0	10,653	160	1,704,400
2.5	8,522	180	1,533,960
3.0	7,102	180	1,278,300
3.5	6,087	180	1,095,686
4.0	5,326	180	958,725
4.5	4,734	220	1,041,578
5.0	4,261	220	937,420

2. Secondary settling tank costs

$$A_{SST} = 1,000 \cdot f_q \cdot (Q_{i,\,ADWF} / 24) / [0.80 \cdot v_o \exp(-r_{hin} \cdot X_{TSS}) \cdot N_{AS} \cdot N_{SST}] \qquad (4.31)$$

Where:

f_q	3.5	
$Q_{i,ADWF}$	15	MLD
v_o	8.6	m/h
r_{hin}	0.486	m^3/kg
N_{AS}	2	No. main treatment lines
N_{SST}	2	No. settling tanks per treatment line

Thus:

A_{SST} = 1,000 · (3.5 ·15 / 24) / [(0.80 · 8.6 exp(−0.486 ·X_{TSS}) · 2 · 2]

As a function of X_{TSS}:

X_{TSS} (kgTSS/m³)	A_{SST} (m²) (of 1 tank)	Diameter (m)	Unit price (€/m²)	Cost (€) (for 1 tank)	Costs of all tanks (€)
1.0	129	13	790	102,093	408,370
1.5	165	14	790	130,175	520,700
2.0	210	16	790	165,982	663,928
2.5	268	18	790	211,638	846,554
3.0	342	21	710	242,527	970,106
3.5	436	24	710	309,238	1,236,952
4.0	555	27	650	360,979	1,443,914
4.5	708	30	650	460,272	1,841,090
5.0	903	34	610	550,763	2,203,054

3. Total construction costs of the aerobic reactors and secondary settling tanks

X_{TSS} (kgTSS/m³)	Aeration tank cost (€)	Settling tank costs (€)	Total costs (€)
1.0	3,408,800	408,370	3,817,170
1.5	2,272,533	520,700	2,793,233
2.0	1,704,400	663,928	2,368,328
2.5	1,533,960	846,554	2,380,514
3.0	1,278,300	970,106	2,248,406
3.5	1,095,686	1,236,952	2,332,638
4.0	958,725	1,443,914	2,402,639
4.5	1,041,578	1,841,090	2,882,668
5.0	937,420	2,203,054	3,140,474

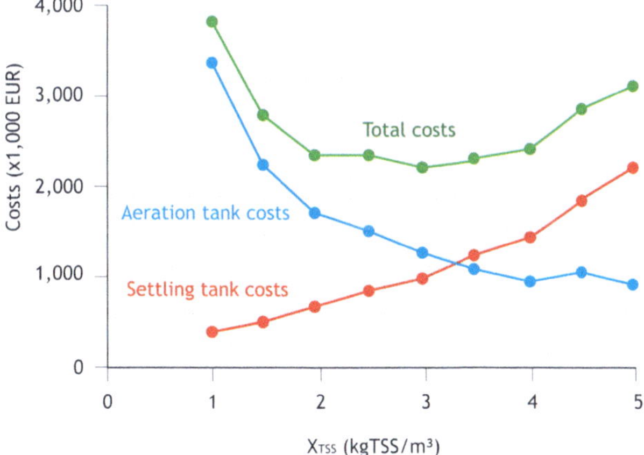

Figure 4.3 Construction costs of the activated sludge plant: aeration tank costs, settling tank costs, and total costs.

As observed in Table 4.8 and Figure 4.3 (Total costs line), the lowest cost is obtained with the X_{TSS} of 3.0 kgTSS/m^3, followed by the costs obtained at the concentrations of 3.5, 2.0, 2.5 and 4.0 kg TSS/m^3. This indicates that although the X_{TSS} of 4 kgTSS/m^3 leads to one of the most economical options, it is not the most economical. While the higher X_{TSS} concentration decreases the volume of the aerobic reactor, it tends to increase A_{SST} because the settling tanks are more loaded which increases the construction costs of the plant.

4.4 EXERCISES

Exercise 4.4.1
Explain the fate of the biodegradable soluble and particulate organics in a CAS system.

Exercise 4.4.2
Explain the fate of the unbiodegradable soluble and particulate organics in a CAS system.

Exercise 4.4.3
Explain what the sources of inorganic or fixed suspended solids and the fate of the unbiodegradable soluble and particulate organics are in a conventional activated sludge system.

Exercise 4.4.4
Explain why a certain percentage of the influent nutrient concentrations are removed in a fully aerobic activated sludge plant even if it is not designed for the removal of nutrients.

Exercise 4.4.5
Given the following wastewater characteristics and COD fractionation, determine the minimum total COD and soluble COD that would be present in the effluent, assuming that the CAS and its secondary settling tank operate under ideal conditions.

COD fraction	COD (mg/l)	COD fraction	COD (mg/l)
$S_{S,i}$	153	$S_{U,i}$	32
$X_{S,i}$	67	$X_{U,i}$	74

Exercise 4.4.6
For the previous COD fractionation, what would be the COD fluxes flowing into the aerobic reactor if the influent flow rate Q_i is 12.5 MLD?

Exercise 4.4.7
Given the following wastewater characteristics and COD fractionation, and considering an effluent solids concentration of 32 mg TSS/l with a f_{VT} ratio of 0.85, will the COD_e be able to meet a standard COD_e of 125 mgCOD/l?

COD fraction	COD (mg/l)	COD fraction	COD (mg/l)
$S_{S,i}$	174	$S_{U,i}$	123
$X_{S,i}$	215	$X_{U,i}$	156

Exercise 4.4.8

For the COD fractionation from Exercise 4.4.7, an influent flow rate (Q_i) of 25 MLD and assuming that a primary settling tank (PST) is installed before the aerobic reactor and operates under ideal conditions, what would be the COD fluxes flowing into the aerobic reactor?

Exercise 4.4.9

For the following COD fractionation (COD_i = 624 gCOD/m^3), a CAS has been designed to treat an influent flow rate (Q_i) of 25 MLD with an SRT of 10 d to operate at a temperature of 18 °C. The proposed s-recycle ratio is 0.75 and an average X_{TSS} in the aerobic reactor of 4 kgTSS/m^3. The influent contains 21 mgFSS/l and the expected $X_{TSS,e}$ is 20 mg/l. If the plant operators decide to minimize costs, what SRT can you suggest and what would be the potential savings?

COD fraction	COD (mg/l)	COD fraction	COD (mg/l)
$S_{S,i}$	187	$S_{U,i}$	96
$X_{S,i}$	205	$X_{U,i}$	136

Exercise 4.4.10

If the plant from Exercise 4.4.9 was designed with SRT of 4 days and with a primary settling tank (PST) that removes 0.5% of Q_i as settled sludge, what will be the implications for the design?

Exercise 4.4.11

For the plant in Exercise 4.4.9, what design considerations need to be made to achieve a stable WAS that does not require a sophisticated sludge handling line?

Exercise 4.4.12

If the plant from Exercise 4.4.11 was designed with SRT of 22 days and an average total solids concentration in the reactor (X_{TSS}) of 4 kgTSS/m^3, a reactor tank volume (V_R) of 24,231 m^3 will be needed to treat the influent wastewater of 25 MLD (25,000 m^3/d) that has a peak factor (f_q) of 4.0. However, a secondary settling tank (SST) has not been designed. Furthermore, the unit price to construct the aeration tanks are (as a function of the volume range):

Volume range (m^3)	Unit price [a] (€/m^3)
< 1,250	360
1,251-2,500	270
2,501-5,000	220
5,001-10,000	180
10,001-25,000	160

[a] Estimated unit price in The Netherlands in the year 2020 for prefabricated concrete elements.

To design the aeration tanks, assume that 3 mainstream treatment lines (N_{AS}) and 2 secondary settling tanks per treatment line (N_{SST}) are constructed. For the secondary settling tanks, consider a circular configuration and the following design parameters: a maximum settling velocity (v_o) of 8.1 m/h and a hindrance coefficient (r_{hin}) of 0.395 m^3/kg.

The estimated costs of the circular settling tanks, as a function of their diameter, are the following:

Diameter (m)	Unit price[b] ($€/m^2$)
< 20	790
21-25	710
26-30	650
31-35	610
36-40	580
41-50	560
> 50	540

[b] Estimated unit price in The Netherlands in the year 2020 for prefabricated concrete elements.

What would be the optimal X_{TSS} concentration that could minimize the construction costs of the activated sludge system?

Exercise 4.4.13

Let us assume that the plant designed in Example 4.3.1 with the SRT of 6 days was finally built with 2 mainstream treatment lines. Each treatment line has 1 aerobic tank reactor of 2,663 m^3 at an average X_{TSS} of 4 kgTSS/m^3 and 2 secondary settling tanks with a diameter of 27 m. In total, the total volume of the aeration tanks is 5,326 m^3 (of the two aerobic tanks) and a total settling area of 2,290 m^2 (of the four final settlers, Example 4.3.4, Table 4.8). However, in the coming year, the plant will receive a discharge of 1.5 MLD of an additional highly biodegradable industrial effluent from a food and processing industry with a COD_i of 1,378 mgCOD/l. The industrial effluent does not contain any toxic compounds and its COD fractionation is:

COD fraction	COD (mg/l)	COD fraction	COD (mg/l)
$S_{S,i}$	1,253	$S_{U,i}$	60
$X_{S,i}$	50	$X_{U,i}$	305

The designers of the plant would like to know if the activated sludge plant is robust enough to treat the additional COD load.

ANNEX 1: SOLUTIONS TO EXERCISES

Solution 4.4.1
Both the biodegradable soluble and the biodegradable particulate organics will be mostly converted into new biomass cells. The difference is that the biodegradable soluble organics are rapidly converted whereas the biodegradable particulate organics get enmeshed in the sludge flocs, being slowly hydrolysed prior to their conversion into new biomass.

Solution 4.4.2
Soluble unbiodegradable organics cannot be removed either biologically (e.g., for the synthesis of new cell biomass) or physically (e.g., because they are soluble and do not get enmeshed in the sludge flocs). Thus, soluble unbiodegradable organics leave through the effluent of the activated sludge plant. Unbiodegradable particulate organics cannot be removed biologically but are enmeshed in the sludge flocs, accumulating in the system and ultimately being removed via the waste of activated sludge.

Solution 4.4.3
Inorganic suspended solids can be present in the influent wastewater depending on the origin and quality of the wastewater (e.g., if hard water is supplied to the water network or if saline water intrusion or groundwater infiltration occurs). In addition, active biomass cells need inorganic compounds (e.g., micronutrients) to carry out their microbial metabolism and therefore they accumulate in the activated sludge plant as a function of the active biomass.

Solution 4.4.4
Ordinary heterotrophic organisms, the dominant organisms in activated sludge systems, require (macro-) nutrients, mostly nitrogen and phosphorus, to synthesize new biomass cells. They cover these biomass nutrient requirements by consuming the nutrients present in the influent wastewater.

Solution 4.4.5
If the CAS and its secondary settling tank operate under ideal conditions: total COD = COD_e = $S_{U,i}$ = 32 mg/l. Otherwise, the minimum soluble COD = $S_{U,i}$ = 32 mg/l and total COD = COD_e = $S_{U,i}$ + $f_{cv} \cdot X_{VSS,e}$ where $X_{VSS,e}$ will depend on the efficiency of the secondary settling tank.

Solution 4.4.6
$FCOD_i$ = 4,075 kgCOD/d
$FCOD_{b,i}$ = 2,750 kgCOD/d
$FX_{U,i}$ = 925 kgCOD/d
$FX_{Uv,i}$ = 625 kgVSS/d

Solution 4.4.7
Total COD = COD_e = $S_{U,i}$ + $f_{cv} \cdot f_{VT} \cdot X_{TSS,e}$ = 123 + (1.48 · 0.85 · 32) = 163.3 mg COD/l. Therefore, the discharge standard of 125 mg COD/l cannot be met.

Solution 4.4.8
Ideally, both particulate organics (biodegradable and unbiodegradable) will be settled out in the PST. Therefore:
$FCOD_i$ = 7,425 kgCOD/d
$FCOD_{b,i}$ = 4,350 kgCOD/d

Solution 4.4.9

Due to the temperature of 18 $^{\circ}$C, the system could be designed with SRT of 4 days (instead of the SRT of 10 days to minimize the construction and operational costs of the plant):

Table 4.9 Summary and comparison between the design carried out with an initial SRT of 10 days and SRT of 4 days.

Description	Parameter	Unit	SRT 10 d	SRT 4 d
1. Influent and bioreactor				
Type of wastewater	raw		raw	raw
Temperature	T	°C	18	18
Influent flow rate	Q_i	MLD	25	25
Influent total COD	COD_i	gCOD/m³	624	624
Influent biodegradable COD	$COD_{b,i}$	gCOD/m³	392	392
Sludge recycle ratio (s)	s	m³.d /m³.d	0.75	0.75
2. $COD_{b,i}$ for OHOs				
Flux of $COD_{b,i}$ for OHOs	$FCOD_{b,OHO}$	kgCOD/d	9,800	9,800
3. System biomass (VSS) equations				
Mass of OHOs	MX_{OHOv}	kgVSS	13,514	9,261
Mass of endogenous residue from OHOs	$MX_{E,OHOv}$	kgVSS	6,126	1,679
Mass of unbiodegradable organics from influent	MX_{Uv}	kgVSS	22,973	9,189
4. Volatile and total suspended solids (VSS and TSS) in system				
Mass of active biomass (=MX_{OHOv})	MX_B	kgVSS	13,514	9,261
Mass of VSS	MX_{VSS}	kgVSS	42,613	20,130
MX_B/VSS	f_{av}	gVSS/gVSS	0.32	0.46
Mass of fixed or inorganic SS	MX_{FSS}	kgFSS	7,277	3,489
Mass of TSS	MX_{TSS}	kgTSS	49,890	23,619
Ratio of VSS/TSS	f_{VT}	gVSS/gTSS	0.85	0.85
5. Reactor volume				
Bioreactor volume	V_R	m³	12,472	5,905
6. Waste of activated sludge				
Activated sludge waste flow rate	Q_w	m³/d	1,247	1,476
7. Nutrient requirement				
N concentration needed for biomass synthesis	N_s	gN/m³	17.0	20.1
P concentration needed for biomass synthesis	P_s	gP/m³	5.1	6.0
8. Oxygen demand [a)]				
Flux of carbonaceous O_2 demand	FO_C	kgO₂/d	6,893	5,752
9. Treated effluent quality				
Effluent total COD	COD_e	gCOD/m³	121	121
10. COD mass balance				
COD output/COD input	COD mass balance	gCOD/gCOD	100 %	100 %

[a)] Due to the temperature (18°C), the system will probably be able to nitrify at SRT of 10 days and maybe even at SRT of 4 days. This needs to be verified since it could affect the oxygen requirements of the plant.

By shortening the SRT from 10 to 4 days, considerable savings can be expected in reactor tank volumes (to less than half the original required volume) and oxygen requirements (17%). However, the activated sludge waste will increase (by 18%) and this implies that the sludge handling line may need to be expanded.

Solution 4.4.10

If a PST is installed, ideally both particulate organics will be settled out. Thus, the design will be as follows:

Table 4.10 Summary of the design carried out with the SRT of 4 days and after installing a PST.

Description	Parameter	Unit	SRT 4 d	SRT 4 d with PST
1. Influent and bioreactor				
Type of wastewater	raw		raw	settled
Temperature	T	°C	18	18
Influent flow rate	Q_i	MLD	25	24.875
Influent total COD	COD_i	$gCOD/m^3$	624	283
Influent biodegradable COD	$COD_{b,i}$	$gCOD/m^3$	392	187
Sludge recycle ratio (s)	s	$m^3.d\,/m^3.d$	0.75	0.75
2. $COD_{b,i}$ for OHOs				
Flux of $COD_{b,i}$ for OHOs	$FCOD_{b,OHO}$	kgCOD/d	9,800	4,652
3. System biomass (VSS) equations				
Mass of OHOs	MX_{OHOv}	kgVSS	9,261	4,396
Mass of endogenous residue from OHOs	$MX_{E,OHOv}$	kgVSS	1,679	797
Mass of unbiodegradable organics from influent	MX_{Uv}	kgVSS	9,189	0
4. Volatile and total suspended solids (VSS and TSS) in system				
Mass of active biomass (=MX_{OHOv})	MX_B	kgVSS	9,261	4,396
Mass of VSS	MX_{VSS}	kgVSS	20,130	5,193
MX_B/VSS	f_{av}	gVSS/gVSS	0.46	0.85
Mass of fixed or inorganic SS	MX_{FSS}	kgFSS	3,489	659 [b]
Mass of TSS	MX_{TSS}	kgTSS	23,619	5,852
Ratio of VSS/TSS	f_{VT}	gVSS/gTSS	0.85	0.89
5. Reactor volume				
Bioreactor volume	V_R	m^3	5,905	1,463
6. Waste of activated sludge				
Activated sludge waste flow rate	Q_w	m^3/d	1,476	366
7. Nutrient requirement				
N requirement concentration for biomass synthesis	N_s	gN/m^3	20.1	5.2
P requirement concentration for biomass synthesis	P_s	gP/m^3	6.0	1.6
8. Oxygen demand [a]				
Flux of carbonaceous O_2 demand	FO_C	kgO_2/d	5,752	2,730

9. Treated effluent quality				
Effluent total COD	COD_e	$gCOD/m^3$	121	122
10. COD mass balance				
COD output/COD input	COD mass balance	$gCOD/gCOD$	100 %	100 %

[a] Due to the temperature (18°C), the system will probably be able to nitrify at SRT of 10 days and maybe even at SRT of 4 days. This needs to be verified since it could affect the oxygen requirements of the plant.
[b] Also, due to the PST the concentration of influent ISS may be removed decreasing the MX_{FSS}.

By introducing the PST, considerable savings can be expected in reactor tank volumes (to approximately 25 % of the volume required without PST) and oxygen requirements (to almost 47 %). However, the main implication in the design will be the need to treat the sludge settled out in the PST. Assuming that the underflow rate is of approximately 0.5 % Q_i, the COD flux to be treated in the sludge handling line will be around 8,560 kgCOD/d. This flux and its volume (of 125 m³/d) needs to be considered to design the sludge treatment line in addition to the Q_W.

Solution 4.4.11
To produce a stabilized digested WAS, the minimum applied SRT should be 22 days.

Table 4.11 Summary of the design carried out with the SRT of 22 days and comparison against the designs performed at SRT of 4 and 10 days.

Description	Parameter	Unit	SRT 4 d	SRT 10 d	SRT 22 d
1. Influent and bioreactor					
Type of wastewater	raw		raw	raw	raw
Temperature	T	°C	18	18	18
Influent flow rate	Q_i	MLD	25	25	25
Influent total COD	COD_i	$gCOD/m^3$	624	624	624
Influent biodegradable COD	$COD_{b,i}$	$gCOD/m^3$	392	392	392
Sludge recycle ratio (s)	s	$m^3.d\,/m^3.d$	0.75	0.75	0.75
2. $COD_{b,i}$ for OHOs					
Flux of $COD_{b,i}$ for OHOs	$FCOD_{b,OHO}$	kgCOD/d	9,800	9,800	9,800
3. System biomass (VSS) equations					
Mass of OHOs	MX_{OHOv}	kgVSS	9,261	13,514	16,222
Mass of endogenous residue from OHOs	$MX_{E,OHOv}$	kgVSS	1,679	6,126	16,179
Mass of unbiodegradable organics from influent	MX_{Uv}	kgVSS	9,189	22,973	50,541
4. Volatile and total suspended solids (VSS and TSS) in system					
Mass of active biomass (=MX_{OHOv})	MX_B	kgVSS	9,261	13,514	16,222
Mass of VSS	MX_{VSS}	kgVSS	20,130	42,613	82,942
MX_B/VSS	f_{av}	gVSS/gVSS	0.46	0.32	0.20
Mass of fixed or inorganic SS	MX_{FSS}	kgFSS	3,489	7,277	13,983
Mass of TSS	MX_{TSS}	kgTSS	23,619	49,890	96,925
Ratio of VSS/TSS	f_{VT}	gVSS/gTSS	0.85	0.85	0.86

5. Reactor volume

Bioreactor volume	V_R	m^3	5,905	12,472	24,231
6. Waste of activated sludge					
Activated sludge waste flow rate	Q_W	m^3/d	1,476	1,247	1,101
7. Nutrient requirement					
N requirement concentration for biomass synthesis	N_s	gN/m^3	20.1	17.0	15.1
P requirement concentration for biomass synthesis	P_s	gP/m^3	6.0	5.1	4.5
8. Oxygen demand [a]					
Flux of carbonaceous O_2 demand	FO_C	kgO_2/d	5,752	6,893	7,620
9. Treated effluent quality					
Effluent total COD	COD_e	$gCOD/m^3$	121	121	121
10. COD mass balance					
COD output/COD input	COD mass balance	$gCOD/gCOD$	100 %	100 %	100 %

[a] Due to the temperature (18°C), the system will be able to nitrify at SRT of 10 days and SRT of 22 days. This needs to be verified since it could affect the oxygen requirements of the plant.

By increasing the SRT to 22d, a stable Q_W of 1,101 m^3/d (up to 25 % lower with regard to the SRT of 4 days) with an active biomass fraction (MX_B / M_{VSS}) of 0.20 can be generated. This daily Q_W could be treated in a relatively easier manner (e.g., using sludge drying beds) compared to the Q_W produced at shorter SRT. However, it should be noticed that the both the tank volumes and oxygen requirements will increase, impacting the capital investment and running costs of the system.

Solution 4.4.12

The lowest cost of approximately €6,315,971 could be obtained with a X_{TSS} of 5.0 $kgTSS/m^3$, followed rather closely by the costs obtained at the concentrations of 4.5 and 4.0 kg TSS/m^3 (of €6,349,211 and €6,390,767, respectively). However, these costs refer to capital investment and the X_{TSS} of 5.0 $kgTSS/m^3$ may decrease the oxygen transfer efficiency in the aerobic reactor which could increase the running costs of the plant. Thus, it can be more convenient to design and operate the plant at the X_{TSS} of 4.0 or 4.5 $kgTSS/m^3$, concentrations at which the oxygen transfer will be higher.

Solution 4.4.13

First, the effect of the discharge of the industrial effluent on the existing plant is evaluated. However, since, as explained below, certain parameters are exceeded, the calculations are also done shortening the SRT from 6 to 4.5 days as an alternative operational strategy. A summary of the calculations is displayed in Table 4.12.

Table 4.12 Summary of the design carried out with the SRT of 22 days and comparison against the designs performed at SRT of 4 and 10 days.

Description	Parameter	Unit	SRT 6 d	SRT 6 d	SRT 4.5 d
1. Influent and bioreactor					
Type of wastewater	raw		raw	raw plus industrial	raw plus industrial
Temperature	T	°C	14	14	14
Influent flow rate	Q_i	MLD	15	16.5	16.5
Influent total COD	COD_i	$gCOD/m^3$	585	684	684
Influent biodegradable COD	$COD_{b,i}$	$gCOD/m^3$	364	449	449
Sludge recycle ratio (s)	s	$m^3.d\,/m^3.d$	0.67	0.67	0.67
2. $COD_{b,i}$ for OHOs					
Flux of $COD_{b,i}$ for OHOs	$FCOD_{b,OHO}$	kgCOD/d	5,460	7,415	7,415
3. System biomass (VSS) equations					
Mass of OHOs	MX_{OHOv}	kgVSS	6,661	9,046	7,862
Mass of endogenous residue from OHOs	$MX_{E,OHOv}$	kgVSS	1,616	2,195	1,430
Mass of unbiodegradable organics from influent	MX_{Uv}	kgVSS	8,878	10,733	8,050
4. Volatile and total suspended solids (VSS and TSS) in system					
Mass of active biomass (=MX_{OHOv})	MX_B	kgVSS	6,661	9,046	7,862
Mass of VSS	MX_{VSS}	kgVSS	17,156	21,974	17,342
MX_B/VSS	f_{av}	gVSS/gVSS	0.39	0.41	0.45
Mass of fixed or inorganic SS	MX_{FSS}	kgFSS	4,149	4,507	3,542
Mass of TSS	MX_{TSS}	kgTSS	21,305	26,481	20,884
Ratio of VSS/TSS	f_{VT}	gVSS/gTSS	0.81	0.83	0.83
5. Reactor volume and X_{TSS}					
Bioreactor volume	V_R	m^3	5,326	5,326	5,326
Average total solids concentration	X_{TSS}	g/m^3	4	4.97	3.92
6. Waste of activated sludge					
Activated sludge waste flow rate	Q_W	m^3/d	888	888	1,160
7. Nutrient requirement					
N requirement concentration for biomass synthesis	N_s	gN/m^3	19.1	22.2	23.4
P requirement concentration for biomass synthesis	P_s	gP/m^3	5.7	6.7	7.0
8. Oxygen demand					
Flux of carbonaceous O_2 demand	FO_C	kgO_2/d	3,418	4,642	4,358
9. Treated effluent quality					
Effluent total COD	COD_e	$gCOD/m^3$	93	92	92
10. COD mass balance					
COD output/COD input	COD mass balance	gCOD/gCOD	100 %	100 %	100 %

If the plant receives the additional industrial effluent COD load operating at SRT of 6 days, the X_{TSS} would increase to 4.97 kgTSS/m^3 and the FO_c to 4,642 kgO$_2$/d. These can be seen as minor increases and the aerobic tank would probably be able to cope with the increased COD load. However, the increased combined Q_i of 16.5 MLD combined with the X_{TSS} of 4.97 kgTSS/m^3 have a major impact on the required area of the secondary settling tanks. The total required settling area would increase from 555 m^2 to 979 m^2. This is an increase of about 75 % which will take the 4 secondary settling tanks above their capacity. Alternatively, the plant could decrease their SRT to 4.5 d and, since the changes are minimal compared to the operation before the industrial effluent addition (last column from Table 4.12), both the aerobic and the final setting stages could work well. However, the Q_W would increase by 20 % from 888 to 1,160 m^3/d. This alternative could be feasible if the sludge handling facilities could cope with the increased Q_W. Still, the aeration capacity needs to be expanded to match the increased requirements for oxygen (of 4,358 kgO$_2$/d).

REFERENCE

Chen GH., van Loosdrecht M.C.M., Ekama G.A. and Brdjanovic D. Ed. (2020) Biological Wastewater Treatment: Principles, Design and Modelling. 2nd edition. IWA Publishing, pg. 850. 9781789060355.

NOMENCLATURE

Symbol	Description	Units
$(F/M)_{COD,TSS}$	Food-to-microorganisms ratio calculated based on COD fractionation	kgCOD.d/kgTSS
A_{ND}	Total mass to COD ratio for activated sludge systems performing only nitrification and denitrification processes	kgTSS/kgCOD.d
A_{SST}	Area of secondary settling tank	m^2
$b_{ANO,20}$	ANO-specific endogenous mass loss rate constant at 20 °C	gVSS/gVSS.d
$b_{ANO,T}$	ANO-specific endogenous mass loss rate constant at temperature T	gVSS/gVSS.d
b_{OHO}	Specific endogenous mass loss rate of the OHOs	gVSS/gVSS.d
$b_{OHO,20}$	Specific endogenous mass loss rate of the OHOs at 20 °C	gVSS/gVSS.d
$b_{OHO,T}$	OHO specific endogenous mass loss rate at temperature T	gVSS/gVSS.d
$COD_{b,i}$	influent concentration of biodegradable COD ($S_{S,i} + X_{S,i}$)	gCOD/m^3
COD_e	Efluent concentration of total COD	gCOD/m^3
COD_i	Concentration of COD in the influent	gCOD/m^3
F/M	Food-to-microorganisms ratio	kgCOD.d/kgTSS
f_{av}	Fraction of active biomass with the regard to the mass of volatile suspended solids	kgVSS/kgVSS
$FCOD_{b,i}$	Daily mass of influent biodegradable organics ($S_{S,i} + X_{S,i}$)	kgCOD/d
$FCOD_{b,OHO}$	Daily mass of biodegradable substrate available to OHOs	kgCOD/d
$FCOD_i$	Daily mass of influent COD	kgCOD/d
f_{cv}	COD/VSS ratio of the sludge	gCOD/gVSS
$f_{FSS,OHO}$	Fraction of fixed (inorganic) suspended solids of OHOs	gFSS/gVSS
f_n	Nitrogen content of the sludge	gN/gVSS
FO_C	Daily mass of carbonaceous oxygen demand	kgO$_2$/d
FO_{NIT}	Daily mass of oxygen consumed by ANO	kgO$_2$/d
FO_{OHO}	Daily mass of oxygen consumed by OHOs	kgO$_2$/d

$FO_{OHO,e}$	Daily mass of oxygen consumed by OHOs for endogenous respiration	kgO_2/d
$FO_{OHO,s}$	Daily mass of oxygen consumed by OHOs for synthesis	kgO_2/d
FO_t	Daily mass of total oxygen demand	kgO_2/d
f_P	Fraction of P in the active OHO mass, endogenous mass and unbiodegradable mass	$gP/gVSS$
fq	Factor for peak wet weather flow	-
$FS_{S,i}$	Influent daily flux of rbCOD	$kgCOD/d$
$f_{SU,CODi}$	Influent unbiodegradable soluble COD fraction	$gCOD/gCOD$
$FS_{U,i}$	Influent daily flux of soluble unbiodegradable COD	$kgCOD/d$
f_{VT}	VSS to TSS ratio of the sludge	$kgVSS/kgTSS$
$FX_{,FSS,i}$	Flux of inorganic or fixed suspended solids in the influent	$kgFSS/d$
$f_{XE,OHO}$	Fraction of endogenous residue of the OHOs	$gEVSS/gAVSS$
$FX_{FSS,i}$	Daily mass of influent inorganics	$gFSS/d$
$f_{XU,CODi}$	Influent unbiodegradable particulate COD fraction	$gCOD/gCOD$
$FX_{U,i}$	Influent daily flux of particulate unbiodegradable COD	$kgCOD/d$
$FX_{Uv,i}$	Influent daily flux of particulate unbiodegradable organics	$kgVSS/d$
MX_{ANOv}	Mass of ANOs in the system	$gAVSS$
MX_{bio}	Sum of all active biomasses in the system	$gVSS$
$MX_{E,OHOv}$	Mass of OHO endogenous residue in the system	$gEVSS$
MX_{FSS}	Mass of fixed (inorganic) suspended solids in the system	$gFSS$
MX_{OHOv}	Mass of OHOs in the system	$gAVSS$
MX_{TSS}	TSS mass in the system	$gTSS$
MX_{Uv}	Mass of unbiodegradable organic matter in the system, coming from the influent	$gVSS$
MX_{VSS}	Mass of volatile suspended solids in the system	$gTSS$
N_{AS}	Number of activated sludge treatment modules in a system	-
N_e	Effluent total nitrogen concentration	gN/m^3
NIT_c	Nitrification capacity of the bioreactor	$gNO_3\text{-}N/m^3$
$N_{ous,i}$	Influent unbiodegradable soluble organic nitrogen	gN/m^3
N_s	Nitrogen required for biomass growth	gN/m^3
N_{SST}	Number of secondary settling tanks per activated sludge treatment modules	-
P_e	Effluent total phosphorus concentration	gP/m^3
P_i	Influent total phosphorus concentration	gP/m^3
P_s	Phosphorus required for biomass growth	gP/m^3
Q_i	Daily average influent flow rate	m^3/d
$Q_{i,ADWF}$	Average dry weather flow	m^3/d
Q_s	Sludge recycle flow rate	m^3/d
r_{hin}	Settling hindrance factor	$m^3/kgTSS$
s	Return activated sludge recycle ratio based on influent flow	$m^3.d/m^3.d$
S_f	Safety factor for the nitrification process	-
$S_{NHx,e}$	Effluent free and saline ammonia	gN/m^3
$S_{NO3,e}$	Effluent nitrate concentration	$gNO_3\text{-}N/m^3$
SRT	Sludge age	d
SRT_{min}	Minimum SRT required for nitrification	d
$S_{S,i}$	Influent readily biodegradable COD concentration	$gCOD/m^3$
$S_{U,i}$	Influent concentration of soluble unbiodegradable COD	$gCOD/m^3$

T	Temperature	$^{\circ}$C
TKN	Total Kjeldahl nitrogen concentration	gN/m^3
TKN_e	Effluent total Kjeldahl nitrogen	gN/m^3
TKN_i	Influent total Kjeldahl nitrogen concentration	gN/m^3
v_o	Maximum initial settling velocity	m/h
V_R	Volume of biological process (bioreactor)	m^3
$X_{FSS,i}$	Influent fixed suspended solids (FSS) concentration	$gFSS/m^3$
$X_{TSS,s}$	Total suspended-solids concentration in the sludge recycle flowrate	$gTSS/m^3$
X_S	Slowly biodegradable organics concentration	$gCOD/m^3$
$X_{S,i}$	Influent concentration of particulate biodegradable COD	$gCOD/m^3$
X_{TSS}	Reactor total suspended-solids concentration	$gTSS/m^3$
$X_{TSS,e}$	Total suspended-solids concentration in the effluent	$gTSS/m^3$
X_U	Unbiodegradable particulate COD	$gCOD/m^3$
$X_{U,i}$	Influent concentration of unbiodegradable particulate COD	$gCOD/m^3$
X_{VSS}	Reactor volatile suspended solids concentration	$gVSS/m^3$
Y_{OHOv}	Biomass yield of OHOs	gVSS/gCOD
$Y_{OHOv,obs}$	Observed biomass yield of OHOs	gVSS/gCOD
$\theta_{b,ANO}$	Temperature coefficient for $b_{ANO,T}$	-
$\theta_{b,OHO}$	Temperature coefficient for $b_{OHO,T}$	-
θ_{NIT}	Temperature coefficient for $\mu_{ANO,T}$ and $K_{ANO,T}$	-
$\mu_{ANOmax,20}$	ANO maximum specific growth rate constant at 20 °C	gVSS/gVSS.d
$\mu_{ANOmax,T}$	Maximum specific biomass growth rate of nitrifiers at temperature T	gVSS/gVSS.d

Abbreviation	Description
ANO	Ammonia nitrifying organisms
COD	Chemical oxygen demand
E	Effluent
FSS	Fixed (inorganic) suspended solids
i	Influent
MLSS	Mixed-liquor suspended solids
MLVSS	Mixed-liquor volatile suspended solids
NIT	Nitrifying organisms
OHO	Ordinary heterotrophic organism
RBCOD	Readily biodegradable COD
SRT	Sludge retention time
SST	Secondary settling tank
TKN	Total Kjeldahl nitrogen
TSS	Total suspended solids
VSS	Volatile suspended solids

Aerobic lagoons and ponds are in many low- and medium-income countries prevailing technologies for the removal of organic matter from sewage, and faecal and septic sludge, often in combination with anaerobic and facultative ponds (photo: M. von Sperling).

doi: 10.2166/9781789062304_0141

5

Nitrogen removal

Carlos M. Lopez-Vazquez, Eveline I.P. Volcke and Mark C.M. van Loosdrecht

5.1 INTRODUCTION

Chapter 5 on biological nitrogen removal in the textbook *Biological Wastewater Treatment: Principles, Modelling and Design* (Chen *et al.*, 2020) introduces the principles, fundamentals and key factors influencing conventional BNR systems as well as innovative nitrogen removal processes. This provides the basis for understanding the design and operation of full-scale wastewater treatment plants that perform the biological removal of nitrogen, through nitrification and denitrification processes. In addition, a steady-state model is introduced and used for the design and evaluation of biological nitrogen removal wastewater treatment plants. Furthermore, the Chapter 5 in the textbook introduces and discusses the main characteristics, features and benefits of innovative nitrogen removal processes, as well as aspects that are taken into account to select the most appropriate innovative nitrogen removal processes such as SHARON, ANAMMOX, and the BABE systems. All the equations are in Chapter 5 on nitrogen removal in the textbook.

5.2 LEARNING OBJECTIVES

After the successful completion of this chapter, the reader will be able to:

- Describe the basic transformations of organic and inorganic nitrogen compounds that take place in biological nitrogen removal systems.
- Describe the principles and microbial mechanisms involved in the conventional nitrification and denitrification processes.
- Discuss and assess the effects of key environmental and operational factors and design parameters affecting the conventional nitrification and denitrification processes in biological nutrient removal wastewater treatment systems.

- Discuss and apply the fundamentals for the selection, design and control of the sludge age in activated sludge wastewater treatment systems performing the conventional nitrification and denitrification processes.
- Identify and describe different process configurations and discuss their advantages and disadvantages with regard to the process performance and efficiency of the biological nitrogen removal process.
- Apply a steady-state model for the design and evaluation of activated sludge wastewater treatment plants performing the biological nitrogen removal process via conventional nitrification and denitrification.
- Describe the principles and microbial mechanisms involved in the application of innovative nitrogen removal processes in wastewater treatment plants.
- Discuss and apply the main key environmental and operational factors and design parameters for the selection of innovative nitrogen removal processes in wastewater treatment plants.
- Describe the main characteristics, effects, advantages and disadvantages of innovative nitrogen removal processes applied in wastewater treatment plants.

5.3 EXAMPLES
Example 5.3.1
N fractionation

The COD fractionation of an influent flow is:

COD fraction	COD (mg/l)	COD fraction	COD (mg/l)
$S_{S,i}$	165	$S_{U,i}$	75
$X_{S,i}$	199	$X_{U,i}$	146

Given the COD fractionation displayed above and considering a TKN_i of 44.5 mgN/l composed of 29.1 mgNH$_4^+$-N/l ($S_{NHx,i}$), 33.1 mg/l of soluble TKN ($TKN_{s,i}$) and 0.49 mg/l of $N_{ous,i}$, answer the following:

a) What N fractions will be present in the influent flow rate?
b) What will be the N concentration that could be available to cover the nitrogen requirements for biomass growth?
c) If the influent were treated under ideal conditions, what would be the minimum TKN_e that could be achieved?

Solution[1]
a) What N fractions will be present in the influent flow rate?

Considering that:

$$TKN_i = N_{o,i} + S_{NHx,i}$$

And, since:

$$TKN_i = 44.5 \text{ mg N/l and } S_{NHx,i} = 29.1 \text{ mg/l, then:}$$

---

[1] Refer to Chapter 3 on wastewater characteristics in Chen *et al.* (2020) and in this book for further details.

$N_{o,i} = 44.5 - 29.1 = 15.4$ mg N/l.

Also:

$TKN_{s,i} = S_{NHx,i} + N_{obs,i} + N_{ous,i}$

Solving for $N_{obs,i}$:

$N_{obs,i} = TKN_{s,i} - S_{NHx,i} - N_{ous,i}$
$N_{obs,i} = 33.1 - 29.1 - 0.49$

And,

$N_{obs,i} = 3.51$ mg N/l

Also:

$TKN_i = TKN_{s,i} + TKN_{p,i}$;

As such:

$TKN_{p,i} = TKN_i - TKN_{s,i}$; thus: $TKN_{p,i} = 44.5 - 33.1 = 11.4$ mg N/l.

In addition:

$TKN_{p,i} = N_{obp,i} + N_{oup,i}$;

where $N_{oup,i}$ is the organic nitrogen associated with the unbiodegradable particulate organics and, therefore, proportional to $X_{U,i}$ (being $N_{oup,i} = f_n \cdot X_{U,i} / f_{cv}$), where: $f_n = 0.10$ mgN/mgVSS and $f_{cv} = 1.48$ mgCOD/mgVSS.

Consequently, for this example:

$N_{oup,i} = (0.10$ mgN/mgVSS$) \cdot (146$ mgCOD/l$) / (1.48$ mgCOD/mgVSS$)$
$N_{oup,i} = 9.86$ mg N/l

Therefore and since:

$TKN_{p,i} = N_{obp,i} + N_{oup,i}$
$N_{obp,i} = TKN_{p,i} - N_{oup,i}$
$N_{obp,i} = 11.4 - 9.86$
$N_{obp,i} = 1.54$ mgN/l

Thereby, the nitrogen fractions are: $S_{NHx,i}$ is 29.1 mgN/l while the rest of the N fractions will be:

N fraction	N (mg/l)	N fraction	N (mg/l)
$N_{obs,i}$	3.51	$N_{ous,i}$	0.49
$N_{obp,i}$	1.54	$N_{oup,i}$	9.86

The total sum of the fractions ($S_{NHx,i} + N_{obs,i} + N_{ous,i} + N_{obp,i} + N_{oup,i} = 44.5$ mgN/l) is equal to $TKN_i = 44.5$ mgN/l, confirming that the N fractions have been correctly determined.

b) What will be the N concentration that could be available to cover the nitrogen requirements for biomass growth?

The nitrogen available will be the sum of the concentrations of the organic biodegradable nitrogen species ($N_{obs,i} + N_{obp,i}$) plus that of ammonia ($S_{NHX,i}$): $29.1 + 3.51 + 1.54 = 34.15$ mgN/l.

c) If the influent were treated under ideal conditions, what would be the minimum TKN_e that could be achieved?

Under ideal conditions, all the nitrogen that can be used for biomass growth will be consumed for biomass synthesis or nitrified while the unbiodegradable particulate nitrogen ($N_{oup,i}$) will be enmeshed in the activated sludge flocs and leave through the activated sludge waste. Therefore, only $N_{ous,i}$ will remain in the effluent of the plant, being equal to the minimum TKN_e that the treatment system can achieve. Thus, under ideal conditions: $TKN_e = N_{ous,i} = 0.49$ mg N/l.

Example 5.3.2

WWTP expanded to achieve nitrification

An activated sludge system has been designed with a SRT of 4 days to treat an influent wastewater (Q_i) with a flow rate of 20 MLD (20,000 m³/d). The plant needs to comply with an effluent total COD concentration (COD_e) of 125 mg/l. The plant does not have a primary settling tank. The raw wastewater contains a total influent COD concentration (COD_i) of 475 mg/l that contains an influent total Kjeldahl nitrogen concentration (TKN_i) and total phosphorus concentration (P_i) of 73.5 mgN/l and 14.5 mgP/l, respectively. In addition, the influent flow rate contains 25 mg/l of inorganic or fixed suspended solids. The minimum yearly wastewater temperature is 14 °C.

The plant has been designed with an average MLSS concentration of 4,000 mg/l, a TSS concentration in the effluent of the secondary settling tank of 15 mg TSS/l and a TSS concentration in the sludge recycle line of 10,000 mg TSS/l.

Table 5.1 presents a summary of the process design results.

Table 5.1 Summary of the process design results of the activated sludge plant that performs the aerobic organic matter removal.

Description	Parameter	Unit	Value
1. Influent and bioreactor			
Type of wastewater	Raw		Raw
Temperature	T	°C	14
Influent flow rate	Q_i	MLD	20
Influent total COD	COD_i	$gCOD/m^3$	475
Influent biodegradable COD	$COD_{b,i}$	$gCOD/m^3$	326
Influent concentration of readily biodegradable COD	$S_{S,i}$	$gCOD/m^3$	149
Influent concentration of slowly biodegradable COD	$X_{S,i}$	$gCOD/m^3$	177
Influent concentration of soluble unbiodegradable COD	$S_{U,i}$	$gCOD/m^3$	89
Influent concentration of particulate unbiodegradable COD	$X_{U,i}$	$gCOD/m^3$	60
Influent concentration of fixed (inorganic) suspended solids	$XF_{SS,i}$	$gFSS/m^3$	25
Influent TKN (= N_i since it is assumed that the influent does not contain any nitrate or nitrite)	TKN_i	gN/m^3	73.5
Influent concentration of free and saline ammonia	$S_{NHx,i}$	gN/m^3	62.7
Influent concentration of biodegradable soluble nitrogen	$N_{obs,i}$	gN/m^3	3.51
Influent concentration of biodegradable particulate nitrogen	$N_{obp,i}$	gN/m^3	2.75
Influent concentration of unbiodegradable soluble nitrogen	$N_{ous,i}$	gN/m^3	0.49
Influent concentration of unbiodegradable particulate nitrogen	$N_{oup,i}$	gN/m^3	4.05
Influent total P	P_i	gP/m^3	14.5
Influent concentration of orthophosphate	$S_{PO4,i}$	gP/m^3	7.9
Sludge retention time	SRT	d	4
Sludge recycle ratio (s)	s	$m^3.d/m^3.d$	0.67
2. $COD_{b,i}$ for OHOs			
Flux of $COD_{b,i}$ for OHOs	$FCOD_{b,OHO}$	kgCOD/d	6,520
3. System biomass (VSS) equations			
Mass of OHOs	MX_{OHOv}	kgVSS	6,489
Mass of endogenous residue from OHOs	$MX_{E,OHOv}$	kgVSS	1,049
Mass of unbiodegradable organics from influent	MX_{Uv}	kgVSS	3,243
4. Volatile and total suspended solids (VSS and TSS) in the system			
Mass of active biomass (= MX_{OHOv})	MX_B	kgVSS	6,489
Mass of VSS	MX_{VSS}	kgVSS	10,781
MX_B/VSS	f_{av}	gVSS/gVSS	0.60
Mass of fixed or inorganic SS	MX_{FSS}	kgFSS	2,973
Mass of TSS	MX_{TSS}	kgTSS	13,755
Ratio of VSS/TSS	f_{VT}	gVSS/gTSS	0.78
5. Reactor volume			
Total suspended solids concentration of design in the aeration tank	X_{TSS}	$kgTSS/m^3$	4.0
Bioreactor volume	V_R	m^3	3,439
6. Waste of activated sludge			
Activated sludge waste flow rate	Q_W	m^3/d	860

7. Nutrient requirement			
Required N concentration for biomass synthesis	N_s	gN/m^3	13.5
Required P concentration for biomass synthesis	P_s	gP/m^3	4.0
8. Oxygen demand			
Flux of carbonaceous O_2 demand	FO_C	kgO_2/d	3,731
9. Treated effluent quality[a]			
Effluent total COD	COD_e	$gCOD/m^3$	106
Effluent total Kjeldahl Nitrogen	TKN_e	gN/m^3	61.2
Effluent total phosphorus	P_e	gP/m^3	10.9
10. Secondary settling tanks			
Maximum peak factor	f_q	-	3.5
Maximum initial settling velocity	v_0	m/h	7.8
Hindrance settling coefficient	r_{hin}	$m^3/kgTSS$	0.426
Number of treatment lines	N_{AS}	-	2
Number of secondary settling tanks per treatment line	N_{SST}	-	2
Diameter of each secondary settling tank	ϕ_{SST}	m	30 m
Total installed area of the secondary settling tanks	A_{SST}	m^2	2,827

[a] Considering an effluent total solids concentration of 15 mgTSS/l ($X_{TSS,e}$), the f_{VT} of 0.78, f_{CV} of 1.48 mgCOD/mgVSS, f_n of 0.10 mgN/mgVSS and f_p of 0.03 mgP/mgVSS. Thus: $COD_e = S_{u,i} + f_{VT} \cdot X_{TSS,e} \cdot f_{CV}$; $TKN_e = N_i - N_s + f_{VT} \cdot X_{TSS,e} \cdot f_n$; and $P_e = P_i - P_s + f_{VT} \cdot X_{TSS,e} \cdot f_p$.

However, the activated sludge wastewater treatment plant needs to comply with the maximum allowable effluent TKN concentration (TKN_e) of 10 mg N/l. Thus, the plant needs to be upgraded to meet the upcoming discharge standard. Assuming that the same design conditions and characteristics remain, what design modifications have to be made to comply with a TKN_e lower than 10 mg N/l?

Solution

One of the possible solutions to meet the discharge standard of 10 mgTKN$_e$/l is shown in Table 5.3. However, first the kinetic and stoichiometric parameters are shown in Table 5.2.

Table 5.2 Kinetic and stoichiometric parameters for the design example to expand an organic matter removal plant to also perform the nitrification process.

Parameter		Symbol	Value	Unit
	OHO			
	Specific endogenous mass loss rate of the OHOs at 20 °C	$b_{OHO,20}$	0.24	gVSS/gVSS.d
	Temperature coefficient for $b_{OHO,T}$	$\theta_{b,OHO}$	1.029	
	Specific endogenous mass loss rate of the OHOs at temperature T	$b_{OHO,T}$	0.202	gVSS/gVSS.d
	ANO			
Kinetic	ANO maximum specific growth rate constant at 20 °C	$\mu_{ANOmax,20}$	0.45	gVSS/gVSS.d
	Temperature coefficient for $\mu_{ANOmax,T}$	θ_{NIT}	1.123	
	ANO maximum specific growth rate constant at temperature T	$\mu_{ANOmax,T}$	0.224	gVSS/gVSS.d
	ANO half-saturation constant at 20 °C	$K_{ANO,20}$	1.0	gN/m^3
	Temperature coefficient for $K_{ANO,T}$	θ_{NIT}	1.123	
	ANO half-saturation constant at temperature T	$K_{ANO,T}$	0.50	gN/m^3
	ANO-specific endogenous mass loss rate constant at 20 °C	$b_{ANO,20}$	0.04	gVSS /gVSS.d
	Temperature coefficient for $b_{ANO,T}$	$\theta_{b,ANO}$	1.029	
	ANO-specific endogenous mass loss rate constant at temperature T	$b_{ANO,T}$	0.034	gVSS /gVSS.d
	OHO			
	Biomass yield of OHOs	Y_{OHOv}	0.45	gVSS/gCOD
	Fraction of endogenous residue of the OHOs	$f_{XE,OHO}$	0.20	gVSS/gVSS
Stoichiometric	Fraction of fixed (inorganic) suspended solids of OHOs	$f_{FSS,OHO}$	0.15	gFSS/gVSS
	ANO			
	Biomass yield of ANOs	Y_{ANOv}	0.10	gVSS/gCOD
	General			
	COD/VSS ratio of the sludge	f_{CV}	1.48	gCOD/gVSS
	Nitrogen content in volatile suspended solids biomass	f_n	0.10	gN/gVSS
	Phosphorus content in volatile suspended solids biomass	f_p	0.03	gP/gVSS

Table 5.3 Process design of the activated sludge plant to perform the aerobic removal of organic matter and the nitrification process.

1. Verify the required SRT$_{MIN}$ for nitrification

First, the minimum required SRT for nitrification (SRT$_{MIN}$) needs to be determined and compared to the SRT of 4 d to assess whether the system can nitrify:

$$\text{SRT}_{min} = 1 / [(\mu_{ANOmax,T} / S_f) - b_{ANO,T}] \qquad \text{(Eq. 5.16)}$$

If:

$\mu_{ANOmax,20}$ = 0.45 1/d

θ_{NIT} = 1.123

$\mu_{ANOmax,T}$ = $\mu_{ANOmax,20} \cdot \theta_{NIT}^{(14-20)}$
= $0.45 \cdot 1.123^{(14-20)}$

$\mu_{ANOmax,T}$ 0.22 1/d

$b_{ANO,20}$ = 0.04 1/d

$\theta_{b,ANO}$		$= 1.029$
$b_{ANO,T}$		$= b_{ANO,20} \cdot \theta_{b,ANO}^{(T-20)}$
		$= 0.04 \cdot 1.09^{(14-20)}$
$b_{ANO,T}$		$0.034 \; 1/d$
And,		
S_f		$= 1.25$
SRT_{min}		$= 1 / [(0.224 / 1.25) - (0.034)]$
		$6.9 \; d$

The applied SRT of 4 d is shorter than SRT_{MIN}. Thus, nitrification cannot take place. In order to achieve nitrification, a SRT longer than 6.9 days needs to be applied and the process design of the plant needs to be recalculated. This is carried out in the following calculations using a newly selected SRT of 7 days.

2. Process design of the activated sludge plant with a SRT of 7 days to achieve organic matter removal and nitrification.

2.1. System configuration

Aerobic completely-stirred tank reactor configuration process configuration operated at 14 °C.

2.2 Influent and sludge recycle composition (from previous tables)

Q_i	20	MLD	(20,000 m³/d) average influent flow rate
COD_i	475	gCOD/m³	Influent concentration of total COD
$S_{S,i}$	149	gCOD/m³	Influent concentration of readily biodegradable COD
$X_{S,i}$	177	gCOD/m³	Influent concentration of slowly biodegradable COD
$COD_{b,i}$	326	gCOD/m³	Influent concentration of biodegradable COD ($S_{S,i} + X_{S,i}$)
$S_{U,i}$	89	gCOD/m³	Influent concentration of soluble unbiodegradable COD
$X_{U,i}$	60	gCOD/m³	Influent concentration of particulate unbiodegradable COD
$X_{FSS,i}$	25	gFSS/m³	Influent concentration of fixed (inorganic) suspended solids
TKN_i	73.5	gN/m³	Influent concentration of total Kjeldahl N
$S_{NHx,i}$	62.7	gN/m³	Influent concentration of free and saline ammonia
$N_{obs,i}$	3.51	gN/m³	Influent concentration of biodegradable soluble nitrogen
$N_{obp,i}$	2.75	gN/m³	Influent concentration of biodegradable particulate nitrogen
$N_{ous,i}$	0.49	gN/m³	Influent concentration of unbiodegradable soluble nitrogen
$N_{oup,i}$	4.05	gN/m³	Influent concentration of unbiodegradable particulate nitrogen
P_i	14.5	gP/m³	Influent concentration of total P
$S_{PO4,i}$	7.9	gP/m³	Influent concentration of orthophosphate
SRT	7	d	Sludge retention time

2.3 Influent fluxes used for calculations ($= Q_i \cdot$ influent concentration of component)

$FCOD_i$	9,500	kgCOD/d	Influent daily flux of total COD
$FS_{S,i}$	2,980	kgCOD/d	Influent daily flux of RBCOD
$FCOD_{b,i}$	6,520	kgCOD/d	Influent daily flux of biodegradable COD ($S_{S,i} + X_{S,i}$)
$FX_{U,i}$	1,200	kgCOD/d	Influent daily flux of particulate unbiodegradable COD
$FS_{U,i}$	1,780	kgCOD/d	Influent daily flux of soluble unbiodegradable COD
$FX_{FSS,i}$	500	kgFSS/d	Influent daily flux of fixed (inorganic) suspended solids

2.4 Sludge recycle characteristics

s		m³.d/m³.d	$s = X_{TSS} / (X_{TSS,s} - X_{TSS})$
		m³.d/m³.d	$s = 4,000 / (10,000 - 4,000)$
s	0.67	m³.d/m³.d	Sludge recycle ratio (s) with regard to influent flow rate

2.5. Biomass equations

Corresponds to the biological mass present in the system as synthesized from the influent COD (in g/d), taking into account the cumulative effect of SRT [(g/d) · d = g in the system]

2.5.1 OHOs
Active mass

Y_{OHOv}	0.45 gVSS/gCOD
$Y_{OHOv,obs}$	$= Y_{OHOv} / (1 + b_{OHO,T} \cdot SRT)$
	$= 0.45 / (1 + 0.202 \cdot 7)$
$Y_{OHOv,obs}$	0.186 gVSS/gCOD
MX_{OHOv}	$= Y_{OHOv,obs} \cdot FCOD_{b,i} \cdot SRT$
	$= 0.186 \cdot 6,520 \cdot 7$
MX_{OHOv}	8,504 kgVSS

Endogenous mass

$MX_{E,OHOv}$	$= f_{XE,OHO} \cdot b_{OHO,T} \cdot MX_{OHOv} \cdot SRT$	(Eq. 4.10)
	$= 0.20 \cdot 0.202 \cdot 8,504 \cdot 7$	
$MX_{E,OHOv}$	2,407 kgVSS	

2.5.2 Unbiodegradable particulate organics

MX_{Uv}	$= FX_{U,i} \cdot SRT / f_{cv}$	(Eq. 4.11)
	$= 1,200 \cdot 7 / 1.48$	
MX_{Uv}	5,676 kgVSS	

2.6 Nutrient requirements
2.6.1. Nitrogen requirements for biomass synthesis

N_s	$= f_n \cdot MX_{VSS} / (Q_i \cdot SRT)$	(Eq. 5.25)
	$= f_n \cdot (MX_{OHOv} + MX_{E,OHOv} + MX_{Uv}) / (Q_i \cdot SRT)$	
	$= (0.10 \cdot (8,504 + 2,407 + 5,676)) / (20 \cdot 7)$	
N_s	11.8 gN/m^3	

2.6.2 Phosphorus requirements for biomass synthesis

P_s	$= f_p \cdot MX_{VSS} / (Q_i \, SRT)$	(Eq. 5.31)
	$= f_p \cdot (MX_{OHOv} + MX_{E,OHOv} + MX_{Uv}) / (Q_i \cdot SRT)$	
	$= (0.03 \cdot (8,504 + 2,407 + 5,676)) / (20 \cdot 7)$	
P_s	3.6 gP/m^3	

It is important to notice that, strictly speaking, the mass of nitrifying organisms (MX_{ANOv}) also needs to be known to calculate MX_{VSS}. However, to calculate MX_{ANOv} it is necessary to know N_s because N_s affects the concentration of nitrogen available for nitrification (NIT_c, also identified as the nitrification capacity of the system) since $NIT_c = TKN_i - TKN_e - N_s$ (Eq. 5.35). As such, N_s is estimated without yet knowing MX_{ANOv} in order to be able to carry out the design of the nitrification process. Nevertheless, the contribution of MX_{ANOv} to MX_{VSS} usually tends to be considerably low (lower than 5 % of MX_{VSS}) due to the lower biomass yield of X_{ANOv} and the lower concentrations of N_i compared to those of COD_i (which contributes to practically all MX_{VSS} in CAS systems).

2.7 Design of the nitrification process

Since the system is being designed with a SRT of 7 days (> SRT_{MIN} of 6.9 d), the plant can nitrify. Thus, the nitrification process can be designed as described below.

2.7.1. Nitrification capacity and nitrate generation

NIT_c	$= TKN_i - TKN_e - N_s$	(Eq. 5.35)

Where:

TKN_e	$= S_{NHx,e} + N_{ous,i}$	(Eq. 5.33)

And,

$S_{NHx,e}$	$= [K_{ANO,T} (b_{ANO,T} + 1 / SRT)] / [\mu_{ANOmax,T} - (b_{ANO,T} + 1 / SRT)]$	(Eq. 5.11)
	$= [0.50 \cdot (0.034 + 1/7)] / [0.224 - (0.034 + 1/7)]$	
$S_{NHx,e}$	$1.84 \; gN/m^3$	

From the N fractionation:

$N_{ous,i}$ $\quad$ $0.49 \; gN/m^3$
Thus:
TKN_e $\quad$ $= 1.84 + 0.49$
TKN_e $\quad$ $2.33 \; gN/m^3$

As observed, the plant is able to meet the new effluent standard of 10 mgN/l of TKN with the new SRT of 7 days.

And,

NIT_c $\quad$ $= 73.5 - 2.33 - 11.8$
NIT_c $\quad$ $59.3 \; gN/m^3$

Since the system has been designed with the SRT_{min} of 7 d, nitrification takes place and consequently nitrate, S_{NO3}, is generated (which is equal to NIT_c) which corresponds to the effluent nitrate concentration.

NIT_c	$= S_{NO3} = TKN_i - TKN_e - N_s$	(Eq. 5.35)

So,
$\quad\quad\quad\quad\quad\quad = 59.3 \; gN/m^3$
S_{NO3}

2.7.2. ANO biomass

Combining equations 5.35 and 5.42b, the ANO biomass (MX_{ANOv}) can be estimated as:

$MX_{ANO,v}$ $\quad$ $= Q_i \cdot NIT_c \cdot Y_{ANOv} \cdot SRT / (1 + b_{ANO,T} \cdot SRT)$
$\quad\quad\quad\quad\quad$ $= (20 \cdot 59.3 \cdot 0.10 \cdot 7) / [1 + (0.034) \cdot (7)]$
$MX_{ANO,v}$ $\quad$ $= 672 \; kgVSS$

2.8 VSS and TSS

2.8.1 VSS and active fraction

MX_B $\quad$ $= MX_{OHOv} + MX_{ANO,v}$
$\quad\quad\quad$ $= 8,504 + 672$
MX_B $\quad$ $9,176 \; kgVSS$
MX_{VSS} $\quad$ $= MX_{OHOv} + MX_{ANO,v} + MX_{E,OHOv} + MX_{Uv}$
$\quad\quad\quad$ $= 8,504 + 672 + 2,407 + 5,676$
MX_{VSS} $\quad$ $17,259 \; kgVSS$
f_{av} $\quad$ $= MX_B / MX_{VSS}$
$\quad\quad$ $= 9,176 / 17,259$
f_{av} $\quad$ 0.53

Compared to the MX_{VSS}, M_{XANOv} comprises a minor fraction in the system:

$f_{ANO,VSS}$ $\quad$ $= MX_{ANOv} / MX_{VSS}$
$\quad\quad\quad$ $= 672 / 17,259$
$f_{ANO,VSS}$ $\quad$ $= 0.038 \; kgVSS/kgVSS$

The $f_{ANO,VSS}$ ratio indicates that $MX_{ANO,v}$ composes approximately 3.8 % of the MX_{VSS} in the system. This can be considered to be negligible when estimating N_s (*e.g.*, if $MX_{ANO,v}$ is considered to calculate N_s, it will increase approximately 0.4 mg N/l from 11.8 to around 12.2 mgN/l).

2.8.2 FSS

MX_{FSS} $= f_{FSS,OHO} \cdot MX_{OHOv} + FX_{FSS,i} \cdot SRT$ (Eq. 4.14a)

$= (0.15 \cdot 8{,}504) + (500 \cdot 7)$

MX_{FSS} 4,776 kgFSS

2.8.3 TSS

MX_{TSS} $= MX_{VSS} + MX_{FSS}$ (Eq. 4.15)

$= 17{,}259 + 4{,}776$

MX_{TSS} 22,035 kgTSS

2.8.4 f_{VT}

f_{VT} $= MX_{VSS} / MX_{TSS}$

$= 17{,}259 / 22{,}035$

f_{VT} 0.78 gVSS/gTSS

Also, compared to the MX_{TSS}, M_{XANOv} comprises a minimum fraction in the system:

$f_{ANO,TSS}$ $= MX_{ANOv} / MX_{TSS}$

$= 672 / 22{,}035$

$f_{ANO,TSS}$ $= 0.03$ kgVSS/kgVSS

The $f_{ANO,TSS}$ ratio indicates that $MX_{ANO,v}$ composes approximately 3 % of MX_{TSS} in the system. This can be considered to be negligible when estimating the volume of the reactor (V_R).

2.9 Process volume

X_{TSS} 4 kgTSS/m^3

V_R $= MX_{TSS} / X_{TSS}$ (Eq. 4.20)

$= 22{,}035/4$

V_R 5,509 m^3

The V_R required for the SRT of 7 days needs to be more than 60 % larger than the previous tank to work at the same X_{TSS} of 4 kgTSS/m^3 (3,439 m^3 vs 5,509 m^3) (see Table 5.1). If the extra aeration tank volume is not added, the X_{TSS} will increase to approximately 6.40 kgTSS/m^3 (since based on Eq. 4.20: $X_{TSS} = MX_{TSS} / V_R = 22{,}035 / 3{,}439$). This concentration is higher than the maximum X_{TSS} recommended for the operation of activated sludge systems (of approximately 5 kgTSS/m^3) because it will decrease the oxygen transfer efficiency (increasing the aeration costs). The latter may be even more critical since the nitrification process and the extended SRT will increase the oxygen requirements of the plant. Also, the X_{TSS} of 6.40 kgTSS/m^3 may overload the capacity of the secondary settling tanks (whose maximum capacity also needs to be checked).

2.10 Effluent quality

2.10.1 COD_e

COD_e $= S_{U,i} + f_{CV} \cdot f_{VT} \cdot X_{TSS,e}$ (non-filtered)

$= 89 + 1.48 \cdot 0.78 \cdot 15$

106.3 gCOD/m^3

2.10.2 TKN_e and N_e

TKN_e $= S_{NHx,i} + N_{ous,i}$

TKN_e $= 1.84 + 0.49$

TKN_e 2.33 gN/m^3

Which, as discussed, meets the TKN_e limit of 10 mgN/l (10 gN/m^3). The total nitrogen in the effluent can be computed as displayed below:

N_e $= TKN_e + S_{NO3,e} + f_n \cdot f_{VT} \cdot X_{TSS,e}$

Where, since the plant can nitrify, $S_{NO3,e} = NIT_c$ and needs to be accounted for in the effluent of the plant. Thus:

$$N_e = 2.33 + 59.3 + 0.10 \cdot 0.78 \cdot 15$$
$$N_e = 62.8 \text{ gN/m}^3$$

While the plant is able to nitrify and, therefore, removes the required TKN to meet the effluent standard of 10 mg N/l, it needs to be underlined that it does not remove a significant amount of total nitrogen; only the nitrogen used for biomass synthesis (N_s) is the total N concentration removed. Actually, the plant removes less nitrogen at the SRT of 7 d than at 4 d due to the lower active biomass fraction (reflected in a lower N_s requirement at 7 d SRT compared to that at 4 d SRT). For the plant to remove total nitrogen, a denitrification step would be required.

2.10.3 P_e

$$P_e = P_i - P_s + f_p \cdot f_{VT} \cdot X_{TSS,e} \text{ (not filtered)}$$
$$= 14.5 - 3.6 + 0.03 \cdot 0.78 \cdot 15$$
$$= 11.3 \text{ gP/m}^3$$

Similar to N_e, the plant removes less phosphorus due to the longer SRT and, thereby, the lower P_s requirements at the 7 d SRT compared to those at 4 d SRT (Table 5.1).

2.11 Oxygen demand (OD)
2.11.1 OD for synthesis and endogenous respiration

$$FO_{OHO} = FO_{OHO,s} + FO_{OHO,e}$$
$$FO_{OHO,s} = FCOD_{b,OHO} \cdot (1 - f_{CV} \cdot Y_{OHOv})$$
$$= 6,520 \cdot (1 - 1.48 \cdot 0.45)$$
$$FO_{OHO,s} \quad 2,178 \text{ kgO}_2/\text{d}$$
$$FO_{OHO,e} = FCOD_{b,OHO} \cdot f_{CV} \cdot (1 - f_{XE,OHO}) \cdot b_{OHO,T} \cdot Y_{OHOv,obs} \cdot SRT$$
$$= 6,520 \cdot 1.48 \cdot (1 - 0.20) \cdot 0.202 \cdot 0.186 \cdot 7$$
$$FO_{OHO,e} \quad 2,036 \text{ kgO}_2/\text{d}$$
$$FO_{OHO} \quad 4,213 \text{ kgO}_2/\text{d}$$

2.11.2 OD for nitrification

$$FO_{NIT} = 4.57 \cdot NIT_c \cdot Q_i \qquad \text{(Based on eqs. 5.35, 5.42b and 5.43b)}$$
$$= 4.57 \cdot 59.3 \cdot 20$$
$$5,420 \text{ kgO}_2/\text{d}$$

2.11.3 Total OD of the system

$$FO_t = FO_{OHO} + FO_{NIT}$$
$$= 4,213 + 5,420$$
$$9,633 \text{ kgO}_2/\text{d}$$

The longer SRT (of 7 days) increases the endogenous respiration rate, increasing the oxygen requirements from 3,731 to 4,213 kgO$_2$/d (approximately 13 % more). However, since nitrification also occurs, the oxygen requirements increase by 5,420 kgO$_2$/d. In total, the system needs 9,633 kgO$_2$/d instead of 3,731 kgO$_2$/d when only the removal of organic matter at the SRT of 4 days is performed (Table 5.1). This implies that the installed aeration capacity of the treatment system will likely need to be increased by a multiplier of more than 2.6.

2.12 Waste of activated sludge

$$Q_W = V_R / SRT \qquad \text{(Eq. 4.1)}$$
$$= 5,509 / 7$$
$$Q_W \quad 787 \text{ m}^3/\text{d}$$

Q_W at 7 d SRT is lower than at 4 d because less sludge is discharged to achieve the longer SRT.

2.13 COD mass balance verification

Input

FCOD$_i$ 9,500 kgCOD/d   IN

Outputs

O_2 demand for synthesis and endogenous respiration

FOc   4,213 kgCOD/d (44.3 % of total COD)

Soluble unbiodegradable organics leaving via the effluent

FS$_{U,i}$ 1,780 kgCOD/d (18.7 % of total COD)

Sludge waste COD flux (excluding MX$_{ANOv}$)

FCOD$_{VSS,w}$ = (MX$_{VSS}$ – MX$_{ANO,v}$) / SRT) · f$_{CV}$
                         = (16,586 / 7) · 1.48
                         3,507 kgCOD/d (36.9 % of total COD)

Sum IN: 9,500 kgCOD/d (100 % of total COD)

Delta (OUT-IN):   0 kgCOD/d (0 %)

The 100 % mass balance for COD indicates that all the influent COD is accounted for in the calculated outputs (carbonaceous oxygen demand, sludge waste and unbiodegradable soluble organics).

2.14 Evaluation of the capacity of the secondary settling tanks

If the aeration tank volume of the plant is not expanded from 3,439 to 5,509 m^2 to cope with the increased mass of solids, X$_{TSS}$ will increase to 6.4 kgTSS/m^3 overloading the secondary settling tanks, as shown in the following calculations.

Area of each secondary settling tank required:

A$_{SST}$ = 1,000 f$_q$ · (Q$_{i, ADWF}$ / 24) / [0.80 v$_o$ exp(–r$_{hin}$ · X$_{TSS}$) · N$_{AS}$ · N$_{SST}$]   (Eq. 4.31)

Where:

f$_q$	3.5	
Q$_{i,ADWF}$	20	MLD
v$_o$	7.8	m/h
r$_{hin}$	0.426	m^3/kg
N$_{AS}$	2	No. of main treatment lines
N$_{SST}$	2	No. of settling tanks per treatment line

Thus, with X$_{TSS}$ = 6.4 kg TSS/m^3

A$_{SST}$ = 1,000 · 3.5 · (20/24) / [0.80 · 7.8 exp(–0.426 ·6.4) · 2 · 2]

A$_{SST}$ = 1,791 m^2

Considering that there are 2 main treatment lines each with 2 secondary settling tanks, this leads in a total settling area of 7,165 m^2 (= 4 · 1,791 m^2). This is approximately 2.5 times the current installed area (of 2,827 m^2) (Table 5.1), implying that if the volume of the aerobic tanks is not expanded (to keep X$_{TSS}$ at approximately 4 kg TSS/m^3) then more secondary settling tanks need to be built. However, it is preferable to expand the volume of the aeration tanks since that can also contribute to maintaining a more efficient aeration due to the lower X$_{TSS}$ concentration.

Table 5.4 presents a summary of the process design results and a comparison with the initial conditions at SRT of 4 days when the system performed only the removal of organic matter (Table 5.1) and after the SRT was extended to 7 days to also perform the nitrification process.

Table 5.4 Summary and comparison between the design carried out with a SRT of 4 days (performing only the removal of organic matter) and SRT of 7 days (achieving also nitrification).

Description	Parameter	Unit	Value @ 4 d SRT	Value @ 7 d SRT
1. Influent and bioreactor				
Type of wastewater	Raw		Raw	Raw
Temperature	T	°C	14	14
Influent flow rate	Q_i	MLD	20	20
Influent total COD	COD_i	$gCOD/m^3$	475	475
Influent biodegradable COD	$COD_{b,i}$	$gCOD/m^3$	326	326
Influent TKN	TKN_i	gN/m^3	73.5	73.5
Influent total P	P_i	gP/m^3	14.5	14.5
Sludge recycle ratio (s)	s	$m^3.d/m^3.d$	0.67	0.67
2. $COD_{b,i}$ for OHOs				
Flux of $COD_{b,i}$ for OHOs	$FCOD_{b,OHO}$	$kgCOD/d$	6,520	6,520
3. System biomass (VSS) equations				
Mass of OHOs	MX_{OHOv}	$kgVSS$	6,489	8,504
Mass of endogenous residue from OHOs	$MX_{E,OHOv}$	$kgVSS$	1,049	2,407
Mass of nitrifying organisms	MX_{ANOv}	$kgVSS$	0	672
Mass of unbiodegradable organics from influent	MX_{Uv}	$kgVSS$	3,243	5,676
4. Volatile and total suspended solids (VSS and TSS) in the system				
Mass of active biomass (= $MX_{OHOv} + MX_{ANO,v}$)	MX_B	$kgVSS$	6,489	9,176
Mass of VSS	MX_{VSS}	$kgVSS$	10,781	17,259
MX_B/VSS	f_{av}	$gVSS/gVSS$	0.60	0.53
Mass of fixed or inorganic SS	MX_{FSS}	$kgFSS$	2,973	4,776
Mass of TSS	MX_{TSS}	$kgTSS$	13,755	22,035
Ratio of VSS/TSS	f_{VT}	$gVSS/gTSS$	0.78	0.78
5. Reactor volume				
Bioreactor volume	V_R	m^3	3,439	5,509
6. Activated sludge waste				
Activated sludge waste flow rate	Q_w	m^3/d	860	787
7. Nutrient requirement				
Required N concentration for biomass synthesis	N_s	gN/m^3	13.5	11.8
Required P concentration for biomass synthesis	P_s	gP/m^3	4.0	3.6
8. Oxygen demand				
Flux of carbonaceous O_2 demand	FO_C	kgO_2/d	3,731	4,213
Flux of O_2 demand for nitrification	FO_{NIT}	kgO_2/d	0	5,420
Total O_2 demand of the system	FO_t	kgO_2/d	3,731	9,633

9. Treated effluent quality

Effluent total COD	COD_e	$gCOD/m^3$	106	106.3
Effluent total nitrogen	N_e	gN/m^3	61.2	62.8
Effluent total Kjeldahl Nitrogen	TKN_e	gN/m^3	61.2	2.33
Effluent nitrate concentration	$S_{NO3,e}$	gN/m^3	0	59.3
Effluent total phosphorus	P_e	gP/m^3	10.9	10.5
10. COD mass balance				
COD output/COD input	COD mass balance	gCOD/gCOD	100 %	100 %

Example 5.3.3

Extended aeration with denitrification

The activated sludge system designed in Example 4.3.3 with a SRT of 25 days to treat an influent wastewater (Q_i) with a flow rate of 15 MLD (15,000 m^3/d) (Figure 5.1) needs to be modified to comply with an effluent total nitrogen concentration of less than 10 mgN/l. The plant does not have a primary settling tank. The raw wastewater contains a total influent COD concentration (COD_i) of 585 mg/l that contains an influent total Kjeldahl nitrogen concentration (TKN_i) and total phosphorus concentration (P_i) of 44.5 mgN/l and 14.5 mgP/l, respectively. In addition, the influent flow rate contains 35 mg/l of inorganic or fixed suspended solids. The minimum yearly wastewater temperature is 14 °C.

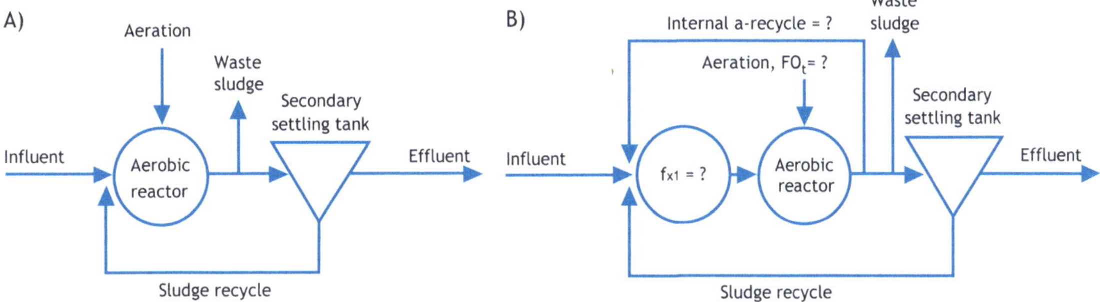

Figure 5.1 Expansion of an extended aeration activated sludge system to perform the denitrification process in addition to the organic matter removal and nitrification processes: A) simplified flow scheme before the expansion and B) after the introduction of an anoxic mass fraction (f_{x1}) and of the internal a-recycle flow rate ratio for denitrification.

Given the previous conditions and suggested configuration (Figure 5.1):

a) What should the unaerated mass fraction f_{x1} be?
b) What will the required internal a-recycle ratio be?
c) What will the total oxygen consumption (FO_t) be after introducing the unaerated mass fraction f_{x1}?

Solution

One of the possible solutions to meet the discharge standard of 10 mgN$_e$/l is shown in Table 5.5.

Table 5.5 Process design of the activated sludge plant to introduce an anoxic mass fraction f_{x1} (pre-denitrification configuration).

1. Estimate the minimum and maximum primary anoxic mass fractions

First, the minimum required primary anoxic mass fraction, $f_{x1,min}$, needs to be determined:

$f_{x1,min}$	$= f_{ss} (1 - Y_{OHO}) (1 + b_{OHO,T} \cdot SRT) / (2.86\ K_{1,T} \cdot Y_{OHOv} \cdot SRT)$	(Eq. 5.51)

If:

f_{ss}	$= S_{S,i} / (S_{S,i} + X_{S,i})$	Table 4.6
	$= 165 / (165 + 199)$	
	0.45	
Y_{OHO}	$0.67\ kgCOD/kgCOD$	
$b_{OHO,20}$	$= 0.24\ 1/d$	
$\theta_{b,OHO}$	$= 1.029$	
$b_{OHO,T}$	$= b_{OHO,20} \cdot \theta_{b,OHO}^{(T-20)}$	
	$= 0.24 \cdot 1.029^{(14-20)}$	
$b_{OHO,14}$	$0.202\ 1/d$	
SRT	$25\ d$	
$K_{1,T}$	$= K_{1,20} \cdot \theta_{d,K1}^{(T-20)}$	Table 5.5 (textbook)
$K_{1,20}$	$= 0.72\ gN/gVSS.d$	
$\theta_{d,K1}$	$= 1.20$	
$K_{1,14}$	$= 0.72 \cdot 0.42^{(14-20)}$	
$K_{1,14}$	$0.241\ gN/gVSS.d$	
Y_{OHOv}	$0.45\ kgVSS/kgCOD$	

Thus:

$f_{x1,min}$	$= 0.45 \cdot (1 - 0.67) \cdot (1 + 0.202 \cdot 25)\ / (2.86 \cdot 0.241 \cdot 0.45 \cdot 25)$
	0.12

Second, the maximum allowable primary anoxic mass fraction needs to be estimated:

$f_{x,max}$	$= 1 - S_f [(b_{ANO,T} + 1 / SRT) / \mu_{ANOmax,T}]$	(Eq. 5.19b)

If:

S_f	$= 1.25$
$b_{ANO,T}$	$= b_{ANO,20} \cdot \theta_{b,ANO}^{(T-20)}$
$b_{ANO,20}$	$= 0.04\ 1/d$
$\theta_{b,ANO}$	$= 1.029$
$b_{ANO,T}$	$= 0.04 \cdot 1.09^{(14-20)}$
	$0.034\ 1/d$
$\mu_{ANOmax,T}$	$= \mu_{ANOmax,20} \cdot \theta_{NIT}^{(14-20)}$
$\mu_{ANOmax,20}$	$= 0.45\ 1/d$
θ_{NIT}	$= 1.123$
$\mu_{ANOmax,T}$	$= 0.45 \cdot 1.123^{(14-20)}$
$\mu_{ANOmax,T}$	$0.22\ 1/d$

Thus:

$f_{x,max}$	$= 1 - 1.25 \cdot [(0.034 + 1 / 25) / 0.22]$
$f_{x,max}$	0.59

At the applied SRT of 25 days, the minimum primary anoxic mass fraction required to use up all the rapidly biodegradable organics ($S_{S,i}$) to denitrify is 0.12, while the maximum to avoid inhibiting the nitrification process is 0.59. Therefore, the primary anoxic mass fraction needs to be selected within this interval. Given these values and considering that it is an extended aeration system, a primary anoxic mass fraction of 0.15 is selected.

2. Design of the nitrification process
2.1. SRT_{MIN} required for nitrification
Since the system is being designed with a SRT of 25 days the plant can nitrify. This can be confirmed by calculating the SRT_{MIN} after including the primary anoxic mass fraction of 0.15:

SRT_{min}	$= 1 / [(\mu_{ANOmax,T} / S_f) \cdot (1 - f_{xt}) - b_{ANO,T}]$	(Eq. 5.16)

If:

$\mu_{ANOmax,T}$	$= 0.22 \ 1/d$
$b_{ANO,T}$	$= 0.034 \ 1/d$
f_{xt}	$= 0.15$

Then:

SRT_{min}	$= 1 / [(0.22 / 1.25) \cdot (1 - 0.15) - 0.034]$
SRT_{min}	$= 8.4 \ d$

2.2. Nitrification capacity and nitrate generation

NIT_c	$= TKN_i - TKN_e - N_s$	(Eq. 5.35)

Where:

TKN_e	$= S_{NHx,e} + N_{ous,i}$	(Eq. 5.33)

And,

$S_{NHx,e}$	$= [K_{ANO,T} \cdot (b_{ANO,T} + 1 / SRT)] /$	(Eq. 5.11)
	$[\mu_{ANOmax,T} \cdot (1 - f_{xt}) - (b_{ANO,T} + 1/SRT)]$	

Where:

$K_{ANO,T}$	$= K_{ANO,20} \ \theta_{NIT}^{(T-20)}$
$K_{ANO,20}$	$= 1 \ g/m^3$
θ_{NIT}	$= 1.123$
	$= 1 \cdot 1.123^{(14-20)}$
$K_{ANO,T}$	$= 0.50 \ gN/m^3$

Then:

$S_{NHx,e}$	$= [0.50 \cdot (0.034 + 1 / 25)] / [0.22 \cdot (1 - 0.15) - (0.034 + 1 / 25)]$
$S_{NHx,e}$	$0.31 \ gN/m^3$

From the N fractionation (data needs to be provided):

$N_{ous,i}$	$0.49 \ gN/m^3$

Thus:

TKN_e	$= 0.31 + 0.49$
TKN_e	$0.80 \ gN/m^3$

The plant is able to meet the effluent standard of 10 mgN/l of TKN, but not a N_e concentration of 10 mg/l.

And:

NIT_c	$= TKN_i - TKN_e - N_s$	(Eq. 5.35)
N_s	$= 15.3 \ gN/m^3$	(Table 4.6)

The introduction of the anoxic mass fraction does not affect the volatile suspended solids mass (MX_{VSS}), thus the N_s calculated in Example 4.3.3 remains. Therefore:

NIT_c	$= 44.5 - 0.80 - 15.3$
NIT_c	28.4 gN/m^3

Since nitrification takes place, the effluent nitrate concentration is:

NIT_c	$= S_{NO3} = TKN_i - TKN_e - N_s$	(Eq.5.35)

So:

S_{NO3}	$= 28.4 \text{ gN/m}^3$

2.3. ANO biomass

Combining equations 5.35 and 5.42b, the ANO biomass (MX_{ANOv}) can be estimated as:

$MX_{ANO,v}$	$= Q_i \cdot NIT_c \cdot Y_{ANOv} \cdot SRT / (1 + b_{ANO,T} \cdot SRT)$
	$= 15 \cdot 28.4 \cdot 0.10 \cdot 25 / (1 + 0.034 \cdot 25)$
$MX_{ANO,v}$	$= 577.9 \text{ kgVSS}$

3. Design of the denitrification process

3.1. Denitrification potential

D_{p1}	$= S_{s,i} \cdot (1 - Y_{OHO}) / 2.86 + f_{x1} \cdot K_{2,T} \cdot (S_{s,i} + X_{s,i}) \cdot Y_{OHO,v} \cdot SRT / (1 + b_{OHO,T} \cdot SRT)$	
		(eqs. 5.46 and 5.47)

Where:

$S_{s,i}$	165 gCOD/m^3	
$X_{s,i}$	199 gCOD/m^3	
Y_{OHO}	0.67 kgCOD/kgCOD	
f_{x1}	0.15	
$K_{2,T}$	$= K_{2,20} \cdot \theta_{d,K2}^{(T-20)}$	Table 5.5 (textbook)
$K_{2,20}$	$= 0.101 \text{ gN/gVSS.d}$	
$\theta_{d,K2}$	$= 1.080$	
$K_{2,14}$	$= 0.101 \cdot 1.080^{(14-20)}$	
$K_{2,14}$	0.064 gN/gVSS.d	
Y_{OHOv}	0.45 kgVSS/kgCOD	
$b_{OHO,T}$	0.202 1/d	

As such:

D_{p1}	$= 165 \cdot (1 - 0.67) / 2.86 + 0.15 \cdot 0.064 \cdot (165+199) \cdot 0.45 \cdot 25 / (1 + 0.202 \cdot 25)$
D_{p1}	25.5 gN/m^3

3.2 Internal a-recycle flow rate ratio

Optimum a-recycle flow rate ratio (a_{opt})

a_{opt}	$= \{-B + (B^2 + 4AC)^{1/2}\} / 2A$	(Eq. 5.56)

Where:

A	$= S_{O2,a} / 2.86$
B	$= NIT_c - D_{p1} + [(s + 1) \cdot S_{O2,a} + s \cdot S_{O2,s}] / 2.86$
C	$= (s + 1) \cdot (D_{p1} - s \cdot S_{O2,s} / 2.86) - s \cdot NIT_c$

Assuming dissolved oxygen concentrations of 2 and 1 mg/l in the internal a-recycle and s-recycle flows, respectively (O_a and O_s) and an s-recycle of 0.67 (Table 4.6), thus:

A	$= 2 / 2.86$	
A	0.70	
B	$= 28.4 - 25.5 + [(0.67 + 1) \cdot 2 + (0.67 \cdot 1)] / 2.86$	
B	4.45	
C	$= (0.67 + 1) \cdot (25.5 - (0.67 \cdot 1) / 2.86) - (0.67 \cdot 28.4)$	
C	23.17	

Thus:

a_{opt}	$= \{-B + (B^2 + 4AC)^{1/2}\} / 2A$	(Eq. 5.56)
a_{opt}	$= \{-(4.45) + [4.45^2 + 4 \cdot (0.70) \cdot (23.17)]^{1/2}\} / (2 \cdot 0.70)$	
a_{opt}	3.4	
Selected a	$= \min \{a_{opt} ; a_{prac}\}$	
	$= \min \{3.4 ; 5.0\}$	
Selected a	3.4	

3.3 Effluent nitrogen concentrations

$S_{NO3,e,min}$	$= NIT_c / (a_{opt} + s + 1)$	(Eq. 5.57)
	$= 28.4 / (3.4 + 0.67 + 1)$	
$S_{NO3,e,min}$	$= S_{NO3,e}$	
$S_{NO3,e}$	5.61 gN/m^3	
TKN_e	$= S_{NHx,e} + N_{ous,i}$	
	$= 0.31 + 0.49$	
TKN_e	0.80 gN/m^3	
$S_{NO3,e} + TKN_e$	$= 5.61 + 0.80$	
	6.41 gN/m^3	

The previous concentration only considers the presence of soluble N compounds. If the contribution of the solids lost through the effluent is calculated to estimate the particulate N:

$X_{N,e}$	$= f_n \cdot f_{VT} \cdot X_{TSS,e}$	

Where,

f_n	0.10 gN/gVSS	
f_{VT}	0.80	(Table 4.6)
$X_{TSS,e}$	15 gTSS/m^3	(Table 4.6)
	$= 0.10 \cdot 0.80 \cdot 15$	
$X_{N,e}$	1.20 gN/m^3	

Thus, total effluent N:

N_e	$= S_{NHx,e} + N_{ous,i} + S_{NO3,e} + X_{N,e}$	
	$= TKN_e + S_{NO3,e} + X_{N,e}$	
	$= 0.80 + 6.41 + 1.20$	
N_e	8.91 gN/m^3	

Thus, with a primary anoxic mass fraction of 0.15 and an internal a-recycle flow ratio of 3.4, the plant is able to comply with the required effluent discharge standard of less than 10 gN/m^3. A lower primary anoxic mass fraction of 0.12 could be used, but the effluent total nitrogen concentration would still have approximately the same value.

3.4 Oxygen demand of the system after the introduction of the primary anoxic mass fraction

3.4.1 OD for synthesis and endogenous respiration is:

From Table 4.6:

FO_{OHO}	$= FO_{OHO,s} + FO_{OHO,e}$
FO_{OHO}	$4,252\ kgO_2/d$

3.4.2 OD for nitrification

FO_{NIT} $= 4.57 \cdot NIT_c \cdot Q_i$ (based on eqs. 5.35, 5.42b and 5.43b)

$= 4.57 \cdot 28.4 \cdot 15$

$1,947\ kgO_2/d$

3.4.3 OD recovered from denitrification

FO_{DENIT} $= 2.86 \cdot Q_i \cdot (NIT_c - S_{NO3,e})$

$= 2.86 \cdot 15 \cdot (28.4 - 5.61)$

FO_{DENIT} $978\ kgO_2/d$

3.4.4 Total OD of the system

$FO_{t,DENIT}$ $= FO_{OHO} + FO_{NIT} - FO_{DENIT}$

$= 4,252 + 1,947 - 978$

$5,221\ kgO_2/d$

The introduction of the anoxic mass fraction (f_{x1}) contributes to reaching the required effluent discharge standard while helping to decrease the oxygen requirements by approximately 16 % from 6,199 to 5,221 kgO_2/d. In addition, it can help to recover part of the alkalinity consumed (approximately $3.57 \cdot (28.4 - 5.61) = 81.4$ mg alkalinity/l as $CaCO_3$) during nitrification (which had roughly consumed $7.14 \cdot 28.4 = 202$ mg alkalinity/l as $CaCO_3$). Thus, the alkalinity consumption would be of 120 mg alkalinity/l as $CaCO_3$.

5.4 EXERCISES

Principles of nitrification (exercises 5.4.1-5.4.10)

Exercise 5.4.1

Given the COD fractionation displayed below and considering a TKN_i of 39 mgN/l composed of 27.5 $mgNH_4^+$-N/l, 31.2 mg/l of soluble TKN and 0.50 mg/l of $N_{ous,i}$, what will the expected N fractions present in the influent flow be?

COD fraction	COD (mg/l)	COD fraction	COD (mg/l)
$S_{S,i}$	153	$S_{U,i}$	32
$X_{S,i}$	67	$X_{U,i}$	74

Exercise 5.4.2

The COD fractionation of an influent flow is:

COD fraction	COD (mg/l)	COD fraction	COD (mg/l)
$S_{S,i}$	174	$S_{U,i}$	123
$X_{S,i}$	215	$X_{U,i}$	156

Given the COD fractionation displayed above and considering a TKN_i of 59 mgN/l composed of 37.5 $mgNH_4^+$-N/l, 45.2 mg/l of soluble TKN and 1.2 mg/l of $N_{ous,i}$, answer the following:

a) What will the N fractions present in the influent flow rate be?
b) What will be the N concentration that could be available to cover the nitrogen requirements for biomass growth?
c) If the influent were treated under ideal conditions, what would be the minimum TKN_e that could be achieved?

Exercise 5.4.3
In the design and operation of nitrification systems, what is the meaning of the SRT_{min}?

Exercise 5.4.4
Why is a safety factor used to calculate the SRT_{min}?

Exercise 5.4.5
In an activated sludge plant that performs the nitrification process, what is the fate of the free and saline ammonia (FSA) concentrations (NH_4^+ and NH_3) and that of the biodegradable organic N compounds?

Exercise 5.4.6
Even if the applied SRT is longer than the SRT_{min}, why cannot full nitrification (a zero effluent FSA concentration) be achieved in a fully aerated activated sludge system?

Exercise 5.4.7
If a fully aerated WWTP does not nitrify, what would be the expected N_e and TKN_e concentrations?

Exercise 5.4.8
If a fully aerated WWTP is able to nitrify, what would be the expected N_e and TKN_e concentrations?

Exercise 5.4.9
What is the meaning of the $f_{x,max}$?

Exercise 5.4.10
What is the influence of temperature on the nitrification process?

Principles of denitrification (exercises 5.4.11-5.4.19)
Exercise 5.4.11
What are the advantages of implementing a pre-denitrification instead of a post-denitrification configuration?

Exercise 5.4.12
What is the meaning of the $f_{x1,min}$?

Exercise 5.4.13
What is the a-optimum internal recycle ratio?

Exercise 5.4.14
What is the a-practical internal recycle ratio?

Exercise 5.4.15

From a process performance perspective, what are the reasons why the value of the a-optimum internal recycle ratio could be higher than that of the a-practical internal recycle ratio?

Exercise 5.4.16

What are the disadvantages of applying an internal a-recycle ratio higher or lower than the a-optimum ratio?

Exercise 5.4.17

Why is the oxygen required to oxidize ammonium to nitrate higher than the oxygen equivalents that can be recovered from the denitrification process?

Exercise 5.4.18

Why cannot all the alkalinity consumed in the nitrification process be recovered during the denitrification process?

Exercise 5.4.19

What is a balanced MLE system?

Process design and evaluation of biological nitrogen removal plants (exercises 5.4.20-5.4.21)

Exercise 5.4.20

Retrofitting a CAS system with a short SRT to achieve nitrification.

A CAS system has been designed with a SRT of 4 days to treat an influent wastewater (Q_i) with a flow rate of 25 MLD (25,000 m^3/d). Originally, the plant needed only to comply with an effluent total COD concentration (COD_e) of 125 mg/l. The plant does not have a primary settling tank. The raw wastewater contains a total influent COD concentration (COD_i) of 624 mg/l that includes an influent total Kjeldahl nitrogen concentration (TKN_i) and total phosphorus concentration (P_i) of 42.5 mgN/l and 9.7 mgP/l, respectively. In addition, the influent flow rate contains 21 mg/l of inorganic or fixed suspended solids. The minimum yearly wastewater temperature is 12 °C.

The plant has been designed with an average MLSS concentration of 3,500 mg/l, a TSS concentration in the effluent of the secondary settling tank of 15 mg TSS/l and a TSS concentration in the sludge recycle line of 10,000 mgTSS/l. Table 5.6 presents a summary of the process design results.

Table 5.6 Summary of the process design results of the activated sludge plant that performs the aerobic removal of organic matter.

Description	Parameter	Unit	Value
1. Influent and bioreactor			
Type of wastewater	Raw		Raw
Temperature	T	°C	12
Influent flow rate	Q_i	MLD	25
Influent total COD	COD_i	$gCOD/m^3$	624
Influent biodegradable COD	$COD_{b,i}$	$gCOD/m^3$	392
Influent concentration of readily biodegradable COD	$S_{S,i}$	$gCOD/m^3$	187
Influent concentration of slowly biodegradable COD	$X_{S,i}$	$gCOD/m^3$	205
Influent concentration of soluble unbiodegradable COD	$S_{U,i}$	$gCOD/m^3$	96

Influent concentration of particulate unbiodegradable COD	$X_{U,i}$	$gCOD/m^3$	136
Influent TKN (=N_i since it is assumed that the influent does not contain any nitrate or nitrite)	TKN_i	gN/m^3	42.5
Influent concentration of free and saline ammonia	$S_{NHx,i}$	gN/m^3	27.5
Influent concentration of biodegradable soluble nitrogen	$N_{obs,i}$	gN/m^3	2.41
Influent concentration of biodegradable particulate nitrogen	$N_{obp,i}$	gN/m^3	2.10
Influent concentration of unbiodegradable soluble nitrogen	$N_{ous,i}$	gN/m^3	1.29
Influent concentration of unbiodegradable particulate nitrogen	$N_{oup,i}$	gN/m^3	9.19
Influent total P	P_i	gP/m^3	9.7
Influent concentration of orthophosphate	$S_{PO4,i}$	gP/m^3	6.5
Sludge retention time	SRT	d	4
Sludge recycle ratio (s)	s	$m^3.d\,/m^3.d$	0.54
2. $COD_{b,i}$ for OHOs			
Flux of $COD_{b,i}$ for OHOs	$FCOD_{b,OHO}$	$kgCOD/d$	9,800
3. System biomass (VSS) equations			
Mass of OHOs	MX_{OHOv}	kgVSS	10,011
Mass of endogenous residue from OHOs	$MX_{E,OHOv}$	kgVSS	1,529
Mass of unbiodegradable organics from influent	MX_{Uv}	kgVSS	9,189
4. Volatile and total suspended solids (VSS and TSS) in the system			
Mass of active biomass (= MX_{OHOv})	MX_B	kgVSS	10,011
Mass of VSS	MX_{VSS}	kgVSS	20,730
MX_B/VSS	f_{av}	gVSS/gVSS	0.48
Mass of fixed or inorganic SS	MX_{FSS}	kgFSS	3,602
Mass of TSS	MX_{TSS}	kgTSS	24,332
Ratio of VSS/TSS	f_{VT}	gVSS/gTSS	0.85
5. Reactor volume			
Total suspended solids design concentration in the aeration tank	X_{TSS}	$kgTSS/m^3$	3.5
Bioreactor volume	V_R	m^3	6,952
6. Waste of activated sludge			
Activated sludge waste flow rate	Q_W	m^3/d	1,738
7. Nutrient requirement			
Required N concentration for biomass synthesis	N_s	gN/m^3	20.7
Required P concentration for biomass synthesis	P_s	gP/m^3	6.2
8. Oxygen demand			
Flux of carbonaceous O_2 demand	FO_C	kgO_2/d	5,530
9. Treated effluent quality [a]			
Effluent total COD	COD_e	$gCOD/m^3$	115
Effluent total Kjeldahl Nitrogen	TKN_e	gN/m^3	23.1
Effluent total phosphorus	P_e	gP/m^3	3.9
10. Secondary settling tanks			
Maximum peak factor	f_q	-	3.5
Maximum initial settling velocity	V_0	m/h	9.3
Hindrance settling coefficient	r_{hin}	$m^3/kgTSS$	0.433

Number of treatment lines	N_{AS}	-	2
Number of secondary settling tanks per treatment line	N_{SST}	-	2
Diameter of each secondary settling tank	ϕ_{SST}	m	27
Total installed area of the secondary settling tanks	A_{SST}	m^2	2,290

[a] Considering an effluent total solids concentration of 15 mgTSS/l ($X_{TSS,e}$), the f_{VT} of 0.85, f_{CV} of 1.48 mgCOD/mgVSS, f_n of 0.10 mgN/mgVSS and f_p of 0.03 mgP/mgVSS. Thus: $COD_e = S_{u,i} + f_{VT} \cdot X_{TSS,e} \cdot f_{CV}$; $TKN_e = N_i - N_s + f_{VT} \cdot X_{TSS,e} \cdot f_n$; and $P_e = P_i - P_s + f_{VT} \cdot X_{TSS,e} \cdot f_p$.

However, the activated sludge wastewater treatment plant needs to comply with a maximum allowable effluent TKN concentration (TKN_e) of 5 mg N/l. Thus, the plant needs to be upgraded to meet the upcoming discharge standard. Assuming that the same design conditions and characteristics remain, what design modifications have to be made to comply with a TKN_e lower than 5 mg N/l?

Consider the following kinetic parameters:
- $\mu_{ANOmax,20} = 0.50$ 1/d
- Safety factor (S_f) = 1.30

Exercise 5.4.21

Retrofitting a CAS system to achieve denitrification with a MLE system.

If the plant designed in Exercise 5.4.20 needs to be retrofitted to meet an N_e of less than 10 mg/l, what modifications need to be made to upgrade it to a MLE system?

Assume:
- $O_a = 2$ mgO$_2$/l
- $O_s = 1$ mgO$_2$/l

Innovative nitrogen removal systems (exercises 5.4.22-5.4.27)

Exercise 5.4.22

What is the SHARON process?

Exercise 5.4.23

What is the ANAMMOX process?

Exercise 5.4.24

What is the BABE process?

Exercise 5.4.25

An activated sludge plant treats a municipal wastewater that has an influent COD_i/N_i ratio higher than 12. The plant has a 4-stage Bardenpho configuration. A gravity thickener is used to thicken the surplus sludge which is later treated in an anaerobic digester. Sludge waste from the anaerobic digester is dewatered in a centrifuge, where a poly-electrolyte is added to enhance the dewatering process, prior to disposal for incineration. The plant struggles to meet the N_e standard of 10 mg/l, particularly during winter when the sewage temperature drops below 10 °C and free and saline ammonia (FSA) comprises most of the nitrogen present in the effluent. Land/space is limited. Supernatant and reject water from the sidestream processes (the thickener and centrifuge) is recirculated to the anoxic stage of the plant. SRT was extended until the maximum capacity of the secondary clarifiers was reached, resulting in a MLSS concentration of around 6,500 mg/l in the aerobic tank, but no improvement was observed in the nitrogen removal efficiency. Explain briefly which sidestream process you would select (out of SHARON, SHARON-ANAMMOX and BABE) to achieve satisfactory nitrogen removal.

Exercise 5.4.26

After certain industrial plants started to discharge their effluent into the sewer network, the COD_i/N_i ratio of a municipal wastewater treated in an activated sludge system decreased to around 7 and the N_e standard of 10 mg/l is not met during most of the year (nitrate being the most abundant N compound present in the effluent). The plant has a MLE configuration. Surplus sludge is thickened (in a gravity thickener), treated in an anaerobic digester, and thereafter dewatered in a centrifuge (where a poly-electrolyte is added for dewatering purposes) prior to disposal for incineration. Supernatant and reject water from the sidestream processes (the thickener and centrifuge) is recirculated to the anoxic stage of the plant. Secondary clarifiers are at their normal capacity contributing to keep a MLSS concentration of around 4,500 mg/l in the aerobic tank. Which sidestream process would you select (out of SHARON, SHARON-ANAMMOX and BABE) to achieve satisfactory nitrogen removal in this plant?

Exercise 5.4.27

After the expansion of the sewer network, a MLE plant struggles to meet the N_e standard of 10 mg/l during most of the year. Ammonium is the major N compound present in the effluent. The plant treats municipal wastewater with an influent COD_i/N_i ratio of approximately 10. A gravity thickener is used to thicken the surplus sludge which is later dewatered on sludge drying beds and used for agricultural purposes. Supernatant from the thickener and sludge drying beds is recirculated to the anoxic section of the MLE system. As a potential solution, the SRT was extended without observing any improvement in the nitrogen removal efficiency. Given the previous conditions, which sidestream process would you select (out of SHARON, SHARON-ANAMMOX and BABE) to achieve satisfactory nitrogen removal in this system?

ANNEX 1: SOLUTIONS TO EXERCISES

Principles of nitrification (solutions 5.4.1-5.4.10)

Solution 5.4.1

Based on the provided data, the $S_{NHx,i}$ concentration is 27.5 mgN/l while the rest of the N fractions are:

N fraction	N (mg/l)	N fraction	N (mg/l)
$N_{obs,i}$	3.2	$N_{ous,i}$	0.5
$N_{obp,i}$	2.8	$N_{oup,i}$	5.0

Solution 5.4.2

a) The $S_{NHx,i}$ concentration is 37.5 mgN/l while the rest of the N fractions will be:

N fraction	N (mg/l)	N fraction	N (mg/l)
$N_{obs,i}$	6,5	$N_{ous,i}$	1.2
$N_{obp,i}$	3.3	$N_{oup,i}$	10.5

b) 47.3 mgN/l

c) 1.2 mgN/l

Solution 5.4.3 It is the minimum SRT required to allow the growth of nitrifying organisms and, consequently, the minimum SRT needed to achieve nitrification.

Solution 5.4.4 A safety factor is used to ensure that nitrification is satisfactorily achieved; otherwise, the wastewater treatment plant may be designed and operated at the edge of the nitrification process where a minor variation in the applied SRT could limit it. Also, it compensates for potential load variations during the day as well as potential variations when controlling the SRT.

Solution 5.4.5

The biodegradable organic N compounds are hydrolysed and their nitrogen contents are released becoming, together with the FSA, bioavailable for biomass synthesis and nitrification.

Solution 5.4.6

The nitrification process is highly dependent on the kinetics; as such, the half-saturation concentration on FSA ($K_{ANO,T}$) plays a major role at low concentrations (< 4 mgN/l) slowing down and limiting the FSA oxidation as the FSA concentration decreases, avoiding a full FSA oxidation being reached.

Solution 5.4.7

The expected N_e will be identical to TKN_e and therefore composed of the total influent N (N_i) minus the N concentration used for biomass synthesis ($N_e = N_i - N_s$).

Solution 5.4.8

Since there is no denitrification (the plant is fully aerated), the expected N_e will be composed of the total influent N (N_i) minus the N concentration used for biomass synthesis ($N_e = N_i - N_s$) which is identical to the sum of $S_{NHx,e}$ plus $N_{ous,e}$ and the nitrate generated $S_{NO3,e}$ ($N_e = S_{NHx,e} + N_{ous,e} + S_{NO3,e}$). This happens because the nitrate generated is not removed via the denitrification process. However, TKN_e will be equal to the N fractions that cannot be oxidized (either because of kinetic limitations or because they are unbiodegradable soluble): $TKN_e = S_{NHx,e} + N_{ous,e}$.

Solution 5.4.9
It is the maximum unaerated mass zone that can be implemented in an activated sludge nutrient removal plant without hampering the nitrification process.

Solution 5.4.10
Temperature plays a major role in nitrification; every increase of 6 °C doubles the maximum growth rate and every decrease of 6 °C halves it.

Principles of denitrification (solutions 5.4.11-5.4.19)
Solution 5.4.11
A pre-denitrification configuration can have a higher denitrification potential than a post-denitrification system. This happens because it can utilize both the readily and the slowly biodegradable organics for denitrification (as well as of the endogenous denitrification), while a post-denitrification can only utilize the slowly biodegradable organics and the endogenous denitrification. Furthermore, the denitrification rate on the readily biodegradable organics is more than 7 times faster than the other two (0.720 compared to 0.101 and 0.072 mg NO_3^--N/mgVSS.d, respectively).

Solution 5.4.12
$f_{x1,min}$ is the minimum unaerated mass fraction required to utilize all the readily biodegradable organics for denitrification in a pre-denitrification configuration.

Solution 5.4.13
In an activated sludge pre-denitrification configuration, it is the internal recycle ratio from the aerobic to the anoxic mass fraction that maximizes the denitrification process by recirculating the optimal nitrate load that can be denitrified.

Solution 5.4.14
Mathematically, an a-optimal recycle ratio can have values higher than 5.0. However, these may lead to high operational costs and, from a practical perspective, may not be applicable. Consequently, an a-recycle ratio with a value of 5.0 is suggested as the maximum or practical a-recycle ratio that can provide a satisfactory denitrification removal while remaining applicable and feasible to implement.

Solution 5.4.15
During the design process, a-optimum values higher than the a-practical value of 5.0 are obtained when the denitrification potential (D_{p1}) is considerably higher than the nitrification capacity (NIT_c). For instance, to decrease an a-optimum value, the D_{p1} can be decreased by shortening the anoxic mass fraction f_{x1} (provided it remains higher than $f_{x1,min}$).

Solution 5.4.16
To apply an internal a-recycle ratio either higher or lower than the a-optimum results in a suboptimal denitrification efficiency. An a-recycle higher than the optimal will increase the intrusion of oxygen into the anoxic mass fraction, decreasing the availability of biodegradable organics for denitrification (and also increasing the operational costs). Meanwhile, an a-recycle ratio lower than the optimal will recycle a lower

nitrate load than the one that could be denitrified, making a suboptimal use of the denitrification potential of the system.

Solution 5.4.17

Because in the nitrification process the oxidation of FSA loses 8 electrons which are passed to oxygen, while the denitrification process only recovers 5 electrons donated from the biodegradable organics. As such, the nitrification process consumes 4.57 $mgO_2/mgFSA$ oxidized while the denitrification process recovers 2.86 $mgO_2/mgNO_3^-$ denitrified. In addition, activated sludge systems rarely achieve denitrification efficiencies higher than 90 %, decreasing even further the 'recovery of oxygen'.

Solution 5.4.18

Because the nitrification process generates two protons per mol of FSA oxidized (resulting in an alkalinity consumption of 7.14 mg alkalinity as $CaCO_3$/mg FSA) while the denitrification process only recovers 1 proton per mol of nitrate denitrified (recovering 3.57 mg alkalinity as $CaCO_3$/mg NO_3-N).

Solution 5.4.19

A balanced MLE system is a system whose unaerated mass fraction f_{x1} is equal to $f_{x,max}$ and the a-optimum ratio is equal to the a-practical ratio. These conditions optimize the design and operation of the nitrogen removal activated sludge plant. This can be achieved because:

1) The applied SRT is the minimum required to achieve nitrification, avoiding the necessity to have larger tanks and higher oxygen consumption requirements.
2) The maximum denitrification potential is used, avoiding any unused anoxic mass fraction, which also contributes to decreasing the volume of the system.
3) It promotes the recycle of an optimal nitrate load (due to a-optimal), making an optimal use of the denitrification potential.
4) It applies the a-practical internal recycle ratio which makes it possible to reach the minimum nitrate concentration in the effluent of the plant.

Process design and evaluation of biological nitrogen removal plants (solutions 5.4.20-5.4.21)
Solution 5.4.20
Retrofitting a CAS system with short SRT to achieve nitrification.
To meet the TKN_e standard of 5 mg/l, an SRT of 8.5 days will need to be applied to comply with an SRT_{MIN} of 8.3 days. If so:

- TKN_e will be around 2.52 gN/m^3.
- The MX_{TSS} will increase to around 45,448 kgTSS, requiring an increase in the volume of the aerobic reactors to 12,985 m^3 to cope with the increased solids mass.
- The aeration requirements will increase to 8,989 kgO_2/d.
- In addition, the nitrification process will consume around 155 mg alkalinity/l.

Solution 5.4.21
Retrofitting a CAS system to achieve denitrification with a MLE system.
In order to upgrade the WWTP designed in Exercise 5.4.20 into an MLE system that meets an N_e standard of 10 mg/l, one solution is to extend the SRT to 11 days. If so:

- A primary anoxic mass fraction of 0.15 could be created together with an internal a-recycle flow ratio of 5.0.
- Thus, N_e will be around 7.2 gN/m^3, being composed of 3.5 gN/m^3 of $S_{NO3,e}$, 1.07 gN/m^3 of $S_{NHX,e}$, 1.29 gN/m^3 of $N_{ous,i}$ and 1.3 gN/m^3 of organic nitrogen present in the solids lost through the effluent of the secondary settling tank.
- The MX_{TSS} will increase to around 56,097 kgTSS, requiring an increase in the volume of the aerobic reactors to keep a X_{TSS} concentration of 3.5 kgTSS/m^3. If the volume of the aeration tanks is not increased, the X_{TSS} concentration will increase to about 4.4 kgTSS/m^3 which may not be detrimental for the system. Nevertheless, the capacity and performance of the secondary settling tanks need to be verified.
- The aeration requirements will increase to 9,390 kgO$_2$/d (for carbon removal and nitrification) but the denitrification process will help to recover approximately 1,384 kgO$_2$/d. This will result in a net oxygen consumption of 8,006 kgO$_2$/d.

Innovative nitrogen removal systems (solutions 5.4.22-5.4.27)

Solution 5.4.22

SHARON stands for Single reactor High Ammonium Removal Over Nitrite. Briefly, it refers to the nitritation process or oxidation of ammonium to nitrite (and not to nitrate). This can be achieved in effluents that contain high ammonium concentrations at temperatures higher than 25-30 °C by applying short solid and hydraulic retention times (of approximately 1-1.5 d). High salinity can also contribute to this. These conditions tend to favour the activity of ammonium oxidizing organisms while limiting that of nitrite oxidizing organisms, leading to nitrite accumulation. If the denitrification (or denitritation process) takes place on nitrite, considerable savings in oxygen, carbon requirements and sludge generation can be achieved.

Solution 5.4.23

The ANAMMOX process stands for anaerobic ammonium oxidation. It is an autotrophic removal process of ammonium by anammox organisms using nitrite. The main advantages are that it does not require organic carbon to drive this process, that only 50% of the ammonium has to be oxidized to nitrite, and that there is a low biomass yield. These features make it a sustainable process for ammonium removal from wastewater.

Solution 5.4.24

The BABE process is a bioaugmentation process that can be implemented in plants that treat their wastewater aerobically in activated sludge systems and their excess of sludge in anaerobic digesters. Thus, a fraction of the sludge recycled from the secondary settling tanks back to the aerobic reactors can be mixed and aerated with a flow of (warm) reject water (*i.e.* in a 1:1 ratio) generated by the anaerobic digesters (usually operated at mesophilic temperatures ~30-35 °C). The advantages of this application are that it boosts the growth of indigenous nitrifiers, contributing to an increase in the oxidation of ammonium when the nitrifiers return to the mainstream treatment line. This makes it possible to improve the nitrification process even at lower environmental temperatures.

Solution 5.4.25

Nitrification is limited because of the high free and saline ammonia concentration observed in the effluent. The SRT was extended at the maximum without observing any results. Thus, to install the BABE process in the recirculation line could be an option to augment the concentration of nitrifiers in the system, helping to improve the nitrification process.

Solution 5.4.26

Nitrate is produced so nitrification is not limiting, but the denitrification appears to be the bottleneck. The relatively low COD_i/N_i ratio (~7.0) suggests that there is not enough carbon available for denitrification, supporting the initial hypothesis that denitrification is limited. Furthermore, the supernatant from the thickener and reject water are recirculated to the anoxic stage, meaning that any COD present in this internal stream is not sufficient.

An adequate alternative could be to implement either the SHARON process or the SHARON-ANAMMOX process. If the SHARON process is implemented, the carbon requirements could decrease since the denitrification over nitrite requires less carbon source (but an external carbon source may still be needed). Alternatively, the SHARON-ANAMMOX process can help to achieve full nitrogen removal without requiring any addition of organic carbon.

Solution 5.4.27

Nitrification is the limiting step due to the high ammonium concentration. The COD_i/N_i ratio of 10 indicates that there is sufficient COD available for denitrification. Because NH_3 is the counterion in the sludge reject water (since it is produced in the sludge drying beds), then an option could be the SHARON process to achieve partial nitrification and denitrify on the nitrite concentration available. This will favour the nitrification process and also reduce the aeration and carbon requirements.

REFERENCE

Chen G.H., Van Loosdrecht M.C.M., Ekama G.A. and Brdjanovic D. (eds.) (2020). Biological Wastewater Treatment: Principles, Modelling and Design. ISBN: 9781789060355. IWA publishing, London, UK.

NOMENCLATURE

Abbreviation	Description
BNR	Biological nutrient removal
WWTP	Wastewater treatment plant
SHARON	Single reactor high ammonium removal rate over nitrite
ANAMMOX	Anaerobic ammonium oxidation
RBCOD	Readily biodegradable COD
BABE	Bioaugmentation batch enhanced
COD	Chemical oxygen demand
TKN	Total Kjeldahl nitrogen
CAS	Conventional activated sludge
MLSS	Mixed liquor suspended solids
MLVSS	Mixed liquor volatile suspended solids
TSS	Total suspended solids
VSS	Volatile suspended solids
SRT	Solids retention time
a	Internal a-recycle flow rate ratio from the aerobic to primary anoxic mass fraction with regard to the influent flow rate ($a = Q_a / Q_i$)
s	Sludge s-recycle flow rate ratio from the bottom of the secondary settling tank to the main wastewater treatment line with regard to the influent flow rate ($s = Q_s / Q_i$)

FSS	Fixed (inorganic) suspended solids
MLE	Modified Ludzack-Ettinger
BARDENPHO	Barnard denitrification and phosphorus removal
FSA	Free saline ammonia
MLD	Millions of litres per day
WAS	Waste of activated sludge

Symbol	Description	Unit
a	Internal a-recycle ratio	$m^3.d/m^3.d$
a_{opt}	Optimum internal a-recycle ratio	$m^3.d/m^3.d$
a_{prac}	Practical a-recycle ratio	$m^3.d/m^3.d$
A_{SST}	Total installed area of the secondary settling tanks	m^2
$b_{ANO,20}$	ANO-specific endogenous mass loss rate constant at 20 °C	$gVSS/gVSS.d$
$b_{ANO,T}$	ANO-specific endogenous mass loss rate constant at temperature T	$gVSS/gVSS.d$
$b_{OHO,20}$	Specific endogenous mass loss rate of the OHOs at 20 °C	$gVSS/gVSS.d$
$b_{OHO,T}$	Specific endogenous mass loss rate of the OHOs at temperature T	$gVSS/gVSS.d$
$COD_{b,i}$	Influent biodegradable COD	$gCOD/m^3$
COD_e	Effluent total COD	$gCOD/m^3$
COD_i	Influent total COD	$gCOD/m^3$
DP_1	Denitrification potential	gN/m^3
$f_{ANO,TSS}$	Ratio of active ANO biomass to TSS in the system	$gVSS/gTSS$
$f_{ANO,VSS}$	Ratio of ANO biomass to VSS in the system	$gVSS/gVSS$
f_{AV}	Active biomass to volatile suspended solids ratio	$gVSS/gVSS$
f_{av}	Fraction of active biomass with the regard to the mass of volatile suspended solids	$kgAVSS/kgVSS$
$FCOD_{b,i}$	Influent daily flux of biodegradable COD ($S_{S,i} + X_{S,i}$)	$kgCOD/d$
$FCOD_{b,OHO}$	Flux of $COD_{b,i}$ for OHOs	$kgCOD/d$
f_{CV}	COD/VSS ratio of the sludge	$gCOD/gVSS$
$f_{FSS,OHO}$	Fraction of fixed (inorganic) suspended solids of OHOs	$gFSS/gVSS$
f_n	Nitrogen content in volatile suspended solids biomass	$gN/gVSS$
FO_C	Flux of carbonaceous O_2 demand	kgO_2/d
FO_{DENIT}	Flux of O_2 demand recovered by denitrification	kgO_2/d
FO_{NIT}	Flux of O_2 demand for nitrification	kgO_2/d
FO_{OHO}	Flux of carbonaceous O_2 demand by OHO	kgO_2/d
$FO_{OHO,e}$	Flux of carbonaceous O_2 demand for endogenous respiration of OHO	kgO_2/d
$FO_{OHO,s}$	Flux of carbonaceous O_2 demand for OHO synthesis	kgO_2/d
$FO_{t,DENIT}$	Total mass of O_2 required less that recovered by denitrification	kgO_2/d
FO_{TOT}	Total O_2 demand of the system	kgO_2/d
f_p	Phosphorus content in volatile suspended solids biomass	$gP/gVSS$
f_q	Maximum peak factor	-
f_{ss}	Fraction of soluble biodegradable organics to biodegradable organics	$gCOD/gCOD$
$FS_{S,i}$	Influent daily flux of RBCOD	$kgCOD/d$
$FS_{U,i}$	Influent daily flux of soluble unbiodegradable COD	$kgCOD/d$

f_{VT}	VSS/TSS ratio of the sludge	gVSS/gTSS
f_{x1}	Primary anoxic mass fraction	-
$f_{x1,min}$	Minimum primary anoxic mass fraction	-
$f_{XE,OHO}$	Fraction of endogenous residue of the OHOs	gVSS/gVSS
$FX_{FSS,i}$	Influent daily flux of fixed (inorganic) suspended solids	kgFSS/d
f_{xt}	Maximum unaerated mass fraction	
$FX_{U,i}$	Influent daily flux of particulate unbiodegradable COD	kgCOD/d
$K_{1,20}$	Maximum denitrification rate on readily or soluble biodegradable organics at 20 °C	gN/gVSS.d
$K_{1,T}$	Maximum denitrification rate on readily or soluble biodegradable organics at temperature T	gN/gVSS.d
$K_{2,20}$	Maximum denitrification rate on slowly biodegradable organics at temperature 20 °C	gN/gVSS.d
$K_{2,T}$	Maximum denitrification rate on slowly biodegradable organics at temperature T	gN/gVSS.d
$K_{ANO,20}$	ANO half-saturation constant at 20 °C	gN/m^3
$K_{ANO,T}$	ANO half-saturation constant at temperature T	gN/m^3
MX_{ANOv}	Active mass of ANOs	kgVSS
MX_B	Mass of active biomass	kgVSS
$MX_{E,OHOv}$	Mass of endogenous residue from OHOs	kgVSS
MX_{FSS}	Mass of fixed or inorganic suspended solids	kgFSS
MX_{OHOv}	Active mass of OHOs	kgVSS
MX_{TSS}	Mass of TSS	kgTSS
MX_{Uv}	Mass of unbiodegradable organics from influent	kgVSS
MX_{VSS}	Mass of VSS	kgVSS
N_{AS}	Number of treatment lines	-
N_e	Effluent total N	gN/m^3
N_i	Influent total N	gN/m^3
NIT_c	Nitrification capacity	gN/m^3
$N_{o,i}$	Influent organic nitrogen	gN/m^3
$N_{obp,i}$	Influent concentration of biodegradable particulate nitrogen	gN/m^3
$N_{obs,i}$	Influent concentration of biodegradable soluble nitrogen	gN/m^3
$N_{oup,i}$	Influent concentration of unbiodegradable particulate nitrogen	gN/m^3
$N_{ous,i}$	Influent concentration of unbiodegradable soluble nitrogen	gN/m^3
$N_{ous,e}$	Effluent concentration of unbiodegradable soluble nitrogen	gN/m^3
N_s	N requirement concentration for biomass synthesis	gN/m^3
N_{SST}	Number of secondary settling tanks per treatment line	-
P_e	Effluent total phosphorus	gP/m^3
P_i	Influent total P	gP/m^3
P_s	P requirement concentration for biomass synthesis	gP/m^3
Q_a	Internal a-recycle flowrate from the aerobic to anoxic tank	m^3/d
Q_i	Influent flow rate	m^3/d
$Q_{i,ADWF}$	Average dry weather flow rate influent	m^3/d

Q_s	S-recycle flowrate from the secondary settling tank to the main-stream treatment line	m^3/d
Q_W	Activated sludge waste flow rate	m^3/d
s	Sludge recycle ratio (s)	$m^3.d/m^3.d$
S_f	Safety factor for the nitrification process	-
$S_{NHx,e}$	Effluent concentration of free and saline ammonia	gN/m^3
$S_{NHx,i}$	Influent concentration of free and saline ammonia	gN/m^3
$S_{NO3,e}$	Effluent nitrate concentration	gN/m^3
$S_{NO3,e,min}$	Minimum effluent nitrate concentration	gN/m^3
$S_{PO4,i}$	Influent concentration of orthophosphate	gP/m^3
SRT	Sludge retention time	d
SRT_{MIN}	Minimum solids retention time required for nitrification	d
$S_{S,i}$	Influent concentration of readily biodegradable COD	$gCOD/m^3$
$S_{U,i}$	Influent concentration of soluble unbiodegradable COD	$gCOD/m^3$
T	Temperature	°C
TKN_e	Effluent total Kjeldahl Nitrogen	gN/m^3
TKN_i	Influent TKN	gN/m^3
$TKN_{p,i}$	Influent particulate TKN	gN/m^3
$TKN_{s,i}$	Influent soluble TKN	gN/m^3
V_o	Maximum initial settling velocity	m/h
V_R	Bioreactor volume	m^3
X_{ANOv}	Concentration of ANO biomass	$gVSS/m^3$
$X_{FSS,i}$	Influent concentration of fixed (inorganic) suspended solids	$gFSS/m^3$
$X_{N,e}$	Effluent particulate nitrogen	gN/m^3
$X_{S,i}$	Influent concentration of slowly biodegradable COD	$gCOD/m^3$
X_{TSS}	Total suspended solids concentration of design in the aeration tank	$kgTSS/m^3$
$X_{TSS,e}$	Effluent total solids concentration	$gTSS/m^3$
$X_{TSS,s}$	Total suspended-solids concentration in the sludge recycle flowrate	$gTSS/m^3$
$X_{U,i}$	Influent concentration of particulate unbiodegradable COD	$gCOD/m^3$
Y_{ANOv}	Biomass yield of ANOs	$gVSS/gCOD$
Y_{OHO}	Biomass yield of heterotrophic organisms	$gCOD/gCOD$
$Y_{OHOv,obs}$	Observed biomass yield of OHOs	$gVSS/gCOD$
Y_{OHOv}	Biomass yield of OHOs	$gVSS/gCOD$
$\theta_{b,ANO}$	Temperature coefficient for $b_{ANO,T}$	-
$\theta_{b,OHO}$	Temperature coefficient for $b_{OHO,T}$	-
θ_{NIT}	Temperature coefficient for $\mu_{ANOmax,T}$ and $K_{ANO,T}$	-
$\theta_{d,K1}$	Temperature coefficient for $K_{1,20}$ denitrification rate	-
$\theta_{d,K2}$	Temperature coefficient for $K_{2,20}$ denitrification rate	-
ϕ_{SST}	Diameter of each secondary settling tank	m
$\mu_{ANOmax,T}$	ANO maximum specific growth rate constant at temperature T	$gVSS/gVSS.d$
$\mu_{ANOmax,20}$	ANO maximum specific growth rate constant at 20 °C	$gVSS/gVSS.d$
r_{hin}	Hindrance settling coefficient	$m^3/kgTSS$

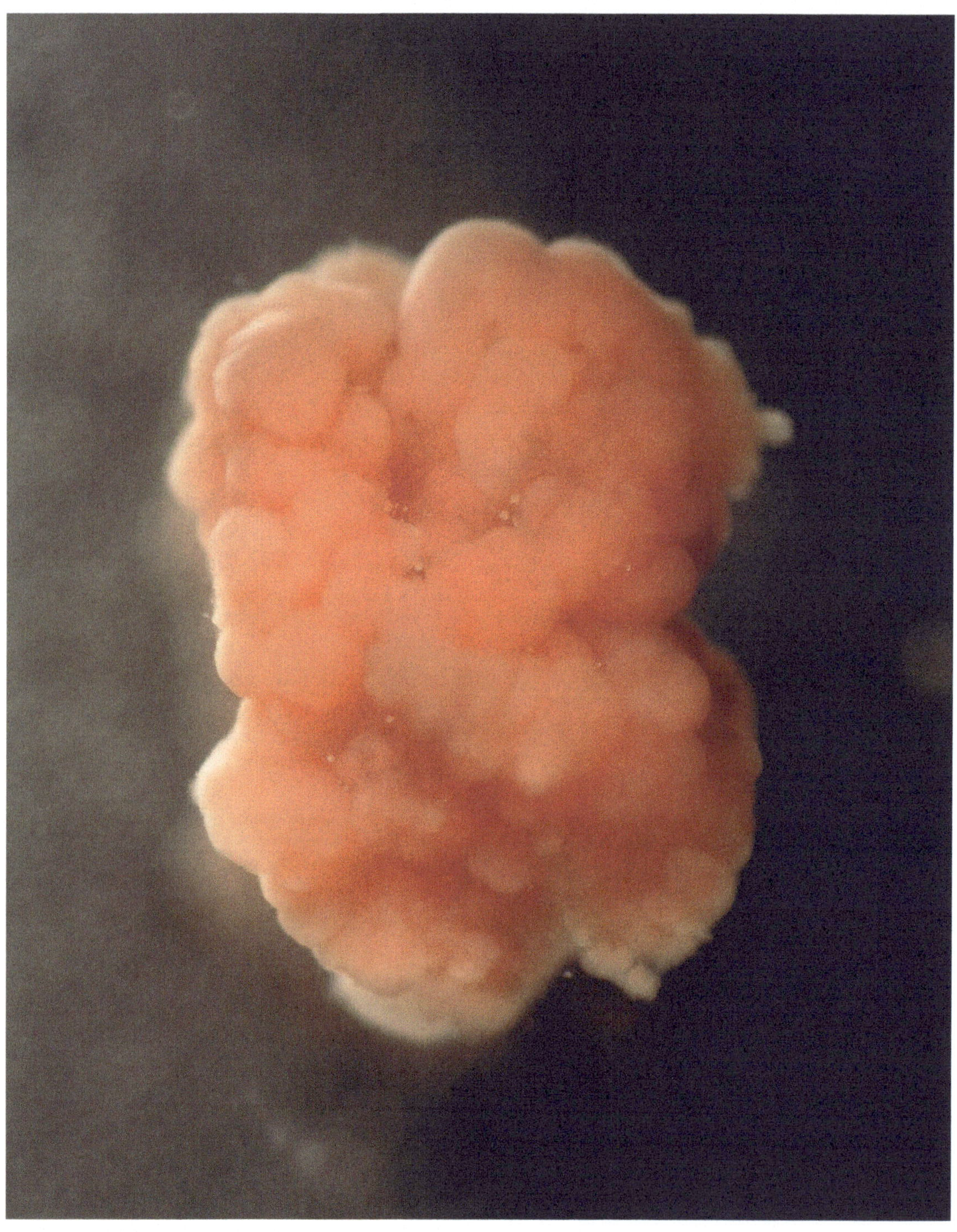

Typical Anammox sludge granule (photo: Water Board Hollandse Delta).

doi: 10.2166/9781789062304_0175

6

Enhanced biological phosphorus removal

Carlos M. Lopez-Vazquez, Francisco Rubio Rincon and Adrian Oehmen

6.1 INTRODUCTION

Chapter 6 on Enhanced biological phosphorus removal (EBPR) in the book *Biological Wastewater Treatment: Principles, Modelling and Design* (Chen *et al.*, 2020) introduces the principles, microbial mechanisms as well as key factors affecting the EBPR process, providing the basis to understand the different full-scale EBPR plant configurations developed since the discovery of the EBPR process. These also serve to present the development of a stoichiometric-based steady-state model for the design and evaluation of activated sludge wastewater treatment plants performing the EBPR process. Last but not least, the steady-state model is used to illustrate the effects of the introduction of the EBPR process in activated sludge systems performing organic matter removal and biological nitrogen removal (via conventional nitrification-denitrification). Therefore, this chapter aims to guide the reader through the principles, microbial mechanisms and the steady-state stoichiometric EBPR model to design, assess and evaluate EBPR wastewater treatment systems. Thus, all the equations belong to Chapter 6 on EBPR from Chen *et al.* (2020).

6.2 LEARNING OBJECTIVES

After the successful completion of this chapter, the reader will be able to:

- Describe the principles, dominant microbial communities and the microbial mechanisms of the EBPR process.
- Identify and describe different EBPR process configurations and discuss their advantages and disadvantages with regard to their process removal efficiency.

- Discuss and assess the effects of key environmental and operational factors and design parameters affecting the EBPR process in biological nutrient removal wastewater treatment systems.
- Apply the EBPR stoichiometric-based steady-state model for the process design and evaluation of activated sludge wastewater treatment plants performing the enhanced biological removal of phosphorus.

6.3 EXAMPLES

Example 6.3.1

Process design of a Phoredox (A/O) EBPR system

By applying the steady-state stoichiometric EBPR model design, design a Phoredox (AO) system (Figure 6.1) to treat an influent wastewater (Q_i) with a flow rate of 15 MLD (15,000 m^3/d) that contains an influent total phosphorus concentration (P_i) of 14.5 mgP/l. The EBPR plant needs to comply with an effluent total phosphorus concentration (P_e) of 1 mg P/l. The plant does not have a primary settling tank. Thus, the raw wastewater contains a total influent COD concentration (COD_i) of 585 mg/l. The influent COD fractionation is shown in Figure 6.2. In addition, the influent flow rate contains 49 mg/l of inorganic or fixed suspended solids.

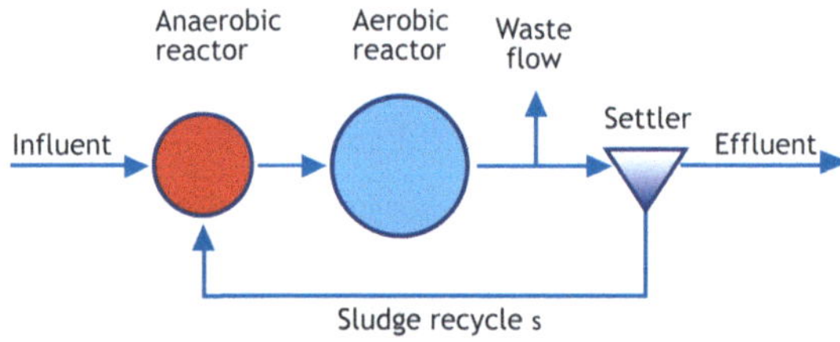

Figure 6.1 A Phoredox (A/O) configuration. The anaerobic reactor is divided into three sections (not illustrated).

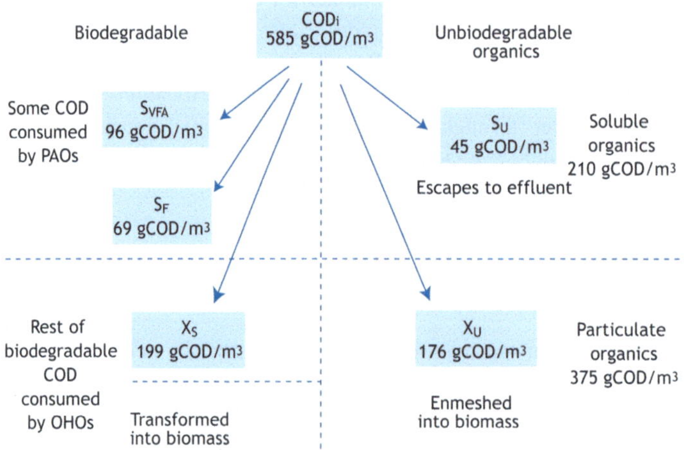

Figure 6.2 Influent COD fractionation for the EBPR design example.

The EBPR plant will be designed with an SRT of 8 days and an anaerobic mass fraction (f_{xa}) of 0.20 divided into 3 different compartments. The plant will operate at an average MLSS concentration of 3,500 mg/l. The average yearly wastewater temperature is 14 °C.

To carry out the design, assume:

a. That nitrification does not take place in this system because of the combination of (*i*) a relatively low wastewater temperature (14 °C), (*ii*) the applied SRT of 8 days; and (*iii*) the net aerobic mass fraction of 0.80 (= 1.0 – f_{xa}),

b. That the sludge recycle flow (Q_r) can convey up to 1 mg/l of dissolved oxygen to the anaerobic reactor ($S_{O2,s}$) but that the influent flow rate does not contain any dissolved oxygen ($S_{O2,i}$).

c. A TSS concentration in the effluent of the secondary settling tank of 10 mg TSS/l.

d. A TSS concentration in the sludge recycle line of 10,000 mg TSS/l.

Based on the above information, determine:

a. The s-recycle ratio required to keep an average MLSS concentration of 3,500 mgTSS/l in the system (s = Q_r / Q_i; where Q_r is the sludge recycle flow rate and Q_i is the influent flow rate).

b. Total COD fluxes consumed by PAO and OHO.

c. P-removal contribution of each biomass present in the EBPR system.

d. Potential and actual P removal by the system.

e. P_e concentration.

f. MX_{VSS} and MX_{TSS} masses.

g. Total volume of the system (V_R) and the volumes of the anaerobic ($V_{R,AN}$) and aerobic stages ($V_{R,OX}$).

h. O_2 and nitrogen requirements of the system (FO_c and FN_s, respectively).

i. The COD mass balance, in order to confirm the reliability of your design.

Solution

Table 6.1 presents a summary of the main design data and Table 6.2 displays the main stoichiometric and kinetic parameters necessary to carry out the EBPR process design.

Table 6.1 Summary of data to carry out the design of the Phoredox (A/O) EBPR plant.

Description	Symbol	Value	Unit
Flow rate	Q_i	15	MLD
Total COD	COD_i	585	$gCOD/m^3$
COD concentrations			
- readily biodegradable COD	$S_{S,i}$	165	$gCOD/m^3$
- volatile fatty acids	$S_{VFA,i}$	96	$gCOD/m^3$
- fermentable COD	$S_{F,i}$	69	$gCOD/m^3$
- slowly biodegradable COD	$X_{S,i}$	199	$gCOD/m^3$
- unbiodegradable soluble COD	$S_{U,i}$	45	$gCOD/m^3$
- unbiodegradable particulate COD	$X_{U,i}$	176	$gCOD/m^3$
Influent total P concentration	P_i	14.5	gP/m^3
Influent inorganic (fixed) suspended solids	$X_{FSS,i}$	49	$gFSS/m^3$
Temperature	T	14	°C
Sludge retention time	SRT	8	d
Anaerobic mass fraction	f_{xa}	0.15	-

Number of anaerobic zones	N	3	reactors
Dissolved O_2 in the influent	$S_{O2,i}$	0	gO_2/m^3
Dissolved O_2 in the sludge recycle	$S_{O2,s}$	1	gO_2/m^3
Influent nitrate concentration	$S_{NO3,i}$	0	$gNO_3\text{-}N/m^3$
Nitrate concentration in the sludge recycle	$S_{NO3,s}$	0	$gNO_3\text{-}N/m^3$
Total suspended solids in the effluent	$X_{TSS,e}$	10	$gTSS/m^3$
Design aerobic TSS concentration	$X_{TSS,OX}$	3,500	$gTSS/m^3$

Table 6.2 Kinetic and stoichiometric parameters for the EBPR design example.

	Parameter	Symbol	Value	Unit
	OHO			
	First-order fermentation rate constant at temperature 20 °C	$k_{F,20}$	0.06	$m^3/gVSS.d$
	Temperature coefficient for $k_{F,T}$	$\theta_{k,F}$	1.029	-
	First-order fermentation rate constant at temperature T	$k_{F,T}$	0.051	$m^3/gVSS.d$
Kinetic	Specific endogenous mass loss rate of the OHOs at 20 °C	$b_{OHO,20}$	0.24	$gVSS/gVSS.d$
	Temperature coefficient for $b_{OHO,T}$	$\theta_{b,OHO}$	1.029	-
	Specific endogenous mass loss rate of the OHOs at temperature T	$b_{OHO,T}$	0.202	$gVSS/gVSS.d$
	PAO			
	PAO-specific endogenous mass loss rate constant at 20 °C	$b_{PAO,20}$	0.04	$gVSS/gVSS.d$
	Temperature coefficient for $b_{PAO,T}$	$\theta_{b,PAO}$	1.029	-
	PAO-specific endogenous mass loss rate constant at temperature T	$b_{PAO,T}$	0.034	$gVSS/gVSS.d$
	OHO			
	Biomass yield of OHOs	Y_{OHOv}	0.45	$gVSS/gCOD$
	Fraction of endogenous residue of the OHOs	$f_{XE,OHO}$	0.20	$gVSS/gVSS$
	Fraction of P in the active OHO mass	f_P	0.03	$gP/gVSS$
	Fraction of P in the endogenous mass (OHO and PAO)	f_P	0.03	$gP/gVSS$
	Fraction of fixed (inorganic) suspended solids of OHOs	$f_{FSS,OHO}$	0.15	$gFSS/gVSS$
	PAO			
Stoichiometric	Biomass yield of PAOs	Y_{PAOv}	0.45	$gVSS/gCOD$
	Fraction of endogenous residue of the PAOs	$f_{XE,PAO}$	0.25	$gVSS/gVSS$
	Fraction of P in the active PAO mass	$f_{P,PAO}$	0.38	$gP/gVSS$
	Fraction of P in the endogenous mass (OHO and PAO)	f_P	0.03	$gP/gVSS$
	VSS/TSS ratio for PAO active mass	$f_{VT,PAO}$	0.44	$gVSS/gTSS$
	Ratio of P release/VFA uptake	$f_{PO4,rel}$	0.50	$gP/gCOD$
	Fraction of fixed (inorganic) suspended solids of PAOs	$f_{FSS,PAO}$	1.30	$gFSS/gVSS$
	Inerts or unbiodegradable mass			
	Fraction of P in the unbiodegradable mass	f_P	0.03	$gP/gVSS$
	Fraction of P in the influent fixed/inorganic mass	$f_{P,FSS,i}$	0.02	$gP/gFSS$
	General			
	COD/VSS ratio of the sludge	f_{CV}	1.48	$gCOD/gVSS$
	Nitrogen content of active biomass	f_n	0.10	$gN/gVSS$

Using the data provided, Table 6.3 shows the detailed calculations to carry out the Phoredox (A/O) EBPR process design to comply with P_e of less than 1 mg/l.

Table 6.3 EBPR system design procedure.

1. System configuration			
Phoredox (A/O) process configuration operated at 14 °C			
2. Influent and sludge recycle composition (from previous tables)			
Q_i	15	MLD	(15,000 m³/d) influent flow rate
2.1 Influent concentrations			
Influent and bioreactor data			
COD_i	585	gCOD/m³	influent concentration of total COD
$S_{S,i}$	165	gCOD/m³	influent concentration of readily biodegradable COD
$S_{VFA,i}$	96	gCOD/m³	influent concentration of VFAs
$S_{F,i}$	69	gCOD/m³	influent concentration of fermentable COD
$X_{S,i}$	199	gCOD/m³	influent concentration of slowly biodegradable COD
$COD_{b,i}$	364	gCOD/m³	influent concentration of biodegradable COD ($S_{S,i}$ + $X_{S,i}$)
$S_{U,i}$	45	gCOD/m³	influent concentration of soluble unbiodegradable COD
$X_{U,i}$	176	gCOD/m³	influent concentration of particulate unbiodegradable COD
$S_{NO3,i}$	0	gNO₃-N/m³	influent concentration of nitrate
$S_{O2,i}$	0	gO₂/m³	influent concentration of dissolved oxygen
$S_{NO3,s}$	0	gNO₃-N/m³	sludge recycle concentration of nitrate
$S_{O2,s}$	1	gO₂/m³	sludge recycle concentration of dissolved oxygen
$X_{FSS,i}$	49	gFSS/m³	influent concentration of fixed (inorganic) suspended solids
P_i	14.5	gP/m³	influent concentration of total P
2.2 Influent fluxes used for calculations (= $Q_i \cdot$ influent concentration of component)			
$FCOD_i$	8,775	kgCOD/d	influent daily flux of total COD
$FS_{S,i}$	2,475	kgCOD/d	influent daily flux of RBCOD
$FS_{VFA,i}$	1,440	kgCOD/d	influent daily flux of VFAs
$FS_{F,i}$	1,035	kgCOD/d	influent daily flux of fermentable COD
$FCOD_{b,i}$	5,460	kgCOD/d	influent daily flux of biodegradable COD ($S_{S,i}$ + $X_{S,i}$)
$FX_{U,i}$	2,640	kgCOD/d	influent daily flux of particulate unbiodegradable COD
$FS_{U,i}$	675	kgCOD/d	influent daily flux of soluble unbiodegradable COD
$FX_{FSS,i}$	735	kgFSS/d	influent daily flux of fixed (inorganic) suspended solids

2.3 Sludge recycle characteristics

s		$m^3.d/m^3.d$	$s = X_{TSS,OX} / (X_{TSS,s} - X_{TSS,OX})$
		$m^3.d/m^3.d$	$s = 3,500 / (10,000 - 3,500)$
s	0.54	$m^3.d/m^3.d$	sludge recycle ratio (s) with regard to influent flow rate
$S_{O2,s}$	1.0	gO_2/m^3	dissolved O_2 in the sludge recycle
$S_{NO3,s}$	0.0	gNO_3^--N/m^3	nitrate concentration in the sludge recycle

3. Division of $S_{S,i}$ between PAOs and OHOs

3.1 Fermentable COD available for conversion into VFAs after denitrification reactor (and O_2 consumption) in the anaerobic reactor (in units of $gCOD/m^3$ of influent)

$$S_{F,i,conv} = S_{F,i} - 8.6 \cdot (s \cdot S_{NO3,s} + S_{NO3,i}) - 3 \cdot (s \cdot S_{O2,s} + S_{O2,i}) \tag{6.8}$$

$$= S_{F,i} - \text{COD for denitrification} - \text{COD for D.O.}$$

$$= 69 - 8.6 \cdot (0.54 \cdot 0 + 0) - 3 \cdot (0.54 \cdot 1 + 0)$$

COD for DN	0.0	$gCOD/m^3$
COD for D.O.	1.6	$gCOD/m^3$
$S_{F,i,conv}$	67	$gCOD/m^3$

3.2 Fermentable COD lost in the effluent of the last anaerobic reactor

N	3		the 3rd compartment of the anaerobic reactor

calculation done by iterations

    a- presume a seed1 $S_{F,ANn}$ value of 0. This value is used to calculate MX_{OHOv}

    b- type the calculated MX_{OHOv} calculated value as seed2 value

    c- repeat steps a and b until the seed2 $S_{F,ANn}$ equals the calculated $S_{F,ANn}$

$$S_{F,ANn} = S_{F,i,conv} / (1 + s) / [1 + (k_{F,T} \cdot (f_{xa} \cdot MX_{OHOv} / (N \cdot Q_i \cdot (1 + s))))]^n \tag{6.9}$$

$$= 67 / (1 + 0.54) / [1 + (0.051 \cdot (0.15 \cdot 4,139 / (3 \cdot 15 \cdot (1 + 0.54))))]^3$$

seed1:

$S_{F,ANn}$	0.0	10.6	$gCOD/m^3$
	↓	↑	

seed1:

MX_{OHOv}	4,139	4,139	kgVSS	(6.14b)

$$= Y_{OHOv} \cdot FCOD_{b,OHO} \cdot SRT / (1 + b_{OHO,T} \cdot SRT) \text{ (note: } FCOD_{b,OHO} \text{ is calculated in steps 3.3 and 3.4)}$$

$$= (0.45 \cdot 3,009 \cdot 8) / (1 + 0.202 \cdot 8)$$

$S_{F,ANn}$	10.6	9.7	$gCOD/m^3$
	↓	↑	

seed2:

MX_{OHOv}	4,475	4,475	kgVSS	(6.14b)

$$= Y_{OHOv} \cdot FCOD_{b,OHO} \cdot SRT / (1 + b_{OHO,T} \cdot SRT) \text{ (note: } FCOD_{b,OHO} \text{ is calculated in steps 3.3 and 3.4)}$$

$$= (0.45 \cdot 3,254 \cdot 8) / (1 + 0.202 \cdot 8)$$

$S_{F,ANn}$ 9.7   9.8 $gCOD/m^3$

↓   ↑

seed3:

MX_{OHOv} 4,447   4,447   kgVSS   (6.14b)

$= Y_{OHOv} \cdot FCOD_{b,OHO} \cdot SRT / (1 + b_{OHO,T} \cdot SRT)$ (note: $FCOD_{b,OHO}$ is calculated in steps 3.3 and 3.4)

$= (0.45 \cdot 3,233 \cdot 8) / (1 + 0.202 \cdot 8)$

$S_{F,ANn}$ 9.8   9.8   gCOD/m3

↓   ↑

seed4:

MX_{OHOv} 4,450   4,450   kgVSS   (6.14b)

$= Y_{OHOv} \cdot FCOD_{b,OHO} \cdot SRT / (1 + b_{OHO,T} \cdot SRT)$ (note: $FCOD_{b,OHO}$ is calculated in steps 3.3 and 3.4)

$= (0.45 \cdot 3,235 \cdot 8) / (1 + 0.202 \cdot 8)$

3.3 VFAs stored by PAOs

$FS_{S,PAO}$ $= Q_i \cdot (S_{F,i,conv} - (1 + s) \cdot S_{F,ANn}) + Q_i \cdot S_{VFA,i}$ (6.12)

$= 15 \cdot (67 - (1 + 0.54) \cdot 9.8) + (15 \cdot 96)$

$FS_{S,PAO}$ 2,225   kgCOD/d

3.4 Remaining biodegradable COD available to OHOs

$FCOD_{b,OHO}$ $= FCOD_{b,i} - FS_{S,PAO}$ (6.13)

$= 5,460 - 2,225$

$FCOD_{b,OHO}$ 3,235   kgCOD/d

4. Biomass (VSS) equations

Corresponds to the biological mass present in the system as synthesized from the influent COD (in g/d), taking into account the cumulative effect of SRT [(g/d) · d = g in the system]

4.1 PAOs

Active mass

Y_{PAOv} 0.45   gVSS/gCOD

$Y_{PAO,obs}$ $= Y_{PAOv} / (1 + b_{PAO,T} \cdot SRT)$

$= 0.45 / (1 + 0.034 \cdot 8)$

$Y_{PAO,obs}$ 0.354   gVSS / gCOD

MX_{PAOv} $= Y_{PAO,obs} \cdot FS_{S,PAO} \cdot SRT$ (6.2)

$= 0.354 \cdot 2,225 \cdot 8$

MX_{PAOv} 6,308   kgVSS in the system

Endogenous mass

$MX_{E,PAOv}$ $= f_{XE,PAO} \cdot b_{PAO,T} \cdot MX_{PAOv} \cdot SRT$ (6.3)

$= 0.25 \cdot 0.034 \cdot 6,308 \cdot 8$

$MX_{E,PAOv}$ 425   kgVSS

4.2 OHOs

Active mass

Y_{OHOv}	0.45	gVSS/gCOD

$Y_{OHO,obs} = Y_{OHOv} / (1 + b_{OHO,T} \cdot SRT)$

$\quad\quad = 0.45 / (1 + 0.202 \cdot 8)$

$Y_{OHO,obs}$	0.172	gVSS/gCOD

$MX_{OHOv} = Y_{OHO,obs} \cdot F_{CODb,OHO} \cdot SRT$

$\quad\quad = 0.172 \cdot 3{,}235 \cdot 8$

MX_{OHOv} 4,450 kgVSS (this value is the calculated MX_{OHOv} value in step 3.2)

Endogenous mass

$MX_{E,OHOv} = f_{XE,OHO} \cdot b_{OHO,T} \cdot MX_{OHOv} \cdot SRT$ $\hfill$ (6.5)

$\quad\quad = 0.20 \cdot 0.202 \cdot 4{,}450 \cdot 8$

$MX_{E,OHOv}$ 1,439 kgVSS

4.3 Unbiodegradable particulate organics

$MX_{Uv} = Q_i \cdot X_{U,i} \cdot SRT / f_{cv}$ $\hfill$ (6.6)

$\quad\quad = 15 \cdot 176 \cdot 8 / 1.48$

MX_{Uv} 14,270 kgVSS

5. VSS and TSS

5.1 VSS and active fraction

$MX_B = MX_{PAOv} + MX_{OHOv}$

$\quad\quad = 6{,}308 + 4{,}450$

MX_B 10,758 kgVSS

$MX_{VSS} = MX_{PAOv} + MX_{OHOv} + MX_{E,PAOv} + MX_{E,OHOv} + MX_{Uv}$ $\hfill$ (6.24b)

$\quad\quad = 6{,}308 + 4{,}450 + 425 + 1{,}439 + 14{,}270$

MX_{VSS} 26,893 kgVSS

$f_{av} = MX_B / MX_{VSS}$

$\quad\quad = 10{,}758 / 26{,}893$

f_{av} 40 %

5.2 FSS

$f_{P,PAO,actual} = [(Q_i \cdot SRT \cdot \Delta P_{SYS,actual}) - (f_P \cdot (MX_{VSS} - MX_{PAOv})] / MX_{PAOv}$ $\hfill$ (6.24a)

$\quad\quad = [(15 \cdot 8 \cdot 14.5) - (0.03 \cdot (26{,}893 - 6{,}308)] / 6{,}308$

$f_{P,PAO,actual}$ 0.17 gP/gVSS

$MX_{FSS} = f_{FSS,OHO} \cdot MX_{OHOv} + f_{FSS,PAO} \cdot (f_{P,PAO,actual} / f_{P,PAO}) \cdot MX_{PAOv} + FX_{FSS,i} \cdot SRT$ $\hfill$ (6.24d)

$\quad\quad = (0.15 \cdot 4{,}450) + (1.3 \cdot (0.17 / 0.38) \cdot 6{,}308) + (735 \cdot 8)$

MX_{FSS} 10,196 kgFSS

5.3 TSS

$$MX_{TSS} = MX_{VSS} + MX_{FSS} \qquad (6.25a)$$
$$= 26,893 + 10,196$$

MX_{TSS} 37,089   kgTSS

5.4 f_{VT}

$$f_{VT} = MX_{VSS} / MX_{TSS} \qquad (6.25c)$$
$$= 26,893 / 37,089$$

f_{VT} 0.73   gVSS/gTSS

$$f_{VT,PAO} = MX_{PAOv} / MX_{TSS}$$
$$= 6,308 / 37,089$$

$f_{VT,PAO}$ 0.17   gVSS/gTSS

5.5 P content of TSS

$$f_{P,TSS} = [(f_P \cdot MX_{OHOv}) + f_P \cdot (MX_{E,OHOv} + MX_{E,PAOv}) + (f_P \cdot MX_{Uv})$$
$$+ (f_{P,PAO,actual} \cdot MX_{PAOv}) + (f_{P,FSS,i} \cdot MX_{FSS})] / MX_{TSS} \qquad (6.26)$$
$$= [(0.03 \cdot 4,450) + 0.03 \cdot (1,439 + 425) + (0.03 \cdot 14,270) + (0.17 \cdot 6,308) + (0.02 \cdot 10,196)] / 37,089$$

$f_{P,TSS}$ 0.051   gP/gTSS

6. P removal

6.0 P release

$$S_{PO4,rel} = f_{PO4,rel} \cdot FS_{S,PAO} / Q_i \qquad (6.15)$$
$$= 0.5 \cdot 2,225 / 15$$

S_{PO4_rel} 74.2   gP/m^3 gP/m^3 of influent, not gP/m^3 of AN reactor

6.1 ΔP by PAOs

$$\Delta P_{PAO} = f_{P,PAO} \cdot MX_{PAOv} / (SRT \cdot Q_i) \qquad (6.16)$$
$$= 0.38 \cdot 6,308 / (8 \cdot 15)$$

ΔP_{PAO} 20.0   gP/m^3

6.2 ΔP by OHOs

$$\Delta P_{OHO} = f_P \cdot MX_{OHOv} / (SRT \cdot Q_i) \qquad (6.17)$$
$$= 0.03 \cdot 4,450 / (8 \cdot 15)$$

ΔP_{OHO} 1.11   gP/m^3

6.3 ΔP by endogenous mass

$$\Delta P_{XE} = \Delta P_{XE,PAO} + \Delta P_{XE,OHO} \qquad (6.18)$$

$$\Delta P_{XE,PAO} = f_P \cdot MX_{E,PAOv} / (SRT \cdot Q_i)$$
$$= 0.03 \cdot 425 / (8 \cdot 15)$$

$\Delta P_{XE,PAO}$ 0.11   gP/m^3

$$\Delta P_{XE,OHO} \quad = f_P \cdot MX_{E,OHOv} / (SRT \cdot Q_i)$$

$$= 0.03 \cdot 1{,}349 / (8 \cdot 15)$$

$$\Delta P_{XE,OHO} \quad 0.36 \quad gP/m^3$$

$$\Delta P_{XE} \quad 0.47 \quad gP/m^3$$

6.4 ΔP by influent unbiodegradable organic mass

$$\Delta P_{XU} \quad = f_P \cdot MX_{Uv} / (SRT \cdot Q_i) \tag{6.19}$$

$$= 0.03 \cdot 14{,}270 / (8 \cdot 15)$$

$$\Delta P_{XU} \quad 3.57 \quad gP/m^3$$

6.5 Potential total P removal

$$\Delta P_{SYS,pot} \quad = \Delta P_{PAO} + \Delta P_{OHO} + \Delta P_{XE} + \Delta P_{XU} \tag{6.20}$$

$$= 20.0 + 1.11 + 0.47 + 3.57$$

$$\Delta P_{SYS,pot} \quad 25.1 \quad gP/m^3$$

6.6 Actual total P removal

$$P_i \quad 14.5 \quad gP/m^3$$

$$\Delta P_{SYS,actual} \quad = \min (\Delta P_{SYS,pot}; P_i) \tag{6.21}$$

$$= \min (25.1;\ 14.5)$$

$$\Delta P_{SYS,actual} \quad 14.5 \quad gP/m^3$$

6.7 Particulate P in the effluent

To calculate after step 6.5 where the P content of TSS is calculated

$$X_{P,e} \quad = f_{P,TSS} \cdot X_{TSS,e} \tag{6.22}$$

$$= 0.051 \cdot 10$$

$$X_{P,e} \quad 0.51 \quad gP/m^3$$

6.8 Effluent total P

$$P_e \quad = P_i - \Delta P_{SYS,actual} + X_{P,e} \tag{6.23}$$

$$= 14.5 - 14.5 + 0.51$$

$$P_e \quad 0.51 \quad gP/m^3$$

With the proposed design, the EBPR plant is able to comply with the effluent discharge standard of less than 1 mgP/l (= 1 gP/m³). Actually, the system has the potential to remove up to 25.1 mgP/l, and thus all the phosphorus present in the influent can be removed (14.5 mgP/l). However, the loss of solids through the effluent of the secondary settling tank (10 mgTSS/l that contain approximately 0.051 mgP/mgTSS) leads to the effluent total P concentration of 0.51 gP/m³ (= mgP/l). Arguably, it cannot be expected to observe a full removal of total phosphorus because the phosphorus intracellularly stored in the solids is lost through the effluent, but effluent discharge standards of less than 1 mgP/l can be satisfactorily met.

7. Process volume (based on TSS)

X_{TSS}	3,500	$gTSS / m^3$
V_R	$= MX_{TSS} / X_{TSS,OX}$	(6.27a)
	$= 37,089/3,500$	
V_R	10,597	m^3

The volume of the anaerobic zone (divided in three sections) depends on the anaerobic mass fraction.

$V_{R,AN}$	$= f_{xa} \cdot V_R$	
	$= 0.20 \cdot 10,597$	
$V_{R,AN}$	2,119	m^3

8. Nitrogen requirements

FN_s	$= f_n \cdot MX_{VSS} / SRT$		(6.28a)
	$= (0.10 \cdot 26,893) / 8$		
FN_s	336	kgN/d	(6.28b)
$TKN_{i,s}$	$= FN_s / Q_i$		
	$= 336 / 15$		
$TKN_{i,s}$	22.4	gN/m^3	

9. Oxygen demand (OD)

OD by PAOs: for synthesis and endogenous respiration

FO_{PAO}	$= FO_{PAO,s} + FO_{PAO,e}$	(6.29a)
$FO_{PAO,s}$	$= FS_{S,PAO} \cdot (1 - f_{CV} \cdot Y_{PAOv})$	
	$= 2,225 \cdot (1 - 1.48 \cdot 0.45)$	
$FO_{PAO,s}$	743	kgO_2/d
$FO_{PAO,e}$	$= FS_{S,PAO} \cdot f_{CV} \cdot (1 - f_{XE,PAO}) \cdot b_{PAO,T} \cdot Y_{PAO,obs} \cdot SRT$	
	$= 2,225 \cdot 1.48 \cdot (1 - 0.25) \cdot 0.034 \cdot 0.354 \cdot 8$	
$FO_{PAO,e}$	236	kgO_2/d
FO_{PAO}	979	kgO_2/d

OD by OHOs: for synthesis and endogenous respiration

FO_{OHO}	$= FO_{OHO,s} + FO_{OHO,e}$	(6.30a)
$FO_{OHO,s}$	$= FCOD_{b,OHO} \cdot (1 - f_{CV} \cdot Y_{OHOv})$	
	$= 3,235 \cdot (1 - 1.48 \cdot 0.45)$	
$FO_{OHO,s}$	1,081	
$FO_{OHO,e}$	$= FCOD_{b,OHO} \cdot f_{CV} \cdot (1 - f_{XE,OHO}) \cdot b_{OHO,T} \cdot Y_{OHO,obs} \cdot SRT$	
	$= 3,235 \cdot 1.48 \cdot (1 - 0.20) \cdot 0.202 \cdot 0.172 \cdot 8$	
$FO_{OHO,e}$	1,065	
FO_{OHO}	2,146	kgO_2/d

OD total (carbonaceous)

$$FO_c = FO_{PAO} + FO_{OHO} \qquad (6.31a)$$
$$= 979 + 2,146$$

| FO_c | 3,125 | kgO_2/d |

COD mass balance verification

Input

| $FCOD_i$ | | 8,775 | kgCOD/d | | 100 % | | IN |

Output

O_2 demand for synthesis and endogenous respiration

| FOc | | 3,125 | kgCOD/d | | 35.6 % |

Soluble unbiodegradable organics leaving via the effluent

$FS_{U,i}$		675	kgCOD/d		7.7 %
Sludge	gVSS	gCOD/d	(= gVSS · f_{CV} / SRT = gVSS · 1.48 / 8 = gVSS · 0.185)		
MX_{PAOv}	6,308	1,167	kgCOD/d		13.3 %
MX_{OHOv}	4,450	823	kgCOD/d		9.4 %
MX_B	10,758	1,990			22.7 %
$MX_{E,PAOv}$	425	79	kgCOD/d		0.9 %
$MX_{E,OHOv}$	1,439	266	kgCOD/d		3.0 %
MX_{Uv}	14,270	2,640	kgCOD/d		30.1 %
$MX_{Ev} + MX_{Uv}$	16,135	2,985			34.0 %
MX_{VSS}	26,893	4,975	kgCOD/d		56.7 %
	Sum:	8,775	kgCOD/d		100 %		OUT
	Delta (OUT-IN): 0		kgCOD/d		0 %

The 100 % mass balance for COD indicates that all the influent COD is accounted for in the calculated values of oxygen demand and sludge production. From the COD mass balance, and for the conditions of the design example, the fate of the influent COD is as follows: 35.6 % is oxidized with oxygen, 7.7 % escapes in the effluent as soluble unbiodegradable organics and 56.7 % becomes activated sludge. The sludge is composed of 40.0 % (10,758 / 26,893) active biomass and 60.0 % (16,135 / 26,893) inactive particulate matter of which 88.4 % (14,270 / 16,135) is influent unbiodegradable particulate organics and 11.6 % ((425 + 1,439) / 16,135) endogenous residue). A summary of the EBPR system design results is presented in Table 6.4.

Table 6.4 Summary of EBPR system design results.

Description	Parameter	Unit	Value
1. Influent and bioreactor			
Type of wastewater	raw		raw
Temperature	T	°C	14
Influent flow rate	Q_i	MLD	15
Influent total COD	COD_i	gCOD/m³	585
Influent soluble biodegradable COD	$S_{S,i}$	gCOD/m³	165
Influent biodegradable COD	$COD_{b,i}$	gCOD/m³	364
Influent total P	P_i	gP/m³	14.5
Sludge retention time	SRT	d	8
Sludge recycle ratio (s)	s	m³.d/m³.d	0.54
Oxygen concentration in sludge recycle	$S_{O2,s}$	gO₂/m³	1
2. Portion of $S_{S,i}$ for PAOs and of $COD_{b,i}$ for OHOs			
Concentration of fermentable COD in the last AN reactor	$S_{F,ANn}$	gCOD/m³	9.8
Flux of $S_{S,i}$ for PAOs	$FS_{S,PAO}$	kgCOD/d	2,225
Flux of $COD_{b,i}$ for OHOs	$FCOD_{b,OHO}$	kgCOD/d	3,235
3. System biomass (VSS) equations			
Mass of PAOs	MX_{PAOv}	kgVSS	6,308
Mass of endogenous residue from PAOs	$MX_{E,PAOv}$	kgVSS	425
Mass of OHOs	MX_{OHOv}	kgVSS	4,450
Mass of endogenous residue from OHOs	$MX_{E,OHOv}$	kgVSS	1,439
Mass of unbiodegradable organics from influent	MX_{Uv}	kgVSS	14,270
4. P removal			
PO₄ release	S_{PO4_rel}	gP/m³	74.2
Maximum P removal by PAOs	ΔP_{PAO}	gP/m³	20.0
Actual P removal by PAOs	$\Delta P_{PAO,actual}$	gP/m³	9.3
P removal by OHOs	ΔP_{OHO}	gP/m³	1.1
P removal by endogenous residue	ΔP_{XE}	gP/m³	0.5
P removal by X_U	ΔP_{XU}	gP/m³	3.6
Potential P removal by system	$\Delta P_{SYS.pot}$	gP/m³	25.1
Actual P removal by system	$\Delta P_{SYS,actual}$	gP/m³	14.5
Effluent particulate P (from $X_{TSS,e}$)	$X_{P,e}$	gP/m³	0.5
Influent total P	P_i	gP/m³	14.5
Effluent total P	P_e	gP/m³	0.5
5. Volatile and total suspended solids (VSS and TSS) in system			
Mass of active biomass	MX_B	kgVSS	10,758
Mass of VSS	MX_{VSS}	kgVSS	26,893
Ratio of AVSS/VSS	f_{av}	gAVSS/gVSS	0.40

Mass of fixed SS	MX_{FSS}	kgFSS	10,196
Mass of TSS	MX_{TSS}	kgTSS	37,089
Ratio of VSS/TSS	f_{VT}	gVSS/gTSS	0.73
Fraction of P in TSS	$f_{P,TSS}$	gP/gTSS	0.05
6. Reactor volume			
Anaerobic reactor volume	$V_{R,AN}$	m^3	2,119
Aerobic reactor volume	$V_{R,OX}$	m^3	8,478
Bioreactor volume	V_R	m^3	10,597
7. N requirement			
N requirement flux for synthesis	FN_s	kgN/d	336
N requirement concentration for synthesis	$TKN_{i,s}$	gN/m^3	22.4
8. Oxygen demand			
Flux of O_2 demand by PAOs	FO_{PAO}	kgO_2/d	979
Flux of O2 demand by OHOs	FO_{OHO}	kgO_2/d	2,146
Flux of carbonaceous O_2 demand	FO_c	kgO_2/d	3,125
COD output/COD input	COD mass balance	gCOD/gCOD	100.0 %

Example 6.3.2

Determination of additional VFA supply to a Phoredox (A/O) EBPR system

A Phoredox (AO) system (Figure 6.3) will be designed to treat an influent wastewater (Q_i) that has a flow rate of 20 MLD (20,000 m^3/d). As originally planned, the EBPR plant will be designed with an SRT of 8 days and an anaerobic mass fraction (f_{xa}) of 0.20 divided into 4 different compartments. The plant will operate at an average MLSS concentration of 4,000 mg/l and the average yearly wastewater temperature is 16 °C. The wastewater influent contains a total phosphorus concentration (P_i) of 12.5 mgP/l and the EBPR plant needs to comply with an effluent total phosphorus concentration (P_e) of 1 mg P/l. The plant does not have a primary settling tank. Thus, the raw wastewater contains a total influent COD concentration (COD_i) of 415 mg/l and 35 mg/l of inorganic or fixed suspended solids. However, as observed in the COD fractionation displayed in Figure 6.4, the influent does not have any VFA.

First, carry out the design to assess what the potential phosphorus removal of the system will be. If the plant does not meet the P_e limit of 1 mg P/l, determine what the additional VFA load should be that needs to be supplied to meet the required standard.

Assume:

a. That the influent flow rate does not contain any dissolved oxygen ($S_{O2,i}$) but that the sludge recycle flow (Q_r) can convey up to 1 mg/l of dissolved oxygen to the anaerobic reactor ($S_{O2,s}$).
b. That the influent flow rate does not contain any nitrate ($S_{NO3,i}$) but the sludge recycle flow (Q_r) contains up to 3 mg/l of nitrate that is discharged into the anaerobic reactor ($S_{NO3,s}$).
c. A TSS concentration in the effluent of the secondary settling tank of 15 mg TSS/l.
d. A TSS concentration in the sludge recycle line of 10,000 mg TSS/l.

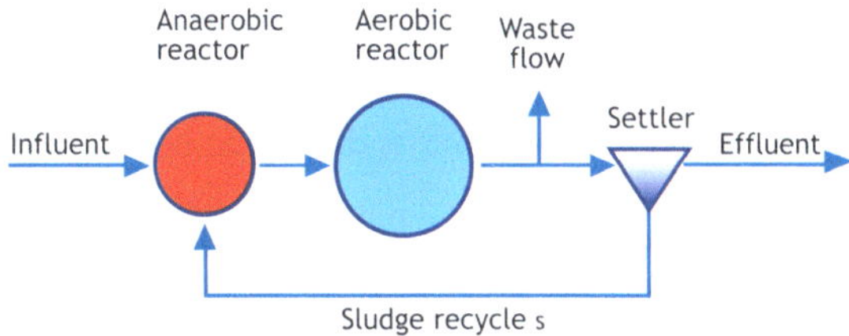

Figure 6.3 A Phoredox (A/O) configuration. The anaerobic reactor is divided into four compartments (not illustrated).

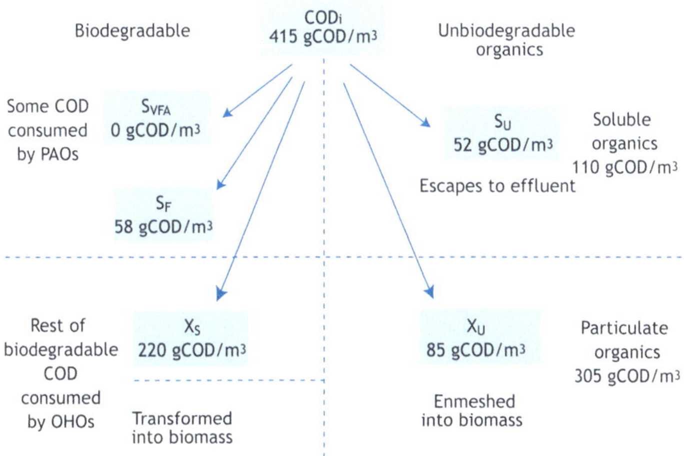

Figure 6.4 Influent COD fractionation for the EBPR design example.

To support your answer and based on the previous information, determine:

a. Estimate the s-recycle ratio (s-ratio between the sludge recycle flow rate, Q_r, to influent flow rate, Q_i : s = Q_r / Q_i) to maintain an average MLSS concentration of 4,000 mgTSS/l in the system.
b. Total COD fluxes consumed by PAO and OHO.
c. P-removal contribution of each biomass present in the EBPR system.
d. Potential and actual P removal by the system.
e. P_e concentration.
f. If the P_e concentration is higher than 1 mgP/l, estimate the additional VFA needs of the plant.
g. Once the additional VFA requirements are determined, estimate:

   - The MX_{VSS} and MX_{TSS} masses.
   - Total volume of the system (V_R) and the volumes of the anaerobic ($V_{R,AN}$) and aerobic stages ($V_{R,OX}$).
   - O_2 and nitrogen requirements of the system (FO_c and FN_s, respectively).
   - A COD mass balance to confirm the reliability of your design.

Solution

Table 6.5 presents a summary of the main design data and Table 6.6 displays the main stoichiometric and kinetic parameters necessary to carry out the EBPR process design.

Table 6.5 Summary of data to carry out the design of the Phoredox (A/O) EBPR plant.

Description	Symbol	Value	Unit
Flow rate	Q_i	20	MLD
Total COD	COD_i	415	$gCOD/m^3$
COD concentrations			
- readily biodegradable COD	$S_{S,i}$	58	$gCOD/m^3$
- volatile fatty acids	$S_{VFA,i}$	0	$gCOD/m^3$
- fermentable COD	$S_{F,i}$	58	$gCOD/m^3$
- slowly biodegradable COD	$X_{S,i}$	220	$gCOD/m^3$
- unbiodegradable soluble COD	$S_{U,i}$	52	$gCOD/m^3$
- unbiodegradable particulate COD	$X_{U,i}$	85	$gCOD/m^3$
Influent total P concentration	P_i	12.5	gP/m^3
Influent inorganic (fixed) suspended solids	$X_{FSS,i}$	35	$gFSS/m^3$
Temperature	T	16	$°C$
Sludge retention time	SRT	8	d
Anaerobic mass fraction	f_{xa}	0.20	-
Number of anaerobic zones	N	4	reactors
Dissolved O_2 in the influent	$S_{O2,i}$	0	gO_2/m^3
Dissolved O_2 in the sludge recycle	$S_{O2,s}$	1	gO_2/m^3
Influent nitrate concentration	$S_{NO3,i}$	0	$gNO_3\text{-}N/m^3$
Nitrate concentration in the sludge recycle	$S_{NO3,s}$	3	$gNO_3\text{-}N/m^3$
Total suspended solids in the effluent	$X_{TSS,e}$	15	$gTSS/m^3$
Design aerobic TSS concentration	$X_{TSS,OX}$	4,000	$gTSS/m^3$

Table 6.6 Kinetic and stoichiometric parameters for the EBPR design example.

	Parameter	Symbol	Value	Unit
Kinetic	*OHO*			
	First-order fermentation rate constant at temperature 20 °C	$k_{F,20}$	0.06	m³/gVSS.d
	Temperature coefficient for $k_{F,T}$	$\theta_{k,F}$	1.029	
	First-order fermentation rate constant at temperature T	$k_{F,T}$	0.054	m³/gVSS.d
	Specific endogenous mass loss rate of the OHOs at 20 °C	$b_{OHO,20}$	0.24	gVSS/gVSS.d
	Temperature coefficient for $b_{OHO,T}$	$\theta_{b,OHO}$	1.029	
	Specific endogenous mass loss rate of the OHOs at temperature T	$b_{OHO,T}$	0.214	gVSS/gVSS.d
	PAO			
	PAO-specific endogenous mass loss rate constant at temperature 20 °C	$b_{PAO,20}$	0.04	gVSS/gVSS.d
	Temperature coefficient for $b_{PAO,T}$	$\theta_{b,PAO}$	1.029	
	PAO-specific endogenous mass loss rate constant at temperature T	$b_{PAO,T}$	0.036	gVSS/gVSS.d
Stoichiometric	*OHO*			
	Biomass yield of OHOs	Y_{OHOv}	0.45	gVSS/gCOD
	Fraction of endogenous residue of the OHOs	$f_{XE,OHO}$	0.20	gVSS/gVSS
	Fraction of P in the active OHO mass	f_P	0.03	gP/gVSS
	Fraction of P in the endogenous mass (OHO and PAO)	f_P	0.03	gP/gVSS
	Fraction of fixed (inorganic) suspended solids of OHOs	$f_{FSS,OHO}$	0.15	gFSS/gVSS
	PAO			
	Biomass yield of PAOs	Y_{PAOv}	0.45	gVSS/gCOD
	Fraction of endogenous residue of the PAOs	$f_{XE,PAO}$	0.25	gVSS/gVSS
	Fraction of P in the active PAO mass	$f_{P,PAO}$	0.38	gP/gVSS
	Fraction of P in the endogenous mass (OHO and PAO)	f_P	0.03	gP/gVSS
	VSS/TSS ratio for PAO active mass	$f_{VT,PAO}$	0.46	gVSS/gTSS
	Ratio of P release/VFA uptake	$f_{PO4,rel}$	0.50	gP/gCOD
	Fraction of fixed (inorganic) suspended solids of PAOs	$f_{FSS,PAO}$	1.30	gFSS/gVSS
	Inerts or unbiodegradable mass			
	Fraction of P in the unbiodegradable mass	f_P	0.03	gP/gVSS
	Fraction of P in the influent fixed/inorganic mass	$f_{P,FSS,i}$	0.02	gP/gFSS
	General			
	COD/VSS ratio of the sludge	f_{CV}	1.48	gCOD/gVSS
	Nitrogen content of active biomass	f_n	0.10	gN/gVSS

Using the data provided, Table 6.7 shows the detailed calculations to carry out the Phoredox (A/O) EBPR process design to determine its potential P removal without the presence of VFA in the influent.

Table 6.7 EBPR system design procedure.

1. System configuration			
Phoredox (A/O) process configuration operated at 16 °C			
2. Influent and sludge recycle composition (from previous tables)			
Q_i	20	MLD	(20,000 m³/d) influent flow rate
2.1 Influent concentrations			
Influent and bioreactor data			
COD_i	415	$gCOD/m^3$	influent concentration of total COD
$S_{S,i}$	58	$gCOD/m^3$	influent concentration of readily biodegradable COD
$S_{VFA,i}$	0	$gCOD/m^3$	influent concentration of VFAs
$S_{F,i}$	58	$gCOD/m^3$	influent concentration of fermentable COD
$X_{S,i}$	220	$gCOD/m^3$	influent concentration of slowly biodegradable COD
$COD_{b,i}$	278	$gCOD/m^3$	influent concentration of biodegradable COD ($S_{S,i}$ + $X_{S,i}$)
$S_{U,i}$	52	$gCOD/m^3$	influent concentration of soluble unbiodegradable COD
$X_{U,i}$	85	$gCOD/m^3$	influent concentration of particulate unbiodegradable COD
$S_{NO3,i}$	0	$gNO_3\text{-}N/m^3$	influent concentration of nitrate
$S_{O2,i}$	0	gO_2/m^3	influent concentration of dissolved oxygen
$S_{NO3,s}$	3	$gNO_3\text{-}N/m^3$	sludge recycle concentration of nitrate
$S_{O2,s}$	1	gO_2/m^3	sludge recycle concentration of dissolved oxygen
$X_{FSS,i}$	35	$gFSS/m^3$	influent concentration of fixed (inorganic) suspended solids
P_i	12.5	gP/m^3	influent concentration of total P
2.2 Influent fluxes used for calculations (= Q_i · influent concentration of component)			
$FCOD_i$	8,300	kgCOD/d	influent daily flux of total COD
$FS_{S,i}$	1,160	kgCOD/d	influent daily flux of rbCOD
$FS_{VFA,i}$	0	kgCOD/d	influent daily flux of VFAs
$FS_{F,i}$	1,160	kgCOD/d	influent daily flux of fermentable COD
$FCOD_{b,i}$	5,560	kgCOD/d	influent daily flux of biodegradable COD ($S_{S,i}$ + $X_{S,i}$)
$FX_{U,i}$	1,700	kgCOD/d	influent daily flux of particulate unbiodegradable COD
$FS_{U,i}$	1,040	kgCOD/d	influent daily flux of soluble unbiodegradable COD
$FX_{FSS,i}$	700	kgFSS/d	influent daily flux of fixed (inorganic) suspended solids
2.3 Sludge recycle characteristics			
s		$m^3.d/m^3.d$	s = $X_{TSS,OX}$ / ($X_{TSS,s}$ − $X_{TSS,OX}$) = 4,000 / (10,000 − 4,000)
s	0.67	$m^3.d/m^3.d$	sludge recycle ratio (s) with regard to influent flow rate
$S_{O2,s}$	1.0	gO_2/m^3	dissolved O_2 in the sludge recycle
$S_{NO3,s}$	3.0	$gNO_3\text{-}N/m^3$	nitrate concentration in the sludge recycle

3. Division of $S_{S,i}$ between PAOs and OHOs

3.1 Fermentable COD available for conversion into VFAs after denitrification reactor (and O_2 consumption) in the anaerobic reactor (in units of $gCOD/m^3$ of influent)

$$S_{F,i,conv} = S_{F,i} - 8.6 \cdot (s \cdot S_{NO3,s} + S_{NO3,i}) - 3 \cdot (s \cdot S_{O2,s} + S_{O2,i}) \qquad (6.8)$$

$$= S_{F,i} - \text{COD for denitrification} - \text{COD for D.O.}$$

$$= 58 - 8.6 \cdot (0.67 \cdot 3.0 + 0) - 3 \cdot (0.67 \cdot 1 + 0)$$

COD for DN	17.2	$gCOD/m^3$
COD for D.O.	2.0	$gCOD/m^3$
$S_{F,i,conv}$	39	$gCOD/m^3$

3.2 Fermentable COD lost in the effluent of the last anaerobic reactor

N   4   the 4th compartment of the anaerobic reactor

calculation done by iterations

   a- assume a seed1 $S_{F,ANn}$ value of 0. This value is used to calculate MX_{OHOv}

   b- type the calculated MX_{OHOv} calculated value as seed2 value

   c- repeat steps a and b until the seed2 $S_{F,ANn}$ equals the calculated $S_{F,ANn}$

$$S_{F,ANn} = S_{F,i,conv}/(1+s) / [1 + (k_{F,T} \cdot (f_{xa} \cdot MX_{OHOv} / (N \cdot Q_i \cdot (1 + s))))]^n \qquad (6.9)$$

$$= 39 / (1 + 0.67) / [1 + (0.054 \cdot (0.20 \cdot 6,349 / (4 \cdot 20 \cdot (1 + 0.67))))]^4$$

seed1:

$S_{F,ANn}$	0.0	4.5	$gCOD/m^3$
	↓	↑	

seed1:

MX_{OHOv}	6,349	6,349	kgVSS	(6.14b)

$$= Y_{OHOv} \cdot FCOD_{b,OHO} \cdot SRT / (1 + b_{OHO,T} \cdot SRT)$$
(note: $FCOD_{b,OHO}$ is calculated in steps 3.3. and 3.4)

$$= (0.45 \cdot 4,784 \cdot 8) / (1 + 0.214 \cdot 8)$$

$S_{F,ANn}$	4.5	4.3	$gCOD/m^3$
	↓	↑	

seed2:

MX_{OHOv}	6,548	6,548	kgVSS	(6.14b)

$$= Y_{OHOv} \cdot FCOD_{b,OHO} \cdot SRT / (1 + b_{OHO,T} \cdot SRT)$$
(note: $FCOD_{b,OHO}$ is calculated in steps 3.3 and 3.4)

$$= (0.45 \cdot 4,934 \cdot 8) / (1 + 0.214 \cdot 8)$$

$S_{F,ANn}$	4.3	4.3	$gCOD/m^3$
	↓	↑	

seed3:

MX_{OHOv}	6,539	6,539	kgVSS	(6.14b)

$$= Y_{OHOv} \cdot FCOD_{b,OHO} \cdot SRT / (1 + b_{OHO,T} \cdot SRT)$$
(note: $FCOD_{b,OHO}$ is calculated in steps 3.3 and 3.4)

$$= (0.45 \cdot 4,927 \cdot 8) / (1 + 0.214 \cdot 8)$$

3.3 VFAs stored by PAOs from the fermentation of $S_{F,i,conv}$

$FS_{S,PAO}$	$= Q_i \cdot (S_{F,i,conv} - (1 + s) \cdot S_{F,ANn}) + Q_i \cdot S_{VFA,i}$		(6.12)
	$= 20 \cdot (39 - (1 + 0.67) \cdot 4.3) + (20 \cdot 0)$		
$FS_{S,PAO}$	633	kgCOD/d	

3.4 Remaining biodegradable COD available to OHOs

$FCOD_{b,OHO}$	$= FCOD_{b,i} - FS_{S,PAO}$		(6.13)
	$= 5,560 - 633$		
$FCOD_{b,OHO}$	4,927	kgCOD/d	

4. Biomass (VSS) equations

Corresponds to the biological mass present in the system as synthesized from the influent COD (in g/d), taking into account the cumulative effect of SRT $[(g/d) \cdot d = g$ in the system]

4.1 PAOs

Active mass

Y_{PAOv}	0.45	gVSS/gCOD
$Y_{PAO,obs}$	$= Y_{PAOv} / (1 + b_{PAO,T} \cdot SRT)$	
	$= 0.45 / (1 + 0.036 \cdot 8)$	
$Y_{PAO,obs}$	0.350	gVSS / gCOD
MX_{PAOv}	$= Y_{PAO,obs} \cdot FS_{S,PAO} \cdot SRT$	(6.2)
	$= 0.350 \cdot 633 \cdot 8$	
MX_{PAOv}	1,772	kgVSS in the system

Endogenous mass

$MX_{E,PAOv}$	$= f_{XE,PAO} \cdot b_{PAO,T} \cdot MX_{PAOv} \cdot SRT$	(6.3)
	$= 0.25 \cdot 0.036 \cdot 1,772 \cdot 8$	
$MX_{E,PAOv}$	126	kgVSS

4.2 OHOs

Active mass

Y_{OHOv}	0.45	gVSS/gCOD
$Y_{OHO,obs}$	$= Y_{OHOv} / (1 + b_{OHO,T} \cdot SRT)$	
	$= 0.45 / (1 + 0.214 \cdot 8)$	
$Y_{OHO,obs}$	0.166	gVSS/gCOD
MX_{OHOv}	$= Y_{OHO,obs} \cdot F_{CODb,OHO} \cdot SRT$	
	$= 0.166 \cdot 4,927 \cdot 8$	
MX_{OHOv}	6,539	kgVSS

(note: this value is the calculated MX_{OHOv} value of step 3.2)

Endogenous mass

$MX_{E,OHOv}$ $= f_{XE,OHO} \cdot b_{OHO,T} \cdot MX_{OHOv} \cdot SRT$ (6.5)

                    $= 0.20 \cdot 0.214 \cdot 6{,}539 \cdot 8$

$MX_{E,OHOv}$ 2,240   kgVSS

4.3 Unbiodegradable particulate organics

MX_{Uv} $= Q_i \cdot X_{U,i} \cdot SRT / f_{CV}$ (6.6)

                    $= 20 \cdot 85 \cdot 8 / 1.48$

MX_{Uv} 9,189   kgVSS

5. VSS and TSS

5.1 VSS and active fraction

MX_B $= MX_{PAOv} + MX_{OHOv}$

                    $= 1{,}772 + 6{,}539$

MX_B 8,311   kgVSS

MX_{VSS} $= MX_{PAOv} + MX_{OHOv} + MX_{E,PAOv} + MX_{E,OHOv} + MX_{Uv}$ (6.24b)

                    $= 1{,}772 + 6{,}539 + 126 + 2{,}240 + 9{,}189$

MX_{VSS} 19,867   kgVSS

f_{av} $= MX_B / MX_{VSS}$

                    $= 8{,}311 / 19{,}867$

f_{av} 42 %

5.2 FSS

$f_{PAO,actual}$ $= [(Q_i \cdot SRT \cdot \Delta P_{SYS,actual}) - (f_P \cdot (MX_{VSS} - MX_{PAOv}))] / MX_{PAOv}$ (6.24a)

                    $= [(20 \cdot 8 \cdot 7.6) - (0.03 \cdot (19{,}867 - 1{,}772)] / 1{,}772$

$f_{PAO,actual}$ 0.38 gP/gVSS

MX_{FSS} $= f_{FSS,OHO} \cdot MX_{OHOv} + f_{FSS,PAO} \cdot (f_{P,PAO,actual} / f_{P,PAO}) \cdot MX_{PAOv} + FX_{FSS,i} \cdot SRT$ (6.24d)

                    $= (0.15 \cdot 6{,}539) + (1.3 \cdot (0.38/0.38) \cdot 1{,}772) + (700 \cdot 8)$

MX_{FSS} 8,884   kgFSS

5.3 TSS

MX_{TSS} $= MX_{VSS} + MX_{FSS}$ (6.25a)

                    $= 19{,}867 + 8{,}884$

MX_{TSS} 28,751   kgTSS

5.4 f_{VT}

f_{VT} $= MX_{VSS} / MX_{TSS}$ (6.25c)

                    $= 19{,}867 / 28{,}751$

f_{VT} 0.69   gVSS/gTSS

$f_{VT,PAO}$ $= MX_{PAOv} / MX_{TSS}$

$$= 1{,}772 / 28{,}751$$

$f_{VT,PAO}$	0.06	gVSS/gTSS

5.5 P content of TSS

$$f_{P,TSS} = [(f_P \cdot MX_{OHOv}) + f_P \cdot (MX_{E,OHOv} + MX_{E,PAOv}) + (f_P \cdot MX_{Uv})$$
$$+ (f_{P,PAO,actual} \cdot MX_{PAOv}) + (f_{P,FSS,i} \cdot MX_{FSS})\,] / MX_{TSS} \tag{6.26}$$
$$= [(0.03 \cdot 6{,}539) + 0.03 \cdot (2{,}240 + 126) + (0.03 \cdot 9{,}189) + (0.38 \cdot 1{,}772) + (0.02 \cdot 8{,}864)] / 28{,}751$$

$f_{P,TSS}$	0.048	gP/gTSS

6. P removal

6.0 P release

$$S_{PO4,rel} = f_{PO4,rel} \cdot FS_{S,PAO} / Q_i \tag{6.15}$$
$$= 0.5 \cdot 633 / 20$$

S_{PO4_rel}	15.8	gP/m^3	(note: gP/m^3 of influent, not gP/m^3 of AN reactor)

6.1 ΔP by PAOs

$$\Delta P_{PAO} = f_{P,PAO} \cdot MX_{PAOv} / (SRT \cdot Q_i) \tag{6.16}$$
$$= 0.38 \cdot 1{,}772 / (8 \cdot 20)$$

ΔP_{PAO}	4.2	gP/m^3

6.2 ΔP by OHOs

$$\Delta P_{OHO} = f_P \cdot MX_{OHOv} / (SRT \cdot Q_i) \tag{6.17}$$
$$= 0.03 \cdot 6{,}539 / (8 \cdot 20)$$

ΔP_{OHO}	1.23	gP/m^3

6.3 ΔP by endogenous mass

$$\Delta P_{XE} = \Delta P_{XE,PAO} + \Delta P_{XE,OHO} \tag{6.18}$$
$$\Delta P_{XE,PAO} = f_P \cdot MX_{E,PAOv} / (SRT \cdot Q_i)$$
$$= 0.03 \cdot 126 / (8 \cdot 20)$$

$\Delta P_{XE,PAO}$	0.02	gP/m^3

$$\Delta P_{XE,OHO} = f_P \cdot MX_{E,OHOv} / (SRT \cdot Q_i)$$
$$= 0.03 \cdot 2{,}240 / (8 \cdot 20)$$

$\Delta P_{XE,OHO}$	0.42	gP/m^3
ΔP_{XE}	0.44	gP/m^3

6.4 ΔP by influent unbiodegradable organic mass

$$\Delta P_{XU} = f_P \cdot MX_{Uv} / (SRT \cdot Q_i) \tag{6.19}$$
$$= 0.03 \cdot 9{,}189 / (8 \cdot 20)$$

ΔP_{XU}	1.72	gP/m^3

6.5 Potential total P removal

$$\Delta P_{SYS,pot} = \Delta P_{PAO} + \Delta P_{OHO} + \Delta P_{XE} + \Delta P_{XU} \tag{6.20}$$

$$= 4.2 + 1.23 + 0.44 + 1.72$$

$\Delta P_{SYS,pot}$ 7.6 gP/m^3

6.6 Actual total P removal

P_i 12.5 gP/m^3

$$\Delta P_{SYS,actual} = \min (\Delta P_{SYS,pot}; P_i) \tag{6.21}$$

$$= \min (7.6; 12.5)$$

$\Delta P_{SYS,actual}$ 7.6 gP/m^3

6.7 Particulate P in the effluent

To calculate after step 6.5 where the P content of TSS is calculated

$$X_{P,e} = f_{P,TSS} \cdot X_{TSS,e} \tag{6.22}$$

$$= 0.048 \cdot 15$$

$X_{P,e}$ 0.72 gP/m^3

6.8 Effluent total P

$$P_e = P_i - \Delta P_{SYS,actual} + X_{P,e} \tag{6.23}$$

$$= 12.5 - 7.6 + 0.72$$

P_e 5.6 gP/m^3

The current proposed design is unable to meet the effluent total P concentration of less than 1 gP/m^3 (= mg/l). The main reason is the low COD flux consumed by PAO ($FS_{S,PAO}$) due to the absence of VFA in the influent. Consequently, the PAO biomass (MX_{PAOv}) is too low and only constitutes approximately 6 % of the total mass of solids present in the system ($f_{VT,PAO}$), contributing to the removal of only 4.2 gP/m^3, Consequently, the effluent contains 5.6 gP/m^3 of which 0.72 gP/m^3 are present in the solids lost through the effluent of the secondary settling tank. As such and without taking into account the loss of solids through the effluent (since they depend on the performance of the secondary settling tank), the actual P-removal capacity of the system ($\Delta P_{SYS,actual}$) needs to be 4.9 gP/m^3 higher to remove all the influent total P concentration of 12.5 gP/m^3: $P_i - \Delta P_{SYS,actual} = 12.5 - 7.6 = 4.9 \ gP/m^3$.

Because the influent wastewater does not contain sufficient VFA, it is necessary to add external VFA to promote the growth of the required PAO biomass to remove 4.9 gP/m^3 more. This means that ΔP_{PAO} needs to increase by 4.9 gP/m^3 from 4.2 gP/m^3 to 9.1 gP/m^3 ($\Delta P_{PAO} = 4.2 + 4.9 = 9.1 \ gP/m^3$). This ΔP_{PAO} can be therefore used to estimate the flux of external VFA that need to be added to meet the effluent total P standard of 1 gP/m^3. This is shown in the following calculations.

7. Determination of additional VFA addition

From the equation to determine the PAO biomass:

$$MX_{PAOv} = Y_{PAO,obs} \cdot FS_{S,PAO} \cdot SRT \tag{6.1}$$

And from the determination of the P-removal potential of PAO:

$$\Delta P_{PAO} = f_{P,PAO} \cdot MX_{PAOv} / (SRT \cdot Q_i) \tag{6.16}$$

Solving Eq. 6.16 for MX_{PAOv}:

$$MX_{PAOv} = \Delta P_{PAO} \cdot SRT \cdot Q_i / f_{P,PAO} \tag{6.16i}$$

Matching Eq. 6.1 and Eq. 6.16i ($MX_{PAOv} = MX_{PAOv}$) and solving for $FS_{S,PAO}$:

$$FS_{S,PAO} = \Delta P_{PAO} \cdot Q_i / (f_{P,PAO} \cdot Y_{PAO,obs}) \tag{6.16ii}$$

From Eq. 6.12:

$$FS_{S,PAO} = Q_i \cdot (S_{F,i,conv} - (1 + s) \cdot S_{F,ANn}) + Q_i \cdot S_{VFA,i}$$

Thus, matching Eq. 6.16ii and Eq. 6.12, and solving for $Q_i \cdot S_{VFA,i}$:

$$Q_i \cdot S_{VFA,i} = [\Delta P_{PAO} \cdot Q_i / (f_{P,PAO} \cdot Y_{PAO,obs})] - [Q_i \cdot (S_{F,i,conv} - (1 + s) \cdot S_{F,ANn})]$$

As previously described, the influent wastewater does not contain VFA. As such, this expression can be used to calculate the additional VFA that needs to be dosed: $Q_{VFA,add} \cdot S_{VFA,add}$. Therefore:

$$Q_{VFA,add} \cdot S_{VFA,add} = [\Delta P_{PAO} \cdot Q_i / (f_{P,PAO} \cdot Y_{PAO,obs})] - [Q_i \cdot (S_{F,i,conv} - (1 + s) \cdot S_{F,ANn})]$$

Substituting the corresponding known values:

$$\begin{aligned} Q_{VFA,add} \cdot S_{VFA,add} &= [9.1 \cdot 20 / (0.38 \cdot 0.35)] - [20 \cdot (39 - (1 + 0.67) \cdot 4.3)] \\ &= 1{,}367 - 633 \\ &= 734 \qquad kgCOD/d \end{aligned}$$

To remove the 12.5 gP/m^3 present in the influent wastewater, the system requires 1,367 kgCOD/d of readily biodegradable organics to promote the growth of the PAO biomass. The system already favours the uptake of 633 kgCOD/d by the PAO biomass. Therefore, 734 kgCOD/d needs to be added using an external carbon source. It is suggested to add 10 m^3/d ($Q_{VFA,add}$) of a highly concentrated acetate solution equivalent to 73.5 gCOD/l. The relatively low $Q_{VFA,add}$ flow rate of 10 m^3/d corresponds to 0.05 % of the influent flow rate received at the plant (20 MLD = 20,000 m^3/d). Consequently, from a hydraulic perspective it will not lead to a major effect, but the additional COD load of 734 kgCOD/d ($Q_{VFA,add} \cdot S_{VFA,add}$) will increase the PAO biomasses (MX_{PAOv} and $MX_{E,PAOv}$) as well as the accumulation of fixed or inorganic suspended solids due to the P removal. Thus, the biomasses need to be recalculated (following a similar procedure like in points 4, 5 and 6).

8. Recalculating the biomass

$FS_{S,PAO}$	= 1,367	kgCOD/d
$FCOD_{b,OHO}$	= 4,927	kgCOD/d

8.1 PAOs

Active mass

Y_{PAOv}	0.45	gVSS/gCOD
$Y_{PAO,obs}$	= $Y_{PAOv} / (1 + b_{PAO,T} \cdot SRT)$	
	= $0.45 / (1 + 0.036 \cdot 8)$	
$Y_{PAO,obs}$	0.350	gVSS/gCOD

MX_{PAOv}	$= Y_{PAO,obs} \cdot FS_{S,PAO} \cdot SRT$		(6.2)
	$= 0.350 \cdot 1{,}367 \cdot 8$		
MX_{PAOv}	3,829	kgVSS in the system	

Endogenous mass

$MX_{E,PAOv}$	$= f_{XE,PAO} \cdot b_{PAO,T} \cdot MX_{PAOv} \cdot SRT$		(6.3)
	$= 0.25 \cdot 0.036 \cdot 3{,}829 \cdot 8$		
$MX_{E,PAOv}$	273	kgVSS	

8.2 OHOs

Active mass

Y_{OHOv}	0.45	gVSS/gCOD
$Y_{OHO,obs}$	$= Y_{OHOv} / (1 + b_{OHO,T} \cdot SRT)$	
	$= 0.45 / (1 + 0.214 \cdot 8)$	
$Y_{OHO,obs}$	0.166	gVSS/gCOD
MX_{OHOv}	$= Y_{OHO,obs} \cdot F_{CODb,OHO} \cdot SRT$	
	$= 0.166 \cdot 4{,}927 \cdot 8$	
MX_{OHOv}	6,539	kgVSS

(this value is the calculated MX_{OHOv} value of step 3.2)

Endogenous mass

$MX_{E,OHOv}$	$= f_{XE,OHO} \cdot b_{OHO,T} \cdot MX_{OHOv} \cdot SRT$		(6.5)
	$= 0.20 \cdot 0.214 \cdot 6{,}539 \cdot 8$		
$MX_{E,OHOv}$	2,240	kgVSS	

8.3 Unbiodegradable particulate organics

MX_{Uv}	$= Q_i \cdot X_{U,i} \cdot SRT / f_{CV}$		(6.6)
	$= 20 \cdot 85 \cdot 8 / 1.48$		
MX_{Uv}	9,189	kgVSS	

9. VSS and TSS

9.1 VSS and active fraction

MX_B	$= MX_{PAOv} + MX_{OHOv}$		
	$= 3{,}829 + 6{,}539$		
MX_B	10,368	kgVSS	
MX_{VSS}	$= MX_{PAOv} + MX_{OHOv} + MX_{E,PAOv} + MX_{E,OHOv} + MX_{Uv}$		(6.24b)
	$= 3{,}829 + 6{,}539 + 273 + 2{,}240 + 9{,}189$		
MX_{VSS}	22,071	kgVSS	
f_{av}	$= MX_B / MX_{VSS}$		
	$= 10{,}368 / 22{,}071$		
f_{av}	47 %		

9.2 FSS

$f_{P,PAO,actual}$ $= [(Q_i \cdot SRT \cdot \Delta P_{SYS,actual}) - (f_P \cdot (MX_{VSS} - MX_{PAOv}))] / MX_{PAOv}$ (6.24a)

$= [(20 \cdot 8 \cdot 12.5) - (0.03 \cdot (22{,}071 - 3{,}829))] / 3{,}829$

$f_{P,PAO,actual}$ 0.36 gP/gVSS

MX_{FSS} $= f_{FSS,OHO} \cdot MX_{OHOv} + f_{FSS,PAO} \cdot (f_{P,PAO,actual} / f_{P,PAO}) \cdot MX_{PAOv} + FX_{FSS,i} \cdot SRT$ (6.24d)

$= (0.15 \cdot 6{,}539) + (1.3 \cdot (0.36 / 0.38) \cdot 3{,}829) + (700 \cdot 8)$

MX_{FSS} 11,293   kgFSS

9.3 TSS

MX_{TSS} $= MX_{VSS} + MX_{FSS}$ (6.25a)

$= 22{,}071 + 11{,}293$

MX_{TSS} 33,363   kgTSS

9.4 f_{VT}

f_{VT} $= MX_{VSS} / MX_{TSS}$ (6.25c)

$= 22{,}071 / 33{,}363$

f_{VT} 0.66   gVSS/gTSS

$f_{VT,PAO}$ $= MX_{PAOv} / MX_{TSS}$

$= 3{,}829 / 33{,}363$

$f_{VT,PAO}$ 0.11   gVSS/gTSS

9.5 P content of TSS

$f_{P,TSS}$ $= [(f_P \cdot MX_{OHOv}) + f_P \cdot (MX_{E,OHOv} + MX_{E,PAOv}) + (f_P \cdot MX_{Uv})$

$+ (f_{P,PAO,actual} \cdot MX_{PAOv}) + (f_{P,FSS,i} \cdot MX_{FSS})] / MX_{TSS}$ (6.26)

$= [(0.03 \cdot 6{,}539) + 0.03 \cdot (2{,}240 + 273) + (0.03 \cdot 9{,}189) + (0.36 \cdot 3{,}829) + (0.02 \cdot 11{,}293)] / 33{,}363$

$f_{P,TSS}$ 0.064   gP/gTSS

10. P removal

10.0 P release

$S_{PO4,rel}$ $= f_{PO4,rel} \cdot FS_{S,PAO} / Q_i$ (6.15)

$= 0.5 \cdot 1{,}367 / 20$

S_{PO4_rel} 34.2   gP/m^3 (note: gP/m^3 of influent, not gP/m^3 of AN reactor)

10.1 ΔP by PAOs

ΔP_{PAO} $= f_{P,PAO} \cdot MX_{PAOv} / (SRT \cdot Q_i)$ (6.16)

$= 0.38 \cdot 3{,}829 / (8 \cdot 20)$

ΔP_{PAO} 9.1   gP/m^3

10.2 ΔP by OHOs

ΔP_{OHO} $= f_P \cdot MX_{OHOv} / (SRT \cdot Q_i)$ (6.17)

$= 0.03 \cdot 6{,}539 / (8 \cdot 20)$

ΔP_{OHO} 1.23   gP/m^3

10.3 ΔP by endogenous mass

ΔP_{XE}	$= \Delta P_{XE,PAO} + \Delta P_{XE,OHO}$	(6.18)
$\Delta P_{XE,PAO}$	$= f_P \cdot MX_{E,PAOv} / (SRT \cdot Q_i)$	
	$= 0.03 \cdot 273 / (8 \cdot 20)$	
$\Delta P_{XE,PAO}$	$0.05 \qquad gP/m^3$	
$\Delta P_{XE,OHO}$	$= f_P \cdot MX_{E,OHOv} / (SRT \cdot Q_i)$	
	$= 0.03 \cdot 2,240 / (8 \cdot 20)$	
$\Delta P_{XE,OHO}$	$0.42 \qquad gP/m^3$	
ΔP_{XE}	$0.47 \qquad gP/m^3$	

10.4 ΔP by influent unbiodegradable organic mass

ΔP_{XU}	$= f_P \cdot MX_{Uv} / (SRT \cdot Q_i)$	(6.19)
	$= 0.03 \cdot 9,189 / (8 \cdot 20)$	
ΔP_{XU}	$1.72 \qquad gP/m^3$	

10.5 Potential total P removal

$\Delta P_{SYS,pot}$	$= \Delta P_{PAO} + \Delta P_{OHO} + \Delta P_{XE} + \Delta P_{XU}$	(6.20)
	$= 9.1 + 1.23 + 0.47 + 1.72$	
$\Delta P_{SYS,pot}$	$12.5 \qquad gP/m^3$	

10.6 Actual total P removal

P_i	$12.5 \qquad gP/m^3$	
$\Delta P_{SYS,actual}$	$= min (\Delta P_{SYS,pot}; P_i)$	(6.21)
	$= min (12.5; 12.5)$	
$\Delta P_{SYS,actual}$	$12.5 \qquad gP/m^3$	

10.7 Particulate P in the effluent

To calculate after step 6.5 where the P content of TSS is calculated

$X_{P,e}$	$= f_{P,TSS} \cdot X_{TSS,e}$	(6.22)
	$= 0.064 \cdot 15$	
$X_{P,e}$	$0.97 \qquad gP/m^3$	

10.8 Effluent total P

P_e	$= P_i - \Delta P_{SYS,actual} + X_{P,e}$	(6.23)
	$= 12.5 - 12.5 + 0.97$	
P_e	$0.97 \qquad gP/m^3$	

As observed, the addition of 1,367 kg COD/d as VFA favours the growth of PAO and consequently leads to a higher P-removal capacity of the plant contributing to meet the discharge limit of 1 mg P/l.

Example 6.3.3

Design of a biological nutrient removal activated sludge wastewater treatment plant

By applying the steady-state stoichiometric design models for nitrification, denitrification and EBPR, design a 3-stage modified Bardenpho system (also known as A2O) (Figure 6.5) to treat an influent wastewater (Q_i) that has a flow rate of 30 MLD (30,000 m^3/d). The wastewater contains a total influent COD concentration (COD$_i$) of 585 mg/l, an influent total nitrogen concentration (N$_i$) of 38.5 mgP/l and influent total phosphorus concentration (P$_i$) of 11.5 mgP/l. The influent COD, N and P fractionations are shown in figures 6.6-6.8. In addition, the influent flow rate contains 25 mg/l of inorganic or fixed suspended solids.

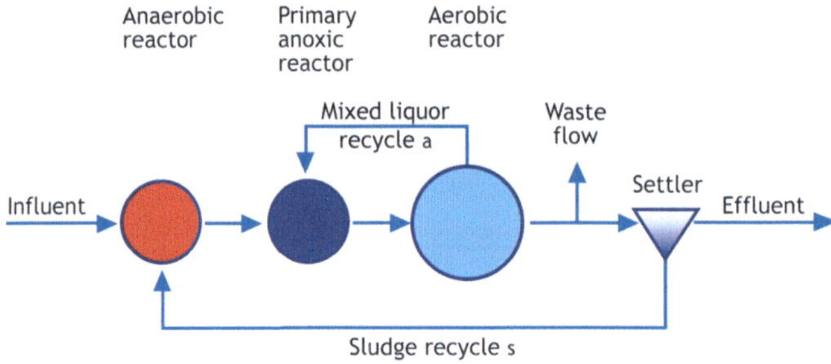

Figure 6.5 A 3-stage modified Bardenpho (a.k.a. A2O) configuration. The anaerobic reactor is divided into three compartments (not illustrated).

The biological nutrient removal plant needs to comply with an effluent total COD concentration of 125 mg/l, effluent total nitrogen concentration of 10 mg N/l, and effluent total phosphorus concentration of 1 mg/l. The plant will operate at an average MLSS concentration of 3,500 mg/l and the lowest yearly wastewater temperature is 14 °C.

To carry out the design, assume:

a. A TSS concentration in the effluent of the secondary settling tank of 15 mg TSS/l.
b. A TSS concentration in the sludge recycle line of 8,000 mg TSS/l with a s-recycle ratio of 0.78.
c. A maximum specific biomass growth rate of nitrifiers ($\mu_{ANO,max,20}$) of 0.40 1/d.
d. A safety factor for nitrification of 1.25.
e. Absence of dissolved oxygen and nitrate in the influent flow rate.
f. A dissolved oxygen concentration of 2 mg/l in the a-recycle ratio and 1 mg/l in the s-recycle ratio.
g. A nitrate concentration in the s-recycle ratio of 4.5 mg/l (approximately similar to the expected effluent nitrate concentration, $S_{NO3,e}$).

Based on the previous information, determine:

a. The required solids retention time (SRT) for the whole plant.
b. The COD fluxes between PAOs and OHOs.
c. The nitrification activity of the system.
d. The effluent total nitrogen and total phosphorus concentrations (N$_e$ and P$_e$, respectively).

e. MX_{VSS} and MX_{TSS} masses.
f.   Total volume of the system (V_R) and the volumes of the anaerobic ($V_{R,AN}$) and aerobic stages ($V_{R,OX}$).
g. O_2 requirements of the system (FO_c and FO_{NIT}).
h.   A COD mass balance to confirm the reliability of your design.

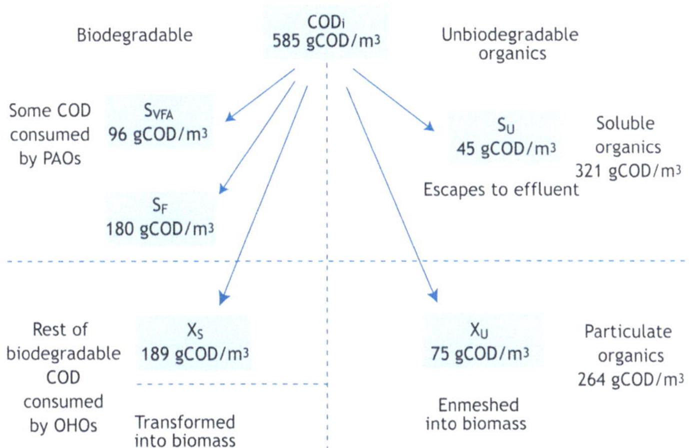

Figure 6.6 Influent COD fractionation for the biological nutrient removal design example.

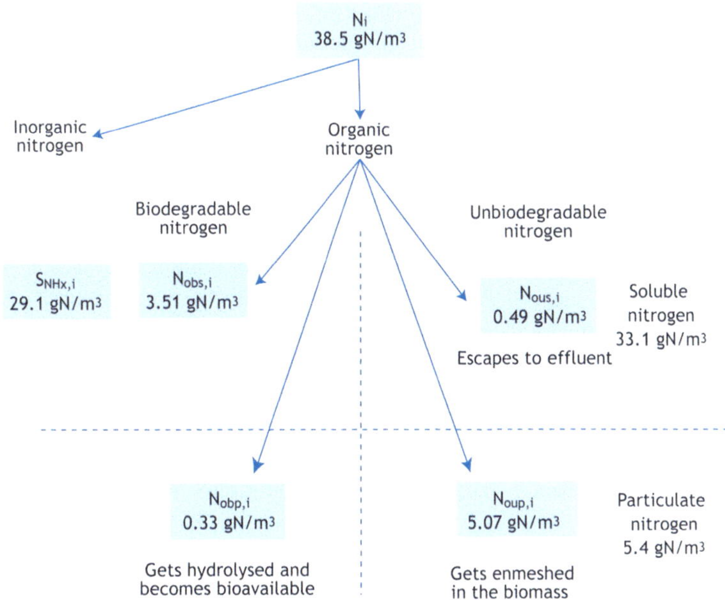

Figure 6.7 Influent nitrogen fractionation for the biological nutrient removal design example.

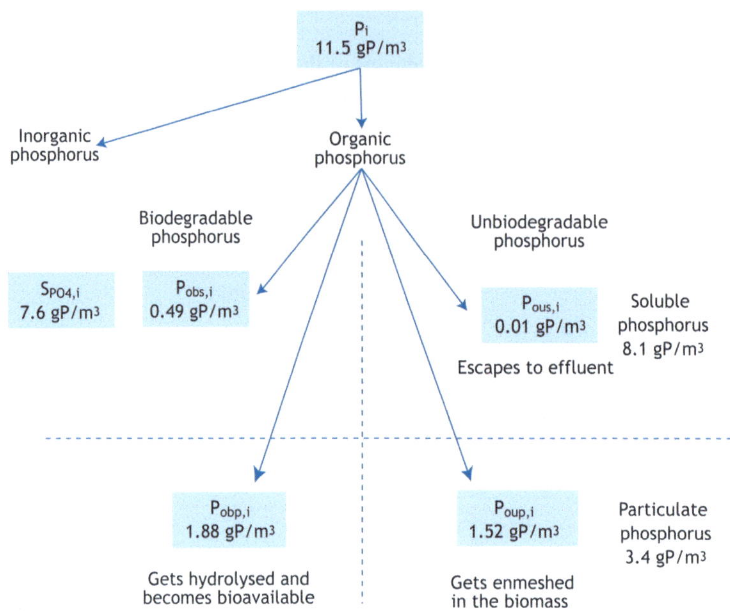

Figure 6.8 Influent phosphorus fractionation for the biological nutrient removal design example.

Solution

To carry out the design of the biological nutrient removal plant, the design algorithm presented in Figure 6.9 can be followed. According to design algorithm, first the configuration of the wastewater treatment plant needs to be defined and the COD, N and P fractionations need to be calculated. For this example, these have been already defined. The selected configuration is a 3-stage modified Bardenpho (Figure 6.5) and the COD, N and P fractionations are presented in figures 6.6, 6.7 and 6.8. Thereafter, the kinetic and stoichiometric parameters are defined considering the COD, N and P bioprocesses and the lowest average temperature of 14 °C. Thus, in the next step, the minimum SRT required for nitrification (SRT_{min}) is calculated suggesting an initial maximum unaerated mass fraction (f_{xt}) to implement the anoxic stage for the pre-denitrification process (f_{x1}) and the anaerobic stage (f_{xa}) to support the EBPR process. Since nitrification is required, the SRT_{min} defines the SRT of the system. Once the SRT_{min} is estimated, the EBPR model is followed to design the P and COD conversions. Thereafter, the nitrification conversions are computed (taking into account the data generated from the EBPR process design, such as the nitrogen consumed for biomass synthesis, N_s, to compute the nitrification capacity, NIT_c) as well as the denitrification process (e.g., using the $S_{F,ANn}$ value from the EBPR design to calculate the denitrification potential, D_{p1}). It can be a good indication that the system will be able to satisfactorily denitrify the nitrate generated, if at this step D_{p1} is comparable to N_c ($D_{p1} \sim NIT_c$). However, a D_{p1} considerably lower than NIT_c ($D_{p1} << NIT_c$) suggests that the anoxic mass fraction (f_{x1}) is too short and needs to be extended. To extend f_{x1} without modifying the SRT_{min} and f_{xt}, f_{xa} could be shortened ($f_{x1} = f_{xt} - f_{xa}$) as far as the P removal is able to meet the corresponding effluent discharge limit. Otherwise, to extend f_{x1} the total unaerated mass fraction (f_{xt}) may need to be extended and, for that purpose, the SRT_{min} may need to increase and the EBPR and N-conversions recalculated. On the other hand, if a_{opt} is too high (e.g., > 5) due to a D_{p1} considerably higher than NIT_c ($D_{p1} >> NIT_c$) then f_{x1} and also f_{xt} could be shortened which could lead to a shorter SRT_{min}. This has promising implications and benefits because a shorter SRT can decrease the tank volumes, the oxygen requirements (both contributing to decrease the capital investments and operational costs) and even lead to an improved nutrient removal efficiency due to higher nutrient biomass requirements. The design process can be done iteratively until the effluent discharge standards are met. The design process is exemplified in the following calculations, starting with the summary of the main design data (Table 6.8) and of the main stoichiometric and kinetic parameters (Table 6.9) necessary to carry out the process design.

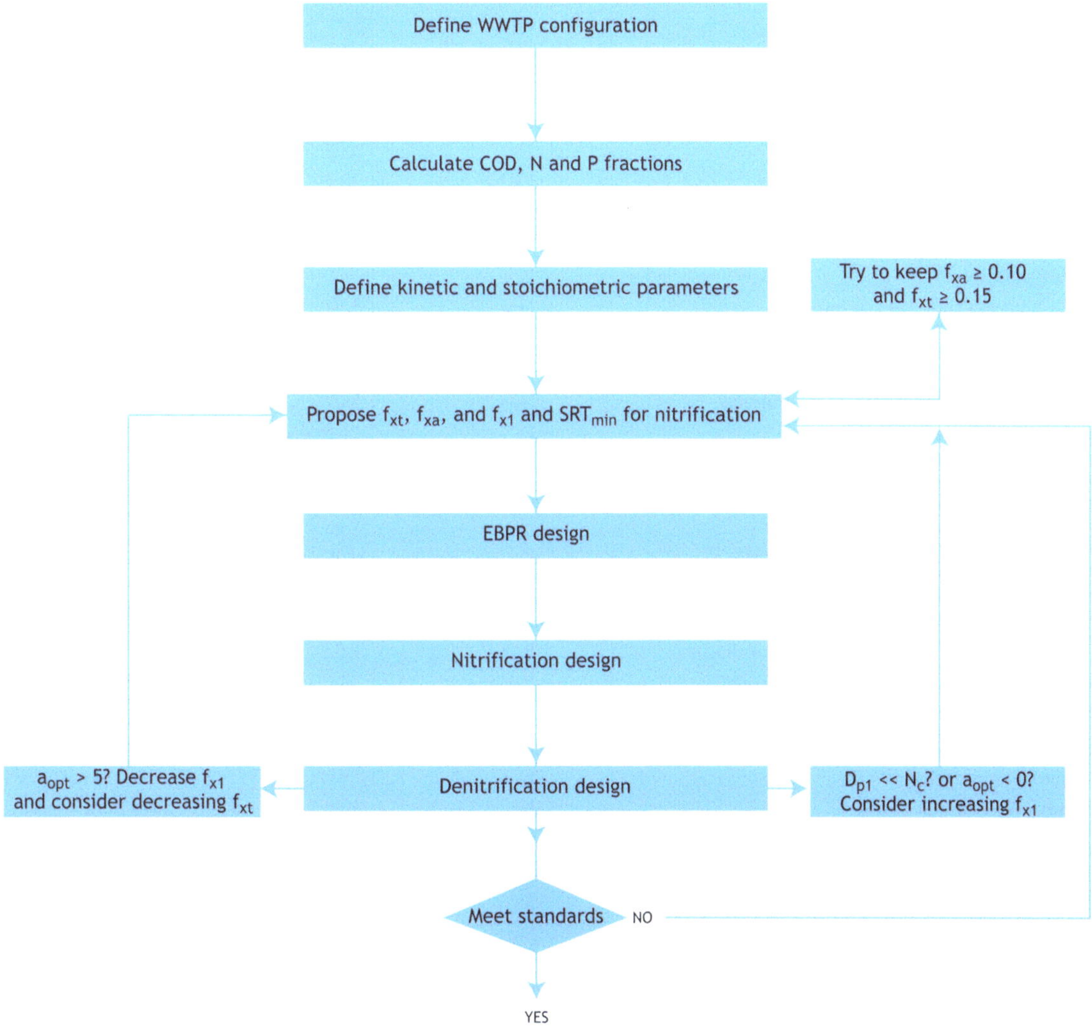

Figure 6.9 Suggested design algorithm to design the biological nutrient removal system.

Table 6.8 Summary of data to carry out the design of the 3-stage modified Bardenpho (A2O) plant.

Description	Symbol	Value	Unit
Flow rate	Q_i	30	MLD
Total COD	COD_i	585	$gCOD/m^3$
COD concentrations			
- readily biodegradable COD	$S_{S,i}$	276	$gCOD/m^3$
- volatile fatty acids	$S_{VFA,i}$	96	$gCOD/m^3$
- fermentable COD	$S_{F,i}$	180	$gCOD/m^3$
- slowly biodegradable COD	$X_{S,i}$	189	$gCOD/m^3$
- unbiodegradable soluble COD	$S_{U,i}$	45	$gCOD/m^3$
- unbiodegradable particulate COD	$X_{U,i}$	75	$gCOD/m^3$
Total nitrogen concentration	N_i	38.5	gN/m^3
- Total Kjeldahl nitrogen (TKN)	TKN_i	38.5[a]	gN/m^3
Inorganic nitrogen			
- free and saline ammonia	$S_{NHx,i}$	29.1	gN/m^3
Organic nitrogen			
- biodegradable soluble nitrogen	$N_{obs,i}$	3.51	gN/m^3
- biodegradable particulate nitrogen	$N_{obp,i}$	0.33	gN/m^3
- unbiodegradable soluble nitrogen	$N_{ous,i}$	0.49	gN/m^3
- unbiodegradable particulate nitrogen	$N_{oup,i}$	5.07	gN/m^3
Influent total P concentration	P_i	11.5	gP/m^3
Inorganic phosphorus			
- orthophosphate	$S_{PO4,i}$	7.6	gP/m^3
Organic phosphorus			
- biodegradable soluble phosphorus	$P_{obs,i}$	0.49	gP/m^3
- biodegradable particulate phosphorus	$P_{obp,i}$	1.88	gP/m^3
- unbiodegradable soluble phosphorus	$P_{ous,i}$	0.01	gP/m^3
- unbiodegradable particulate phosphorus	$P_{oup,i}$	1.52	gP/m^3
Influent inorganic (fixed) suspended solids	$X_{FSS,i}$	25	$gFSS/m^3$
Temperature	T	14	°C
Sludge retention time	SRT	[b]	d
Anaerobic mass fraction	f_{xa}	[c]	-
Number of anaerobic zones	N	3[c]	reactors
Anoxic mass fraction	f_{x1}	[d]	
Dissolved O_2 in the influent	$S_{O2,i}$	0	gO_2/m^3
Dissolved O_2 in the a recycle	$S_{O2,a}$	2	gO_2/m^3
Dissolved O_2 in the sludge recycle	$S_{O2,s}$	1	gO_2/m^3
Influent nitrate concentration	$S_{NO3,i}$	0	$gNO_3\text{-}N/m^3$
Nitrate concentration in the sludge recycle	$S_{NO3,s}$	4.5	$gNO_3\text{-}N/m^3$
Total suspended solids in the effluent	$X_{TSS,e}$	15	$gTSS/m^3$
Design aerobic TSS concentration	$X_{TSS,OX}$	3,500	$gTSS/m^3$
s-recycle ratio	s	0.78	-

[a] In this example, $N_i = TKN_i$ because the TKN analytical technique measures the organically bound nitrogen plus $S_{NHx,i}$ (free and saline ammonia) even though the latter is inorganic. Had the influent contained nitrate or nitrite ($S_{NO3,i}$ or $S_{NO2,i}$, respectively), then N_i would be higher: $N_i = TKN_i + S_{NO3,i} + S_{NO2,i}$.

[b] The SRT will be defined after calculating SRT_{min}.

[c] An initial f_{xa} fraction should be suggested (in this example divided into 3 compartments, N) to carry out the first iteration.

[d] An initial f_{x1} fraction will be suggested to carry out the first iteration.

Table 6.9 Kinetic and stoichiometric parameters for the EBPR design example.

	Parameter	Symbol	Value	Unit
Kinetic	*OHO*			
	First-order fermentation rate constant at temperature 20 °C	$k_{F,20}$	0.06	m³/gVSS.d
	Temperature coefficient for $k_{F,T}$	$\theta_{k,F}$	1.029	
	First-order fermentation rate constant at temperature T	$k_{F,T}$	0.051	m³/gVSS.d
	Specific endogenous mass loss rate of the OHOs at 20 °C	$b_{OHO,20}$	0.24	gVSS/gVSS.d
	Temperature coefficient for $b_{OHO,T}$	$\theta_{b,OHO}$	1.029	
	Specific endogenous mass loss rate of the OHOs at temperature T	$b_{OHO,T}$	0.202	gVSS/gVSS.d
	Specific denitrification rate for NDEBPR pre-denitrification at 20 °C	$K_{2,20}$	0.23	gNO₃/gVSS.d
	Temperature coefficient for $K_{2,T}$	$\theta_{K2,DENIT}$	1.08	
	Specific denitrification rate for NDEBPR pre-denitrification at temperature T	$K_{2,T}$	0.14	gNO₃/gVSS.d
	ANO			
	ANO maximum specific growth rate constant at 20 °C	$\mu_{ANO,max,20}$	0.45	gVSS/gVSS.d
	Temperature coefficient for $\mu_{ANO,max,T}$	θ_{NIT}	1.123	
	ANO maximum specific growth rate constant at temperature T	$\mu_{ANO,max,T}$	0.22	gVSS/gVSS.d
	ANO half-saturation constant at 20 °C	$K_{ANO,20}$	1	gN/m³
	Temperature coefficient for $K_{ANO,T}$	θ_{NIT}	1.123	
	ANO maximum specific growth rate constant at temperature T	$K_{ANO,T}$	0.50	gN/m³
	ANO-specific endogenous mass loss rate constant at 20 °C	$b_{ANO,20}$	0.04	gVSS/gVSS.d
	Temperature coefficient for $b_{ANO,T}$	$\theta_{b,ANO}$	1.029	
	ANO-specific endogenous mass loss rate constant at temperature T	$b_{ANO,T}$	0.034	gVSS/gVSS.d
	PAO			
	PAO-specific endogenous mass loss rate constant at 20 °C	$b_{PAO,20}$	0.04	gVSS/gVSS.d
	Temperature coefficient for $b_{PAO,T}$	$\theta_{b,PAO}$	1.029	
	PAO-specific endogenous mass loss rate constant at temperature T	$b_{PAO,T}$	0.034	gVSS/gVSS.d
Stoichiometric	*OHO*			
	Biomass yield of OHOs	Y_{OHOv}	0.45	gVSS/gCOD
	Fraction of endogenous residue of the OHOs	$f_{XE,OHO}$	0.20	gVSS/gVSS
	Fraction of P in the active OHO mass	f_P	0.03	gP/gVSS
	Fraction of P in the endogenous mass (OHO and PAO)	f_P	0.03	gP/gVSS
	Fraction of fixed (inorganic) suspended solids of OHOs	$f_{FSS,OHO}$	0.15	gFSS/gVSS
	ANO			
	Biomass yield of ANOs	Y_{ANOv}	0.10	gVSS/gCOD
	PAO			
	Biomass yield of PAOs	Y_{PAOv}	0.45	gVSS/gCOD
	Fraction of endogenous residue of the PAOs	$f_{XE,PAO}$	0.25	gVSS/gVSS
	Fraction of P in the active PAO mass	$f_{P,PAO}$	0.38	gP/gVSS
	Fraction of P in the endogenous mass (OHO and PAO)	f_P	0.03	gP/gVSS
	VSS/TSS ratio for PAO active mass	$f_{VT,PAO}$	0.44	gVSS/gTSS
	Ratio of P release/VFA uptake	$f_{PO4,rel}$	0.50	gP/gCOD
	Fraction of fixed (inorganic) suspended solids of PAOs	$f_{FSS,PAO}$	1.30	gFSS/gVSS
	Inerts or unbiodegradable mass			
	Fraction of P in the unbiodegradable mass	f_P	0.03	gP/gVSS
	Fraction of P in the influent fixed/inorganic mass	$f_{P,FSS,i}$	0.02	gP/gFSS
	General			
	COD/VSS ratio of the sludge	f_{CV}	1.48	gCOD/gVSS
	Nitrogen content of active biomass	f_n	0.10	gN/gVSS

Using the data provided, Table 6.10 shows the detailed calculations to carry out the 3-stage modified Bardenpho (A2O) process design to comply with the maximum allowable discharge standards.

Table 6.10 System design procedure.

1. System configuration			
3-stage modified Bardenpho (A2O) process configuration operated at 14 °C			
2. Influent and sludge recycle composition (from previous tables)			
Q_i	30	MLD	(30,000 m^3/d) influent flow rate
2.1 Influent concentrations - *Influent and bioreactor data*			
COD_i	585	gCOD/m^3	influent concentration of total COD
$S_{S,i}$	276	gCOD/m^3	influent concentration of readily biodegradable COD
$S_{VFA,i}$	96	gCOD/m^3	influent concentration of VFAs
$S_{F,i}$	180	gCOD/m^3	influent concentration of fermentable COD
$X_{S,i}$	189	gCOD/m^3	influent concentration of slowly biodegradable COD
$COD_{b,i}$	465	gCOD/m^3	influent concentration of biodegradable COD ($S_{S,i} + X_{S,i}$)
$S_{U,i}$	45	gCOD/m^3	influent concentration of soluble unbiodegradable COD
$X_{U,i}$	75	gCOD/m^3	influent concentration of particulate unbiodegradable COD
N_i	38.5	gN/m^3	influent concentration of total nitrogen
TKN_i	38.5	gN/m^3	influent concentration of total nitrogen
$S_{NHx,i}$	29.1	gN/m^3	influent concentration of free and saline ammonia
$N_{obs,i}$	3.51	gN/m^3	influent concentration of biodegradable soluble nitrogen
$N_{obp,i}$	0.33	gN/m^3	influent concentration of biodegradable particulate nitrogen
$N_{ous,i}$	0.49	gN/m^3	influent concentration of unbiodegradable soluble nitrogen
$N_{oup,i}$	5.07	gN/m^3	influent concentration of unbiodegradable particulate nitrogen
P_i	14.5	gP/m^3	influent concentration of total P
$S_{PO4,i}$	7.6	gP/m^3	influent concentration of orthophosphate
$P_{obs,i}$	0.49	gP/m^3	influent concentration of biodegradable soluble phosphorus
$P_{obp,i}$	1.88	gP/m^3	influent concentration of biodegradable particulate phosphorus
$P_{ous,i}$	0.01	gP/m^3	influent concentration of unbiodegradable soluble phosphorus
$P_{oup,i}$	1.52	gP/m^3	influent concentration of unbiodegradable particulate phosphorus
$S_{NO3,i}$	0	gNO$_3$-N/m^3	influent concentration of nitrate
$S_{O2,i}$	0	gO$_2$/m^3	influent concentration of dissolved oxygen
$S_{O2,a}$	2	gO$_2$/m^3	a-recycle concentration of dissolved oxygen
$S_{NO3,s}$	4.5	gNO$_3$-N/m^3	sludge recycle concentration of nitrate
$S_{O2,s}$	1	gO$_2$/m^3	sludge recycle concentration of dissolved oxygen
$X_{FSS,i}$	25	gFSS/m^3	influent concentration of fixed (inorganic) suspended solids

2.2 Influent fluxes used for calculations (= Qi · influent concentration of component)

$FCOD_i$	17,550	kgCOD/d	influent daily flux of total COD
$FS_{S,i}$	8,280	kgCOD/d	influent daily flux of rbCOD
$FS_{VFA,i}$	2,880	kgCOD/d	influent daily flux of VFAs
$FS_{F,i}$	5,400	kgCOD/d	influent daily flux of fermentable COD
$FCOD_{b,i}$	13,950	kgCOD/d	influent daily flux of biodegradable COD ($S_{S,i}$ + $X_{S,i}$)
$FX_{U,i}$	2,250	kgCOD/d	influent daily flux of particulate unbiodegradable COD
$FS_{U,i}$	1,350	kgCOD/d	influent daily flux of soluble unbiodegradable COD
$FX_{FSS,i}$	750	kgFSS/d	influent daily flux of fixed (inorganic) suspended solids

2.3 Sludge recycle characteristics

s	0.78	$m^3.d/m^3.d$	sludge recycle ratio (s) with regard to influent flow rate
$S_{O2,s}$	1.0	gO_2/m^3	dissolved O_2 in the sludge recycle
$S_{NO3,s}$	4.5	gNO_3^--N/m^3	nitrate concentration in the sludge recycle

3. Minimum required SRT for nitrification

Assumptions for the 1st iteration:

f_{xt} $= 0.30$

Being

f_{xa} $= 0.10$

f_{xl} $= 0.20$

S_f $= 1.25$

Given the previous data:

$$SRT_{min} = 1 / [(\mu_{ANO,max,T} / S_f) \cdot (1 - f_{xt}) - b_{ANO,T}] \tag{5.16}$$

Thus:

$$SRT_{min} = 1 / [(0.224 / 1.25) \cdot (1 - 0.30) - (0.034)]$$

10.9 d

Rounding up the SRT_{min}:

SRT_{min} 11 d

4. Division of $S_{S,i}$ between PAOs and OHOs

4.1 Fermentable COD available for conversion into VFAs after denitrification reactor (and O_2 consumption) in the anaerobic reactor (in units of $gCOD/m^3$ of influent)

$$S_{F,i,conv} = S_{F,i} - 8.6 \cdot (s \cdot S_{NO3,s} + S_{NO3,i}) - 3 \cdot (s \cdot S_{O2,s} + S_{O2,i}) \tag{6.8}$$

$$= S_{F,i} - COD \text{ for denitrification} - COD \text{ for D.O.}$$

$$= 180 - 8.6 \cdot (0.78 \cdot 4.5 + 0) - 3 \cdot (0.78 \cdot 1 + 0)$$

COD for DN	30.1	$gCOD/m^3$
COD for D.O.	2.3	$gCOD/m^3$
$S_{F,i,conv}$	148	$gCOD/m^3$

4.2 Fermentable COD lost in the effluent of the last anaerobic reactor

N   3   the 3rd compartment of the anaerobic reactor

calculation done by iterations

    a- assume a seed1 $S_{F,ANn}$ value of 0. This value is used to calculate MX_{OHOv}

    b- type the calculated MX_{OHOv} calculated value as seed2 value

    c- repeat steps a and b until the seed2 $S_{F,ANn}$ equals the calculated $S_{F,ANn}$

$S_{F,ANn}$ $\quad = S_{F,i,conv}/(1+s) / [1 + (k_{F,T} \cdot (f_{xa} \cdot MX_{OHOv} / (N \cdot Q_i \cdot (1 + s))))]^n$ (6.9)

$\quad\quad\quad = 148 / (1 + 0.78) / [1 + (0.051 \cdot (0.10 \cdot 10{,}200 / (3 \cdot 30 \cdot (1 + 0.78))))]^3$

Seed1:

$S_{F,ANn}$ 0.0   35.9   gCOD/m^3

    ↓   ↑

    seed1:

MX_{OHOv} 10,200   10,200   kgVSS   (6.14b)

$\quad\quad\quad = Y_{OHOv} \cdot FCOD_{b,OHO} \cdot SRT / (1 + b_{OHO,T} \cdot SRT)$ (note that $FCOD_{b,OHO}$ is calculated in step 3.4)

$\quad\quad\quad = (0.45 \cdot 6{,}643 \cdot 11) / (1 + 0.202 \cdot 11)$

$S_{F,ANn}$ 35.9   29.3   gCOD/m^3

    ↓   ↑

    seed2:

MX_{OHOv} 13,140   13,140   kgVSS   (6.14b)

$\quad\quad\quad = Y_{OHOv} \cdot FCOD_{b,OHO} \cdot SRT / (1 + b_{OHO,T} \cdot SRT)$ (note: $FCOD_{b,OHO}$ is calculated in steps 3.3 and 3.4)

$\quad\quad\quad = (0.45 \cdot 8{,}558 \cdot 11) / (1 + 0.202 \cdot 11)$

$S_{F,ANn}$ 29.3   30.2   gCOD/m^3

    ↓   ↑

    seed3:

MX_{OHOv} 12,599   12,599   kgVSS   (6.14b)

$\quad\quad\quad = Y_{OHOv} \cdot FCOD_{b,OHO} \cdot SRT / (1 + b_{OHO,T} \cdot SRT)$ (note: $FCOD_{b,OHO}$ is calculated in steps 3.3. and 3.4)

$\quad\quad\quad = (0.45 \cdot 8{,}206 \cdot 11) / (1 + 0.202 \cdot 11)$

$S_{F,ANn}$ 30.2   30.2   gCOD/m^3

    ↓   ↑

    seed4:

MX_{OHOv} 12,673   12,673   kgVSS   (6.14b)

$\quad\quad\quad = Y_{OHOv} \cdot FCOD_{b,OHO} \cdot SRT / (1 + b_{OHO,T} \cdot SRT)$
(note: $FCOD_{b,OHO}$ is calculated in steps 3.3 and 3.4)
$\quad\quad\quad = (0.45 \cdot 8{,}254 \cdot 11) / (1 + 0.202 \cdot 11)$

4.3 VFAs stored by PAOs

$$FS_{S,PAO} = Q_i \cdot (S_{F,i,conv} - (1 + s) \cdot S_{F,ANn}) + Q_i \cdot S_{VFA,i} \tag{6.12}$$

$$= 30 \cdot (148 - (1 + 0.0.78) \cdot 30.2) + (30 \cdot 96)$$

$FS_{S,PAO}$ 5,696 kgCOD/d

4.4 Remaining biodegradable COD available to OHOs

$$FCOD_{b,OHO} = FCOD_{b,i} - FS_{S,PAO} \tag{6.13}$$

$$= 13,950 - 5,696$$

$FCOD_{b,OHO}$ 8,254 kgCOD/d

5. Biomass (VSS) equations

Corresponds to the biological mass present in the system as synthesized from the influent COD (in g/d), taking into account the cumulative effect of SRT [(g/d) $\cdot$ d = g in the system]

5.1 PAOs

Active mass

Y_{PAOv} 0.45 gVSS/gCOD

$$Y_{PAO,obs} = Y_{PAOv} / (1 + b_{PAO,T} \cdot SRT)$$

$$= 0.45 / (1 + 0.034 \cdot 11)$$

$Y_{PAO,obs}$ 0.328 gVSS / gCOD

$$MX_{PAOv} = Y_{PAO,obs} \cdot FS_{S,PAO} \cdot SRT \tag{6.2}$$

$$= 0.328 \cdot 5,696 \cdot 11$$

MX_{PAOv} 20,572 kgVSS in the system

Endogenous mass

$$MX_{E,PAOv} = f_{XE,PAO} \cdot b_{PAO,T} \cdot MX_{PAOv} \cdot SRT \tag{6.3}$$

$$= 0.25 \cdot 0.034 \cdot 20,572 \cdot 11$$

$MX_{E,PAOv}$ 1,906 kgVSS

5.2 OHOs

Active mass

Y_{OHOv} 0.45 gVSS/gCOD

$$Y_{OHO,obs} = Y_{OHO} / (1 + b_{OHO,T} \cdot SRT)$$

$$= 0.45 / (1 + 0.202 \cdot 11)$$

$Y_{OHO,obs}$ 0.140 gVSS/gCOD

$$MX_{OHOv} = Y_{OHO,obs} \cdot FCOD_{b,OHO} \cdot SRT$$

$$= 0.140 \cdot 8,254 \cdot 11$$

MX_{OHOv} 12,673 kgVSS (this value is the calculated MX_{OHOv} value of step 3.2)

Endogenous mass

$$MX_{E,OHOv} = f_{XE,OHO} \cdot b_{OHO,T} \cdot MX_{OHOv} \cdot SRT \tag{6.5}$$

$$= 0.20 \cdot 0.202 \cdot 12{,}673 \cdot 11$$

$MX_{E,OHOv}$ 5,637   kgVSS

5.3 Unbiodegradable particulate organics

MX_{Uv} $= Q_i \cdot X_{U,i} \cdot SRT / f_{CV}$ (6.6)

$$= 30 \cdot 75 \cdot 11 / 1.48$$

MX_{Uv} 16,723   kgVSS

6. VSS and TSS

6.1 VSS and active fraction

MX_B $= MX_{PAOv} + MX_{OHOv}$

$$= 20{,}572 + 12{,}673$$

MX_B 33,245   kgVSS

MX_{VSS} $= MX_{PAOv} + MX_{OHOv} + MX_{E,PAOv} + MX_{E,OHOv} + MX_{Uv}$ (6.24b)

$$= 20{,}572 + 12{,}673 + 1{,}906 + 5{,}637 + 16{,}723$$

MX_{VSS} 57,511   kgVSS

f_{av} $= MX_B / MX_{VSS}$

$$= 33{,}245 / 57{,}511$$

f_{av} 58 %

6.2 FSS

$f_{P,PAO,actual}$ $= [(Q_i \cdot SRT \cdot \Delta P_{SYS,actual}) - (f_P \cdot (MX_{VSS} - MX_{PAOv}))] / MX_{PAOv}$ (6.24a)

$$= [(30 \cdot 11 \cdot 11.5) - (0.03 \cdot (57{,}511 - 20{,}572))] / 20{,}572$$

$f_{P,PAO,actual}$ 0.12 gP/gVSS

MX_{FSS} $= f_{FSS,OHO} \cdot MX_{OHOv} + f_{FSS,PAO} \cdot (f_{P,PAO,actual} / f_{P,PAO}) \cdot MX_{PAOv} + FX_{FSS,i} \cdot SRT$ (6.24d)

$$= (0.15 \cdot 12{,}673) + (1.3 \cdot (0.12 / 0.38) \cdot 20{,}572) + (750 \cdot 11)$$

MX_{FSS} 18,596   kgFSS

6.3 TSS

MX_{TSS} $= MX_{VSS} + MX_{FSS}$ (6.25a)

$$= 57{,}511 + 18{,}596$$

MX_{TSS} 76,107   kgTSS

6.4 f_{VT}

f_{VT} $= MX_{VSS} / MX_{TSS}$ (6.25c)

$$= 57{,}511 / 76{,}107$$

f_{VT} 0.76   gVSS/gTSS

$f_{VT,PAO}$ $= MX_{PAOv} / MX_{TSS}$

$$= 20{,}572 / 76{,}107$$

$f_{VT,PAO}$ 0.27   gVSS/gTSS

6.5 P content of TSS

$f_{P,TSS}$ $= [(f_P \cdot MX_{OHOv}) + f_P \cdot (MX_{E,OHOv} + MX_{E,PAOv}) + (f_P \cdot MX_{Uv})$

$+ (f_{P,PAO,actual} \cdot MX_{PAOv}) + (f_{P,FSS,i} \cdot MX_{FSS})] / MX_{TSS}$ (6.26)

$= [(0.03 \cdot 12{,}673) + 0.03 \cdot (5{,}637 + 1{,}906) + (0.03 \cdot 16{,}723) + (0.12 \cdot 20{,}572) + (0.02 \cdot 18{,}569)] / 76{,}107$

$f_{P,TSS}$ 0.052 gP/gTSS

7. P removal

7.0 P release

$S_{PO4,rel}$ $= f_{PO4,rel} \cdot FS_{S,PAO} / Q_i$ (6.15)

$= 0.5 \cdot 5{,}696 / 30$

S_{PO4_rel} 94.9 gP/m^3 (note: gP/m^3 of influent, not gP/m^3 of AN reactor)

7.1 ΔP by PAOs

ΔP_{PAO} $= f_{P,PAO} \cdot MX_{PAOv} / (SRT \cdot Q_i)$ (6.16)

$= 0.38 \cdot 20{,}572 / (11 \cdot 30)$

ΔP_{PAO} 23.7 gP/m^3

7.2 ΔP by OHOs

ΔP_{OHO} $= f_P \cdot MX_{OHOv} / (SRT \cdot Q_i)$ (6.17)

$= 0.03 \cdot 12{,}673 / (11 \cdot 30)$

ΔP_{OHO} 1.15 gP/m^3

7.3 ΔP by endogenous mass

ΔP_{XE} $= \Delta P_{XE,PAO} + \Delta P_{XE,OHO}$ (6.18)

$\Delta P_{XE,PAO}$ $= f_P \cdot MX_{E,PAOv} / (SRT \cdot Q_i)$

$= 0.03 \cdot 1{,}906 / (11 \cdot 30)$

$\Delta P_{XE,PAO}$ 0.17 gP/m^3

$\Delta P_{XE,OHO}$ $= f_P \cdot MX_{E,OHOv} / (SRT \cdot Q_i)$

$= 0.03 \cdot 5{,}637 / (11 \cdot 30)$

$\Delta P_{XE,OHO}$ 0.51 gP/m^3

ΔP_{XE} 0.69 gP/m^3

7.4 ΔP by influent unbiodegradable organic mass

ΔP_{XU} $= f_P \cdot MX_{Uv} / (SRT \cdot Q_i)$ (6.19)

$= 0.03 \cdot 16{,}723 / (11 \cdot 30)$

ΔP_{XU} 1.52 gP/m^3

7.5 Potential total P removal

$\Delta P_{SYS,pot}$ $= \Delta P_{PAO} + \Delta P_{OHO} + \Delta P_{XE} + \Delta P_{XU}$ (6.20)

$= 23.7 + 1.15 + 0.69 + 1.52$

$\Delta P_{SYS,pot}$ 27.0 gP/m^3

7.6 Actual total P removal

P_i 	 11.5 	 gP/m^3

$\Delta P_{SYS,actual}$ 	 $= \min(\Delta P_{SYS,pot}; P_i)$ 	 (6.21)

$= \min(27.0; 11.5)$

$\Delta P_{SYS,actual}$ 	 11.5 	 gP/m^3

7.7 Particulate P in the effluent

To calculate after step 6.5 where the P content of TSS is calculated

$X_{P,e}$ 	 $= f_{P,TSS} \cdot X_{TSS,e}$ 	 (6.22)

$= 0.052 \cdot 15$

$X_{P,e}$ 	 0.78 	 gP/m^3

7.8 Effluent total P

P_e 	 $= P_i - \Delta P_{SYS,actual} + X_{P,e}$ 	 (6.23)

$= 11.5 - 11.5 + 0.78$

P_e 	 0.78 	 gP/m^3

With the proposed design, the EBPR WWTP is able to comply with the effluent discharge standard of less than 1 mgP/l ($= 1$ gP/m^3) with an effluent total P concentration of 0.78 gP/m^3 ($=$ mgP/l). Arguably, it cannot be expected to observe a full removal of total phosphorus due to the phosphorus intracellularly stored in the solids being lost through the effluent, but effluent discharge standards of less than 1 mgP/l can be met.

8. Effluent total COD

COD_e 	 $= S_{U,e} + f_{cv} \cdot X_{VSS,e}$ 	 (6.23)

$= S_{U,e} + f_{cv} \cdot f_{VT} \cdot X_{TSS,e}$

$= 45 + (1.48) \cdot (0.76) \cdot (15)$

COD_e 	 61.9 	 $gCOD/m^3$

Also, the effluent discharge standard of less than 125 mgCOD/l can be met.

9. Oxygen requirements for COD and EBPR processes

The 100 % mass balance for COD indicates that all the influent COD is accounted for in the calculated values of oxygen demand and sludge production. From the COD mass balance, and for the conditions of the design example, the fate of the influent COD is as follows: 48.2 % is oxidized with oxygen, 7.7 % escapes in the effluent as soluble unbiodegradable organics and 44.1 % becomes activated sludge. 57.8 % of the sludge is composed of active biomass (33,245 / 57,511) and 42.2 % (24,266 / 57,511) of inactive particulate matter of which 68.9 % (16,723 / 24,266) is influent unbiodegradable particulate organics and 31.1 % is endogenous residues ((1,906 + 5,637) / 24,266).

OD by PAOs: for synthesis and endogenous respiration

FO_{PAO} 	 $= FO_{PAO,s} + FO_{PAO,e}$ 	 (6.29a)

$FO_{PAO,s}$ 	 $= FS_{S,PAO} \cdot (1 - f_{cv} \cdot Y_{PAOv})$

$= 5,696 \cdot (1 - 1.48 \cdot 0.45)$

$FO_{PAO,s}$ 	 1,903 	 kgO_2/d

$FO_{PAO,e}$ $= FS_{S,PAO} \cdot f_{CV} \cdot (1 - f_{XE,PAO}) \cdot b_{PAO,T} \cdot Y_{PAO,obs} \cdot SRT$

$= 5{,}696 \cdot 1.48 \cdot (1 - 0.25) \cdot 0.034 \cdot 0.328 \cdot 11$

$FO_{PAO,e}$ 769 kgO_2/d

FO_{PAO} 2,672 kgO_2/d

OD by OHOs: for synthesis and endogenous respiration

FO_{OHO} $= FO_{OHO,s} + FO_{OHO,e}$ (6.30a)

$FO_{OHO,s}$ $= FCOD_{b,OHO} \cdot (1 - f_{CV} \cdot Y_{OHOv})$

$= 8{,}254 \cdot (1 - 1.48 \cdot 0.45)$

$FO_{OHO,s}$ 2,757

$FO_{OHO,e}$ $= FCOD_{b,OHO} \cdot f_{CV} \cdot (1 - f_{XE,OHO}) \cdot b_{OHO,T} \cdot Y_{OHO,obs} \cdot SRT$

$= 8{,}254 \cdot 1.48 \cdot (1 - 0.20) \cdot 0.202 \cdot 0.140 \cdot 11$

$FO_{OHO,e}$ 3,040

FO_{OHO} 5,790 kgO_2/d

OD total (carbonaceous)

FO_c $= FO_{PAO} + FO_{OHO}$ (6.31a)

$= 2{,}672 + 5{,}790$

FO_c 8,462 kgO_2/d

COD mass balance verification

Input

FCOD$_i$ 17,550 kgCOD/d 100 % IN

Output

O_2 demand for synthesis and endogenous respiration

FOc 8,462 kgCOD/d 48.2 %

Unbiodegradable soluble organics leaving via the effluent

FS$_{U,i}$ 1,350 kgCOD/d 7.7 %

Sludge gVSS gCOD/d (= gVSS · f_{CV} / SRT = gVSS · 1.48 / 11 = gVSS · 0.135)

MX_{PAOv} 20,572 2,768 kgCOD/d 15.8 %

MX_{OHOv} 12,673 1,705 kgCOD/d 9.7 %

MX_B 33,245 4,473 kgCOD/d 25.5 %

$MX_{E,PAOv}$ 1,906 256 kgCOD/d 1.5 %

$MX_{E,OHOv}$ 5,637 758 kgCOD/d 4.3 %

MX_{Uv} 16,723 2,250 kgCOD/d 12.8 %

$MX_{Ev} + MX_{Uv}$ 24,266 3,265 kgCOD/d 18.6 %

MX_{VSS} 57,511 7,738 kgCOD/d 44.1 %

Sum: 17,550 kgCOD/d 100 % OUT

Delta (OUT-IN): 0 kgCOD/d 0 %

10. Nitrogen requirements for COD and EBPR processes

FN_s	$= f_n \cdot MX_{VSS} / SRT$		(6.28a)
	$= (0.10 \cdot 57,511) / 11$		
FN_s	523	kgN/d	
N_s	$= FN_s / Q_i$		
	$= 523 / 30$		
N_s	17.4	gN/m^3	

11. Nitrification process

11.1. Nitrification capacity and nitrate generation

NIT_c	$= TKN_i - TKN_e - N_s$		(5.35)

Where:

TKN_e	$= S_{NHx,e} + N_{ous,i}$		(5.33)

and,

$S_{NHx,e}$	$= [K_{ANO,T} (b_{ANO,T} + 1 / SRT)] / [\mu_{ANO,max,T} (1 - f_{xt}) - (b_{ANO,T} + 1 / SRT)]$		(5.15)
	$= [0.50 \cdot (0.034 + 1 / 11)] / [0.224 \cdot (1 - 0.30) - (0.034 + 1/ 11)]$		
$S_{NHx,e}$	1.91	gN/m^3	

From the N-fractionation diagram (Fig. 6.7)

$N_{ous,i}$	0.49	gN/m^3

Thus:

TKN_e	$= 1.91 + 0.49$	
TKN_e	2.40	gN/m^3

And:

NIT_c	$= 38.5 - 2.40 - 17.4$	
NIT_c	18.7	gN/m^3

Since the system has been designed with the SRT_{min} of 11 d, nitrification takes place and consequently nitrate, S_{NO3}, is generated (which is equal to NIT_c). However, the S_{NO3} concentration does not correspond to the actual effluent nitrate concentration since the A2O system will be designed with a pre-denitrification stage. This is discussed in detail below.

NIT_c	$= S_{NO3} = TKN_i - TKN_e - N_s$		(5.35)

So:

S_{NO3}	$= 18.7$	gN/m^3

11.2 ANO biomass

Combining equations 5.35 and 5.42b, the ANO biomass (MX_{ANOv}) can be estimated as:

MX_{ANOv}	$= Q_i NIT_c Y_{ANO} SRT / (1 + b_{ANO,T} \cdot SRT)$	
	$= (30 \cdot 18.7 \cdot 0.10 \cdot 11) / [1 + (0.034) \cdot (11)]$	
MX_{ANOv}	$= 450$	$kg\ VSS$

Compared to the MX_{VSS} and MX_{TSS}, MX_{ANOv} comprises a minimum fraction in the system:

$f_{VT,ANO}$ $\quad = MX_{ANOv} / MX_{TSS}$

$\quad = 450 / 76,107$

$f_{VT,ANO}$ $\quad = 0.006 \quad$ kgVSS/kgTSS

The $f_{VT,ANO}$ ratio indicates that MX_{ANOv} composes of approximately 0.6 % of MX_{TSS} in the system.

11.3. Oxygen demand for nitrification.

Based on eqs. 5.35, 5.42b and 5.43b:

FO_{NIT} $\quad = 4.57 \cdot Q_i \cdot NIT_c$

$\quad = 4.57 \cdot 30 \cdot 18.7$

FO_{NIT} $\quad 2,560 \quad$ kgO$_2$/d

12. Denitrification process

12.1 Denitrification potential in the A2O system (NDEBPR)

D_{p1} $\quad = S_{F,ANn} \cdot (1 + r) \cdot (1 - f_{cv} \cdot Y_{OHOv}) / 2.86 + f_{x1} \cdot K'_{2,T} \cdot (COD_{b,i} - S_{s,PAO}) \cdot Y_{OHO,v} \cdot SRT / (1 + b_{OHO,T} \cdot SRT)$ $\qquad$ (6.34a)

$\quad = (30.2) \cdot (1+0) \cdot (1 - 1.48 \cdot 0.45) / 2.86 + [(0.20 \cdot 0.14)[465 - 189.9](0.45 \cdot 11)] / (1 + 0.202 \cdot 11)$

D_{p1} $\quad 15.73 \quad$ gN/m^3

In this example, $r = 0$ because there is no internal recirculation from the anoxic to the anaerobic reactor. Also, $S_{s,PAO}$ is estimated as $FS_{S,PAO} / Q_i = 5,696 / 30 = 189.9$ mg COD/l.

12.2 a-recycle ratio

Optimum a recycle (a_{opt})

a_{opt} $\quad = \{-B + (B^2 + 4AC)^{1/2}\} / 2A$ $\qquad$ (5.56)

A $\quad = S_{O2,a} / 2.86$

B $\quad = NIT_c - D_{p1} + [(s + 1) \cdot S_{O2,a} + s \cdot S_{O2,s}] / 2.86$

C $\quad = (s + 1) \cdot (D_{p1} - s \, S_{O2,s} / 2.86) - s \, NIT_c$, thus:

A $\quad = 2 / 2.86$

A $\quad 0.70$

B $\quad = 18.67 - 15.73 + [(0.78 + 1) \cdot (2) + (0.78) \cdot (0)] / 2.86$

B $\quad 4.18$

C $\quad = (0.78 + 1) \cdot (15.73 - (0.78) \cdot (0) / 2.86) - (0.78) \cdot (18.67)$

C $\quad 13.44$, thus:

a_{opt} $\quad = \{-(4.18) + [(4.18)^2 + 4 \cdot (0.70) \cdot (13.44)]^{1/2}\} / (2) \cdot (0.70)$

a_{opt} $\quad 2.32$

Selected a $\quad = \min \{a_{opt} \, ; \, a_{prac}\}$

$\quad = \min \{2.32 \, ; \, 5.0\}$

$\quad\quad a_{opt}$

Selected a $\quad 2.32$

Note that, the calculations of B and C, $S_{O2,s}$ is considered zero it this step because it is a A2O configuration and, as such, the s-recycle flowrate discharges in the anaerobic reactor and not in the primary anoxic reactor having, therefore, no influence in the denitrification process. Nevertheless, the actual value of $S_{O2,s}$ (1 mgO$_2$/l) was considered when designing the EBPR process (Table 6.10, step 2.3).

12.3 Effluent nitrogen concentrations

$$S_{NO3,e,min} \qquad = S_{NO3,e,opt} = NIT_e \,/\, (a_{opt} + s + 1) \tag{5.57}$$

$$= 18.67 \,/\, (2.32 + 0.78 + 1)$$

$$S_{NO3,e,min} \qquad = S_{NO3,e} = 4.56$$

$$S_{NO3,e} \qquad 4.56 \qquad gN/m^3$$

$$TKN_e \qquad = S_{NHx,e} + N_{ous,i}$$

$$= 1.91 + 0.49$$

$$TKN_e \qquad 2.4 \qquad gN/m^3$$

$$S_{NO3,e} + TKN_e \; = 4.56 + 2.40$$

$$6.96 \qquad gN/m^3$$

The previous concentration only considers the presence of soluble N compounds. If the contribution of the solids lost through the effluent is calculated to estimate the particulate N:

$$X_{N,e} \qquad = f_n \cdot f_{VT} \cdot X_{TSS,e}$$

$$= (0.10) \cdot (0.76) \cdot (15)$$

$$X_{N,e} \qquad 1.14 \qquad gN/m^3$$

Thus, total effluent N:

$$N_e \qquad = S_{NHx,e} + N_{ous,i} + S_{NO3,e} + X_{N,e}$$

$$= TKN_e + S_{NO3,e} + X_{N,e}$$

$$= 2.40 + 4.56 + 1.14$$

$$N_e \qquad 8.1 \qquad gN/m^3$$

As with the other parameters (e.g., COD and P), the system can comply with the effluent discharge standard of less than 10 mgN/l.

13. Total oxygen consumption

13.1. Oxygen recovery by denitrification

$$FO_{DENIT} \qquad = 2.86 \cdot (NIT_e - S_{NO3,e}) \cdot Q_i \tag{5.62}$$

$$= 2.86 \cdot (18.67 - 4.56) \cdot (30)$$

$$FO_{DENIT} \qquad 1{,}211 \qquad kgO_2/d$$

13.2. Total oxygen consumption

$$FO_{t,DENIT} \qquad = FO_c + FO_{NIT} - FO_{DENIT}$$

$$= 8{,}462 + 2{,}560 - 1{,}211$$

$$FO_{t,DENIT} \qquad 9{,}811 \qquad kgO_2/d$$

The oxygen consumption decreases by approximately 11 % due to the oxygen recovery by denitrification.

14. Reactor volume

V_R $= MX_t / X_{TSS}$

         $= 76{,}529 / 3.5$ (MX$_t$ after the addition of MX_{ANOv} which is sometimes neglected)

         21,873   m^3

And the volumes of the different tanks can be estimated based on the mass fractions of each stage:

$V_{R,AN}$ $= f_{xa} \cdot V_R$

         $= 0.10 \cdot 21{,}873$

$V_{R,AN}$ 2,187   m^3

$V_{R,AX}$ $= f_{x1} \cdot V_R$

         $= 0.20 \cdot 21{,}873$

$V_{R,AX}$ 4,373   m^3

$V_{R,OX}$ $= f_{ox} \cdot V_R$

         $= (1 - 0.10 - 0.20) \cdot 21{,}873$

$V_{R,OX}$ 15,311   m^3

Table 6.11 Summary of the design results of the 3-stage modified Bardenpho (also known as A2O) system.

Description	Parameter	Unit	Value
1. Influent and bioreactor			
Type of wastewater	raw		raw
Temperature	T	°C	14
Influent flow rate	Q_i	MLD	30
Influent total COD	COD_i	gCOD/m^3	585
Influent soluble biodegradable COD	$S_{S,i}$	gCOD/m^3	276
Influent biodegradable COD	$COD_{b,i}$	gCOD/m^3	465
Influent total nitrogen concentration	N_i	gN/m^3	38.5
Influent total Kjeldahl nitrogen (TKN)	TKN_i	gN/m^3	38.5
Influent free and saline ammonia	$S_{NHx,i}$	gN/m^3	29.1
Influent unbiodegradable soluble nitrogen	$N_{ous,i}$	gN/m^3	0.49
Influent total P	P_i	gP/m^3	11.5
Influent orthophosphate	$S_{PO4,i}$	gP/m^3	7.6
Influent inorganic (fixed) suspended solids	$X_{FSS,i}$	gFSS/m^3	25
Sludge retention time	SRT	d	11
Internal recycle ratio (a)	a	m^3.d/m^3.d	2.32
Oxygen concentration in a recycle	$S_{O2,a}$	gO$_2$/m^3	2
Sludge recycle ratio (s)	s	m^3.d/m^3.d	0.78
Oxygen concentration in sludge recycle	$S_{O2,s}$	gO$_2$/m^3	1
Nitrate concentration in sludge recycle	$S_{NO3,s}$	gN/m^3	4.5

2. Portion of $S_{S,i}$ for PAOs and of $COD_{b,i}$ for OHOs

Concentration of fermentable COD in the last AN reactor	$S_{F,ANn}$	$gCOD/m^3$	30.2
Flux of $S_{S,i}$ for PAOs	$FS_{S,PAO}$	$kgCOD/d$	5,696
Flux of $COD_{b,i}$ for OHOs	$FCOD_{b,OHO}$	$kgCOD/d$	8,254

3. System biomass (VSS) equations

Mass of PAOs	MX_{PAOv}	$kgVSS$	20,572
Mass of endogenous residue from PAOs	$MX_{E,PAOv}$	$kgVSS$	1,906
Mass of OHOs	MX_{OHOv}	$kgVSS$	12,673
Mass of endogenous residue from OHOs	$MX_{E,OHOv}$	$kgVSS$	5,637
Mass of ANO	MX_{ANOv}	$kgVSS$	450
Mass of unbiodegradable organics from influent	MX_{Uv}	$kgVSS$	16,723

4. COD removal

Effluent soluble COD concentration	$S_{U,e}$	$gCOD/m^3$	45.0
Effluent total COD concentration	COD_e	$gCOD/m^3$	61.9

5. Nitrogen removal

Nitrogen requirements for biomass synthesis	N_s	gN/m^3	17.4
Nitrification capacity	NIT_c	gN/m^3	18.7
Denitrification potential	D_{p1}	gN/m^3	15.7
Effluent free and saline ammonia concentration	$S_{NHx,e}$	gN/m^3	1.9
Effluent total Kjeldahl concentration	TKN_e	gN/m^3	2.4
Effluent nitrate concentration	$S_{NO3,e}$	gN/m^3	4.6
Effluent particulate nitrogen	$X_{N,e}$	gN/m^3	1.1
Effluent total nitrogen concentration	N_e	gN/m^3	8.1

6. P removal

PO4 release	S_{PO4_rel}	gP/m^3	94.9
Maximum P removal by PAOs	ΔP_{PAO}	gP/m^3	23.7
Actual P removal by PAOs	$\Delta P_{PAO,actual}$	gP/m^3	7.46
P removal by OHOs	ΔP_{OHO}	gP/m^3	1.15
P removal by endogenous residue	ΔP_{XE}	gP/m^3	0.69
P removal by X_U	ΔP_{XU}	gP/m^3	1.52
Potential P removal by system	$\Delta P_{SYS,pot}$	gP/m^3	27.1
Actual P removal by system	$\Delta P_{SYS,actual}$	gP/m^3	11.5
Effluent particulate P (from X_e)	$X_{P,e}$	gP/m^3	0.78
Influent total P	P_i	gP/m^3	11.5
Effluent total P	P_e	gP/m^3	0.78

7. Volatile and total suspended solids (VSS and TSS) in system

Mass of active biomass	MX_B	kgVSS	33,695
Mass of VSS	MX_{VSS}	kgVSS	57,961
Ratio of AVSS/VSS	f_{av}	gAVSS/gVSS	0.58
Mass of fixed SS	MX_{FSS}	kgFSS	18,596
Mass of TSS	MX_{TSS}	kgTSS	76,556
Ratio of VSS/TSS	f_{VT}	gVSS/gTSS	0.76
Fraction of P in TSS	$f_{P,TSS}$	gP/gTSS	0.05

8. Reactor volume

Anaerobic reactor volume	$V_{R,AN}$	m^3	2,187
Anoxic reactor volume	$V_{R,AX}$	m^3	4,373
Aerobic reactor volume	$V_{R,OX}$	m^3	15,311
Bioreactor volume	V_R	m^3	21,873

9. Oxygen demand

Flux of O_2 demand by PAOs	FO_{PAO}	kgO_2/d	2,672
Flux of O2 demand by OHOs	FO_{OHO}	kgO_2/d	5,790
Flux of carbonaceous O_2 demand	FO_c	kgO_2/d	8,462
Flux of O_2 demand by ANO	FO_{NIT}	kgO_2/d	2,560
Flux of O_2 recovered by denitrification	FO_{DENIT}	kgO_2/d	1,211
Total O_2 flux	$FO_{t,DENIT}$	kgO_2/d	9,811

6.4 EXERCISES

EBPR principles (exercises 6.4.1-6.4.5)

Exercise 6.4.1

Using a chart and graph, describe the most basic process configuration that can lead to the proliferation and enrichment of PAO in activated sludge wastewater treatment systems.

Exercise 6.4.2

In what form do PAO store internally the phosphorus that they remove?

Exercise 6.4.3

Which factors are known to trigger the competition between PAO and GAO?

Exercise 6.4.4

What is the main selection pressure that favours the proliferation of PAO in an EBPR system?

Exercise 6.4.5

What are the main characteristics of EBPR systems?

EBPR microbiology (exercises 6.4.6-6.4.10)

Exercise 6.4.6

List the two most common PAO found in full-scale EBPR WWTPs.

Exercise 6.4.7

Explain the classical definition for the PAO phenotype.

Exercise 6.4.8

Explain the general metabolism of a PAO phenotype.

Exercise 6.4.9

Why is phosphorus the key element to remove from wastewater to minimize the occurrence of eutrophication in surface water bodies?

EBPR microbial metabolism (exercises 6.4.10-6.4.15)

Exercise 6.4.10

What is the principal source of ATP for anaerobic organic carbon uptake by PAO?

Exercise 6.4.11

Name two ways that NADH can be produced anaerobically within PAO.

Exercise 6.4.12

Why is more phosphorus stored aerobically than released anaerobically by PAO?

Exercise 6.4.13

What determines the type of PHA stored by PAOs?

Exercise 6.4.14

How does pH affect the VFA uptake by PAOs?

Exercise 6.4.15

What is the main difference in terms of ATP and reducing power generation between the glycogen degradation and polyphosphate hydrolysis processes?

EBPR process configuration (exercises 6.4.16-6.4.18)

Exercise 6.4.16

What are the advantages and disadvantages of the modified UCT configuration compared to the 5-stage Bardenpho configuration?

Exercise 6.4.17

What is the source of organic carbon metabolized by PAO in S2EBPR systems?

Exercise 6.4.18

What is the function of the anaerobic zone in EBPR systems?

Effects of key environmental, design and operational factors on EBPR (exercises 6.4.19-6.4.25)

Exercise 6.4.19

Explain why recycling nitrate or oxygen into the anaerobic reactor can destabilize the EBPR process.

Exercise 6.4.20

Name two conditions that can promote the growth of glycogen-accumulating organisms.

Exercise 6.4.21

How does a primary settler affect the EBPR process?

Exercise 6.4.22

Discuss the effect of long SRT (>20 d) on EBPR systems.

Exercise 6.4.23

If nitrification could take place in the EBPR Phoredox A/O system designed in example 6.3.1:

    a.   What could be the maximum nitrate concentration ($S_{NO3,s}$) that could get into the anaerobic reactor via the sludge recycle and consume all S_F concentration?

    b.   What will be the expected P_c?

Exercise 6.4.24

An AO EBPR configuration treats a pre-settled wastewater that contains a S_{bi} concentration of 125 mgCOD/l with an influent dissolved oxygen ($S_{O2,i}$) and nitrate ($S_{NO3,i}$) concentrations of 1 mg/l and 3 mgN/l, respectively. If the sludge recycle ratio of 0.75 (s) contains a dissolved oxygen concentration of 1 mg/l ($S_{O2,s}$), what will be the sludge recycle nitrate concentration ($S_{NO3,s}$) that could lead to the complete consumption of the influent S_{bi} concentration and result in the collapse of the EBPR process?

Exercise 6.4.25

A UCT EBPR configuration treats a raw wastewater that contains a S_{VFA} and S_F concentrations of 75 mgCOD/l and 90 mgCOD/l respectively. Also, the influent dissolved oxygen ($S_{O2,i}$) and nitrate ($S_{NO3,i}$) concentrations are 1 mg/l and 2 mgN/l, respectively. If the internal recycle ratio r is 1.0 and assuming that it does not contain oxygen ($S_{O2,r} = 0$), what will be the recycle nitrate concentration ($S_{NO3,r}$) that could lead to the complete consumption of the influent S_F concentration, jeopardizing the efficiency of the EBPR process?

EBPR process design and evaluation (exercises 6.4.26-6.4.29)

Exercise 6.4.26

For Example 4.3.1, convert the first 10 % of the aerobic reactor to an anaerobic reactor to modify the configuration to a Phoredox or AO system for EBPR to meet an effluent total phosphorus concentration lower than 1 mgP/l. For this purpose, also consider that 45 % of $S_{S,i}$ is composed of S_{VFA}. Assume that oxygen is absent in the influent but present in the sludge recycle ($S_{O2,s} = 1$ mg/l). Also, consider the absence of nitrate in the influent and sludge recycle flows.

Exercise 6.4.27

An MLE system has been designed to treat an influent wastewater (Q_i) that has a flow rate of 20 MLD (20,000 m³/d). The wastewater contains a total influent COD concentration (COD_i) of 494 mg/l, an influent total nitrogen concentration (N_i) of 42.5 mgP/l and influent total phosphorus concentration (P_i) of 12.4 mgP/l. In addition, the influent flow rate contains 35 mg/l of inorganic or fixed suspended solids.

The MLE plant was designed to comply with an effluent total COD concentration of 125 mg/l and effluent total nitrogen concentration of 10 mg N/l, but now it has to comply with an effluent total phosphorus concentration of 1 mg/l. The plant operates at an average MLSS concentration of 3,500 mg/l and the lowest yearly wastewater temperature is 12 $^\circ$C. The MLE system was designed, assuming: (a) a maximum specific biomass growth rate of nitrifiers ($\mu_{ANO,max,20}$) of 0.45 1/d, (b) a safety factor for nitrification of 1.25, (c) absence of dissolved oxygen and nitrate in the influent flow rate, (d) a dissolved oxygen concentration of 2 mg/l in the a-recycle ratio and 1 mg/l in the s recycle, and (e) a TSS concentration in the sludge recycle line of 12,000 mg TSS/l with a s-recycle ratio of 0.41. In addition, to upgrade the MLE system to an A2O configuration consider: (a) a TSS concentration in the effluent of the secondary settling tank of 15 mg TSS/l, and (b) a nitrate concentration in the s-recycle ratio of 4.5 mg/l (approximately similar to the expected effluent nitrate concentration, $S_{NO3,e}$).

The main design data and the main stoichiometric and kinetic parameters necessary to carry out the process design are presented in tables 6.12 and 6.13.

Table 6.12 Summary of the MLE system data to upgrade it to an A2O process configuration.

Description	Parameter	Unit	Value
1. Nitrogen removal design and operating conditions			
Nitrogen requirements for biomass synthesis	N_s	gN/m^3	12.5
Nitrification capacity	NIT_c	gN/m^3	28.0
Denitrification potential	D_{p1}	gN/m^3	38.7
Internal recycle ratio (a)	a	m^3.d/m^3.d	5.0
Effluent free and saline ammonia concentration	$S_{NHx,e}$	gN/m^3	1.5
Effluent total Kjeldahl concentration	TKN_e	gN/m^3	2.0
Effluent nitrate concentration	$S_{NO3,e}$	gN/m^3	4.4
Effluent total nitrogen concentration	N_e	gN/m^3	7.5
2. Volatile and total suspended solids (VSS and TSS) in system			
Mass of active biomass	MX_B	kgVSS	13,051
Mass of VSS	MX_{VSS}	kgVSS	65,622
Ratio of AVSS/VSS	f_{av}	gAVSS/gVSS	0.20
Mass of fixed SS	MX_{FSS}	kgFSS	20,038
Mass of TSS	MX_{TSS}	kgTSS	85,560
Ratio of VSS/TSS	f_{VT}	gVSS/gTSS	0.76
3. Reactor volume			
Anoxic reactor volume	$V_{R,AX}$	m^3	12,237
Aerobic reactor volume	$V_{R,OX}$	m^3	12,237
Bioreactor volume	V_R	m^3	24,474
4. Oxygen demand			
Flux of carbonaceous O$_2$ demand	FO_c	kgO$_2$/d	4,850
Flux of O$_2$ demand by ANO	FO_{NIT}	kgO$_2$/d	2,564
Flux of O$_2$ recovered by denitrification	FO_{DENIT}	kgO$_2$/d	1,354
Total O$_2$ flux	$FO_{t,DENIT}$	kgO$_2$/d	6,060

Table 6.13 Summary of the MLE system data to carry out the upgrade to an A2O process.

Description	Symbol	Value	Unit
Flow rate	Q_i	20	MLD
Total COD	COD_i	494	$gCOD/m^3$
COD concentrations			
- readily biodegradable COD	$S_{S,i}$	189	$gCOD/m^3$
- volatile fatty acids	$S_{VFA,i}$	96	$gCOD/m^3$
- fermentable COD	$S_{F,i}$	93	$gCOD/m^3$
- slowly biodegradable COD	$X_{S,i}$	123	$gCOD/m^3$
- unbiodegradable soluble COD	$S_{U,i}$	67	$gCOD/m^3$
- unbiodegradable particulate COD	$X_{U,i}$	115	$gCOD/m^3$
Total nitrogen concentration	N_i	42.5	gN/m^3
- Total Kjeldahl nitrogen (TKN)	TKN_i	42.5[a]	gN/m^3
Inorganic nitrogen			
- free and saline ammonia	$S_{NHx,i}$	29.1	gN/m^3
Organic nitrogen			
- biodegradable soluble nitrogen	$N_{obs,i}$	4.51	gN/m^3
- biodegradable particulate nitrogen	$N_{obp,i}$	0.63	gN/m^3
- unbiodegradable soluble nitrogen	$N_{ous,i}$	0.49	gN/m^3
- unbiodegradable particulate nitrogen	$N_{oup,i}$	7.77	gN/m^3
Influent total P concentration	P_i	12.4	gP/m^3
Inorganic phosphorus			
- orthophosphate	$S_{PO4,i}$	7.6	gP/m^3
Organic phosphorus			
- biodegradable soluble phosphorus	$P_{obs,i}$	0.49	gP/m^3
- biodegradable particulate phosphorus	$P_{obp,i}$	1.97	gP/m^3
- unbiodegradable soluble phosphorus	$P_{ous,i}$	0.01	gP/m^3
- unbiodegradable particulate phosphorus	$P_{oup,i}$	2.33	gP/m^3
Influent inorganic (fixed) suspended solids	$X_{FSS,i}$	35	$gFSS/m^3$
Temperature	T	12	$°C$
Sludge retention time of MLE system	SRT	26	d
Anaerobic mass fraction	f_{xa}	0.10[b]	-
Number of anaerobic zones	N	3[b]	reactors
Anoxic mass fraction of MLE system	f_{x1}	0.35[c]	-
Dissolved O_2 in the influent	$S_{O2,i}$	0	gO_2/m^3
Dissolved O_2 in the a recycle	$S_{O2,a}$	2	gO_2/m^3
Dissolved O_2 in the sludge recycle	$S_{O2,s}$	1	gO_2/m^3
Influent nitrate concentration	$S_{NO3,i}$	0	$gNO_3\text{-}N/m^3$
Nitrate concentration in the sludge recycle	$S_{NO3,s}$	4.5	$gNO_3\text{-}N/m^3$
Total suspended solids in the effluent	$X_{TSS,e}$	15	$gTSS/m^3$
Design aerobic TSS concentration	$X_{TSS,OX}$	3,500	$gTSS/m^3$
s-recycle ratio	s	0.41	-

[a]Assuming that the influent contains neither nitrate nor nitrite ($S_{NO3,i}$ or $S_{NO2,i}$, respectively). [b]An initial f_{xa} fraction should be suggested (in this example divided into 3 compartments, N) to carry out the first iteration. [c]An initial f_{x1} fraction will be suggested to carry out the first iteration.

Exercise 6.4.28

For Example 6.3.3, instead of designing a 3-stage modified Bardenpho system, design a UCT process configuration to comply with an effluent total COD concentration of 125 mg/l, effluent total nitrogen concentration of 10 mg N/l, and effluent total phosphorus concentration of 1 mg/l. Assume the absence of nitrate and oxygen in the r-recycle ratio.

Exercise 6.4.29

For Example 6.3.3 and Exercise 6.4.28 design the UCT process configuration considering a r-recycle ratio of 1.0, the presence of oxygen in the influent flow rate ($S_{O2,i}$ = 1 mg/l) and of nitrate in the r-recycle ratio ($S_{NO3,r}$ = 3 mgN/l), while keeping the same volumes determined in Example 6.3.3 (V_R = 21,866 m^3 composed of $V_{R,AN}$ = 2,187 m^3, $V_{R,AX}$ = 4,373 m^3 and $V_{R,OX}$ = 15,306 m^3).

WWTP plant Gama in Brazil – an example of a 5-stage Bardenpho process for the removal of organic matter, nitrogen and phosphorus (photo: G. Da Luz Lima Júnior).

ANNEX 1: SOLUTIONS TO EXERCISES
EBPR principles (solutions 6.4.1-6.4.5)
Solution 6.4.1

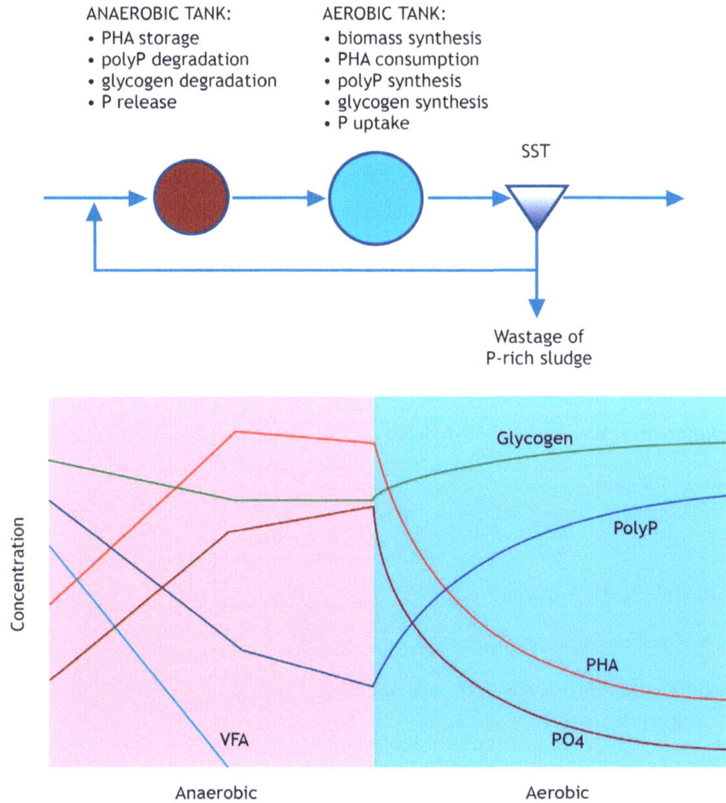

Figure 6.10. Most basic configuration that can favour the proliferation of PAO for the implementation of EBPR systems.

Solution 6.4.2
Polyphosphate.

Solution 6.4.3
Several factors exist that affect the PAO-GAO competition, including: (*i*) type of VFA, (*ii*) dissolved oxygen concentration, (*iii*) pH, and (*iv*) temperature.

Solution 6.4.4
There are several factors, but the most accepted is the P/VFA ratio fed. A higher ratio (e.g., P/VFA>0.5) would give rise to PAOs whereas a lower one (i.e., P/VFA<0.5) to GAOs.

Solution 6.4.5
The selective growth of biomass that is capable of removing phosphorus beyond its anabolic requirements by accumulating intracellular polyphosphate reserves.

EBPR microbiology (solutions 6.4.6-6.4.9)
Solution 6.4.6
Candidatus Accumulibacter and *Tetrasphaera.*

Solution 6.4.7
Anaerobic P release with PHA storage, aerobic (or anoxic) P uptake and growth from PHA oxidation.

Solution 6.4.8
In the absence of any electron acceptor, PAO sequesters and intracellularly stores RBCOD as PHA. The energy and redox sources are provided by utilizing intracellular polymers (poly-P and glycogen). Thereafter, when exposed to the presence of electron acceptors (e.g., nitrate or oxygen), PAO oxidizes the previously stored PHA to replenish their intracellular polymer pools (poly-P and glycogen) and to grow.

Solution 6.4.9
Unlike nitrogen (which can be fixed from the atmosphere), phosphorus can only be available for the microorganisms when it is dissolved in water.

EBPR microbial metabolism (exercises 6.4.10-6.4.15)
Solution 6.4.10
Polyphosphate hydrolysis.

Solution 6.4.11
Glycogen degradation and the TCA cycle.

Solution 6.4.12
PAO cells grow aerobically; the increase in cell quantity (growth) leads to higher levels of P taken up aerobically than released anaerobically.

Solution 6.4.13
The type of VFA stored determines the types of PHA stored: a higher consumption of acetate (HAc) leads to a higher accumulation of poly-β-hydroxybutyrate (PHB), whereas a higher consumption and storage of propionate (HPr) leads to a higher generation and accumulation of poly-β-hydroxyvalerate (PHV) and poly-β-hydroxy-2-methylvalerate (PH2MV).

Solution 6.4.14
VFAs require energy to be transported through the cell membrane, which is often provided by polyphosphate hydrolysis and glycogen consumption. As such, the energy requirements for VFA uptake increase when the pH increases.

Solution 6.4.15
During glycogen degradation ATP and NADH are formed, while during the hydrolysis of poly-P only ATP is generated.

EBPR process configuration (solutions (6.4.16-6.4.18)

Solution 6.4.16

The modified UCT configuration tends to minimize the intrusion of electron acceptors into the anaerobic stage, which can result in a stable and robust EBPR process. However, compared to the 5-stage Bardenpo configuration, it has a lower anaerobic solids concentration and tends to have a more complex operation. On the other hand, the 5-stage Bardenpho has a relatively simpler configuration that can also lead to lower effluent nitrate concentrations (e.g., by adding external carbon sources in the secondary anoxic reactor). However, the 5-stage Bardenpho may have higher construction and operation costs (since the system requires 5 different tanks or stages and due to the purchase of the external carbon source).

Solution 6.4.17

Hydrolysis and fermentation of RAS producing soluble organic carbon (e.g., VFA).

Solution 6.4.18

Conversion of fermentable COD into VFA, allowing PAOs to intracellularly store the influent and generated VFA.

Effects of key environmental, design and operational factors on EBPR (solutions 6.4.19-6.4.25)

Solution 6.4.19

With nitrate present in the anaerobic zone, denitrifiers compete for organic carbon uptake with PAO and fermenters, leading to less carbon source taken up by PAO anaerobically for PHA production, and less aerobic P uptake from PHA oxidation.

Solution 6.4.20

(A combination of) high temperature, low pH, high influent COD/P ratio.

Solution 6.4.21

The sedimentation and removal of influent biodegradable particulate COD ($X_{S,i}$) in the primary settling tank (PST) would reduce the flux of biodegradable COD entering into the activated sludge system. This decreases the concentration of fermentable of COD converted, while the influent orthophosphate concentration (the most abundant P compound) will not be affected by the PST. Thus, less active biomass grows in the system and less fermented COD can be generated in the anaerobic reactor ($S_{F,i,conv}$). Consequently, the COD flux to PAOs will tend to be lower ($FS_{s,PAO}$), decreasing the PAO biomass, and meanwhile the influent orthophosphate concentration will be similar with or without the PST.

Solution 6.4.22

Long SRT leads to lower active biomass. Thus, the poly-P content in PAO biomass would be higher and could reach their maximum storage capacity, limiting their P-removal capacity.

Solution 6.4.23

Based on Eq. 6.43, maximum $S_{NO3,s}$ concentration is 14.6 mgN/l. Keeping the rest of the design conditions presented in Example 6.3.1, the expected P_e will be approximately 0.52 mg/l and the system will still comply with the effluent discharge standard of 1 mgP/l.

Solution 6.4.24

It can be demonstrated that the COD consumed by O_2 and NO_3 intrusion into the anaerobic stage can be estimated as:

$$S_{bi} = 3\,(S_{O2,i} + r\,S_{O2,r} + s\,S_{O2,s}) + 8.6\,(S_{NO3,i} + r\,S_{NO3,r} + s\,S_{NO3,s})$$

Thus, solving for $S_{NO3,s}$, its maximum concentration that can lead full S_{bi} consumption is: 14.1 mgN/l.

Solution 6.4.25

$S_{NO3,r} = 8.1$ mgN/l.

EBPR process design and evaluation (solutions (6.4.26-6.4.29)

Solution 6.4.26

Suggesting a $f_{xa} = 0.10$ divided into 3 compartments:

Table 6.12 Summary of the retrofit of an aerobic system to achieve C and P removal.

Description	Parameter	Unit	Value
1. Influent and bioreactor			
Type of wastewater	raw		raw
Temperature	T	°C	14
Influent flow rate	Q_i	MLD	15
Influent total COD	COD_i	gCOD/m^3	585
Influent soluble biodegradable COD	$S_{S,i}$	gCOD/m^3	165
Influent fermentable COD	$S_{F,i}$	gCOD/m^3	91
Influent volatile fatty acids	$S_{VFA,i}$	gCOD/m^3	74
Influent biodegradable COD	$COD_{b,i}$	gCOD/m^3	364
Influent TKN	TKN_i	gN/m^3	44.5
Influent total P	P_i	gP/m^3	14.5
Influent inorganic (fixed) suspended solids	$X_{FSS,i}$	gFSS/m^3	35
Sludge retention time	SRT	d	6
Sludge recycle ratio (s)	s	m^3.d/m^3.d	0.67
Oxygen concentration in sludge recycle	$S_{O2,s}$	gO$_2$/m^3	1
Nitrate concentration in s-recycle	$S_{NO3,s}$	gN/m^3	0
2. Portion of $S_{S,i}$ for PAOs and of $COD_{b,i}$ for OHOs			
Concentration of fermentable COD in the last AN reactor	$S_{F,ANn}$	gCOD/m^3	24.4
Flux of $S_{S,i}$ for PAOs	$FS_{S,PAO}$	kgCOD/d	1,835
Flux of $COD_{b,i}$ for OHOs	$FCOD_{b,OHO}$	kgCOD/d	3,625
3. System biomass (VSS) equations			
Mass of PAOs	MX_{PAOv}	kgVSS	4,125
Mass of endogenous residue from PAOs	$MX_{E,PAOv}$	kgVSS	209
Mass of OHOs	MX_{OHOv}	kgVSS	4,423
Mass of endogenous residue from OHOs	$MX_{E,OHOv}$	kgVSS	1,073
Mass of ANO	MX_{ANOv}	kgVSS	0
Mass of unbiodegradable organics from influent	MX_{Uv}	kgVSS	8,878

4. COD removal			
Effluent soluble COD concentration	$S_{U,e}$	$gCOD/m^3$	75.0
Effluent total COD concentration	COD_e	$gCOD/m^3$	91.4
5. Nitrogen removal			
Nitrogen requirements for biomass synthesis	N_s	gN/m^3	20.8
Effluent total Kjeldahl concentration (soluble)	TKN_e	gN/m^3	23.7
Effluent nitrate concentration	$S_{NO3,e}$	gN/m^3	0
Effluent particulate nitrogen	$X_{N,e}$	gN/m^3	1.1
Effluent total nitrogen concentration	N_e	gN/m^3	24.8
6. P removal			
PO$_4$ release	S_{PO4_rel}	gP/m^3	61.2
Maximum P removal by PAOs	ΔP_{PAO}	gP/m^3	17.4
Actual P removal by PAOs	$\Delta P_{PAO,actual}$	gP/m^3	9.21
P removal by OHOs	ΔP_{OHO}	gP/m^3	1.47
P removal by endogenous residue	ΔP_{XE}	gP/m^3	0.43
P removal by X_U	ΔP_{XU}	gP/m^3	2.96
Potential P removal by system	$\Delta P_{SYS,pot}$	gP/m^3	22.3
Actual P removal by system	$\Delta P_{SYS,actual}$	gP/m^3	14.5
Effluent particulate P (from X_e)	$X_{P,e}$	gP/m^3	0.83
Influent total P	P_i	gP/m^3	14.5
Effluent total P	P_e	gP/m^3	0.83
7. Volatile and total suspended solids (VSS and TSS) in system			
Mass of active biomass	MX_B	$kgVSS$	8,548
Mass of VSS	MX_{VSS}	$kgVSS$	18,708
Ratio of AVSS/VSS	f_{av}	$gAVSS/gVSS$	0.46
Mass of fixed SS	MX_{FSS}	$kgFSS$	6,650
Mass of TSS	MX_{TSS}	$kgTSS$	25,354
Ratio of VSS/TSS	f_{VT}	$gVSS/gTSS$	0.74
Fraction of P in TSS	$f_{P,TSS}$	$gP/gTSS$	0.055
8. Reactor volume			
Anaerobic reactor volume	$V_{R,AN}$	m^3	533
Aerobic reactor volume	$V_{R,OX}$	m^3	4,793
Bioreactor volume	V_R	m^3	5,326
9. Biomass concentration[a)]			
Total solids concentration in the aerobic reactor	$X_{TSS,OX}$	$gTSS/m^3$	4.76
Total solids concentration in the anaerobic reactor	$X_{TSS,AN}$	$gTSS/m^3$	4.76
10. Oxygen demand			
Flux of O$_2$ demand by PAOs	FO_{PAO}	kgO_2/d	768
Flux of O$_2$ demand by OHOs	FO_{OHO}	kgO_2/d	2,269
Flux of carbonaceous O$_2$ demand	FO_c	kgO_2/d	3,036

[a)] It is recommendable to verify the oxygen transfer efficiency and operation of the secondary settling tanks at this increased biomass concentration.

Solution 6.4.27

Table 6.15 Summary of the design results of upgrading the MLE to an A2O configuration.

Description	Parameter	Unit	Value
1. Influent and bioreactor			
Sludge retention time	SRT	d	40
Internal recycle ratio (a)	a	$m^3.d/m^3.d$	3.7
Oxygen concentration in a recycle	$S_{O2,a}$	gO_2/m^3	2
Sludge recycle ratio (s)	s	$m^3.d/m^3.d$	0.41
Oxygen concentration in sludge recycle	$S_{O2,s}$	gO_2/m^3	1
Nitrate concentration in sludge recycle	$S_{NO3,s}$	gN/m^3	4.5
2. Portion of $S_{S,i}$ for PAOs and of $COD_{b,i}$ for OHOs			
Concentration of fermentable COD in the last AN reactor	$S_{F,ANn}$	$gCOD/m^3$	19.8
Flux of $S_{S,i}$ for PAOs	$FS_{S,PAO}$	kgCOD/d	2,878
Flux of $COD_{b,i}$ for OHOs	$FCOD_{b,OHO}$	kgCOD/d	3,362
3. System biomass (VSS) equations			
Mass of PAOs	MX_{PAOv}	kgVSS	22,816
Mass of endogenous residue from PAOs	$MX_{E,PAOv}$	kgVSS	7,261
Mass of OHOs	MX_{OHOv}	kgVSS	7,013
Mass of endogenous residue from OHOs	$MX_{E,OHOv}$	kgVSS	10,712
Mass of ANO	MX_{ANOv}	kgVSS	940
Mass of unbiodegradable organics from influent	MX_{Uv}	kgVSS	62,162
4. COD removal			
Effluent soluble COD concentration	$S_{U,e}$	$gCOD/m^3$	67.0
Effluent total COD concentration	COD_e	$gCOD/m^3$	82.7
5. Nitrogen removal			
Nitrogen requirements for biomass synthesis	N_s	gN/m^3	13.7
Nitrification capacity	NIT_c	gN/m^3	26.7
Denitrification potential	D_{p1}	gN/m^3	24.0
Effluent free and saline ammonia concentration	$S_{NHx,e}$	gN/m^3	1.6
Effluent total Kjeldahl concentration	TKN_e	gN/m^3	2.1
Effluent nitrate concentration	$S_{NO3,e}$	gN/m^3	5.2
Effluent particulate nitrogen	$X_{N,e}$	gN/m^3	1.0
Effluent total nitrogen concentration	N_e	gN/m^3	8.3
6. P removal			
PO_4 release	S_{PO4_rel}	gP/m^3	71.9
Maximum P removal by PAOs	ΔP_{PAO}	gP/m^3	10.8
Actual P removal by PAOs	$\Delta P_{PAO,actual}$	gP/m^3	8.5
P removal by OHOs	ΔP_{OHO}	gP/m^3	0.3
P removal by endogenous residue	ΔP_{XE}	gP/m^3	0.7
P removal by X_U	ΔP_{XU}	gP/m^3	2.3
Potential P removal by system	$\Delta P_{SYS,pot}$	gP/m^3	14.1
Actual P removal by system	$\Delta P_{SYS,actual}$	gP/m^3	12.4

Effluent particulate P (from X_e)	$X_{P,e}$	gP/m^3	1.0
Influent total P	P_i	gP/m^3	12.4
Effluent total P	P_e	gP/m^3	1.0
7. Volatile and total suspended solids (VSS and TSS) in system			
Mass of active biomass	MX_B	kgVSS	30,768
Mass of VSS	MX_{VSS}	kgVSS	110,903
Ratio of AVSS/VSS	f_{av}	gAVSS/gVSS	0.28
Mass of fixed SS	MX_{FSS}	kgFSS	52,200
Mass of TSS	MX_{TSS}	kgTSS	163,103
Ratio of VSS/TSS	f_{VT}	gVSS/gTSS	0.68
Fraction of P in TSS	$f_{P,TSS}$	gP/gTSS	0.064
8. Reactor volume			
Anaerobic reactor volume (divided in 3 compartments)	$V_{R,AN}$	m^3	4,660
Anoxic reactor volume	$V_{R,AX}$	m^3	23,300
Aerobic reactor volume	$V_{R,OX}$	m^3	18,640
Bioreactor volume	V_R	m^3	46,600
9. Oxygen demand			
Flux of O_2 demand by PAOs	FO_{PAO}	kgO_2/d	1,766
Flux of O_2 demand by OHOs	FO_{OHO}	kgO_2/d	2,707
Flux of carbonaceous O_2 demand	FO_c	kgO_2/d	4,473
Flux of O_2 demand by ANO	FO_{NIT}	kgO_2/d	2,441
Flux of O_2 recovered by denitrification	FO_{DENIT}	kgO_2/d	1,222
Total O_2 flux	$FO_{t,\,DENIT}$	kgO_2/d	5,692

Solution 6.4.28

Suggestion: To keep the same f_{xa} of 0.10, a r-recycle ratio of 1.0 ($r = Q_r / Q_i$) can be selected and the net volume of the anaerobic tank will need to be doubled to compensate for the expected $X_{TSS,AN}$, which will be lower than $X_{TSS,OX}$. This is suggested since, based on mass balances, the anaerobic biomass concentration can be estimated as: $X_{TSS,AN} = r \cdot X_{TSS,OX} / (1 + r)$.

Since, $f_{xa} = V_{R,AN} \cdot X_{TSS,AN} / MX_{TSS}$,

and: $X_{TSS,AN} = (1) \cdot X_{TSS,OX} / (1+1) = X_{TSS,OX} / 2$.

Then, $f_{xa} = V_{R,AN} \cdot X_{TSS,AN} / MX_{TSS}$.

Solving for $V_{R,AN}$: $V_{R,AN} = f_{xa} \cdot V_R \cdot X_{TSS,OX} / (X_{TSS,OX} / 2) = 2 \cdot f_{xa} \cdot V_R$.

Thus, f_{xa} will remain the same (0.10) but $V_{R,AN}$ will be twice as much, increasing the whole volume of the plant (V_R). As such and assuming that there is neither oxygen nor nitrate in the r-recycle, the summary of the calculations can be the following:

Table 6.16 Summary of the main design results of the design of a UCT process system.

Description	Parameter	Unit	Value
1. Influent and bioreactor			
Type of wastewater	raw		raw
Temperature	T	°C	14
Influent flow rate	Q_i	MLD	30
Influent total COD	COD_i	$gCOD/m^3$	585
Influent soluble biodegradable COD	$S_{S,i}$	$gCOD/m^3$	276
Influent biodegradable COD	$COD_{b,i}$	$gCOD/m^3$	465
Influent total nitrogen concentration	N_i	gN/m^3	38.5
Influent total Kjeldahl nitrogen (TKN)	TKN_i	gN/m^3	38.5
Influent free and saline ammonia	$S_{NHx,i}$	gN/m^3	29.1
Influent unbiodegradable soluble nitrogen	$N_{ous,i}$	gN/m^3	0.49
Influent total P	P_i	gP/m^3	11.5
Influent orthophosphate	$S_{PO4,i}$	gP/m^3	7.6
Influent inorganic (fixed) suspended solids	$X_{FSS,i}$	$gFSS/m^3$	25
Sludge retention time	SRT	d	11
Internal recycle ratio (a)	a	$m^3.d/m^3.d$	2.35
Oxygen concentration in a recycle	$S_{O2,a}$	gO_2/m^3	2
Sludge recycle ratio (s)	s	$m^3.d/m^3.d$	0.78
Oxygen concentration in sludge recycle	$S_{O2,s}$	gO_2/m^3	1
Anoxic to anaerobic recycle ratio (r)	r	$m^3.d/m^3.d$	1.0
Oxygen concentration in r-recycle	$S_{O2,r}$	gO_2/m^3	0
Nitrate concentration in r-recycle	$S_{NO3,r}$	gN/m^3	0
2. Portion of $S_{S,i}$ for PAOs and of $COD_{b,i}$ for OHOs			
Concentration of fermentable COD in the last AN reactor	$S_{F,ANn}$	$gCOD/m^3$	37.3
Flux of $S_{S,i}$ for PAOs	$FS_{S,PAO}$	$kgCOD/d$	6,041
Flux of $COD_{b,i}$ for OHOs	$FCOD_{b,OHO}$	$kgCOD/d$	7,909
3. System biomass (VSS) equations			
Mass of PAOs	MX_{PAOv}	kgVSS	21,816
Mass of endogenous residue from PAOs	$MX_{E,PAOv}$	kgVSS	2,022
Mass of OHOs	MX_{OHOv}	kgVSS	12,144
Mass of endogenous residue from OHOs n	$MX_{E,OHOv}$	kgVSS	5,401
Mass of ANO	MX_{ANOv}	kgVSS	445
Mass of unbiodegradable organics from influent	MX_{Uv}	kgVSS	16,723
4. COD removal			
Effluent soluble COD concentration	$S_{U,e}$	$gCOD/m^3$	45.0
Effluent total COD concentration	COD_e	$gCOD/m^3$	61.9
5. Nitrogen removal			
Nitrogen requirements for biomass synthesis	N_s	gN/m^3	17.6
Nitrification capacity	NIT_c	gN/m^3	18.5
Denitrification potential	D_{p1}	gN/m^3	16.0

Effluent free and saline ammonia concentration	$S_{NHx,e}$	gN/m^3	1.9
Effluent total Kjeldahl concentration	TKN_e	gN/m^3	2.4
Effluent nitrate concentration	$S_{NO3,e}$	gN/m^3	4.5
Effluent particulate nitrogen	$X_{N,e}$	gN/m^3	1.1
Effluent total nitrogen concentration	N_e	gN/m^3	8.1
6. P removal			
PO_4 release	S_{PO4_rel}	gP/m^3	100.1
Maximum P removal by PAOs	ΔP_{PAO}	gP/m^3	25.1
Actual P removal by PAOs	$\Delta P_{PAO,actual}$	gP/m^3	7.53
P removal by OHOs	ΔP_{OHO}	gP/m^3	1.10
P removal by endogenous residue	ΔP_{XE}	gP/m^3	0.67
P removal by X_U	ΔP_{XU}	gP/m^3	1.52
Potential P removal by system	$\Delta P_{SYS,pot}$	gP/m^3	28.4
Actual P removal by system	$\Delta P_{SYS,actual}$	gP/m^3	11.5
Effluent particulate P (from X_e)	$X_{P,e}$	gP/m^3	0.77
Influent total P	P_i	gP/m^3	11.5
Effluent total P	P_e	gP/m^3	0.77
7. Volatile and total suspended solids (VSS and TSS) in system			
Mass of active biomass	MX_B	kgVSS	34,405
Mass of VSS	MX_{VSS}	kgVSS	58,551
Ratio of AVSS/VSS	f_{av}	gAVSS/gVSS	0.59
Mass of fixed SS	MX_{FSS}	kgFSS	18,568
Mass of TSS	MX_{TSS}	kgTSS	77,119
Ratio of VSS/TSS	f_{VT}	gVSS/gTSS	0.76
Fraction of P in TSS	$f_{P,TSS}$	gP/gTSS	0.05
8. Reactor volume			
Anaerobic reactor volume	$V_{R,AN}$	m^3	4,407
Anoxic reactor volume	$V_{R,AX}$	m^3	4,407
Aerobic reactor volume	$V_{R,OX}$	m^3	15,424
Bioreactor volume	V_R	m^3	24,237
9. Biomass concentration			
Total solids concentration in the anaerobic reactor	$X_{TSS,AN}$	$gTSS/m^3$	1.75
Total solids concentration in the anoxic reactor	$X_{TSS,AX}$	$gTSS/m^3$	3.50
Total solids concentration in the aerobic reactor	$X_{TSS,OX}$	$gTSS/m^3$	3.50
10. Oxygen demand			
Flux of O_2 demand by PAOs	FO_{PAO}	kgO_2/d	2,834
Flux of O2 demand by OHOs	FO_{OHO}	kgO_2/d	5,549
Flux of carbonaceous O_2 demand	FO_c	kgO_2/d	8,382
Flux of O_2 demand by ANO	FO_{NIT}	kgO_2/d	2,535
Flux of O_2 recovered by denitrification	FO_{DENIT}	kgO_2/d	1,202
Total O_2 flux	$FO_{t,DENIT}$	kgO_2/d	9,715

Solution 6.4.29

From Example 6.3.3, the A2O process is converted to an UCT process configuration considering a r-recycle ratio of 0.9, the presence of oxygen in the influent flow rate ($S_{O2,i}$ = 1 mg/l) and of nitrate in the r-recycle ratio ($S_{NO3,r}$ = 3 mgN/l), while keeping the same volumes determined in Example 6.3.3 (V_R = 21,866 m^3 composed of $V_{R,AN}$ = 2,187 m^3, $V_{R,AX}$ = 4,373 m^3 and $V_{R,OX}$ = 15,311 m^3).

Consider that:

$$X_{TSS,AN} = r \cdot X_{TSS,OX} / (1 + r),$$

And:

$$f_{xa} = V_{R,AN} \cdot X_{TSS,AN} / MX_{TSS}.$$

Thus, f_{xa} = 0.05 with r = 0.9. This will modify the unaerated mass fraction f_{xt} (to 0.21) and also the aerobic mass fraction (to 0.74). The summary of the calculations can be the following:

Table 6.17 Summary of the main design results of the modification of an existing A2O system to a UCT process configuration.

Description	Parameter	Unit	Value
1. Influent and bioreactor			
Type of wastewater	raw		raw
Temperature	T	°C	14
Influent flow rate	Q_i	MLD	30
Influent total COD	COD_i	gCOD/m^3	585
Influent soluble biodegradable COD	$S_{S,i}$	gCOD/m^3	276
Influent biodegradable COD	$COD_{b,i}$	gCOD/m^3	465
Influent total nitrogen concentration	N_i	gN/m^3	38.5
Influent total Kjeldahl nitrogen (TKN)	TKN_i	gN/m^3	38.5
Influent free and saline ammonia	$S_{NHx,i}$	gN/m^3	29.1
Influent unbiodegradable soluble nitrogen	$N_{ous,i}$	gN/m^3	0.49
Influent total P	P_i	gP/m^3	11.5
Influent orthophosphate	$S_{PO4,i}$	gP/m^3	7.6
Influent inorganic (fixed) suspended solids	$X_{FSS,i}$	gFSS/m^3	25
Influent dissolved oxygen concentration	$S_{O2,i}$	gO$_2$/m^3	1.0
Sludge retention time	SRT	d	11
Internal recycle ratio (a)	a	m^3.d/m^3.d	4
Oxygen concentration in a recycle	$S_{O2,a}$	gO$_2$/m^3	2
Sludge recycle ratio (s)	s	m^3.d/m^3.d	0.83
Oxygen concentration in sludge recycle	$S_{O2,s}$	gO$_2$/m^3	1
Anoxic to anaerobic recycle ratio (r)	r	m^3.d/m^3.d	0.9
Oxygen concentration in r-recycle	$S_{O2,r}$	gO$_2$/m^3	0
Nitrate concentration in r-recycle	$S_{NO3,r}$	gN/m^3	3.0

2. Portion of $S_{S,i}$ for PAOs and of $COD_{b,i}$ for OHOs

Concentration of fermentable COD in the last AN reactor	$S_{F,ANn}$	$gCOD/m^3$	46.1
Flux of $S_{S,i}$ for PAOs	$FS_{S,PAO}$	kgCOD/d	4,865
Flux of $COD_{b,i}$ for OHOs	$FCOD_{b,OHO}$	kgCOD/d	9,085

3. System biomass (VSS) equations

Mass of PAOs	MX_{PAOv}	kgVSS	17,568
Mass of endogenous residue from PAOs	$MX_{E,PAOv}$	kgVSS	1,628
Mass of OHOs	MX_{OHOv}	kgVSS	13,950
Mass of endogenous residue from OHOs	$MX_{E,OHOv}$	kgVSS	6,205
Mass of ANO	MX_{ANOv}	kgVSS	470
Mass of unbiodegradable organics from influent	MX_{Uv}	kgVSS	16,723

4. COD removal

Effluent soluble COD concentration	$S_{U,e}$	$gCOD/m^3$	45.0
Effluent total COD concentration	COD_e	$gCOD/m^3$	61.7

5. Nitrogen removal

Nitrogen requirements for biomass synthesis	N_s	gN/m^3	17.0
Nitrification capacity	NIT_c	gN/m^3	19.5
Denitrification potential	D_{p1}	gN/m^3	19.5
Effluent free and saline ammonia concentration	$S_{NHx,e}$	gN/m^3	1.5
Effluent total Kjeldahl concentration	TKN_e	gN/m^3	1.99
Effluent nitrate concentration	$S_{NO3,e}$	gN/m^3	3.3
Effluent particulate nitrogen	$X_{N,e}$	gN/m^3	1.1
Effluent total nitrogen concentration	N_e	gN/m^3	6.4

6. P removal

PO_4 release	S_{PO4_rel}	gP/m^3	81.1
Maximum P removal by PAOs	ΔP_{PAO}	gP/m^3	20.2
Actual P removal by PAOs	$\Delta P_{PAO,actual}$	gP/m^3	7.29
P removal by OHOs	ΔP_{OHO}	gP/m^3	1.27
P removal by endogenous residue	ΔP_{XE}	gP/m^3	0.71
P removal by X_U	ΔP_{XU}	gP/m^3	1.52
Potential P removal by system	$\Delta P_{SYS,pot}$	gP/m^3	23.7
Actual P removal by system	$\Delta P_{SYS,actual}$	gP/m^3	11.5
Effluent particulate P (from X_e)	$X_{P,e}$	gP/m^3	0.79
Influent total P	P_i	gP/m^3	11.5
Effluent total P	P_e	gP/m^3	0.79

7. Volatile and total suspended solids (VSS and TSS) in system

Mass of active biomass	MX_B	kgVSS	31,998
Mass of VSS	MX_{VSS}	kgVSS	56,543
Ratio of AVSS/VSS	f_{av}	gAVSS/gVSS	0.56
Mass of fixed SS	MX_{FSS}	kgFSS	18,570
Mass of TSS	MX_{TSS}	kgTSS	75,113
Ratio of VSS/TSS	f_{VT}	gVSS/gTSS	0.75
Fraction of P in TSS	$f_{P,TSS}$	gP/gTSS	0.05

8. Reactor volume			
Anaerobic reactor volume	$V_{R,AN}$	m^3	2,187
Anoxic reactor volume	$V_{R,AX}$	m^3	4,373
Aerobic reactor volume	$V_{R,OX}$	m^3	15,311
Bioreactor volume	V_R	m^3	21,871
9. Biomass concentrations			
Total solids concentration in the anaerobic reactor	$X_{TSS,AN}$	$gTSS/m^3$	1.71
Total solids concentration in the anoxic reactor	$X_{TSS,AX}$	$gTSS/m^3$	3.62
Total solids concentration in the aerobic reactor	$X_{TSS,OX}$	$gTSS/m^3$	3.62
10. Oxygen demand			
Flux of O_2 demand by PAOs	FO_{PAO}	kgO_2/d	2,282
Flux of O_2 demand by OHOs	FO_{OHO}	kgO_2/d	6,374
Flux of carbonaceous O_2 demand	FO_c	kgO_2/d	8,656
Flux of O_2 demand by ANO	FO_{NIT}	kgO_2/d	2,676
Flux of O_2 recovered by denitrification	FO_{DENIT}	kgO_2/d	1,389
Total O_2 flux	$FO_{t,DENIT}$	kgO_2/d	9,942

REFERENCE

Chen GH., van Loosdrecht M.C.M., Ekama G.A. and Brdjanovic D. Ed. (2020) Biological Wastewater Treatment: Principles, Design and Modelling. 2nd edition. IWA Publishing, pg. 850. 9781789060355.

NOMENCLATURE

Symbol	Description	Unit
a	Internal mixed liquor recycle ratio based on influent flow	$m^3.d/m^3.d$
a_{opt}	Optimal a-recycle ratio (it gives a minimum N_{ne})	$m^3.d/m^3.d$
a_{prac}	Practical a-recycle flow ratio	
$b_{ANO,20}$	ANO-specific endogenous mass loss rate constant at 20 °C	gVSS /gVSS.d
$b_{ANO,T}$	ANO-specific endogenous mass loss rate constant at temperature T	gVSS /gVSS.d
b_{OHO}	Specific endogenous mass loss rate of the OHOs	gEVSS/gVSS.d
$b_{OHO,20}$	Specific endogenous mass loss rate of the OHOs at 20 °C	gVSS /gVSS.d
$b_{OHO,T}$	OHO specific endogenous mass loss rate at temperature T	gEVSS/gVSS.d
b_{PAO}	Specific endogenous mass loss rate of the PAOs	gEVSS/gVSS.d
$b_{PAO,20}$	PAO-specific endogenous mass loss rate constant at 20 °C	gVSS /gVSS.d
$b_{PAO,T}$	PAO specific endogenous mass loss rate at temperature T	gEVSS/gVSS.d
COD_b	Concentration of biodegradable COD	$gCOD/m^3$
$COD_{b,i}$	influent concentration of biodegradable COD ($S_{S,i} + X_{S,i}$)	$gCOD/m^3$
$COD_{b,OHO}$	Concentration of biodegradable COD available to the OHOs	$gCOD/m^3$
COD_e	Effluent concentration of total COD	$gCOD/m^3$
COD_i	Concentration of COD in the influent	$gCOD/m^3$
D_{P1}	Denitrification potential of the primary anoxic reactor	$gNO_3^-\text{-}N/m^3$
D_{P3}	Denitrification potential of the secondary anoxic reactor	$gNO_3^-\text{-}N/m^3$
f_{ox}	Aerobic mass fraction	

Symbol	Description	Units
f_{ax1}	Primary anoxic reactor mass fraction	gVSS/gVSS
f_{av}	Fraction of active biomass with the regard to the mass of volatile suspended solids	kgAVSS/kgVSS
$FCOD_{b,i}$	Daily mass of influent biodegradable organics ($S_{S,i} + X_{S,i}$)	gCOD/d
$FCOD_{b,OHO}$	Daily mass of biodegradable substrate available to OHOs	gCOD/gCOD
$FCOD_i$	Daily mass of influent COD	gCOD/d
f_{cv}	COD/VSS ratio of the sludge	gCOD/gVSS
$f_{FSS,OHO}$	Fraction of fixed (inorganic) suspended solids of OHOs	gFSS/gVSS
$f_{FSS,PAO}$	Fraction of fixed (inorganic) suspended solids of PAOs	gFSS/gVSS
f_n	Nitrogen content of active biomass	gN/gVSS
FN_s	Daily mass of nitrogen required for sludge production	kgN/d
FO_c	Daily mass of carbonaceous oxygen demand	kgO$_2$/d
FO_{DENIT}	Oxygen recovered from dentrification	kgO$_2$/d
FO_{OHO}	Daily mass of oxygen consumed by OHOs	kgO$_2$/d
$FO_{OHO,e}$	Daily mass of oxygen consumed by OHOs for endogenous respiration	kgO$_2$/d
$FO_{OHO,s}$	Daily mass of oxygen consumed by OHOs for synthesis	kgO$_2$/d
$FO_{PAO,e}$	Daily mass of oxygen consumed by PAOs for endogenous respiration	kgO$_2$/d
$FO_{PAO,s}$	Daily mass of oxygen consumed by PAOs for synthesis	kgO$_2$/d
$FO_{t,DENIT}$	Total mass per day (flux) of oxygen required less that recovered by denitrification	kgO$_2$/d
f_P	Fraction of P in the active OHO mass, endogenous mass (OHO and PAO) and unbiodegradable mass	gP/gVSS
$f_{P,FSS,i}$	Fraction of P in the influent FSS	gP/gFSS
$f_{P,OHO}$	Fraction of P in the active OHO mass	gP/gAVSS
$f_{P,PAO}$	Fraction of P in the active PAO mass	gP/gAVSS
$f_{P,TSS}$	P content with respect to TSS	gP/gTSS
$f_{P,PAO,actual}$	Actual phosphorus fraction stored in PAO biomass	kgP/kgVSS
$f_{PO4,rel}$	Ratio of P release/VFA uptake	gP/gCOD
$FS_{F,i}$	Influent daily flux of fermentable COD	kgCOD/d
$FS_{S,i}$	Influent daily flux of RBCOD	
$FS_{S,PAO}$	Daily mass of S_S stored by PAOs in the anaerobic reactor	kgCOD/d
$FS_{U,i}$	Influent daily flux of soluble unbiodegradable COD	kgCOD/d
$FS_{VFA,i}$	Daily mass of influent VFAs	kgCOD/d
f_{vT}	VSS/TSS ratio for OHO active and endogenous masses, PAO endogenous mass and inert mass	gVSS/gTSS
$f_{vT,ANO}$	VSS/TSS ration for ANO active mass	gVSS/gVSS
$f_{vT,PAO}$	VSS/TSS ratio for PAO active mass	gVSS/gTSS
f_{xa}	Anaerobic mass fraction	-
$f_{XE,OHO}$	Fraction of endogenous residue of the OHOs	gEVSS/gAVSS
$f_{XE,PAO}$	Fraction of endogenous residue of the PAOs	gEVSS/gAVSS
$FX_{FSS,i}$	Daily mass of influent inorganics	gFSS/d
f_{xt}	Total unaerated mass fraction (sum of all unaerated mass fractions)	-
$FX_{U,i}$	influent daily flux of particulate unbiodegradable COD	kgCOD/d

$K_{2,20}$	Specific denitrification rate for NDEBPR pre-denitrification at 20 °C	$gNO_3^--N/gVSS.d$
K_{2T}	Specific denitrification rate in primary anoxic reactor of NDEBPR system on SBCOD at temperature T	$gNO_3^--N/gOHOVSS.d$
$K_{ANO,20}$	ANO half-saturation constant at 20 °C	gN/m^3
$K_{ANO,T}$	ANO maximum specific growth rate constant at temperature T	gN/m^3
$k_{F,20}$	First-order fermentation rate constant at temperature 20 °C	$m^3/gVSS.d$
$k_{F,T}$	First-order fermentation rate constant at temperature T	$m^3/gVSS.d$
K'_T	Specific denitrification rate of OHOs for an NDEBPR system (') at temperature T	$gNO_3^--N/gVSS.d$
MX_B	Sum of all active biomasses in the system	$gVSS$
$MX_{E,OHOv}$	Mass of OHO endogenous residue in the system	$gEVSS$
$MX_{E,PAOv}$	Mass of PAO endogenous residue in the system	$gEVSS$
MX_{Ev}	Mass of unbiodegradable endogenous residue	$gVSS$
MX_{FSS}	Mass of fixed (inorganic) suspended solids in the system	$gFSS$
MX_{OHOv}	Mass of OHOs in the system	$gAVSS$
MX_{PAOv}	Mass of PAO in the system	$gAVSS$
MX_{TSS}	TSS mass in the system	$gTSS$
MX_{Uv}	Mass of unbiodegradable organic matter in the system, coming from the influent	$gVSS$
MX_{VSS}	Mass of volatile suspended solids in the system	$gVSS$
n	Number of the anaerobic reactor from a series	-
N	Total number of anaerobic reactors of equal volume in the series n = 1,2...N	-
NIT_c	Nitrification capacity of the bioreactor	gNO_3^--N/m^3
N_e	Effluent total nitrogen concentration	gN/m^3
$N_{obp,i}$	Influent biodegradable particulate organic nitrogen	gN/m^3
$N_{obs,i}$	Influent biodegradable soluble organic nitrogen	gN/m^3
$N_{oup,i}$	Influent unbiodegradable particulate organic nitrogen	gN/m^3
$N_{ous,i}$	Influent unbiodegradable soluble organic nitrogen	gN/m^3
N_s	Nitrogen required for biomass growth	gN/m^3
P_e	Effluent total phosphorus concentration	gP/m^3
P_i	Influent total phosphorus concentration	gP/m^3
$P_{obp,i}$	Influent biodegradable particulate organic phosphorus	gP/m^3
$P_{obs,i}$	Influent biodegradable soluble organic phosphorus	gP/m^3
$P_{oup,i}$	Influent unbiodegradable particulate organic phosphorus	gP/m^3
$P_{ous,i}$	Influent unbiodegradable soluble organic phosphorus	gP/m^3
P_s	Phosphorus required for biomass growth	gP/m^3
Q_i	Daily average influent flow rate	MLD, m^3/d
Q_r	Sludge recycle flow rate	m^3/d
$Q_{VFA,add}$	Additional flowrate of VFA	m^3/d
r	Mixed-liquor recycle ratio from the aerobic to anoxic (or anaerobic) reactor based on influent flow	$m^3.d/m^3.d$
s	Return activated sludge recycle ratio based on influent flow	$m^3.d/m^3.d$
S_{bi}	Soluble biodegradable organics	$gCOD/m^3$

S_F	Fermentable organic matter concentration	$gCOD/m^3$
S_f	Safety factor for the nitrification process	
$S_{F,ANn}$	Fermentable organic matter conc. in the nth AN reactor	$gCOD/m^3$
$S_{F,i}$	Fermentable organic matter concentration in the influent	$gCOD/m^3$
$S_{F,i,conv}$	$S_{F,i}$ available for conversion into VFAs per volume of influent	$gCOD/m^3$
$S_{NHx,e}$	Effluent free and saline ammonia	gN/m^3
$S_{NHx,i}$	Influent inorganic free and saline ammonia concentration	gN/m^3
$S_{NO2,i}$	Influent nitrite concentration	gN/m^3
$S_{NO3,e}$	Effluent nitrate concentration	gNO_3^--N/m^3
$S_{NO3,e,min}$	Minimum effluent nitrate concentration	gN/m^3
$S_{NO3,i}$	Influent nitrate concentration	gNO_3^--N/m^3
$S_{NO3,r}$	Nitrate concentration in the r-recycle flow	gN/m^3
$S_{NO3,s}$	Nitrate concentration in the sludge recycle	gNO_3^--N/m^3
S_{O2}	Oxygen concentration	gO_2/m^3
$S_{O2,a}$	Oxygen concentration in the anoxic recycle to the AN reactor	gO_2/m^3
$S_{O2,i}$	Influent oxygen concentration	gO_2/m^3
$S_{O2,s}$	Oxygen concentration in the sludge recycle to the AN reactor	gO_2/m^3
$S_{PO4,i}$	Influent orthophosphate concentration	gP/m^3
$S_{PO4,rel}$	Concentration of P released	gP/m^3
SRT	Sludge age (sludge retention time)	d
SRT_{min}	Minimum SRT required for nitrification	d
$S_{S,i}$	Influent readily biodegradable COD concentration	$gCOD/m^3$
$S_{S,PAO}$	Concentration of S_S stored by PAOs	$gCOD/m^3$
S_U	Soluble unbiodegradable COD	$gCOD/m^3$
$S_{U,e}$	Effluent concentration of soluble unbiodegradable COD	$gCOD/m^3$
$S_{U,i}$	Influent concentration of soluble unbiodegradable COD	$gCOD/m^3$
S_{VFA}	Volatile fatty acids concentration	$gCOD/m^3$
$S_{VFA,add}$	VFA concentration to be dosed	$kgCOD/m^3$
$S_{VFA,i}$	VFA concentration in the influent	$gCOD/m^3$
T	Temperature	$^{\circ}C$
TKN	Total Kjeldahl nitrogen concentration	gN/m^3
TKN_e	Effluent total Kjeldahl nitrogen	gN/m^3
TKN_i	Influent total Kjeldahl nitrogen concentration	gN/m^3
$TKN_{i,s}$	Influent TKN required for biomass synthesis	gN/m^3
TSS	Total suspended solids	$gTSS/m^3$
$X_{TSS,e}$	Total suspended solids in the effluent	$gTSS/m^3$
V_R	Volume of biological process (bioreactor)	m^3
$V_{R,OX}$	Aerobic volume of the biological process (bioreactor)	m^3
$V_{R,AN}$	Anaerobic volume of the biological process (bioreactor)	m^3
$V_{R,AX}$	Anoxic volume of the biological process (bioreactor)	m^3
VSS	VSS concentration	$gVSS/m^3$
$X_{FSS,i}$	Influent fixed suspended solids (FSS) concentration	$g\ FSS/m^3$
$X_{N,e}$	Effluent particulate nitrogen	gN/m^3

X_{OHO}	Ordinary heterotrophic organism concentration	$gCOD/m^3$
$X_{P,e}$	Effluent particulate phosphorus concentration	gP/m^3
X_{PAO}	Phosphorus-accumulating organisms concentration	$gCOD/m^3$
$X_{TSS,s}$	Total suspended-solids concentration in the sludge recycle flowrate	$gTSS/m^3$
X_S	Slowly biodegradable organics concentration	$gCOD/m^3$
$X_{S,i}$	Influent concentration of particulate biodegradable COD	$gCOD/m^3$
X_{TSS}	Reactor total suspended solids concentration	$gTSS/m^3$
$X_{TSS,AN}$	Anaerobic total suspended solids concentration	$gTSS/m^3$
$X_{TSS,AX}$	Anoxic total suspended solids concentration	$gTSS/m^3$
$X_{TSS,OX}$	Aerobic total suspended solids concentration	$gTSS/m^3$
$X_{U,i}$	Influent concentration of particulate unbiodegradable COD	$gCOD/m^3$
X_{VSS}	Volatile suspended solids concentration	$gVSS/m^3$
$X_{VSS,e}$	Volatile suspended solids concentration in the effluent	$gVSS/m^3$
$Y_{OHO,obs}$	Observed biomass yield of OHOs	$gVSS/gCOD$
Y_{OHOv}	Biomass yield of OHOs	$gVSS/gCOD$
$Y_{PAO,obs}$	Observed biomass yield of PAOs	$gVSS/gCOD$
Y_{PAOv}	Biomass yield of PAOs	$gVSS/gCOD$
ΔP_{OHO}	P removal due to OHOs	gP/m^3
ΔP_{PAO}	P removal due to PAOs	gP/m^3
$\Delta P_{SYS,pot}$	Total P potential removal by the system	gP/m^3
$\Delta P_{SYS.actual}$	Total P actual removal by the system	gP/m^3
ΔP_{XE}	P removal due to endogenous residue mass	gP/m^3
$\Delta P_{XE,OHO}$	P removal due to endogenous residue mass of OHO	gP/m^3
$\Delta P_{XE,PAO}$	P removal due to endogenous residue mass of PAO	gP/m^3
ΔP_{XU}	P removal due to inert mass	gP/m^3
$\theta_{b,ANO}$	Temperature coefficient for $b_{ANO,T}$	-
$\theta_{b,OHO}$	Temperature coefficient for $b_{OHO,T}$	-
$\theta_{b,PAO}$	Temperature coefficient for $b_{PAO,T}$	-
$\theta_{k,F}$	Temperature coefficient for $k_{F,T}$	-
$\theta_{K2,DENIT}$	Temperature coefficient for $K_{2,T}$	-
θ_{NIT}	Temperature coefficient for $\mu_{ANO,max,T}$ and $K_{ANO,T}$	-
$\mu_{ANO,max,T}$	Maximum specific biomass growth rate of nitrifiers at temperature T	$gVSS/gVSS.d$
$\mu_{ANO,max,20}$	ANO maximum specific growth rate constant at 20 °C	$gVSS/gVSS.d$

Abbreviation	Description
A/O	Anaerobic/oxic process
ANO	Ammonia nitrifying organisms
A^2O	Anaerobic, anoxic, aerobic process
PHOREDOX	Anaerobic/oxic process
AN	Anaerobic
AX	Anoxic
AVSS	Active volatile suspended solids
BNR	Biological nitrogen removal

COD	Chemical oxygen demand
e	Effluent
EBPR	Enhanced biological phosphorus removal
EVSS	Endogenous residue as volatile suspended solids
FSS	Fixed (inorganic) suspended solids
IVSS	Inert volatile suspended solids
i	Influent
JHB	Johannesburg process
MLE	Modified Ludzack-Ettinger process
MLSS	Mixed-liquor suspended solids
MLVSS	Mixed-liquor volatile suspended solids
MUCT	Modified UCT process
NIT	Nitrifying organisms
ND	Nitrification-denitrification
NDEBPR	Nitrification-denitrification EBPR
ATP	Adenosin triphosphate
NADH	Nicotinamide-adenine-dinucleotide
HAc	Acetic acid
HPr	Propionic acid
OHO	Ordinary heterotrophic organism
OUR	Oxygen uptake rate
OX	Aerobic
PAO	Phosphate-accumulating organism
GAO	Glycogen-accumulating organism
PHA	Poly-β-hydroxyalkanoates
PHB	Poly-β-hydroxybutyrate
PHV	Poly-β-hydroxyvalerate
PO_4	Phosphate
RAS	Return activated sludge
RBCOD	Readily biodegradable COD
SBCOD	Slowly biodegradable particulate organic matter
SRT	Sludge retention time
SST	Secondary settling tank
TCA	Tricarboxylic acid cycle
TKN	Total Kjeldahl nitrogen
TSS	Total suspended solids
UCT	University of Cape Town process
VFA	Volatile fatty acids
VSS	Volatile suspended solids

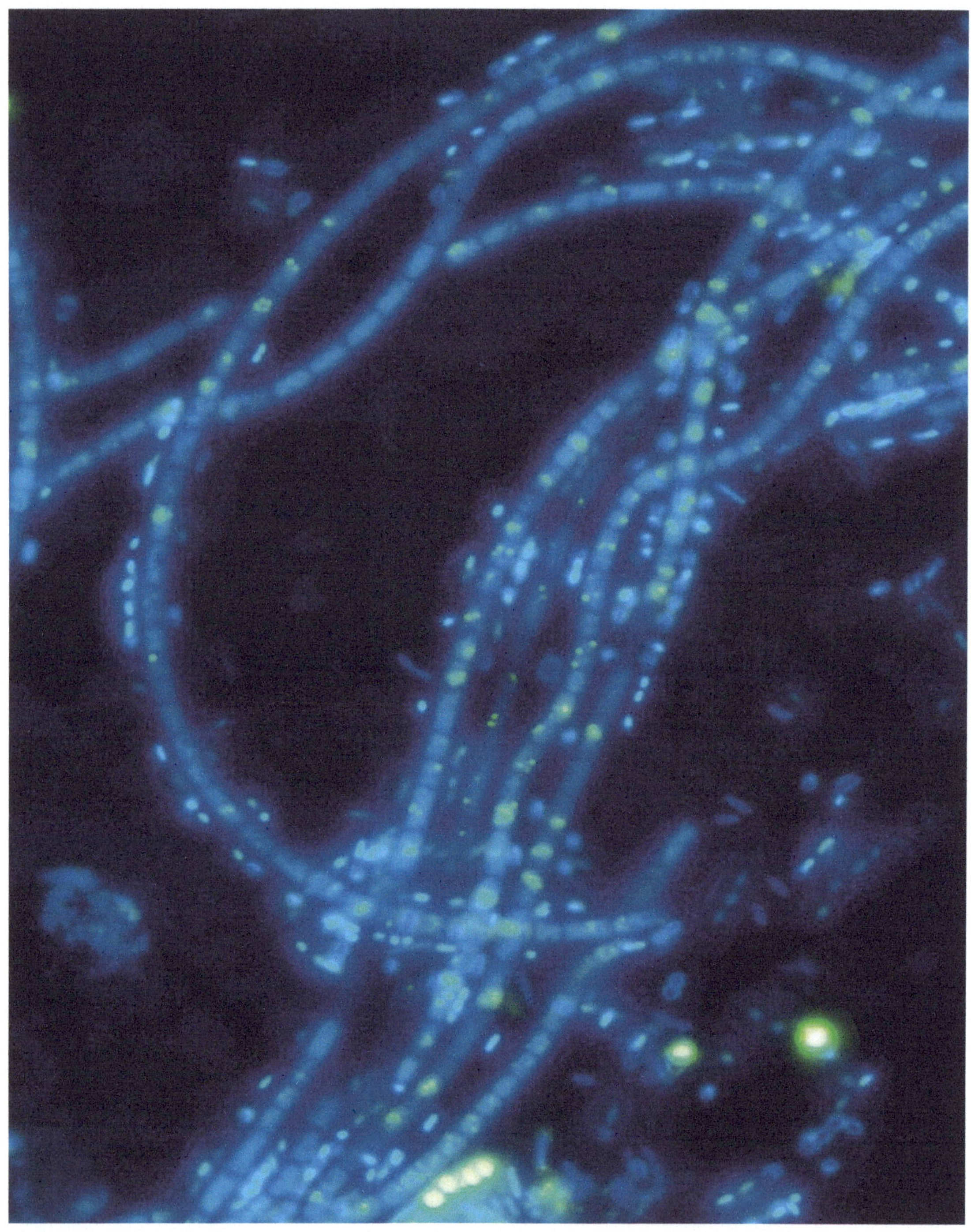

In the laboratory-enriched culture of *Thiothrix caldifontis* capable of removing carbon, nitrogen, phosphorus and sulphur from wastewater (photo: Francisco Rubio-Rincon).

doi: 10.2166/9781789062304_0245

7

Innovative sulphur-based wastewater treatment

Di Wu, Xu Zou, Hui Lu, Tianwei Hao, Ho Kwong Chui, George A. Ekama[†], Mark C.M. van Loosdrecht and Guanghao Chen

7.1 INTRODUCTION

Chapter 7 on innovative sulphur-based wastewater treatment in the textbook *Biological Wastewater Treatment: Principles, Modelling and Design* (Chen *et al.*, 2020) introduces the microbial mechanisms and principles of sulphur-based bioprocess(es), providing the basis to scale them up to full-scale application. These also serve as the fundamentals to develop a stoichiometric-based steady-state model for the design and evaluation of sulphur-based bioprocesses such as sulphate reduction and sulphur-driven autotrophic denitrification, as well as their integrated process (named SANI®). This chapter aims to help the reader to understand the principles, microbial mechanisms, and the steady-state stoichiometric model in order to design, assess and evaluate sulphur-based wastewater treatment systems.

7.2 LEARNING OBJECTIVES

After the successful completion of this chapter, the reader should be able to:

- Describe the microbiological pathways, key microorganisms and relevant factors of biological sulphate reduction (BSR) and design a typical BSR process.
- Describe the biochemical reactions, key microorganisms and governing parameters involved in sulphur-driven autotrophic denitrification (SdAD) and design a typical SdAD process.
- Describe the resources that can be recovered from sulphur-based processes and design typical sulphur-based resource recovery technology.
- Describe the application of sulphur biotechnology for wastewater treatment and apply steady-state modelling tools to design a SANI process.

7.3 EXAMPLES

Design a SANI process to treat 10,000 m^3/d domestic wastewater containing a total COD of 431 mg/l and a total Kjeldahl nitrogen (TKN) of 57.5 mgN/l. The annual average wastewater temperature is 22 °C.

Example 7.3.1

Without considering the hydrolysis of biodegradable particles, determine:

   a. Total volume of the SRUSB.
   b. Daily demand for sulphate.

Example 7.3.2

Considering the hydrolysis of biodegradable particles, determine:

   a. Total volume of the SRUSB and daily demand for sulphate.
   b. Daily organic removal and total dissolved sulphide (TDSd) production in the SRUSB.
   c. Daily oxygen demand in the aeration tank.
   d. Volume of autotrophic nitrification and the denitrification biofilm reactors.

Prior to practising the design, the reader should refresh their knowledge on stoichiometric-based steady-state modelling introduced in Chapter 4 (organic removal) and Chapter 5 (nitrogen removal), as well as sulphur-based bioprocess design in Chapter 7, in Chen *et al.*, 2020.

A SANI plant is typically designed with a sulphate-reducing up-flow sludge bed (SRUSB) anaerobic reactor as well as anoxic and aerobic reactors based on the moving-bed biofilm reactor (MBBR), followed by post-treatment (see Figure 7.12 in Chen *et al.*, 2020). For the SRUSB, two levels of design, with and without considering hydrolysis of biodegradable particles, are demonstrated in examples 7.3.1 and 7.3.2, respectively. For the design of MBBR-type biofilm reactors, the simplified method can be adopted by using a specific surface loading rate as the key factor to calculate the reactor volume and so on. The post-treatment to polish SANI effluent is mature technology, so it is not included in this example.

The governing equations for sulphate reduction, denitrification and nitrification as well as sulphide oxidation are established as below (reactions 1-4, respectively). The influent organic substrate used in Reaction (1) is defined as $C_1H_{1.13}O_{0.441}N_{0.091}P_{0.0051}$ in this design example. The method to determine the formula of influent organic substrate will be described in Exercise 7.4.2. The organic composition of the produced sludge $C_kH_lO_mN_nP_p$ is much less important for steady-state model design calculations, so its specific formula can be neglected.

- Dominant reaction in an anaerobic reactor:

Reaction (1): biological sulphate reduction

$$C_1H_{1.13}O_{0.441}N_{0.0911}P_{0.00516} + 0.473SO_4^{2-} + [3x - z + 4b - \frac{E\gamma_S}{\gamma_B}(3k - m + 4p) - 4 - E\frac{\gamma_S}{8})]H_2O$$

$$\rightarrow [(1 - E\frac{\gamma_S}{8})]TDSd + \left[\frac{E\gamma_S}{\gamma_B}\right]C_kH_lO_mN_nP_p + [a - n\frac{E\gamma_S}{\gamma_B}]NH_4^+ + \left[x - k\frac{E\gamma_S}{\gamma_B}\right]HCO_3^-$$

$$+ \left[b - \frac{pE\gamma_S}{\gamma_B}\right]H_2PO_4^-$$

Also see Eq. 7.21 in Chen *et al.*, 2020.

- Dominant reaction in an anoxic reactor:

Reaction (2): sulphur-driven autotrophic denitrification

$$0.093NH_4^+ + 1.23NO_3^- + 0.438HCO_3^- + HS^- + 0.573H^+$$
$$\rightarrow 0.614N_2 + 1SO_4^{2-} + 0.093C_5H_7O_2N + 0.866H_2O$$

Also see Eq. 7.13 in Chen *et al.,* 2020.

- Dominant reactions in an aerobic reactor:

Reaction (3): nitrification

$$1NH_4^+ + 1.99HCO_3^- + 1.8O_2 + 0.00037H_2PO_4^-$$
$$\rightarrow 0.99NO_3^- + 0.0064C_{3.98}H_7O_{1.72}N_{0.82}P_{0.059} + 1.97CO_2 + 3.03H_2O$$

Also see Eq. 7.25 in Chen *et al.,* 2020.

Reaction (4): sulphide oxidation
$$S^{2-} + 2O_2 \rightarrow SO_4^{2-}$$

In the examples 7.3.1 and 7.3.2, the SRUSB will be designed with a sludge retention time (SRT) of 30 days and an average mixed liquor suspended solid (MLSS) of 5.9 gMLSS/l. Notably, these two critical design factors are case-based and are therefore adjusted according to the requirements of each case.

In the steady-state model, it is assumed that after acidogenic hydrolysis, particulate biodegradable COD is directly transferred into biomass. To simplify the steady-state model, SRB biomass, which has the capability of particulate biodegradable COD hydrolysis and biological sulphate reduction, is used to represent the dominant bacteria in the SRUSB. The hydrolysis process can be described by a Monod-type kinetic model (see Section 7.4.4.1 in Chen *et al.,* 2020).

Furthermore, to carry out the design calculation, some assumptions need to be made such as: 1) the particulate organics and filtered organics (*i.e.* soluble) in the SRUSB effluent are unbiodegradable, and 2) the influent wastewater is stable. Peak factors in the practical design procedure are not considered in this example.

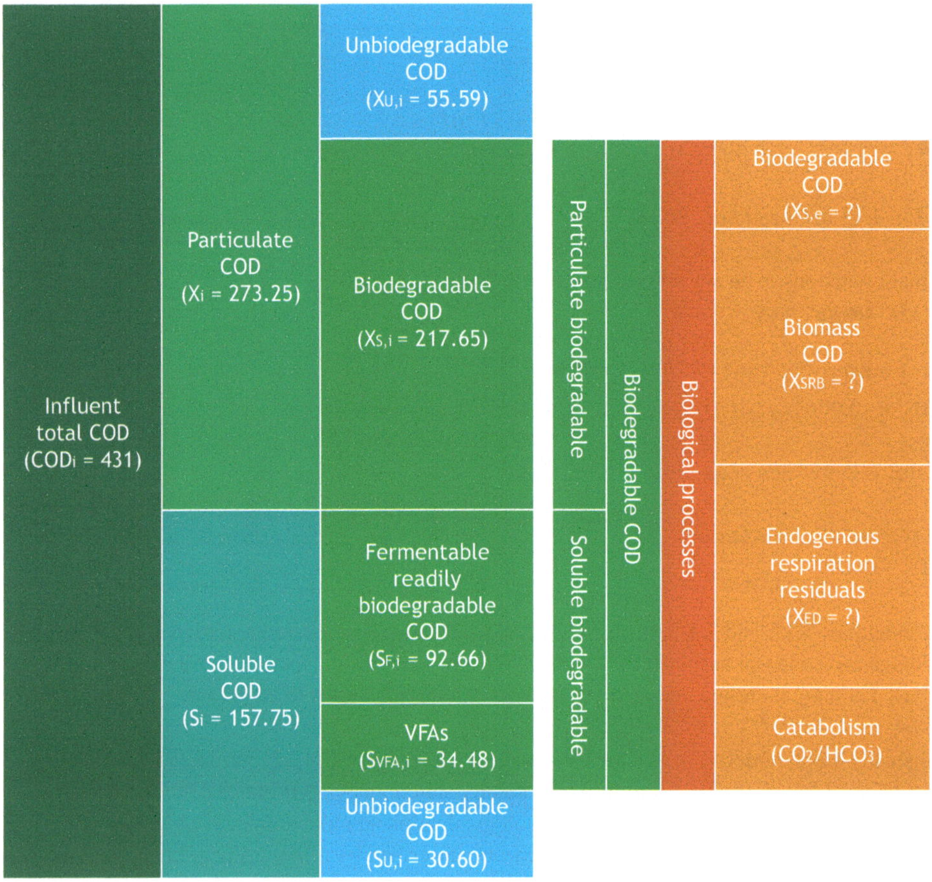

Figure 7.1 SRUSB influent and effluent COD fractionations.

Table 7.1 Summary of the data needed to carry out the design of the SANI® plant.

Description	Symbol	Value	Unit
Influent flow rate for the SRUSB	Q_i	10,000	m^3/d
Influent COD concentration	COD_i	431	$gCOD/m^3$
Influent COD fractions			
- fermentable COD fraction	f_{FBSO}	0.22	-
- volatile fatty acid fraction	f_{VFA}	0.08	-
- biodegradable particulate COD fraction	f_{BPO}	0.51	-
- unbiodegradable soluble COD fraction	f_{USO}	0.071	-
- unbiodegradable particulate COD fraction	f_{UPO}	0.13	-
Temperature	T	22	°C
Sludge retention time	SRT	30	d
Non-biodegradable particulate COD/influent total COD	$f_{XU,CODi}$	0.13	-
Non-biodegradable soluble COD/influent total COD	$f_{SU,CODi}$	0.071	-
Influent soluble total Kjeldahl nitrogen	TKN_i	57.5	gN/m^3
Design effluent soluble total Kjeldahl nitrogen	TKN_e	5.7	gN/m^3
Influent soluble unbiodegradable organic nitrogen	$N_{ous,i}$	0.7	gN/m^3
Nitrogen required for sludge production	N_s	5.0	gN/m^3
Design average TSS concentration in the SRUSB	X_{TSS}	5,900	$gTSS/m^3$
Influent inorganic suspended solid concentration	$X_{FSS,i}$	10	$gISS/m^3$

Table 7.2 Kinetic and stoichiometric parameters for the SANI® design example.

	Parameter	Symbol	Value	Unit
Kinetic	*Hydrolysis*			
	Maximum specific hydrolysis rate constant of acidogens	$k_{SRB,max}$	3.25	$1/d$
	Half saturation constant of acidogens	K_S	557	$gCOD/m^3$
	Endogenous respiration rate of the SRB biomass	b_{SRB}	0.04	$1/d$
Stoichiometric	*SRB & acidogenic biomass (assumed to belong to SRB biomass)*			
	Biomass yield of SRB (from FBSO and BPO)	Y_{SRB}	0.11	gCOD/gCOD
	Biomass yield of SRB (from VFA)	Y_{VFA}	0.023	gCOD/gCOD
	Endogenous residual fraction of SRB	f_{SRB}	0.08	gVSS/gCOD
	Surface specific conversion rate of ammonia	$B_{A,NH4}$	1	gNH_4^+-$N/m^2.d$
	Surface specific conversion rate of nitrate/nitrite	$B_{A,NOx}$	1	gNO_x^--$N/m^2.d$
	General			
	Mixed liquor recycle ratio from the aerobic to the anoxic reactor	a	2.5	-
	COD/VSS ratio of the sludge	f_{CV}	1.48	gCOD/gVSS
	Specific surface area of the biofilm	a_F	500	m^2/m^3
	Filling ratio	f_{fill}	60 %	-

Solution 7.3.1

Design calculation without considering hydrolysis of biodegradable particles (equation numbers refer to Chen *et al.*, 2020):

7.3.1.1 System configuration

SANI® full-scale plant operated at 22 °C.
Raw wastewater characteristics are shown in Table 7.1.

7.3.1.2 Influent concentrations

Influent COD concentrations are calculated based on the influent COD fractions in Table 7.1

(also shown in Figure 7.1).

COD_i	431	$gCOD/m^3$	influent concentration of total COD
$S_{U,i}$	30.60	$gCOD/m^3$	influent concentration of soluble unbiodegradable COD
$X_{U,i}$	55.59	$gCOD/m^3$	influent concentration of particulate unbiodegradable COD
$S_{VFA,i}$	34.48	$gCOD/m^3$	influent concentration of VFAs
$S_{F,i}$	92.66	$gCOD/m^3$	influent concentration of fermentable COD
$X_{S,i}$	217.65	$gCOD/m^3$	influent particulate biodegradable COD

7.3.1.3 Design calculation of the SRUSB

Load of total COD: (7.32)

$$FCOD_i = Q_i \cdot COD_i = 10{,}000 \cdot 431 = 4{,}310 \quad kgCOD/d$$

Load of biodegradable COD: (7.33)

$$FCOD_{b,i} = \left(1 - f_{SU,CODi} - f_{XU,CODi}\right) \cdot FCOD_i$$

$$= (1 - 0.071 - 0.129) \cdot 4{,}310 = 3{,}448.05 \quad kgCOD/d$$

Load of unbiodegradable particulate organics: (7.34)

$$FX_{Uv,i} = FCOD_i \cdot \frac{f_{XU,CODi}}{f_{cv}} = 4{,}310 \cdot \frac{0.13}{1.48} = 375.66 \quad kgVSS/d$$

Mass of VSS in the SRUSB at steady state: (7.36)

$$MX_{VSS} = FCOD_{b,i} \cdot \frac{Y_{SRB} \cdot SRT}{1 + b_{SRB}SRT} \cdot (1 + f_{SRB} \cdot b_{SRB}SRT) + FX_{Uv,i}SRT$$

$$= 4{,}310 \cdot \frac{0.113 \cdot 30}{1 + 0.04 \cdot 30} \cdot \frac{1 + 0.08 \cdot 0.04 \cdot 30}{1.48} + 375.66 \cdot 30$$

$$= 1.52 \cdot 10^4 \quad kgVSS$$

(note: $MX_{VSS} = MX_{SRB} + MX_{ED} + FX_{Uv,i} \cdot SRT$)

Load of inorganic suspended solids: (7.35)

$$FX_{FSS,i} = Q_i \cdot X_{FSS,i} = 10{,}000 \cdot \frac{10}{1{,}000} = 100 \quad kgISS/d$$

Active fraction of SRB: (7.38)

$$f_{av,SRB} = \frac{1}{[1 + f_{SRB} \cdot b_{SRB}SRT + f_{cv}(1 + b_{SRB} \cdot SRT)]}$$

$$= \frac{1}{[1 + 0.08 \cdot 0.04 \cdot 30 + 1.48 \cdot (1 + 1.48 \cdot 30)]} = 0.014$$

Mass of FSS in the SRUSB at steady state: (7.37)

$$MX_{FSS} = FX_{FSS,i} \cdot SRT + f_{XU,CODi} \cdot f_{av,SRB} \cdot MX_{VSS}$$

$$= 100 \cdot 30 + 0.129 \cdot 0.014 \cdot 1.52 \cdot 10^4$$

$$= 3.03 \cdot 10^3 \qquad\qquad\qquad\qquad kgISS$$

Mass of TSS in the SRUSB at steady state: (7.39)

$$MX_{TSS} = MX_{VSS} + MX_{FSS} = 1.52 \cdot 10^4 + 3.03 \cdot 10^3$$

$$= 1.82 \cdot 10^4 \qquad\qquad\qquad\qquad kgTSS$$

Volume of SRUSB:

$$V_{SRUSB} = \frac{MX_{TSS}}{X_{TSS}} = \frac{1.82 \cdot 10^4}{5,900} \cdot 1,000 = 3,090.4 \qquad\qquad m^3$$

Demand for sulphate: (7.41)

$$FSO_4 = FCOD_i \cdot \left(1 - f_{SU.CODi} - f_{XU,CODi}\right) \cdot [(1 - f_{cv} \cdot Y_{SRBv})$$

$$+ (1 - f_{SRB}) \cdot b_{SRB} \cdot \frac{(Y_{SRBv} \cdot f_{cv} \cdot SRT)}{(1 + b_{SRB} \cdot SRT)}]$$

$$= 4,310 \cdot (1 - 0.071 - 0.13) \cdot [(1 - 0.11)$$

$$+ (1 - 0.08) \cdot 0.04 \cdot 0.11 \cdot \frac{30}{(1 + 0.04 \cdot 30})]$$

$$= 3,254 \qquad\qquad\qquad\qquad kgCOD/d$$

$$= 1,627 \qquad\qquad\qquad\qquad kgS/d$$

Solution 7.3.2
Design calculation considering the hydrolysis of biodegradable particles:

7.3.2.1 SRUSB design calculation

The calculation of the steady-state concentration of particulate biodegradable COD in the SRUSB is done by iteration, via steps 1-7:

1. Assume an initial $X_{S,e}$ (effluent biodegradable COD) concentration of 0 g/m^3, and an initial SRUSB volume (V_R) value of 0.001 m^3. These values are used to calculate X_{SRB}.

2. Apply the calculated X_{SRB} to calculate X_{ED}.

3. Apply the calculated X_{SRB} to calculate r_h.

4. Apply the calculated X_{SRB} and X_{ED} to calculate MX_{TSS}.

5. Apply the calculated r_h to calculate $X_{S,e}$.

6. Apply the calculated MX_{TSS} to calculate V_R.

7. Substitute the calculated $X_{S,e}$ and V_R for the initial $X_{S,e}$ and V_R values in Step 1, then repeat steps 1 to 7 until the calculated $X_{S,e}$ and V_R values stabilize.

Step 1. Calculate the produced SRB biomass concentration (X_{SRB}):

$$
\begin{aligned}
X_{SRB} \quad &= biomass_{BPO} + biomass_{FBSO} + biomass_{VFA} \\
&= Y_{SRB} \cdot [(X_{OS,i} + S_{F,i}) \cdot SRT \cdot Q_i / V_R - X_{S,e}] / (1 + b_{SRB} \cdot SRT) \\
&\quad + Y_{VFA} \cdot Q_i / V_R \cdot S_{VFA,i} \cdot SRT / (1 + b_{SRB} \cdot SRT) \\
&= 0.113 \cdot [(217.65 + 92.66) \cdot 30 \cdot 1 \cdot 10^4 / 0.001 - 0] / (1 + 0.04 \cdot 30) \\
&\quad + 0.023 \cdot 1 \cdot 10^4 / 0.001 \cdot 34.48 \cdot 30 / (1 + 0.04 \cdot 30) \\
&= 4.78 \cdot 10^9 \qquad\qquad\qquad\qquad\qquad\qquad gCOD/m^3
\end{aligned}
$$

Step 2. Calculate the produced endogenous respiration residues (X_{ED}):

$$
\begin{aligned}
X_{ED} \quad &= X_{SRB} \cdot f_{SRB} \cdot b_{SRB} \cdot SRT \\
&= 4.78 \cdot 10^9 \cdot 0.08 \cdot 0.04 \cdot 30 \\
&= 4.69 \cdot 10^8 \qquad\qquad\qquad\qquad\qquad\qquad gCOD/m^3
\end{aligned}
$$

Step 3. Combine the X_S mass balance equation with the hydrolysis rate expression (Eq. 7.27b), and calculate the volumetric hydrolysis rate at steady state (r_h):

$$
\begin{aligned}
r_h \quad &= Q_i \cdot X_{S,i} / V_R - X_{S,e} / SRT + b_{SRB} \cdot X_{SRB} \\
&= 1 \cdot 10^4 \cdot 217.65 / 0.001 - 0/30 + 0.04 \cdot 4.78 \cdot 10^9 \\
&= 2.37 \cdot 10^9 \qquad\qquad\qquad\qquad\qquad\qquad gCOD/m^3.d
\end{aligned}
$$

Step 4. Calculate the mass of TSS in the SRUSB at steady state:

$$
\begin{aligned}
MX_{VSS} \quad &= (X_{SRB} + X_{ED} + X_{S,e}) \cdot V_R / f_{cv} + FX_{Uv,I} \cdot SRT \\
&= (4.78 \cdot 10^9 + 4.69 \cdot 10^8 + 0) \cdot 0.001 / 1.48 + 3.76 \cdot 10^2 \cdot 30 \\
&= 1.49 \cdot 10^4 \qquad\qquad\qquad\qquad\qquad\qquad kgVSS
\end{aligned}
$$

$$
\begin{aligned}
MX_{FSS} \quad &= FX_{FSS,i} \cdot SRT + f_{XU,CODi} \cdot f_{av,SRB} \cdot MX_{VSS} \\
&= 1 \cdot 10^4 \cdot 10 / 1,000 \cdot 30 + 0.13 \cdot 0.014 \cdot 1.49 \cdot 10^4 \\
&= 3.03 \cdot 10^3 \qquad\qquad\qquad\qquad\qquad\qquad kgISS
\end{aligned}
$$

MX_{TSS} $= MX_{VSS} + MX_{FSS}$

              $= 1.49 \cdot 10^4 + 3.03 \cdot 10^3$

              $= 1.79 \cdot 10^4$ kgTSS

Step 5. According to Eq. 7.29, calculate the particulate biodegradable COD retained in the SRUSB.

Note: in a Monod-type equation, substrate concentrations are not affected by particle accumulation.

$X_{S,c}$ $= r_h \cdot K_S / (K_{SRB,max} \cdot X_{SRB} - r_h)$

              $= 2.37 \cdot 10^9 \cdot 557 / (3.25 \cdot 4.78 \cdot 10^9 - 2.37 \cdot 10^9)$

              $= 100.33$ gCOD/m^3

Step 6. Calculate reactor volume V_R:

V_R $= MX_{TSS} / X_{TSS}$

              $= 1.79 \cdot 10^4 \cdot 10^3 / 5,900$

              $= 3.04 \cdot 10^3$ m^3

Step 7. Repeat V_R, $X_{S,c} \rightarrow X_{SRB} \rightarrow MX_{TSS}$, $r_h \rightarrow V_R$, $X_{S,c}$ until $X_{S,c}$ and V_R stabilize.

The calculation results after iteration are:

$X_{S,c}$ $= 100.18$ gCOD/m^3

X_{SRB} $= 1,587.42$ gCOD/m^3

X_{ED} $= 152.39$ gCOD/m^3

r_h $= 769$ gCOD/m^3.d

MX_{FSS} $= 3.03 \cdot 10^3$ kgTSS

MX_{VSS} $= 1.51 \cdot 10^4$ kgTSS

MX_{TSS} $= 1.81 \cdot 10^4$ kgTSS

V_R $= 3,070.49$ m^3

Mass of SRB in the SRUSB at steady state:

MX_{SRB} $= X_{SRB} \cdot V_R / f_{cv}$

              $= 3.29 \cdot 10^3$ kgVSS

Mass of endogenous residual from SRB at steady state:

MX_{ED} $= X_{ED} \cdot V_R / f_{cv}$

              $= 3.16 \cdot 10^2$ kgVSS

254

Calculate the waste sludge flow rate (Q_w):

$$Q_w = V_R / SRT$$

$$= 3{,}070.49 / 30$$

$$= 102.34 \qquad \text{m}^3/\text{d}$$

Calculate the hydrolysed BPO fraction:

$$f_{BPO,h} = (Q_i \cdot X_{S,i} - Q_w \cdot X_{S,e}) / (Q_i \cdot X_{S,i})$$

$$= (1 \cdot 10^4 \cdot 217.65 - 102.34 \cdot 100.18) / (1 \cdot 10^4 \cdot 217.65)$$

$$= 99.53\ \%$$

When acidogenic hydrolysis is not taken into account in the calculation, it is assumed that the biodegradable COD in the SRUSB will be hydrolysed completely and transferred into biomass sludge afterwards. However, when acidogenic hydrolysis is considered in the calculation, the model results indicate that most of the biodegradable COD (99.53 %) is hydrolysed in the SRUSB reactor, and the unhydrolysed biodegradable particulate COD ($X_{S,e}$) which is approximately 100.18 g/m^3 is enmeshed into the biomass and finally discharged as the waste sludge Q_w. When comparing the results obtained from the above calculations, the volume of SRUSB is almost the same (<1 % difference) in both scenarios, demonstrating that both methods are acceptable for process design.

7.3.2.2 Daily organic removal and TDSd production in the SRUSB

Based on the stoichiometric ratio of reaction (1), if 1 mol organic (32 gCOD) were consumed, there would be 0.47 mol (15.04 gS) sulphate consumed and 0.47 mol (15.04 gS) TDSd produced.

$$FX_S = Q_i \cdot X_{S,I} - Q_w \cdot X_{S,e}$$

$$= 1 \cdot 10^4 \cdot 217.65 - 101.84 \cdot 68.01$$

$$= 2.17 \cdot 10^6 \qquad \text{gCOD/d}$$

$$FS_{F,i} = S_{F,i} \cdot Q_{in}$$

$$= 92.66 \cdot 10{,}000$$

$$= 9.05 \cdot 10^5 \qquad \text{gCOD/d}$$

$$FS_{VFA} = S_{VFA,I} \cdot Q_{in}$$

$$= 34.48 \cdot 10{,}000$$

$$= 3.45 \cdot 10^5 \qquad \text{gCOD/d}$$

Daily organic COD removal:

$$FCOD_o = FX_S + FS_{F,i}$$

$$= 2.17 \cdot 10^6 + 9.05 \cdot 10^5$$

$$= 3.08 \cdot 10^6 \qquad \text{gCOD/d}$$

COD removal in the SRUSB:

$$\Delta COD = (FCOD_o + FS_{VFA}) / Q_i$$

$$= (3.08 \cdot 10^6 + 3.45 \cdot 10^5) / 10{,}000$$

$$= 342.5 \qquad gCOD/m^3$$

TDSd produced in the SRUSB:

$$TDSd = \Delta COD \cdot (15.04 / 32)$$

$$= 342.5 \cdot 15.04 / 32$$

$$= 161 \qquad gS/m^3$$

Daily demand for sulphate:

$$FSO_4 = TDSd \text{ produced} \cdot Q_i$$

$$= 161 / 1{,}000 \cdot 1 \cdot 10^4$$

$$= 1{,}610 \qquad kgS/d$$

Influent sulphate concentration to the SRUSB:

$$S_{SO4} = FSO_4 / Q_i$$

$$= 161 \qquad gS/m^3$$

7.3.2.3 Dimension of aerobic bioreactor and daily oxygen demand

Based on the stoichiometric ratio of reaction (3), if 1 mol NH_4-N (14 gN) were consumed, there would be 1.8 mol of oxygen consumed (57.6 gO_2). If 1 mol TDSd (32 gS) were oxidized, according to reaction (4), there would be 64 gO_2 consumed.

Nitrification capacity:

$$NIT_c = TKN_i - TKN_e - N_s$$

$$= 57.5 - 5.7 - 5$$

$$= 46.8 \qquad gN/m^3$$

The necessary surface area for the nitrifying reactor:

$$A_{F,NH4} = Q_i \cdot NITc / B_{A,NH4}$$

$$= 1 \cdot 10^4 \cdot 46.8 / 1$$

$$= 468{,}000 \qquad m^2$$

Volume of aerobic reactor:

V_{OX} $\quad = A_{F,NH4} / (a_F \cdot f_{fill})$

$\quad = 468{,}000 / 300$

$\quad = 1{,}560 \qquad \qquad m^3$

Daily oxygen consumption by nitrification:

FO_N $\quad = Q_i \cdot NITc \cdot$ (reaction mass ratio between oxygen and NH_4^+-N)

$\quad = 1 \cdot 10^4 \cdot 46.8 \cdot 57.6 / 14 \qquad \qquad gO_2/d$

$\quad = 1{,}920 \qquad \qquad kgO_2/d$

TDSd oxidized by aeration (note that ΔTDSd-ax is calculated in Section 7.3.2.4):

ΔTDSd-ox $\quad = $ ΔTDSd-SR $-$ ΔTDSd-ax

$\quad = 161 - 81.75$

$\quad = 79.3 \qquad \qquad gS/m^3$

Daily oxygen consumption by TDSd oxidation:

FO_S $\quad = Q_i \cdot$ ΔTDSd-ox $\cdot$ (reaction mass ratio between oxygen and TDSd)

$\quad = 1 \cdot 10^4 \cdot 79.3 \cdot (64 / 64) \qquad \qquad gO_2/d$

$\quad = 793 \qquad \qquad kgO_2/d$

Note that the extra oxygen consumed for TDSd oxidation can be positively reduced, for instance by applying primary sedimentation prior to the SRUSB to reduce COD and TDSd production.

7.3.2.4 Dimensioning and calculation of TDSd consumption in the anoxic tank

If 0.16 mol NO_3^--N (2.29 gN) were consumed, according to reaction (2), there would be 4 gS TDSd consumed; 4.77 (gCaCO₃) HCO_3^- consumed and 1 gVSS biomass produced.

Effluent nitrate concentration:

$S_{NO3,e}$ $\quad = NIT_c / (a + 1)$

$\quad = 46.8 / (2.5 + 1)$

$\quad = 13.37 \qquad \qquad gN/m^3$

TDSd consumed in the anoxic tank:

ΔTDSd-ax $\quad = NITc \cdot$ (reaction mass ratio between TDSd and NO_3-N)

$\quad = 46.8 \cdot (4 / 2.29)$

$\quad = 81.75 \qquad \qquad gS/m^3$

SO_4^{2-} produced $\quad$ = TDSd consumed

$\qquad$ = 81.75 $\qquad\qquad\qquad\qquad$ gS/m^3

The necessary surface area for the anoxic reactor:

$A_{F,NO3}$ $\qquad$ = $Q_i \cdot (NIT_c - S_{NO3,e}) / B_{A,NH4}$

$\qquad$ = 10,000 · (46.8 −13.37) / 1

$\qquad$ = 334,300 $\qquad\qquad\qquad\qquad$ m^2

The volume of the anoxic reactor:

V_{AX} $\qquad$ = $A_{F,NOx} / (a_F \cdot f_{fill})$

$\qquad$ = 334,300 / 300

$\qquad$ = 1,114.3 $\qquad\qquad\qquad\qquad$ m^3

Table 7.3 Summary of the SANI system design results.

Description	Parameter	Unit	Value
1. Influent and SRUSB			
Type of wastewater	Raw/settled		Raw
Temperature	T	°C	16
Influent flow rate	Q_i	m^3/d	10,000
Influent total COD	COD_i	gCOD/m^3	431
Influent soluble unbiodegradable COD	$S_{U,i}$	gCOD/m^3	30.60
Influent particulate unbiodegradable COD	$X_{U,i}$	gCOD/m^3	55.59
Influent VFAs	$S_{VFA,i}$	gCOD/m^3	34.48
Influent fermentable COD	$S_{F,i}$	gCOD/m^3	92.66
Influent particulate biodegradable COD	$X_{S,i}$	gCOD/m^3	217.65
Sludge retention time	SRT	d	30
Recirculation ratio	a	-	2.5
Waste sludge flow rate	Q_w	m^3/d	96.16
Particulate biodegradable COD left in the SRUSB	$X_{S,e}$	gCOD/m^3	100.18
Volumetric hydrolysis rate	r_h	gCOD/m^3.d	769
2. System biomass (VSS) equations			
Incomplete hydrolysis			
Mass of SRB	MX_{SRB}	kgVSS	3,290
Mass of endogenous residue from SRB	MX_{ED}	kgVSS	316

Complete hydrolysis			
Mass of SRB	MX_{SRB}	kgVSS	3,590
Mass of endogenous residue from SRB	MX_{ED}	kgVSS	345
3. Sulphur compounds			
Complete hydrolysis			
Daily sulphate demand in the SRUSB	FSO_4	kgS/d	1,627
Effluent TDSd concentration in the SRUSB	$\Delta TDSd\text{-}SR$	gS/m^3	162.7
Incomplete hydrolysis			
Daily sulphate demand in the SRUSB	FSO_4	kgS/d	1,610
Effluent TDSd concentration in the SRUSB	$\Delta TDSd\text{-}SR$	gS/m^3	161
TDSd consumed in the anoxic tank	$\Delta TDSd\text{-}ax$	gS/m^3	81.75
TDSd oxidized in the aerobic tank	$\Delta TDSd\text{-}ox$	gS/m^3	79.3
4. Volatile and total suspended solids (VSS and TSS) in the system			
Incomplete hydrolysis			
Mass of VSS	MX_{VSS}	kgVSS	15,100
Mass of FSS	MX_{FSS}	kgISS	3,030
Mass of TSS	MX_{TSS}	kgTSS	18,100
Complete hydrolysis			
Mass of VSS	MX_{VSS}	kgVSS	15,200
Mass of FSS	MX_{FSS}	kgISS	3,030
Mass of TSS	MX_{TSS}	kgTSS	18,200
5. Reactor volumes			
SRUSB reactor volume (complete hydrolysis)	V_{SRUSB}	m^3	3,090.4
SRUSB reactor volume (incomplete hydrolysis)	V_{SRUSB}	m^3	3,070.5
Aerobic reactor volume	V_{OX}	m^3	1,560
Anoxic reactor volume	V_{AX}	m^3	1,114.3
6. N compounds			
Effluent nitrate concentration	$S_{NO3,e}$	gN/m^3	13.37
7. Oxygen demand			
Flux of O_2 demand by nitrification	FO_N	kgO_2/d	1,920
Flux of O_2 demand by TDSd oxidation	FO_S	kgO_2/d	793

7.4. EXERCISES

Biological sulphate reduction (exercises 7.4.1-7.4.5)

Exercise 7.4.1

Briefly describe the following aspects of the biological sulphate reduction (BSR) process:

a. Sulphate-reducing pathways.
b. The key microorganisms driving sulphate reduction.
c. The electron donors for the sulphate-reducing bioprocess.
d. Applications of the BSR process and influencing factors.

Exercise 7.4.2

Determine the influent biodegradable composition and its electron-donating capacity for biological sulphate reduction.

Exercise 7.4.3

Assuming that the biodegradable particulate COD is hydrolysed completely in the SRUSB, examine the effect of sludge age on (*i*) the mass of TSS and VSS sludge in the SRUSB tank (MX_{TSS}, kg TSS; MX_{VSS}, kg VSS); (*ii*) the active fractions of the sludge with respect to VSS ($f_{av,SRB}$); (*iii*) the mass of TSS sludge wasted per day (FX_{TSS}, kgTSS/d); and (*iv*) the average daily sulphate demand (FSO_4, kgS/d).

Exercise 7.4.4

Assuming that the influent biodegradable particulate COD is hydrolysed completely in the SRUSB, calculate the different fractions of COD flux in the SRUSB for different sludge ages at steady state, *i.e.* (*i*) degraded COD; (*ii*) particulate unbiodegradable COD; and (*iii*) particulate biodegradable COD (biomass). Use the influent flow rate Q_i and influent COD fractions in Table 7.1.

Exercise 7.4.5

Calculate the relationship between sludge age with reactor volume requirements at different average reactor TSS concentrations for raw wastewater X_{TSS}: 3,000 gTSS/m^3, 4,500 gTSS/m^3 and 5,900 gTSS/m^3.

Sulphur-driven autotrophic denitrification (SdAD) (exercises 7.4.6-7.4.9)

Exercise 7.4.6

Briefly describe the following aspects of the SdAD process:

a. The advantages of SdAD.
b. The electron donor and acceptor of SdAD.
c. Biochemical reactions involved in SdAD.
d. Microorganisms involved with SdAD.
e. Factors affecting the SdAD process.

Exercise 7.4.7

Develop a kinetic model for the SdAD process, using the activated sludge modelling method introduced in Chapter 14 in Chen *et al.*, 2020.

Exercise 7.4.8

Predict the performance of SdAD in an MBBR reactor in batch mode (without either influent or effluent). The given initial concentrations of $S_{TDSd,ini}$, $S_{S0,ini}$, $S_{NO3,ini}$, $S_{NO2,ini}$, $S_{N2O,ini}$, $X_{SOB,ini}$ and $X_{I,ini}$ are equal to 90 gS/m^3, 0 gS/m^3, 30 gN/m^3, 0 gN/m^3, 0 gN/m^3, 500 gCOD/m^3 and 10 gVSS/m^3, respectively, using the given stoichiometric and kinetic parameters in Table 7.4. Plot the substrates over time.

Table 7.4 Stoichiometric and kinetic parameters of the SdAD model.

Parameter	Definition	Value	Unit	Source
Stoichiometric parameters				
Y_{SOB}	Yield coefficient for SOB	0.28	gCOD/gS	Xu *et al.*, 2016
f_I	Fraction of X_U in biomass decay	0.1	gCOD/gCOD	Henze *et al.*, 2000
Kinetic parameters				
$\mu_{TDSd,NO3\text{-}NO2}$	Maximum reaction rate of Process 1	0.24	1/h	Xu *et al.*, 2014
$\mu_{TDSd,NO2\text{-}N2O}$	Maximum reaction rate of Process 2	0.31	1/h	Liu *et al.*, 2016
$\mu_{TDSd,N2O\text{-}N2}$	Maximum reaction rate of Process 3	0.076	1/h	Liu *et al.*, 2016
$\mu_{S,NO3\text{-}NO2}$	Maximum reaction rate of Process 4	0.020	1/h	Xu *et al.*, 2014
$\mu_{S,NO2\text{-}N2O}$	Maximum reaction rate of Process 5	0.021	1/h	Liu *et al.*, 2016
$\mu_{S,N2O\text{-}N2}$	Maximum reaction rate of Process 6	0.017	1/h	Liu *et al.*, 2016
b_{SOB}	Decay rate coefficient of SOB	0.002	1/h	Xu *et al.*, 2014
K_1^{TDSd}	S_{H2S} affinity constant for Process 1	1.36	gS/m^3	Liu *et al.*, 2016
K_{NO3}^{TDSd}	S_{NO3} affinity constant for Process 1	0.2	gN/m^3	Liu *et al.*, 2016
K_2^{TDSd}	S_{H2S} affinity constant for Process 2	1.8	gS/m^3	Wang *et al.*, 2010
K_{NO2}^{TDSd}	S_{NO2} affinity constant for Process 2	0.21	gN/m^3	Xu *et al.*, 2014
K_3^{TDSd}	S_{H2S} affinity constant for Process 3	1.48	gS/m^3	Xu *et al.*, 2014
K_{N2O}^{TDSd}	S_{N2O} affinity constant for Process 3	0.21	gN/m^3	Xu *et al.*, 2014
K_4^{S0}	S_{S0} affinity constant for Process 4	0.21	gS/m^3	Xu *et al.*, 2014
K_{NO3}^{S0}	S_{NO3} affinity constant for Process 4	0.18	gN/m^3	Xu *et al.* 2014
K_5^{S0}	S_{S0} affinity constant for Process 5	175	gS/m^3	Xu *et al.*, 2014
K_{NO2}^{S0}	S_{NO2} affinity constant for Process 5	0.10	gN/m^3	Xu *et al.*, 2014
K_6^{S0}	S_{S0} affinity constant for Process 6	175	gS/m^3	Xu *et al.*, 2014
K_{N2O}^{S0}	S_{N2O} affinity constant for Process 6	0.10	gN/m^3	Xu *et al.*, 2014

Exercise 7.4.9

Use the kinetic model established in Exercise 7.4.7 to simulate the performance of the SdAD process in a continuous-flow MBBR. Assuming a stable SOB concentration in this anoxic MBBR tank, then (*i*) calculate the steady-state concentrations of S^0, NO_2-N and N_2O-N in the anoxic tank; (*ii*) find out the relationship between the S/N ratio and the accumulation of S^0 and the other end products; and (*iii*) find out the relationship between HRT and the accumulation of S^0 and the other end products.

Sulphur conversion-based resource recovery (exercises 7.4.10-7.4.13)
Exercise 7.4.10

Briefly describe sulphur-based resource recovery technologies.

Exercise 7.4.11

Develop a kinetic model to calculate the metal sulphide formation and recovery.

Metal recovery from a wastewater stream can be performed in an integrated system, where biological conversion of the sulphates into sulphides and precipitation of metal ions as metal sulphides takes place in a sulphate-reducing bioreactor and chemical precipitation reactor, respectively (see Figure 7.2). Taking the recovery of zinc as an example, in a zinc precipitation process, the pH value in the precipitator is controlled because Zn^{2+} is extracted from the liquid phase as ZnS in the precipitator (see Figure 7.2), and the next step in the separation process will take place in a filter or a sedimentation tank. Use the following mass balance equations and information given in Table 7.5 to calculate:

a. The steady-state concentrations of TDSd, $[Zn^{2+}]$ and $[S^{2-}]$.
b. The value of $[Zn^{2+}]$ removed each day and the removal efficiency in the precipitator.
c. The pS value at steady state (pS = $-\log([S^{2-}])$).

The mass balance of total dissolved sulphide (TDSd, concentrations of H_2S, HS^- and S^{2-}) and zinc (Zn^{2+}) in a CSTR (König et $al.$, 2006) are described below.

$$\frac{dTDSd}{dt} = \frac{Q_b}{V_R} \cdot TDSd_{SRUSB} - \frac{Q_b + Q_i}{V_R} \cdot TDSd - k \cdot [Zn^{2+}] \cdot [S^{2-}], \qquad TS > 0$$

$$\frac{d[Zn^{2+}]}{dt} = \frac{Q_i}{V_R} \cdot [Zn^{2+}]_i - \frac{Q_b + Q_i}{V_R} \cdot [Zn^{2+}] - k \cdot [Zn^{2+}] \cdot [S^{2-}], \qquad [Zn^{2+}] > 0$$

The relationship between the concentration of sulphide and the amount of total sulphide is given by:

$$[S^{2-}] = \frac{TDSd}{\dfrac{[H^+]^2}{K_{a1} \cdot K_{a2}} + \dfrac{[H^+]}{K_{a2}} + 1}$$

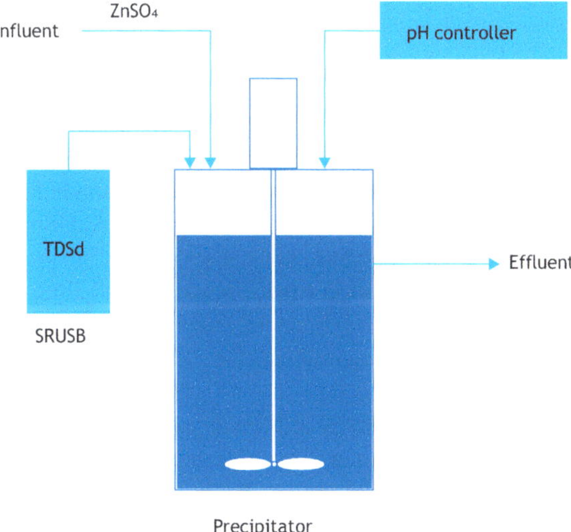

Figure 7.2 Precipitation of zinc with sulphide from the SRUSB in a lab-scale process set-up.

Table 7.5 Parameters needed to calculate the steady-state concentrations.

Symbol	Description	Value	Unit
Q_i	Influent flow rate	2	l/h
$[Zn^{2+}]_i$	Influent $[Zn^{2+}]$ concentration	$75.0 \cdot 10^{-3}$	mol/l
Q_{SRUSB}	Flow rate from the SRUSB to the precipitator	5	l/h
$TDSd_{SRUSB}$	TDSd in the SRUSB	162.7	mg/l
	$=162.7(\text{mg/l}) / 32(\text{g/mol}) / 1,000$	$5.08 \cdot 10^{-3}$	mol/l
k	Precipitation kinetic constant	$2.5 \cdot 10^{20}$	l/mol.h
K_{a1}	Equilibrium constant for the reaction $H_2S \leftrightarrow HS^- + H^+$	10^{-7}	mol/l
K_{a2}	Equilibrium constant for the reaction $HS^- \leftrightarrow S^{2-} + H^+$	$10^{-13.9}$	mol/l
V_R	Volume of the reactor	0.6	l
$[H^+]$	Concentration of protons	10^{-6}	mol/l

Exercise 7.4.12

Apply the steady-state model in Exercise 7.4.11 to further investigate the ZnS precipitation process.

a. Calculate the minimal TDSd concentration to remove 99 % of the zinc ($Zn^{2+} = 0.075$ mol/l).
b. Assuming that the TDSd concentration in the SRUSB is stable (162.5 mgS/l), calculate the minimal flow rate (Q_{SRUSB}) to make the $[Zn^{2+}]$ removal efficiency higher than 99 %.
c. Simulate the ZnS precipitation with the initial $[Zn^{2+}]$ at 0.025 mol/l, 0.05 mol/l, and 0.075 mol/l.

Exercise 7.4.13

Elemental sulphur can be recovered from sulphide-laden wastewater such as SRUSB effluent. In this exercise, a mathematical model is developed to simulate the sulphur recovery in a sulphide oxidation bioreactor (Figure 7.3).

Assume that the aerobic tank is well mixed and that the element sulphur concentration in the effluent is equal to the element sulphur concentration in the reactor. The chemical sulphide oxidation process is ignored. Sulphide is oxidized to element sulphur or sulphate by chemolithotrophic sulphide oxidation bacteria, *e.g.* SOB. The SOB biomass concentration in the aerobic reactor remains stable at 400 g/m^3. The influent is designed as fixed at a constant level of 500 m^3/d with three different TDSd concentrations of 100, 200, and 300 gS/m^3.

Develop the model to answer the following two questions:

a. Assuming the hydraulic retention time (HRT) is 2 hours, calculate the optimum dissolved oxygen (DO) concentration which can result in the highest sulphur production efficiency.
b. Assuming the DO has a constant value of 1 g/m^3, calculate the HRT required to achieve the highest sulphur production efficiency.

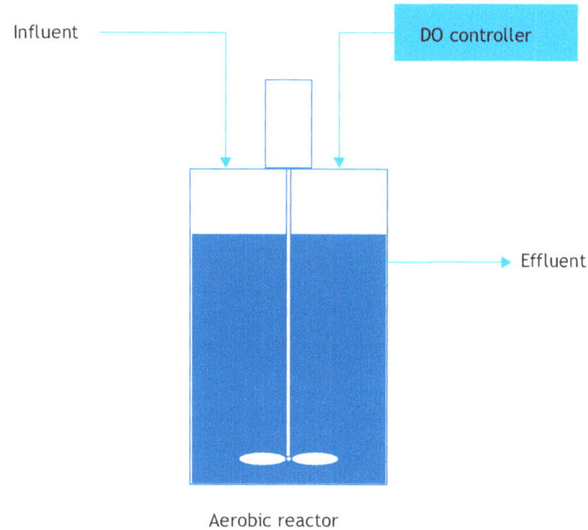

Figure 7.3 Element sulphur recovery in a sulphide oxidation reactor.

ANNEX 1: SOLUTIONS TO EXERCISES

Biological sulphate reduction (solutions 7.4.1-7.4.5)

Solution 7.4.1

a. In both natural and engineered systems, biological sulphur conversion can be classified into two categories, *i.e.* 'assimilatory sulphate reduction' and 'dissimilatory sulphate reduction (DSR)' pathways, for the purpose of synthesizing sulphur-containing molecules and cellular energy production, respectively. The main reactions associated with DSR pathways are closely related to the sulphur and carbon cycles and play a crucial role for wastewater organics degradation in the sulphur reduction/SANI process.

b. Sulphate-reducing bacteria (SRB) and sulphate-reducing archaea (SRA), generally called SRB, can perform respiration with oxidized sulphur compounds in their energy metabolism. To date, more than 120 species and 40 genera belonging to four bacterial classes and two archaeal classes have been documented. Refer to Table 7.2 in Chen *et al.,* 2020 for a list of SRB genera, their electron acceptor and the optimum temperature for their growth.

c. According to the type of SRB growth (autotrophic or heterotrophic), the different electron donors of SRB reactions are inorganic substrates (H_2 and CO) and organic substrates. For more details on SRB reduction rates, related reactions and benefits/drawbacks, please refer to Table 7.3 in Chen *et al.,* 2020. It should be noted that, stoichiometrically, 0.67 mol of COD is needed for the complete reduction of 1 mol sulphate.

d. SRB-based technologies can be applied in different types of wastewater treatment, such as the removal and reuse of sulphur compounds from sulphur-laden wastewaters (*e.g.* leather tanning, kraft pulping and food processing), and for flue-gas desulphurization wastewater as well as domestic wastewater with sufficient sulphate compounds contained/added. These applications are significantly affected by temperature, pH, sulphide concentration and COD/SO_4^{2-} mass ratio (for details see Table 7.5 in Chen *et al.,* 2020). It should be also noted that there are different dominant kinetic principles in different SRB applications (refer to Table 7.4 in Chen *et al.,* 2020).

Solution 7.4.2

The influent stoichiometric composition with x, y, z, a and b values in $C_xH_yO_zN_aP_b$, are assigned to each of the main influent wastewater organic fractions. Thus the organics composition ($C_xH_yO_zN_aP_b$) can be expressed as $C_{f_c/12}H_{f_h/1}O_{f_o/16}N_{f_n/14}P_{f_p/31}$, where f_c, f_h, f_o, f_n, f_p are the mass ratios (g element/g compound) of each element in the organics, and $f_c + f_h + f_o + f_n + f_p = 1$.

The parameters of COD, VSS, TOC, OrgN and OrgP concentrations are measured (Table 7.6) to conduct element analysis for BPO_i and UPO_i components and mass ratios of f_c, f_h, f_o, f_n, f_p . The obtained results are shown in Table 7.6:

f_c = TOC / VSS; f_n = OrgN / VSS; f_p = OrgP / VSS; f_{cv} = COD / VSS.

Table 7.6 Calculation procedure to determine the mass ratios of the influent biodegradable particulate organic (BPOi) and influent unbiodegradable particulate organic (UPOi).

		COD (gCOD/m³)	VSS (gVSS/m³)	TOC (gC/m³)	OrgN (gN/m³)	OrgP (gP/m³)
BPO_i	Measured concentration	217.65	147.06	81.62	8.67	1.08
	Mass ratio	f_{cv} 1.48	-	f_c 0.55	f_n 0.059	f_p 0.0074
UPO_i	Measured concentration	COD (gCOD/m³) 55.6	VSS (gVSS/m³) 37.57	TOC (gC/m³) 20.06	OrgN (gN/m³) 8.26	OrgP (gP/m³) 0.9
	Mass ratio	f_{cv} 1.48	-	f_c 0.53	f_n 0.22	f_p 0.024

$$C_xH_yO_zN_aP_b \equiv C_{\frac{f_c}{12}}H_{\frac{f_h}{1}}O_{\frac{f_o}{16}}N_{\frac{f_n}{14}}P_{\frac{f_p}{31}} \equiv C_1H_{\frac{y}{x}}O_{\frac{z}{x}}N_{\frac{a}{x}}P_{\frac{b}{x}}$$

BPO_i elemental analysis: (Ekama, 2009)

$$f_o = \frac{16}{18} \cdot \left(1 - \frac{f_{cv}}{8} - \frac{8 \cdot f_c}{12} - \frac{17 \cdot f_n}{14} - \frac{26 \cdot f_p}{31}\right)$$

$$= 0.33$$

$$f_h = \frac{2}{18} \cdot \left(1 + f_{cv} - \frac{44 \cdot f_c}{12} + \frac{10 \cdot f_n}{14} - \frac{71 \cdot f_p}{31}\right)$$

$$f_h = 0.052$$

$$x = f_c/12 \qquad\qquad y = f_h/1$$

$$= 0.55/12 \qquad\qquad\qquad = 0.052$$

$$= 0.046$$

z	$= f_o / 16$	a	$= f_n / 14$
	$= 0.33 / 16$		$= 0.059 / 14$
	$= 0.02$		$= 0.0042$
b	$= f_p / 31$		
	$= 0.0074 / 31$		
	$= 0.00024$		

BPO_i elemental composition: $C_1 H_{1.13} O_{0.44} N_{0.091} P_{0.0052}$

UPO_i elemental analysis:

$$f_o = \frac{16}{18} \cdot \left(1 - \frac{f_{cv}}{8} - \frac{8 \cdot f_c}{12} - \frac{17 \cdot f_n}{14} - \frac{26 \cdot f_p}{31} \right)$$

$$= 0.15$$

$$f_h = \frac{2}{18} \cdot \left(1 + f_{cv} - \frac{44 \cdot f_c}{12} + \frac{10 \cdot f_n}{14} - \frac{71 \cdot f_p}{31} \right)$$

$$f_h = 0.069$$

x	$= f_c / 12$	y	$= f_h / 1$
	$= 0.534 / 12$		$= 0.069$
	$= 0.044$		
z	$= f_o / 16$	a	$= f_n / 14$
	$= 0.15 / 16$		$= 0.22 / 14$
	$= 0.0095$		$= 0.016$
b	$= f_p / 31$		
	$= 0.024 / 31$		
	$= 0.00077$		

UPO_i elemental composition: $C_1 H_{1.56} O_{0.21} N_{0.35} P_{0.017}$

The elemental compositions of UPO_i and BPO_i are determined from the mass ratios:

BPO_i: $C_1 H_{1.13} O_{0.44} N_{0.091} P_{0.0052}$

UPO_i: $C_1 H_{1.56} O_{0.21} N_{0.35} P_{0.017}$

From Section 7.4.4.1 in Chapter 7 in Chen *et al.*, 2020, the mass of COD produced (as active SRB biomass and endogenous sludge) per day can be calculated as follows:

$$E = \frac{Y_{SRB}}{[1 + b_{SRB} \cdot SRT \cdot (1 - Y_{SRB})]} = \frac{0.113}{[1 + 0.04 \cdot 30 \cdot (1 - 0.113)]} = 0.054$$

Electrons available for redox reactions per mol of the biodegradable organics (γ_S, Eq. 7.22 in Chen *et al.*, 2020):

$$\gamma_S = 4x + y - 2z - 3a + 5b = 4 \cdot 1 + 1.13 - 2 \cdot 0.44 - 3 \cdot 0.091 + 5 \cdot 0.0052 = 4$$

Note: x, y, z, a, b here refers to the elemental composition of BPO_i.

The stoichiometric coefficient of sulphate in the sulphate reduction equation (Eq. 7.21 Chen *et al.*, 2020):

$$1 - E \cdot \frac{\gamma_S}{8} = 1 - 0.0547 \cdot \frac{4}{8} = 0.47$$

Therefore, this means that for 1 mol of $C_1H_{1.13}O_{0.44}N_{0.091}P_{0.0052}$ (32g COD) X_S consumed, there would be 0.47 mol (15.04 gS) of SO_4-S being reduced.

Solution 7.4.3

Using the equations listed below, MX_{VSS}, MX_{TSS}, $f_{av,SRB}$, FX_{TSS}, and FSO_4 can be calculated for different sludge ages, and the obtained results are illustrated in Table 7.7 and Figure 7.4.

$$MX_{VSS} = FCOD_{b,i} \cdot \frac{Y_{SRB} \cdot SRT}{1 + b_{SRB} \cdot SRT} \cdot \frac{1 + f_{SRB} \cdot b_{SRB}SRT}{f_{cv}} + FX_{Uv,i} \cdot SRT \tag{7.36}$$

$$MX_{TSS} = MX_{VSS} + FX_{FSS,i} \cdot SRT + f_{XU,CODi} \cdot f_{av,SRB} \cdot MX_{VSS} \tag{7.39}$$

$$f_{av,SRB} = 1/[1 + f_{SRB}b_{SRB}SRT + f_{cv}(1 + b_{SRB}SRT)] \tag{7.38}$$

$$FX_{TSS} = Q_wX_{TSS} = MX_{TSS}/SRT \tag{7.35}$$

$$FSO_4 = FCOD_i \cdot (1 - f_{SU.CODi} - f_{XU,CODi}) \cdot [(1 - f_{cv} \cdot Y_{SRBv}) \tag{7.41}$$

Table 7.7 Steady-state calculation results for SRT from 5 days to 60 days.

SRT (d)	MX_{VSS} (kgVSS)	MX_{TSS} (kgTSS)	$f_{av,SRB}$	FX_{TSS} (kgVSS/d)	FSO_4 (kgS/d)
5	2,992	3,498	0.074	699.7	1,559
10	5,697	6,708	0.041	670.8	1,580
20	10,625	12,645	0.022	632.3	1,608
30	15,204	18,233	0.015	607.8	1,626
40	19,595	23,632	0.011	590.8	1,639
50	23,873	28,918	0.0089	578.4	1,648
60	28,078	34,131	0.0075	568.8	1,655

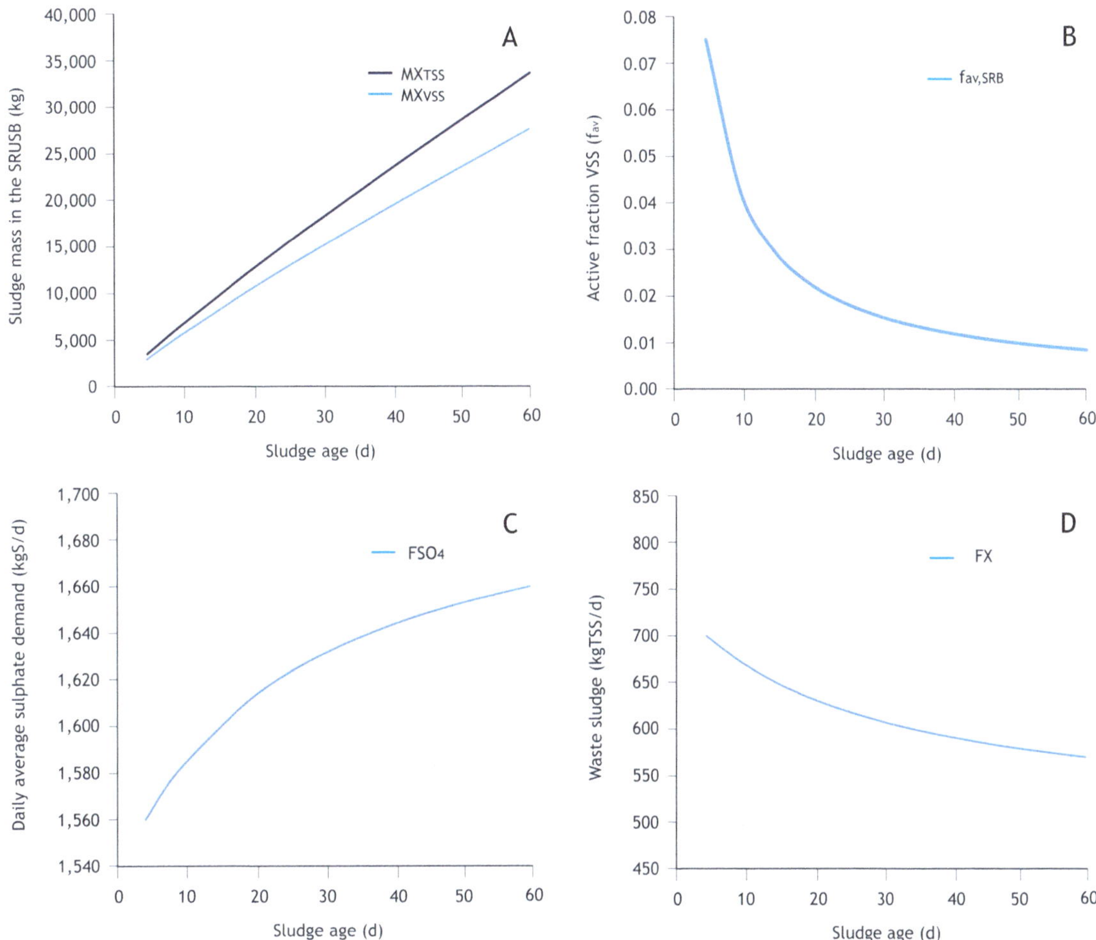

Figure 7.4 Mass of sludge MX$_{TSS}$ (kgTSS) and MX$_{VSS}$ (kgVSS) (A); active fractions of the sludge with respect to VSS (f$_{av,SRB}$) (B); average sulphate demand FSO$_4$ (kgS/d) (C); mass of VSS sludge produced per day (kgVSS/d) (D), versus sludge age for the example raw sewage.

In summary, the results show that longer sludge retention times will proportionally increase the mass of sludge MX$_{TSS}$ and MX$_{VSS}$ in the reactor (Figure 7.4 A). However, SRB sludge systems operating at very long sludge ages allow anaerobic sludge reduction, thereby leading to a lower active fraction (Figure 7.4 B). Meanwhile, an increase in the sludge age decreases the sludge production (Figure 7.4 D) and causes more demand for sulphate in the SRUSB (Figure 7.4 C).

Solution 7.4.4

According to the equations below, calculate the COD flux fractions for different SRT (the calculation equations are listed as below). The different COD flux fractions are plotted in diagrams in Figure 7.5; some key results are shown in Table 7.8. Based on these model simulations, the effect of SRT can be further elaborated: 1) high COD degradation can be achieved in short SRT conditions, while the degradation will increase as long as the SRT increases, and 2) concomitantly the remaining biodegradable COD and unbiodegradable COD in the SRUSB decreases. It is also noteworthy that the UPO fraction of COD can significantly affect sludge production.

$$FCOD_{VSS} = MX_{VSS} \cdot \frac{f_{cv}}{SRT}$$

$$FCOD_{biomass} = FX_{u,i} - FCOD_{VSS}$$

$$\frac{FS_{u,i}}{FCOD_i} = f_{uso} \text{ (see Table 7.1)}$$

$$\frac{FX_{u,i}}{FCOD_i} = f_{upo} \text{ (see Table 7.1)}$$

Table 7.8 Fraction COD flux in the SRUSB for different SRT.

SRT (d)	$\dfrac{FCOD_{VSS}}{FCOD_i}$	$\dfrac{FCOD_{biomass}}{FCOD_i}$	$\dfrac{FS_{u,i}}{FCOD_i}$	$\dfrac{FX_{u,i}}{FCOD_i}$	$\dfrac{FCOD_i}{FCOD_i}$
5	0.21	0.077	0.071	0.13	1
10	0.20	0.067	0.071	0.13	1
20	0.18	0.053	0.071	0.13	1
30	0.17	0.045	0.071	0.13	1
40	0.17	0.039	0.071	0.13	1
50	0.16	0.035	0.071	0.13	1
60	0.16	0.032	0.071	0.13	1

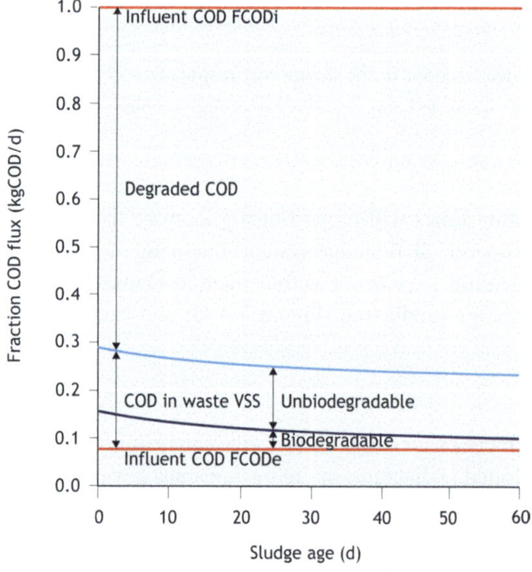

Figure 7.5 Proportion of influent COD flux exiting as COD in the effluent, as degraded COD, and COD in waste sludge solids (WAS) and the residual biodegradable COD in the WAS versus sludge age for the example raw wastewaters.

Solution 7.4.5

Design volumes are calculated for different TSS concentrations and different sludge retention times. The calculation values are shown in Table 7.9. The results reveal two basic principles for the design of the SRUSB reactor: the higher the TSS concentrations obtained in the reactor, the smaller the SRUSB reactor volume required; and the longer the sludge age designed, the larger the SRUSB volume required (see Figure 7.6).

$$Volume = \frac{MX_{VSS} + MX_{FSS}}{X_{TSS}}$$

Table 7.9 Various design volumes for the reactors with SRT under different TSS.

| SRT (d) | 3,000 mg/l | | 4,500 mg/l | | 5,900 mg/l | |
	Volume (m³)	V/FCOD$_i$ (m³/kg.d)	Volume (m³)	V/FCOD$_i$ (m³/kg.d)	Volume (m³)	V/FCOD$_i$ (m³/kg.d)
5	1,166.15	0.27	777.43	0.18	592.96	0.14
10	2,236.02	0.52	1,490.68	0.35	1,136.96	0.26
20	4,215.26	0.98	2,810.17	0.65	2,143.35	0.54
30	6,077.79	1.41	4,051.86	0.94	3,090.40	0.72
40	7,877.46	1.83	5,251.645	1.22	4,005.49	0.93
50	9,639.43	2.24	6,426.28	1.49	4,901.40	1.14
60	11,376.99	2.64	7,584.66	1.76	5,784.91	1.34

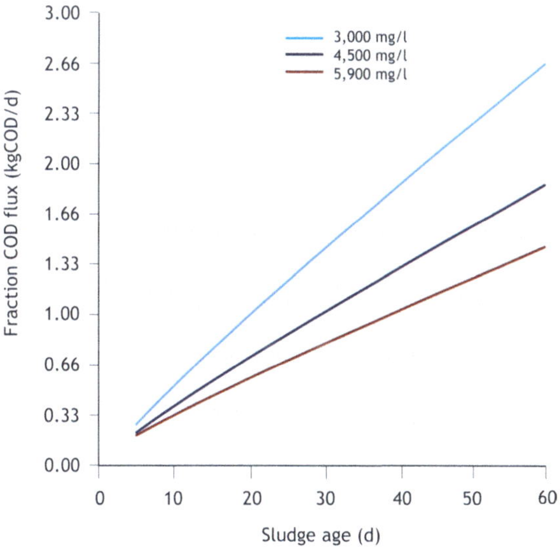

Figure 7.6 Reactor volume requirements in m³ per kg COD per day versus sludge age at different average reactor TSS concentrations for raw wastewater.

Sulphur-driven autotrophic denitrification (SdAD) (solutions 7.4.6-7.4.9)

Solution 7.4.6

a. Compared with the conventional heterotrophic denitrification process, the SdAD process does not require an exogenous organic carbon source, it eliminates problems associated with residual organics, and it produces much less biomass.

b. In SdAD, sulphur-oxidizing bacteria (tables 7.6 and 7.7 in Chen *et al.*, 2020) utilize the reduced sulphur compounds (mainly sulphide, thiosulphate and elemental sulphur) as electron donors to reduce nitrogen oxides (mainly nitrate, nitrite, nitrous oxide, and nitric oxide) as electron acceptor(s). The end products are commonly sulphate and dinitrogen gas, while the intermediates (mainly NO_2^-, N_2O, and S^0_{bio}) can also be produced due to the influent sulphur-to-nitrogen ratio.

c. Some typical biochemical reaction equations of SdAD are described by equations 7.7-7.12 in Chen *et al.*, 2020, but these equations are only catabolism reactions, showing stoichiometric ratios and neglecting the biomass growth. When considering anabolism (biomass growth) in the process, the SdAD reaction equations can be re-written as equations 7.13-7.15 in Chen *et al.*, 2020.

d. Microorganisms involved in the SdAD process are phylogenetically diverse and mainly consist of sulphur-oxidizing bacteria (SOB) and nitrate-reducing bacteria. The various types of electron donor, and the diverse communities of SOB and their optimal growth conditions are listed in Table 7.6 and Table 7.7 in Chen *et al.*, 2020, respectively.

e. The reaction rates of SdAD processes are affected by the following factors: 1) the types of sulphur substrates (total dissolved sulphide (TDSd), S^0 and $S_2O_3^{2-}$) and nitrogen oxides (NO_3^-, NO_2^-, N_2O, NO); 2) the mass ratio of sulphur-to-nitrogen (1.4 gS/gN theoretically) and alkalinity-to-removed nitrogen (maintained at a minimum of at least 4.0-4.6 gCaCO$_3$/gN); and 3) operational parameters including pH (6-8), DO concentration (0.1-0.3 gO$_2$/m^3) and temperatures.

Note: the SdAD process is complex with multiple reactions and electron acceptors/donors involved. Therefore, in order to further investigate this process, mathematical models need to be established; these are introduced in Exercise 7.4.7 and practiced in exercise 7.4.8 and 7.4.9.

Solution 7.4.7

This SdAD model is developed in three steps, as described below.

<u>Step 1</u>

Firstly, the SdAD process is described in the following 7 sub-processes. In this example the dissolved sulphide and elemental sulphur are considered to be electron donors while nitrate, nitrite, and nitrous oxide are considered as electron acceptors. Nitric oxide is neglected because it can react rapidly without accumulation. The endogenous respiration is also calculated.

Process 1: $\dfrac{1}{Y_{SOB}}[g] \cdot HS^- +? [g]NO_3^- +? [g]HCO_3^- \rightarrow 1[g] \cdot X_{SOB} + \dfrac{1}{Y_{SOB}}[g] \cdot S^0 +? [g]NO_2^-$

Process 2: $\dfrac{1}{Y_{SOB}}[g] \cdot HS^- +? [g]NO_2^- +? [g]HCO_3^- \rightarrow 1[g] \cdot X_{SOB} + \dfrac{1}{Y_{SOB}}[g] \cdot S^0 +? [g]N_2O$

Process 3: $\dfrac{1}{Y_{SOB}}[g] \cdot HS^- +? [g]N_2O +? [g]HCO_3^- \rightarrow 1[g] \cdot X_{SOB} + \dfrac{1}{Y_{SOB}}[g] \cdot S^0 +? [g]N_2$

Process 4: $\dfrac{1}{Y_{SOB}}[g] \cdot S^0 +? [g]NO_3^- +? [g]HCO_3^- \rightarrow 1[g] \cdot X_{SOB} + \dfrac{1}{Y_{SOB}}[g] \cdot SO_4^{2-} +? [g]NO_2^-$

Process 5: $\dfrac{1}{Y_{SOB}}[g] \cdot S^0 +? [g]NO_2^- +? [g]HCO_3^- \rightarrow 1[g] \cdot X_{SOB} + \dfrac{1}{Y_{SOB}}[g] \cdot SO_4^{2-} +? [g]N_2O$

Process 6: $\dfrac{1}{Y_{SOB}}[g] \cdot S^0 +? [g]N_2O +? [g]HCO_3^- \rightarrow 1[g] \cdot X_{SOB} + \dfrac{1}{Y_{SOB}}[g] \cdot SO_4^{2-} +? [g]N_2$

Process 7: Endogenous respiration of SOB (X_{SOB}), production of respiration residuals (X_I)

Secondly, the stoichiometric coefficients for NO_X (NO_3^-, NO_2^-, N_2O, N_2) are calculated in these sub-processes based on the electron balance analysis. In each sub-process, the electrons donated by electron donors should be equal to the electrons accepted by electron acceptors.

Note: $v_{j,i}$ represents the stoichiometric coefficient of target substrates in the stoichiometric matrix (Table 7.10), where j means the process number and i means the number of substrates in the stoichiometric matrix. For example, in Process 1, the stoichiometric coefficient of the fourth substrate NO_3^- can be represented by $v_{1,4}$.

For Process 1, nitrate reduction by sulphide:

The electron-donating reaction can be described as:

$$\frac{1}{32 \cdot Y_{SOB}}[mol] \cdot HS^- - (\frac{2}{32 \cdot Y_{SOB}} - \frac{4}{32})[mol] \cdot e^- \rightarrow \frac{1}{32 \cdot Y_{SOB}}[mol] \cdot S^0 + \frac{1}{32}[mol] \cdot X_{SOB}$$

The electron-accepting reaction can be described as:

$$(\frac{2}{32 \cdot Y_{SOB}} - \frac{4}{32})/2[mol] \cdot NO_3^- + (\frac{2}{32 \cdot Y_{SOB}} - \frac{4}{32})[mol] \cdot e^- \rightarrow (\frac{2}{32 \cdot Y_{SOB}} - \frac{4}{32})/2[mol]NO_2^-$$

Thus, the stoichiometric coefficient for NO_3^- in Process 1 is:

$$v_{1,4} = \frac{\frac{2}{32 \cdot Y_{SOB}} - \frac{4}{32}}{2} \cdot 14 = \frac{1 - 2 \cdot Y_{SOB}}{2.28 \cdot Y_{SOB}}$$

The stoichiometric coefficient for NO_2^- in Process 1 is:

$$v_{1,5} = \frac{\frac{2}{32 \cdot Y_{SOB}} - \frac{4}{32}}{2} \cdot 14 = \frac{1 - 2 \cdot Y_{SOB}}{2.28 \cdot Y_{SOB}}$$

For Process 2, nitrite reduction by sulphide:

The electron-donating reaction of Process 2 is the same as that of Process 1.

The electron-accepting reaction can be described as:

$$(\frac{2}{32 \cdot Y_{SOB}} - \frac{4}{32})/2[mol] \cdot NO_2^- + (\frac{2}{32 \cdot Y_{SOB}} - \frac{4}{32})[mol] \cdot e^- \rightarrow \frac{\frac{2}{32 \cdot Y_{SOB}} - \frac{4}{32}}{2[mol]\frac{1}{2}N_2O}$$

Thus, the stoichiometric coefficient of NO_2^- in Process 2 is:

$$v_{2,5} = \frac{\frac{2}{32 \cdot Y_{SOB}} - \frac{4}{32}}{2} \cdot 14 = \frac{1 - 2 \cdot Y_{SOB}}{2.28 \cdot Y_{SOB}}$$

The stoichiometric coefficient of N_2O in Process 2 is:

$$v_{2,6} = \frac{\dfrac{2}{32 \cdot Y_{SOB}} - \dfrac{4}{32}}{2} \cdot \frac{1}{2} \cdot 28 = \frac{1 - 2 \cdot Y_{SOB}}{2.28 \cdot Y_{SOB}}$$

For Process 3, nitrous oxide reduction by sulphide:

Electron donation: same as Process 1.

Electron acceptance:

$$\left(\frac{2}{32 \cdot Y_{SOB}} - \frac{4}{32}\right)[mol] \cdot N_2O + \left(\frac{2}{32 \cdot Y_{SOB}} - \frac{4}{32}\right)[mol] \cdot e^- \rightarrow \left(\frac{2}{32 \cdot Y_{SOB}} - \frac{4}{32}\right)[mol]N_2$$

Stoichiometric coefficient of N_2O:

$$v_{3,6} = \left(\frac{2}{32 \cdot Y_{SOB}} - \frac{4}{32}\right) \cdot 14 = \frac{1 - 2 \cdot Y_{SOB}}{1.14 \cdot Y_{SOB}}$$

Stoichiometric value of N_2:

$$v_{3,7} = \left(\frac{2}{32 \cdot Y_{SOB}} - \frac{4}{32}\right) \cdot 14 = \frac{1 - 2 \cdot Y_{SOB}}{1.14 \cdot Y_{SOB}}$$

For Process 4, nitrate reduction by elemental sulphur:

Electron donation:

$$\frac{1}{32 \cdot Y_{SOB}}[mol] \cdot S^0 - \left(\frac{6}{32 \cdot Y_{SOB}} - \frac{4}{32}\right)[mol] \cdot e^- \rightarrow \frac{1}{32 \cdot Y_{SOB}}[mol] \cdot SO_4^{2-} + \frac{1}{32}[mol] \cdot X_{SOB}$$

Electron acceptance:

$$\left(\frac{6}{32 \cdot Y_{SOB}} - \frac{4}{32}\right)/2[mol] \cdot NO_3^- + \left(\frac{6}{32 \cdot Y_{SOB}} - \frac{4}{32}\right)[mol] \cdot e^- \rightarrow \left(\frac{2}{32 \cdot Y_{SOB}} - \frac{4}{32}\right)/2[mol]NO_2^-$$

Stoichiometric coefficient of NO_3^-:

$$v_{4,4} = \frac{\dfrac{6}{32 \cdot Y_{SOB}} - \dfrac{4}{32}}{2} \cdot 14 = \frac{3 - 2 \cdot Y_{SOB}}{2.28 \cdot Y_{SOB}}$$

Stoichiometric coefficient of NO_2^-:

$$v_{4,5} = \frac{\dfrac{2}{32 \cdot Y_{SOB}} - \dfrac{4}{32}}{2} \cdot 14 = \frac{1 - 2 \cdot Y_{SOB}}{2.28 \cdot Y_{SOB}}$$

For Process 5, nitrite reduction by elemental sulphur:

Electron donation: same as Process 4.

Electron acceptance:

$$(\frac{6}{32 \cdot Y_{SOB}} - \frac{4}{32})/2[mol] \cdot NO_2^- + (\frac{6}{32 \cdot Y_{SOB}} - \frac{4}{32})[mol] \cdot e^- \rightarrow \frac{\dfrac{6}{32 \cdot Y_{SOB}} - \dfrac{4}{32}}{2[mol]\dfrac{1}{2}N_2O}$$

Stoichiometric coefficient of NO_2^-:

$$v_{5,5} = \frac{\dfrac{6}{32 \cdot Y_{SOB}} - \dfrac{4}{32}}{2} \cdot 14 = \frac{3 - 2 \cdot Y_{SOB}}{2.28 \cdot Y_{SOB}}$$

Stoichiometric coefficient of N_2O:

$$v_{5,6} = \frac{\dfrac{6}{32 \cdot Y_{SOB}} - \dfrac{4}{32}}{2} \cdot \frac{1}{2} \cdot 28 = \frac{3 - 2 \cdot Y_{SOB}}{2.28 \cdot Y_{SOB}}$$

For Process 6, nitrous oxide reduction by elemental sulphur:

Electron donation: same as Process 4.

Electron acceptance:

$$(\frac{6}{32 \cdot Y_{SOB}} - \frac{4}{32})[mol] \cdot N_2O + (\frac{6}{32 \cdot Y_{SOB}} - \frac{4}{32})[mol] \cdot e^- \rightarrow (\frac{6}{32 \cdot Y_{SOB}} - \frac{4}{32})[mol]N_2$$

Stoichiometric coefficient of N_2O:

$$v_{6,6} = \left(\frac{6}{32 \cdot Y_{SOB}} - \frac{4}{32}\right) \cdot 14 = \frac{3 - 2 \cdot Y_{SOB}}{1.14 \cdot Y_{SOB}}$$

Stoichiometric value of N_2:

$$v_{6,7} = \left(\frac{6}{32 \cdot Y_{SOB}} - \frac{4}{32}\right) \cdot 14 = \frac{3 - 2 \cdot Y_{SOB}}{1.14 \cdot Y_{SOB}}$$

For Process 7, decay of X_{SOB}:

After endogenous respiration, the X_{SOB} will be degraded to X_I:

$$1[mol]X_{SOB} \rightarrow f_I[mol]X_I$$

Stoichiometric coefficient of X_{SOB}:

$$v_{7,8} = -1$$

Stoichiometric coefficient of X_I:

$$v_{7,9} = f_I$$

<u>Step 2</u>
Develop the tables for the stoichiometric matrix and rate expressions for the SdAD model.

Table 7.10 Stoichiometric matrix of the SdAD process.

Component (i) Process (j)	S_{TDSd} gS/m^3	S_{S0} gS/m^3	S_{SO42-} g/m^3	S_{NO3-} gN/m^3	S_{NO2-} gN/m^3	S_{N2O} gN/m^3	S_{N2} gN/m^3	X_{SOB} gCOD/m^3	X_I gCOD/m^3
Process 1	$\dfrac{-1}{Y_{SOB}}$	$\dfrac{1}{Y_{SOB}}$		$\dfrac{-(0.5 - Y_{SOB})}{1.14Y_{SOB}}$	$\dfrac{(0.5 - Y_{SOB})}{1.14Y_{SOB}}$			1	
Process 2	$\dfrac{-1}{Y_{SOB}}$	$\dfrac{1}{Y_{SOB}}$			$-\dfrac{(0.5 - Y_{SOB})}{1.15Y_{SOB}}$	$\dfrac{(0.5 - Y_{SOB})}{1.15Y_{SOB}}$		1	
Process 3	$\dfrac{-1}{Y_{SOB}}$	$\dfrac{1}{Y_{SOB}}$				$\dfrac{-(0.5 - Y_{SOB})}{0.571Y_{SOB}}$	$\dfrac{0.5 - Y_{SOB}}{0.571Y_{SOB}}$	1	
Process 4		$\dfrac{-1}{Y_{SOB}}$	$\dfrac{-1}{Y_{SOB}}$	$\dfrac{-(1.5 - Y_{SOB})}{1.14Y_{SOB}}$	$\dfrac{(1.5 - Y_{SOB})}{1.14Y_{SOB}}$			1	
Process 5		$\dfrac{-1}{Y_{SOB}}$	$\dfrac{-1}{Y_{SOB}}$		$-\dfrac{(1.5 - Y_{SOB})}{1.15Y_{SOB}}$	$\dfrac{(1.5 - Y_{SOB})}{1.15Y_{SOB}}$		1	
Process 6		$\dfrac{-1}{Y_{SOB}}$	$\dfrac{-1}{Y_{SOB}}$			$-\dfrac{(1.5 - Y_{SOB})}{0.57Y_{S0}}$	$\dfrac{1.5 - Y_{SOB}}{0.57Y_{SOB}}$	1	
Process 7								-1	f_I

Table 7.11 Rate expressions of the SdAD processes.

Process	Notation	Rate
Process 1	r_1	$\mu_{TDSd,NO3-NO2} \cdot \dfrac{S_{TDSd}}{K_1^{TDSd} + S_{TDSd}} \cdot \dfrac{S_{NO3}}{K_{NO3}^{TDSd} + S_{NO3}} \cdot X_{SOB}$
Process 2	r_2	$\mu_{TDSd,NO2-N2O} \cdot \dfrac{S_{TDSd}}{K_2^{TDSd} + S_{TDSd}} \cdot \dfrac{S_{NO2}}{K_{NO2}^{TDSd} + S_{NO2}} \cdot X_{SOB}$
Process 3	r_3	$\mu_{TDSd,N2O-N2} \cdot \dfrac{S_{TDSd}}{K_3^{TDSd} + S_{TDSd}} \cdot \dfrac{S_{N2O}}{K_{N2O}^{TDSd} + S_{N2O}} \cdot X_{SOB}$
Process 4	r_4	$\mu_{S,NO3-NO2} \cdot \dfrac{S_{S0}}{K_4^{S0} + S_{S0}} \cdot \dfrac{S_{NO3}}{K_{NO3}^{S0} + S_{NO3}} \cdot X_{SOB}$
Process 5	r_5	$\mu_{S,NO2-N2O} \cdot \dfrac{S_{S0}}{K_5^{S0} + S_{S0}} \cdot \dfrac{S_{NO2}}{K_{NO2}^{S0} + S_{NO2}} \cdot X_{SOB}$
Process 6	r_6	$\mu_{S,N2O-N2} \cdot \dfrac{S_{S0}}{K_6^{S0} + S_{S0}} \cdot \dfrac{S_{N2O}}{K_{N2O}^{S0} + S_{N2O}} \cdot X_{SOB}$
Process 7	r_7	$b_{SOB} \cdot X_{SOB}$

<u>Step 3</u>
The general mass balance equation for determining a conversion rate can be written as:

$$\frac{dS_i}{dt} = \sum_j v_{j,i} \cdot r_j + \Delta input(i)/\Delta t - \Delta output(i)/\Delta t$$

With the information in Table 7.10 and Table 7.11, the conversion rate of S_{TDSd}, S_{S0}, S_{NO3}, S_{NO2}, S_{N2O}, X_{SOB} and X_I can be calculated via the following equations.

S_{TDSd}: $\quad \dfrac{dS_{TDSd}}{dt} = \dfrac{-1}{Y_{SOB}} \cdot (r1 + r2 + r3) + \dfrac{Q_i \cdot S_{TDSd,inp}}{V_{AX}} - \dfrac{Q_e \cdot S_{TDSd}}{V_{AX}}$

S_{S0}: $\quad \dfrac{dS_{S0}}{dt} = \dfrac{1}{Y_{SOB}} \cdot (r1 + r2 + r3 - r4 - r5 - r6) + \dfrac{Q_i \cdot S_{S0,inp}}{V_{AX}} - \dfrac{Q_e \cdot S_{S0}}{V_{AX}}$

S_{NO3}: $\quad \dfrac{dS_{NO3}}{dt} = \dfrac{-(0.5-Y_{SOB})}{1.14Y_{SOB}} \cdot r1 - \dfrac{1.5-Y_{SOB}}{1.14Y_{SOB}} \cdot r4 + \dfrac{Q_i \cdot S_{NO3,inp}}{V_{AX}} - \dfrac{Q_e \cdot S_{NO3}}{V_{AX}}$

S_{NO2}: $\quad \dfrac{dS_{NO2}}{dt} = \dfrac{0.5-Y_{SOB}}{1.14Y_{SOB}} \cdot r1 - \dfrac{(0.5-Y_{SOB})}{1.15Y_{SOB}} \cdot r2 + \dfrac{1.5-Y_{SOB}}{1.14Y_{SOB}} \cdot r4 - \dfrac{(1.5-Y_{SOB})}{1.15Y_{SOB}} \cdot r5 + \dfrac{Q_i \cdot S_{NO2,inp}}{V_{AX}} - \dfrac{Q_e \cdot S_{NO2}}{V_{AX}}$

S_{N2O}: $\quad \dfrac{dS_{N2O}}{dt} = \dfrac{(0.5-Y_{SOB})}{1.15Y_{SOB}} \cdot r2 - \dfrac{(0.5-Y_{SOB})}{0.571Y_{SOB}} \cdot r3 + \dfrac{1.5-Y_{SOB}}{1.15Y_{SOB}} \cdot r5 - \dfrac{(1.5-Y_{SOB})}{0.57Y_{SOB}} \cdot r6 + \dfrac{Q_i \cdot S_{N2O,inp}}{V_{AX}} - \dfrac{Q_e \cdot S_{N2O}}{V_{AX}}$

X_{SOB}: $\quad \dfrac{dX_{SOB}}{dt} = r1 + r2 + r3 + r4 + r5 + r6 - r7 + \dfrac{Q_i \cdot X_{SOB,inp}}{V_{AX}} - \dfrac{Q_e \cdot X_{SOB}}{V_{AX}}$

X_I: $\quad \dfrac{dX_I}{dt} = f_I \cdot r7 + \dfrac{Q_i \cdot X_{I,inp}}{V_{AX}} - \dfrac{Q_e \cdot X_I}{V_{AX}}$

A kinetic model of the SdAD process can be developed by following the steps above. Two examples of this model (solutions 7.4.8 and 7.4.9) are provided below for exercise purposes.

Solution 7.4.8

Assume that the biofilm thickness is low and ignore the mass transfer limitation of the biofilm. Simulate the SdAD process in a batch experiment for three hours by solving the model equations (described in Step 3 in Exercise 7.4.7). In order to solve these equations numerically, it is necessary to generate a column of time values using small increments of dt (in this example, consider dt = 0.01 hours). Then, columns are required to be created for S_{TDSd}, S_{S0}, S_{NO3}, S_{NO2}, S_{N2O}, X_{SOB}, X_I and for r1, r2, r3, r4, r5, r6 and r7.

The first values in the S_{H2S} column will be $S_{TDSd}(0) = S_{TDSd,ini} = 90$ [gS/m³], $S_{NO3}(0) = S_{NO3,ini} = 30$ [gN/m³], $S_{NO2}(0) = S_{NO2,ini} = 0$ [gN/m³], $S_{N2O}(0) = S_{N2O,ini} = 0$ [gN/m³], $S_{S0}(0) = S_{S0,ini} = 0$ [gS/m³], $X_{SOB}(0) = X_{SOB,ini} = 500$ [gCOD/m³].

The calculation procedure follows these four steps:

<u>Step 1</u>
Using the initial substrate concentrations, calculate the reaction rates of the different processes.

$$r1(0) = \mu_{TDSd,NO3-NO2} \cdot \frac{S_{TDSd}(0)}{K_1^{H2S} + S_{TDSd}(0)} \cdot \frac{S_{NO3}(0)}{K_{NO3}^{TDSd} + S_{NO3}(0)} \cdot X_{SOB}(0) = 119.88 \text{ gCOD/m}^3.\text{h}$$

$$r2(0) = \mu_{TDSd,NO2-N2O} \cdot \frac{S_{TDSd}(0)}{K_2^{H2S} + S_{TDSd}(0)} \cdot \frac{S_{NO2}(0)}{K_{NO2}^{TDSd} + S_{NO2}(0)} \cdot X_{SOB}(0) = 0 \text{ gCOD/m}^3.\text{h}$$

$$r3(0) = \mu_{TDSd,N2O-N2} \cdot \frac{S_{TDSd}(0)}{K_3^{TDSd} + S_{TDSd}(0)} \cdot \frac{S_{N2O}(0)}{K_{N2O}^{TDSd} + S_{N2O}(0)} \cdot X_{SOB}(0) = 0 \text{ gCOD/m}^3.\text{h}$$

$$r4(0) = \mu_{S,NO3-NO2} \cdot \frac{S_{S0}(0)}{K_4^{S0} + S_{S0}(0)} \cdot \frac{S_{NO3}(0)}{K_{NO3}^{S0} + S_{NO3}(0)} \cdot X_{SOB}(0) = 0 \text{ gCOD/m}^3.\text{h}$$

$$r5(0) = \mu_{S,NO2-N2O} \cdot \frac{S_{S0}(0)}{K_5^{S0} + S_{S0}(0)} \cdot \frac{S_{NO2}(0)}{K_{NO2}^{S0} + S_{NO2}} \cdot X_{SOB}(0) = 0 \text{ gCOD/m}^3.\text{h}$$

$$r6(0) = \mu_{S,N2O-N2} \cdot \frac{S_{S0}(0)}{K_6^{S0} + S_{S0}(0)} \cdot \frac{S_{N2O}(0)}{K_{N2O}^{S0} + S_{N2O}(0)} \cdot X_{SOB}(0) = 0 \text{ gCOD/m}^3.\text{h}$$

<u>Step 2</u>
Calculate the conversion rates of the substrates in their initial state. [Note: in a batch reactor (this example), input and output flows are equal to zero. However, in a continuous flow reactor, the input and output flow rates and substrate concentrations should be considered, as shown in Exercise 7.4.9.]

$$\frac{dTDSd}{dt}|_0 = \frac{-1}{Y_{SOB}} \cdot \left(r1(0) + r2(0) + r3(0)\right) + 0 - 0 = -428.13 \text{ gS/m}^3.\text{h}$$

$$\frac{dS_{S0}}{dt}|_0 = \frac{1}{Y_{SOB}} \cdot (r1 + r2 + r3 - r4 - r5 - r6) + 0 - 0 = -428.13 \text{ gS/m}^3.\text{h}$$

$$\frac{dS_{NO3}}{dt}\Big|_0 = \frac{-(0.5 - Y_{SOB})}{1.14Y_{SOB}} \cdot r1 - \frac{1.5 - Y_{SOB}}{1.14Y_{SOB}} \cdot r4 + 0 - 0 = -82.62 \text{ gS/m}^3.\text{h}$$

$$\frac{dS_{NO2}}{dt}\Big|_0 = \frac{0.5 - Y_{SOB}}{1.14Y_{SOB}} \cdot r1 - \frac{(0.5 - Y_{SOB})}{1.15Y_{SOB}} \cdot r2 + \frac{1.5 - -Y_{SOB}}{1.14Y_{SOB}} \cdot r4 - \frac{(1.5 - Y_{SOB})}{1.15Y_{SOB}} \cdot r5 + 0 - 0 = 82.62 \text{ gS/m}^3.\text{h}$$

$$\frac{dS_{N2O}}{dt}\Big|_0 = \frac{(0.5 - Y_{SOB})}{1.15Y_{SOB}} \cdot r2 - \frac{(0.5 - Y_{SOB})}{0.571Y_{SOB}} \cdot r3 + \frac{1.5 - Y_{SOB}}{1.15Y_{SOB}} \cdot r5 - \frac{(1.5 - Y_{SOB})}{0.57Y_{SOB}} \cdot r6 + 0 - 0 = 0 \text{ gS/m}^3.\text{h}$$

$$\frac{dX_{SOB}}{dt}\Big|_0 = r1 + r2 + r3 + r4 + r5 + r6 - r7 = 118.88 \text{ gCOD/m}^3.\text{h}$$

<u>Step 3</u>
Calculate the substrate concentrations at the next time interval.

$$S_{TDSd}(0.01) = S_{TDSd}(0) + \frac{dTDSd}{dt}\Big|_0 \cdot 0.01 = 85.72 \text{ gS/m}^3$$

$$S_{S0}(0.01) = S_{S0}(0) + \frac{dS_{S0}}{dt}\Big|_0 \cdot 0.01 = 4.28 \text{ gS/m}^3$$

$$S_{NO3}(0.01) = S_{NO3}(0) + \frac{dS_{NO3}}{dt}\Big|_0 \cdot 0.01 = 29.17 \text{ gN/m}^3$$

$$S_{NO2}(0.01) = S_{NO2}(0) + \frac{dS_{NO2}}{dt}\Big|_0 \cdot 0.01 = 0.83 \text{ gN/m}^3$$

$$S_{N2O}(0.01) = S_{N2O}(0) + \frac{dS_{N2O}}{dt}\Big|_0 \cdot 0.01 = 0 \text{ gN/m}^3$$

$$X_{SOB}(0.01) = X_{SOB}(0) + \frac{dX_{SOB}}{dt}\Big|_0 \cdot 0.01 = 501.19 \text{ gCOD/m}^3$$

<u>Step 4</u>
Using the updated substrate concentrations obtained from Step 3 to repeat Step 1 through Step 3, calculate the new numerical values of the process rates, conversion rates and substrate concentrations.

For example:

$$S_{TDSd}(0.02) = S_{TDSd}(0.01) + \frac{dTDSd}{dt}\Big|_{0.01} \cdot 0.01 = 77 \text{ gS/m}^3$$

When the running time reaches the target (3h in this example), the calculation processes stop.

Through the modelling analysis, the concentrations of TDSd, S^0 and SO_4-S as well as NO_3-N, NO_2-N and N_2O-N can be simulated as well as the batch reaction carried out, as shown in Figure 7.7.

The results imply that most of the TDSd is transferred into S^0 within a very short time at the beginning of the reaction. When TDSd is consumed completely, the intermediate compound of S^0 starts to be consumed, but at a much low reaction rate. On the other hand, for nitrogen conversion, the nitrate concentration is reduced rapidly when using TDSd as the electron donor. In the meantime, the intermediate compounds of NO_2-N and N_2O-N are produced and accumulated. However, it should be noted that the accumulation of N_2O-N ceases once all the TDSd has been consumed, and decreases to very low concentrations within 30 min by using S^0 as the electron donor. Nitrite is continuously accumulated until all the nitrate is consumed; then it starts to decrease as long as sufficient S^0 exists. Note that in this example, the sulphur load is relatively higher than the theoretical stoichiometric S/N ratio, leading to the conversion of all the nitrogen oxidants to dinitrogen gas.

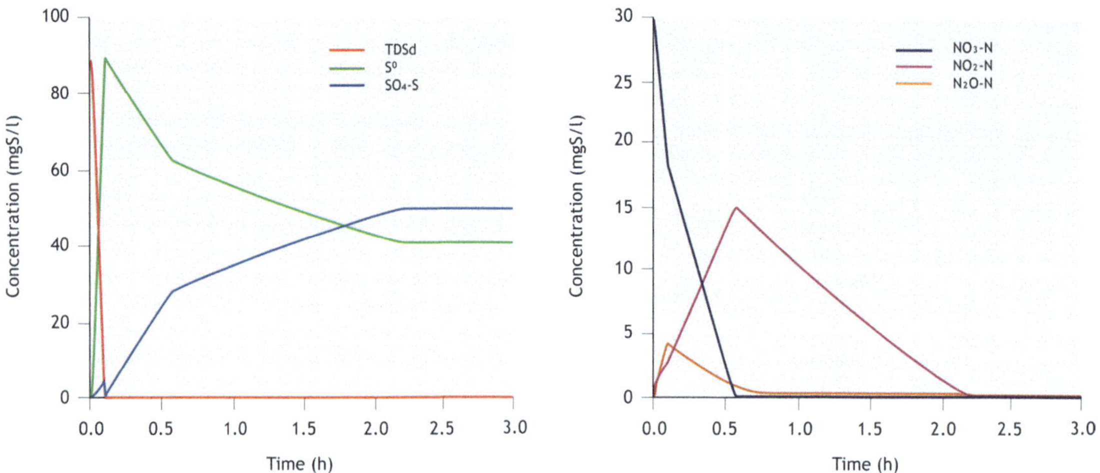

Figure 7.7 Simulated TDSd, S^0, sulphate concentrations (left) and NO_3-N, NO_2-N, N_2O-N concentrations (right).

Solution 7.4.9

(1) Calculation of the bulk liquid concentrations (= effluent) under steady-state conditions

Calculate the nitrate concentration entering into the anoxic tank.

The nitrate generated in the aerobic tank ($S_{NO3,e}$, see Table 7.3) is recycled back as the input nitrate of the anoxic tank ($S_{NO3,inp}$).

$$S_{NO3,inp} = S_{NO3,e} \text{ (effluent nitrate concentration in the aerobic tank)} \cdot a \cdot Q_I / [(a + 1) \cdot Q_i]$$

$$= 13.37 \cdot 2.5 / (2.5 + 1)$$

$$= 9.55 \qquad\qquad gN/m^3$$

Calculate TDSd entering into the anoxic tank:

The effluent TDSd from the SRUSB is used as input TDSd to the anoxic tank.

$$S_{TDSd,inp} = S_{TDSd,SRUSB} \text{ (effluent TDSd concentration in the SRUSB)} \cdot Q_I / [(1 + a) \cdot Q_i]$$

$$= 168.38 / (1 + 2.5)$$

$$= 48.1 \qquad\qquad gS/m^3$$

Calculate the volume of the anoxic tank:

V_{AX} =1,114.3 m^3

Calculate the influent flow rate of the anoxic tank (taking the recirculation rate into account):

$Q_{i,ax}$ $= Q_i \cdot (1 + a)$

                          $= 10{,}000 \cdot (1 + 2.5)$

                          $= 35{,}000$ m^3/d

                          $= 1{,}458.33$ m^3/h

Similar to the four steps in Solution 7.4.8, maintain the SOB biomass concentration XSOB = 500 gCOD/m³. Add the impact of the input concentrations, reactor volume V_{AX} and influent flow rate $Q_{i,ax}$ to Step 2 in Solution 7.4.7 and calculate the substrate concentrations in the next step. The calculation procedures are shown in Figure 7.8. After 2,500 calculations, the substrate concentrations stabilize at t = 2.5 [h].

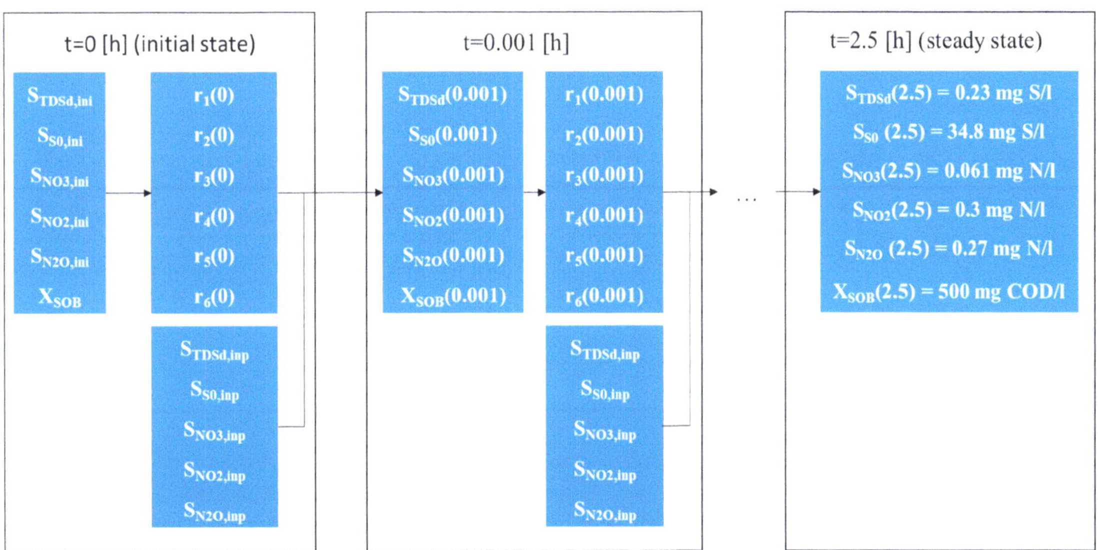

Figure 7.8 Calculation procedure for steady-state substrate concentrations.

Assume that the influent flow rate, HRT and S/N ratio are kept stable. The influent substrate concentrations are set from 10 to 190 gS/m³ for TDSd and from 5 to 95 gN/m³ for nitrate, but with a constant S/N ratio at 2. The simulated results show that the steady-state element sulphur and nitrite concentrations (*i.e.* the intermediate compounds of the SdAD process) are proportional to the loading rates, as shown in Figure 7.9.

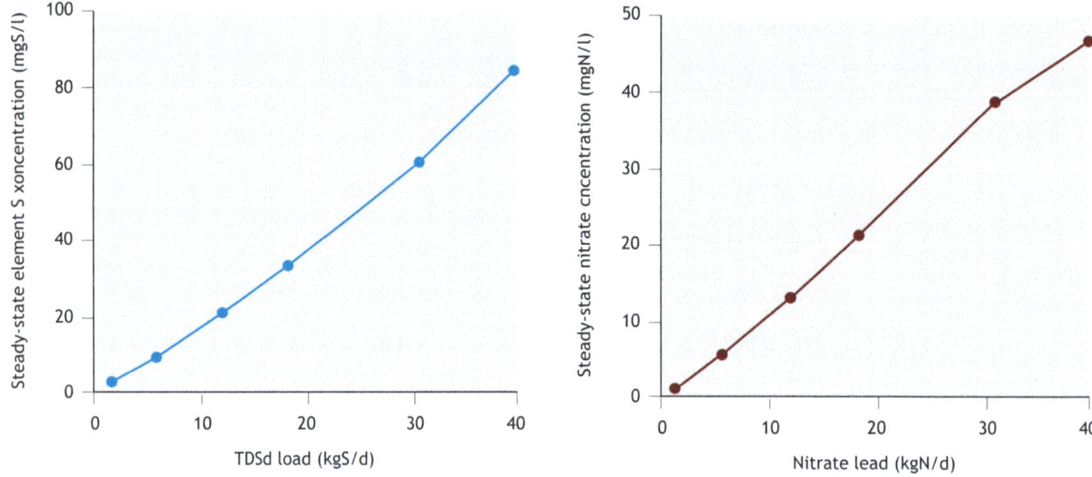

Figure 7.9 Simulated steady-state S° at different TDSd loading rates (left) and steady-state NO₂-N concentrations at different NO₃-N loading rates (right).

(2) Simulation for different S/N conditions

Influent nitrate concentrations are kept the same as before while TDSd concentrations are changed to achieve different influent S/N ratios (see Table 7.12).

Table 7.12 Simulation of different influent S/N ratio

$S_{TDSd,inp}$ (gS/m^3)	$S_{NO3-N,inp}$ (gN/m^3)	X_{SOB} (gCOD/m^3)	S/N
48.1	9.55	500	5.04
28.5	9.55	500	3
19.1	9.55	500	2
9.55	9.55	500	1

Following the same procedure as mentioned above in Figure 7.8, all the substrate concentrations are calculated and listed in Table 7.13. In addition, Figure 7.10 shows the relationships between intermediate concentrations with the S/N ratio. Under steady-state conditions, the element sulphur increases if the S/N ratio increases while the nitrite concentration will decrease if the S/N ratio increases. The results also show that the N₂O concentrations is not significantly influenced by the S/N ratio.

Table 7.13 Steady state concentrations at different S/N ratios

S/N	$S_{TDSd,steady}$	$S_{S0,steady}$	$S_{NO3,steady}$	$S_{NO2,steady}$	$S_{N2O,steady}$
5.04	0.30	37.42	0.16	1.78	1.04
3	0.15	19.12	0.21	4.90	0.86
2	0.10	10.44	0.24	6.48	0.70
1	0.04	1.68	0.35	8.06	0.49

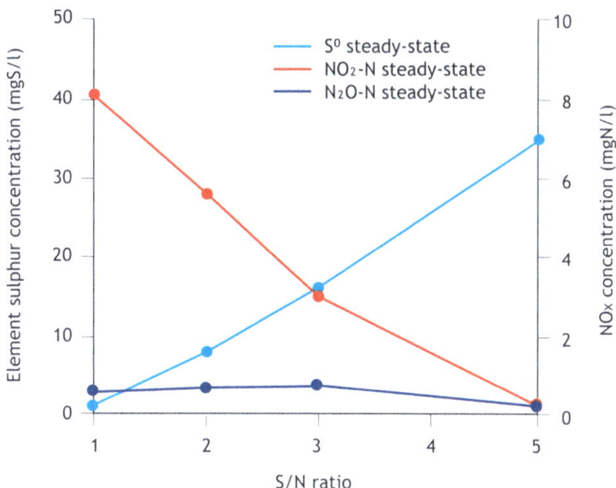

Figure 7.10 Steady-state concentration of intermediate substrates at different S/N ratios.

(3) Simulation for different HRT conditions

The hydraulic retention time (HRT) is changed from 0.25 to 3.06 hours by altering the influent flow rate. Table 7.14 shows the steady-state concentrations of different substrates in the reactor. Figure 7.11 shows that the NO_x rapidly decreases within the first hours, hence controlling the HRT at 1 hours is sufficient.

Table 7.14 Steady state concentrations at different HRT

HRT (h)	$S_{TDSd,steady}$ (mgS/l)	$S_{S0,steady}$ (mgS/l)	$S_{NO3,steady}$ (mgN/l)	$S_{NO2,steady}$ (mgN/l)	$S_{N2O,steady}$ (mgN/l)
0.25	0.42	38.97	0.32	2.65	1.29
0.38	0.30	37.24	0.16	1.67	1.01
0.76	0.23	34.80	0.06	0.30	0.27
1.53	0.26	34.27	0.03	0.07	0.07
2.29	0.27	34.18	0.02	0.04	0.04
3.06	0.27	34.14	0.01	0.03	0.03

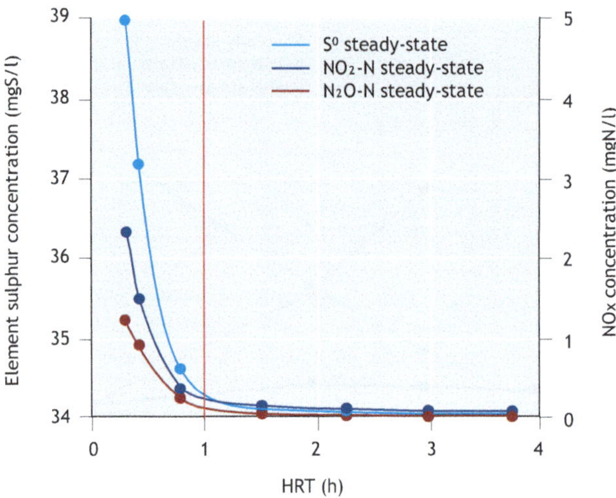

Figure 7.11 Steady-state concentration of intermediate substrates at different HRT.

Sulphur conversion-based resource recovery (solutions 7.4.10-7.4.13)

Solution 7.4.10

   a. Sulphur-based technologies should be considered as an emerging solution for transitioning from passive wastewater treatment to carbon-/energy-neutral treatment with concomitant resource recovery.

   b. Selective metabolic products and/or intermediates of the sulphate-reducing process, such as biohydrogen (H_2), hydrocarbons and poly-hydroxy-alkanoates (PHA) are valuable and recoverable resources.

   c. Elemental sulphur is also worth recovering and reusing with the assistance of a group of sulphur conversion-based bacteria, namely chemolithotrophic SOB. SOB can use O_2/NO_x as the electron acceptor to oxidize sulphide to elemental sulphur for resource recovery. This sulphur recovery technology (such as THIOPAQ®) has been successfully applied in industrial wastewater and off-gas treatment. A kinetic model for elemental sulphur production via sulphide oxidation can be developed by using a similar kinetic model for the SdAD process (see Exercise 7.4.8).

   d. In addition, sulphide is the product of sulphate reduction mediated by SRB, which can extracellularly bound to or precipitate with metal ions in solution, which can be applied for valuable metal sulphide (MeS) recovery (see Table 7.16 in Chen *et al.*, 2020 for examples of MeS compounds). There are two strategies when SRB are applied in metal sulphide production: 1) the metal reacts directly with the sulphide in the sulphidogenic (in-line) system and 2) sulphide is produced and separated as a metal precipitation agent (in the offline system). This (biogenic-)metal sulphide recovery system can be designed based on the following design examples (exercises 7.4.11 and 7.4.12).

Solution 7.4.11

At steady state, both $\frac{dTDSd}{dt}$ and $\frac{d[Zn^{2+}]}{dt}$ are equal to zero:

$$\begin{cases} 0 = \dfrac{Q_{SRUSB}}{V_R} \cdot TDSd_{SRUSB} - \dfrac{Q_{SRUSB} + Q_i}{V_R} \cdot TDSd - k \cdot [Zn^{2+}] \cdot [S^{2-}] \\[2ex] 0 = \dfrac{Q_i}{V_R} \cdot [Zn^{2+}]_i - \dfrac{Q_{SRUSB} + Q_i}{V_R} \cdot [Zn^{2+}] - k \cdot [Zn^{2+}] \cdot [S^{2-}] \\[2ex] [S^{2-}] = \dfrac{TDSd}{[H^+]^2/(K_{a1} \cdot K_{a2}) + [H^+]/K_{a2} + 1} \end{cases}$$

Substitute the given values in Table 7.15 into the equations above:

$$\begin{cases} 0 = \dfrac{5}{0.6} \cdot 5.08 \cdot 10^{-3} - \dfrac{5+2}{0.6} \cdot \text{TDSd} - 2.5 \cdot 10^{20} \cdot [\text{Zn}^{2+}] \cdot [\text{S}^{2-}] \\[2mm] 0 = \dfrac{2}{0.6} \cdot 0.075 - \dfrac{5+2}{0.6} \cdot [\text{Zn}^{2+}] - 2.5 \cdot 10^{20} \cdot [\text{Zn}^{2+}] \cdot [\text{S}^{2-}] \\[2mm] [\text{S}^{2-}] = \dfrac{\text{TDSd}}{(10^{-6})^2/(10^{-7} \cdot 10^{-13.9}) + 10^{-6}/10^{-13.9} + 1} \end{cases}$$

Solve the equation set and calculate the steady-state concentrations in mol/l:

$$\begin{cases} \text{TDSd} = 8.31 \cdot 10^{-12} \\[2mm] [\text{Zn}^{2+}] = 0.017 \\[2mm] [\text{S}^{2-}] = 9.51 \cdot 10^{-21} \end{cases}$$

$[\text{Zn}^{2+}]$ removed daily:

$$\begin{aligned} \Delta \text{FZn} \quad &= [\text{Zn}^{2+}]_i \cdot Q_i - [\text{Zn}^{2+}] \cdot (Q_i + Q_{\text{SRUSB}}) \\ &= 0.075 \cdot 2 - 0.017 \cdot (2 + 5) \\ &= 0.025 \qquad\qquad\qquad\qquad \text{mol/d} \end{aligned}$$

pS value:

$$\begin{aligned} \text{pS} \quad &= -\log([\text{S}^{2-}]) \\ &= -\log(9.51 \cdot 10^{-21}) \\ &= 20.02 \end{aligned}$$

$[\text{Zn}^{2+}]$ removal efficiency:

$$\begin{aligned} \gamma_{\text{Zn}} \quad &= [\text{Zn}^{2+}] / [\text{Zn}^{2+}] \text{ load removed daily} \\ &= 0.025 / (0.075 \cdot 2) \\ &= 16.93 \ \% \end{aligned}$$

Solution 7.4.12

(1) Calculate the minimal TDSd concentration

Maintain the Q_{SRUSB} at 5 l/h and change the TDSd concentration in the SRUSB from 0 to 0.04 mol/l. Apply numerical calculation tools (such as MATLAB) to calculate the steady-state concentrations $[\text{Zn}^{2+}]$ and $[\text{S}^{2-}]$ for each TDSd concentration (see MATLAB Code-I in Annex 2). Find the $[\text{Zn}^{2+}]$ concentration in the precipitator at steady state without considering the precipitation process. The calculated $[\text{Zn}^{2+}]$ concentration is the maximum $[\text{Zn}^{2+}]$ concentration ($[\text{Zn}^{2+}]_{\text{max}}$) at steady state no matter how the $\text{TDSd}_{\text{SRUSB}}$ concentration changes.

$$\frac{d[\text{Zn}^{2+}]}{dt} = \frac{Q_i}{V_R} \cdot [\text{Zn}^{2+}]_i - \frac{Q_{\text{SRUSB}} + Q_i}{V_R} \cdot [\text{Zn}^{2+}]$$

$$0 = \tfrac{2}{0.6} \cdot 0.075 - \tfrac{5+2}{0.6} \cdot [\text{Zn}^{2+}]_{\text{max}} \quad \text{and} \quad [\text{Zn}^{2+}]_{\text{max}} = 0.021 \text{ mol/l}$$

Calculate the $[Zn^{2+}]$ removal efficiency $(1 - [Zn^{2+}] / [Zn^{2+}]_{max})$ and pS value for each TDSd concentration (see Figure 7.12).

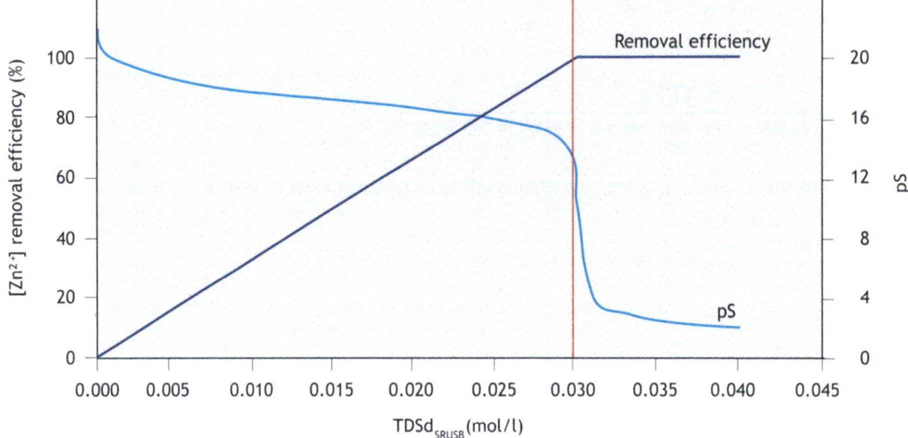

Figure 7.12 $[Zn^{2+}]$ removal efficiency and pS vs. TDSd$_{SRUSB}$ concentrations.

With the fixed Q_{SRUSB} /Q_i equal to 2.5, it can be seen from Figure 7.12 that to increase the Zn removal efficiency to more than 99 %, the TDSd$_{SRUSB}$ concentrations in the SRUSB should be higher than 0.03 mol/l. From Figure 7.12, it can be seen that the pS value is very sensitive to the TDSd$_{SRUSB}$ concentrations if TDSd$_{SRUSB}$ concentrations are approximately 0.03 mol/l. Therefore, it is easy to control the TDSd$_{SRUSB}$ concentration at 0.03 mol/l by monitoring the pS value in the reactor.

(2)   Calculate the minimal influent flow rate
Maintain TDSd = 162.7 mg/l and change the flow rate Q_{SRUSB} from 0 to 40 l/h. Apply numerical calculation tools (*e.g.* MATLAB) to calculate the steady-state concentrations $[Zn^{2+}]$ and $[S^{2-}]$ for each Q_{SRUSB} (see MATLAB Code-II). Calculate the $[Zn^{2+}]$ removal efficiency and pS value for each Q_{SRUSB} (see Figure 7.13).

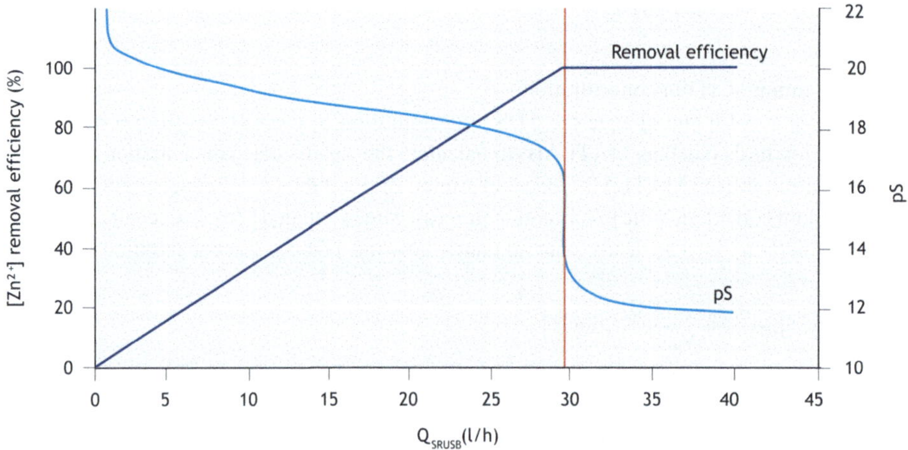

Figure 7.13 $[Zn^{2+}]$ removal efficiency and pS vs. Q_{SRUSB}.

When the TDSd concentration in the SRUSB equals 162.7 mg/l, it can be seen from Figure 7.13 that to increase the Zn removal efficiency to more than 99 %, the Q_{SRUSB} should be higher than 29.5 l/h.

(3) Simulate ZnS precipitation

Calculate the zinc removal efficiency under different Zn^{2+} concentrations including 0.075 mol/l, 0.05 mol/l and 0.025 mol/l, and compare the pS vs. $[Zn^{2+}]$ removal efficiency curve for each $[Zn^{2+}]_i$.

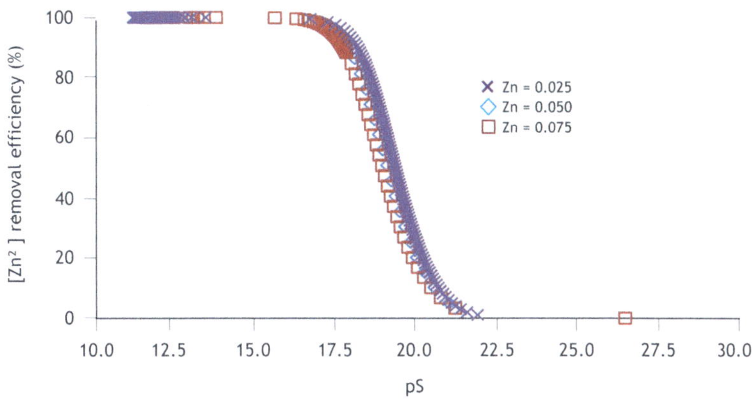

Figure 7.14 $[Zn^{2+}]$ removal efficiency vs. pS for different $[Zn^{2+}]_i$.

Figure 7.14 shows that, no matter how the $[Zn^{2+}]_i$ concentration and the load of TDSd changes, the relationship between $[Zn^{2+}]$ removal efficiency and pS remains the same. Moreover, the removal of $[Zn^{2+}]$ can be maintained at high levels (> 99 %) when the pS value is lower than 15-16.

Therefore, in this precipitation, pS can be considered as the key parameter for process control. For example, the control setpoint of pS can be set to 15, when pH equals 6. In practice, the pS can be controlled by two different control strategies: 1) controlling the influent TDSd concentration, and/or 2) controlling the feed TDSd loading by adjusting the influent flow rate.

Note: the MATLAB codes are provided in the Annex 2 to help readers become familiar with using a numerical tool to calculate steady-state concentrations $[Zn^{2+}]$ under different operational conditions.

Solution 7.4.13

(1) The model development

There are two biochemical reactions that should be considered for elemental sulphur recovery:

Reaction 1: sulphide is oxidized to elemental sulphur:

$$H_2S + 0.5O_2 \rightarrow S^0 + H_2O$$

Reaction 2: elemental sulphur is further oxidized to sulphate:

$$S^0 + 1.5O_2 + H_2O \rightarrow SO_4^{2-} + 2H^+$$

A Monod-type mathematical model can be developed. Table 7.15 summarizes the model matrix for the stoichiometric matrix [TDSd (S_{H_2S}), dissolved oxygen (S_{O_2}), element sulphur (X_{S^0}), sulphate (S_{SO4}) and SOB (X_{SOB})] and the process rates of two processes (r1 and r2). The kinetic and stoichiometric constants are listed in Table 7.16.

Table 7.15 The stoichiometric matrix and process rate of the element sulphur recovery model.

	S_{H_2S} (gS/m^3)	S_{O_2} (gO$_2$/m^3)	X_{S^0} (gS/m^3)	S_{SO4} (gS/m^3)	X_{SOB} (gCOD/m^3)
Process 1	$\dfrac{-1}{Y_{SOB1}}$	$-\dfrac{1 - 2Y_{SOB1}}{2Y_{SOB1}}$	$\dfrac{1}{Y_{SOB1}}$		1
Process 2		$-\dfrac{3 - 2Y_{SOB2}}{2Y_{SOB2}}$	$\dfrac{-1}{Y_{SOB2}}$	$\dfrac{1}{Y_{SOB2}}$	1
r1	$\mu_{max,1} \cdot \dfrac{S_{H2S}}{K_{XSOB,H2S} + S_{H2S}} \cdot \dfrac{S_{O_2}}{K_{XSOB,H2S,0} + S_{O_2}} \cdot X_{SOB}$				
r2	$\mu_{max,2} \cdot \dfrac{X_{S0}}{K_{XSOB,S} + X_{S0}} \cdot \dfrac{S_{O_2}}{K_{XSOB,S,0} + S_{O_2}} \cdot X_{SOB}$				

Table 7.16 Kinetic and stoichiometric parameters of the sulphur recovery process model.

Parameter	Description	Unit	Value	Reference
$\mu_{max,1}$	Maximum growth rate of X_{SOB} on sulphide	1/d	4	Sun *et al.*, 2017
$\mu_{max,2}$	Maximum growth rate of X_{SOB} on sulphur	1/d	1.4	Sun *et al.*, 2017
$K_{XSOB,H2S}$	Half-saturation constant of X_{SOB} for sulphide	gS/m^3	0.0013	Jensen *et al.*, 2009
$K_{XSOB,H2S,O}$	Half-saturation constant of X_{SOB} for S$_O$ for sulphide oxidation	gO$_2$/m^3	0.1	Jensen *et al.*, 2009
$K_{XSOB,S}$	Half-saturation constant of X_{SOB} for S$_O$ for sulphide oxidation	gS/m^3	0.9	Jensen *et al.*, 2009
$K_{XSOB,S,O}$	Half-saturation constant of X_{SOB} for S$_O$ for sulphur oxidation	gO$_2$/m^3	0.45	Jensen *et al.*, 2009
Y_{SOB1}	Yield of X_{SOB} on sulphide	gCOD/gCOD	0.09	Xu *et al.*, 2013
Y_{SOB2}	Yield of X_{SOB} on sulphur	gCOD/gCOD	0.25	Buisman *et al.*, 1991

The steady-state TDSd and element sulphur concentrations in the effluent of aerobic reactor can be calculated via Step 1 and Step 2, and the respective element sulphur production efficiency can be calculated via Step 3.

Step 1

Calculate the steady-state TDSd concentration by substituting the influent TDSd concentrations $S_{H_2S,influent}$ and setpoints of dissolved oxygen concentrations S_{O_2} into the following equation:

$$\frac{dS_{H_2S}}{dt} = \frac{-1}{Y_{SOB1}} \cdot r1 + \frac{Q_{in} \cdot S_{H_2S,influent}}{V} - \frac{Q_{eff} \cdot S_{H_2S}}{V} = 0$$

Step 2

Calculate the steady-state element sulphur concentration in the aerobic tank by substituting both the setpoints of dissolved oxygen concentrations S_{O_2} and calculated effluent TDSd concentrations S_{H_2S} (from Step 1) into the following equation:

$$\frac{dX_{S^0}}{dt} = \frac{1}{Y_{SOB1}} \cdot r1 + \frac{-1}{Y_{SOB2}} \cdot r2 - \frac{Q_{eff} \cdot X_{S^0}}{V} = 0$$

Step 3

The sulphur production efficiency (γ_S) can be calculated for different $S_{H_2S,influent}$ and S_{O_2} conditions:

$$\gamma_S = \frac{FS^0}{FTDSd} \cdot 100 \text{ \%}$$

Where the TDSd loading rate (FTDSd) and element sulphur production rate (FS^0) can be calculated according to the following equations:

$$FTDSd = Q_{in} \cdot S_{H_2S,influent}$$

$$FS^0 = Q_{eff} \cdot X_{S^0,effluent}$$

Calculation example: TDSd equals 100 g/m^3, DO equals to 1 g/m^3 and HRT equals to 2 h:

$$
\begin{aligned}
V \quad &= Q_{in} \cdot HRT \\
&= 20.83 \qquad m^3
\end{aligned}
$$

According to Step 1:

$$0 = \frac{-1}{0.09} \cdot 4 \cdot \frac{S_{H2S}}{0.0013 + S_{H2S}} \cdot \frac{1}{0.1 + 1} \cdot 400 + \frac{500 \cdot 100}{20.83} - \frac{500 \cdot S_{H_2S}}{20.83}$$

$$S_{H2S} = 0.0019 \qquad gS/m^3$$

According to Step 2:

$$
\begin{aligned}
0 = {}& \frac{1}{0.09} \cdot 4 \cdot \frac{0.0019}{0.0013 + 0.0019} \cdot \frac{1}{0.1 + 1} \cdot 400 + \frac{-1}{0.25} \cdot 1.4 \\
& \cdot \frac{X_{S0}}{0.9 + X_{S0}} \cdot \frac{1}{0.45 + 1} \cdot 400 - \frac{500 \cdot X_{S^0}}{20.83}
\end{aligned}
$$

$$X_{S0} = 84.07 \qquad gS/m^3$$

According to Step 3:

$$Y_S = \frac{500 \cdot 84.07}{500 \cdot 100} \cdot 100\,\% $$

$$= 84.07\,\%$$

(2) Analysis of the performance

a) Calculating the optimum DO concentration

In order to calculate the optimum dissolved oxygen (DO) concentration which can result in the highest sulphur production efficiency, the HRT is fixed at 2 h, the dissolved oxygen concentrations are set at a range of 0 to 1.5 g/m^3, and three different TDSd concentrations (100 g/m^3, 200 g/m^3 and 300 g/m^3) are investigated. The sulphur-recovery efficiencies correspond to different DO concentrations and TDSd concentrations are calculated through the model and the calculation process shown in part 1 of solution 7.4.13. The results are plotted in Figure 7.15.

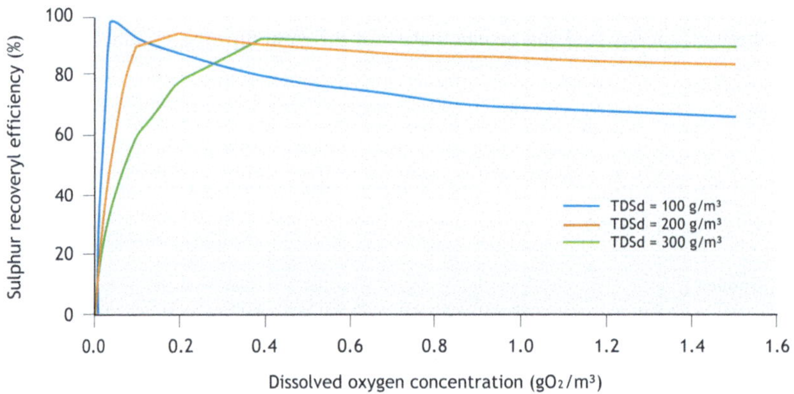

Figure 7.15 Element sulphur production efficiency vs. DO concentration at three influent TDSd concentrations (100, 200 and 300 gS/m³) when HRT equals 2 h.

Figure 7.15 shows that for the influent TDSd concentration 100 g/m^3 with HRT equal to 2 h, the highest sulphur production efficiency of 95 % is obtained at the dissolved oxygen concentration 0.05 g/m^3. In addition, it can be seen that for the influent TDSd concentration 200 g/m^3 with HRT equal to 2 h, the highest sulphur production efficiency of 93 % is obtained at the dissolved oxygen concentration 0.2 g/m^3, and for the influent TDSd concentration 300 g/m^3 with HRT equal to 2 h, the highest sulphur production efficiency of 91 % is obtained at the dissolved oxygen concentration 0.4 g/m^3. The graph shows that the higher the TDSd loading rate, the lower the maximum sulphur production efficiency it can reach. However, a high TDSd loading rate (*e.g.* 300 g/m^3) can achieve high and stable sulphur production efficiency in a wider range of DO concentrations (0.4-1.5 g/m^3) compared to a low TDSd loading rate (*e.g.* 100 g/m^3).

b) Calculating the optimum HRT value

In order to calculate the optimum HRT which results in the highest sulphur production efficiency, the DO is fixed at 1 g/m3, the HRT are set at a range of 1 to 5 h, and three different TDSd concentrations (100 g/m^3, 200 g/m^3 and 300 g/m^3) are investigated. Sulphur-recovery efficiencies corresponding to different HRT and TDSd concentrations are calculated through the model and the calculation process shown in part 1 of solution 7.4.13.

The results are plotted in Figure 7.16.

Figure 7.16 shows that for the influent TDSd concentration of 100 g/m^3 with the dissolved oxygen setpoint 1 g/m^3, the highest sulphur production efficiency of 84 % is obtained at HRT equal to 1 h. It can also be seen that for the influent TDSd concentration 200 g/m^3 with the dissolved oxygen setpoint 1 g/m^3, the highest sulphur production efficiency of 88 % is obtained at HRT equal to 1.5 h. and for the influent TDSd concentration 300 g/m^3 with the dissolved oxygen setpoint 1 g/m^3, the highest sulphur production efficiency of 89 % is obtained at HRT equal to 2 h.

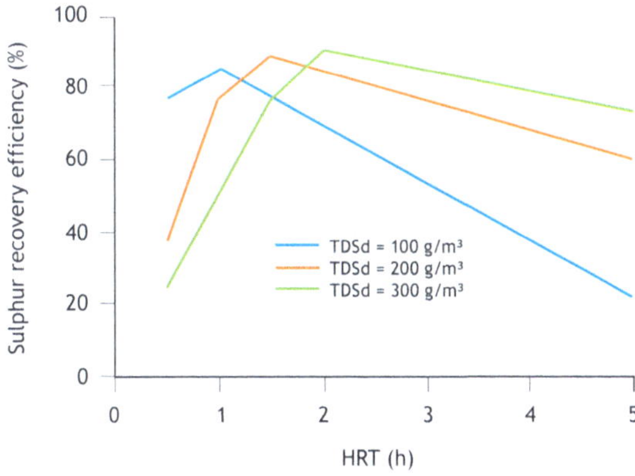

Figure 7.16 Element sulphur production efficiency vs. HRT concentration at three influent TDSd concentration (100, 200 and 300 gS/m³) with a DO setpoint of 1 g/m³.

REFERENCES

Buisman C.J.N., Jspeert P.l., Hof A., Janssen A.J.H., Hagen R.T. and Lettinga G. (1991). Kinetic parameters of a mixed culture oxidizing sulfide and sulfur with oxygen. *Biotechnol. Bioeng.*, 38, 813–820.

Ekama G.A. (2009). Using bioprocess stoichiometry to build a plant-wide mass balance based steady-state WWTP model. *Water Research,* 43, 2101–2120.

Henze, M., Gujer, W., Mino, T., and van Loosdrecht, M. C. M. (2000). Activated sludge models ASM1, ASM2, ASM2d and ASM3. ISBN: 1900222248. IWA publishing, London, UK.

Jensen H.S., Nielsen A.H., Hvitved-Jacobsen T. and Vollertsen J. (2009). Modeling of Hydrogen Sulfide Oxidation in Concrete Corrosion Products from Sewer Pipes. *Water Environment Research,* 81, 365–373.

König J., Keesman K.J., Veeken A. and Lens P.N.L. (2006). Dynamic Modelling and Process Control of ZnS Precipitation. *Separation Science and Technology,* 41, 1025–1042.

Sun J., Dai X., Liu Y., Peng L. and Ni B.-J. (2017). Sulfide removal and sulfur production in a membrane aerated biofilm reactor: Model evaluation. *Chemical Engineering Journal,* 309, 454–462.

Xu X., Chen C., Lee D.-J., Wang A., Guo W., Zhou X., Guo H., Yuan Y., Ren N. and Chang J.-S. (2013). Sulfate-reduction, sulfide-oxidation and elemental sulfur bioreduction process: Modeling and experimental validation. *Bioresource Technology,* 147, 202–211.

NOMENCLATURE

Symbol	Description	Unit
a	Mixed liquor recycle ratio from the aerobic to the anoxic reactor	
a_F	Specific surface area of the biofilm	
$B_{A,NH4}$	Surface specific conversion rate of ammonia	
$B_{A,NOx}$	Surface specific conversion rate of nitrate/nitrite	
b_{SOB}	Decay rate coefficient of SOB	
b_{SRB}	Endogenous respiration rate of the SRB biomass	
COD_i	Influent COD concentration	
f_{BPO}	Biodegradable particulate COD fraction	
f_{CV}	COD/VSS ratio of the sludge	
f_{FBSO}	Fermentable COD fraction	
f_{fill}	Filling ratio	
f_I	Fraction of XU in biomass decay	
FO_N	Flux of O_2 demand by nitrification	
FO_S	Flux of O_2 demand by TDSd oxidation	
FSO_4	Daily sulphate demand in the SRUSB	
f_{SRB}	Endogenous residual fraction of SRB	
$f_{SU,CODi}$	Non-biodegradable soluble COD/influent total COD	
f_{UPO}	Unbiodegradable particulate COD fraction	
f_{USO}	Unbiodegradable soluble COD fraction	
f_{VFA}	Volatile fatty acid fraction	
$f_{XU,CODi}$	Non-biodegradable particulate COD/influent total COD	
$[H^+]$	Concentration of protons	
k	Precipitation kinetic constant	
K_{a1}	Equilibrium constant for the reaction $H_2S \leftrightarrow HS^- + H^+$	
K_{a2}	Equilibrium constant for the reaction $HS^- \leftrightarrow S^{2-} + H^+$	
K_S	Half saturation constant of acidogens	
$k_{SRB,max}$	Maximum specific hydrolysis rate constant of acidogens	
K_1^{TDSd}	S_{H2S} affinity constant for Process 1	
K_2^{TDSd}	S_{H2S} affinity constant for Process 2	
K_3^{TDSd}	S_{H2S} affinity constant for Process 3	
K_4^{S0}	S_{S0} affinity constant for Process 4	
K_5^{S0}	S_{S0} affinity constant for Process 5	
K_6^{S0}	S_{S0} affinity constant for Process 6	
K_{N2O}^{S0}	S_{N2O} affinity constant for Process 6	
K_{N2O}^{TDSd}	S_{N2O} affinity constant for Process 3	
K_{NO2}^{S0}	S_{NO2} affinity constant for Process 5	
K_{NO2}^{TDSd}	S_{NO2} affinity constant for Process 2	
K_{NO3}^{S0}	S_{NO3} affinity constant for Process 4	
K_{NO3}^{TDSd}	S_{NO3} affinity constant for Process 1	

$K_{XSOB,H2S,O}$	Half-saturation constant of X_{SOB} for S_O for sulphide oxidation
$K_{XSOB,H2S}$	Half-saturation constant of X_{SOB} for sulphide
$K_{XSOB,S,O}$	Half-saturation constant of X_{SOB} for S_O for sulphur oxidation
$K_{XSOB,S}$	Half-saturation constant of X_{SOB} for S_O for sulphide oxidation
MX_{ED}	Mass of endogenous residue from SRB
MX_{FSS}	Mass of FSS
MX_{SRB}	Mass of SRB
MX_{TSS}	Mass of TSS
MX_{VSS}	Mass of VSS
$N_{ous,i}$	Influent soluble unbiodegradable organic nitrogen
N_s	Nitrogen required for sludge production
Q_i	Influent flow rate for the SRUSB
Q_{SRUSB}	Flow rate from the SRUSB to the precipitator
Q_w	Waste sludge flow rate
r_h	Volumetric hydrolysis rate
$S_{F,i}$	Influent concentration of fermentable COD
$S_{NO3,c}$	Effluent nitrate concentration
SRT	Sludge retention time
$S_{U,i}$	Influent concentration of soluble unbiodegradable COD
$S_{VFA,i}$	Influent concentration of VFAs
T	Temperature
$TDSd_{SRUSB}$	TDSd in the SRUSB
TKN_c	Design effluent soluble total Kjeldahl nitrogen
TKN_i	Influent soluble total Kjeldahl nitrogen
V_{AX}	Anoxic reactor volume
V_{OX}	Aerobic reactor volume
V_R	Volume of the reactor
V_{SRUSB}	SRUSB reactor volume (complete hydrolysis)
$X_{FSS,i}$	Influent inorganic suspended solid concentration
$X_{S,c}$	Particulate biodegradable COD left in the SRUSB
$X_{S,i}$	Influent particulate biodegradable COD
X_{TSS}	Design average TSS concentration in the SRUSB
$X_{U,i}$	influent concentration of particulate unbiodegradable COD
Y_{SOB}	Yield coefficient for SOB
Y_{SOB1}	Yield of X_{SOB} on sulphide
Y_{SOB2}	Yield of X_{SOB} on sulphur
Y_{SRB}	Biomass yield of SRB (from FBSO and BPO)
Y_{VFA}	Biomass yield of SRB (from VFA)
$[Zn^{2+}]_i$	Influent $[Zn^{2+}]$ concentration

Greek symbols	Description	Unit
$\Delta TDSd\text{-}ax$	TDSd consumed in the anoxic tank	
$\Delta TDSd\text{-}ox$	TDSd oxidized in the aerobic tank	
$\Delta TDSd\text{-}SR$	Effluent TDSd concentration in the SRUSB	
$\mu_{max,1}$	Maximum growth rate of X_{SOB} on sulphide	
$\mu_{max,2}$	Maximum growth rate of X_{SOB} on sulphur	
$\mu_{S,N2O\text{-}N2}$	Maximum reaction rate of Process 6	
$\mu_{S,NO2\text{-}N2O}$	Maximum reaction rate of Process 5	
$\mu_{S,NO3\text{-}NO2}$	Maximum reaction rate of Process 4	
$\mu_{TDSd,N2O\text{-}N2}$	Maximum reaction rate of Process 3	
$\mu_{TDSd,NO2\text{-}N2O}$	Maximum reaction rate of Process 2	
$\mu_{TDSd,NO3\text{-}NO2}$	Maximum reaction rate of Process 1	

Acronym	Description
BSR	Biological sulphate reduction
DO	Dissolved oxygen
DSR	Dissimilatory sulphate reduction
HRT	Hydraulic retention time
MBBR	Moving-bed biofilm reactor
MeS	Metal sulphide
MLSS	Mixed liquor suspended solid
PHA	Poly-hydroxy-alkanoates
SdAD	Sulphur-driven autotrophic denitrification
SOB	Sulphate oxidizing bacteria
SRA	Sulphate-reducing archaea
SRB	Sulphate reducing bacteria
SRT	Sludge retention time
SRUSB	Sulfate reducing upflow sludge blanket
TDSd	Total dissolved sulphide
TKN	Total Kjeldahl nitrogen
VFAs	Volatile fatty acids

ANNEX 2: MATLAB CODE
MATLAB Code I - at a fixed flow rate Q and with different TDSd concentrations in SRUSB.

Matlab code:

```matlab
syms Zn S TDSd

Qi=2;           %influent flowrate
V_R=0.6;        %volume of the reactor
k=2.5E20;       %precipitation kinetic constant
Zni=0.075;      %influent [Zn2+] concentration
Ka1=1E-7;       %equilibrium constant for the reaction H2S?HS- +H+
Ka2=10^(-13.9); %equilibrium constant for the reaction HS-?S2-+H+
H=1E-6;         %concentration of protons
n=40;           %iteration steps
Q_SRUSB = 5;       %assume the Q_SRSUB equals to 5 h/l

M=zeros(n,5);   %define an zero matrix to load the calculated data

for i=1:n
    TDSd_SRUSB=0.001*i;
    eq1=Q_SRUSB/V_R*TDSd_SRUSB-(Q_SRUSB+Qi)/V_R*TDSd-k*Zn*S;
    eq2=Qi/V_R*Zni-(Q_SRUSB+Qi)/V_R*Zn-k*Zn*S;
    eq3=TDSd/(H^2/(Ka1*Ka2)+H/Ka2+1)-S;
    solution=solve(eq1,eq2,eq3);
    S_steady=double(solution.S(2,1));
    TDSd_steady=double(solution.TDSd(2,1));
    Zn_steady=double(solution.Zn(2,1));
    pS=-log10(S_steady);
    % write the TDSd_SRUSB, pS and steady state substrate concentrations into the matrix M
    M(i,1)=TDSd_SRUSB;
    M(i,2)=S_steady;
    M(i,3)=TDSd_steady;
    M(i,4)=Zn_steady;
    M(i,5)=pS;
end

figure,
plot(M(:,1),M(:,4));
title('[Zn2+] removal efficiency vs. TDSd_SRUSB concentration')
xlabel('TDSd_SRUSB concentration [mgS/l]')
ylabel('[Zn2+] removal efficiency')
figure,
plot(M(:,1),M(:,5));
title('pS vs. TDSd_SRUSB concentration')
xlabel('TDSd_SRUSB concentration [mgS/l]')
ylabel('pS')
```

MATLAB Code II – at a constant TDSd concentration and different flow-rates for the TDSd solution.

Matlab code:

```matlab
syms Zn S TDSd

Qi=2;          %influent flowrate
V_R=0.6;       %volume of the reactor
TDSdi=5.08E-3; %initial TDSd from the SRUSB
k=2.5E20;      %precipitation kinetic constant
Zni=0.075;     %influent [Zn2+] concentration
Ka1=1E-7;      %equilibrium constant for the reaction H2S?HS- +H+
Ka2=10^(-13.9); %equilibrium constant for the reaction HS-?S2-+H+
H=1E-6;        %concentration of protons
n=40;          %iteration steps

M=zeros(n,5);  %define an zero matrix to load the calculated data

for i=1:n
  Q_SRUSB=1*i;
  eq1=Q_SRUSB/V_R*TDSdi-(Q_SRUSB+Qi)/V_R*TDSd-k*Zn*S;
  eq2=Qi/V_R*Zni-(Q_SRUSB+Qi)/V_R*Zn-k*Zn*S;
  eq3=TDSd/(H^2/(Ka1*Ka2)+H/Ka2+1)-S;
  solution=solve(eq1,eq2,eq3);
  S_steady=double(solution.S(2,1));
  TDSd_steady=double(solution.TDSd(2,1));
  Zn_steady=double(solution.Zn(2,1));
  pS=-log10(S_steady);
  % write the Q_SRUSB, pS and steady state substrate concentrations into the matrix M
  M(i,1)=Q_SRUSB;
  M(i,2)=S_steady;
  M(i,3)=TDSd_steady;
  M(i,4)=Zn_steady;
  M(i,5)=pS;
end

figure,
plot(M(:,1),M(:,4));
title('[Zn2+] removal efficiency vs. Q_SRUSB')
xlabel('Q_SRUSB [l/h]')
ylabel('[Zn2+] removal efficiency')
figure,
plot(M(:,1),M(:,5));
title('pS vs. Q_SRUSB')
xlabel('Q_SRUSB [l/h]')
ylabel('pS')
```

doi: 10.2166/9781789062304_0295

8

Wastewater disinfection

Ernest R. Blatchley III

8.1 INTRODUCTION

Chapter 8 introduces the mechanisms and key factors that affect disinfection processes, providing the basis to understand full-scale wastewater disinfection. These principles form the basis of models that are used to describe the intrinsic kinetics of disinfection processes, as well as deterministic/probabilistic models to describe overall disinfection process dynamics (*i.e.* reactor behaviour). The approaches used for design of chemical and photochemical disinfection processes are described as well. Overall, this chapter aims to guide the reader through the principles, mechanisms, and model(s) used to design, assess, and evaluate commonly applied wastewater disinfection processes.

8.2 LEARNING OBJECTIVES

Having studied Chapter 8, the reader should be able to:

- Describe the indicator organism concept and the characteristics of an ideal indicator organism.
- Define the physical chemistry of chemical disinfectants used in wastewater applications (chlorine and peracids).
- Define the mechanisms of disinfection associated with chemicals used in wastewater applications (chlorine and peracids).
- Describe the laws of photochemistry.
- Define the mechanisms of disinfection by UV irradiation.
- Define the models used to describe disinfection kinetics.
- Describe the deterministic and probabilistic process models of disinfection process performance.
- Describe approaches used in design of chemical and photochemical (*i.e.* UV) disinfection process.

8.3 EXAMPLES

Example 8.3.1

List and briefly describe the desirable characteristics of microbial or viral indicators. Give examples of common bacterial and viral indicators as part of your answer.

Solution

Indicator organisms (or viruses) are applied because it is generally impractical (or impossible) to identify all the potential microbial and viral pathogens that could be present in a wastewater sample. As such, indicators are used as a surrogate measure of the pathogen burden in wastewater. Desirable characteristics of indicators include:

- Ubiquity in undisinfected effluent samples – the indicator is intended for use as a surrogate measure of the pathogen burden in a sample; therefore, the indicator chosen for a given application should be common in ambient samples.
- Non-pathogenic toward humans – culture-based methods are the gold standard for quantification of the concentration of viable microbes and viruses. Therefore, the methods that are used to measure the concentrations of indicators will require that they be cultured (grown) in the laboratory that conducts these tests. Selection of a non-pathogenic indicator has the benefit of minimizing the risk of disease transmission among those who collect and analyse samples.
- Simple, inexpensive, and rapid to quantify using culture-based methods – measurement of the concentration of an indicator is likely to be conducted as a routine. Therefore, culture-based methods for quantification of viability or infectivity that are simple, inexpensive, and rapid will be beneficial.
- Their presence (or absence) should correlate strongly with the presence (or absence) of microbial or viral pathogens – indicators are intended to provide an *indication* of the presence or absence of microbial or viral pathogens, so a strong (positive) correlation between viable or infective concentrations of indicators and pathogens will be beneficial.
- They should be at least as resistant to disinfectants as microbial pathogens that are likely to be present in an effluent sample – for an indicator to meet the requirement of strongly correlating with pathogens (see above), it will need to be at least as resistant to inactivation by a disinfectant as the target pathogens. This is because measurements of the concentrations of indicators are generally conducted on a treated (disinfected) sample.

As indicated in the text, common bacterial indicators include *E. coli* and faecal coliforms. For effluents that are discharged to marine waters, *Enterococcus* is often used as an indicator. Common viral (phage) indicators include F-specific and somatic coliphages.

Example 8.3.2

For compounds that demonstrate pH-dependent speciation, it is often desirable to be able to illustrate pH-dependence of speciation in a graphical form. In the case of free chlorine, we may consider the following definition to describe speciation:

$$C_{T,Cl} = \text{free chlorine} = \left[Cl_2\right] + \left[HOCl\right] + \left[OCl^-\right] + 2\left[Cl_2O\right] + \left[Cl_3^-\right]$$

The distribution of free chlorine among these various forms is defined by the following reactions, each of which is sufficiently fast in both the forward and reverse directions that we can (almost) always assume that they are described by their respective equilibria.

Table 8.1 Equilibrium constants for free chlorine species (at 25 °C). Values of equilibrium constants are from Odeh *et al.* (2004) and Deborde and Von Gunten (2008).

Reaction	Equilibrium constant
$Cl_2 + H_2O \rightleftharpoons HOCl + H^+ + Cl^-$	$K_h = \dfrac{[HOCl] \cdot [H^+] \cdot [Cl^-]}{[Cl_2]} = 1.04 \cdot 10^{-3} \ M^2$
$HOCl \rightleftharpoons H^+ + OCl^-$	$K_a = \dfrac{[H^+] \cdot [OCl^-]}{[HOCl]} = 3.39 \cdot 10^{-8} \ M$
$HOCl + HOCl \rightleftharpoons Cl_2O + H_2O$	$K_c = \dfrac{[Cl_2O]}{[HOCl]^2} = 8.74 \cdot 10^{-3} \ M^{-1}$
$Cl_2 + Cl^- \rightleftharpoons Cl_3^-$	$K_t = \dfrac{[Cl_3^-]}{[Cl_2] \cdot [Cl^-]} = 0.18 \ M^{-1}$

In this problem, you will use these equilibria to calculate the distribution of free chlorine among its five forms and prepare graphs to illustrate this pH-dependence. However, to do this, we will use a two-step process and a few simplifying assumptions. It is helpful to apply these assumptions because they allow us to work around the problem of the squared dependence of Cl_2O concentration on the concentration of HOCl.

Our first step will be to assume a total free chlorine concentration of 5.0 mg/l as Cl_2 (*i.e.* a molar concentration of $7.04 \cdot 10^{-5}$ M). For the first part of the calculation, assume that the contributions of Cl_2O and Cl_3^- to the total free chlorine concentration are negligible; therefore, you will start by excluding these forms from your calculations. After you have calculated the concentrations of HOCl, OCl^-, and Cl_2 that exist in solution as a function of pH for the range $0 \leq pH \leq 14$ (you should be able to do this using the first two equilibrium expressions from the table above), then calculate the concentrations of Cl_2O and Cl_3^- for this same pH range using the last two equations from the table together with your calculated values of [HOCl] and [Cl_2]. Confirm the assumption that the contributions of Cl_2O and Cl_3^- to the total free chlorine concentration are negligible. If you are satisfied that these assumptions are correct, then prepare a graph of $-\log_{10}[X]$ vs. pH, where [X] represents the molar concentration of each of the five forms of free chlorine. Perform this set of calculations and prepare the graph for two conditions of chloride ion concentration:

a. $[Cl^-] = 10^{-4}$ M (representative of tap water)
b. $[Cl^-] = 0.545$ M (representative of seawater)

Solution

The approach used in developing estimates of the pH-dependent distribution of free chlorine among its various forms follows the logic described in the problem statement. Specifically, we use the equilibrium expressions, the definition of free chlorine, and the total free chlorine concentration to conduct these calculations.

298

To start, we simplify the definition of free chlorine, as described in the problem statement:

$$C_{T,Cl} = [Cl_2] + [HOCl] + [OCl^-]$$

(8.1)

From here, we need to substitute information from the equilibrium expressions to allow development of a mathematical expression that relates the molar concentration of each form of free chlorine to the total free chlorine concentration. These algebraic manipulations are presented below.

$[Cl_2]$:

To define $[Cl_2]$ as a function of pH, each of the terms on the right-hand-side of the definition of free chlorine are expressed in terms of $[Cl_2]$. The molar concentration of Cl_2 is calculated as:

$$C_{T,Cl} = [Cl_2] + \frac{K_h[Cl_2]}{[H^+][Cl^-]} + \frac{K_a K_h[Cl_2]}{[H^+]^2[Cl^-]}$$

(8.2)

Rearranging, we find:

$$[Cl_2] = \frac{C_{T,Cl}}{\left(1 + \dfrac{K_h}{[H^+]\cdot[Cl^-]} + \dfrac{K_a \cdot K_h}{[H^+]^2 \cdot[Cl^-]}\right)}$$

(8.3)

$[HOCl]$:

Following similar logic, we develop an equation to describe $[HOCl]$ by using the equilibria to define the concentrations of all forms of free chlorine as functions of $[HOCl]$.

$$C_{T,Cl} = \frac{[HOCl]\cdot[H^+]\cdot[Cl^-]}{K_h} + [HOCl] + \frac{K_a \cdot[HOCl]}{[H^+]}$$

(8.4)

Rearranging we find:

$$[HOCl] = \frac{C_{T,Cl}}{\left(\dfrac{[H^+]\cdot[Cl^-]}{K_h} + 1 + \dfrac{K_a}{[H^+]}\right)}$$

(8.5)

$[OCl^-]$:

Again following similar logic, we develop an equation to describe $[OCl^-]$ by using the equilibria to define the concentrations of all forms of free chlorine as functions of $[OCl^-]$.

$$C_{T,Cl} = \frac{\left[H^+\right]^2 \cdot \left[Cl^-\right] \cdot \left[OCl^-\right]}{K_a \cdot K_h} + \frac{\left[H^+\right] \cdot \left[OCl^-\right]}{K_a} + \left[OCl^-\right] \tag{8.6}$$

Rearranging we find:

$$\left[OCl^-\right] = \frac{C_{T,Cl}}{\left(\dfrac{\left[H^+\right]^2 \cdot \left[Cl^-\right]}{K_a \cdot K_h} + \dfrac{\left[H^+\right]}{K_a} + 1\right)} \tag{8.7}$$

From here, we can use the remaining equilibria to define $[Cl_2O]$ and $[Cl_3{}^-]$. Specifically:

$$\left[Cl_2O\right] = K_c \cdot \left[HOCl\right]^2 \tag{8.8}$$

$$\left[Cl_3^-\right] = K_t \cdot \left[Cl_2\right] \cdot \left[Cl^-\right] \tag{8.9}$$

The equations above were used to calculate the concentrations of the various forms of free chlorine as functions of pH for tap water and seawater (a future communication will be made to indicate to the reader where to find a spreadsheet to ease the calculations). These results were used to prepare graphs of the two conditions described in the problem statement (see graphs below).

In the spreadsheet that was used to conduct these calculations, a column was added to check the validity of the assumption that the concentrations of Cl_2O and $Cl_3{}^-$ were negligible. This was done by summing the concentrations of the five compounds that comprise free chlorine and dividing by $C_{T,Cl}$.

For tap water, free chlorine is dominated by HOCl and OCl⁻ for virtually the entire pH range included in this graph. It is only when pH approaches zero that significant contributions from molecular chlorine (Cl_2) become evident. Trichloride ($Cl_3{}^-$) and chlorine monoxide (Cl_2O) are always present in trace concentrations. The check of the validity of the assumption that the concentrations of Cl_2O and $Cl_3{}^-$ were negligible appears to be valid for the case of tap water.

For the case of seawater, HOCl and OCl⁻ again dominated over most of the pH range. However, because the chloride ion concentration is orders of magnitude higher than in tap water, Cl_2 and $Cl_3{}^-$ contribute significantly to total free chlorine at pH values as high as 4−5. Cl_2O is always present at trace concentration, relative to the other forms of free chlorine. Therefore, this method of calculating the distribution of free chlorine among its various forms is subject to error for conditions of high chloride ion concentration and low pH. The check of the validity of the assumption that the concentrations of Cl_2O and $Cl_3{}^-$ were negligible indicated error on the order of 10 % at low pH for this assumption. Therefore, if accurate estimates of the concentrations of Cl_2O and $Cl_3{}^-$ are required, another method of calculation should be applied.

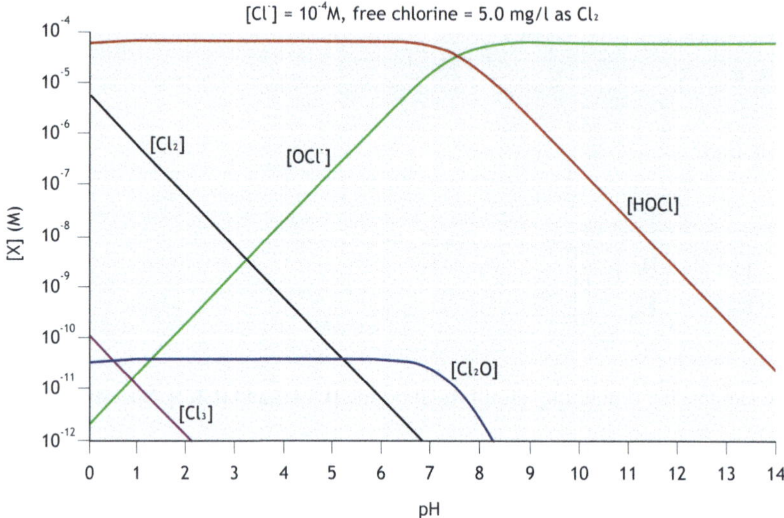

Figure 8.1 Chlorine distribution in tap water.

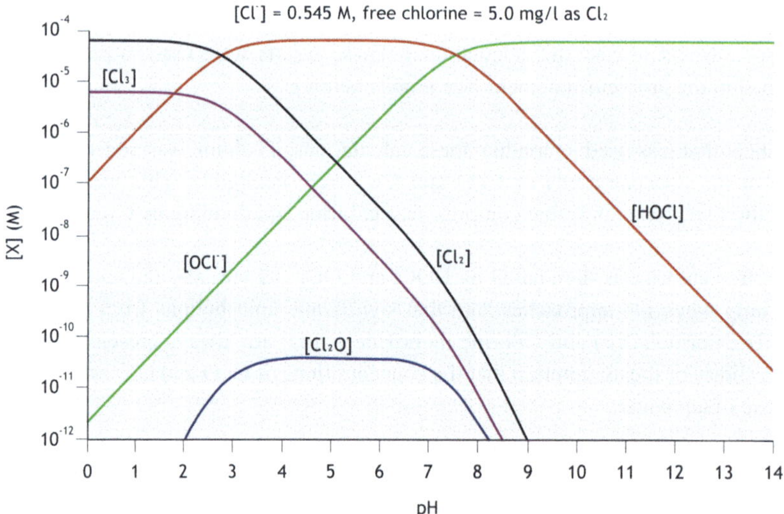

Figure 8.2 Chlorine distribution in seawater.

Example 8.3.3

Table 8.2 provides a summary of disinfectant doses reported to yield 3.0 $\log_{10}$ units of inactivation for common disinfection target microbes and viruses. For the purposes of this problem, assume that inactivation responses of the microbes and viruses listed in Table 8.2 to disinfectants can be described by Chick-Watson (1st-order) kinetics. For chemical disinfectants and UV_{254} disinfection, this implies the following kinetic expressions, respectively:

$$\frac{dN}{dt} = -\Lambda CN \qquad (8.10)$$

$$\frac{dN}{dt} = -kE_iN \qquad (8.11)$$

For the purposes of this problem, you may assume that each of the disinfectants listed in Table 8.2 is applied to a well-mixed batch sample under conditions that accomplish 3.0 $\log_{10}$ units of inactivation of *E. coli*, a treatment endpoint that is often implied by discharge permit limitations. Using this information, define the number of $\log_{10}$ units of inactivation that will be accomplished for *C. parvum* oocycts, murine norovirus, and coliphage MS2.

Given that a large fraction of waterborne disease transmission is associated with protozoan parasites (*e.g. C. parvum*) and viruses (especially noroviruses), what do these estimates of inactivation imply about the appropriateness of using *E. coli* as an indicator for disinfection systems based on HOCl, NH$_2$Cl, peracetic acid (PAA), and UV$_{254}$? Justify your answers for each disinfectant.

Solution

Table 8.2 from Chen *et al.*, 2020 is copied below for reference. Bibliographic citations have been omitted for brevity but can be found in Chapter 8 in Chen *et al.*, 2020.

Table 8.2 Summary of disinfectant doses reported to yield 3.0 $\log_{10}$ units of inactivation for common disinfection target microorganisms. Chemical disinfectant doses are reported in units of mg·min/l. UV$_{254}$ doses are presented in units of mJ/cm^2.

Disinfectant	Microorganism			
	Escherichia coli	*Cryptosporidium parvum* oocysts	Murine norovirus	Coliphage MS2
HOCl	0.10	5,300[1]	0.179	0.142
NH$_2$Cl	6.9	14,000	11	655
PAA	80	N/A[2]	73	609
UV$_{254}$	5.5	5.2	15	97

[1] Inactivation responses for HOCl were estimated by assuming that HOCl was the only form of free chlorine that contributed to inactivation of *C. parvum* oocsysts.

[2] Reports of the efficacy of peracetic acid for inactivation of *C. parvum* oocycts are widely variable in the literature. Peracetic acid is often reported to be similar to free chlorine for inactivation of *C. parvum* oocysts.

We will address this problem in a stepwise manner. First, the kinetic expressions can be integrated then rearranged (algebraically) to produce equations that allow estimation of the rate constants ('coefficient of specific lethality' or 'inactivation constant') for chemical disinfection and UV$_{254}$ irradiation, respectively. The resulting equations are:

$$k = -\frac{\ln\left(\frac{N}{N_0}\right)}{E_i t} \qquad (8.12)$$

and,

$$\Lambda = -\frac{\ln\left(\dfrac{N}{N_0}\right)}{Ct} \qquad (8.13)$$

For both equations, the terms in the denominator represent the disinfectant dose required to achieve 3 $\log_{10}$ units of inactivation, while the ratio in the numerator represents the inactivation extent (*i.e.* $N/N_0 = 10^{-3}$). Applying these equations, we find the following values for the rate constants (units are l/mg.min for chemical disinfectants and cm^2/mJ for UV_{254} irradiation):

Disinfectant	*Escherichia coli*	*Cryptosporidium parvum* oocysts	Murine norovirus	Coliphage MS2
HOCl	69.1	0.00130	38.6	48.6
NH$_2$Cl	1.00	0.000493	0.628	0.0105
PAA	0.086	N/A	0.094626785	0.0113
UV$_{254}$	1.26	1.33	0.461	0.0712

Next, the integrated kinetic expressions are rearranged again to allow calculation of the extent of inactivation of each microbe or virus for the disinfectant exposures (doses) defined in Table 8.2. These expressions take the following forms for chemical and UV_{254} disinfection, respectively:

$$\frac{N}{N_0} = \exp\left(-\Lambda Ct\right) \qquad (8.14)$$

$$\frac{N}{N_0} = \exp\left(-kE_i t\right) \qquad (8.15)$$

Applying the disinfectant doses from Table 8.2 and the rate constants from the table above, we find values of N/N_0 for each disinfectant/microbe (virus) pair:

Disinfectant	*Escherichia coli*	*Cryptosporidium parvum* oocysts	Murine norovirus	Coliphage MS2
HOCl	0.001	0.999	0.0211	0.00771
NH$_2$Cl	0.001	0.997	0.0131	0.930
PAA	0.001	N/A	0.000516	0.404
UV$_{254}$	0.001	0.000671	0.0794	0.676

Notice that the values in the *E. coli* column provide a check of the math; specifically, these values confirm that the disinfectant doses described in Table 8.2 result in 3 $\log_{10}$ units of inactivation of *E. coli*.

Lastly, the '$\log_{10}$ inactivation' values are calculated by taking the $-\log_{10}$ of each value reported in the table above:

Disinfectant	*Escherichia coli*	*Cryptosporidium parvum* oocysts	Murine norovirus	Coliphage MS2
HOCl	3.00	$5.66 \cdot 10^{-5}$	1.68	2.11
NH$_2$Cl	3.00	0.00148	1.88	0.0316
PAA	3.00	N/A	3.29	0.394
UV$_{254}$	3.00	3.17	1.10	0.170

These results indicate that the ability of *E. coli* to serve as an indicator of inactivation of true microbial and viral pathogens is highly variable. Specifically, doses of HOCl and NH$_2$Cl that are effective for inactivation of *E. coli* have essentially no effect on *C. parvum*, but will yield relevant inactivation of some viruses. Similarly, the dose of PAA required to achieve effective inactivation of *E. coli* will also yield relevant inactivation of some viruses, but little or no apparent change in *C. parvum*. On the other hand, the dose of UV$_{254}$ radiation required to achieve acceptable inactivation of *E. coli* will also achieve effective inactivation of *C. parvum*, but somewhat more modest inactivation of viruses.

Example 8.3.4

Figure 8.3 below (Figure 8.4 from the Chen *et al.*, 2020) illustrates normalized absorption spectra of DNA that has been extracted from cultures of three distinct microbes. Because the composition of RNA is similar to that of DNA, the absorption spectra illustrated below are generally representative of nucleic acids (DNA and RNA). Also included below is an absorption spectrum that illustrates absorbance characteristics of proteins.

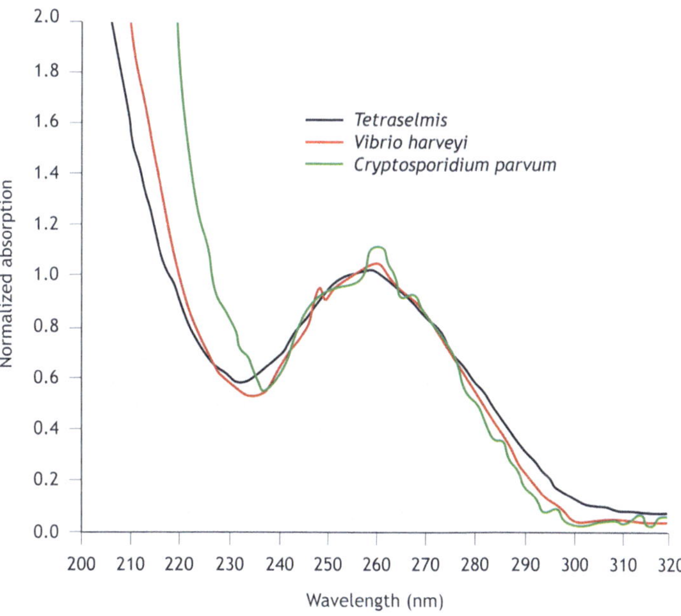

Figure 8.3 (Figure 8.4 in Chen *et al.*, 2020) Absorption spectra for DNA extracted from three aquatic organisms: *Tetraselmis* (a marine alga), *Vibrio harveyi* (a bacterium), and *Cryptosporidium parvum* (a common protozoan parasite). These absorption spectra were normalized to their values measured at 254 nm.

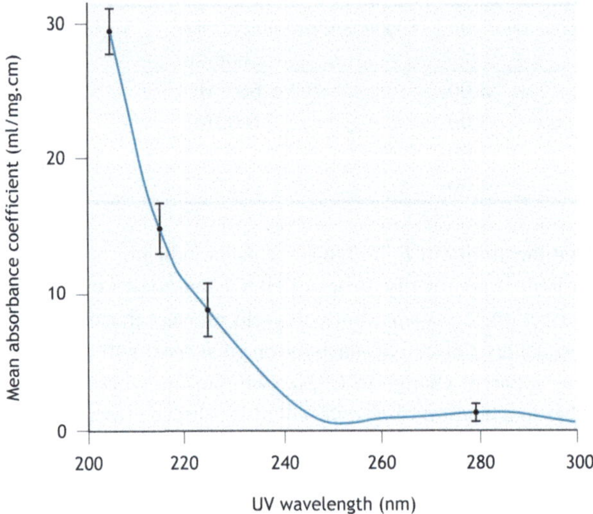

Figure 8.4 Mean wavelength-dependent UV absorbance coefficients, averaged over published measurements for eight common proteins (Image: Buonanno *et al.*, 2013).

Below are two additional figures that illustrate the wavelength-dependence of the inactivation of two waterborne viruses: PBCV-1 and adenovirus type 2. The wavelength-dependence of the response of a microbe or virus to UV exposure is defined as an 'action spectrum.' The image on the left illustrates the action spectrum for an algal virus PBCV-1. The action spectra for this double-stranded DNA virus are indicated both for infectivity and for specific damage to its DNA. The image on the right provides a similar illustration of the action spectra for adenovirus type 2 (also a double-stranded DNA virus), again with separate indications of the wavelength-dependence of loss of infectivity and DNA damage.

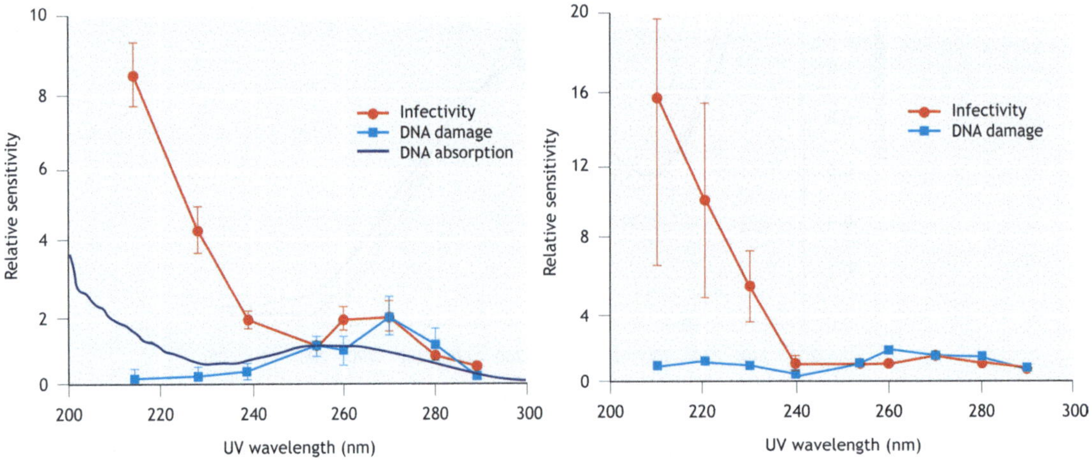

Figure 8.5 Action spectra of two double-stranded DNA viruses. Left panel illustrates action spectra for inactivation and DNA damage within the algal virus PBCV-1 (Image: Sun *et al.*, 2020). Right panel illustrates action spectra for inactivation and DNA damage within adenovirus type 2 (data: Beck *et al.*, 2014).

a. Describe similarities and differences between the absorbance spectra of nucleic acids and proteins.

b. Based on the laws of photochemistry, explain why the shapes of the action spectra for DNA damage are similar to their respective absorbance spectra.

c. For both sets of the action spectra illustrated above, there are substantial deviations between the inactivation responses and the DNA damage responses below about 240 nm. Using the laws of photochemistry and the information presented in the figures above, explain how this can be true.

Solution

a. Nucleic acids (DNA and RNA) show locally strong absorbance in the range 240 nm $\lesssim \lambda \lesssim$ 280 nm. A local minimum in absorbance is evident at about 240 nm. At wavelengths below 240 nm, absorbance by nucleic acids increases sharply as wavelength decreases.

Proteins demonstrate strong absorbance at wavelengths below about 240 nm, with absorbance increasing sharply as wavelength decreases below 240 nm. At longer wavelengths, proteins show relatively weak absorbance.

b. The first law of photochemistry states that electromagnetic radiation (photons) must be absorbed by a target molecule or compound to bring about a photochemical change. In other words, a target molecule (in this case DNA) must absorb radiation for it to be possible for a photochemical reaction to take place. With all other factors being similar, the rate of a photochemical reaction will be directly related to the efficiency of photon absorption, which is described by its absorbance. Therefore, we expect the rate of DNA damage to be related to DNA absorbance, and because of this we expect the DNA action spectrum to have a similar shape to that of its absorption spectrum.

c. The deviations between the inactivation action spectra of these viruses and their corresponding DNA damage action spectra are evident at wavelengths below about 240 nm. For these wavelengths, proteins become increasingly effective absorbers as wavelength decreases. Since it is known that viruses (and microbes) can be inactivated by damage to their genome and/or their proteins, this suggests that the inactivation responses of both viruses become increasingly affected by protein damage at wavelengths below about 240 nm.

8.4 EXERCISES

Exercise 8.4.1

If we confine our calculations to HOCl, OCl⁻, and Cl_2, then it is possible to develop exact solutions to describe the fraction of free chlorine that is represented by each of these forms.

$$\alpha_{Cl_2} = \frac{[Cl_2]}{C_{T,Cl}}$$

$$\alpha_{HOCl} = \frac{[HOCl]}{C_{T,Cl}}$$

$$\alpha_{OCl^-} = \frac{[OCl^-]}{C_{T,Cl}}$$

The parameters α_{Cl_2}, α_{HOCl}, and α_{OCl^-} are known as 'distribution fractions,' which in this case represent the fraction of free chlorine that exists in each form (as a reminder, for this problem we will ignore the contributions and behaviour of Cl_2O and Cl_3^-). From the definition of free chlorine provided above, we know:

$$\alpha_{Cl_2} + \alpha_{HOCl} + \alpha_{OCl^-} \equiv 1$$

a. Using this information, prepare a spreadsheet to calculate α_{Cl_2}, α_{HOCl}, and α_{OCl^-} as functions of pH for $0 \leq pH \leq 14$ and $[Cl^-] = 10^{-4}$ M (an approximation of the chloride concentration that is likely to be present in tap water and most municipal wastewater samples) in increments of 0.1 pH units. The reader should also include a column in your spreadsheet to check the validity of these calculations based on the definition of free chlorine listed above (*i.e.* the sum of the distribution fractions must equal 1). Use the information from this spreadsheet to prepare a graph that includes plots of α_{Cl_2} vs. pH, α_{HOCl} vs. pH, and α_{OCl^-} vs. pH for $0 \leq pH \leq 14$ and $[Cl^-] = 10^{-4}$ M.

b. Prepare a similar graph like that shown in Figure 8.2 for the same pH range, but now using a chloride concentration that is representative of seawater (0.545 M).

c. Provide a brief discussion of the effects of an increase in $[Cl^-]$ from tap water to seawater in terms of speciation of free chlorine.

Exercise 8.4.2

Chemicals that demonstrate acid/base behaviour will be affected by solution pH. The pH-dependent distribution of these chemicals among their various forms can affect many aspects of their physical chemistry. In the case of free chlorine, an example of this is the volatility of chlorine as a function of pH. The 'effective Henry's constant' was developed as a concept to illustrate the pH-dependence of free chlorine, but can be applied to other chemicals as well (Blatchley *et al.,* 1992). For free chlorine, the effective Henry's law constant (H_{Cl}) is defined as the weighted average of the Henry's law constants of the chemicals that comprise free chlorine; the weighting function is the distribution fraction, as calculated in Example 8.3.2. In most practical applications involving free chlorine, we need to consider the pH-dependent behaviours of Cl_2, $HOCl$, and OCl^-. In mathematical terms, H_{Cl} can be defined as follows:

$$H_{Cl} = \alpha_{Cl_2} H_{Cl_2} + \alpha_{HOCl} H_{HOCl} + \alpha_{OCl^-} H_{OCl^-}$$

Using the solutions from example 8.3.2, prepare a graph to illustrate the effective Henry's law constant for free chlorine (H_{Cl}, as defined above) vs. pH for $0 \leq pH \leq 10$ under the following conditions:

a. $[Cl^-] = 10^{-4}$ M (an approximation of freshwater)
b. In seawater

For this graph, use the following Henry's law constants:

$$H_{Cl_2} = 767\,atm$$

$$H_{HOCl} = 0.060\,atm$$

$$H_{OCl^-} = 0$$

Given the range of H_{Cl} that will be evident in this graph, you should use a $\log_{10}$ scale for H_{Cl}. The graph you prepare should allow comparisons between these two conditions. Provide a paragraph or two of text to describe the similarities and differences between calculated H_{Cl} values for tap water and seawater and describe the effects of changes in pH and $[Cl^-]$ on the volatility of free chlorine.

Exercise 8.4.3

A pair of web-based executable programs has been made available by a researcher at the US EPA, Dr. David Wahman (Wahman, 2018). These programs represent effective tools for simulation of the reactions between chlorine and ammonia-N that govern so-called 'chloramination' and 'breakpoint chlorination' processes. They also complement the papers of Jafvert and Valentine (1992) and Vikesland *et al.* (2001) which provided the first comprehensive summaries of the chemistry that defines reactions between free chlorine and ammonia, including common applications of chloramination and breakpoint chlorination. Here we will explore the use of both programs, as well as the implications of these model simulations.

a. Use the program found at https://usepaord.shinyapps.io/Unified-Combo/) to simulate the dynamic behaviour of free chlorine and the inorganic chloramines over a period of 60 minutes at pH = 8.0, total alkalinity of 150 mg/l (as $CaCO_3$), water temperature of 25 °C, TOC = 0, and free ammonia initial concentration of 1.0 mg/l (as N). Conduct these simulations under the assumption of 'simultaneous addition' of free chlorine and ammonia; do these simulations for initial free chlorine concentrations of 4 and 8 mg/l (as Cl_2). Note that based on the molecular/atomic weights of chlorine (Cl_2, MW = 70.9 g/mole) and nitrogen (NH_3-N, MW = 14.0 g/mole), these mass-based concentrations correspond to molar ratios (Cl_2:N) of 0.79 and 1.6, respectively. Download the results of these simulations to .csv files, then use these results to prepare graphs of the concentrations of free chlorine, monochloramine, dichloramine, trichloramine, and (free) ammonia in the software of your choice. Choose axis scales to facilitate comparisons between these simulation results for the two initial chlorine concentration conditions. Briefly comment on and explain the similarities and differences you observe in the predicted time-course behaviours of each of the following:

   - Total chlorine
   - Monochloramine
   - Dichloramine
   - Trichloramine
   - Free chlorine
   - (Free) ammonia.

b. Use the program found at: https://usepaord.shinyapps.io/Breakpoint-Curve/ to simulate breakpoint chlorination at pH = 7.0 and pH = 8.0. For these simulations, use the following input parameter values: initial 'Free Ammonia' concentration fixed at 1.0 mg/l (as N); total alkalinity = 150 mg/l (as $CaCO_3$); water temperature = 25 °C; reaction time of 30 minutes. Use the 'printscreen' function to download the graphical summary of this simulation for t=30 minutes; copy this graphical image into your solution of the question. Briefly comment on the similarities and differences you observe in the predicted behaviours of each of the following and include a brief explanation for each comparison.

   - Total chlorine
   - Monochloramine
   - Dichloramine

- Trichloramine
- Free chlorine
- (Free) ammonia.

Exercise 8.4.4

A municipal wastewater treatment facility operates at a flow rate of 1,500 m^3/hr. Residual chlorine reaches the end of the chlorine contact chamber at a concentration of 1.2 mg/l as Cl_2. At what mass rate must $NaHSO_3$ be added to the effluent to accomplish complete dechlorination. Express your answer in kg/d.

Exercise 8.4.5

Hypochlorous acid (HOCl) and peracetic acid (CH_3COOOH) are both weak acids that function as disinfectants. For both compounds, the neutral (acid) form is far more effective as a disinfectant than its conjugate base. For temperature conditions that are commonly applied in municipal wastewater disinfection, representative values of pK_a for HOCl and CH_3COOOH are 7.5 and 8.1, respectively. Given this information:

a. Comment on the efficacy of these compounds at pH values above and below their respective pK_a values. In other words, would you expect them to function better at pH above the pK_a value or at pH below the pK_a value. Justify your answer.
b. Discharge permit limitations for municipal wastewaters often require that treated wastewater have a pH between 6-9; typical effluent pH is often in the vicinity of 7.5. Comment on the relative efficacy of free chlorine and PAA at low, medium, and high values of effluent pH (*i.e.* pH = 6, pH = 7.5, and pH = 9). Justify your answer.

ANNEX 1: SOLUTIONS TO EXCERSISES
Solution 8.4.1

The calculations for this problem are similar to those developed above for Example 8.3.2. Specifically, the distribution fractions can be described as functions of pH and chloride ion concentration using algebraically manipulated forms of the equilibrium expressions:

$$\alpha_{Cl_2} = \frac{\left[Cl_2\right]}{C_{T,Cl}} = \left(1 + \frac{K_h}{\left[H^+\right]\cdot\left[Cl^-\right]} + \frac{K_a \cdot K_h}{\left[H^+\right]^2}\right)^{-1}$$

$$\alpha_{HOCl} = \frac{\left[HOCl\right]}{C_{T,Cl}} = \left(\frac{\left[H^+\right]\cdot\left[Cl^-\right]}{K_h} + 1 + \frac{K_a}{\left[H^+\right]}\right)^{-1}$$

$$\alpha_{OCl^-} = \frac{\left[OCl^-\right]}{C_{T,Cl}} = \left(\frac{\left[H^+\right]^2\cdot\left[Cl^-\right]}{K_h \cdot K_a} + \frac{\left[H^+\right]}{K_a} + 1\right)^{-1}$$

The logic and algebraic manipulations used to develop these equations are almost identical to those used to solve Example 8.3.2.

A spreadsheet[1] has been developed to summarize these calculations. In that file, note that 'check' columns are included to confirm that the distribution fractions at any pH value added up to 1 (unity), identically. In other words:

$$\alpha_{Cl_2} + \alpha_{HOCl} + \alpha_{OCl^-} \equiv 1$$

Graphs of the distribution fractions as functions of pH are presented below for tap water (and municipal wastewater) and seawater. As with Example 8.3.2, we see that the increase in chloride ion concentration in seawater (relative to tap water) results in a shift of free chlorine speciation toward molecular chlorine (Cl_2). This is most evident at low pH values.

---

[1] https://www.iwapublishing.com/books/9781789062298/biological-wastewater-treatment

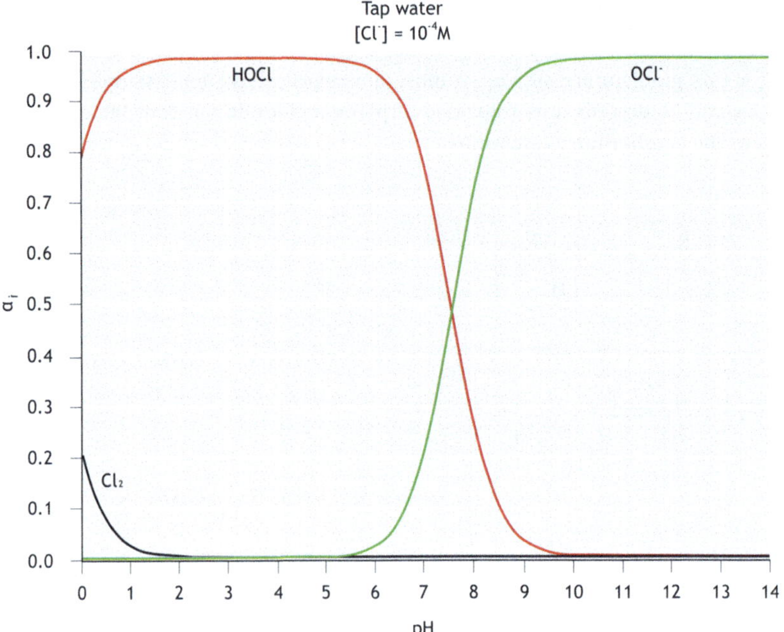

Figure 8.6 Chlorine distribution in tap water for Exercise 8.4.1.

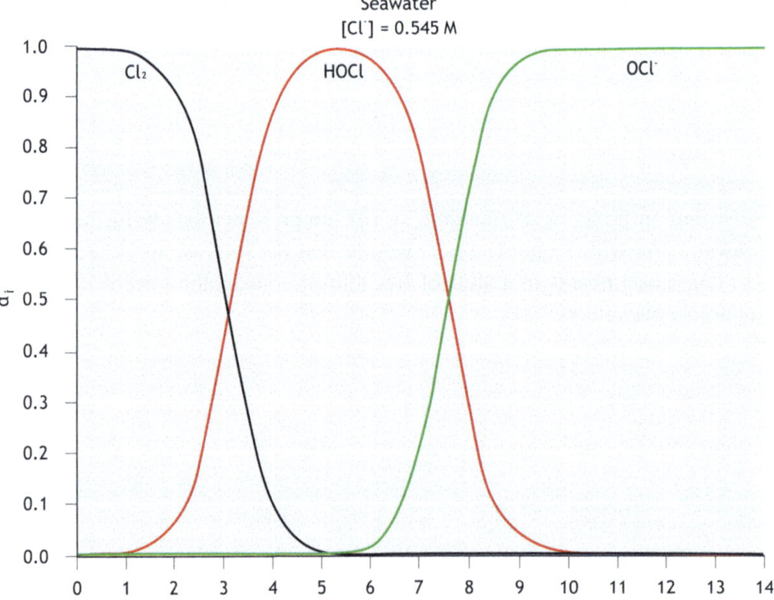

Figure 8.7 Chlorine distribution in seawater for Exercise 8.4.1.

Solution 8.4.2

The spreadsheet for Exercise 8.4.1 was extended to develop estimates of the effective Henry's law constant for free chlorine (H_{Cl}) as a function of pH. These calculations were conducted as follows (note that OCl^- is excluded from this calculation because it is non-volatile):

$$H_{Cl} = H_{Cl_2} \alpha_{Cl_2} + H_{HOCl} \, \alpha_{HOCl}$$

The results of these calculations are summarized in the graph below. The graph includes values of the effective Henry's law constant for free chlorine (H_{Cl}) for $0 \leq pH \leq 10$; values of H_{Cl} are essentially zero at $pH \gtrsim 10$ because virtually all free chlorine is in the form of OCl^- under these conditions. For the case of seawater, free chlorine becomes much more volatile than it is with tap water. This change in behaviour is attributable to the shift of speciation toward Cl_2 that takes place in seawater because of the elevated chloride ion concentration.

Chlorination of municipal wastewater is not likely to be conducted in water that contains a chloride ion concentration that is in the order of that observed in seawater. However, there are applications of chlorine (*e.g.* cooling towers) in which high chloride ion concentrations may be observed (Holzwarth *et al.*, 1984).

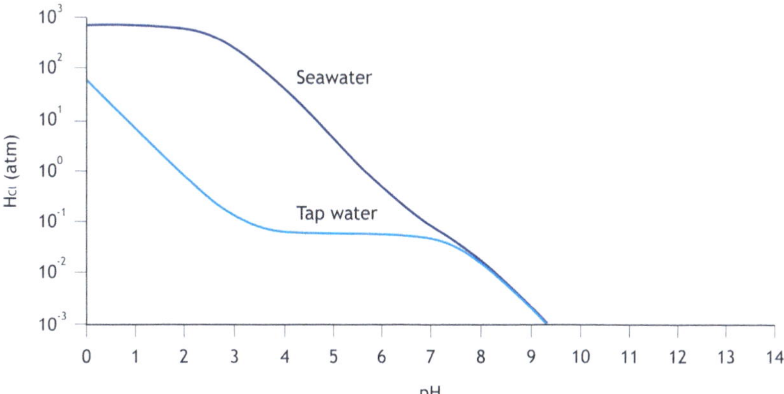

Figure 8.8 Effective Henry's law constant as a function of pH.

Solution 8.4.3

a. As described in the problem statement, the Cl:N mass ratios of 4 and 8 can be converted to molar ratios as follows:

$$\frac{4 \text{ mg Cl}_2}{1 \text{ mg N}} \cdot \frac{14{,}000 \text{ mg N}}{\text{mole N}} \cdot \frac{\text{mole Cl}_2}{70{,}900 \text{ mg Cl}_2} = 0.79$$

$$\frac{8 \text{ mg Cl}_2}{1 \text{ mg N}} \cdot \frac{14{,}000 \text{ mg N}}{\text{mole N}} \cdot \frac{\text{mole Cl}_2}{70{,}900 \text{ mg Cl}_2} = 1.6$$

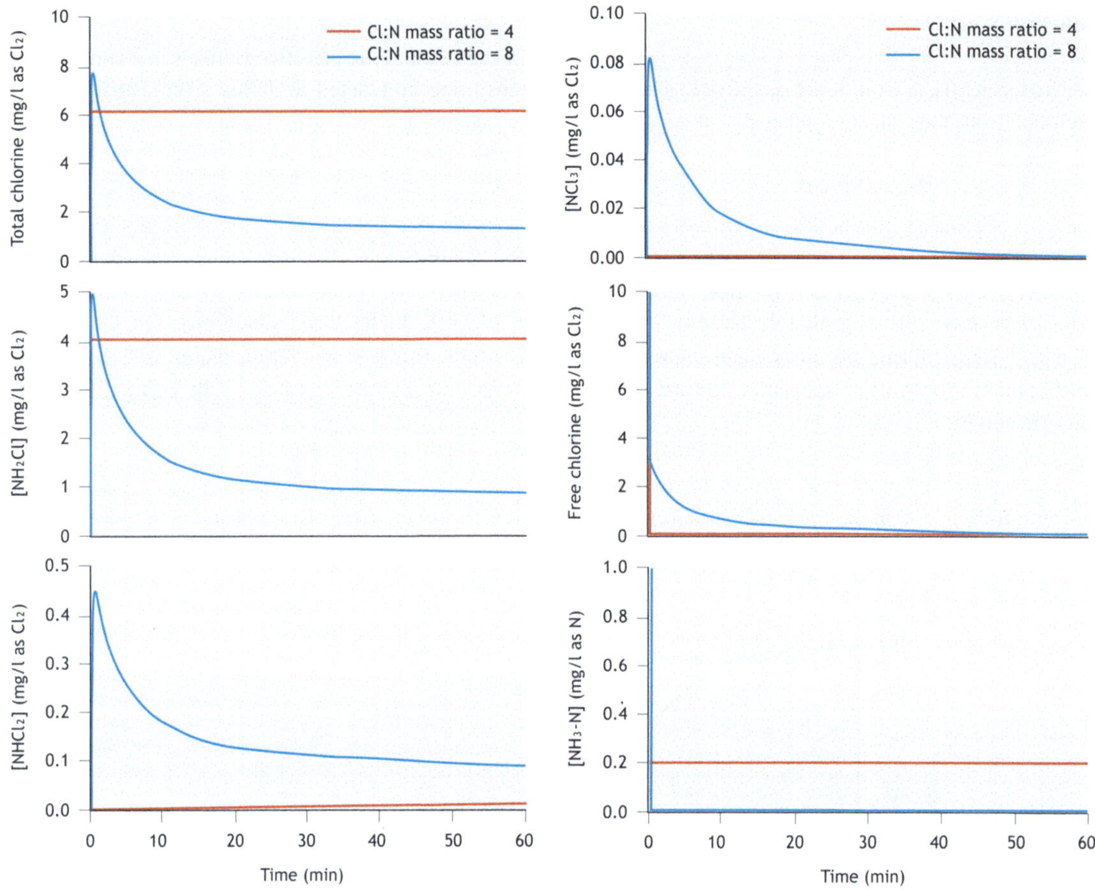

Figure 8.9 Dynamic behaviour of free chlorine and the inorganic chloramines over a period of 60 minutes at pH = 8.0, total alkalinity of 150 mg/l (as $CaCO_3$), water temperature of 25 °C, TOC = 0, and free ammonia initial concentration of 1.0 mg/l (as N): (a) total chlorine for the case of Cl:N mass ratios of 4 and 8; (b) monochloramine; (c) dichloramine; (d) trichloramine; (e) free chlorine; and (f) (free) ammonia.

Note that the first condition is the 'chloramination' region, where we would expect most free chlorine to be converted to NH_2Cl, and some ammonia will remain unreacted. The second condition is very close to the breakpoint, where we should expect both residual chlorine and ammonia to be unstable. Moreover, since the breakpoint defines the stoichiometric condition at which free chlorine added to solution is (just) sufficient to satisfy chlorine demand imposed by ammonia-N, this condition will result in very low concentrations of free chlorine and ammonia (*i.e.* both will be 'consumed' in the breakpoint reactions).

- Total chlorine: for the case of a Cl:N mass ratio of 4, total residual chlorine is stable at a value of about 4 mg/l (as Cl_2) for a timescale of 60 minutes. This is because the free chlorine that has been added to the solution has been converted almost entirely to NH_2Cl. For the Cl:N = 8 case, which is close to the breakpoint, residual chlorine is unstable.

- Monochloramine: for the case of a Cl:N mass ratio of 4, NH_2Cl comprises essentially the entire residual. As described above, under this condition NH_2Cl is stable on this timescale. For Cl:N = 8, NH_2Cl is also formed, but demonstrates steady decay over a 60-minute period.
- Dichloramine: very little $NHCl_2$ is formed at Cl:N = 4, but it is formed at Cl:N = 8, and because all of the redox chemistry in this system proceeds through $NHCl_2$, it behaves as an unstable intermediate.
- Trichloramine: essentially no NCl_3 is formed at Cl:N = 4, but it is formed at Cl:N = 8 and behaves as an unstable intermediate. Loss of NCl_3 over time is probably attributable to hydrolysis (back to $NHCl_2$), which in turn allows decay via redox reactions.
- Free chlorine: essentially no free chlorine is present at Cl:N = 4. At Cl:N = 8, free chlorine concentration decreases almost immediately from its initial value of 8 mg/l (as Cl_2) to about 3 mg/l (as Cl_2), after which it displays steady decay. The initial loss is attributable to substitution reactions that result in formation of the inorganic chloramines. The decay that is observed over the 60-minute period is attributable to redox reactions that involve the inorganic chloramines and free chlorine.
- (Free) ammonia: roughly 80 % of ammonia is converted to NH_2Cl at t = 0 for the case of Cl:N = 4. At Cl:N = 8, the ammonia is entirely consumed at the time of chlorine addition.

b. See Figure 8.10 and Figure 8.11.

- Total chlorine: at the lower pH condition (top panel, next page), there is a slightly higher total residual chlorine concentration than at the higher pH at t=30 min. This is most evident at Cl:N molar ratios that are slightly above 1 up to the breakpoint (Cl:N mass ratios between 5-9), and largely attributable to the formation of $NHCl_2$.
- Monochloramine: the concentration of $NHCl_2$ is higher at pH 7 than at pH 8. This is because of the formation of $NHCl_2$ by disproportionation of NH_2Cl, which is general-acid catalysed.
- Dichloramine: the concentration of NH_2Cl is lower at pH 7 than at pH 8. This is because of the formation of $NHCl_2$ by disproportionation, which is general-acid catalysed.
- Trichloramine: NCl_3 is evident at Cl:N ratios that exceed the breakpoint. The concentration of NCl_3 is higher in the pH 7 solution than in the pH 8 solution.
- Free chlorine: like NCl_3, free chlorine is evident at Cl:N ratios that exceed the breakpoint. The concentrations of free chlorine in solution are similar in both solutions, but slightly higher at pH 8 than at pH 7.
- (Free) ammonia: free NH_3 is consumed when the Cl:N ratio reaches 1 on a molar basis (roughly at a Cl:N mass ratio of 5). The pattern of NH_3 consumption is similar between the two conditions.

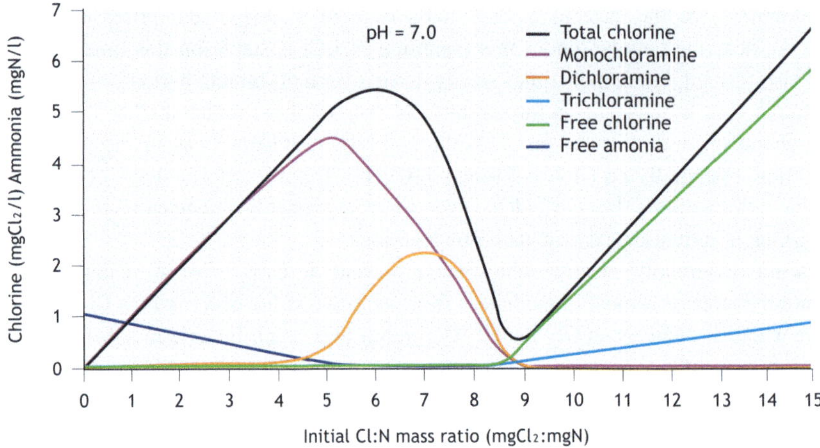

Figure 8.10 Breakpoint curves of free chlorine and the inorganic chloramines over a period of 30 minutes at pH 7.0.

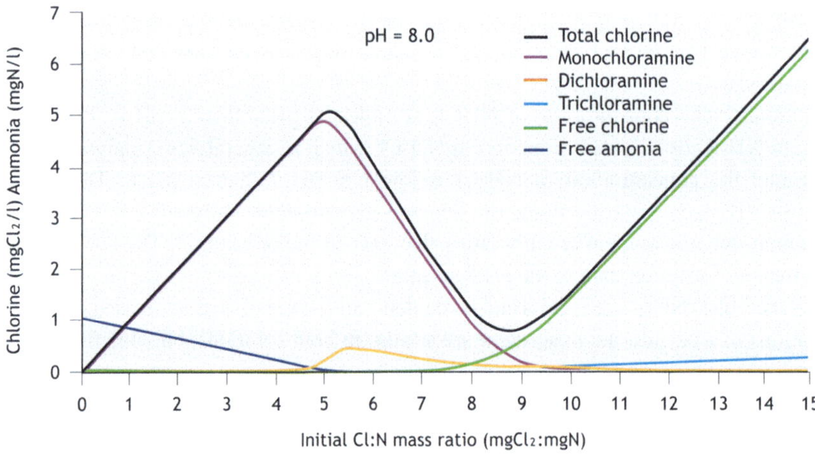

Figure 8.11 Breakpoint curves of free chlorine and the inorganic chloramines over a period of 30 minutes at pH 8.0.

Solution 8.4.4

The stoichiometry of dechlorination (of NH_2Cl) by $NaHSO_3$ is defined as follows (note that the sodium ion plays no role in this reaction and has been excluded from the stoichiometric expression):

$$NH_2Cl + HSO_3^- + H_2O \rightleftharpoons SO_4^{2-} + H^+ + NH_4^+$$

This expression indicates a 1:1 molar ratio between NH_2Cl and $NaHSO_3$ to complete this reaction. Therefore, the mass concentration of $NaHSO_3$ required for dechlorination can be defined as:

$$1.2 \frac{\text{mg Cl}_2}{1} \cdot \frac{\text{mmole Cl}_2}{70.9 \text{ mg Cl}_2} \cdot \frac{\text{mmole NaHSO}_3}{\text{mmole Cl}_2} \cdot \frac{104 \text{ mg}}{\text{mmole NaHSO}_3} = 1.76 \frac{\text{mg NaHSO}_3}{1}$$

We can calculate the mass rate of $NaHSO_3$ use by multiplying this mass concentration by the flow rate, with the application of appropriate unit conversions:

$$\dot{M}_{NaHSO_3} = 1,500 \frac{\text{m}^3}{\text{hr}} \cdot 1.76 \frac{\text{mg NaHSO}_3}{1} \cdot \frac{\text{kg}}{10^6 \text{ mg}} \cdot \frac{24 \text{ hr}}{\text{d}} = 0.063 \frac{\text{kg NaHSO}_3}{\text{d}}$$

Solution 8.4.5

a.   The acid/base reactions that apply for HOCl and PAA are as follows:

HOCl	$HOCl \rightleftharpoons H^+ + OCl^-$	$K_a = \dfrac{[H^+] \cdot [OCl^-]}{[HOCl]}$
PAA	$CH_3COOOH \rightleftharpoons H^+ + CH_3COO^-$	$K_a = \dfrac{[H^+] \cdot [CH_3COO^-]}{[CH_3COOOH]}$

When the pH is below the pK_a of an acid, the protonated form will be present at relatively high concentration. The farther the pH is depressed below the pK_a, the larger the fraction of the protonated form. In the case of HOCl and PAA, the protonated forms are the most effective disinfectants. Therefore, both compounds will function better as disinfectants when they are present in solution at pH below their respective pK_a values.

b.
- At low pH (*i.e.* pH = 6), both compounds will be present (largely) in their respective protonated forms. Therefore, HOCl and PAA are both likely to be relatively effective at this pH value.
- At medium pH (*i.e.* pH = 7.5) HOCl will be roughly evenly split between HOCl and its conjugate acid, OCl^-. PAA will also be present as a mixture of its neutral (protonated) form and its conjugate acid, but a larger fraction of PAA will exist in solution as its protonated form than HOCl. In other words, it is likely that PAA will be less affected by this pH condition than HOCl.
- At high pH (*i.e.* pH = 9), both compounds will be dominated by their deprotonated forms (*i.e.* their conjugate bases, both of which are anions). As such, the performance of both compounds will be adversely affected by this pH condition, relative to lower pH conditions. However, since the pK_a of PAA is roughly 0.6 pH units higher than that of HOCl, it is likely that PAA will be less adversely affected by this relatively high pH condition than will HOCl.

REFERENCES

Beck S.E., Rodriguez R.A., Linden K.G., Hargy T.M., Larason T.C. and Wright H.B. (2014). Wavelength Dependent UV Inactivation and DNA Damage of Adenovirus as Measured by Cell Culture Infectivity and Long Range Quantitative PCR. *Environmental Science & Technology*, *48*, (1), 591-598.

Blatchley III E.R., Johnson R.W., Alleman J.E. and McCoy W.F. (1992). Effective Henry's law constants for free chlorine and free bromine. *Water Research*, *26*, (1), 99-106.

Buonanno M., Randers-Pehrson G., Bigelow A.W., Trivedi S., Lowy F.D., Spotnitz H.M., Hammer S.M. and Brenner D.J. (2013). 207-nm UV Light - A Promising Tool for Safe Low-Cost Reduction of Surgical Site Infections. I: *In Vitro* Studies. *Plos One*, *8*, (10).

Chen GH., van Loosdrecht M.C.M., Ekama G.A. and Brdjanovic D. Ed. (2020) Biological Wastewater Treatment: Principles, Design and Modelling. 2nd edition. IWA Publishing, pg. 850. 9781789060355.

Holzwarth G., Balmer R.G. and Soni L. (1984). The fate of chlorine and chloramines in cooling towers; Henry's law constants for flashoff. *Water Research*, *18*, (11), 1421-1427.

Jafvert C.T. and Valentine R.L. (1992). Reaction Scheme for the Chlorination of Ammoniacal Water. *Environmental Science & Technology*, *26*, (3), 577-586.

Sun Z., Fu J.N., Li X., Blatchley E.R. and Zhou Z. (2020). Using Algal Virus Paramecium bursaria Chlorella Virus as a Human Adenovirus Surrogate for Validation of UV Treatment Systems. *Environmental Science & Technology*, *54*, (23), 15507-15515.

Vikesland P.J., Ozekin K. and Valentine R.L. (2001). Monochloramine decay in model and distribution system waters. *Water Research, 35*, (7), 1766-1776.

Wahman D.G. (2018). Web-Based Applications to Simulate Drinking Water Inorganic Chloramine Chemistry. *Journal American Water Works Association, 110*, (11), E43-E61.

NOMENCLATURE

Symbol/Abbreviation	Description	Unit
C	(Chemical) disinfectant concentration	mg/l
$C_{T,Cl}$	Free chlorine concentration $[Cl_2] + [HOCl] + [OCl^-] + 2[Cl_2O] + [Cl_3^-]$	mole/l
$[Cl_2]$	Concentration of molecular chlorine	mole/l
$[HOCl]$	Concentration of hypochlorous acid	mole/l
$[OCl^-]$	Concentration of hypochlorite ion	mole/l
$[Cl_2O]$	Concentration of chlorine monoxide	mole/l
$[Cl_3^-]$	Concentration of trichloride ion	mole/l
E_i	Incident fluence rate (or irradiance) of UV radiation	mW/cm^2
H_{Cl}	Effective Henry's law constant for free chlorine	atm
k	Inactivation constant for UV disinfection	cm^2/mJ
N	Concentration of viable or infective microbes or viruses	number/l
t	Time of exposure to disinfectant	s
UV_{254}	Ultraviolet radiation	254 nm
PAA	Peracetic acid	
Λ	Coefficient of specific lethality	l/mg.min

doi: 10.2166/9781789062304_0317

9

Aeration and mixing

Diego Rosso, Eveline I.P. Volcke, Manel Garrido-Baserba, Coenraad Pretorius and Michael K. Stenstrom

9.1 INTRODUCTION

Chapter 9 Aeration and Mixing in the textbook *Biological Wastewater Treatment: Principles, Modelling and Design* (Chen *et al.*, 2020) introduces the fundamental quantities related to oxygen transfer and aeration, their dynamics in relation to the biological process dynamics, the equipment required to provide aeration, and the relation between mixing and aeration. Here we also present the energy implications of aeration and mixing. This chapter applies all of this content through examples, questions and exercises.

9.2 LEARNING OBJECTIVES

After the successful completion of this chapter, the reader will be able to:

- Describe the equipment for aeration and mixing and its functioning.
- Size the equipment based on average and peak process conditions.
- Specify the number of aeration diffusers necessary to meet the oxygen requirement of a biological process.
- Quantify the mixing effectiveness in a suspended-growth process.

9.3 EXAMPLES
Example 9.3.1

Aeration system design

A wastewater treatment plant requires an average mass of oxygen per day $R_{O2} = 8{,}640\ kgO_2/d$. The aeration tanks are 5 m deep and the diffusers are to be installed 30 cm above the floor. The diffusers utilize 9 inch

(22.86 cm) membrane disks (each with the area = 0.0373 m^2) with a desired average specific airflow rate of 9.44 $\cdot 10^{-4}$ m^3/s (corresponding to 2 SCFM, *i.e.,* standard cubic feet per minute) per diffuser; for new diffusers, this is specified by the manufacturer. The plant receives a daily peak flow rate of 1.3 times the average. The alpha factor is assumed α = 0.35, which is a conservative (low) value typical for processes with only BOD removal (for which α ranges between 0.25 - 0.45).

Once in operation, diffusers are characterized by an increased pressure drop (increasing DWP: dynamic wet pressure) as well as by a reduced oxygen transfer efficiency, described by the parameters Ψ (pressure factor, Eq. 9.1) and F (fouling factor, Eq. 9.2), respectively. For new diffusers, Ψ =1.0 and F =1.0; upon usage Ψ > 1.0 and F < 1.0.

$$\Psi = \frac{DWP_{new\_diffuser}}{DWP_{used\_diffuser}} \tag{9.1}$$

$$F = \frac{\alpha SOTE_{new\_diffuser}}{\alpha SOTE_{used\_diffuser}} \tag{9.2}$$

The blower has an efficiency of 75 % and is operated at an inlet temperature of 20 °C and an inlet pressure of 0.9 atm.

a) Consider new diffusers. What would be the number of diffusers necessary to meet the aeration requirements in process water? Calculate the number of diffusors based on *average* conditions as well as based on *peak* conditions. Also, calculate the blower discharge pressure at peak airflow and the corresponding peak blower power.

b) Consider fouled diffusers (Ψ = 1.3 and F = 0.8) but now keep the same number of diffusers as calculated from new diffusers under peak conditions. What is the total airflow rate (AFR) and the airflow rate per diffuser (AFR$_{diff}$) under peak conditions? Calculate again the blower discharge pressure at peak airflow and the corresponding peak blower power.

Additional data: At standard conditions (20 °C and 1 atm) assume an air density of 1.225 kg/m^3 and a ponderal oxygen concentration in air of 23 %, *i.e.,* 0.23 kg O$_2$ per kg air.

Hint
Size the blower airflow rate (AFR, in m^3/h) for the average load and for the peak load. Assume the diffusers are new. The number of diffusers will be based on the peak load.

Additional information A1: estimation of SOTE from manufacturers' clean water curves.
As an alternative to the design algorithm in Figure 9.35 in the textbook, the oxygen transfer efficiency at standard conditions, SOTE, can be estimated from manufacturers' clean water curves, when available. An example is displayed in Fig. 9.1.

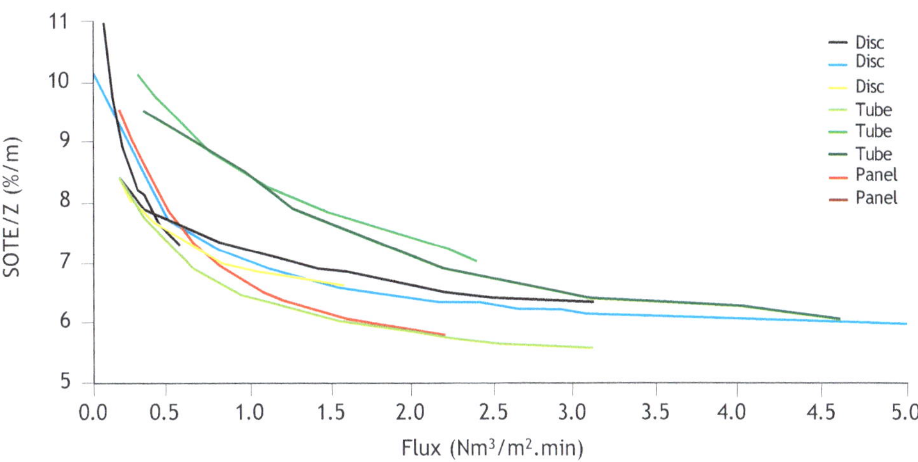

Figure 9.1 Example of efficiency curves (expressed in % per unit depth) for various diffuser geometries over a range of air flux (expressed as airflow per unit diffuser area).

The manufacturers' clean water curves express the SOTE (%) per unit diffuser submergence Z (m) as a function of the airflow rate per unit diffuser area (or air flux). The latter is expressed by Eq. 9.3:

$$\frac{AFR}{N_D \cdot A_{diff}} = \frac{AFR_{diff}}{A_{diff}} \tag{9.3}$$

in which AFR is the total airflow rate (m^3/s), A_{diff} is the specific area of each diffuser (m^2) and N_D is the total diffuser number (dimensionless). AFR_{diff} denotes the airflow rate per diffuser.

Using the clean water curves thus makes it possible to estimate the SOTE, given the desired airflow rate specified by the manufacturer (AFR_{diff}, in m^3/s), the diffuser specific area (A_{diff}, in m^2) and the diffuser submergence Z (m).

Additional information A2: blower curve, system curve, operating point
Each type of blower has a different blower curve, which is provided by the manufacturer and specifies the blower discharge pressure as a function of airflow rate (see Figure 9.14 in the textbook Chen *et al.*, 2020).

The discharge pressure of the blower (p_{disch}, relative to atmospheric pressure, in Pa) must always equal or exceed the sum of the pressure losses in the system, to guarantee that the air is released (Eq. 9.4):

$$p_{disch} \geq \rho \cdot g \cdot Z + h_L\,(AFR) + DWP(AFR) \tag{9.4}$$

in which g denotes the gravitational constant (g = 9.81 m/s^2) and ρ the water density (kg/m^3).

The pressure losses on the right-hand side of Eq. 9.4 indicate the hydrostatic head loss due to the diffuser submergence, the head loss of the air distribution line (friction head h_L, in Pa) and the dynamic wet pressure (DWP, in Pa), *i.e.*, the diffuser head loss, which is a function of the diffuser specific airflow rate and needs to be provided by the manufacturer. Figure 9.2 summarizes some DWP(AFR) curves from manufacturers, which are typically provided for new diffusers.

For practical purposes, the line head loss h_L may be considered constant, such that Eq. 9.4 simplifies to:

$$p_{disch} \geq \rho \cdot g \cdot Z + h_L + DWP(AFR) \tag{9.5}$$

When diffusers foul, the DWP in Eq. 9.5 increases by a factor Ψ (>1, see Eq. 9.1) compared to new diffusers.

The sum of the dynamic wet pressure, static head loss and friction losses in the air distribution line (*i.e.* the right-hand-side of Eq. 9.5), make up the system curve (see Figure 9.14 in Chen *et al.*, 2020). When the aeration system is in operation, Eq. 9.5 becomes an equality and the pressure in excess of the equality results in bubbles detaching from the diffuser with initial velocity > 0. The system curve indicates what the requirement of the system is in terms of aeration pressure and needs to be compared against the blower curve representing the pressure supply.

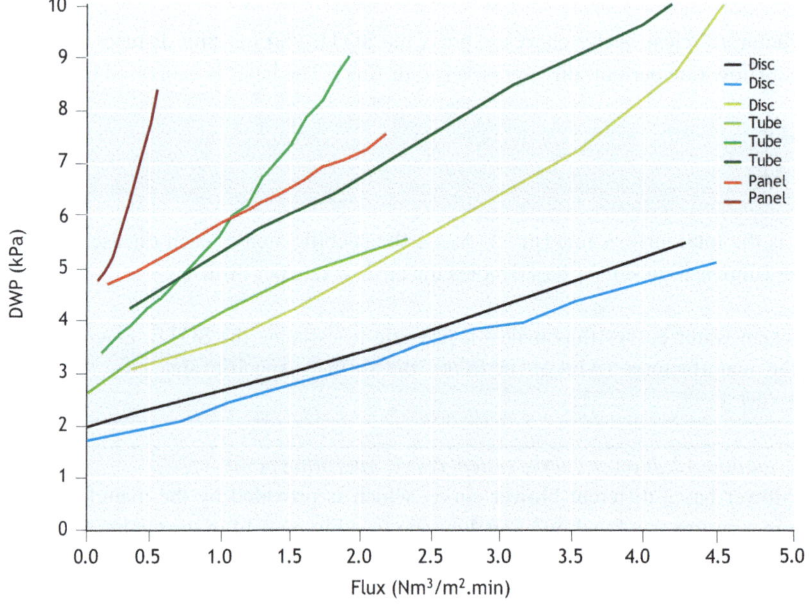

Figure 9.2 Examples of diffuser pressure drop (DWP, *i.e.*, dynamic wet pressure) for a range of air fluxes.

The operating point is where the system curve and blower curve meet and should correspond with the required airflow rate in the system. Should the airflow range of a certain manufacturer be insufficient, at a given pressure requirement, a battery of identical blowers arranged in parallel can be selected. Indeed, the AFR of parallel blowers is cumulative, while the pressure is the same.

In fact, it is always necessary to make sure that the actual discharge pressure of the blower is higher than the system needs. This is because when the pressure requirements of the system approach the maximum discharge pressure of the blowers, the blowers begin to surge; this is manifested through vibrations of the blowers, which may cause structural failure. The surge zone is an area of operation which therefore must be avoided. An automated system shuts down the blowers when surging begins.

Additional information A3: blower brake horsepower (BHP)
The blower brake horsepower (BHP), in short 'blower power', represents the force needed to brake or stop the blower motor. This is the minimum amount of power needed to operate the blower. It equals the mechanical work that the blower performs to impart velocity to the air. It differs from the total wire power, for it does not include the inefficiency of the electrical motor moving the blower.

The blower brake horsepower (BHP, in kW) is a function of the required air mass flow rate and is calculated using the following adiabatic compression formula (Metcalf & Eddy, 2014):

$$BHP = \frac{W_{air} \cdot R \cdot T_{in}}{29.7 \cdot n \cdot e} \left[\left(\frac{p_{disch}}{p_{in}} \right)^n - 1 \right] \tag{9.6}$$

where:
W_{air} = air mass flow rate (kg/s)
R = universal gas constant (8.314 J/mol.K)
T_{in} = absolute inlet temperature (K)
p_{in} = absolute inlet pressure (Pa)
$p_{disch,abs}$ = absolute discharge pressure (Pa)
n = 0.283 for air
e = blower efficiency

As an alternative to Eq. 9.6, it is also possible to use manufacturer curves expressing BHP in terms of the airflow rate (examples can be found on blower manufacturers' websites).

Solution

Note: The calculations are provided in the spreadsheet 'Chapter 9 Design examples.xlsx' on sheet 'Example 9.3.1'. The results are summarized in Table 9.1.

a) New diffusers

The number of diffusers needed to meet the aeration requirements is determined by the ratio between the required airflow rate and the diffuser manufacturer's recommended airflow rate per diffuser, the latter being specified by the manufacturer (Eq. 9.7).

$$N_D = \frac{AFR}{AFR_{diff}} \tag{9.7}$$

The required airflow rate (AFR) can be derived from the amount of oxygen that needs to be fed to the aeration tank, which is characterized by the mass flow of oxygen through the blowers, W_{O2} (kgO$_2$/s) (Eq. 9.8):

$$AFR = W_{O2} \cdot \frac{1}{\rho_{air}} \cdot \frac{1}{\hat{y}_{O2}} \qquad (9.8)$$

where ρ_{air} is the air density (= 1.225 kg/m^3) and $\hat{y}_{O2}$ is the weight fraction of oxygen in air (= 23 wt%).

The required oxygen mass flow rate, W_{O2}, in its turn follows from the required oxygen transfer rate and the oxygen transfer efficiency (Eq. 9.9):

$$W_{O2} = \frac{SOTR}{\alpha FSOTE} \qquad (9.9)$$

The required average oxygen transfer rate under standard conditions, SOTR, is calculated from the average required oxygen mass flow, $R_{O2} = 8{,}640$ kgO$_2$/d, taking into account a safety factor S_F (Eq. 9.10):

$$SOTR = S_F \cdot R_{O2} \qquad (9.10)$$

The safety factor S_F (*e.g.*, 1.2 - 1.5) conservatively addresses the need for higher OTR at peak loading times or seasons. As an alternative to calculating the SOTR with a safety factor, it would also be possible to use the distribution of R_{O2} over time, as can be calculated by a dynamic simulator.

The average SOTR being 360 kgO$_2$/h, applying a safety factor S_F of 1.3 in this example results in a peak oxygen SOTR of 468 kgO$_2$/h.

αFSOTE is the oxygen transfer efficiency in standard conditions in process water, taking into account the time the diffuser has been in operation. It is proportional to the oxygen transfer efficiency in standard conditions in process water (SOTE) through the factors α and F, and can also be written in terms of the SOTE per unit diffuser submergence Z (m) (Eq. 9.11):

$$\alpha FSOTE = \alpha \cdot F \cdot SOTE = \alpha \cdot F \cdot SOTE/Z \cdot Z \qquad (9.11)$$

SOTE/Z is estimated from the clean water curves (see *Additional information A1*). The disk diffuser proposed in this example has an area of $A_{diff} = 0.0373$ m^2 and a desired average specific airflow rate AFR_{diff}^{avg} 9.44$\cdot$10^{-4} m^3/s, which corresponds to an air flux of 1.52 m^3/m^2.min. From the given manufacturers' clean water curve (Figure 9.1; blue line), we find a corresponding value of SOTE/Z^{avg} of approximately 6.20 %/m. The aeration tanks are 5 m deep and the diffusers are to be installed 30 cm above the floor, so the diffuser submergence Z equals 5.0 - 0.30 = 4.70 m. Taking into account the given alpha factor $\alpha = 0.35$ and assuming new diffusers (F = 1.0), the corresponding αFSOTE^{avg} (= αSOTE^{avg}) is calculated from Eq. 9.11 as 10.2 %.

Taking into account SOTR^{avg} = 360 kgO$_2$/h and αFSOTE^{avg} 10.2 %, the required oxygen mass flow for average load conditions is calculated (Eq. 9.9) as W_{O2}^{avg} = 3,530 kgO$_2$/h, which corresponds to a required airflow rate (Eq. 9.8) of AFR^{avg} = 12,528 m^3/h.

The required number of diffusers based on average flow conditions is calculated using Eq. 9.7 as:

$$N_D{}^{avg} = \frac{AFR^{avg}}{AFR_{diff}^{avg}} = \frac{12,528}{9.44 \cdot 10^{-4} \cdot 3,600} = 3,687 \tag{9.12}$$

Under peak flow conditions, the airflow rate per diffuser will increase according to the peak factor, resulting in a specific airflow rate $AFR_{diff}{}^{peak} = 1.23 \cdot 10^{-3}$ m^3/s, which corresponds to an air flux of 1.97 m^3/m^2.min. This implies a decreased SOTE compared to average flow conditions; from the given manufacturers' clean water curve (Figure 9.1; blue line), we find a corresponding value of SOTE/Z^{peak} of approximately 6.0 %/m. The associated αFSOTE^{peak} is calculated from Eq. 9.11 as 9.9 %.

Taking into account SOTR^{peak} = 468 kgO$_2$/h and αFSOTE^{peak} 9.9 %, the required oxygen mass flow (Eq. 9.9) under peak load becomes $W_{O_2}{}^{peak}$ = 4,742 kgO$_2$/h. This corresponds to a required airflow rate (Eq. 9.8) of AFR = 16,829 m^3/h.

The required number of diffusers (Eq. 9.7) based on peak flow conditions becomes:

$$N_D{}^{peak} = \frac{AFR^{peak}}{AFR_{diff}^{peak}} = \frac{16,829}{1.23 \cdot 10^{-3} \cdot 3,600} = 3,810 \tag{9.13}$$

Note that the lower oxygen transfer efficiency (αFSOTE) under peak load conditions results in a higher number of diffusers required. It is important to recognize that the number of diffusers calculated at peak conditions, N_D = 3,810, will be the total we will use from now on, so that a sufficient number of diffusers will be installed in the tank for <u>all</u> operating conditions.

The procedure for calculating the number of diffusers is visualized in Figure 9.3; the resulting numerical values for average load and peak load are summarized in Table 9.1.

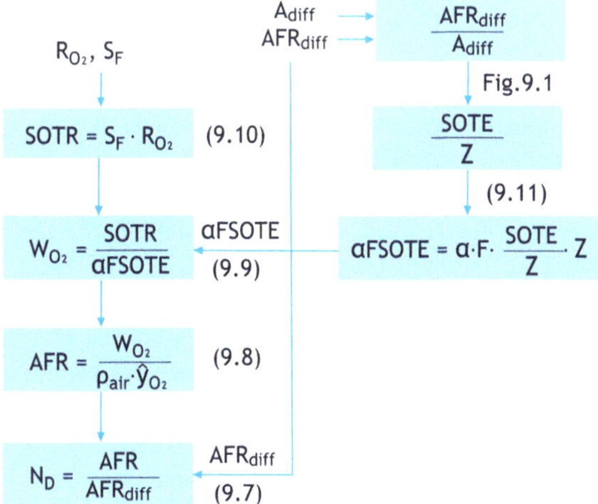

Figure 9.3 Procedure for calculating the number of diffusers based on the load to be treated (oxygen demand R_{O_2}) and the desired airflow rate per diffuser specified by the manufacturer (AFR_{diff}).

The blower discharge pressure at peak airflow is calculated from Eq. 9.5, taking into account that the blower discharge pressure should equal the pressure losses in the system (equality):

$$p_{disch}{}^{peak} = \rho \cdot g \cdot Z + h_L + DWP_{new,\ diffuser}(AFR^{peak}) \tag{9.14}$$

From the given dynamic wet pressure (DWP) curves in Figure 9.2 (blue line), we can identify a DWP (AFR^{peak}) value of approximately 3.2 kPa at the peak air flux of 1.97 $m^3/m^2.min$.

The diffuser submergence here is given as 4.70 m, which corresponds to 47.6 kPa (a 10-meter water column corresponding to the atmospheric pressure 101.325 kPa). The line head loss (h_L) can be assumed conservatively to be 0.3 m of water column (3.04 kPa). In reality, there is a detailed procedure to calculate the sum of the head loss from all the piping and valve elements, but because these are not known we must assume a conservative round number.

As a result, the discharge pressure to specify a blower (*i.e.*, the peak pressure requirements) is calculated from Eq. 9.14 as:

$$p_{disch}{}^{peak} = 47.6 + 3.04 + 3.20 = 53.84\ kPa \tag{9.15}$$

The blower brake horsepower (BHP) corresponding with the peak AFR is calculated using the adiabatic compression formula (Eq. 9.6):

$$BHP^{peak} = \frac{W_{air} \cdot R \cdot T_{in}}{29.7 \cdot n \cdot e}\left[\left(\frac{p_{disch,abs}^{peak}}{p_{in}}\right)^{n} - 1\right] = \frac{\dfrac{W_{O2}}{\hat{y}_{O2}} \cdot R \cdot T_{in}}{29.7 \cdot n \cdot e}\left[\left(\frac{p_{disch,abs}^{peak}}{p_{in}}\right)^{n} - 1\right] = \tag{9.16}$$

$$\frac{\dfrac{4{,}742}{3{,}600} \cdot 0.23 \cdot 8{,}314 \cdot 293.15}{29.7 \cdot 0.283 \cdot 0.75}\left[\left(\frac{53.86 + 101.325}{0.9 \cdot 101.325}\right)^{0.283} - 1\right]\ kW = 359\ kW$$

<u>b) Fouled diffusers – fixed number of diffusers</u>
In the case of fouled diffusers, in order to keep the same desired airflow rate per diffuser (AFR_{diff}) as specified by the manufacturer, it is necessary to recalculate the total number of diffusers based on the actual number needed under peak conditions after fouling. The procedure for calculating the number of diffusers in the case of fouling is completely analogous as for new diffusers (Eq. 9.7 and Figure 9.3); only a reduced value for the oxygen transfer efficiency αFSOTE needs to be taken into account.

In the case of fouled diffusers, the oxygen transfer efficiency αFSOTE is decreased proportionally to the fouling factor F = 0.8, resulting in $\alpha FSOTE^{avg}$ = 8.2 % for average flow conditions and $\alpha FSOTE^{peak}$ = 7.9 % for peak flow conditions. As a result, the airflow requirements are higher. The required oxygen mass flow for *average load* conditions is now calculated (Eq. 9.9) as $W_{O2}{}^{avg}$ = 4,412 kgO_2/h, which corresponds to a required airflow rate (Eq. 9.8) of AFR^{avg} = 15,660 m^3/h. The required oxygen mass flow for *peak load* conditions (Eq. 9.9) therefore becomes $W_{O2}{}^{peak}$ = 5,927 kgO_2/h, which corresponds to a required airflow rate (Eq. 9.8) of AFR^{peak} = 21,037 m^3/h. Note that the standard oxygen transfer rate SOTR, required for the calculation of W_{O2} (Eq. 9.9), remains unchanged compared to the case of new diffusers.

However, in practice it is not possible to change the number of diffusers once installed, so the number of diffusers is not changed with time. At best, the number of installed diffusers that discharge air could be changed, in a so-called swing zone, could be changed. However, in this example, we will keep the number of diffusers constant and we will consider that all diffusers are active at all times. The airflow rate and the corresponding air flux per diffuser should be recalculated, since the number of diffusers was kept constant for the peak conditions of new diffusers. In figures 9.1 and 9.2 we need to look for the SOTE/Z and DWP, respectively, at this recalculated air flux per diffuser. The calculations are detailed below.

When keeping the number of diffusers constant, more air per diffuser must be discharged to compensate for fouling. This means that, in addition to a decreased α (*i.e.*, αF instead of α), there will be a decreased SOTE/Z and thus a further decreased αFSOTE. The procedure for calculating the total airflow rate (AFR) and the airflow rate per diffuser (AFR_{diff}) given the number of diffusers (N_D) is depicted in Figure 9.4. In contrast to the procedure for a fixed AFR_{diff}, the calculation needs to be performed iteratively. It will be illustrated here for peak conditions, performing a single iteration.

In our example the number of diffusers under peak conditions was calculated as 3,810. The oxygen transfer rate (SOTR) at peak conditions amounts to 468 kgO_2/h. In order to calculate the corresponding oxygen mass flow (Eq. 9.9), a value needs to be assumed for the oxygen transfer efficiency αFSOTE. As a first guess, we take its value determined for fouled diffusers under peak conditions, $\alpha FSOTE^{peak}$ = 7.9 %, corresponding with a required oxygen mass flow of $W_{O2}{}^{peak}$ = 5,927 kgO_2/h (Eq. 9.9) and a required airflow rate (Eq. 9.8) of AFR^{peak} = 21,037 m^3/h. The airflow rate per diffuser then follows from rearranging Eq. 9.7:

$$AFR_{diff} = \frac{AFR}{N_D} = \frac{21,037/60}{3,810} = 0.092 \ m^3/min \tag{9.17}$$

which corresponds to an air flux per diffuser of AFR_{diff}/A_{diff} = 2.47 m^3/m^2.min for peak flow conditions. From the given manufacturers' clean water curve (Figure 9.1), we find a corresponding value of $SOTE/Z^{peak}$ of approximately 5.9 %/m. The associated $\alpha FSOTE^{peak}$ is calculated from Eq. 9.11 as 7.76 %, which is lower than the value $\alpha FSOTE^{peak}$ = 7.9 % assumed initially (the 'first guess').

Consequently, the required oxygen mass flow is recalculated as $W_{O2}{}^{peak}$ = 6,028 kgO_2/h (Eq. 9.9) and the corresponding required airflow rate (Eq. 9.8) as AFR^{peak} = 21,393 m^3/h, which are higher than based on the initial assumption for αFSOTE. This is because more air needs to be released by the same number of diffusers, and therefore the diffusers will be operating further on the right of the efficiency and head loss curves.

The higher airflow rate per diffuser also results in a higher required blower discharge pressure. The dynamic wet pressure (DWP) can still be read from Figure 9.2, even though it needs to be compensated for fouling. Indeed, from the given dynamic wet pressure (DWP) curves in Figure 9.2, we can identify a DWP (AFR^{peak}) value of approximately 3.8 kPa at the peak air flux of 2.47 m^3/m^2.min. The resulting blower peak discharge pressure is calculated from Eq. 9.14 as 55.60 kPa.

$$p_{disch}{}^{peak} = \rho \cdot g \cdot Z + h_L + \Psi \cdot DWP_{new, \ diffuser}(AFR^{peak}) = 47.6 + 3.04 + 1.3 \cdot 3.80 = 55.58 \ Pa \tag{9.18}$$

The higher oxygen requirement W_{O2} and the increased peak blower discharge pressure $p_{disch}{}^{peak}$ further result in a higher required blower power BHP^{peak} compared to the case with a higher number of diffusers, namely (Eq. 9.6):

$$BHP^{peak} = \frac{W_{air} \cdot R \cdot T_{in}}{29.7 \cdot n \cdot e}\left[\left(\frac{p^{peak}_{disch,abs}}{p_{in}}\right)^{n} - 1\right] = \frac{\dfrac{W_{O2}}{\hat{y}_{O2}} \cdot R \cdot T_{in}}{29.7 \cdot n \cdot e}\left[\left(\frac{p^{peak}_{disch,abs}}{p_{in}}\right)^{n} - 1\right] =$$

$$\frac{\dfrac{6,028}{3,600} \cdot 0.23 \cdot 8,314 \cdot 293.15}{29.7 \cdot 0.283 \cdot 0.75}\left[\left(\frac{55.58 + 101.325}{0.9 \cdot 101.325}\right)^{0.283} - 1\right] kW = 467\ kW \qquad (9.19)$$

The final note on diffuser flexing: flexing, also known as purging, is the practice of inflating diffusers by feeding airflow higher than the usual maximum range (*e.g.*, 130 – 150 % of maximum operating airflow), as applied in this example. Manufacturers of membrane diffusers recommend this practice as prophylaxis to dilate the pores and delay the ensuing fouling. To carry out flexing, the blower needs to be able to discharge a larger airflow, with the corresponding DWP(AFR). Hence, if this practice is envisioned, the higher DWP(AFR) and its corresponding p_{disch} and BHP can be calculated using the procedure above with the adjusted AFR value. When the total required airflow rate increases because of fouling, it is always necessary to verify that the airflow per diffuser does not fall outside the range of data guaranteed by the manufacturer. In this example, if the daily peak conditions were, for example, 1.5 times the average, it can be quickly verified that the air flux per diffuser would fall outside figs. 9.1 and 9.2 after fouling, if the number of diffusers used were those from the peak conditions for new diffusers. The remedy is to reset the desired airflow per diffuser to a lower value and repeat the calculation. A spreadsheet is an adequate way to perform these calculations; however, when there is a desire to couple them with a dynamic biokinetic model, it is necessary to employ a simulator.

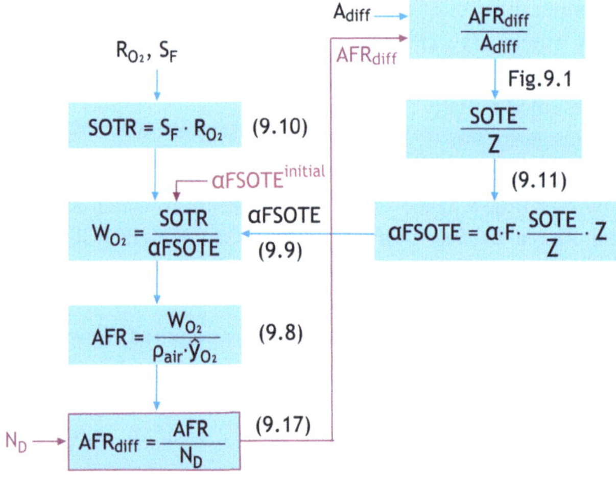

Figure 9.4 Procedure for calculating the total airflow rate (AFR) and the airflow rate per diffuser (AFR$_{diff}$) given the load to be treated (oxygen demand R$_{O2}$) and the number of diffusers (ND). The differences with the procedure to calculate ND for given (AFR$_{diff}$) are indicated in pink.

The results are summarized in Table 9.1.

Table 9.1 Summary of the calculations and results for Example 9.3.1. The calculations are provided in the spreadsheet 'Chapter 9 Design examples.xlsx' on sheet 'Example 9.3.1'.

	New diffusers $F = 1.0$ $\Psi = 1.0$	New diffusers $F = 1.0$ $\Psi = 1.0$	Fouled diffusers $F = 0.8$ $\Psi = 1.3$ N_D constant	Fouled diffusers $F = 0.8$ $\Psi = 1.3$ N_D constant	Unit
Parameter / Conditions	Average	Peak	Average	Peak	
Required oxygen (R_{O2})	8,640	11,232	8,640	11,232	kgO_2/d
Peak factor (S_F)		1.30		1.30	-
SOTR	360	468	360	468	kgO_2/d
α	0.35	0.35	0.35	0.35	-
F	1.00	1.00	0.80	0.80	-
Desired AFR per diffuser (AFR_{diff})	0.06	0.07	0.06	0.07	m^3/min
Diffuser specific area (A_{diff})	0.04	0.04	0.04	0.04	m^2
Air flux per diffuser (AFR_{diff} / A_{diff})	1.52	1.97	1.52	1.97	$m^3/m^2.min$
SOTE/Z	6.20	6.00	6.20	6.00	% / m
αSOTE/Z^{avg}	5.00	5.00	5.00	5.00	%/ft
αSOTE/Z^{avg}	0.30	0.30	0.30	0.30	% / m
Side water depth (SWD)	5.00	5.00	5.00	5.00	m
Diffuser height above floor	0.30	0.30	0.30	0.30	m
Diffuser submergence Z	4.70	4.70	4.70	4.70	m
αFSOTE	10.20	9.87	8.16	7.90	%
Oxygen supplied (W_{O2})	3,530	4,742	4,412	5,927	kgO_2/h
Air density (ρ_{air})	1.23	1.23	1.23	1.23	kg/m^3
$\hat{y}_{O2}$	0.23	0.23	0.23	0.23	kg/kg
Air flow rate (AFR)	12,528	16,829	15,660	21,037	m^3/h
Number of diffusers (N_D)	3,687	3,810	3,810	3,810	-
Actual AFR_{diff}			0.069	0.092	m^3/min
Actual air flux per diffuser			1.84	2.47	$m^3/m^2.min$
Actual SOTE/Z			6.05	5.90	%/m
αFSOTE			7.96	7.76	
Actual W_{O2}			4,522	6,028	kgO_2/h
Actual AFR			16,048	21,393	m^3/h
Hydrostatic head		47.6		47.6	kPa
Air piping head loss		3.04		3.04	kPa
$DWP_{new.diffuser}^{peak}$		3.20		3.80	kPa
Ψ		1.00		1.30	
Discharge pressure required (p_{disch}^{peak})		53.86		55.60	kPa
Blower efficiency e		0.75		0.75	-
Blower inlet temperature (T_{in})		293.15		293.15	K
Blower inlet pressure (p_{in})		91.19		91.19	kPa
Absolute blower discharge pressure ($p_{disch,abs}$)		155.19		156.93	kPa
BHP^{peak}		359		467	kW

Example 9.3.2

Diffuser specification

A wastewater treatment plant consisting of three compartments (3 CSTRs) in series needs new fine-pore diffusers. The OUR breakdown in the three compartments is as follows: 50 %, 35 %, 15 %, corresponding with a progressively lower oxygen requirement along the treatment train. The total airflow in the series of 3 reactors spans the range 6,800 - 13,600 m^3/h.

Calculate the number of diffusers for the three aeration grids for the following cases:

a) Silicone tubes, 1,000 mm long, 50 mm in diameter, with an operating range of 7 - 14 m/min.
b) EPDM discs, 9 in in diameter (229 mm), with an operating range of 3.5 - 5.3 m/min.
c) Polyurethane panels, 200 mm wide, 3.6 m long, with an operating range of 1.75 - 3.50 m/min.

Solution

Since the required airflow rate is given, the number of diffusers can be calculated from Eq. 9.7. The number of diffusers corresponding with the minimum airflow rate per diffuser at minimum airflow conditions is calculated as:

$$N_D^{MIN} = \frac{AFR^{MIN}}{AFR_{diff}^{MIN}} \qquad (9.20)$$

while the number of diffusers corresponding with the maximum airflow rate per diffuser at maximum airflow conditions is calculated as:

$$N_D^{MAX} = \frac{AFR^{MAX}}{AFR_{diff}^{MAX}} \qquad (9.21)$$

The actual number of diffusers, N_D, should be lower than or equal to N_D^{MIN} in order to ensure that the airflow per diffuser remains higher than the minimum boundary of its operating range; at the same time N_D, should be higher than or equal to N_D^{MAX} such that the airflow per diffuser remains lower than the maximum boundary of its operating range:

$$N_D^{MAX} \leq N_D \leq N_D^{MIN} \qquad (9.22)$$

The minimum and maximum airflow are given as AFR^{MIN} = 6,800 m^3/h and AFR^{MAX} = 13,600 m^3/h. The minimum and maximum airflow rate per diffuser can be calculated from the specified minimum and maximum air flux for each type of diffuser, taking into account the diffuser-specific area A_{diff}, which in its turn is determined by the specified diffuser geometry and dimensions. The results are summarized in Table 9.2. The calculations are provided in the spreadsheet 'Chapter 9 Design examples.xlsx' on sheet 'Example 9.3.2'.

Given that the minimum and maximum airflow rate per diffuser for the silicone tubes (case a) and the polyurethane panels (case c) are perfectly aligned with the range of total airflow rates that needs to be provided, the number of diffusers N_D^{MIN} and N_D^{MAX} is the same. The number of diffusers is found to be

$N_D = 324$ for the silicone tubes and $N_D = 90$ for the polyurethane panes. Their distribution over the three compartments is performed proportionally to the required OUR. As for the EPDM discs, it is found that $N_D^{MIN} = 780$ and $N_D^{MAX} = 1{,}040$. As $N_D^{MAX} > N_D^{MIN}$ for this case, it is not possible to satisfy both the minimum and maximum airflow rate per diffuser requirements for the given range of airflow rates that needs to be provided. As a result, this type of diffuser is not an option for the WWTP under study.

Table 9.2 Summary of the results for Example 9.3.2. The calculations are provided in the spreadsheet 'Chapter 9 Design examples.xlsx' on sheet 'Example 9.3.2'.

	Silicone tubes	EPDM discs	Polyurethane panels	Unit
AFR^{MIN}	6,800	6,800	6,800	m³/h
AFR^{MAX}	13,600	13,600	13,600	m³/h
Air flux^{MIN}	7.0	3.5	1.75	m/min
Air flux^{MAX}	14.0	5.3	3.50	m/min
A_{diff}	0.05	0.042	0.72	m²
AFR_{diff}^{MIN}	21.00	8.7	75.6	m³/h
AFR_{diff}^{MAX}	42.00	13.1	151	m³/h
N_{diff}^{MIN}	324	780	90	
N_{diff}^{MAX}	324	1,040	90	
$N_{diff}\,1$	162		45	
$N_{diff}\,2$	114		32	
$N_{diff}\,3$	49		14	

Example 9.3.3

Aeration energy dynamics and costs

The costs for electrical energy (power) to operate the blowers varies during the day (Figure 9.5), as does the energy demand of the WWTP itself, the latter corresponding to the load variations.

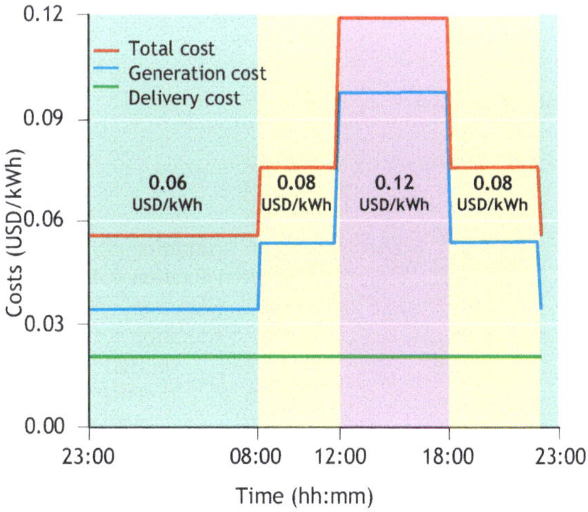

Figure 9.5 Daily variation in power generation and delivery costs, and the resulting total energy cost.

Reconsider the case from Example 9.3.1, for new diffusers. Calculate the power demand and associated power costs during the day, as well as the annual power cost to operate the aeration system. Use the case of new diffusers. To calculate costs, use the tariff structure from Figure 9.5. Assume that the period with the highest total power cost (USD0.12/kWh, 12:00-18:00) requires peak aeration flow rate, the periods with medium power cost (USD0.08/kWh, 8:00-12:00 and 18:00-23:00) corresponds to the average aeration flow rate, and the period with the lowest total power cost (USD0.06/kWh, 23:00-8:00) is also the one in which only a low aeration flow rate is required, *i.e.*, during the night. As for the low (nightly) aeration flow rate, assume that it equals the average airflow divided by the safety factor $S_F = 1.3$).

Solution

The blower power demand (BHP) is calculated using the adiabatic compression formula (Eq. 9.6). The BHP corresponding with the peak AFR has already been calculated in Example 9.3.1 as $BHP^{peak} = 359$ kW (Eq. 9.16); the BHP for low and average flow conditions is calculated analogously as $BHP^{low} = 206$ kW and $BHP^{avg} = 268$ kW, respectively. The power costs for each time interval are obtained by multiplying the duration of each time interval (h) with the corresponding BHP value (kW) and the corresponding energy cost (USD/kWh); the summation gives the power cost per day. The results are summarized in Table 9.3. The calculations are provided in the spreadsheet 'Chapter 9 Design examples.xlsx' on sheet 'Example 9.3.3'. It is striking how the high-power cost during the peak period (12:00-18:00) substantially increases the overall power costs: this period counts for a quarter of the day, but half of the aeration costs.

Table 9.3 Summary of the results for Example 9.3.3. The calculations are provided in the spreadsheet 'Chapter 9 Design examples.xlsx' on sheet 'Example 9.3.3'.

Parameter / Conditions	New diffusers F=1.0 Ψ=1.0	New diffusers F=1.0 Ψ=1.0	Unit
	Low-load	Average	Peak	
Peak factor (S_F)			1.30	-
Diffuser submergence Z	4.70	4.70	4.70	m
Oxygen supplied (W_{O2})	2,715	3,530	4,742	kgO_2/h
Air density (ρ_{air})	1.225	1.225	1.225	kg/m^3
$\hat{y}_{O2}$	0.23	0.23	0.23	kg O_2/kg air
Air flow rate (AFR)	9,637	12,528	16,829	m^3/h
Discharge pressure required (p_{disch})	53.86	53.86	53.86	kPa
Absolute blower discharge pressure ($p_{disch,abs}$)	155.19	155.19	155.19	kPa
Blower efficiency e	0.75	0.75	0.75	
Blower inlet temperature (T_{in})	293.15	293.15	293.15	K
Blower inlet pressure (p_{in})	91.19	91.19	91.19	kPa
BHP	206	268	360	kW
Unit cost (23:00 - 08:00)	0.06	(duration = 9 h)		USD/kWh
Unit cost (08:00 - 12:00)	0.08	(duration = 4 h)		USD/kWh
Unit cost (12:00 - 18:00)	0.12	(duration = 6 h)		USD/kWh
Unit cost (18:00 - 23:00)	0.08	(duration = 5 h)		USD/kWh
Cost (23:00 - 08:00)	111.2	($@BHP^{min}$)		USD/d
Cost (08:00 - 12:00)	85.6	($@BHP^{avg}$)		USD/d
Cost (12:00 - 18:00)	258.9	($@BHP^{peak}$)*		USD/d
Cost (18:00 - 23:00)	107.1	($@BHP^{avg}$)		USD/d
Total daily cost	562.7			USD/d
Total yearly cost	205,390			USD/yr

*Note that peak cost is roughly half of total

Example 9.3.4

Mixing and energy

Quantify the minimum power requirement to maintain mixed liquor suspended with a mechanical mixer, and the minimum airflow requirement to maintain the same mixed liquor suspended with bubbles.

Additional information A4: mixer design

Mixer design is typically based on a target power level, which may be expressed in terms of W/m^3 for mechanical mixers or air flux, $Nm^3/h.m^2$, for air-based mixing. The power level method quantifies mixing from the mixing power input, regardless of the distribution of velocity gradients. A disadvantage of the power level or air flux approach is that designers often treat homogeneity as the *only* goal of mixing, so that power levels are often increased to achieve ever higher levels of homogeneity. However, overmixing also causes disadvantages, from shearing biological flocs to using too much air, because mixing results in high DO concentrations and this can compromise BNR process performance. Another consideration is that the power level method makes no distinction between different impellers. This means that specifying a mixer in this way creates a strong incentive for the equipment vendor to supply a cheap and inefficient impeller. Refer to Grenville *et al.* (2017) for comprehensive descriptions of the unique characteristics of each impeller.

As an alternative to the power level method, mixer design can be based on flocculation, characterized by the velocity gradient G as a key design parameter. Wahlberg *et al.* (1994) developed a method for this approach and provided a set of experimental data, which can be used to determine best, average and worst-case design requirements. More specifically, for any given HRT there is an optimum G, where flocculation is just completed, and the minimum possible effluent turbidity is attained. Combining this method with the dataset, an average, minimum (for the best performing flocculating system) and maximum (for the worst flocculating system) G value can be determined, as given in Table 9.4.

Table 9.4 Example of optimum velocity gradient G for flocculation (adapted from Wahlberg *et al.*, 1994).

| HRT (min) | Velocity gradient G (1/s) | | | | | |
| | Plug flow | | | Completely mixed | | |
	Min.	Avg.	Max.	Min.	Avg.	Max.
5	4.32	10.4	16.3	4.6	12.3	22.8
10	2.89	7	10.7	3.5	9.4	16.8
20	1.82	4.5	9	2.7	7.1	12.2
30	1.35	3.4	7.7	2.2	6	10.1
45	1	2.5	6.4	1.9	5	8.4
60	0.8	2	5.5	1.6	4.4	7.3

It is clear from Table 9.4 that there is considerable variation between the best, average and worst flocculation systems in terms of optimum G value, suggesting that there is considerable benefit in doing site-specific testing to determine the flocculation characteristics of a specific system prior to design. The plug flow system has a lower optimum G value compared to the completely mixed system. Further optimization of the design is possible if a tapered velocity gradient is employed in a plug flow system, or by placing a number of completely mixed zones in series with lower G values in the downstream zones.

From the optimal G values (1/s) for a given system (Table 9.4), the corresponding power levels at 20 °C for mechanical mixers can be calculated using the definition of G for the mixing power P_{mix} (W) per unit liquid volume V (m^3) (Metcalf and Eddy, 2014):

$$G = \left(\frac{P_{mix}}{\mu_w V}\right)^{1/2} \tag{9.23}$$

$$\frac{P_{mix}}{V} = G^2 \cdot \mu_w \tag{9.24}$$

in which μ_w is the dynamic viscosity of water ($\mu_w = 1.0005 \cdot 10^{-3}$ N.s./m^2).

The G values from Table 9.4 can also be used to calculate the air flux required for mixing with air-powered mixers. The G values correspond to mixing power P_{mix} (kW), calculated through Eq. 9.24, which in the case of air mixing is the power corresponding to the isothermal work of expansion of the rising bubbles (Schroeder, 2021):

$$\frac{P_{mix}}{V} = K \cdot AFR \cdot \ln\left(\frac{Z + 10.33}{10.33}\right) \tag{9.25}$$

$$AFR = \frac{\dfrac{P_{mix}}{V}}{K \cdot \ln\left(\dfrac{Z + 10.33}{10.33}\right)} \tag{9.26}$$

Where K is a constant (1.689 in SI units) and AFR (Nm^3_{air}/ $m^3_{reactor}$.min) is the required airflow through the diffusers installed at a submergence Z (m).

Solution

The minimum power requirements P_{mix}/V to maintain mixed liquor suspended with a mechanical mixer correspond to the optimum G values for flocculation from Table 9.4 and are calculated from Eq. 9.24. The results are summarized in Table 9.5. Please note that the units are W/m^3.

Table 9.5 Power levels for optimum flocculation.

HRT (min)	Power level (W/$m^3_{reactor}$)					
	Plug flow			Completely mixed		
	Min.	Avg.	Max.	Min.	Avg.	Max.
5	1.87E-02	1.08E-01	2.66E-01	2.12E-02	1.51E-01	5.20E-01
10	8.36E-03	4.90E-02	1.15E-01	1.23E-02	8.84E-02	2.82E-01
20	3.31E-03	2.03E-02	8.10E-02	7.29E-03	5.04E-02	1.49E-01
30	1.82E-03	1.16E-02	5.93E-02	4.84E-03	3.60E-02	1.02E-01
45	1.00E-03	6.25E-03	4.10E-02	3.61E-03	2.50E-02	7.06E-02
60	6.40E-04	4.00E-03	3.03E-02	2.56E-03	1.94E-02	5.33E-02

Note: A key difference between the laboratory-scale flocculation test and full-scale applications is that at the smaller scale the G value can be assumed to be constant for the entire test volume, but in a full-scale application, local G values would be much higher in the vicinity of the impeller and much lower in corners and at other extremes (Wahlberg *et al.*, 1994; Pretorius *et al.*, 2015). The designer needs to make some allowance for this variation in G value and power level. One approach would be to use computational fluid dynamics to gain an understanding of how much these parameters can vary throughout the reactor zone.

The minimum airflow rate per unit reactor volume required to maintain the same liquid suspended with bubbles follows from Eq. 9.26;   Table 9.6 summarizes the results for a typical diffuser submergence Z of 5 m.

Table 9.6 Optimum airflow for air-powered mixers corresponding to the same mixing as the mechanical mixers from Table 9.5.

HRT (min)	Air flow ($Nm^3_{air}/m^3_{reactor}.min$)					
	Plug flow			Completely mixed		
	Min.	Avg.	Max.	Min.	Avg.	Max.
5	6.45E-02	3.74E-01	9.18E-01	7.31E-02	5.23E-01	1.80E+00
10	2.89E-02	1.69E-01	3.96E-01	4.23E-02	3.05E-01	9.75E-01
20	1.14E-02	7.00E-02	2.80E-01	2.52E-02	1.74E-01	5.14E-01
30	6.30E-03	3.99E-02	2.05E-01	1.67E-02	1.24E-01	3.52E-01
45	3.46E-03	2.16E-02	1.42E-01	1.25E-02	8.64E-02	2.44E-01
60	2.21E-03	1.38E-02	1.05E-01	8.85E-03	6.69E-02	1.84E-01

Note: Air-powered mixing has the potential to approach uniform distribution of mixer power throughout the reactor zone. However, at these low air flows, the challenge would be to design a diffuser layout that would achieve full floor coverage, without dead zones, while adhering to equipment limitations. In fact, when designing an aeration system for oxygen transfer, it is immediately clear that the minimum mixing requirements are far exceeded using the airflow rates necessary to aerate the process. This is because oxygen transfer is very inefficient and only a small aliquot of each bubble is transferred during the bubble rise, while the mixing benefits from the entire bubble volume expanding.

9.4 EXERCISES
Blower sizing (exercises 9.4.1-9.4.3)
Exercise 9.4.1
What is the required pressure to guarantee air discharge during operations?

Exercise 9.4.2
What happens when the pressure requirements approach the maximum discharge pressure of the blowers?

Exercise 9.4.3
Explain what diffuser flexing is. What is the required pressure to effectively perform diffuser flexing at 130 % of the maximum process airflow?

Oxygen transfer rate (exercises 9.4.4-9.4.6)
Exercise 9.4.4
What is the difference between the oxygen transfer rate (OTR) and the standard oxygen transfer rate (SOTR) if the test is conducted at 15 °C and 700m AMSL (*i.e.,* height above mean sea level)?

Exercise 9.4.5
What happens to the k_La if we use excessively slow DO sensors?

Exercise 9.4.6
Can we use plant effluent to perform a clean water test in order to determine the k_La?

Aeration system specification (exercises 9.4.7-9.4.9)
Exercise 9.4.7
What are the main types of aerators and what are their main differences?

Exercise 9.4.8
If cleaning is not an option at a given facility, what are the challenges that the operators will face?

Exercise 9.4.9
What is the cleaning frequency for fine-pore diffusers?

Aeration efficiency (exercises 9.4.10-9.4.12)
Exercise 9.4.10
Define aeration efficiency.

Exercise 9.4.11
With regard to the effect of process design on aeration efficiency, briefly describe two favourable design criteria or configurations useful for improving the aeration efficiency in wastewater treatment plants.

Exercise 9.4.12
With regard to the effect of process operation on aeration efficiency, describe three operating factors that adversely affect (reduce or deteriorate) the aeration efficiency in wastewater treatment plants.

Aeration energy – oxygen transfer efficiency (exercises 9.4.13-9.4.14)
Exercise 9.4.13
What is the process performance for an aeration system operating at constant airflow and variable process loading (hence, variable DO)?

Exercise 9.4.14
What should the diurnal curves for process load, airflow rate, and blower power demand look like?

ANNEX 1: SOLUTIONS TO EXCERSISES
Blower sizing (solutions 9.4.1-9.4.3)
Solution 9.4.1
To guarantee that the air is released, the discharge pressure of the blower (p_{disch}, relative to atmospheric pressure) must exceed the pressure requirements of the system, consisting of the sum of the pressure losses in the system:

$$p_{disch} \geq \rho \cdot g \cdot Z + h_L + DWP(AFR) \quad (Eq.9.5)$$

The pressure losses on the right-hand side of Eq. 9.5 are the following:
- the hydrostatic head loss $\rho \cdot g \cdot Z$ due to the diffuser submergence Z;
- the head loss of the air distribution line (friction head h_L), which for practical purposes may be considered independent of the airflow rate;
- the dynamic wet pressure (DWP), *i.e.,* the diffuser head loss, which is a function of the airflow rate (AFR) and needs to be provided by the manufacturer.

Solution 9.4.2
When the pressure requirements approach the maximum discharge pressure of the blowers, the blowers begin to surge. The surge zone is an area of operation and must be avoided because the vibrations that the blowers are subject to may cause structural failure. An automated system shuts down the blowers when the search begins.

Solution 9.4.3
Diffuser flexing, also known as purging, is the practice of inflating membrane diffusers by feeding airflow higher than the usual maximum range (*e.g.,* 130-150 % of maximum operating airflow), in order to dilate the pores of the membrane diffusers and therefore delay the occurrence of fouling. The required pressure for flexing is calculated from Eq. 9.5 using the DWP related to the envisaged airflow, in this case, 130 % of the maximum process airflow.

Oxygen transfer rate (solutions 9.4.4-9.4.6)
Solution 9.4.4
The OTR is calculated from the dissolved oxygen time series of reiteration in field conditions. In this case field conditions are 15 °C and 700 m above sea level. The standardized OTR, SOTR, is the value corrected for 20 °C and 1 atm. For more information on this procedure, refer to *e.g.,* ASCE (2018).

Solution 9.4.5
Answer 9.4.2.2 Using excessively slow DO sensors causes a delay between the actual dissolved oxygen and the reading of that value by the sensor. The delay is very visible at the beginning of the re-aeration process, *i.e.,* when the slope of the re-aeration curve (DO vs. time) is very high and any error on the horizontal axis corresponds to large errors on the vertical axis. This can be visibly observed because the fitting curve, which is an ideal exponential, will differ from the experimental data, which will be more of a sigmoid. At the end of re-aeration, when the curve is almost horizontal because it is approaching the asymptote (DO saturation), the speed of the DO sensor is immaterial. Hence, a slow DO sensor results in underestimation of $k_L a$.

Solution 9.4.6
No, we cannot, because plant effluent still contains residual organics that would alter the result and produce lower $k_L a$ than tap water.

Aeration system specification (exercises 9.4.7-9.4.9)

Solution 9.4.7

There are two main types of aeration technologies: the first is surface aeration and the second is submerged aeration.

- Surface aeration is mechanical and relies on impellers.

Subtypes: High-speed versus low-speed surface aerators, with horizontal or vertical shafts.

- Submerged aeration relies on bubbles released throughout the floor by coarse-bubble or fine-bubble diffusers, or by mechanical devices (turbines or jets).

Subtypes: Coarse-bubble versus fine-bubble aeration systems.

Solution 9.4.8

When cleaning is not practised or not possible, fine-pore diffusers foul and the pressure drop of these diffusers increases. At the same time, the diffusers experience a decline in performance, measured as αFSOTE so αF decreases over time. The consequence is an increase in the requirements for airflow, blower energy and pressure, with potential blower failure if pressure requirements become excessive. Cleaning is always recommended but when it cannot be practised, technologies immune from fouling, such as coarse-bubble diffusers or mechanical aeration systems, are recommended.

Solution 9.4.9

Experience and documented measurements in the field suggest that the diffusers should be cleaned at least once every 24 months, and preferably once a year.

Aeration efficiency (exercises 9.4.10-9.4.12)

Solution 9.4.10

The aeration efficiency (AE, expressed in kgO_2/kWh) is defined by the oxygen transfer rate OTR (in kg/h) relative to the power P (in kW) drawn by the aeration system (Eq. 9.4 in Chen *et al.*, 2020): AE = OTP/R.

The AE is a measure for the efficiency of the aeration system, in contrast to the OTR, which only defines the capacity of the aeration system

Solution 9.4.11

The aeration efficiency in wastewater treatment plants can be improved by the following design measures:

- Use of anoxic or anaerobic selectors: they result in improved (higher) alpha factors, probably by uptake of soluble contaminants (surfactants) into the biomass.
- Relatively high sludge retention time (SRT): higher SRT systems, which operate with higher biomass concentrations, remove or sorb the surfactants early in the process, improving the average oxygen transfer efficiency and outweighing the increasing oxygen requirement for increasing SRT.

Solution 9.4.12

- Type of aerators used: fine-pore diffusers are the most efficient, coarse bubble diffusers are much less efficient; surface aerators are also less efficient.
- Fouling: the oxygen transfer efficiency of fine-pore diffusers decreases over time. Cleaning fine-pore diffusers is almost always required and restores process efficiency and reduces power costs. Other aerator types are less prone to fouling.
- Surfactants: surface active agents accumulate at the air-water interface of rising bubbles, increasing the rigidity of the interface and reducing internal gas circulation and overall transfer rate.

Aeration energy – oxygen transfer efficiency (exercises 9.4.13-9.4.14)

Solution 9.4.13

An aeration system operating a constant airflow will over-deliver air when process loading is below average and under-deliver air when the process loading is above average. Although the power demand and energy requirement for this aeration system is constant because the airflow is constant, the process performs poorly at peak loading (*i.e.*, when it is most needed) and excessively well at minimum loading. This is not the correct way to operate a process and should be avoided

Solution 9.4.14

Usually, a plant experiences a peak in process loading during the day. Responding to the peak loading, the aeration control system will discharge more airflow, consequently demanding more blower power. However, at peak loading, the plant receives a higher concentration of contaminants that depress the alpha factor, requiring more airflow in proportion to reach the same oxygen transfer. Moreover, at higher airflows, the oxygen transfer efficiency (*i.e.*, OTE or SOTE, %), which is the ratio between the oxygen transfer rate and the mass flow delivered to the system (OTE = OTR/WO$_2$, Eq. 9.6 in Chen *et al.*, 2020) is lower. The compounding effect is such that at peak loading the process requires proportionally much more airflow and much more power demand. This phenomenon is called peak loading amplification and is illustrated below.

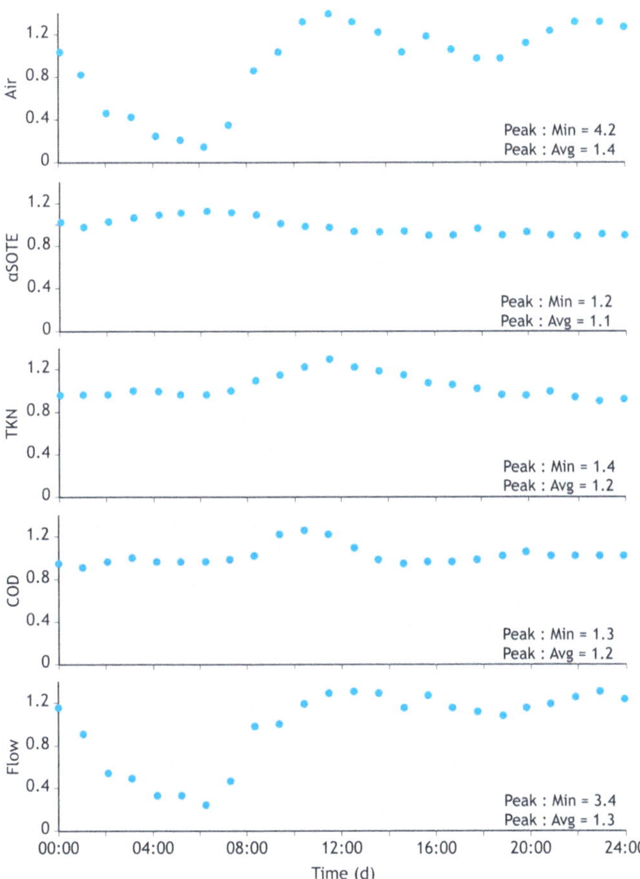

Figure 9.6 Example of daily patterns of influent flow, BOD, TKN, oxygen transfer efficiency (αSOTE) and airflow *vs.* time, all normalized against their daily averages. Note the amplification of air requirements, due to the compounding of flow and load with αSOTE variations. Image from Emami *et al.* (2018).

REFERENCES

ASCE - American Society of Civil Engineers (2018). ASCE/EWRI 18-18 Standard Guidelines for In-Process Oxygen Transfer Testing, 2nd ed. Reston, VA: American Society of Civil Engineers.

Chen G.H., van Loosdrecht M.C.M., Ekama G.A. and Brdjanovic D. (eds.). (2020). Biological Wastewater Treatment: Principles, Modelling and Design. ISBN: 9781789060355. IWA Publishing, London, UK.

Emami N., Sobhani R. and Rosso D. (2018). Diurnal variations of the energy intensity and associated greenhouse gas emissions for activated sludge processes. *Water Science & Technology,* 77(7) 1838-1850. https://doi.org/10.2166/wst.2018.054

Grenville R.K., Giacomelli J.J., Padron G. and Brown D.A.R. (2017). Impeller performance in stirred tanks, *Chemical Engineering,* April 2017, pp.42-51.

Metcalf & Eddy (2014). Wastewater Engineering: Treatment and Resource Recovery. 5th Edition, McGraw-Hill, New York.

Pretorius C., Wicklein E., Rauch-Williams T., Samstag R. and Sigmon C. (2015). How oversized mixers became an industry standard, in: *Proceedings of the Water Environment Federation.* pp. 4379–4411.

Schroeder D.V. (2021). An Introduction to Thermal Physics, Oxford University Press, ISBN 0192895540.

Wahlberg E.J., Keinath T.M. and Parker D.S. (1994). Influence of activated sludge flocculation time on secondary clarification, *Water Environment Research*, pp. 779-786

NOMENCLATURE

Symbol or abbreviation	Description	Unit
A_{diff}	Specific area of each diffuser	m^2
AE	Aeration efficiency in clean water	kgO_2/kWh
AFR	Airflow rate	m^3/s
AFR_{diff}	Desired airflow rate per diffuser – specified by manufacturer	m^3/s
BHP	Blower brake horsepower	kW
DO	Dissolved oxygen in water	kgO_2/m^3
DWP	Dynamic wet pressure	Pa
e	Blower efficiency	-
F	Fouling factor	-
g	Gravitational constant (= 9.81)	m/s^2
G	Velocity gradient	1/s
h_L	Hydrostatic head loss	Pa
k_La	Liquid side mass transfer coefficient	1/h
n	Empirical constant (= 0.283 for air)	-
N_D	Number of diffusers	-
OTE	Oxygen transfer efficiency in clean water	%
OTR	Oxygen transfer rate in clean water	kgO_2/h
K	Empirical constant	S.I. units
p_{in}	Absolute inlet pressure	Pa
p_{disch}	Discharge pressure of the blower (relative to atmospheric pressure)	Pa
$p_{disch,abs}$	Absolute discharge pressure of the blower	Pa
P_{mix}	Mixing power	W
P	Power drawn by the aeration system	kW
R	Universal gas constant (= 8.314)	J/mol.K
R_{O2}	Average required oxygen mass flow	kgO_2/s

S_F	Safety factor	-
SOTE	Oxygen transfer efficiency in standard conditions in clean water	%
SOTR	Oxygen transfer rate in standard conditions in clean water	kgO_2/h
SRT	Sludge retention time	days
T_{in}	Absolute inlet temperature	K
V	Water volume	m^3
W_{air}	Air mass flow rate	kg/s
W_{O2}	Oxygen mass flow fed to aeration tank	kgO_2/s
$\hat{y}_{O2}$	Weight fraction of oxygen in air	wt%
Z	Diffuser submergence	m

Superscripts	Description
avg	Average
MAX	Maximum
MIN	Minimum

Greek symbol	Description	Unit
α	Ratio of process- to clean-water mass transfer	-
αFSOTE	Oxygen transfer efficiency in standard conditions in process water for used diffusers	%
αSOTE	Oxygen transfer efficiency in process water at standard conditions in process water	%
μ_w	Dynamic viscosity of water	$N.s/m^2.$
ρ	Water density	kg/m^3
ρ_{air}	Air density	kg/m^3
Ψ	Pressure factor	-

Aeration has been used for centuries in wastewater treatment. Still, scientist and engineers are continuously searching for new ways to increase its efficiency, save energy, reduce costs, minimise emissions, and protect the environment (photo: D. Brdjanovic).

doi: 10.2166/9781789062304_0341

10

Bulking sludge

Eveline I.P. Volcke, Laurence Strubbe, Carlos M. Lopez-Vazquez and Mark C.M. van Loosdrecht

10.1 INTRODUCTION

Overall, a good separation (settling) and compaction (thickening) of activated sludge in the secondary clarifier are necessary conditions to guarantee a good and efficient operation of the wastewater treatment plant. When there is excessive growth of filamentous bacteria, the settling properties of the sludge are strongly reduced. This phenomenon, described as bulking sludge, is a common and long-standing problem for activated sludge processes where suspended solids cannot be retained in the settler and compaction is hampered affecting the overall efficiency of the plant.

Chapter 10 on bulking sludge in the textbook *Biological Wastewater Treatment: Principles, Modelling and Design* (Chen *et al.*, 2020) presents an overview of relevant historical aspects of the development of activated sludge systems and their relationship with the occurrence of filamentous bulking sludge. Fundamentals on the relationship between morphology and ecophysiology are analysed as well as the identification of the filamentous bacteria involved. Different theories formulated to explain the filamentous bulking sludge are discussed on the basis of lab- and full-scale observations and remedial actions to control and suppress the growth of filamentous organisms are presented. Finally, recent developments in mathematical modelling and aerobic granular sludge are introduced.

10.2 LEARNING OBJECTIVES

After the successful completion of this chapter, the reader will be able to:

- Understand how the sludge volume index is related to the settler area.
- Understand which factors govern the sludge volume index value and how to minimize sludge volume index by process design and operation.

- Calculate the sludge loading rate and contact time of a given selector and apply the selector design guidelines for typical operating conditions and influent flow characteristics of a conventional activated sludge wastewater treatment plant.
- Relate control of bulking sludge to aerobic granular sludge formation.

10.3 EXAMPLES

Wastewater treatment plant under study

Consider a conventional activated sludge wastewater treatment plant (WWTP), named WWTP1. The influent characteristics and the WWTP design and operating conditions are summarized in Table 10.1.

Table 10.1 Influent characteristics and WWTP design and operating conditions. Note that the subscript 'in' here refers to the influent, while the subscript 'i' refers to the selector compartment (i=1,2,3).

Description	Symbol	Value	Unit
Influent flow rate	Q_{in}	14,500	m^3/d
Influent total COD concentration	COD_{in}	816	10^{-3} kgCOD/m
Influent biodegradable COD concentration	$COD_{b,in}$	636	-
Readily biodegradable COD fraction of the influent biodegradable COD	f_{SS}	0.3	-
Influent total Kjeldahl nitrogen concentration	TKN_{in}	50	$gN.m^{-3}$
Influent orthophosphate concentration	$P_{IN,in}$	13	$gP.m^{-3}$
Total reactor volume	V_R	14,650	m^3
Volume of the 1st compartment in the selector	V_{S1}	250	m^3
Volume of the 2nd compartment in the selector	V_{S2}	500	m^3
Volume of the 3rd compartment in the selector	V_{S3}	1,000	m^3
Sludge retention time	SRT	8	d
Total suspended solids mass	MX_{TSS}	51,000	kgTSS
Sludge VSS:TSS ratio	f_{VT}	0.85	kgVSS/kgTSS
Sludge underflow recycle ratio	s	1:1	-

In what follows, the selector design will be assessed for a plant with organic matter removal only (Example 10.3.1), a plant where biological nitrogen removal via nitrification-denitrification takes place in a Modified Ludzack-Ettinger (MLE) process configuration (Example 10.3.2) and a plant where the aerobic selector is replaced by an anaerobic selector (Example 10.3.3).

Aerobic selector design

Example 10.3.1

The WWTP under study has been designed for organic matter removal (only). To this end, all tanks are aerobic, including the three selector compartments (Figure 10.1). A minimum oxygen concentration of 2 g/m^3 is maintained along the aerobic tanks.

1. What is the sludge loading rate (SLR) in each of the three compartments of the aerobic selector?
2. What is the actual contact time in the aerobic selector?

3. Based on the previous operational data and on the selector design guidelines recommended for aerobic selectors in municipal wastewater treatment systems (Table 10.2 in the textbook Chen *et al.*, 2020), does the aerobic selector provide favourable conditions to reduce the occurrence of filamentous bulking sludge? If not, (how) can this be remedied?

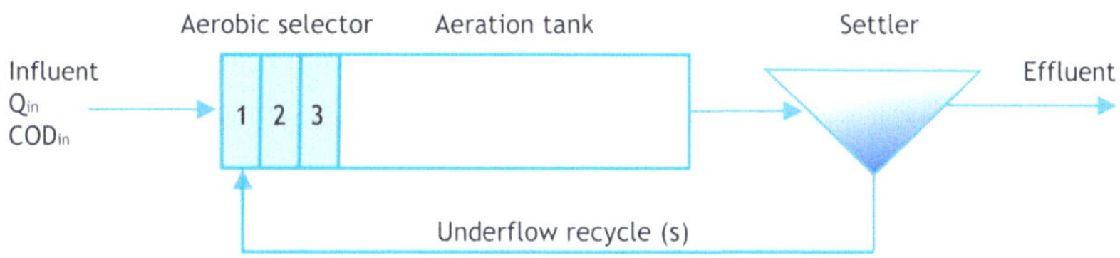

Figure 10.1 Layout of a WWTP with aerobic selector (Example 10.3.1). 'Comp.' refers to 'compartment'. The same configuration holds for the WWTP with anaerobic selector from Example 10.3.3, only then the selector compartments are not aerated, resulting in anaerobic conditions.

Solution

All calculations discussed below are provided in the spreadsheet 'Chapter 10 Design examples.xlsx' on the sheet 'aerobic selector'.

1. The sludge loading rate in selector compartment i (SLR_i, in kgCOD/kgTSS.d) is calculated as the ratio between the incoming COD load ($FCOD_i$, in kgCOD/d) and the TSS mass in that selector compartment (MX_{TSSi}, in kg TSS):

$$SLR_i = \frac{FCOD_i}{MX_{TSSi}} \tag{10.1}$$

The amount of TSS in selector compartment i is calculated according to Eq. 10.2, assuming an even distribution of the TSS over all the reactor compartments, *i.e.*, the three selector compartments (with volume V_{Si}, i=1,2,3) and the bioreactor (with volume V_R):

$$MX_{TSS,i} = V_{Si} / V_R \cdot MX_{TSS} \tag{10.2}$$

which yields:

$$MX_{TSS,1}: 250 \text{ m}^3 / 14{,}650 \text{ m}^3 \cdot 51{,}000 \text{ kgTSS} = 870 \text{ kgTSS} \tag{10.3}$$

$$MX_{TSS,2}: 500 \text{ m}^3 / 14{,}650 \text{ m}^3 \cdot 51{,}000 \text{ kgTSS} = 1{,}741 \text{ kgTSS} \tag{10.4}$$

$$MX_{TSS,3}: 1{,}000 \text{ m}^3 / 14{,}650 \text{ m}^3 \cdot 51{,}000 \text{ kgTSS} = 3{,}481 \text{ kgTSS} \tag{10.5}$$

Note: the TSS concentration is the same everywhere, in the reactor as well as in all the selector compartments, and is calculated as:

$$X_{TSS} = MX_{TSS} / V_R \tag{10.6}$$

$$X_{TSS} = 51,000 \text{ kgTSS} / 14,650 \text{ m}^3 = 3.5 \text{ kgTSS/m}^3 \tag{10.7}$$

The COD load on the first selector compartment equals the influent COD load:

$$FCOD_1 = FCOD_{in} = Q_{in} \cdot COD_{in} \tag{10.8}$$

which amounts to:

$$FCOD_1 = 14.5 \cdot 10^6 \text{ l/d} \cdot 816 \text{ mgCOD/l} = 11,832 \text{ kgCOD/d} \tag{10.9}$$

The COD load on the second selector compartment should strictly speaking be calculated as the influent COD load minus the load of COD which has been degraded in the first selector compartment. However, it is not exactly known how much COD is degraded in each selector compartment. In fact, the selector design guidelines (Table 10.2 in the textbook) are such that only the readily biodegradable COD is converted. In this example, the readily biodegradable COD load amounts to:

$$FRBCODin = Q_{in} \cdot f_{SS} \cdot COD_{b,in} \tag{10.10}$$

$$FRBCODin = 2,769 \text{ kgCOD/d} \tag{10.11}$$

which represents 23 % of the influent load. Given that the fraction of readily biodegradable COD only represents a relatively small fraction of the influent COD load, the COD load on the second and third selector compartment will be approximated by the influent COD load:

$$FCOD_3 = FCOD_2 = FCOD_1 = FCOD_{in} \tag{10.12}$$

The sludge loading rates in the three selector compartments are thus calculated from Eq. 10.1 as:

$$SLR_1 = \frac{FCOD_1}{MX_{TSS1}} = \frac{11,832 \text{ kgCOD/d}}{870 \text{ kg/TSS}} = 13.6 \text{ kgCOD/kgTSS.d} \tag{10.13}$$

$$SLR_2 = \frac{FCOD_2}{MX_{TSS2}} = \frac{11,832 \text{ kgCOD/d}}{1,741 \text{ kg/TSS}} = 6.8 \text{ kgCOD/kgTSS.d} \tag{10.14}$$

$$SLR_3 = \frac{FCOD_3}{MX_{TSS3}} = \frac{11,832 \text{ kgCOD/d}}{3.481 \text{ kg/TSS}} = 3.4 \text{ kgCOD/kgTSS.d} \tag{10.15}$$

2. The actual contact time in the ith aerobic selector compartment is equal to the actual hydraulic residence time, $HRT_{a,i}$ (d) in that compartment:

$$HRT_{a,i} = V_{Si} / (Q_{in} \cdot (1 + s)) \tag{10.16}$$

which is calculated as:

$$HRT_{a,1} = 250 \text{ m}^3 / 14{,}500 \text{ m}^3/\text{d} / (1 + 1) \cdot 24 \text{ h/d} \cdot 60 \text{ min/h} = 12 \text{ min} \qquad (10.17)$$

$$HRT_{a,2} = 500 \text{ m}^3 / 14{,}500 \text{ m}^3/\text{d} / (1 + 1) \cdot 24 \text{ h/d} \cdot 60 \text{ min/h} = 25 \text{ min} \qquad (10.18)$$

$$HRT_{a,3} = 1{,}000 \text{ m}^3 / 14{,}500 \text{ m}^3/\text{d} / (1 + 1) \cdot 24 \text{ h/d} \cdot 60 \text{ min/h} = 60 \text{ min} \qquad (10.19)$$

Thus, the total contact time in the aerobic selector is 87 min.

3. The selector design guidelines for aerobic selectors in municipal wastewater treatment systems are listed in Table 10.2. The single parameter that remains to be calculated is the floc loading in the first compartment. The floc loading expresses the mass of organic matter per mass of TSS and is calculated as the total COD concentration entering the first selector compartment, taking into account the dilution by the recycle flow, divided by the total suspended solids concentration, according to Eq. 10.20:

$$\text{Floc loading in first selector compartment} = COD_{in} / (1 + s) / X_{TSS} \qquad (10.20)$$

In this case, Eq. 10.20 is evaluated as:

$$\text{Floc loading in first selector compartment} =$$
$$816 \text{ g COD/m}^3 / (1 + 1) / 3.5 \text{ kgTSS/m}^3 = 117 \text{ gCOD/kgTSS} \qquad (10.21)$$

Table 10.2 Comparison between operational data of the wastewater treatment plant under study and the selector design guidelines recommended for aerobic selectors. Calculations are provided in the spreadsheet 'Chapter 10 Design examples.xlsx' on the sheet 'aerobic selector'.

Parameter	Design value (guideline)	WWTP1		WWTP1b (adjusted selector volume)	
Number of compartments	≥ 3	3	✔	3	✔
Contact time (min)					
total	10-15	87		14	
- compartment 1		12	✖	2	✔
- compartment 2		25		3	
- compartment 3		50		8	
SLR (kgCOD.kg/TSS.d)					
- compartment 1	12	13.6	✔	85	✖
- compartment 2	6	6.8		43	
- compartment 3	3	3.4		21	
Floc loading in first compartment (gCOD/kgTSS)	50-150	117	✔	117	✔
DO concentration (gO_2/m^3)	≥ 2	Not specified => should be ≥ 2	✔	Not specified => should be ≥ 2	✔

Conclusion:

Comparing the design guidelines recommended for aerobic selectors in municipal wastewater treatment systems with the values for the WWTP1 under study, it is clear that several requirements are met: the number of compartments is satisfactory and so are the sludge loading rates in the selector compartments and the floc loading in the first compartment. One can also assume that the DO concentration will be kept sufficiently high $(DO > 2 \text{ gO}_2/\text{m}^3)$.

However, the total contact time in the selector is much too high (6-8 times): 87 minutes whereas the recommended time is 10-15 minutes. As such, the aerobic selector is not likely to provide favourable conditions to reduce the occurrence of filamentous bulking sludge.

How can this be remedied?

The total contact time in the selector should be reduced by a factor 6-8, which means that the total volume of the three selector compartments should be reduced accordingly, to 200-300 m³ instead of the current 1,750 m³. Assume for instance V_{S1}=40 m³, V_{S2}=80 m³ and V_{S3}=160 m³ (total selector volume 160 m³); however, then the sludge loading rate becomes 6-8 times too high (see Table 10.2 and the Excel file, case WWTP1b). Since it is not an option to increase the TSS concentration by the same factor, it seems that this cannot be remedied.

Anoxic selector design

Example 10.3.2

The WWTP under study is upgraded to achieve biological nitrogen removal via nitrification-denitrification in a MLE process configuration (WWTP2). An average nitrate concentration (S_{NO3-}) of 25 mg N/l is reached in the beginning of the selector through an internal recirculation flow (a) of 2:1 with respect to the influent flow rate.

Would the selector comply with the design guidelines for anoxic selectors (Table 10.2 in the textbook)? If not, what additional design and operational modifications would be required to meet the guidelines?

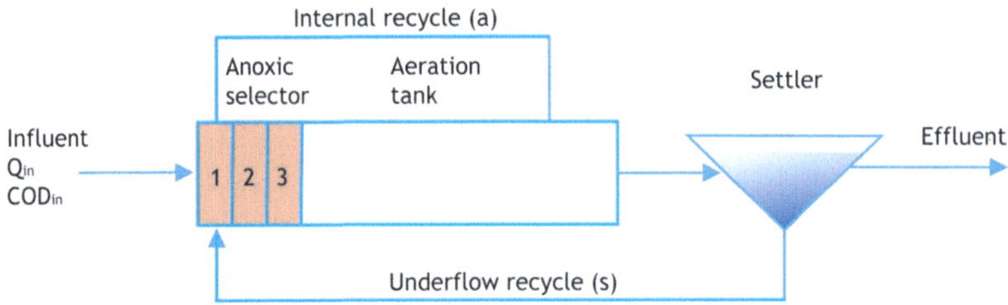

Figure 10.2 Layout of a WWTP with anoxic selector (Example 10.3.2). 'Comp.' refers to 'compartment'.

Solution

The selector design guidelines recommended for aerobic selectors in municipal wastewater treatment systems are listed in Table 10.3. The sludge loading rate (SLR) remains unchanged compared to Example 10.3.1 as the COD load and X_{TSS} remain constant. All the calculations are provided in the spreadsheet 'Chapter 10 Design examples.xlsx' on the sheet 'anoxic selector'.

The contact time in the ith selector compartment $HRT_{a,i}$ (d) is calculated through Eq. 10.22, taking into account the internal recycle ratio a:

$$HRT_{a,i} = V_{Si} / (Q_{in} \cdot (1 + s + a)) \tag{10.22}$$

which is calculated as:

$$HRT_{a,1} = 250 \ m^3 / 14{,}500 \ m^3/d / (1 + 1 + 2) \cdot 24 \ h/d \cdot 60 \ min/h = 6 \ min \tag{10.23}$$

$$HRT_{a,2} = 500 \ m^3 / 14{,}500 \ m^3/d / (1 + 1 + 2) \cdot 24 \ h/d \cdot 60 \ min/h = 12 \ min \tag{10.24}$$

$$HRT_{a,3} = 1{,}000 \ m^3 / 14{,}500 \ m^3/d / (1 + 1 + 2) \cdot 24 \ h/d \cdot 60 \ min/h = 25 \ min \tag{10.25}$$

Thus, the total contact time in the anoxic selector is 43 min. It is half of the contact time compared to the case of the aerobic selector, because of the internal recycle.

The ratio between the amount of readily biodegradable COD consumed and the amount of nitrate consumed is calculated as:

$$(RBCOD / NO_3^--N)_{consumed} = COD_{b,in} \cdot f_{SS} / S_{NO3} \tag{10.26}$$

$$(RBCOD / NO_3^--N)_{consumed} = 636 \ gCOD/m^3 \cdot 0.3 / 25 \ gN/m^3 = 8 \ gCOD/gN \tag{10.27}$$

Table 10.3 Comparison between operational data of the wastewater treatment plant under study and the selector design guidelines recommended for anoxic selectors. Calculations are provided in the spreadsheet 'Chapter 10 Design examples.xlsx'on the sheet 'anoxic selector'.

Parameter	Design value (guideline)	WWTP2		WWTP2c	
Number of compartments	≥ 3	3	✔	3	✔
Contact time (min) total	45-60	43		56	
- compartment 1		6	✔	8	✔
- compartment 2		12		16	
- compartment 3		24		32	
SLR (kgCOD.kg/TSS.d) - compartment 1	6	13.6	✘	6.1	(✘)
- compartment 2	3	6.8		5.2	
- compartment 3	1.5	3.4		2.6	
$(RBCOD / NO_3^--N)_{consumed}$ (gCOD/gN)	7-9	8	✔	8	✔

Comparing the obtained values with the design guidelines for anoxic selectors, it appears that the only criterion which is not met is the one for the sludge loading rates in the selector compartments, which are approximately two times too high. The solution would be to double the reactor volume or to double the MLSS concentration or a combination of both (see Eq. 10.1).

Since the contact time in the selectors is at the low end, the selector compartment values could be increased by 30 % and still meet the guideline, obtaining a contact time of approximately 60 min (43 min $\cdot$ 1.30 = 56 min). The resulting volume of the three compartments is then V_{S1} = 325 m^3, V_{S2} = 650 m^3, V_{S3} = 1,300 m^3. In order to further decrease the SLR to the recommended values of 6, 3 and 1.5 kgCOD/kgTSS.d respectively, the MLSS concentration needs to be increased as well. Increasing the MLSS concentration to X_{TSS} = 6 kgTSS/m^3 (corresponding with a total suspended solids mass MX_{TSS} = 87,500 kgTSS) results in an adequate sludge loading rate in the first selector compartment; however, the sludge loading rates in the second and third reactor compartments are still too high. This could theoretically be solved by further increasing the MLSS concentration; however, this does not seem realistic, a value of X_{TSS} = 6 kg/TSS.m^3 already being very (or even too) high.

Anaerobic selector design
Example 10.3.3
The WWTP under study (WWTP1) is modified in order to incorporate an anaerobic selector instead of an aerobic one (Figure 10.1). It can be assumed that there is no internal recirculation rate sent to the selector, so neither oxygen nor nitrate enters the selector.

1. Which modifications must be made to meet the guidelines for anaerobic selectors (Table 10.2 in the textbook).
2. What will be the unaerated mass fraction?

Solution
1. Guidelines for anaerobic selectors

Table 10.4 compares the operational data and the selector design guidelines recommended for anaerobic selectors in municipal wastewater treatment systems. The contact time remains unchanged compared to Example 10.3.1 *i.e.,* the case without internal recycle to the selector, and meets the corresponding guideline for anoxic selectors. All the calculations are provided in the spreadsheet 'Chapter 10 Design examples.xlsx' on the sheet 'anaerobic selector'.

The only additional criterion that needs to be checked is the ratio between influent readily biodegradable COD (*i.e.,* the sum of VFAs and fermentable COD) and influent phosphate (Eq. 10.28):

$$(COD_{VFA+fermentable} / PO_4^{-3}\text{-}P)_{in} = COD_{b,in} \cdot f_{SS} / P_{IN,in} \qquad (10.28)$$

which is calculated as:

$$(COD_{VFA+fermentable} / PO_4^{-3}\text{-}P)_{in} = 636 \text{ gCOD/m}^3 \cdot 0.3 / 13 \text{ gP/m}^3 = 15 \text{ gCOD/gP} \qquad (10.29)$$

and adheres to the corresponding design guideline.

Table 10.4 Comparison between operational data of the wastewater treatment plant under study and the selector design guidelines recommended for anaerobic selectors. Calculations are provided in the spreadsheet 'Chapter 10 Design examples.xlsx' on the sheet 'anaerobic selector'.

Parameter	Design value (guideline)	WWTP1	
Number of compartments	≥ 3	3	✔
Contact time (min)	60-120	87	✔
$(COD_{VFA+fermentable} / PO_4^{3-}\text{-}P)_{in}$ (gCOD/gP)	9-20	15	✔

Conclusion: the anaerobic selector fulfils the guidelines and thus yields favourable conditions to reduce the occurrence of bulking sludge.

2. Unaerated sludge mass fraction
The unaerated sludge mass fraction amounts to:

$$f_{XT} = V_{S,tot,\ anaerobic} / V_{tot} \tag{10.30}$$

and is calculated as:

$$f_{XT} = (250 + 500 + 1,000)\ m^3 / 14,650\ m^3 = 0.12\ or\ 12\ \% \tag{10.31}$$

Note that it is assumed here that the WWTP does not contain an anoxic zone, otherwise the anoxic tank volume would need to be considered in the calculation of the unaerated sludge mass fraction.

10.4 EXERCISES
Bulking sludge characteristics (exercises 10.4.1-10.4.3)
Exercise 10.4.1
Explain what the sludge volume index (SVI) is and its practical relevance.

Exercise 10.4.2
Explain what bulking sludge is and its consequences for process performance.

Exercise 10.4.3
To what extent is bulking sludge related to the SVI?

Causes and remediation of bulking sludge (exercises 10.4.4-10.4.7)
Exercise 10.4.4
Describe the diffusion-based selection theory and the kinetic selection theory to explain the occurrence of bulking sludge. Discuss how these theories relate to each other.

Exercise 10.4.5
Explain the storage selection theory.

Exercise 10.4.6

What are the main non-specific methods to control bulking sludge? What drawbacks do these methods have?

Exercise 10.4.2.7

Describe the principle of specific methods to control bulking sludge. What is the principle behind these methods? How are they established in practice?

Selectors to control bulking sludge - relation with granular sludge (exercises 10.4.8-10.4.10)

Exercise 10.4.8

Explain the principle of selectors to control bulking sludge.

Exercise 10.4.9

Discuss the implementation of selectors through their main design parameters.

Exercise 10.4.10

Describe how aerobic granular sludge relates to bulking sludge.

ANNEX 1: SOLUTIONS TO EXERCISES
Bulking sludge characteristics (solutions 10.4.1-10.4.3)
Solution 10.4.1
The sludge volume index (SVI, in ml/g) is an empirical measure of the sludge settling characteristics, expressing the volume taken by one gram of sludge after 30 minutes of settling. The SVI is obtained by having a sludge sample settling in a 1-liter measurement cylinder for 30 minutes. The volume of the sludge layer is read and divided by the original suspended solids content of the sludge sample.

Practical relevance: the SVI has a direct and strong effect on the required settler (surface) area. For instance, an increase in the SVI from 100 to 150 ml/g will result in almost double the required settler area (Figure 10.3).

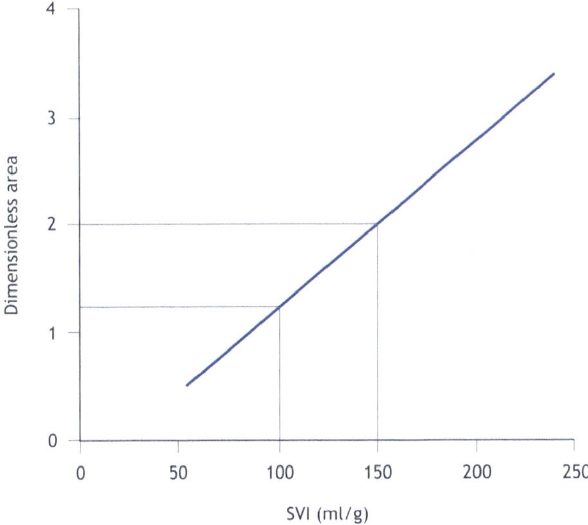

Figure 10.3 Relation between sludge volume index and surface area needed for a settler according to the STOWA design guidelines for settlers (STOWA, 1994). Figure reproduced from Chen *et al.* (2020).

Solution 10.4.2
- 'Bulking sludge' refers to the excessive growth of filamentous bacteria in an activated sludge process, to such an extent that suspended solids cannot be maintained in the settler. Indeed, filamentous bacteria result in open and porous sludge flocs, which experience hindered settling.
- Note 1: bulking sludge is an operational or empirical problem, there is not an exact scientific index to distinguish bulking sludge from non-bulking sludge.
- Note 2: the volume fraction of filamentous bacteria in the activated sludge community which causes settling problems could be minor. Volume fractions of 1-20 % are sufficient to cause bulking sludge.

Solution 10.4.3

- In practice, bulking sludge is associated with a high SVI. However, there is no generally accepted 'critical value' above which bulking sludge occurs, *i.e.* in the sense that it causes settler failure.
- In the Netherlands for instance, an SVI above 120 ml/g is already considered bulking sludge.

Causes and remediation of bulking sludge (solutions 10.4.4-10.4.7)

Solution 10.4.4

- *Diffusion-based selection theory.*
  Filamentous bacteria can easily grow outside the flocs, which means that the filamentous bacteria observe a higher substrate concentration than the floc-formers inside the floc. As a result, diffusion-dominated conditions (*i.e.*, low substrate concentrations) typically result in the growth of open, filamentous structures.
- *Kinetic selection theory.*
  This theory regards filamentous microorganisms as K strategists, *i.e.,* characterized by a higher affinity (lower affinity constant K_S) and a lower maximum growth rate than floc-forming bacteria, the latter being r strategists. In systems where the substrate concentration is low (typically $C_S < K_S$), as in continuously-fed completely mixed systems, filamentous bacteria have a higher specific growth rate than floc-forming bacteria, and thereby win the competition for substrate. In systems where the substrate concentration is high, as in plug-flow reactors and sequencing batch reactor (SBR) systems, floc-forming bacteria will dominate.
- *Relation between diffusion-based selection theory and kinetic selection theory.*
  K_S is typically used in activated sludge processes as an apparent mass transfer parameter, lumping substrate affinity and diffusion resistance. As such, the kinetic selection theory and the diffusion-based selection theory effectively both describe growth limited by diffusion at low substrate concentration. Indeed, the lower mass transfer resistance experienced by filamentous bacteria will translate into a lower K_S, classifying them as K strategists.

Solution 10.4.5

Microorganisms generally store substrate under high substrate concentrations, certainly when they undergo feast/famine conditions as usually occurs in a wastewater treatment plants. Non-filamentous (floc-forming) microorganisms are traditionally supposed to have a higher storage ability than filamentous ones. The storage ability gives them a competitive advantage in highly dynamic activated sludge systems such as plug-flow reactors, SBR and selector systems. Even though storage and regeneration may not be prime selection factors against filamentous bacteria, they do play a key role in selector-like systems because the high loading conditions and plug-flow conditions to prevent diffusion limitations also induce the storage response by the floc-forming bacteria. The prime selection factors for filamentous bacteria causing bulking sludge are generally accepted to be diffusion-based and/or kinetic selection (see Exercise 10.4.8).

Solution 10.4.6

- Non-specific methods comprise the use of oxidants to destroy filamentous bacteria causing bulking sludge. Examples are chlorination, ozonation and the application of hydrogen peroxide. They do not specifically target filamentous bacteria but make use of the fact that filamentous bacteria are placed mostly outside the floc and are therefore more susceptible to oxidants than the floc-forming bacteria.
- Drawbacks: (*i*) non-specific methods do not remove the causes for the excessive growth of filamentous microorganisms and their effect is therefore only transient; (*ii*) when using oxidants, there are environmental and ecotoxicological concerns about the potential formation of undesirable by-products such as halogenated organic components, and (*iii*) other micro-organisms could also be affected by the

oxidants, given that this method is non-specific. If slow-growing bacteria such as nitrifiers were to be affected, they would take a long time to recover, resulting in a suboptimal effluent quality.

Solution 10.4.7

- Specific methods are preventive methods which favour the growth of floc-forming bacteria at the expense of filamentous bacterial structures. Finding the right environmental conditions to achieve this is a challenge in activated sludge plants, but when successful, it enables the permanent control of bulking sludge.
- Preventive actions are based on the idea that the substrate conversion should not be limited by diffusion. This requires that readily biodegradable substrates are consumed at high substrate concentrations, such that the substrate uptake rate is close to its maximum value (at least $q_s/q_{smax} > 0.6$ and preferably $q_s/q_{smax} > 0.8$). In addition, the oxygen concentration should also be non-limiting (typically $O_2 > 1$ mg/l), at least during the period where readily biodegradable substrate is available. This is largely ensured by a properly designed plug flow or compartmentalized selector tank.

Selectors to control bulking sludge - relation with granular sludge (solutions 10.4.8-10.4.10)
Solution 10.4.8

In order to ensure that substrate conversion is not limited by diffusion and thus prevent bulking sludge, a high uptake rate and (almost) complete removal of readily biodegradable organics should be established in the entrance part of the activated sludge process. This initial part of the biological reactor, characterized by a plug-flow hydraulics (*i.e.,* a low dispersion number) and by an adequate macro-gradient of substrate, is termed the selector. The selector can also be a small initial zone of the reactor which receives the influent and sludge return flows, as long as it leads to a high uptake rate and almost complete removal of readily biodegradable organics.

Solution 10.4.9

- Selectors need to consist of at least three compartments to ensure sufficient plug-flow behaviour.
- Selectors can be aerobic, but also anoxic or anaerobic. In an aerobic selector, the oxygen concentration should be sufficiently high ($O_2 > 2$mg/l) as verified with an oxygen sensor in the first compartment. The entrance of oxygen into anoxic or anaerobic selectors needs to be avoided at all times. A low DO in the selector leads to oxygen gradients and growth of filamentous bacteria (low DO bulking).
- The sludge loading rate (SLR, kgCOD/kgTSS.d) needs to be kept close to a prescribed value[1], which depends on the selector type (higher SLR for aerobic than for anoxic selectors, not specified for an anaerobic one) and decreases along the selector path (highest SLR in the first selector compartment). The SLR needs to be high enough to avoid substrate limitation and low enough to enable full conversion of readily biodegradable substrate.
- The contact time needs to be kept within a certain prescribed range[1], which depends on the selector type (increases for aerobic $\geq$ anoxic $\geq$ anaerobic), load, temperature and influent readily biodegradable COD fraction. The contact time needs to be sufficiently long to ensure complete conversion of readily biodegradable substrate. On the other hand, too long contact times may result in a too low concentration of substrate, favouring the growth of filamentous microorganisms.
- For an aerobic selector, the floc loading (in kgCOD/kgTSS) needs to be in a certain range[1].
- The design of an anoxic selector is primarily based on the ratio of readily biodegradable organics to nitrate (RBCOD / NO_3^--N), which needs to be higher than the typical value for direct denitrification, taking into

---

[1] Values and ranges are specified in the examples 10.3.1-10.3.3.

account that an important fraction of readily biodegradable COD is converted into storage products. Nitrate needs to be in surplus in the anoxic reactor, even though periods with temporarily anaerobic conditions are not harmful.

- For anaerobic selector design, the ratio of readily biodegradable COD (VFA and fermentable COD) to phosphate needs to be kept within a certain range to make sure that hardly any readily biodegradable COD enters the main aeration basin.
- For anaerobic selector design, the main criterion is that readily biodegradable COD (VFA and fermentable COD) is fully converted under anaerobic conditions.
- Mixing conditions in anoxic and anaerobic conditions are not critical. Also, carry-over of readily biodegradable organics into the aerated stage is much less detrimental than in aerobic conditions. This is based on the experimental work by Martins (2004).

Solution 10.4.10

Aerobic granular sludge is the opposite phenomenon to bulking sludge. Granular sludge is formed when substrate conversion is not limited by diffusion. For aerobic granules to be formed, the removal of readily biodegradable organics should take place with minimal substrate gradients over the sludge floc or under fully anaerobic conditions. Anaerobic selectors are particularly suitable to promote aerobic granular sludge formation since they select phosphate and glycogen accumulating bacteria (PAO and GAO), given that ordinary heterotrophic filamentous bacteria are unable to grow under anaerobic conditions. The relatively low maximum growth rates of PAO and GAO result in lower substrate and oxygen uptake rates, which means that substrate/oxygen uptake rather than diffusion is the rate-limiting-step. Moreover, diffusion limitation is minimized since growth takes place on substrate stored within the cell.

REFERENCES

Chen G.H., van Loosdrecht M.C.M., Ekama G.A. and Brdjanovic D. (eds.) (2020). Biological Wastewater Treatment: Principles, Modelling and Design. ISBN: 9781789060355. IWA publishing, London, UK.
Martins A.M.P. (2004). Bulking sludge control: kinetics, substrate storage, and process design aspects, Engineering. PhD thesis TUDelft. ISBN 972-9098-07-7 http://resolver.tudelft.nl/uuid:b0cf9088-57b2-4f35-a3da-3b94c37a64e2
STOWA (1994). Selector design: the role of the influent characterization. Report 94-16 (in Dutch).

NOMENCLATURE

Abbreviation	Description
DO	Dissolved oxygen
RBCOD	Readily biodegradable COD
SRT	Sludge retention time (sludge age)
SLR	Sludge loading rate
SVI	Sludge volume index
TSS	Total suspended solids
VFAs	Volatile fatty acids
VSS	Volatile suspended solids
TSS	Total suspended solids

Symbol	Description	Unit
a	Mixed liquor recycle ratio (Q_a / Q_{in})	-
$COD_{b,in}$	Influent biodegradable COD concentration	$kgCOD/m^3$
COD_{in}	Influent total COD concentration	$kgCOD/m^3$
C_S	Substrate concentration	g/m^3
f_{SS}	Influent readily biodegradable fraction of the influent biodegradable COD	-
$f_{SU,CODin}$	Soluble unbiodegradable fraction of total influent COD	-
$f_{XU,CODin}$	Particulate unbiodegradable fraction of total influent COD	-
f_{XT}	Unaerated sludge mass fraction in the reactor	-
f_{VT}	Ratio of VSS over TSS of the sludge	$gVSS/gTSS$
$FCOD_{b,in}$	Daily load of influent biodegradable COD	$kgCOD/d$
$FCOD_{in}$	Daily load of influent COD	$kgCOD/d$
$FCOD_i$	Daily load of COD on the ith selector compartment	$kgCOD/d$
$FRBCOD_{in}$	Daily load of influent readily biodegradable COD	$kgCOD/d$
$FRBCOD_i$	Daily load of readily biodegradable COD on the ith selector compartment	$kgCOD/d$
HRT_a	Actual hydraulic retention time	d
K_S	Half-saturation constant for substrate	g/m^3
MX_{TSS}	Mass of solids in the bioreactor	$kgTSS$
$MX_{TSS,i}$	Mass of solids in the ith selector compartment	$kgTSS$
$P_{IN,in}$	Influent orthophosphate concentration	gP/m^3
Q_a	Mixed liquor recycle flow rate = internal recycle flow rate	m^3/d
Q_{in}	Influent flow rate	m^3/d
Q_s	Sludge recycle flow rate	m^3/d
$RBCOD_{in}$	Influent readily biodegradable COD concentration	$kgCOD/m^3$
SRT	Sludge retention time	d
s	Sludge underflow recycle ratio (Q_s/Q_{in})	-
SLR	Sludge loading rate	$kgCOD/kgTSS.d$
SLR_i	Sludge loading rate of the ith selector compartment	$kgCOD/kgTSS.d$
S_{NO3}	Nitrate concentration	kgN/m^3
TKN_{in}	Influent total Kjeldahl nitrogen (TKN) concentration	kgN/m^3
V_{Si}	Volume of the ith selector compartment	m^3
V_R	Volume of bioreactor	m^3

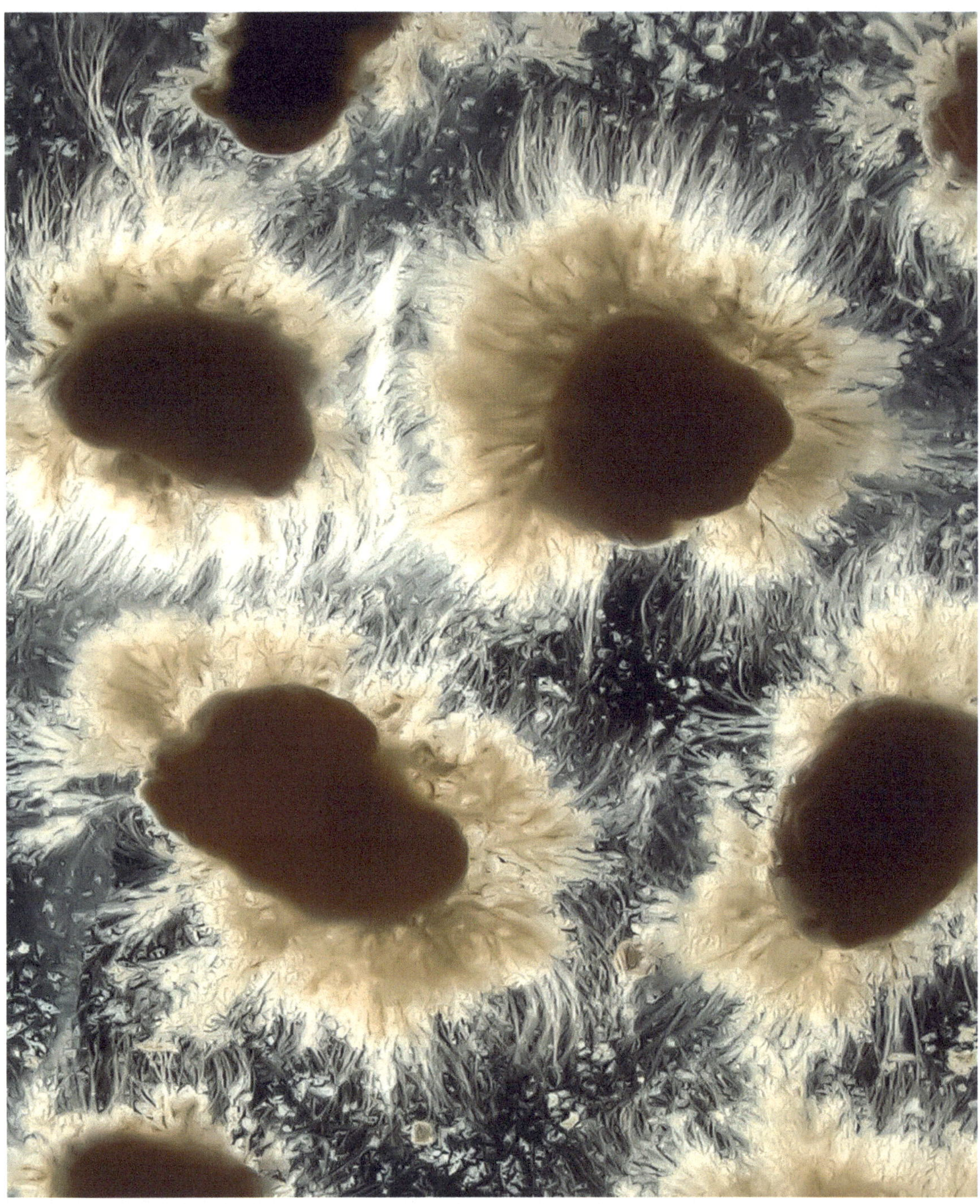

Low concentrations of RBCOD in the influent may even lead to filamentous outgrowth in aerobic granular sludge when also the oxygen transport rate to the biofilm is limiting. This dual oxygen and COD limitation have a negative impact on the settling rate, stability and process performance. This problem has been overcome in Nereda® technology by applying an anaerobic feed of sewage allowing full uptake of readily degradable COD before aeration starts (photo: M. de Kreuk).

doi: 10.2166/9781789062304_0357

11

Aerobic granular sludge

Laurence Strubbe, Merle de Kreuk, Edward J.H. van Dijk, Mark C.M. van Loosdrecht and Eveline I.P. Volcke

11.1 INTRODUCTION

Chapter 11 on aerobic granular sludge (AGS) in the main textbook *Biological Wastewater Treatment: Principles, Modelling and Design* (Chen *et al.,* 2020) introduces the principles, underlying processes, the functionality of the (biological) mechanisms and the implementation of AGS design in different situations. The examples, questions and exercises given here will guide you through the procedures and considerations related to AGS reactor design, operation, and control. An appropriate process control shows how the daily variations in batches are handled by adjusting the batch length and the scheduling.

The questions relate to the general background of the AGS system, the principles of the formation of AGS and how this is translated into the reactor operation. In addition, this chapter relates the dynamics of a typical cycle to the biological conversions in the granules. The influence of process operation conditions and influent characteristics are also discussed in order to achieve a nitrification/denitrification process that is balanced at all times. Fluctuations and control options are identified. Finally, resource recovery aspects are addressed.

11.2 LEARNING OBJECTIVES

After the successful completion of this chapter, the reader will be able to:

- Identify the main characteristics of aerobic granular sludge processes and the differences with respect to conventional activated sludge processes.
- Discuss the prerequisites for the granulation process and the selection of aerobic granular sludge.
- Interpret the dynamics of a typical cycle: aerated/non-aerated periods, corresponding processes, and spatial distribution (substrate, redox zones, etc.) in the granules.

- Quantify conversion rates in aerobic granular sludge processes for typical influent characteristics and reactor operating conditions.
- Quantify and interpret the influence of process operation conditions (dissolved oxygen concentration, temperature) and influent characteristics (*e.g.*, VFA in influent) on organic matter and nutrient removal.
- Identify typical control strategies to account for influent flow and composition dynamics.
- Dimension and compare an aerobic granular sludge plant with and without a buffer tank for given influent characteristics.
- Apply the principle of batch scheduling in a typical example.
- Describe resource recovery options for AGS plants.

11.3 EXAMPLES

Additional information A1: Design of an AGS reactor based on hydraulic constraints
Two hydraulic constraints need to be considered for the design of an aerobic granular sludge (AGS) reactor.

1) The first hydraulic constraint relates to the volume exchange ratio (VER, in %), *i.e.*, the ratio between the volume fed during a new batch cycle (V_{batch}, in m^3) and the total reactor volume ($V_{reactor}$, in m^3):

$$VER = V_{batch} / V_{reactor} \cdot 100 \% \tag{11.1}$$

in which:

$$V_{batch} = Q \cdot t_{feed} \tag{11.2}$$

with

Q = Influent flow rate (m^3/h)
t_{feed} = Feeding phase duration (h)

The VER is limited to a maximum value, $VER_{max} = 65 \%$, to prevent the influent being mixed with effluent during the (plug-flow) feeding phase:

$$VER \leq VER_{max} = 65 \% \tag{11.3}$$

The hydraulic constraint regarding the VER is typically applied to determine the minimum reactor volume or at least to check whether the given reactor volume meets the hydraulic constraint.

2) The second hydraulic constraint deals with the upflow velocity in the reactor (v_{upflow}, in m/h):

$$v_{upflow} = \frac{Q}{\dfrac{V_{reactor}}{H}} \tag{11.4}$$

in which:

H = Reactor height (m)

The upflow velocity in the reactor is limited to a maximum value of 5 m/h (Eq. 11.5) to prevent sludge being spilled into the effluent during plug-flow feeding:

$$v_{upflow} \leq 5 \text{ m/h} \tag{11.5}$$

Eq. 11.5 is typically applied to determine the reactor height for a given reactor volume. This condition needs to be fulfilled under all conditions, so it is evaluated under peak wet weather conditions as the worst-case scenario.

Eq. 11.3 and Eq. 11.5 thus constitute the two hydraulic constraints that need to be fulfilled in the design of an AGS reactor.

Additional information A2: Design of an AGS reactor based on biological constraints
In addition to the hydraulic constraints, there is also a biological constraint which needs to be fulfilled for the design of an AGS reactor, namely regarding the sludge loading rate.

The sludge loading rate (SLR, kgCOD/kgTSS.d) is calculated by Eq. 11.6 (rearrangement of Eq. 11.7 from Chen *et al.*, 2020):

$$SLR = \frac{Q \cdot COD}{\dfrac{t_{react,day}}{24} \cdot X_{TSS} \cdot V_{reactor} \cdot n_{reactor}} \tag{11.6}$$

in which:

COD = Influent COD concentration (kgCOD/m^3)
$t_{react,day}$ = Total reaction time per day (h)
X_{TSS} = Mixed liquor suspended solids or sludge concentration (kgTSS/m^3)
$n_{reactor}$ = Number of reactors (-)

The sludge loading rate needs to be limited to 0.4 kgCOD/kgTSS.d at moderate temperatures to ensure that nitrification takes place:

$$SLR \leq 0.4 \text{ kgCOD/kgTSS.d} \tag{11.7}$$

Example 11.3.1

Design and upgrading of an AGS system without a buffer tank
A new wastewater treatment plant needs to be built to serve 300,000 people equivalent (PE), taking into a wastewater production of 150 L/d.PE, a daily peak flow factor of 1.5 (S_F^{PDWF}) and a wet weather peak flow factor (S_F^{PWWF}) of 3.0. The incoming wastewater has a COD concentration of 500 g/m^3.

It has been decided to build an AGS system without a buffer tank, operated in batch mode. The feeding phase (t_{feed}) takes 60 minutes, the aeration phase ($t_{aeration}$) 140 minutes and the settling phase ($t_{settling}$) 20 minutes. The schedule for the settling phase holds during average dry weather flow (DWF) as well as peak dry weather flow (PDWF) conditions.

The task is to answer the following questions:

a) Calculate the dry weather flow rate (Q_{DWF}), peak dry weather flow rate (Q_{PDWF}) and peak wet weather flow rates (Q_{PWWF}) (all in m³/h). With this information, determine the reactor volume (for one AGS reactor) based on the hydraulic constraint concerning the volume exchange ratio.
b) Calculate the number of AGS reactors required.
c) Calculate the sludge loading rate (SLR, in kgCOD/kgTSS.d) in the reactors, for the average dry weather flow as well as the daily peak flow. To do this, assume a typical total suspended sludge concentration in the reactor.
d) Is this sludge loading rate low enough for nitrification to take place during increased temperatures of 25 °C?
e) Calculate the maximum upflow velocity in the reactor, assuming a reactor height of 8 m. Evaluate the obtained value.
f) Do you recommend any adjustments in the design or operation?

Once the AGS system has been constructed, the situation in the catchment area changes: an additional sewer system, characterised by a very high peak wet weather flow (PWWF) over dry water flow rate (DWF), is connected to this treatment plant. As a result, the average daily flow rate increases to 2,200 m³/h, while the peak wet weather flow increases to 8,800 m³/h.

g) Describe the consequences of these increased flow rates and envisage in a qualitative way what could be done to be able to treat all of the influent?

Solution

a) Reactor volume

The dry weather flow rate (Q^{DWF}) is the average collected sewage flow rate during periods without rain, consisting of wastewater from households as well as industrial effluents. It is calculated by Eq. 11.8.

$$Q^{DWF} = 150 \cdot 10^{-3} \text{ m}^3/\text{d.PE} \cdot 300{,}000 \text{ PE} \cdot (1/24) \text{ d/h} = 1{,}875 \text{ m}^3/\text{h} \tag{11.8}$$

The peak dry weather flow rate or daily peak flow rate (Q^{PDWF}) is the maximum flow rate during one day, which is calculated by multiplying the dry weather flow rate with the peak flow factor (Eq. 11.9):

$$Q^{PDWF} = S_F^{PDWF} \cdot Q^{DWF} \tag{11.9}$$

which in this case becomes:

$$Q^{PDWF} = 1.5 \cdot 1{,}875 \text{ m}^3/\text{h} = 2{,}813 \text{ m}^3/\text{h} \tag{11.10}$$

The peak wet weather flow rate (Q^{PWWF}) is the peak flow rate during rainy weather, which determines the maximum hydraulic load expected to reach the wastewater treatment plant. The peak wet weather flow rate is calculated from the dry weather flow rate and the wet weather peak flow factor (Eq. 11.11):

$$Q^{PWWF} = S_F^{PWWF} \cdot Q^{DWF} \tag{11.11}$$

In this case:

$$Q^{PWWF} = 3 \cdot 1,875 \ m^3/h = 5,625 \ m^3/h \tag{11.12}$$

Taking into account the hydraulic constraint concerning the VER (Eq. 11.3), the definition of VER (Eq. 11.1), the reactor volume is determined by Eq. 11.13:

$$V_{reactor} \geq V_{batch} \cdot 100 \ \% \ / \ 65 \ \% \tag{11.13}$$

which, taking into account Eq.11.2, is equivalent to Eq. 11.14:

$$V_{reactor} \geq Q \cdot t_{feed} \cdot 100 \ \% \ / \ 65 \ \% \tag{11.14}$$

The feeding phase duration (t_{feed}) is fixed at 1 hour during DWF and PDWF conditions. The corresponding volume fed during a cycle (V_{batch}, Eq. 11.2) is the highest under PDWF conditions, and so is the corresponding minimum reactor volume.

As a result, the reactor volume is calculated from Eq. 11.14 for PDWF conditions, as:

$$V_{reactor} = Q^{PDWF} \cdot t_{feed} \cdot 100 \ \% \ / \ 65 \ \% = 2,813 \ m^3/h \cdot 1 \ h \ / \ 0.65 = 4,328 \ m^3 \approx 4,500 \ m^3 \tag{11.15}$$

Setting the reactor volume at 4,500 m³ to comply with the peak dry weather flow, the VER during average dry weather is calculated (from Eq. 11.1 and Eq. 11.2) as:

$$VER = V_{batch}^{DWF} \ / \ V_{reactor} \cdot 100 \ \% = Q^{DWF} \cdot t_{feed} \ / \ V_{reactor} \cdot 100 \ \% = 1,875 \ m^3/h \cdot 1 \ h \ / \ 4,500 \ m^3 = 42 \ \% \tag{11.16}$$

A VER of 42 % during the average dry weather flow would be appropriate and ensures that during the daily peak flow, the effective VER will stay within the hydraulic limits of 65 %. However, during peak wet weather conditions, the feeding time will need to be reduced to maintain the VER below the maximum value of 65 %.

b) <u>Number of reactors</u>
The number of reactors is determined by taking into account that feeding is continuous, so:

$$n_{reactors} \cdot t_{feed} = t \tag{11.17}$$

in which t denotes the total cycle duration (h), which equals to:

$$t = t_{feed} + t_{react} + t_{settle} \tag{11.18}$$

Where:

t_{react} = Reaction phase duration (h)
t_{settle} = Time for sludge settling (h)

By combining Eq. 11.17 and Eq. 11.18, the required number of reactors is determined as:

$$n_{reactors} = \frac{t}{t_{feed}} = \frac{t_{feed} + t_{react} + t_{settle}}{t_{feed}} \qquad (11.19)$$

In this case:

$$n_{reactors} = \frac{(60 + 140 + 20)\,min}{60\,min} = 3.67 \qquad (11.20)$$

The resulting number needs to be rounded to the highest whole number, so at least four reactors are required to ensure the influent can be fed to one of the reactors.

c) <u>Sludge loading rate</u>
The sludge loading rate is calculated by Eq. 11.6.

$$SLR = \frac{Q \cdot COD}{\dfrac{t_{react,day}}{24} \cdot X_{TSS} \cdot V_{reactor} \cdot n_{reactor}} \qquad (11.6)$$

A typical MLSS concentration for granular sludge is:

$$X_{TSS} = 8 \ kgTSS/m^3 \qquad (11.21)$$

The total reaction time per day is calculated by Eq. 11.22:

$$t_{react,day} = n_{cycles} \cdot t_{react} \qquad (11.22)$$

where n_{cycles} denotes the number of cycles per day and per reactor (-), which is determined by Eq. 11.23:

$$n_{cycles} = \frac{24}{t} = \frac{24}{t_{feed} + t_{react} + t_{settle}} \qquad (11.23)$$

In this case:

$$n_{cycles} = \frac{24 \ h/d}{(60 + 140 + 20)\,min} \cdot 60 \ min/h = 6.5 \ cycles/d \qquad (11.24)$$

The total reaction time per day is thus calculated (by substituting Eq. 11.24 in Eq. 11.22) as:

$$t_{react.day} = \frac{6.5 \ cycles/d \cdot 140 \ min/cycle}{60 \ min/h} = 15 \ h/d \qquad (11.25)$$

With this information, the sludge loading rate at average dry weather flow is calculated (by substituting Eq. 11.25 and Eq. 11.21 in Eq. 11.6) as:

$$\text{SLR}^{\text{DWF}} = \frac{1,875 \text{ m}^3/\text{h} \cdot 24 \text{ h/d} \cdot 500 \text{ gCOD/m}^3}{\dfrac{15 \text{ h/d}}{24 \text{ h/d}} \cdot 8,000 \text{ gTSS/m}^3 \cdot 4,500 \text{ m}^3 \cdot 4} = 0.25 \text{ kgCOD/kgTSS.d} \tag{11.26}$$

while the sludge loading rate at daily peak flow becomes:

$$\text{SLR}^{\text{PDWF}} = \frac{2,813 \text{ m}^3/\text{h} \cdot 24 \text{ h/d} \cdot 500 \text{ gCOD/m}^3}{\dfrac{15 \text{ h/d}}{24 \text{ h/d}} \cdot 8,000 \text{ gTSS/m}^3 \cdot 4,500 \text{ m}^3 \cdot 4} = 0.38 \text{ kgCOD/kgTSS.d} \tag{11.27}$$

d) Biological constraint
Nitrification occurs for sludge loading rates up to 0.4 kgCOD/kgTSS.d at moderate temperatures, while higher temperatures can enable nitrification even under higher loading rates. In this example, the SLR in average dry weather conditions and in peak dry weather conditions are both lower than 0.4 kgCOD/kgTSS.d, which means that the biological constraint is definitely fulfilled during elevated temperatures. Therefore, in this example, the hydraulic constraint set by $\text{VER}_{\text{max}} = 0.65$ % (Eq. 11.3) is more limiting than the biological constraint.

e) Maximum upflow velocity
The maximum upflow velocity is reached under peak wet weather conditions and is calculated from Eq. 11.4:

$$v_{\text{upflow}} = \frac{Q_{\text{PWWF}}}{\dfrac{V_{\text{reactor}}}{H}} = \frac{5.625 \text{ m}^3/\text{h}}{\dfrac{4,500 \text{ m}^3}{8 \text{ m}}} = 10 \text{ m/h} \tag{11.28}$$

The upflow velocity exceeds the maximum of 5 m/h imposed by the hydraulic constraint Eq. 11.5.

f) Adjustments in design and operation
In order to meet the hydraulic constraint regarding the maximum upflow velocity during peak wet weather conditions and thus make sure that no sludge is spilled into the effluent during rainy weather, one option would be to increase either the reactor aspect ratio (*i.e.*, the ratio between the reactor height and diameter) or the AGS reactor volume. The maximum allowable reactor height is determined by construction restrictions and typically does not exceed 10-12 meters. Assume that the reactor height needs to be kept at 8 m in this example; therefore, the reactor volume would need to be doubled to meet the hydraulic constraint Eq. 11.5, since the maximum upflow velocity is twice as high as allowed. Alternatively, a buffer tank could be installed. Both these solutions would entail a significant cost increase.

An alternative would be to change the reactor operation instead of the design. During periods of intense rainfall, two reactors could be fed at the same time (lowering Q^{PWWF} in Eq. 11.28 and thus v_{upflow}). Feeding two reactors at the same time would imply shorter reaction times (to ensure that the four AGS reactors in this example remain sufficient to keep a continuous feeding, see Eq. 11.19), which is possible during PWWF conditions because of the lower concentrations.

g) <u>Consequences of increased flow rates</u>

The average daily flow rate has increased from $Q^{DWF} = 1,875$ m³/h to 2,200 m³/h, while the peak wet weather flow rate has increased from $Q^{PWWF} = 5,625$ m³/h to 8,800 m³/h. Given the absence of additional data, we assume that the daily peak flow factor remains at $S_F^{PDWF} = 1.5$, so the daily peak flow has become $Q^{PDWF} = 1.5 \cdot 2,200$ m³/h = 3,300 m³/h. The consequences of these increased flow rates are investigated by checking the hydraulic and biological constraints for this changed situation. The volume exchange ratio under daily peak flow conditions becomes:

$$\text{VER}^{PDWF} = \frac{Q \cdot t_{feed}}{V_{reactor}} \cdot 100 \ \% = \frac{3,300 \text{ m}^3/\text{h} \cdot 1 \text{ h}}{4,500 \text{ m}^3} \cdot 100 \ \% = 73 \ \% \tag{11.29}$$

which exceeds the imposed maximum of 65 %.

However, the upflow velocity in the reactor at peak wet weather flow conditions was already too high for the initial situation (Eq. 11.28) and now becomes even higher:

$$v_{upflow}^{PWWF} = \frac{Q_{PWWF}}{\dfrac{V_{reactor}}{H}} = \frac{8,800 \text{ m}^3/\text{h}}{\dfrac{4,500 \text{ m}^3}{8 \text{ m}}} = 15.6 \text{ m/h} \tag{11.30}$$

This exceeds the maximum of 5 m/h imposed by the hydraulic constraint Eq. 11.5 by a factor of over 3. Even in peak dry weather flow conditions, the upflow velocity in the reactor is too high:

$$v_{upflow}^{PDWF} = \frac{Q_{PDWF}}{\dfrac{V_{reactor}}{H}} = \frac{3,300 \text{ m}^3/\text{h}}{\dfrac{4,500 \text{ m}^3}{8 \text{ m}}} = 5.9 \text{ m/h} \tag{11.31}$$

In addition, the corresponding SLR at peak dry weather conditions is calculated as:

$$\text{SLR}^{PDWF} = \frac{3,300 \text{ m}^3/\text{h} \cdot 24 \text{ h/d} \cdot 500 \text{ gCOD/m}^3}{\dfrac{15 \text{ h/d}}{24 \text{ h/d}} \cdot 8,000 \text{ gTSS/m}^3 \cdot 4,500 \text{ m}^3 \cdot 4} = 0.46 \text{ kgCOD/kgTSS.d} \tag{11.32}$$

which is too high to ensure good nitrification performance in moderate temperatures.

Overall, it is clear that this situation violates both the hydraulic constraints as well as the biological constraint. Even though it would still be possible to feed two AGS reactors at the same time during PWWF conditions, while shortening the reaction phase length, this is not likely to be sufficient to meet all the constraints. Indeed, a shorter reaction phase length makes the relative time spent on feeding longer, which in its turn leads to an increased volume exchange ratio, which was already too high.

Alternatively, a buffer tank or an extra AGS reactor could be added to the plant to overcome all the constraints during PDWF conditions. The installation of a buffer tank, which ensures a good effluent quality at all times, is the preferred option during short intense rainfall when equalisation of the flow rate is required.

Example 11.3.2

Design of an AGS system – influence of a buffer tank

An AGS system is constructed to treat wastewater characterised by a dry weather flow rate of 3,250 m^3/h, a daily peak factor S_F^{PDWF} =1.8, and a peak wet weather flow of 12,000 m^3/h. The design needs to fulfil the following two requirements in average dry weather conditions: (*i*) a maximum volume exchange (VER) ratio of 35 % and (*ii*) the maximum volumetric loading rate needs to be 1.2 m^3/m^3.d. The time schedule during dry weather flow is t_{feed} = 60 min, t_{react} = 300 min and t_{settle} = 20 min.

The task is to perform the following:

a) Determine the minimal volume for each reactor and the total number of AGS reactors (without a buffer tank) based on the given requirements (*i*) and (*ii*). Indicate the relation with the previously defined hydraulic and biological constraints (Eq. 11.3 and Eq. 11.7, respectively).

b) Calculate the resulting VER of the reactor during dry weather flow, daily peak flow and peak wet weather flow conditions. Determine the maximum flow rate to maintain the same cycle time with a maximum VER of 65 %. Also calculate the volumetric loading rate for each case. How do you expect the sludge loading rates to vary?

c) In order to maintain a sufficient quality of the effluent, the VER needs to be kept below 65 % and the volumetric loading rate below 3.5 m^3/m^3.d during wet weather peak flow conditions, while keeping the reactor volume and the number of reactors. Therefore, the batch schedule must be changed. Recalculate the cycle time (t), feeding time (t_{feed}) and reaction time (t_{react}) to meet these PWWF conditions.

d) To use the reactor volume more efficiently, the number of AGS reactors is reduced to four and a buffer tank is built instead. Calculate the required buffer tank volume to be able to deal with PWWF conditions, keeping the batch schedule determined under c).

Solution

a) <u>Reactor volume and number of reactors</u>

Based on the hydraulic constraint of VER = 35 % during DWF conditions, the reactor volume is determined (by substituting Eq. 11.3 in Eq. 11.1, after rearrangement) as:

$$V_{reactor} = \frac{V_{batch}^{DWF}}{VER_{max}} = \frac{Q^{DWF} \cdot t_{feed}}{100\ \%\ /\ 65\ \%} = \frac{3,250}{0.35} = 9,286\ m^3 \tag{11.33}$$

The corresponding VER under daily peak flow conditions amounts to:

$$VER = \frac{V_{batch}^{PDWF}}{V_{reactor}} \cdot 100\ \% = \frac{Q^{PDWF} \cdot t_{feed}}{V_{reactor}} \cdot 100\ \%$$

$$= \frac{S_F^{PDWF} \cdot Q^{DWF} \cdot t_{feed}}{V_{reactor}} \cdot 100\ \% = 1.8 \cdot 0.35 \cdot 100\ \% = 63\ \% \approx 65\ \% \tag{11.34}$$

So the given requirement of a minimal volume exchange ratio of 35 % under average dry weather conditions will ensure a minimum volume exchange ratio of 65 % under daily peak flow conditions, given S_F^{PDWF} = 1.8. Thus in this case, the hydraulic constraint Eq. 11.33 for DWF conditions is equivalent to the hydraulic constraint Eq. 11.3 for PDWF conditions.

The volumetric loading rate (VLR, in $m^3/m^3.d$) is determined by Eq. 11.35:,

$$VLR = \frac{Q}{\frac{t_{react,day}}{24} \cdot V_{reactor} \cdot n_{reactor}} \tag{11.35}$$

and should adhere to:

$$VLR \leq 1.2 \ m^3/m^3.d \tag{11.36}$$

The minimum reactor volume resulting from Eq. 11.35 and Eq. 11.36 is determined by Eq. 11.37:

$$V_{reactor} = \frac{Q}{\frac{t_{react,day}}{24} \cdot VLR \cdot n_{reactor}} \tag{11.37}$$

The total reaction time per day is calculated from Eq. 11.22, which requires the knowledge of the number of cycles per day and per reactor (Eq. 11.23). In this example,

$$n_{cycles} = \frac{24 \ h/d}{(600 + 300 + 20)\,min} \cdot 60 \ min/h = 3.8 \ cycles/d \tag{11.38}$$

$$t_{react,day} = \frac{3,8 \ cycles/d \ \cdot \ 300 \ min/cycle}{60 \ min/h} = 19 \ h/d \tag{11.39}$$

The number of reactors is calculated from Eq. 11.19 as:

$$n_{reactors} = \frac{(600 + 300 + 20)\,min}{60\,min} = 6.3 \tag{11.40}$$

At least 7 reactors are required to maintain a continuous feeding.

The volume of each reactor is calculated from Eq. 11.36 as:

$$V_{reactor} = \frac{3,250 \ m^3/h \cdot 24 \ h/d}{\frac{19 \ h}{24 \ h/d} \cdot 1.2 \ 1/d \cdot 7} = 11,729 \ m^3 \approx 12,000 \ m^3 \tag{11.41}$$

The obtained volume based on the VER_{max} (Eq. 11.41) is larger than the one based on the maximum allowed VLR (Eq. 11.33) which means that in this example, the biological constraint is more limiting than the hydraulic constraint of the applied VER during DWF.

The final layout comprises 7 AGS reactors which each have a minimum volume of 12,000 m^3.

Note that the volumetric loading rate (Eq. 11.35) combines the sludge loading rate (SLR, in kgCOD/kgTSS.d, Eq. 11.6) with the reactor MLSS concentration (X_{TSS}, in kgTSS/m^3) and the influent COD concentration (kgCOD/m^3), according to Eq. 11.42:

$$VLR = SLR \cdot \frac{X_{TSS}}{COD} \tag{11.42}$$

Given the influent COD concentration of 500 g/m^3 and assuming a typical MLSS concentration of $X_{TSS} = 8$ kgTSS/m^3 for an AGS reactor, the requirement to maintain the VLR below 1.2 m^3/m^3.d (Eq. 11.36) is equivalent to:

$$VLR = SLR \cdot \frac{COD}{X_{TSS}} \le 1.2 \; 1/d \cdot \frac{500 \text{ gCOD/m}^3}{8,000 \text{ gTSS/m}^3} = 0.075 \text{ gCOD/gTSS.d} \tag{11.43}$$

which is more stringent than Eq.11.7. However, the requirement VLR $\le$ 1.2 m^3/m^3.d was imposed for average dry weather conditions. The corresponding SLR for peak flow conditions, given $S_F^{PDWF} = 1.8$, amounts to:

$$SLR = VLR \cdot \frac{COD}{X_{TSS}} \le 1.2 \; 1/d \cdot 1.8 \cdot \frac{500 \text{ gCOD/m}^3}{8,000 \text{ gTSS/m}^3} = 0.14 \text{ gCOD/gTSS.d} \tag{11.44}$$

which is still more stringent than Eq. 11.7. So in this example, the biological constraint Eq. 11.36 for DWF conditions is more stringent than the biological constraint Eq. 11.3 for PDWF conditions.

b) <u>Volume exchange ratios and volumetric loading rates</u>
The volume exchange ratio is calculated from Eq. 11.1 and Eq. 11.2 as:

$$VER = \frac{V_{batch}}{V_{reactor}} \cdot 100 \text{ \%} = \frac{Q \cdot t_{feed}}{V_{reactor}} \cdot 100 \text{ \%} \tag{11.45}$$

The results for PWF, PDWF and PWWF conditions are summarized in Table 11.1. It is clear that the VER for DWF and PDWF conditions fulfils the hydraulic constraint Eq. 11.3, while the VER under PWWF is too high, which implies that the feeding time will need to be reduced to maintain the VER below the maximum value VER$_{max}$ = 65 %. The maximum flow rate for which Eq. 11.3 holds is obtained by rearranging Eq. 11.45, as:

$$Q = \frac{VER_{max}}{100 \text{ \%}} \cdot \frac{V_{reactor}}{t_{feed}} \tag{11.46}$$

which is calculated for this example as 7,800 m^3/h and indicated in Table 11.1 as part of the maximum wet weather flow conditions (WWF$_{max}$).

The corresponding volumetric loading rates are calculated from Eq. 11.35; their values are summarized in Table 11.1. The VLR ranges from 1.2 to 4.3 m^3/m^3.d (DWF and PWWF conditions, respectively). The sludge loading rate is not expected to present equally large variations, since the influent will likely be diluted under rainy weather conditions, implying a lower COD concentration, which compensates for the increasing flow rate, resulting in a relatively lower SLR (Eq. 11.6).

Table 11.1 Summary of calculations and results for Example 11.3.2. The values in black are given, and calculated values are denoted in blue. The calculations are provided in the spreadsheet 'Chapter 11 Design example 2.xlsx'.

	DWF	PDWF	PWWF	WWF_{max}	Unit
S_F	1	1.8	3.7		-
Q	3,250	5,850	12,025	7,800	m^3/h
VER	27	49	100	65	%
VLR	1.2	2.1	4.3	2.8	$m^3/m^3.d$

c) Alternative constraints for VER and SLR

Alternatively, it is required that the VER is maximum 65% under PWWF conditions:

$$VER^{PWWF} = \frac{V_{batch}^{PWWF}}{V_{reactor}} \cdot 100\ \% = \frac{Q^{PWWF} \cdot t_{feed}}{V_{reactor}} \cdot 100\ \% = 65\ \% \tag{11.47}$$

From Eq. 11.47, the feeding time is calculated as:

$$t_{feed} = \frac{VER_{max} \cdot V_{reactor}}{100\ \% \cdot Q^{PWWF}} = 0.65 \cdot \frac{12,000\ m^3}{12,025\ m^3/h} \cdot 60\ min/h = 39\ min \tag{11.48}$$

At the same time, the volumetric loading rate needs to be kept below 3.5 $m^3/m^3.d$ during PWWF conditions:

$$VLR^{PWWF} \leq 3.5\ m^3/m^3.d \tag{11.49}$$

The VLR is expressed by Eq. 11.35. However, it can also be alternatively expressed by Eq. 11.37:

$$VLR = \frac{VER_{max}}{100\% \cdot t_{reactor}} \tag{11.50}$$

The equivalence between Eq. 11.35 and Eq. 11.50 can be seen by substitution of Eq. 11.22 in Eq. 11.35:

$$VLR = \frac{Q}{n_{cycles} \cdot \frac{t_{reactor}}{24} \cdot V_{reactor} \cdot t_{reactor}} \tag{11.51}$$

followed by substitution of Eq. 11.23:

$$VLR = \frac{Q \cdot t}{t_{react} \cdot V_{reactor} \cdot n_{reactor}} \tag{11.52}$$

which is equivalent to:

$$VLR = \frac{Q \cdot t_{feed}}{t_{react} \cdot V_{reactor}} \tag{11.53}$$

and thus to Eq. 11.50.

From Eq. 11.50, the reaction time during PWWF for a maximum volumetric loading rate of 3.5 m^3/m^3.d is determined by:

$$t_{react} = \frac{VER^{PWWF}}{100\% \cdot VLR^{PWWF}} \tag{11.54}$$

which is calculated for this example as:

$$t_{react} = \frac{65\ \%}{100\ \% \cdot 3.5\ 1/d} \cdot 24\ h/d \cdot 60\ min = 267\ min \tag{11.54}$$

The total cycle duration is calculated from Eq. 11.18 as:

$$t = 39\ min + 267\ min + 20\ min = 326\ min \tag{11.55}$$

d) <u>Buffer tank</u>

The buffer volume is calculated based on Eq. 11.8 from Chen *et al.* (2020), which in this example is applied for PWWF conditions, since the latter correspond to the highest flow that needs to be buffered:

$$V_{buffer} = \frac{Q^{PWWF}}{n_{reactor} \cdot n_{cycles}} - Q^{PWWF} \cdot \frac{t_{feed}}{24} \tag{11.56}$$

The number of AGS reactors was given as four and the number of cycles during PWWF is calculated from Eq. 11.23 and Eq. 11.55 as:

$$n_{cycles} = \frac{24}{t} = \frac{24\ h/d}{326\ min} \cdot 60\ min/h = 4.4\ per\ day \tag{11.57}$$

The required buffer volume thus becomes:

$$V_{buffer} = \frac{12,025\ m^3/h \cdot 24\ h/d}{4 \cdot 4.4\ m^3/m^3.d} - 12,025\ m^3/h \cdot \frac{39\ min}{60\ min/h} = 8,582\ m^3 \tag{11.58}$$

Example 11.3.3

Buffer tank operation

Consider an AGS plant consisting of 2 reactors of 6,000 m^3 each, with a height of 8 m. The total cycle time in dry weather conditions is 240 minutes, consisting of 60 minutes feeding (t_{feed}), 150 min reaction (t_{react}) and 30 minutes settling (t_{settle}). Each day one of the cycles starts at 08:00 hours. There is also a buffer tank installed to store the influent wastewater while it cannot be fed to one of the two reactors. Table 11.2 summarizes the daily flow variation reaching the AGS plant during the specified 1-hour time intervals.

Table 11.2 Specific flow rates reaching the aerobic granular sludge plant during dry weather (Q^{DWF}) and peak wet weather (Q^{PWWF}) conditions for a time interval of one hour over one day.

Start time	End time	Q^{DWF} (m^3/h)	Q^{PWWF} (m^3/h)
08:00	09:00	1,022	1,022
09:00	10:00	1,181	1,181
10:00	11:00	1,224	1,224
11:00	12:00	1,238	1,238
12:00	13:00	1,224	1,224
13:00	14:00	1,166	1,166
14:00	15:00	1,109	1,931
15:00	16:00	1,008	3,278
16:00	17:00	936	4,421
17:00	18:00	936	4,498
18:00	19:00	950	3,212
19:00	20:00	1,051	1,256
20:00	21:00	1,152	1,152
21:00	22:00	1,152	1,152
22:00	23:00	1,094	1,094
23:00	24:00	994	994
00:00	01:00	792	792
01:00	02:00	634	3,049
02:00	03:00	475	2,874
03:00	04:00	374	1,198
04:00	05:00	302	765
05:00	06:00	288	288
06:00	07:00	346	346
07:00	08:00	590	590

The following tasks should be carried out:

a) Visualize a logical batch schedule for the two reactors under dry weather conditions. Indicate when the flow needs to be stored.
b) What is the volume of the buffer tank that is used, based on the peak flow during dry weather?
c) The wastewater stored in the buffer tank is emptied each time in the subsequent cycle, adding to the influent flow rate. Calculate the feed flow rate in the 1-hour time interval following the largest buffered wastewater volume, as well as the upflow velocity and VER for the corresponding batch. Are they within the hydraulic limits, *i.e.*, do they fulfil the hydraulic constraints?
d) What is the effect of emptying the storage tank on the SLR? Assess in detail for the 1-hour time interval following the largest buffered wastewater volume.

The operator reported a summer thunderstorm event as described in the last column of Table 11.2.

e) Calculate the wastewater volumes fed and stored during the thunderstorm period (1 day), assuming that the wastewater stored in the buffer tank is emptied each time in the subsequent cycle, adding to the influent flow rate. Do the upflow velocity and VER during the thunderstorm event remain within the hydraulic boundaries?

f) How could the system operation be changed in order to meet the hydraulic constraints during these heavy thunderstorms while maintaining the batch schedule from under DWF conditions? How large should the corresponding buffer volume be? Depict the full 24 hours of batch scheduling.

g) Alternatively, could the hydraulic constraints be met by changing the batch schedule? What would be the impact on the required buffer volume? Do you see any other limitations?

Solution

a) <u>Batch schedule under dry weather conditions</u>

A logical batch schedule for the two reactors under dry weather conditions is provided in Figure 11.1 (see the spreadsheet 'Chapter 11 Design example 3.xlsx' on the sheet 'Schedule DWF'). The influent wastewater needs to be stored when one or both reactors are in the reaction or settling phase. Note that the two cycles of the two AGS reactors have been aligned so that the feeding phase of the second AGS reactors starts in the middle of the cycle of the first AGS reactor, *i.e.*, after 120 minutes.

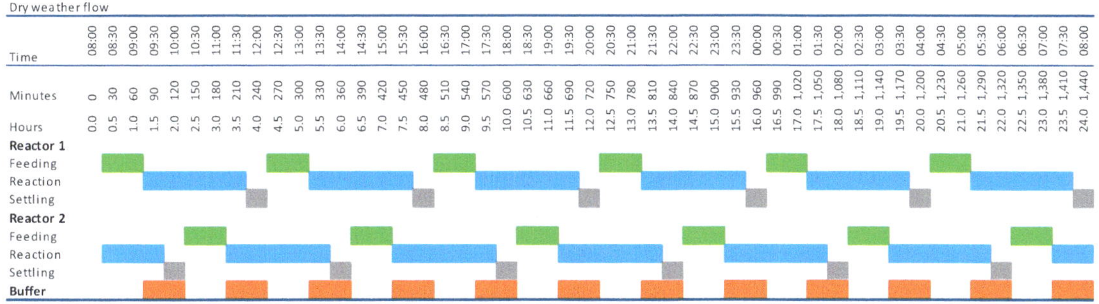

Figure 11.1 Batch schedule for the two AGS reactors under dry weather conditions - Example 11.3.3.

b) <u>Used volume of the buffer tank</u>

The buffered wastewater volume is calculated based on Eq. 11.56 (*i.e.*, Eq. 11.8 from Chen *et al.*, 2020) for the highest flow rate during dry weather conditions. The highest flow rate is identified from Table 11.2 as $Q^{PDWF} = 1,238$ m³/h and takes place between 11:00 and 12:00. The buffered volume becomes:

$$V_{buffer} = \frac{Q^{PWWF}}{n_{reactor} \cdot n_{cycles}} - Q^{PWWF} \cdot \frac{t_{feed}}{24} \tag{11.59}$$

The number of cycles during dry weather conditions is calculated from Eq. 11.23 as:

$$n_{cycles} = \frac{24}{t} = \frac{24 \text{ h/d}}{240 \text{ min}} \cdot 60 \text{ min/h} = 6 \text{ per day} \tag{11.60}$$

With this information, the buffered wastewater volume is calculated from Eq. 11.59 as:

$$V_{buffer} = \frac{1,238 \text{ m}^3/\text{h} \cdot 24 \text{ h/d}}{2 \cdot 6 \text{ m}^3/\text{m}^3 \cdot \text{d}} - 1,238 \text{ m}^3/\text{h} \cdot \frac{60 \text{ min}}{60 \text{ min/h}} = 1,238 \text{ m}^3 \tag{11.61}$$

c) <u>Maximum VER and upflow velocity</u>

An overview of the buffer tank operation under DWF conditions, together with the calculation of the VER and v_{upflow} at every feeding phase, is provided in the spreadsheet 'Chapter 11 Design example 3.xlsx' on the sheet 'Buffer operation DWF'. The volume fed in the 1-hour time interval following the largest stored buffer volume (1,238 m^3, from 11:00 until 12:00) amounts to 2,462 m^3 and is fed between 12:00 and 13:00.

The corresponding VER is calculated as:

$$VER = \frac{V_{batch}}{V_{reactor}} \cdot 100\ \% = \frac{Q^{PDWF} \cdot t_{feed}}{V_{reactor}} \cdot 100\ \% = \frac{2,462 \text{ m}^3/\text{h} \cdot 1 \text{ h}}{6,000 \text{ m}^3} \cdot 100\ \% = 41\ \% \tag{11.62}$$

which fulfils requirement Eq. 11.3.

The upflow velocity in the reactor becomes:

$$v_{upflow} = \frac{2,462 \text{ m}^3/\text{h}}{\dfrac{6,000 \text{ m}^3}{8 \text{ m}}} = 3.3 \text{m/h} \tag{11.63}$$

which adheres to Eq. 11.5. This implies that the hydraulic limits are not exceeded when emptying the buffer tank after the largest volume has been stored.

d) <u>Corresponding sludge loading rate</u>

The SLR is proportional to the flow rate (Eq. 11.6). Emptying the buffer tank during each feeding phase implies an increased feed flow rate compared to only feeding the incoming wastewater flow rate. As a result, the SLR also increases. In this example, no particular information is given on the incoming COD concentration, so it can be assumed that the incoming COD concentration is fairly constant. If the COD concentration is constant, the SLR increases linearly with the feed flow rate (Eq. 11.6).

As for the 1-hour time interval following the largest buffered wastewater volume, the feed flow rate amounts to 2,462 m^3/h, of which the influent flow rate amounts to 1,224 m^3/h. As a result, the SLR increases by a factor:

$$\frac{SLR_{influent+buffer}}{SLR_{influent}} = \frac{2,462 \text{ m}^3/\text{h}}{1,224 \text{ m}^3/\text{h}} = 2 \tag{11.63}$$

i.e., it doubles compared to the case where no stored wastewater is fed. This highlights the impact of processing buffered wastewater on the AGS reactor design.

e) <u>Summer thunderstorm – PWWF conditions</u>

An overview of the buffer tank operation under PWWF conditions, together with the calculation of VER and v_{upflow} at every feeding phase is provided in the spreadsheet 'Chapter 11 Design example 3.xlsx' on the sheet 'Buffer operation PWWF_1'. The calculations indicate that for some of the batch cycles, VER and v_{upflow} exceed the hydraulic limitations when emptying the buffer tank each time in the subsequent cycle, adding to the influent flow rate. Moreover, a buffer volume of 1,238 m³ is no longer sufficient to buffer the flow that needs to be stored when one or both reactors are in the reaction or settling phase.

f) <u>Meeting hydraulic constraints during PWWF conditions</u>

In order to meet the hydraulic constraints, the buffer tank is partially emptied each time in the following cycle, but only to such an extent that the hydraulic constraints are still fulfilled. The spreadsheet 'Chapter 11 Design example 3.xlsx' on the sheet 'Buffer operation WWF_1' shows the 24-hour operation during WWF maintaining the batch schedule in dry weather flow, while taking into account the hydraulic limitations (v_{upflow} limits over the VER). The minimal buffer volume required is 9,165 m³. It is important to note that Eq. 11.56 (Eq. 11.8 from Chen *et al.*, 2020) is no longer valid in this case because the hydraulic limits are not fulfilled for every cycle, which means that the buffer cannot be emptied during feeding.

g) <u>Buffer volume</u>

The reaction time and thus total cycle time during PWWF conditions could be shortened because the influent concentrations (not detailed in this example) are also expected to be lower. A possible alternative batch schedule for PWWF conditions is proposed in the spreadsheet 'Chapter 11 Design example 3.xlsx' on the sheet 'Buffer operation PWWF_2'. The total cycle time is decreased for some batches to 3 hours (60 min fill/draw, 90 min reaction phase and 30 minutes settling and sludge discharge) and in extreme cases to 2 hours (60 min fill/draw, 30 min reaction phase and 30 minutes settling and sludge discharge). In this example, it is opted to shorten the cycles when the buffer tank is not emptied during the feeding phase. This new schedule lowers the required buffer volume to 4,697 m³. However, this alternative batch schedule runs the risk of not meeting the effluent quality. If this was often the case, a larger buffer or extra reaction tank would be required.

Example 11.3.4

Comparison of AGS systems with and without a buffer

Consider a plant with three AGS reactors (without a buffer tank) and a plant with two AGS reactors and one buffer tank. Design both plants based on a Q_{DWF} of 4,800 m³/d, a COD influent concentration of 600 g/m³, an MLSS concentration of 8 kgTSS/m³ and an SLR of 0.3 kgCOD/kgTSS.d. The settling time (t_{settle}) is set at 30 minutes. For the AGS plant with the buffer, consider a constant t_{feed} of 60 minutes.

Tasks in this exercise are to:

a) Calculate and plot for both designs the influence of t_{react} on the total plant volume. Take a range from 60 to 600 minutes for t_{react}. This plot will result in Figure 11.15 (A) from Chen *et al.* (2020).

b) Calculate and plot for both designs the influence of t_{reac} on the efficiency of the AGS reactors, expressed as t_{reac}/t. Take a range from 60 to 600 minutes for t_{reac}. This plot will result in Figure 11.15 (B) from Chen *et al.* (2020).

c) Check if the calculated buffer volume for a t_{reac} of 2.5 hours complies with the hydraulic limits during one day at Q_{DWF}. What is the limiting hydraulic constraint?

d) The buffer volume is always emptied during feeding. Derive the flow rate during feeding (Q_{feed}) based on Eq. 11.8 from Chen *et al.* (2020).

e) Explain how the PDWF and PWWF will influence the total plant volume, efficiency of the AGS reactor and the batch scheduling for both plants.

Solution

a,b) <u>Reactor volumes and buffer volume</u>

The calculations and plots are provided in the spreadsheet 'Chapter 11 Design example 4.xlsx'. The sheet 'Without buffer' provides the calculations and plots related to the plant with three AGS reactors. The sheet 'With buffer' provides the calculations and plots related to the plant with two AGS reactors and one buffer tank. The sheet 'Comparison' shows the plots given in Figure 11.15 from Chen *et al.* (2020).

c) <u>Hydraulic constraints</u>

The hydraulic constraints of the design with a buffer tank were always within the hydraulic limits as can be seen in the sheet 'Buffer check'. The upflow velocity is the limiting hydraulic constraint.

d) <u>Feed flow rate</u>

As the hydraulic constraints are fulfilled for one day at Q_{DWF}, the buffer volume is always emptied during feeding. In this case, Q_{feed} can be derived by rearranging Eq. 11.8 from Chen *et al.* (2020) to Eq. 11.41.

During the feeding phase at a Q_{DWF} of 4,800 m^3/d, V_{buffer} = 0 m^3 or:

$$V_{buffer} = 0 = \frac{Q_{DWF}}{n_{reactor} \cdot n_{cycles}} - Q_{feed} \cdot \frac{t_{feed}}{24} \tag{11.64}$$

In this case, for which specific data can be found in the sheet 'Buffer check':

$$Q_{feed} = \frac{Q_{DWF}}{n_{reactor} \cdot n_{cycles}} \cdot \frac{24}{t_{feed}} = 400 \text{ m}^3/\text{h} \tag{11.65}$$

e) <u>Influence of PDWF and PWWF.</u>

Both plants could be designed for PDWF conditions assuming a certain S_F^{PDWF}. S_F^{PDWF} will increase the total plant volume proportionally. The efficiency remains constant. PWWF conditions will comply with the design for PDWF if the reactor scheduling is changed (*e.g.,* shorter feeding and/or reaction time). The buffer tank will not be emptied during feeding as the hydraulic constraints cannot be fulfilled at all times.

11.4 EXERCISES

Process characteristics (exercises 11.4.1-11.4.4)

Exercise 11.4.1

Reactor configuration

1. What are the main differences concerning reactor configuration between an aerobic granular sludge (batch) system and a conventional (continuous) activated sludge system?
2. What are the main differences between traditional sequencing batch processes and aerobic granular sludge (batch) systems?

Exercise 11.4.2

Advantages of AGS processes

What are the main advantages of an aerobic granular sludge (batch) system compared to a conventional (continuous) activated sludge system?

Exercise 11.4.3

Unit operations and conversion processes

1. Which four unit operations of an activated sludge plant can be compared to the processes in a single aerobic granular sludge tank?
2. How is it possible that all four processes can be performed in one reactor?

Exercise 11.4.4

Primary settling – suspended solids removal

1. What are the advantages of including a primary settling tank in conventional activated sludge systems and to a lesser extent in AGS systems?
2. How are suspended solids removed when a primary settling tank is not included in the AGS system design?
3. What are the advantages of avoiding the construction of a primary settling tank?

Granulation process (exercises 11.4.5-11.4.12)

Exercise 11.4.5

Drivers for granulation

What are the main drivers for granulation and how are these drivers established?

Exercise 11.4.6

Feast-famine regime

How does the feast-famine regime contribute to stable granulation? What happens during the feast-famine regime? Situate this regime in the SBR cycle.

Exercise 11.4.7

CSTR

Why is a continuous stirred tank reactor (CSTR) not preferable for granule formation?

Exercise 11.4.8

Heterotrophs

Will fast-growing heterotrophic bacteria be able to develop in the aerobic granular sludge reactor?

Exercise 11.4.9

Substrate type

Which substrate is suitable for the formation of compact granules? How does the take-up rate of this substrate influence the anaerobic feeding time of a full-scale and lab-scale plant and how does this differ with industrial wastewater?

Exercise 11.4.10

Shear stress

Does aerobic granular sludge coagulate under reduced shear stress? Is shear an important granular selection prerequisite?

Exercise 11.4.11

Sludge selection spill versus excess granular sludge

What is the difference between sludge selection spill and excess granular sludge?

Exercise 11.4.12

Relation between substrate uptake profile, biomass growth and floc/granule structure

The substrate diffusion profile influences the biomass growth pattern and the resulting floc or granule structure. However, the effects are mixed in Figure 11.2. Can you combine the images on the substrate diffusion (A-D) with the corresponding biomass growth pattern (I-IV) and floc or granule structure (1-4)? Which feeding pattern and/or type of influent substrate does this correspond with?

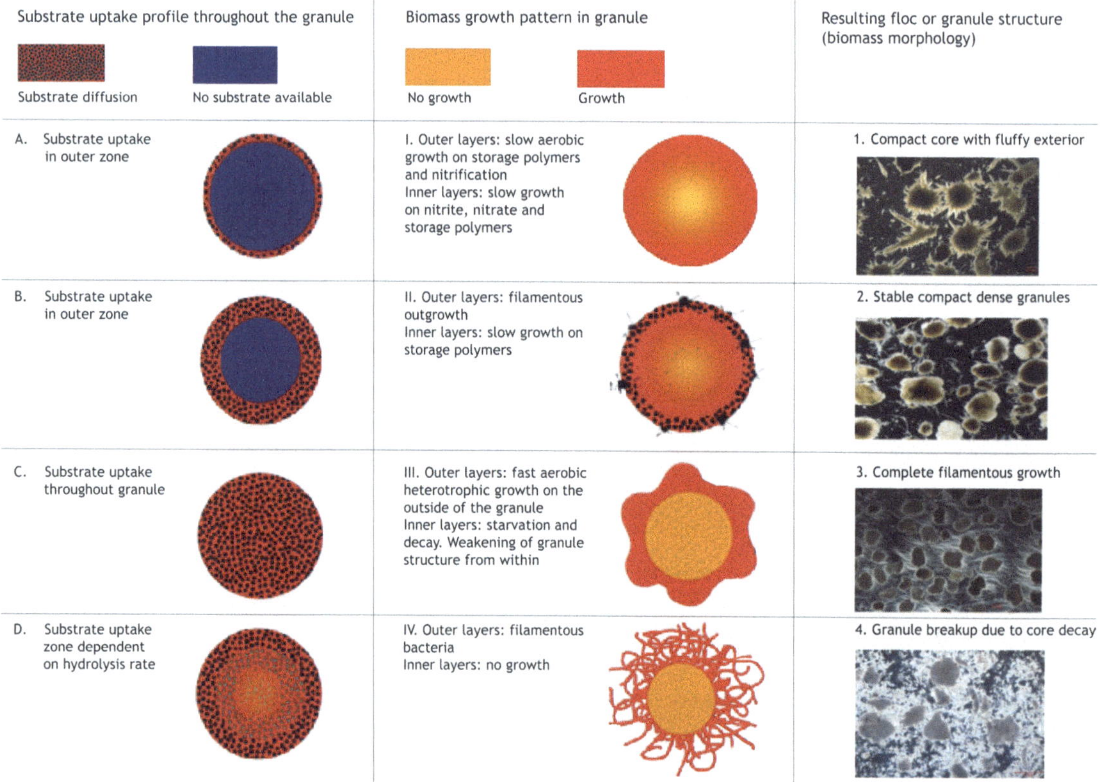

Figure 11.2 Variations in substrate uptake profile, biomass growth pattern and biomass morphology.

Kinetics (exercises 11.4.13-11.4.25)

Exercise 11.4.13

Batch cycle dynamics

The (predicted) concentration profile of several substrates during a batch cycle of an AGS reactor is depicted in Figure 11.3. Indicate which substrates are represented by the curves A to D.

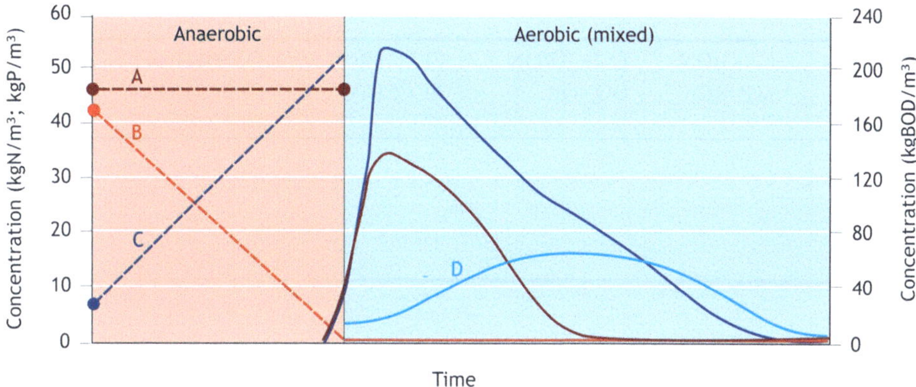

Figure 11.3 Batch cycle dynamics. The solid lines are measured concentrations at the top of the reactor and the dashed lines denote predictions of liquid concentrations during the anaerobic plug-flow regime.

Exercise 11.4.14

Batch cycle dynamics

The AGS process can be divided into two different redox phases: the anaerobic and the aerobic phase, which determine the feast/famine cycle in the reactor. During these phases the granules, composed of different bacteria, complete some reactions of their metabolism, changing the water quality.

Complete the table below, indicating whether the bulk concentration of the specific compound increases or decreases in the specific phase (indicate with 'Increase' or 'Decrease'). If the compound is not involved or remains constant during the specific phase choose —'—'.

	ANAEROBIC PHASE	AEROBIC PHASE
BOD		
PHA		
CO_2		
PO_4^{3-}		
POLY-P		
NH_4^+		
NO_3^-		
N_2		

Exercise 11.4.15

Microbial populations - substrates

Characterize the microbial populations according to their respective carbon source and electron donor and acceptor pairs in the aerobic growth period. Complete the table below with these possible answers (multiple answers are possible per table cell):

BOD, CO_2, N_2, H_2O, NO_3^-, PHA, NH_4^+, O_2

	CARBON SOURCE	ELECTRON DONOR	ELECTRON ACCEPTOR	OXIDIZED ELECTRON DONOR	REDUCED ELECTRON ACCEPTOR
PAO					
GAO					
NITRIFIERS					
DENITRIFIERS					

Exercise 11.4.16

Microbial populations - location in the granule

Complete the table choosing from the options given; multiple answers are possible per table cell:

	ORGANISM TYPE (autotroph or heterotroph?)	METABOLISM (aerobic or anoxic?)	LOCATION IN THE GRANULE (inner, middle or outer?)
PAO			
GAO			
NITRIFIERS			
DENITRIFIERS			

Exercise 11.4.17

Effluent concentrations

Why is the COD, N and P concentration in the effluent of an AGS system usually low enough to be discharged without post-treatment?

Exercise 11.4.18

Substrate conversion rates

The substrate conversion rate in an SBR is depicted in Figure 11.4. The red dashed line represents the growth rate for a CSTR, which is constant over time. The theoretical NH_4-N removal efficiency of the SBR is higher than of a CSTR. Which line represents the substrate conversion rate in the SBR?

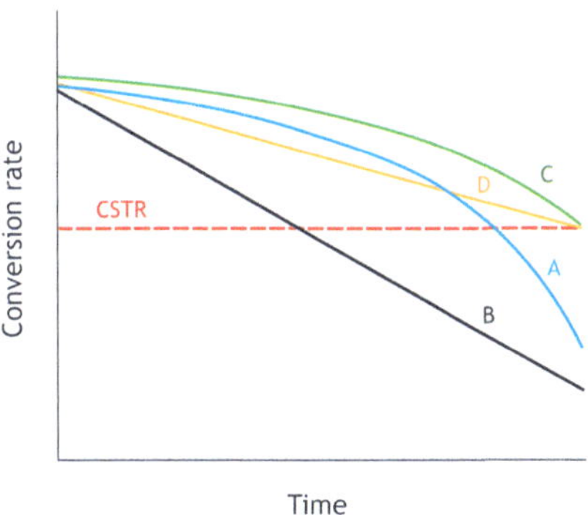

Figure 11.4 Substrate conversion rates in a sequencing batch reactor.

Exercise 11.4.19

Nitrification and denitrification
Is there a need for external dosage of organic carbon during post-denitrification in AGS?

Exercise 11.4.20

Phosphate removal in AGS
One peculiarity of the PAOs is the production of PHA and the capacity of PO_4^{3-} uptake. From literature, it is usually found that PAO can accumulate up to 0.30 gP/gVSS, while in normal organism P uptake is usually around 0.02 gP/gVSS. The growth yield of both ordinary heterotrophs, as well as phosphate accumulating organisms averages 0.4 kgVSS/kgCOD$_{consumed}$.

a) Why is the phosphate uptake higher than the phosphate release?
b) What causes nett P removal in an AGS system?
c) Why is PO_4^{3-} uptake competing with the oxygen consumption of nitrification?
d) Assume an AGS reactor with a biomass concentration of 8 kgVSS/m^3 . This plant receives an influent flow of 2,000 m^3/d, with a VFA concentration of 100 mgCOD/L, a total COD concentration of 400 mg/L and a phosphate concentration of 10 mg PO_4^{3-}-P/ L. Calculate the theoretical P uptake by PAO and the phosphate concentration that can be reached in the effluent. Assume that PAO only use the influent VFA for growth, while the other heterotrophs use the remaining COD for growth.

Exercise 11.4.21

Influence of temperature
The activity of bacteria, and consequently the oxygen uptake rate of micro-organisms is highly dependent on the temperature, which can be expressed by the Arrhenius equation (Eq. 11.66).

$$k(T) = k(20°C) \cdot \theta^{(T-20)} \tag{11.66}$$

k(T) is the maximum growth rate at temperature T (°C) and θ the Arrhenius coefficient (-).

a) Complete the missing words to describe the temperature dependency:
    A lower temperature means that nitrifiers will consume …............. oxygen per time, and thus the thickness
    of the aerobic layer is ……… .
b) Figure 11.5 displays the ammonium consumption rate as a function of temperature (De Kreuk *et al.*, 2005).
    Which line indicates the activity in AGS that is due to the temperatures in the system? The red line or the
    blue one?

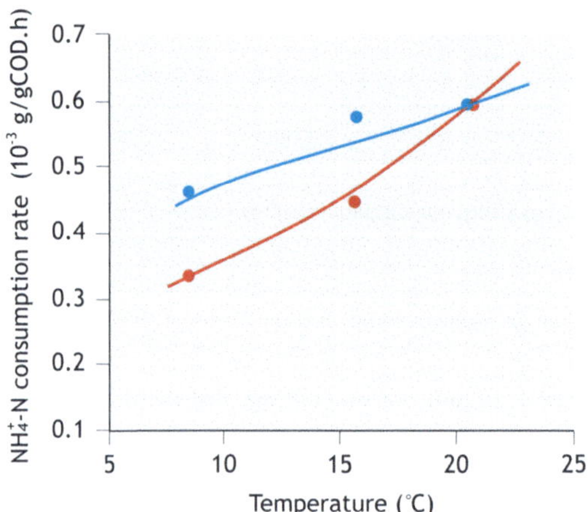

Figure 11.5 Temperature dependency of the ammonium consumption rate.

c) The oxygen uptake rate of nitrifying bacteria is 1.9 $gO_2/gVSS.h$ at 20 °C. What is the conversion rate at
    5 °C, given the Arrhenius coefficient of $\theta = 1.06$?
d) Will PAO adapt their activity at varying temperatures in the long term?

Exercise 11.4.22
Influence of dissolved oxygen concentration
What are the two main parameters determining the thickness of the aerobic layer?

Exercise 11.4.23
Influence of dissolved oxygen concentration
Which conversion process will be affected and what is the effect on the effluent in the case of
a) an increasing DO?
b) a decreasing DO?

Exercise 11.4.24
Influence of granule size
How does the size of the granules influence the ratio of aerobic/anoxic volume of the granules? Complete the
following sentence.

At the same DO, smaller granules have a relatively …… aerobic volume and …… anoxic volume than
larger granules.

Exercise 11.4.25

Oxygen consumption rate, oxygen penetration depth and anoxic volume fraction

Assume an aerobic granular sludge reactor filled with granules of a diameter of 3 mm. The oxygen concentration during the nitrification phase in the bulk liquid is kept at 3 mgO$_2$/L. The nitrification process is the highest oxygen consumer, accounting for 80 % of the oxygen uptake rate. The ammonium conversion rate is 0.4 gNH$_4$-N/gVSS.h at 20 °C. Recall that the stoichiometric oxygen use is 2 moles O$_2$ per mol NH$_4$-N.

a) What is the oxygen consumption rate q$_s^{max}$ in gO$_2$/gVSS.h in this system during aeration?

b) Calculate the oxygen penetration depth in the granules with Eq. 11.67. Take the following assumptions into account: the boundary layer approaches 0 mm (C$_{si}$= C$_b$); the diffusion coefficient D$_{O2}$ = 1.4 ·10^{-9} m^2/s; the biomass concentration in the granule (C$_x$) is 80 kgVSS/m^3.

$$3 \cdot \partial - \frac{2 \cdot \partial^3}{R} = \frac{6 \cdot D_{O2} \cdot C_{si}}{q_s^{max} \cdot C_x} \tag{11.67}$$

With ∂, the penetration depth (m), R, the granule radius (m), D$_{O2}$, the diffusion coefficient of oxygen in the granule (m^2/s), C$_{si}$, the concentration at the granule surface (gO$_2$/m^3), q$_s^{max}$, the maximum uptake rate (gO$_2$/gVSS.h) and C$_x$, biomass concentration in the granule (kgVSS/m^3).

c) What is the fraction (%) of anoxic biomass in the granule, assuming that the nitrate diffuses to the core of the granule?

Process monitoring and control (exercises 11.4.26-11.4.33)

Exercise 11.4.26

Monitoring

1. Why are the measured concentrations low during reactor feeding?
2. Indicate which measurements are usually automated and which ones are manual?

Exercise 11.4.27

Process operation

1. What are the different phases in the operation of an aerobic granular sludge lab-scale SBR?
2. What is the difference in phases with a full-scale aerobic granular sludge process and what are the consequences?
3. How is the sludge spilled in a lab-scale reactor compared to a full-scale reactor?

Exercise 11.4.28

Batch operation

What is the critical difference between process control of a continuous versus a batch process?

Exercise 11.4.29

Upflow velocity and volume exchange ratio

Figure 11.6 displays the upflow velocity in the reactor, v$_{upflow}$ (m/h), and the volume exchange ratio, VER (-), as a function of the wastewater flow rate fed to the AGS reactor.

a) How is the upflow velocity calculated?
b) What does VER represent?
c) Find the maximum flow that can be reached without changing the feeding time. What will happen at higher flow rates?

d) What is the risk when the maximum upflow velocity is exceeded?

e) To reduce the upflow velocity, which is an important constraint during rainy weather flow, one possible solution is the decrease of the height of the reactor. What are the negative impacts related to lowering the design height?

f) Why is cycle shortening a standard solution for very high rain weather flow conditions?

g) When influent reaches the WWTP via a pressurised sewer, why is it not recommended to switch to the shorter rain weather cycle times immediately? What is usually done?

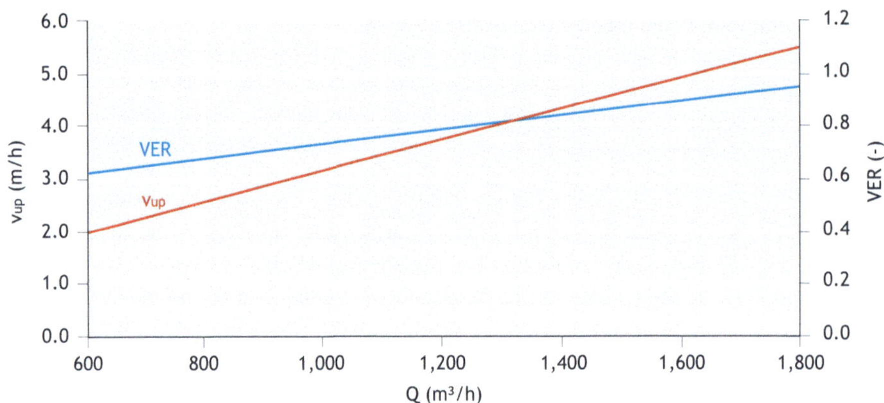

Figure 11.6 Upflow velocity in the reactor v_{up} and volume exchange ratio VER as a function of the wastewater flow rate Q.

Exercise 11.4.30

Variations between cycles

SBR reactors experience fluctuations due to the different influent batches they receive. The following variables may differ between batches: temperature, COD, nitrogen and phosphorus concentration, flow rate and pH. Figure 11.7 shows the concentration profiles of ammonium, nitrate, phosphate and dissolved oxygen over different batch cycles.

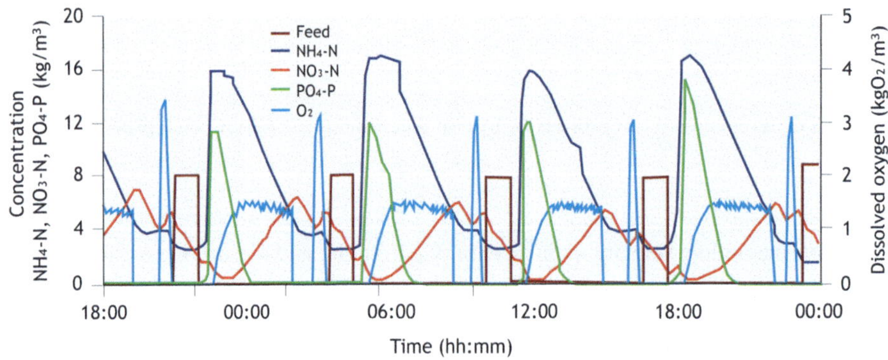

Figure 11.7 Substrate concentration profiles over different batch cycles.

a) What are the two main parameters influencing the concentration peaks?
b) Just before the next feeding period, the DO concentration peaks to a high value. It appears that the aeration is started for a short while at high capacity. Why?
c) Identify the moment at which the operator turns off the aeration. Why is this done?
d) In domestic wastewater treatment the concentration of NH_4^+ is normally approximately 4 to 7 times higher than the concentration of PO_4^{-3}. However, in the graph this is not the case. What happened in the AGS reactor?

Exercise 11.4.31

Sludge loading rate and reactor performance

The concentration peaks per batch vary over the day. Accordingly, the sludge load also varies over the day; this similarly occurs in the conventional activated sludge process. However, to make sure that the sludge loading per batch stays optimal, the concentration peaks in the batch cycle are controlled by adapting the feeding time.

a) When the influent is transported to the WWTP via a combined sewer, the COD concentration in the influent will decrease significantly during a rain event (assume that the COD of rain is equal to zero). How does the sludge loading rate change?
b) Which possible risks occur for the AGS reactor at non-optimal sludge loading rates?

Exercise 11.4.32

Simultaneous nitrification-denitrification

The graph below shows the ammonium and nitrate profile during a cycle. Which arrow represents the N conversion via simultaneous nitrification-denitrification?

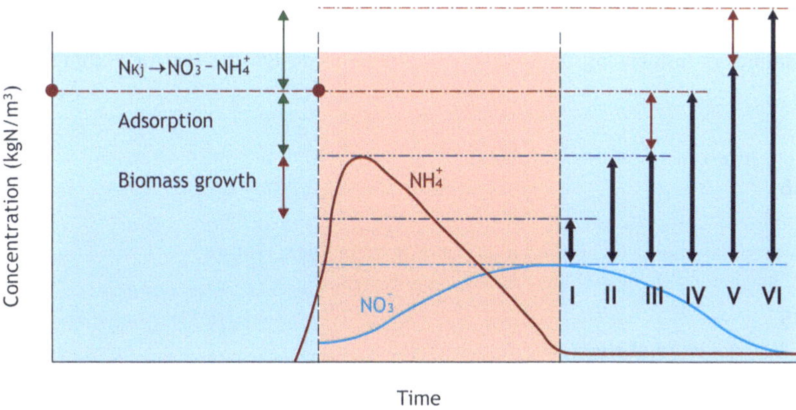

Figure 11.8 Concentration profiles of nitrogen compounds and their relation to conversion processes.

Exercise 11.4.33

Suspended solids

a) Which are the two factors that could hinder the settling of suspended solids and contribute to their presence in the effluent?
b) Which three measures can be taken to marginalise the presence of suspended solids in the effluent?

Process configuration (exercises 11.4.34-11.4.37)

Exercise 11.4.34

System setup

Which three setups ensure processing of a continuous flow of wastewater in an aerobic granular sludge wastewater treatment plant?

Exercise 11.4.35

Influence of a buffer tank

Why can the AGS reactor be designed smaller when a buffer tank is applied?

Exercise 11.4.36

Reactor volume – influencing factors

Indicate in the following table whether the variables will lead to a smaller or larger reactor volume per reactor.

Variable	Does the variable make the volume (per reactor) smaller or larger?
Influent flow increases (m^3/d)	
PWWF/DWF ratio increases (-)	
COD concentration increases (g/m^3)	
Total reaction time per day increases	
Sludge concentration increases (g/m^3)	
Sludge loading rate increases (kgCOD/kg TSS.d)	
Volumetric loading rate increases (m^3/m^3.d)	
Number of reactors increases (-)	

Exercise 11.4.37

Hydraulic constraints

What are the parameters influencing the hydraulic restrains to design the reactor volumes, assuming a fixed Q^{DWF}?

Resource recovery (exercises 11.4.38-11.4.39)

Exercise 11.4.38

Sludge application

How can excess granular sludge be applied today and/or in the (near) future?

Exercise 11.4.39

Methane production potential of sludge

Which sludge type has the highest methane potential (in m^3CH_4/gVSS)? Put the following in order from high to low: primary sludge, activated sludge, aerobic granules, selection spill.

ANNEX 1: SOLUTIONS TO EXERCISES

Process characteristics (solutions 11.4.1-11.4.4)

Solution 11.4.1

Reactor configuration

1. Main differences concerning reactor configuration between an aerobic granular sludge (batch) system and a conventional (continuous) activated sludge system:

    - In contrast to conventional activated sludge systems, AGS systems are either built as minimum three AGS reactors in parallel, or as one or more AGS reactors (usually two) and a buffer tank. Indeed, the batch-wise operation implies that a single reactor cannot receive influent all the time, since influent cannot be fed during the reaction or settling phase. A minimum of three AGS reactors in parallel is required for continuous operation. Alternatively, it is possible to opt to build a buffer tank and as such reduce the number of AGS reactors. The latter may be more economical for smaller plants (requiring a low reaction volume) as well as for larger plants that would require more than three reactors (and where a buffer tank enables a shorter feeding phase with higher flows, and thus a relatively longer reaction time). Note that for large facilities, activated sludge processes are also designed with parallel reactors.

    - Another main difference is that an AGS reactor is always fed in plug-flow mode. The plug-flow regime allows simultaneous feeding and discharge.

2. Main differences between traditional sequencing batch processes and aerobic granular sludge (batch) systems:

    - The feeding in AGS is from the bottom in a plug flow through the reactor. Therefore, the reactor has a constant volume and there is no need for a mechanical decanter for effluent extraction, nor for decanting time within the cycle length. The plug-flow feeding also eliminates the need for a mixer during the anaerobic phase, since the influent is in contact with the granules from the moment it enters the reactor.

    - Due to the fast settling, there is virtually no cycle time required for sludge settling; only a short time to eliminate turbulence after the aeration phase is required.

Solution 11.4.2

Advantages of AGS processes

An aerobic granular sludge batch reactor compared to a continuous activated sludge plant:

- Has a smaller footprint: no need for settling tanks and the AGS reactor is designed to be more compact, because of high biomass concentrations allowing a high volumetric loading rate (Pronk *et al.*, 2015a).
- Needs fewer construction materials, for the same reasons.
- Needs less mechanical equipment, such as recycle pumps, return sludge pumps, clarifier bridges, mixers. This saves on investment, maintenance requirements and energy usage (Pronk *et al.*, 2015a).
- Reaches a high effluent quality more easily: batch-wise operation ensures a relatively high substrate concentration compared to completely mixed reactors, thus minimizing diffusion limitation and therefore allowing higher conversion rates. Also, the reactor content is not continuously fed with wastewater, resulting in a faster drop in concentration in the liquid phase of the reactor. Lastly, the N removal is not dependent on an internal recycle flow from the aeration tank to the pre-denitrification tank[1], but can be controlled by the oxygen set-point and length of the post-denitrification phase.

---

[1] In activated sludge plants, the internal recycle from the nitrification tank to the pre-denitrification tank determines the nitrate concentration in the effluent assuming all ammonium is nitrified and all nitrate returned is denitrified. For example: with a recycle flow of 4 times the influent flow, approximately 20 % of the incoming ammonium will still leave the activated sludge plant in the form of nitrate.

Solution 11.4.3

Unit operations and conversion processes

1. The anaerobic, aerobic, anoxic, and settling tank.
2. The large size of granules compared to activated sludge flocs enables different redox zones (from outside to core: aerobic, anoxic, and (with large granules) anaerobic) in each granule, favouring bacterial communities with different functions (organic carbon removal, phosphorus removal, nitrification, and denitrification). The anaerobic plug-flow feeding through the settled bed followed by the aeration and post-denitrification phase enables biological phosphate removal. The high settling velocity of the granules enables the integration of the settling in the reactor. Moreover, batch scheduling makes it possible to adapt the phases within the cycles as well as the cycle length to reach the desired effluent quality (*e.g.,* by applying a separate denitrification phase to complement simultaneous nitrification-denitrification).

Solution 11.4.4

Primary settling – suspended solids removal

1. A primary settling tank for conventional activated sludge is advantageous to decrease the load on the biological reactor and the associated oxygen consumption by removing suspended solids. Moreover, primary sludge has a very good biogas production potential and is therefore advantageous in view of energy recovery. In an AGS system, suspended solids are removed via the sludge selection spill during settling so there is not a real need for a primary settler for suspended solids removal. Also, the total AGS excess sludge (consisting of the floc fraction with suspended solids on the one hand and granules on the other hand) has a similar biogas production potential as the combined primary and secondary sludge for activated sludge tanks.
2. The suspended solids will be mainly either removed by uptake by protozoa, metazoa or end up in the flocculated sludge fraction. This is removed as excess sludge during the sludge selection spill during settling in the batch cycle. Colloidal material that will also partly be removed as primary sludge could potentially also be hydrolysed in the anaerobic feeding phase and result in substrate for EBPR. The flocculent fraction has a short solid retention time (SRT) in the process (Ali *et al.*, 2019) and particulates are therefore marginally mineralised in the AGS reactor.
3. The advantages of avoiding the construction of a primary settling tank are: a smaller area required for the treatment plant (even though the biological reactor may need to be slightly larger), less odour emission and easier operation.

Granulation process (solutions 11.4.5-11.4.12)

Solution 11.4.5

Drivers for granulation

The formation of compact granules is stimulated by:

- having high substrate concentrations during feeding and making sure that diffusion is not limiting which is established by plug-flow feeding under anaerobic conditions.
- selecting slow-growing organisms which is established by an anaerobic feeding phase in which bacteria take up readily biodegradable substrate and convert it into cell internally stored polymers. These bacteria therefore do not compete on growth rate and do not have a high growth rate.
- selective pressure due to the difference in settling velocity. This makes it possible to selectively remove the flocculent biomass and selectively feed the faster settling (larger) granules.

Solution 11.4.6

Feast-famine regime

The feast-famine regime ensures periods with high substrate concentrations in the reactor alternate with periods without substrate supply. This favours slow-growing organisms. During the feast phase, *i.e.*, the anaerobic feeding phase, there is readily biodegradable BOD (RBCOD) uptake by PAO and GAO and conversion to storage polymers. However, during the famine phase, *i.e.*, the aeration phase, there is consumption of storage polymers by PAO and GAO, since they do not compete on growth rate; they do so at a relatively low specific growth rate.

Solution 11.4.7

CSTR

Substrate concentrations in CSTRs are typically very low, leading to low substrate penetration in immobilised biomass such as granular sludge. This induces flocculent biomass growth since this will have less influence from diffusion limitation.

Solution 11.4.8

Heterotrophs

No, because the RBCOD concentration in the aerobic phase is almost zero and if they grow, they are removed within 0.5-5.0 d via the sludge selection (or excess sludge) spill.

Solution 11.4.9

Substrate type

A suitable substrate for the formation of compact granules is RBCOD. Municipal wastewater does not usually contain a lot of RBCOD (≤ 100 g/m^3, depending on the type of sewer system). An anaerobic feeding phase of 1 hour (determined in general more by the hydraulic design than based on biokinetic parameters) is more than sufficient to take up all the RBCOD by the granular sludge. A part of the slowly biodegradable COD can also be anaerobically hydrolysed, fermented and stored during the 1-hour feeding phase. However, in lab-scale reactors and/or industrial treatment plants, the anaerobic feeding time could differ according to the anaerobic PHA storage. In these reactors the hydraulic considerations are less dominant and the anaerobic time might be designed according to the anaerobic substrate uptake kinetics.

Solution 11.4.10

Shear stress

No, the granule has a stable structure and does not coagulate under reduced hydrodynamic shear. With the selection for PAO and GAO during the anaerobic feed, shear is not an important granular selection prerequisite.

Solution 11.4.11

Sludge selection spill versus excess granular sludge

The sludge selection spill mainly consists of the flocculent fraction and is discharged during settling to establish a hydraulic selection pressure for the sludge that is not settling well (Pronk *et al.*, 2015a). The excess granular sludge is granules that are discharged in order to maintain a constant biomass concentration in the reactor and also to remove large granules that consist of a significant inactive volume fraction. Both these discharges differ in concentration, morphology and SRT of the sludge.

Solution 11.4.12

Relation between substrate uptake profile, biomass growth and floc/granule structure

			Feeding pattern	Influent substrate type
C	I	2	Anaerobic	Readily biodegradable dissolved substrates
B	III	4	Aerobic mixed (pulse)	Readily biodegradable dissolved substrates
D	II	1	Anaerobic	Polymeric substrates
A	IV	3	Aerobic mixed (slow)	Readily biodegradable dissolved substrates

Kinetics (solutions 11.4.13-11.4.25)
Solution 11.4.13
Batch cycle dynamics

$A = NH_4^+$

B = BOD (NOTE: BOD could be higher at the start of the aeration time and will follow a decreasing trend during aeration. BOD ending up in the aerobic phase comes from slowly biodegradable COD. It either partly contributes to flocculent sludge growth (removed via the sludge selection spill indirectly) or is incorporated with the more flocculant sludge and then discharged (removed via the sludge selection spill directly).

$C = PO_4^{3-}$

$D = NO^{3-}$

Solution 11.4.14

Batch cycle dynamics

	ANAEROBIC PHASE	AEROBIC PHASE
BOD	Decrease	--/ Decrease
PHA	Increase	Decrease
CO_2	Small increase[2]	Increase
PO_4^{3-}	Increase	Decrease
POLY-P	Decrease	Increase
NH_4^+	--	Decrease
NO_3^-	---	Increase[3]
N_2	--	Increase

---

[2] Some CO_2 is produced by EBPR and some by fermentation processes.

[3] Due to simultaneous nitrification-denitrification, nitrogen is removed during the aeration phase of the AGS process. Depending on the anoxic volume of the granule, not all nitrate will be removed and therefore nitrate will accumulate during the aeration phase. To increase nitrate removal, the process operation can be adapted to stimulate denitrification. This can be done by switching aeration on and off during the reaction phase or by adding a post-denitrification phase after the aeration phase.

Solution 11.4.15

Microbial populations – substrates

	CARBON SOURCE	ELECTRON DONOR	ELECTRON ACCEPTOR	OXIDIZED ELECTRON DONOR	REDUCED ELECTRON ACCEPTOR
PAO	PHA	PHA	O_2/NO_3^-	CO_2	H_2O/N_2
GAO	PHA	PHA	O_2/NO_3^-	CO_2	H_2O/N_2
NITRIFIERS	CO_2	NH_4^+	O_2	NO_3^-	H_2O
DENITRIFIERS	BOD	BOD	NO_3^-	CO_2	N_2

Solution 11.4.16

Microbial populations - location in the granule.

	ORGANISM TYPE (autotroph or heterotroph?)	METABOLISM (aerobic or anoxic?)	LOCATION IN THE GRANULE[4] (inner, middle or outer?)
PAO	Heterotroph	Substrate uptake: anaerobic Growth: aerobic/anoxic	Middle
GAO	Heterotroph	Substrate uptake: anaerobic Growth: aerobic/anoxic	Middle
NITRIFIERS	Autotroph	Aerobic	Outer
DENITRIFIERS	Heterotroph	Anoxic	Inner

Solution 11.4.17

Effluent concentrations

In a continuous reactor the concentrations are always low, thus giving conversion close to the K_S value for the substrate, *i.e.*, reaction rates are reduced. In a batch operation the concentrations are initially high resulting in high conversion rats. Moreover, in a continuous reactor there is continuous input of pollutants making it more difficult to reach very low values. In addition, the cycle time and different phase lengths can be adapted to reach the optimal effluent quality.

---

[4] A granule can be described by different zones due to the diffusion limitation of oxygen. There is an aerobic outer layer and an anoxic core. Sometimes an anaerobic core can be present although this is not taken into account in this exercise, as an anaerobic core does not have a function. However, it can exist if the oxygen as well as the nitrate does not penetrate to the inner core of the granule. This happens when granules become large.

Solution 11.4.18

Substrate conversion rates

Line A.

Theoretically the rate of conversion is not linear (which eliminates B and D), and also the conversion rate at the end of a AGS reactor cycle is usually lower than for a CSTR reactor. At the end of the batch cycle, the effluent concentrations reached in a batch reactor are generally lower than in a CSTR system.

Solution 11.4.19

Nitrification and denitrification

External carbon dosage is not required because during the anaerobic feeding there is storage of PHA throughout the granule. During aeration, only the PHA in the outer layer is oxidised with O_2. Inside the granule the PHA is already used for denitrification during the aerobic phase. During the post-denitrification phase PHA is used in the entire granule. This is in contrast to activated sludge, where the entire floc is aerobic during aeration and many more storage polymers are oxidised with O_2.

Solution 11.4.20

Phosphate removal in AGS

a) Due to the net biomass growth (which is determined by the biomass yield).
b) Nett P removal is obtained by wasting the excess of biomass accumulating P.
c) PAO will use oxygen as the electron acceptor to oxidize the storage polymer PHA during the aerobic phase. This creates energy for their anabolism and for the uptake of PO_4^{3-}. Since nitrifiers also need oxygen, both populations compete for it.
d) 400 mgCOD/l is consumed, which means that 160 mgVSS/l of heterotrophs are produced (0.4 mgVSS/mgCOD consumed · 400 mgCOD consumed/l). To maintain a biomass concentration of 8 kgVSS/m^3, 160 mgVSS of heterotrophs will be wasted. Roughly, ¼ of the COD (100 mg of the total 400 mgCOD is VFA) is consumed by PAO, the other ¾ is likely consumed by ordinary heterotrophs.

Net PO_4^{3-} uptake:
By PAO: (1/4 · 160 mgVSS/L) · 40 mgVSS/L · 0.30 mgP/mgVSS = 12 mgP/L
By OHO: (3/4 · 160 mgVSS/l) · 120 mgVSS/L · 0.02 mgP/mgVSS = 2.4 mgP/L
Phosphate concentration reached in the effluent: 0 mgP/L.

Solution 11.4.21

Influence of temperature

a) A lower temperature means that nitrifiers will consume less oxygen per time, and thus the thickness of the aerobic layer is extended.
b) The blue line represents the long temperature effect, as the overall nitrification rate is recovered by granule adaptation to the new situation. The red line represents short temperature effects.
c) $k(5\ °C) = 1.9\ gO_2/gVSS.h · 1.06^{(5-20)} = 0.79\ gO_2/gVSS.h$.
d) The overall denitrification and phosphate removal rate stay lower for decreasing temperature in the short and the long term. PAOs grow with oxygen or nitrate as an electron acceptor. An increased aerobic zone means at the same time a decreased anoxic zone. The granule volume in which PAO can grow stays the same, so there will be no extra space for them to grow to increase their concentration and compensate for the lower conversion rates because of the lower temperature.

Solution 11.4.22

Influence of dissolved oxygen concentration

1) Activity of the micro-organisms and dissolved oxygen (DO) concentration.
2) Oxygen is used to oxidize ammonium to nitrate and to oxidise BOD. The aerobic and anoxic layer thickness in the granules will be determined by the amount of oxygen in the bulk and the rate at which it is oxidized at the granule outer layer.

Solution 11.4.23

Influence of dissolved oxygen concentration

a) Increasing DO: denitrification will be affected because the anoxic volume is smaller. NO_3^- will remain in the effluent.
b) Decreasing DO: nitrification will be affected because the aerobic volume is smaller. More NH_4^+ in the effluent.

Solution 11.4.24

Influence of granule size

At the same DO, smaller granules have a relatively higher aerobic volume and lower anoxic volume than larger granules.

Solution 11.4.25

Oxygen consumption rate, oxygen penetration depth and anoxic volume fraction

a) To oxidize 0.4 gNH_4-N/gVSS.h or 0.03 molN/gVSS.h (N = 14 g/mol), 0.06 $molO_2$/gVSS.h or 1.92 gO_2/gVSS.h is consumed. In total 2.4 gO_2/gVSS.h is consumed (1.92/0.8 gO_2/gVSS.h).
b) Wolfram Alpha or Excel Solver can be used to solve this cubic equation. Tutorials of how to solve equations using the Excel solver are readily accessible on YouTube. The oxygen penetration depth in the granules is 443 μm.
c) Anoxic biomass volume fraction = $(1.5 - 0.443)^3$ mm^3 / 1.5^3 mm^3 = 0.35 or 35 %.

Process monitoring and control (solutions 11.4.26-11.4.33)

Solution 11.4.26

Monitoring

1. Due to the plug-flow regime during the feeding phase, the instruments (kept at the top of the reactor) still measure the effluent concentration of the previous batch, thereby measuring low values (Pronk *et al.*, 2015a).
2. Automated: NH_4^+, NO_3^-, ORP, pH. Manual: COD, total N, total P.

Solution 11.4.27

Process operation

1. Fill, react, settle, drain, and idle phase.
2. In the full-scale process the filling and draining phase are combined, which saves cycle time and complex constructions. Smaller reactors can be built. The full-scale plant works at a constant volume.
3. In a lab-scale reactor, withdrawal of liquid at around the mid-column level is used as a selection pressure on fast-settling sludge. In full-scale Nereda© plants however, effluent flows out from the top. The hydraulic selection pressure is established via a separate sludge selection spill from the reactor for the poor settling sludge, which is called the sludge selection spill (Pronk *et al.*, 2015b).

Solution 11.4.28

Batch operation

Both operations try to achieve the desired effluent concentrations. In a continuous process the control is often based on a set-point for the concentration equal to the effluent concentration. In a batch process the control is based on the reaction rate (*i.e.*, the change in concentration over time). The batch operation allows a more flexible operation. A flexible operation can better deal with external conditions, maintaining a high quality of the effluent. The batch operation can change the time schedule based on the influent conditions (flow rate, concentrations), therefore *e.g.*, the time schedule can be shortened during a peak hydraulic load. As in the conventional activated sludge system, there is the necessity to both design and operate the reactors based on the hydraulic and biological conditions.

Solution 11.4.29

Upflow velocity and volume exchange ratio

a) The upflow velocity in the reactor represents the influent (feed) flow rate divided by the cross-section area of the reactor (see Eq. 11.4).

b) The volume exchange ratio represents the ratio between the influent wastewater volume fed during a batch cycle and the total reactor volume (see Eq. 11.1).

c) $Q_{max} = 800$ m^3/h as the maximum VER is 65 %. At higher exchange ratios, breakthrough of influent in the effluent can take place.

d) At an upflow velocity higher than 5 m/h, there is the risk of fluidization of the settled granule bed causing influent and effluent mixing.

e) A larger footprint of the reactors, a larger number of air diffusors and injection points and enhancement of vertical diffusion. Besides, the VER is still 0.65, so the influent volume that can be dosed per batch will not change by changing the height over diameter ratio.

f) Shortening the cycle time, by reducing the reaction time, is a countermeasure to the VER increase when VER is over 0.65.

g) Because of the 'first flush' peak, causing an immediate high BOD loading rate. The first flush can be stored in a rainwater buffer tank and treated during dry weather conditions or it can be divided over the available reactors that are in the feed phase.

Solution 11.4.30

Variations between cycles

a) The influent concentration variations and the VER.

b) This is known as the stripping phase. Denitrification takes place in the settling granule bed, leading to dinitrogen gas formation and lower local NO_3^--N concentrations. The short aeration pulse helps to get rid of the dinitrogen gas that is in solution and would lead to bubble formation during the next feeding phase and by that flotation of the flocculent fraction and suspended solids accompanying the AGS in practice (van Dijk *et al.*, 2018). Moreover, it mixes the bulk liquid and gives a final boost for nitrification to reduce ammonium in the final effluent.

c) 4 gNH_4^+-N/m^3. When ammonium is sufficiently lowered, the remaining nitrate should be converted to N_2 by denitrifiers. Therefore, the anoxic sludge fraction is increased by switching off the aeration.

d) PAOs release PO_4^{-3} in the anaerobic phase, leading to a peak concentration at the moment aeration mixes the reactor completely. Furthermore NH_4^+ is absorbed onto the granule during the feeding phase, leading to a lower concentration than expected by the VER and the influent concentration (Bassin *et al.*, 2011).

Solution 11.4.31

Sludge loading rate and reactor performance

a) The sludge loading rate remains more or less constant. The sludge loading rate is defined by the COD concentration multiplied by the flow rate, divided by the amount of biomass in the reactor (Eq. 11.6). The loading rate of the system does not change too much with a rain event (the flow increases but the total COD load reaching the plant will probably be the same). It is assumed here that rainwater is relatively clean, although it might contain street run-off or solids from the sewer.

b) • More readily biodegradable COD than PAOs and GAOs can take up in a cycle, which reduces selection pressure for PAO and GAO and causes the proliferation of fast-growing heterotrophic bacteria in non-granular shape.

   • Oxygen demand is higher than oxygenation capacity.

   • Peak concentrations of NH_4-N and oxidisable organic N for which the available mass of nitrifiers cannot oxidise within the available aeration time and thus ammonium concentrations exceed effluent demands.

   • The presence of specific degradable compounds that may cause substrate inhibition when their concentration is too high (*e.g.*, for industrial wastewater).

NOTE: If there are a couple of cycles with non-optimal sludge loading, this will not damage the system straight away but it should not last for weeks.

Solution 11.4.32

Simultaneous nitrification-denitrification

Arrow V.

The fraction that is simultaneously converted cannot simply be determined by the difference between the ammonium consumed and the nitrate produced. It is also necessary to take into account the ammonia consumption due to biomass growth, since 12 % of the biomass consists of nitrogen. At the same time, ammonia is adsorbed to the biomass during feeding. This ammonia is still available for bioconversions. This fraction of adsorbed ammonia is not measured by the installed online measurements and can be as high as 25 % of the ammonia that is fed to the reactor. Furthermore, note that the online measurement determines ammonia, but not organic nitrogen. The nitrogen Kjeldahl is also available for nitrification, so that adds another 30 % to the nitrogen that needs to be nitrified. To summarize, if you determine the simultaneous denitrification rate by the difference between the ammonium consumption rate and the nitrate production rate, you underestimate the rate due to the ammonium adsorption and organic nitrogen, and you overestimate it due to the nitrogen that is used for growth.

Solution 11.4.33

Suspended solids

a) The presence of suspended solids in the effluent is caused either by rising sludge due to the degasification of nitrogen gas during the feed and decant phase (when the stripping phase has not been applied or was insufficient) and/or by wash-out of particles that intrinsically do not settle (*i.e.*, fats and foams in the influent) (Van Dijk *et al.*, 2018).

b) • Add an N_2 stripping phase before the feeding phase.

   • Install a baffle in front of the effluent discharge gutter that prevents a possible floating layer from entering the effluent discharge.

   • When effluent demands are very strict, it is possible to add a post-treatment such as sand filtration or membranes.

Process configuration (solutions 11.4.34-11.4.37)

Solution 11.4.34

System setup

- Minimum of three sequencing batch reactors in parallel with always one reactor in feed and draw mode for 1/3 of the total cycle time.
- Buffer tank in front of the sequencing batch reactor (only one batch reactor is sufficient). The influent is stored when the reactor(s) is/are in the aeration or settling phase.
- Combination of a conventional activated sludge system and AGS reactor in parallel where the influent flow is divided over the two systems.

Solution 11.4.35

Influence of a buffer tank

The AGS reactor can always be fed closer to the maximum VER or loading rate. The buffer tank levels the hydraulic peaks (*e.g.*, hydraulic regulation of daily peaks, wet weather flow).

Solution 11.4.36

Reactor volume – influencing factors

Variable	Does the variable make the volume (per reactor) smaller or larger?
Influent flow increases (m^3/d)	Larger
PWWF/DWF ratio increases (-)	Larger
COD concentration increases (g/m^3)	Larger
Total reaction time per day increases	Smaller
Sludge concentration increases (g/m^3)	Smaller
Sludge loading rate increases (kgCOD/kg TSS.d)	Smaller
Volumetric loading rate increases (m^3/m^3.d)	Smaller
Number of reactors increases (-)	Smaller

Solution 11.4.37

Hydraulic constraints

- The peak dry weather flow: this is the largest flow corresponding with the maximum reaction time (because the concentrations will be the highest).
- The length of the feed and draw phase.
- The volume exchange ratio (max. 0.65).
  - These three parameters will influence the hydraulic restraints to design the total reactor volume (m^3).
- The peak dry weather flow.
- The upflow velocity (max. 5m/h)
  - These two parameters will influence the hydraulic restraints to design the reactor area (m^2)

Resource recovery (solutions 11.4.38-11.4.39)

Solution 11.4.38

Sludge application

For agricultural reuse (if legally allowed), energy recovery, inoculation of new treatment plants, resource recovery of Extracellular polysaccharides (EPS) (Kaumera^{TM}).

Solution 11.4.39

Methane production potential of sludge
Primary sludge > Selection spill > Activated sludge > Aerobic granules (Guo *et al.*, 2020).

REFERENCES

Ali M., Wang Z., Salam K.W., Hari A.R., Pronk M., van Loosdrecht M.C.M. and Saikaly P.E. (2019). Importance of Species Sorting and Immigration on the Bacterial Assembly of Different-Sized Aggregates in a Full-Scale Aerobic Granular Sludge Plant. *Environmental Science and Technology,* 53, 8291–8301. https://doi.org/10.1021/ACS.EST.8B07303

Bassin J.P., Pronk M., Kraan, R., Kleerebezem R. and van Loosdrecht M.C.M. (2011). Ammonium adsorption in aerobic granular sludge, activated sludge and anammox granules. *Water Research,* 45, 5257–5265. https://doi.org/10.1016/J.WATRES.2011.07.034

Chen, G. H., van Loosdrecht, M.C.M., Ekama, G. A. and Brdjanovic D. (eds.) (2020). *Biological Wastewater Treatment: Principles, Modelling and Design.* ISBN: 9781789060355. IWA Publishing, London, UK.

De Kreuk M.K., Pronk M. and van Loosdrecht, M.C.M. (2005). Formation of aerobic granules and conversion processes in an aerobic granular sludge reactor at moderate and low temperatures. *Water Research,* https://doi.org/10.1016/j.watres.2005.08.031

Guo H., van Lier J.B. and de Kreuk M. (2020). Digestibility of waste aerobic granular sludge from a full-scale municipal wastewater treatment system. *Water Research,* 173. https://doi.org/10.1016/J.WATRES.2020.115617

Pronk M., de Kreuk M.K., de Bruin B., Kamminga P., Kleerebezem R. and van Loosdrecht M.C.M. (2015a). Full scale performance of the aerobic granular sludge process for sewage treatment. *Water Research,* https://doi.org/10.1016/j.watres.2015.07.011

Pronk M., de Kreuk M.K., de Bruin B., Kamminga P., Kleerebezem R. and van Loosdrecht M.C.M. (2015b). Full scale performance of the aerobic granular sludge process for sewage treatment. *Water Research,* 84, 207–217. https://doi.org/10.1016/j.watres.2015.07.011

van Dijk E.J.H., Pronk M. and van Loosdrecht M.C.M. (2018). Controlling effluent suspended solids in the aerobic granular sludge process. *Water Research,* 147, 50–59. https://doi.org/10.1016/j.watres.2018.09.052

NOMENCLATURE

Symbol	Description	Unit
C_b	Concentration in the boundary layer	gO_2/m^3
C_{si}	Concentration at the granule surface	gO_2/m^3
C_x	Biomass concentration in the granule	$kgVSS/m^3$
COD	Influent COD concentration	$kgCOD/m^3$
D_{O2}	Diffusion coefficient of oxygen in the granule	m^2/s
H	Reactor height	m
X_{TSS}	Mixed liquor suspended solids (sludge) concentration	$kgTSS/m^3$
k(T)	Maximum bacterial growth rate at temperature T	
n_{cycles}	Number of cycles per day per reactor	1/d
$n_{reactor}$	Number of reactors	-
R	Granule radius	m
S_F^{PDWF}	Daily peak flow factor	-
S_F^{PWWF}	Wet weather peak flow factor (= rain weather peak flow factor)	-

SLR	Sludge loading rate	kgCOD/kgTSS.d
T	Temperature	°C
t	Total cycle duration	h
t_{feed}	Feeding phase duration	h
t_{react}	Reaction phase duration	h
$t_{react,day}$	Total reaction time per day	h
t_{settle}	Sludge settling phase duration	h
V_{batch}	Volume fed during a new batch cycle	m^3
V_{buffer}	Total buffer volume	m^3
$V_{reactor}$	Reactor volume (for one reactor)	m^3
V_{upflow}	Upflow velocity	m/h
VER	Volume exchange ratio	%
VER_{max}	Maximum volume exchange ratio	%
VLR	Volumetric loading rate	$m^3/m^3.d$
Q	Influent flow rate	m^3/h
Q^{DWF}	Dry weather flow rate	m^3/h
Q_{feed}	Flow rate during the feeding phase	m^3/h
Q^{PDWF}	Peak dry weather flow rate (= daily peak flow rate)	m^3/h
Q^{PWWF}	Peak wet weather flow rate	m^3/h
q_s^{max}	Maximum oxygen uptake rate	$gO_2/gVSS.h$
∂	Oxygen penetration depth	m
θ	Arrhenius coefficient	-

Abbreviation	Description
AGS	Aerobic granular sludge
CSTR	Continuous stirred tank reactor
DO	Dissolved oxygen
DWF	Dry weather flow
EBPR	Enhanced biological phosphorous removal
EPS	Extracellular polysaccharides
GAO	Glycogen accumulating organism
PAO	Phosphate accumulating organism
PDWF	Peak dry weather flow
PE	Population equivalent
PHA	Poly hydroxy-alkanoates
PWWF	Peak wet weather flow
RBCOD	Readily biodegradable COD
SBR	Sequencing batch reactor
SRT	Solid retention time
WWF_{max}	Maximum wet weather flow

doi: 10.2166/9781789062304_0397

12

Final settling

Ferenc Házi and Eveline I.P. Volcke

12.1 INTRODUCTION

The abbreviations and nomenclature used in this chapter are identical to those in Chapter 12 in the textbook *Biological Wastewater Treatment: Principles, Modelling and Design* (Chen *et al.*, 2020), hereafter referred to as 'the textbook'. Chapter 12 of the textbook introduces the principles and mechanisms of the final settling process as well as the key factors that affect it. It gives an overview of the practical implementation of (gravitational) final settlers and provides the basics of the different design and operation methods for final settler design. These methods lead from understanding the basic measures and the flux theory to represent the observed settling behaviour to the comparison of different standards developed for the design of final settlers. The current chapter focuses on examples and exercises for the design and operational evaluation of final settlers. In particular, the examples help to generate state point analysis and Ekama D&O charts for operational support.

12.2 LEARNING OBJECTIVES

After the successful completion of this chapter, the reader will be able to:

- Describe the principles and objectives of final settling (clarification, thickening, and sludge storage).
- Describe and apply the basic measures of sludge settleability, and understand and choose between the different settling tank configurations.
- Apply flux theory and state point diagrams for the design and operation of secondary settling tanks.
- Apply and compare different methods (empirical, DWA, STOWA, etc) for the design of secondary settling tanks.

12.3 EXAMPLES

Evaluation of sludge settling properties

Example 12.3.1

Determine the Vesilind relationship between settling velocity and the test MLSS concentrations.

System under study – available data

A Stirred Zone Settling Velocity (SZSV) test has been performed with mixed liquor from a biological nutrient removal (BNR) activated sludge wastewater treatment plant (WWTP) at different solids concentrations ranging from 1,000 to 12,000 mg/l. The SZSV test results are summarized in Table 12.1.

Table 12.1 Experimental data from a Stirred Zone Settling Velocity (SZSV) test.

Time (min)	Time (h)	Sludge blanket height (cm) after settling starts in a 2 m-high column for different MLSS concentrations (mg/l)					
		1,000	2,000	4,000	6,000	8,000	12,000
0	0.000	200.0	200.0	200.0	200.0	200.0	200.0
10	0.167	82.9	109.5	158.8	182.9	195.1	197.0
20	0.333	11.0	32.9	115.9	165.9	189.9	193.9
30	0.500	7.9	17.1	72.3	149.1	185.1	190.9
40	0.667	7.0	11.9	43.9	132.0	179.9	188.1
50	0.833	7.0	11.0	33.8	114.9	174.1	186.0
60	1.000	6.1	10.1	28.0	97.9	168.9	184.1

Solution[1]

The SZSV test results from Table 12.1 are represented in Figure 12.1.

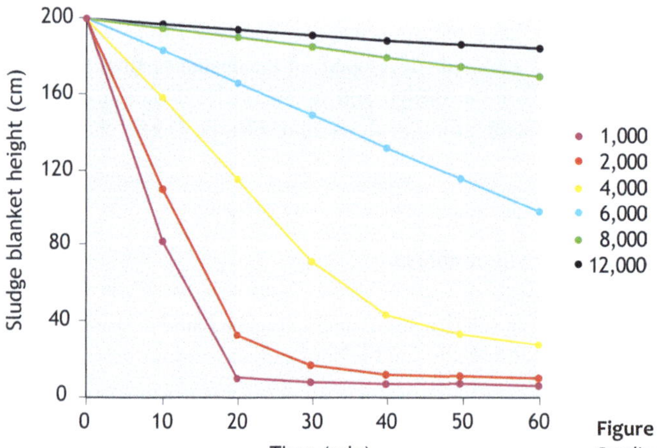

Figure 12.1 Graphical representation of the Stirred Zone Settling Velocity (SZSV) test results from Table 12.1.

---

[1] The solution is described below and made available as supplementary information for this chapter in the spreadsheet Chapter 12 Design example 1.xlsx, in the tabs 'ZSV data' and 'ZSV results'.

The settling velocity v_s for each MLSS concentration is determined as the slope of the linear section on the corresponding settling curve. However, the identification of the linear section of the settling curve is somewhat subjective and prone to uncertainty. The obtained data points are plotted in Figure 12.2 (markers) and show a decreasing exponential behaviour, described by the Vesilind function $v_s = v_0 \cdot e^{-p_{hin} \cdot X}$ (Eq. 12.1 in the textbook).

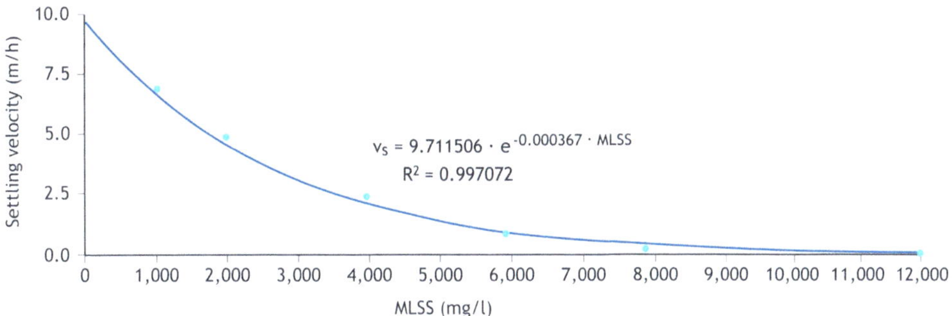

Figure 12.2 Vesilind relationship (solid line) based on ZSV measurements (markers).

By fitting an exponential curve to the data points, the parameter values are determined as v_0 = 9.4 m/h (initial settling velocity) and p_{hin} = 0.36 L/g [or m³/kg] (hindered settling parameter).

Final settling design
Example 12.3.2[2]
<u>System under study – design question</u>
Design a secondary settler for an activated sludge plant where the MLSS is 4.5 kg/m³ and the average dry weather flow rate is 1,000 m³/h. Use the Vesilind function parameters obtained in Example 12.3.1. The additional required data are given in Table 12.2.

The goal of the design is to calculate the settler area and provide information on the underflow pump to handle peak flows. To this end, perform steady-state design according to the different methods from the textbook, based on flux theory, empirical design rules, STOWA, WRC, and ATV standards.

---

[2] This example and its solution are made available as supplementary information in the spreadsheet Chapter 12 Design example 1.xlsx, allowing the reader to check the calculations in detail. The figures provided in this chapter to illustrate the design process are also generated by this spreadsheet.

Table 12.2 Data summary for final settler design – Example 12.3.2.

Description	Symbol	Value	Unit
Plant design parameters			
MLSS	X_F	4.5	kg/m^3
Average dry weather flow (ADWF) rate	Q_{ADWF}	1,000	m^3/h
Diurnal peaking factor	P_{FDW}	1.5	-
Storm peak factor	P_{FWW}	2.5	-
Peak dry weather flow (PDWF)	Q_{PDWF}	1,500	m^3/h
Peak wet weather flow (PWWF)	Q_{PWWF}	3,000	m^3/h
Parameters for the flux theory based on Equation 12.1 and Figure 12.17			
Initial settling velocity	v_0	9.7[a]	m/h
Hindered settling parameter	p_{hin}	0.367[a]	m^3/kg
Safety factor on area	F_{corr}	1.25	-
Empirical design parameters			
Recycle ratio at PDWF	R_{PDWF}	1.0	-
Recycle ratio at PWWF	R_{PWWF}	0.5	-
Maximum dry weather surface overflow rate	SOR_{ADWF}	1	m/h
Maximum wet weather surface overflow rate	SOR_{PWWF}	2.5	m/h
Maximum dry weather solids loading rate	SLR_{ADWF}	6	kg/m^2.h
Maximum wet weather solids loading rate	SLR_{PWWF}	10	kg/m^2.h
WRC design parameters			
Stirred Sludge Volume Index (SSVI) at 3.5 g/l MLSS	$SSVI_{3.5}$	60	mg/l
Safety factor on area	F_{corr}	1.25	-
ATV and STOWA design parameters			
Diluted Sludge Volume Index (DSVI)	DSVI	100	mg/l
Maximum overflow rate	$q_{O,max}$	1.6	m/h
Minimum sludge volume loading	$q_{sv,min}$	300	l/m^2.h
Maximum sludge volume loading	$q_{sv,max}$	400	l/m^2.h
Maximum reduction due to sludge transferred at PWWF	f_{red}	0.7	-

[a] Vesilind function parameters obtained in Section 12.3.1.

The data is summarized in the data sheet of the spreadsheet 'Chapter 12 Design example 1.xlsx' (Figure 12.3).

Input data for all methods

Mixed liquour suspended solids	MLSS	**4.5**	kg/m^3
Average dry weather flow	Q_{ADWF}	**1,000**	m^3/h
Dry weather flow peak factor	PF_{DW}	**1.5**	-
Peak dry weather flow	Q_{PDWF}	1,500	m^3/h
Wet weather peak factor	PF_{WW}	**3**	-
Wet weather peak flow	Q_{PWWF}	3,000	m^3/h

Enter or change red numbers only

Use to check for DWF operation

Design is based on this flow

Data specific to method

Empirical
No settleability input

Flux theory

Initial settling velocity	v_o	**9.7**	m/h
Hindered settling parameter	p_{hin}	**0.367**	m^3/kg
Design safety factor	F_{corr}	1.25	-

Direct information from flux theory input data

X at inflection	5.45 kg/m^3
X at maximum flux	2.72 kg/m^3
Slope at inflection ($q_{R,crit}$)	1.31 m/h

WRC

Stirred Sludge Volume Index at 3.5 g/L	$SSVI_{3.5}$	**60**	ml/g
Design safety factor	F_{corr}	1.25	-

ATV and STOWA

Diluted Sludge Volume Index	DSVI	**100**	ml/g
Settled volume of sludge at MLSS	DSV_{30}	450	ml/l

Figure 12.3 Data sheet for Example 12.3.2.

Solution

<u>Design using flux theory</u>

- Design assumptions

To carry out the design using flux theory, assume the following:

a) Steady-state design conditions.
b) The effluent solids flux is zero.
c) Solids Handling Criteria SHC I and SHC II are satisfied for the peak dry weather flow conditions.
d) SHC I and SHC II are critical for the peak wet weather flow conditions.
e) There is a 25 % safety factor (F_{corr}) in terms of surface area.

- Design procedure

Designing a settler according to the design using flux theory comes down to determining the following information, for both average dry weather flow (ADWF) and peak wet weather flow (PWWF) conditions:

1. Define gravity (J_S, kg/m^2/h) and bulk (J_B, kg/m^2/h) flux curves based on the Vesilind curve.
2. Calculate settling velocity (v_s, m/h) at the design MLSS concentration.
3. Calculate the minimum settler surface area based on PWWF (A, m^2) to meet SHC II.
4. Calculate the overflow line and state point for the state point diagram:
   a. Overflow line parameter: applied overflow rate (hydraulic loading, q_l, m/h).
   b. Overflow flux at feed concentration ($J_{l,F}$, kg/m^2.h), corresponding to the state point.

5. Determine the minimum recycle ratio required to meet the SHC I critical conditions on the state point diagram by aligning the underflow line for both PWWF and PDWF. Start with an initial recycle ratio (R_{init}) of 0.5:
   a. Calculate the underflow line parameters:
      i. Underflow flux (J_R, kg/m^2/h).
      ii. Total applied flux (J_{AP}, kg/m^2/h) or sludge loading rate (SLR).
   b. Repeat while changing the recycle ratio until the underflow line tangents the gravity flux curve:
      - If the underflow line is under the gravity flux curve above the design MLSS, decrease the recycle ratio.
      - If the underflow line intersects with the gravity flux curve above the design MLSS concentrations, increase the recycle ratio.
6. Calculate the corresponding recycle flow rates, underflow fluxes, total applied fluxes and recycle concentrations (X_R, kg/m^3) for both PDWF and PWWF.
7. As an alternative to using the state point diagram (Step 5), the minimum recycle flow can be determined using the Ekama Design & Operation (D&O) chart, as follows.
   a. Draw the SHC II line as the settling velocity at feed concentration ($v_{s,MLSS}$, m/h) (see Step 2).
   b. Determine the SCH I line or overflow rate (Q/A, m/h).
   c. To complete the Ekama D&O chart, the criterion boundary line is drawn.
8. Recalculate the settler surface area and the recycle pump capacity, taking into account the safety factor.

The calculations required to design a settler according to the flux theory, complying with SHC I and SHC II, are detailed below step by step.

1. Define gravity, bulk, and total flux curves for the Vesilind settling velocity function
The gravity flux (J_S, in kg/m^2.h) is obtained from:

$$J_S = v_s \cdot X \hspace{4cm} \text{(Eq. 12.2 in the textbook)}$$

v_s represents the settling velocity at the design MLSS concentration X and is obtained from the Vesilind settling velocity function (Eq. 12.1 in the textbook) with the parameters determined in Example 12.3.1. Plotting the gravity flux as a function of the solids concentration X results in the gravity flux curve in Figure 12.4.

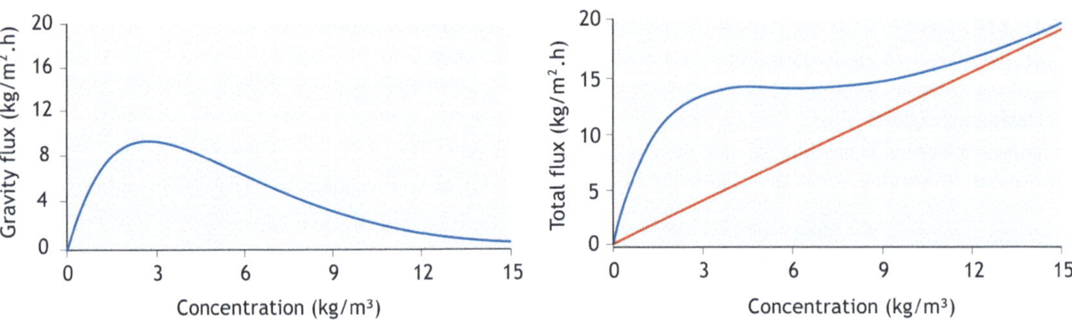

Figure 12.4 Gravity flux curve.　　　　　　　　**Figure 12.5** Bulk flux (red) and total flux curve (blue) at PWWF.

The bulk flux (J_B, in kg/m^2.h) represents the downward motion of solids in the settling tank, generated by the recycle flow (Q_R, in m^3.h). It is calculated from Eq. 12.1:

$$J_B = \frac{Q_R}{A} \cdot X \tag{12.1}$$

in which Q_R represents the recycle flow rate (m^3/h) and A the settler surface area (m^2). The total flux transporting solids to the bottom of the clarifier is the sum of the gravity and bulk fluxes. Figure 12.5 displays the bulk flux and the total flux in PWWF conditions. The calculation of the bulk flux and total flux at PDWF is performed completely analogously.

2. Calculate settling velocity at the design MLSS

The settling velocity at the design MLSS concentration, *i.e.*, the solids concentration fed to the clarifier, is calculated from the Vesilind function (Eq. 12.1 in the textbook) as:

$$v_{s,MLSS} = 9.7 \cdot e^{-0.367 \cdot 4.5} = 1.86 \qquad \text{m/h} \tag{12.2}$$

3. Calculate settler minimum surface (SHC II)

SHC II requires that the overflow rate during PWWF must not exceed the settling velocity at feed concentrations:

$$q_{I,PWWF} = \frac{Q_{PWWF}}{A} \le v_{s,MLSS} \qquad \text{(Eq. 12.5 in the textbook)}$$

The minimum surface of the settler is calculated from SHC II, by setting the maximum overflow rate equal to the settling velocity for peak wet weather conditions:

$$A = \frac{Q_{PWWF}}{v_{s,MLSS}} = \frac{3,000}{1.86} = 1.611 \qquad \text{m}^2 \tag{12.3}$$

4. Calculate overflow for the state point diagram

The overflow line in the state point diagram represents the overflow flux J_I in terms of the solids concentration X (see Eq. 12.6 in the main textbook):

$$J_I = \frac{Q_I}{A} \cdot X = q_I \cdot X \qquad \text{kg/m}^2\text{.h} \tag{12.4}$$

The slope of the overflow line, *i.e.*, the applied overflow rate or hydraulic loading, is calculated for PDWF as:

$$q_{I,PDWF} = \frac{Q_{PDWF}}{A} = \frac{1,500}{1,611} = 0.93 \qquad \text{m/h} \tag{12.5}$$

For PWWF conditions, the overflow rate equals the design settling velocity, as prescribed by SHC II (see Eq. 12.3):

$$q_{I,PWWF} = v_{s,MLSS} = 1.86 \qquad\qquad h \qquad\qquad (12.6)$$

The overflow flux at the design MLSS (*i.e.*, the feed concentration) corresponds to the flux at the state point and is calculated as follows for PDWF and PWWF, respectively:

$$J_{I,PDWF,F} = \frac{Q_{PDWF}}{A} \cdot X_F = q_{I,PDWF} \cdot X_F = 0.93 \cdot 4.5 = 4.19 \qquad\qquad kg/m^2.h \qquad (12.7)$$

$$J_{I,PWWF,F} = \frac{Q_{PWWF}}{A} \cdot X_F = q_{I,PWWF} \cdot X_F = 1.86 \cdot 4.5 = 8.38 \qquad\qquad kg/m^2.h \qquad (12.8)$$

In the state point diagram, SHC II is fulfilled if the state point lies below the gravity flux curve, which is ensured by the calculation of the settler surface area by Eq. 12.3.

5. Determination of minimum recycle ratio using a state point diagram (SHC I)
SHC I requires the applied flux to the clarifier (*i.e.*, the mass of solids applied per unit settler area) to be lower than the minimum total flux. In the state point diagram, SHC I is evaluated by the position of the underflow line, which should be below the descending limb of the gravity curve.

The underflow line is determined by the underflow flux J_R (see Eq. 12.6 in the main textbook):

$$J_R = \frac{Q_R}{A} \cdot X = q_R \cdot X \qquad\qquad kg/m^2.h \qquad (12.9)$$

which is plotted with a negative slope q_R (termed the hydraulic underflow rate) and shifted upwards such that the underflow line intersects the vertical axis at the total applied flux, which is the sum of the overflow flux and the underflow flux at the feed concentration (Eq. 12.8 in the textbook):

$$J_{AP} = \frac{Q_I + Q_R}{A} \cdot X_F = \frac{Q_I}{A} \cdot (1+R) \cdot X_F = J_{I,F} + J_{R,F} \qquad\qquad kg/m^2.h \qquad (12.10)$$

As a result, the underflow line is defined by:

$$\frac{Q_I + Q_R}{A} \cdot X_F - \frac{Q_R}{A} \cdot X = 0 \qquad\qquad kg/m^2.h \qquad (12.11)$$

From Eq. 12.11, it is clear that the position of the underflow line and thus SHC I is influenced by both the settler surface area (A) and the recycle flow rate (Q_R). A being determined by SHC II (see Step 3.), SHC I will now be applied to determine the minimum recycle ratio $R = Q_I / Q_R$.

More specifically, the minimum recycle ratio is determined as the recycle ratio for which the underflow line is tangential to the descending limb of the gravity flux curve. This is done following an iterative procedure, applying different values of R, each time recalculating the position of the underflow line (Eq. 12.11) using updated values for the underflow flux (Eq. 12.9) and the total applied flux (Eq. 12.10).

Applying an initial value for the recycle ratio of R = 0.5 results in the state point diagrams of Figure 12.6 (PWWF) and 12.7 (PDWF). For PWWF conditions, the underflow line intersects with the gravity flux curve above the feed concentration, which means that the recycle ratio is too low and should be increased. A recycle ratio R = 1 is found too high because the corresponding underflow line is under the gravity flux curve above the feed concentration (Figure 12.8). After some trial and error the minimum recycle flow at PWWF conditions is found as $R_{min,PWWF} = 0.69$, resulting in an underflow line that is tangential to the descending limb of the gravity flux curve (Figure 12.10). For PDWF, a recycle ratio of R = 0.5 was found to be too high (the underflow line is under the gravity flux curve above the feed concentration, Figure 12.7), R = 0.3 was found to be too low (the underflow line intersects with the gravity flux curve above the feed concentration) and $R_{min,PDWF} = 0.4$ was found to be the minimum recycle flow at PDWF conditions (Figure 12.11).

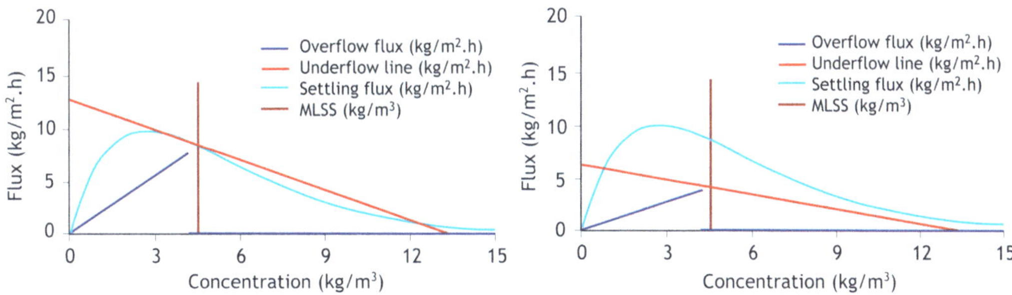

Figure 12.6 SPD at PWWF at R = 0.5.

Figure 12.7 SPD at PDWF at R = 0.5.

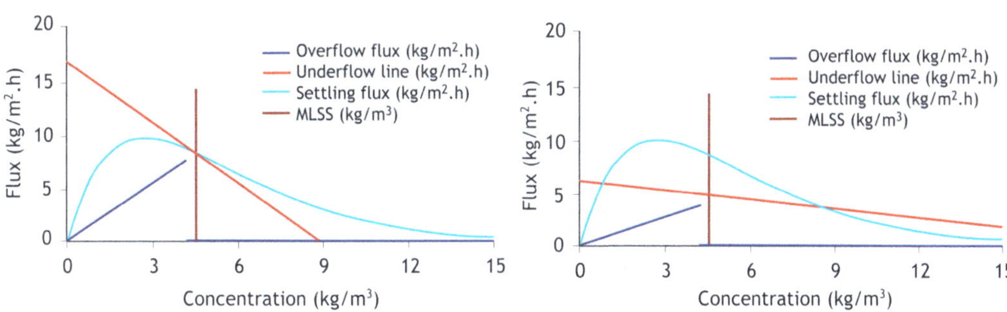

Figure 12.8 SPD at PWWF at R = 1.

Figure 12.9 SPD at PDWF at R = 0.3.

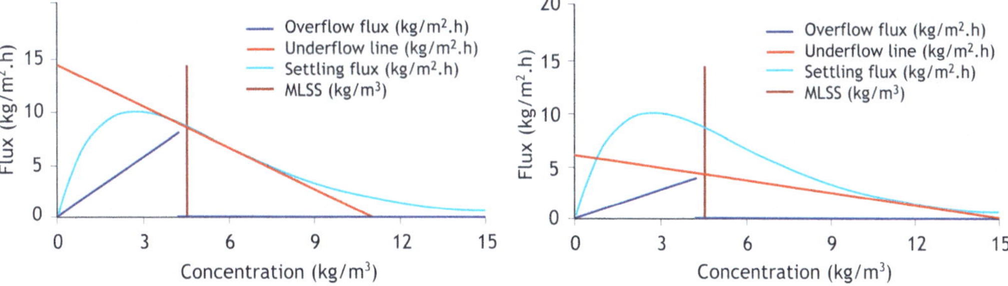

Figure 12.10 SPD at PWWF at $R_{min,PWWF}$ = 0.69.

Figure 12.11 SPD at PDWF at $R_{min,PDWF}$ = 0.4.

6. Calculate recycle flow rates and recycle concentrations

Taking into account $R_{min,PWWF} = 0.69$ and $R_{min,PDWF} = 0.40$, the minimum recycle flow rates required to fulfil SHC I for PDWF and PWWF are calculated as:

$$Q_{R,PDWF} = Q_{PDWF} \cdot R_{min,PDWF} = 1,500 \cdot 0.40 = 600 \qquad \text{m}^3/\text{h} \qquad (12.12)$$

$$Q_{R,PWWF} = Q_{PWWF} \cdot R_{min,FWWF} = 3,000 \cdot 0.69 = 2,073 \qquad \text{m}^3/\text{h} \qquad (12.13)$$

The corresponding hydraulic underflow rates under PDWF and PWWF conditions are obtained (see Eq. 12.9) as:

$$q_{R,PDWF} = \frac{Q_{R,PDWF}}{A} = \frac{1,500 \cdot 0.40}{1,611} = 0.37 \qquad \text{m/h} \qquad (12.14)$$

$$q_{R,PWWF} = \frac{Q_{R,PWWF}}{A} = \frac{3,000 \cdot 0.69}{1,611} = 1.29 \qquad \text{m/h} \qquad (12.15)$$

and the underflow fluxes at design MLSS (*i.e.*, the feed concentration) are calculated from Eq. 12.9 as:

$$J_{R,PDWF,F} = 0.37 \cdot 4.5 = 1.68 \qquad \text{kg/m}^2.\text{h} \qquad (12.16)$$

$$J_{R,PWWF,F} = 1.29 \cdot 4.5 = 5.79 \qquad \text{kg/m}^2.\text{h} \qquad (12.17)$$

The total applied flux at design MLSS under PDWF and PWWF conditions follow from eqs. 12.10, 12.7, 12.8, 12.16 and 12.17 as:

$$J_{AP,PDWF} = J_{I,PDWF,F} + J_{R,PDWF,F} = 4.19 + 1.68 = 5.87 \qquad \text{kg/m}^2.\text{h} \qquad (12.18)$$

$$J_{AP,PWWF} = J_{I,PWWF,F} + J_{R,PWWF,F} = 8.38 + 5.79 = 14.16 \qquad \text{kg/m}^2.\text{h} \qquad (12.19)$$

Assuming no solids in the clarifier effluent, the applied flux equals the flux leaving the settler with the underflow (recycle stream):

$$J_{AP} = \frac{Q_R}{A} \cdot X_R \qquad (12.20)$$

The sludge concentration in the recycle stream can be calculated from rearranging Eq. 12.20:

$$X_R = \frac{J_{AP} \cdot A}{Q_R} \qquad (12.21)$$

which, applied to PDWF and PWWF conditions, results in:

$$X_{R,PDWF} = \frac{J_{AP,PDWF} \cdot A}{Q_{R,ADWF}} = \frac{5.87 \cdot 1,611}{1,500} = 15.75 \qquad \text{kg/m}^3 \qquad (12.22)$$

$$X_{R,PWWF} = \frac{J_{AP,PWWF} \cdot A}{Q_{R,PWWF}} = \frac{14.16 \cdot 1,611}{3,000} = 11.02 \qquad \text{kg/m}^3 \qquad (12.23)$$

7. Iterate R_{min} values to calculate variables for the Ekama D&O chart
As an alternative to using the state point diagram, the minimum recycle ratio can be determined from the Ekama Design & Operation (D&O) chart.

Three lines are drawn to make up the Ekama D&O chart: the SHC II line, the SHC I line and the criterion boundary line. The SHC II line is easily drawn as a straight horizontal line representing the settling velocity at the design MLSS (= feed) concentration, $v_{s,MLSS} = 1.86$ (from Step 2., Eq 12.2). The latter value corresponds to the overflow rate at PWWF conditions: $v_{s,MLSS} = Q_{PWWF} / A$, see Eq. 12.3).

The SHC I line is determined by equations 12.9 and 12.10 in the textbook, for the given Vesilind parameters and the given feed concentration:

$$\frac{Q_I}{A} = \frac{v_0}{R} \cdot \frac{1+\alpha}{1-\alpha} e^{\frac{-p_{hin}(1+R) \cdot X_F \cdot (1+\alpha)}{2R}} = \frac{9.7}{R} \cdot \frac{1+\alpha}{1-\alpha} e^{\frac{-0.367(1+R) \cdot 4.5 \cdot (1+\alpha)}{2R}} \qquad \text{m/h} \qquad (12.24)$$

in which:

$$\alpha = \sqrt{1 - \frac{4R}{0.367 \cdot (1+R) \cdot 4.5}} \qquad (12.25)$$

The value of Eq. 12.24 and Eq. 12.25 are calculated for a range of increasing recycle ratio values R, until the SHC I line intersects with the SHC II line, *i.e.*, until $Q_I / A = v_{s,MLSS}$, Figure 12.12). The minimum recycle ratio for peak wet weather conditions is thus found as $R_{min,PWWF} = 0.69$. In Excel, this is done by using the built-in 'goal seek' function (see the spreadsheet 'Chapter 12 Design example 1.xlsx').

The safe operating zone in the Ekama D&O chart is the zone below SHC II and to the right of SHC I (the green zone in Figure 12.12). For the clarifier not to be overloaded, the overflow rate Q_I / A should be below the SHC I line in the case $R < R_{min,PWWF}$; the overflow rate Q_I / A should be lower than $v_{s,MLSS}$ in the case $R > R_{min,PWWF}$. From the SHC I line, the minimum recycle rate for PDWF conditions (*i.e.*, corresponding with $Q_{PDWF} / A = 1,500 / 1,611 = 0.93$ is found as $R_{min,PDWF} = 0.4$. The values for the minimum recycle ratios at PWWF and PDWF conditions found in the Ekama D&O chart correspond with the values found in the state point diagram (Figure 12.10 and Figure 12.11, respectively).

To complete the Ekama D&O chart, the criterion boundary line is drawn based on Equation 12.11 in the textbook:

$$\frac{Q_I}{A} = \frac{v_0}{e^2 \cdot R} = \frac{9.7}{e^2 \cdot R}$$

The criterion boundary line is the boundary between lower recycle ratios where a critical flux can be found (SHC I applies) and higher recycle ratios where a critical flux cannot be found (only SHC II applies).

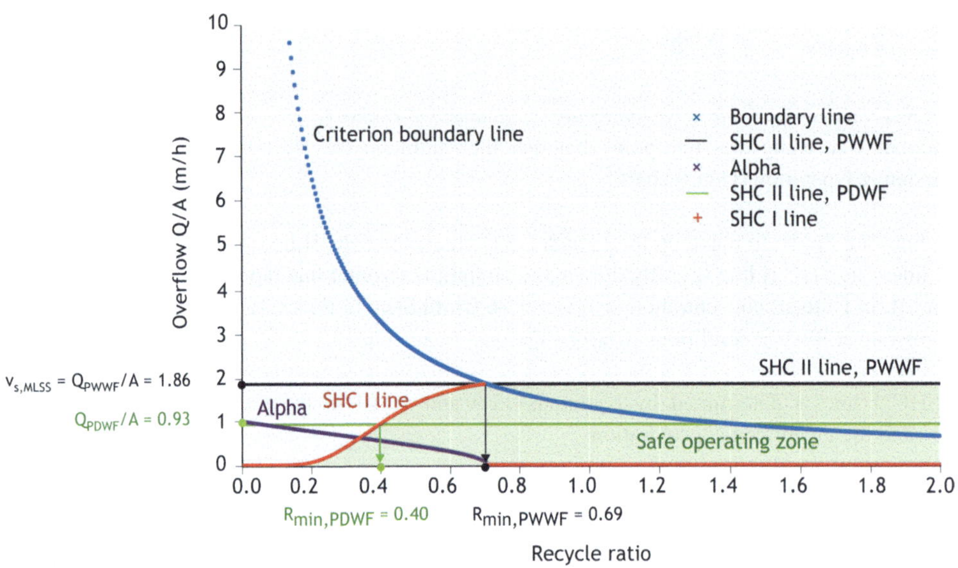

Figure 12.12 Ekama D&O chart for this example. The zone in green indicates the safe operating zone.

8. *Calculate the surface area and recycle pump capacity with a safety threshold*
Taking into account the safety factor F_{corr}, Table 12.2), the settler surface area (Eq. 12.3) is recalculated as:

$$A_{design,flux} = A \cdot F_{corr} = 1,611 \cdot 1.25 = 2,014 \qquad\qquad m^2 \qquad (12.26)$$

Due to practical considerations (standard design), the actual area chosen could be somewhat larger than the area theoretically required (Eq. 12.26), for instance:

$$A_{design,flux} \approx 2,020 \qquad\qquad m^2 \qquad (12.27)$$

The PWWF overflow rate becomes:

$$\frac{Q_{PWWF}}{A} = \frac{3,000}{2,020} = 1,5 \qquad\qquad h \qquad (12.28)$$

Given that the settler area increases by including a safety factor, the settler design will be in the safe operating zone. This can be verified by redrawing the state point diagram: revising the overflow line (Eq. 12.4), the state point (eqs. 12.7 and 12.8) and the underflow line (Eq. 12.11) (results not shown). In the Ekama D&O chart, it is immediately clear that the operating point corresponding with Q_{PWWF} / A = 1.5, Eq. 12.28) and that $R_{min,PWWF} = 0.69$ is in the safe operating zone.

The recycle pump capacity is calculated according to the minimum recycle flow rates required at PDWF (and PWWF) conditions. Given that $Q_{R,PDWF}$ = 600 m³/h (Eq. 12.12) and $Q_{R,PWWF}$ = 2,073 m³/h (Eq. 12.13), say $Q_{R,PWWF}$ = 2,100 m³/h, a total recycle pump capacity of 2,100 m³/h needs to be provided. To optimize the pump operating time at maximum operating conditions (corresponding with maximum efficiency), a logical choice is to provide two pumps: one pump with a capacity of 600 m³/h for dry weather conditions, and an additional pump with a capacity of 1,500 m³/h to work in parallel with the first one in rain weather conditions. Taking into account the settler area including the safety factor (Eq. 12.27), some of the settler specifications need to be recalculated. A summary of the final settler design specifications is given in Table 12.3.

<u>Empirical design</u>
The settler surface area and the recycle pump capacity can also be calculated based on empirical methods. The recycle ratio serves as an empirical design parameter. Recycle ratios between 0.5 and 1.0 are usually satisfactory. In this example, the recycle ratio at PDWF is set at R_{PDWF} =1.0 and the recycle ratio at PWWF is set at R_{PWWF} = 0.5 (from Table 12.2). The recycle pump capacities at PDWF and PWWF are thus calculated as:

$$Q_{R,PDWF} = R_{PDWF} \cdot Q_{PDWF} = 1.0 \cdot 1,500 = 1,500 \qquad\qquad \text{m}^3/\text{h} \qquad (12.29)$$

$$Q_{R,PWWF} = R_{PWWF} \cdot Q_{PWWF} = 0.5 \cdot 3,000 = 1,500 \qquad\qquad \text{m}^3/\text{h} \qquad (12.30)$$

The corresponding recycle pump capacity amounts to:

$$Q_{R,Design} = Max(Q_{R,PDWF}; Q_{R,PWWF}) = 1,500 \qquad\qquad \text{m}^3/\text{h} \qquad (12.31)$$

The selection of the required settler surface area can be based on the maximum hydraulic loading specification and/or maximum solids loading specification (as presented in this example) or other criteria. In this case, the maximum surface overflow rates (hydraulic loading) for average dry weather flow and peak wet weather flow conditions are envisaged as SOR_{ADWF} = 1 m/h and SOR_{PWWF} = 2.5 m/h, respectively. In addition, the clarifier should not be loaded higher than 6 kg/m²/h during average dry weather and 10 kg/m².h during wet weather conditions (*i.e.*, maximum solids loading specification). Each of the above four specifications will lead to a different required clarifier area, the largest of which will be selected, which in this case is 2,025 m² (Eq. 12.36). The calculations are provided in the design sheet in the spreadsheet 'Chapter 12 Design example 1.xlsx' (Figure 12.13).

The settler surface area based on maximum overflow rates at ADWF and PWWF are calculated as:

$$A_{design,SOR,ADWF} = \frac{Q_{ADWF}}{SOR_{ADWF}} = \frac{1,000}{1} = 1,000 \qquad\qquad \text{m}^2 \qquad (12.32)$$

$$A_{design,SOR,PWWF} = \frac{Q_{PWWF}}{SOR_{PWWF}} = \frac{3,000}{2.5} = 1,200 \qquad\qquad \text{m}^2 \qquad (12.33)$$

while the settler surface area based on the maximum solids loading at ADWF and PWWF is found as:

$$A_{design,SLR,ADWF} = X_F \cdot \frac{Q_{ADWF} + Q_{R,PDWF}}{SLR_{ADWF}} = 4.5 \cdot \frac{1,000 + 1,500}{6} = 1,875 \qquad m^2 \qquad (12.34)$$

$$A_{design,SLR,PWWF} = X_F \cdot \frac{Q_{PWWF} + Q_{R,PWWF}}{SLR_{PWWF}} = 4.5 \cdot \frac{3,000 + 1,500}{10} = 2,025 \qquad m^2 \qquad (12.35)$$

The settler surface area based on empirical design thus becomes:

$$A_{design,emp} = Max\left(A_{design,SOR,ADWF}, A_{design,SOR,PWWF}, A_{design,SLR,ADWF}, A_{design,SLR,PWWF}\right) = 2,025 \ m^3 \qquad (12.36)$$

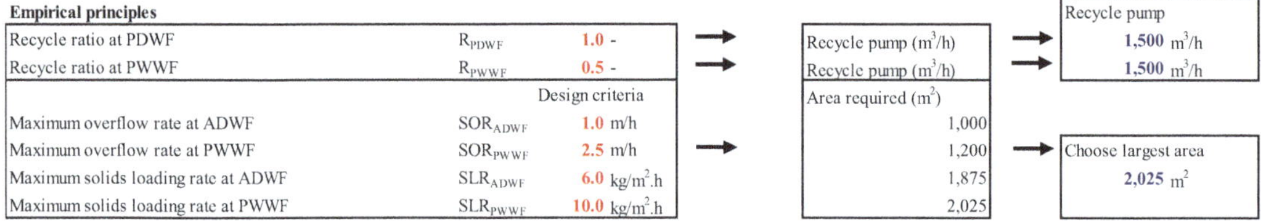

Figure 12.13 Empirical clarifier design. Screenshot from the design sheet in the spreadsheet 'Chapter 12 Design example 1.xlsx'.

A number of additional settler specifications can be calculated and are summarized in Table 12.3.

<u>WRC design</u>
The WRC design method with the extension by Ekama *et al.* (1997) calculates the critical hydraulic underflow rate and the required settler surface area based on an empirical relationship with the SSVI$_{3.5}$ (Eq. 12.12 in the textbook). The Stirred Sludge Volume Index at 3.5 g/l MLSS for the given example is SSVI$_{3.5}$= 60 mg/l (Table 12.2).

The resulting critical hydraulic underflow rate is calculated as:

$$q_{R,crit} = 1.612 - 0.00793 \cdot SSVI3.5 - 0.0015 \cdot \left(MAX\left(0, SSVI3.5 - 125\right)\right)^{1.115} =$$
$$1.612 - 0.00793 \cdot 60 - 0.0015 \cdot \left(MAX\left(0, 60 - 125\right)\right)^{1.115} = 1.14 \qquad h \qquad (12.37)$$

and the required surface area is obtained from:

$$A_{design,PWWF} = X_F \cdot \frac{Q_{PWWF}}{306.86 \cdot SSVI3.5^{-0.77} \cdot q_{R,crit}^{0.68} - X_F \cdot Q_{R,crit}} =$$
$$4.5 \cdot \frac{3000}{306.86 \cdot 60^{-0.77} \cdot 1.14^{0.68} - 4.5 \cdot 1.14} = 1,469 \qquad m^2 \qquad (12.38)$$

which is larger than the area required which would be obtained when applying the same relation for PDWF conditions.

Taking into account a safety factor of 25 % (F_{corr} = 1.25), the surface area becomes:

$$A_{design,WRC} = A_{design,PWWF} \cdot F_{corr} = 1,469 \cdot 1.25 = 1,836 \qquad m^2 \qquad (12.39)$$

The PWWF overflow rate is then:

$$q_{PWWF} = \frac{Q_{PWWF}}{A_{design,WRC}} = \frac{3,000}{1,836} = 1.63 \qquad m/h \qquad (12.40)$$

The recycle pump capacity corresponding with the obtained values for the critical hydraulic underflow rate and the settler surface area is then calculated and, for practical purposes, rounded off as:

$$Q_R = A_{design,WRC} \cdot q_{R,crit} = 1,469 \cdot 1.14 = 2,086 \approx 2,100 \; 2,086 \approx 2,100 \qquad m^3/h \qquad (12.41)$$

The calculations are provided in the design sheet in the spreadsheet 'Chapter 12 Design example 1.xlsx' (Figure 12.14). Additional settler specifications are summarized in Table 12.3.

WRC principles

Calculate Critical Recycle Rate	$q_{R,crit}$	1.14 m/h
Calculate Area based on SSVI$_{3.5}$ for PWWF		1.469 m^2
Calculate Area based on SSVI$_{3.5}$ for PDWF		734 m^2

Choose larger area	1.469 m^2
Area with safety	**1,836** m^2
PWWF recycle flow	2.086 m^3/h
Choose Pump	**2,100** m^3/h
PWWF overflow rate	1.6 m/h

Figure 12.14 WRC principle-based clarifier design. Screenshot from the design sheet in the spreadsheet 'Chapter 12 Design example 1.xlsx'.

ATV design

The ATV design method is based on the Diluted Sludge Volume Index (DSVI) test. In addition, it considers the settled volume of the MLSS under test conditions (DSV$_{30}$, Eq. 12.13 in the textbook), which is calculated for this example as:

$$DSV_{30} = X_{F,DWF} \cdot DSVI = 4.5 \cdot 100 = 450 \qquad ml/l \qquad (12.42)$$

The design steps are as follows:

1) The permissible overflow rate, q_O, is calculated as Eq. 12.15 in the textbook:

$$q_O = 2,400 \cdot DSV_{30}^{-1.34} = 2,400 \cdot 450^{-1.34} = 0.67 \qquad m/h \qquad (12.43)$$

2) q_O must be smaller than $q_{O,max} = 6$ m/h, which is the case.

3) The settler surface area follows from the permissible overflow rate and the influent PWWF flow rate:

$$A_{design,ATV} = \frac{Q_{PWWF}}{Min\left(q_O, q_{O,max}\right)} = \frac{3,000}{0.67} = 4,490 \qquad\qquad m^2 \qquad (12.44)$$

4) For practical considerations, the value obtained is rounded off as:

$$A_{design,ATV} \approx 4,500 \qquad\qquad m^2 \qquad (12.45)$$

5) The maximum recycle solids concentration that can be reached under ADWF and PWWF conditions is calculated according to Eq. 12.16 and Eq. 12.17 in the textbook, respectively:

$$X_{R,ADWF} = \frac{1,200}{DSVI} = \frac{1,200}{100} = 12 \qquad\qquad g/l \qquad (12.46)$$

$$X_{R,PWWF} = X_{RAS,ADWF} + 2 = 12 + 2 = 14 \qquad\qquad g/l \qquad (12.47)$$

6) The necessary recycle flow is calculated from the solids mass balance over the settler (Eq. 12.18 in the main textbook), both for ADWF and PWWF, as:

$$Q_{R,ADWF} = \frac{X_F}{X_{RAS,ADWF} - X_F} \cdot Q_{ADWF} = \frac{4.5}{12 - 4.5} \cdot 1,000 = 600 \qquad\qquad m^3/h \qquad (12.48)$$

$$Q_{R,PWWF} = \frac{X_F}{X_{RAS,PWWF} - X_F} \cdot Q_{PWWF} = \frac{4.5}{14 - 4.5} \cdot 3,000 = 1,421 \qquad\qquad m^3/h \qquad (12.49)$$

The recycle pump capacity is rounded up, to a practical value of:

$$Q_R = 1,500 \qquad\qquad m^3/h \qquad (12.50)$$

for all weather conditions.

The calculations are provided in the design sheet of the spreadsheet 'Chapter 12 Design example 1.xlsx' (Figure 12.15). Additional settler specifications are summarized in Table 12.3.

ATV principles

Permissible overflow rate (depends on DSV30)	q_O	0.67 m/h		Choose area	4,500 m^2
Maximum q_O	$q_{O,max}$	1.60 m/h			
area based on smaller of the two above rates	A	4,490 m^2	$\rightarrow$	ADWF recycle flow	600 m^3/h
Maximum RAS, ADWF	$X_{RAS,\,ADW}$	12 g/l	$\rightarrow$	PWWF recycle flow	1,421 m^3/h
Maximum RAS, PWWF	$X_{RAS,\,PWW}$	14 g/l	$\rightarrow$	Choose pump	1,500 m^3/h

Figure 12.15 ATV principle-based clarifier design for this example; screenshot from the design sheet in the spreadsheet 'Chapter 12 Design example 1.xlsx'.

STOWA design

The STOWA design method is closely related to the ATV design procedure and is performed according to the following steps:

1) The permissible overflow rate is calculated based on the DSV_{30} (Eq. 12.19 in the textbook), as follows:

$$q_O = \frac{1}{3} + \frac{200}{DSV30} = \frac{1}{3} + \frac{200}{450} = 0.78 \qquad\qquad \text{m/h} \qquad\qquad (12.51)$$

2) The sludge volume loading rate is calculated according to Eq. 12.52, taking into account Eq. 12.42 and 12.51.

$$q_{sv} = q_O \cdot DSV30 = 0.78 \cdot 450 = 350 \qquad\qquad \text{m}^2\text{/h} \qquad\qquad (12.52)$$

3) The sludge volume loading rate must be between 300 and 400 $l/m^2.h$ – this condition is fulfilled. As a result, the permissible overflow rate remains q_O =78 m/h (Eq. 12.51).

4) The settler surface area follows from the permissible overflow rate and the influent ADWF flow rate according to Eq. 12.53:

$$A_{design,ADWF} = \frac{Q_{ADWF}}{q_O} = \frac{1,000}{0.78} = 1,286 \qquad\qquad \text{m}^2 \qquad\qquad (12.53)$$

5) The settler surface area based on PWWF conditions, taking into account a maximum reduction on the MLSS concentration of 30 % (corresponding with a reduction factor f_{red} = 0.7), results from Eq. 12.54:

$$A_{design,PWWF} = f_{red} \cdot \frac{Q_{PWWF}}{q_O} = 0.7 \cdot \frac{3,000}{0.78} = 2,700 \qquad\qquad \text{m}^2 \qquad\qquad (12.54)$$

6) The settler surface area is taken as the maximum of Eq. 12.53 and Eq. 12.54, and rounded up:

$$A_{design,STOWA} = \text{above Max}(A_{design,ADW}, A_{design,PWWF}) = 2,700 \text{ m}^2$$

7) The recycle pump design is identical to the ATV method; the selected recycle pump has 1,500 m^3/h capacity.

The calculations are provided in the design sheet in the spreadsheet 'Chapter 12 Design example 1.xlsx' (Figure 12.16). Additional settler specifications are summarized in Table 12.3.

Table 12.3 A summary of the final settler design specifications. The variables in bold have been explicitly prescribed or calculated in the example; the remaining ones can be calculated additionally from the given relations. The calculations are provided in the summary sheet in the spreadsheet 'Chapter 12 Design example 1.xlsx'.

Parameter	Unit	Flux theory	Empirical	WRC	ATV	STOWA
Area, A	m^2	**2,020**	**2,025**	**1,836**	**4,500**	**2,700**
At ADWF						
Overflow rate, Q_{ADWF}	m^3/h			**1,000**		
Surface overflow rate, SOR_{ADWF}	m/h	0.50	0.49	0.54	0.22	0.37
Recycle flow rate, $Q_{R,ADWF}$	m^3/h	600	1,500	**2,100**	1,500	**1,500**
Hydraulic underflow rate, $q_{R,ADWF}$		0.30	0.74	**1.14**	0.33	0.56
Recycle ratio, R	-	0.60	1.50	2.10	1.50	1.50
RAS concentration, $X_{R,ADWF}$[1]	kg/m^3	12.00	7.50	6.64	7.50	7.50
Solids loading rate, SLR_{ADWF}[2]	$kg/m^2.h$	3.56	5.56	7.60	2.50	4.17
At PDWF						
Overflow rate, Q_{PDWF}	m^3/h			**1,500**		
Surface overflow rate, SOR_{PDWF}	m/h	0.74	0.74	0.82	0.33	0.56
Recycle flow rate, $Q_{R,PDWF}$	m3/h	**600**	**1,500**	**2,100**	**1,500**	**1,500**
Hydraulic underflow rate, $q_{R,PDWF}$	m/h	0.30	0.74	**1.14**	0.33	0.56
Recycle ratio, $R_{min,PDWF}$	-	**0.4**	**1.0**	1.4	1.0	1.0
RAS concentration, $X_{R,PDWF}$[1]	kg/m^3	15.75	9.00	7.71	9.00	9.00
Solids loading rate, SLR_{PDWF}[2]	$kg/m^2.h$	4.68	6.67	8.82	3.00	5.00
At PWWF						
Overflow rate, Q_{PWWF}	m^3/h			**3,000**		
Surface overflow rate, SOR_{PWWF}	m/h	1.49	1.48	**1.63**	0.67	1.11
Recycle flow rate, $Q_{R,PWWF}$	m^3/h	**2,100**	1,500	**2,100**	1,500	**1,500**
Hydraulic underflow rate, $q_{R,PWWF}$	m/h	1.04	0.74	**1.14**	0.33	0.56
Recycle ratio, $R_{min,PWWF}$	-	**0.69**	**0.5**	0.7	0.5	0.5
RAS concentration, $X_{R,PWWF}$[1]	kg/m^3	10.93	13.50	10.93	13.5	13.50
Solids loading rate, SLR_{PWWF}[2]	$kg/m^2.h$	11.36	**10**	12.50	4.50	7.50

[1] The sludge concentration in the recycle stream (*i.e.*, the RAS concentration) can be calculated from:

$$\frac{(Q + Q_R)}{Q_R} \cdot X_F \tag{12.55}$$

[2] The solids loading rate is calculated as:

$$SLR = \frac{Q + Q_R}{A} \cdot X_F \tag{12.56}$$

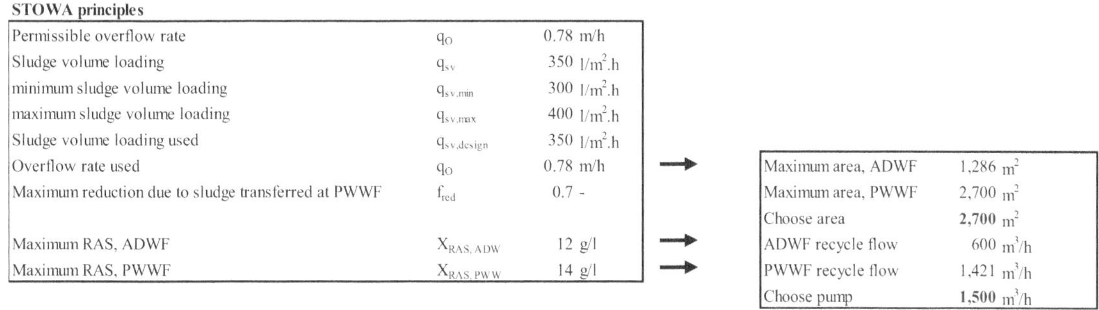

Figure 12.16 STOWA principle-based clarifier design for this example; a screenshot from the design sheet in the spreadsheet 'Chapter 12 Design example 2.xlsx'.

Summary and discussion

An overview of the design specifications obtained with the different methods is given in Table 12.3. The design input indicates a sludge with not particularly good settling properties because DSVI is 100 ml/g. Besides, the settler is loaded with a high MLSS concentration (4.5 kg/m^3) and the peak wet weather flow is high (Q_{PWWF} = 3,000 m^3/h). The combination of these factors results in a high settler area, which is more or less pronounced according to the extent to which they are taken into account in a given design procedure. The empirical design does not consider the settling properties of the sludge at all but does take into account the high MLSS and PWWF flow rates.

Final settling design
Example 12.3.3[3]
System under study - design question

Design a secondary settler for an activated sludge plant where the MLSS is 3.5 kg/m^3 and the average dry weather flow rate is 625 m^3/h. Additional required data is given in Table 12.4, including values for the Vesilind function parameters.

The goal of the design is to calculate the settler area and provide information on the underflow pump to handle peak flows. To this end, perform steady-state design according to the different methods from the textbook, based on flux theory, empirical design rules, STOWA, WRC and ATV standards.

---

[3] This example and its solution are available as supplementary information in the spreadsheet 'Chapter 12 Design example 2.xlsx', allowing the reader to check the calculations in detail. Note that this example is completely the same as Example 1, but with different design parameters (Table 12.2 versus Table 12.4).

Table 12.4 Data summary for final settler design – Example 12.3.3.

Description	Symbol	Value	Units
Plant design parameters			
MLSS	X_F	3.5	kg/m^3
Average dry weather flow rate (ADWF)	Q_{ADWF}	625	m^3/h
Diurnal peaking factor	P_{FDW}	1.5	-
Storm peak factor	P_{FWW}	2.5	-
Peak dry weather flow (PDWF)	Q_{PDWF}	938	m^3/h
Peak wet weather flow (PWWF)	Q_{PWWF}	1,563	m^3/h
Parameters for the flux theory based on Equation 12.1 and Figure 12.17			
Initial settling velocity	v_0	16.8	m/h
Hindered settling parameter	p_{hin}	0.36	m^3/kg
Safety factor on area	F_{corr}	1.25	-
Empirical design parameters			
Recycle ratio at PDWF	R_{PDWF}	0.6	-
Recycle ratio at PWWF	R_{PWWF}	0.4	-
Maximum dry weather surface overflow rate	SOR_{ADWF}	1	m/h
Maximum wet weather surface overflow rate	SOR_{PWWF}	2.5	m/h
Maximum dry weather solids loading rate	SLR_{ADWF}	6	kg/m^2.h
Maximum wet weather solids loading rate	SLR_{PWWF}	15	kg/m^2.h
WRC design parameters			
Stirred Sludge Volume Index at 3.5 g/l MLSS	$SSVI_{3.5}$	48	mg/l
Safety factor on area	F_{corr}	1.25	-
ATV and STOWA design parameters			
Diluted Sludge Volume Index	DSVI	60	mg/l
Maximum overflow rate	$q_{O,max}$	1.6	m/h
Minimum sludge volume loading	$q_{sv,min}$	300	l/m^2.h
Maximum sludge volume loading	$q_{sv,max}$	400	l/m^2.h
Maximum reduction due to sludge transferred at PWWF	f_{red}	0.7	-

The data is summarized in the data sheet of the spreadsheet 'Chapter 12 Design example 2.xlsx' (Figure 12.17).

Input data for all methods

Mixed liquour suspeneded solids	MLSS	**3.5** kg/m^3	
Average dry weather flow	Q_{ADWF}	**625** m^3/h	**Enter or change red numbers only**
Dry weather flow peak factor	PF_{DW}	**1.5** -	
Peak dry weather flow	Q_{PDWF}	938 m^3/h	Use to check for DWF operation
Wet weather peak factor	PF_{WW}	**2.5** -	
Wet weather peak flow	Q_{PWWF}	1562.5 m^3/h	Design is based on this flow

Data specific to method

Empirical
No settleability input

Flux theory				Direct information from flux theory input data	
Initial settling velocity	v_o	**16.8** m/h		X at inflection	5.56 kg/m^3
Hindered settling parameter	p_{hin}	**0.36** m^3/kg		X at maximum flux	2.78 kg/m^3
Design safety factor	F_{corr}	1.25 -		Slope at inflection ($q_{R.crit}$)	2.27 m/h

WRC			
Stirred Sludge Volume Index at 3.5 g/L	$SSVI_{3.5}$	**48** ml/g	
Design safety factor	F_{corr}	1.25 -	

ATV and STOWA			
Diluted Sludge Volume Index	DSVI	**60** ml/g	
Settled volume of sludge at MLSS	DSV_{30}	210 ml/l	

Figure 12.17 Data sheet for Example 12.3.3.

Design assumptions and design procedure

The design assumptions and the design procedure are completely the same as Example 12.3.2.

Solution

Design using flux theory

1. Define gravity, bulk, and total flux curves for the Vesilind settling velocity function

The gravity flux is obtained from Eq. 12.2 in the main textbook using the Vesilind settling velocity function with parameters v_o and p_{hin} from Table 12.4. The gravity flux curve is plotted in Figure 12.18. The bulk flux (J_B, in kg/m^2.h) is calculated from Eq. 12.1. Figure 12.19 displays the bulk flux and the total flux at PWWF conditions.

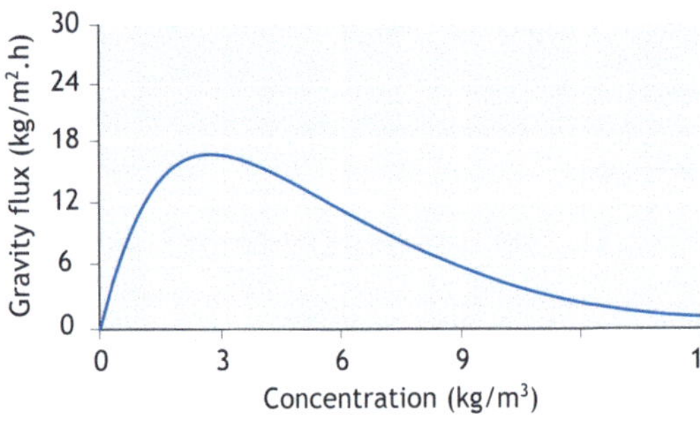

Figure 12.18 Gravity flux curve.

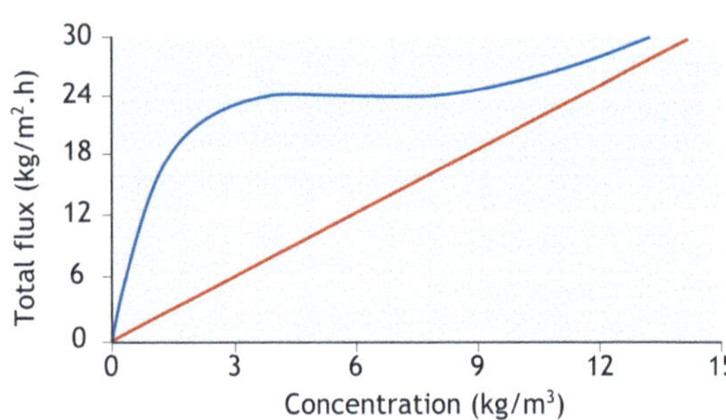

Figure 12.19 Bulk flux (red) and total flux curve (blue) at PWWF.

2. Calculate the settling velocity at the design MLSS

The settling velocity at the design MLSS concentration, *i.e.*, the solids concentration fed to the clarifier, is calculated from the Vesilind function (Eq. 12.1 in the main textbook) as:

$$v_{s,MLSS} = 16.8 \cdot e^{-0.36 \cdot 3.5} = 4.77 \qquad \text{m/h} \qquad (12.57)$$

3. Calculate the settler minimum surface (SHC II)

The minimum surface of the settler is calculated from SHC II, by setting the maximum overflow rate equal to the settling velocity for peak wet weather conditions:

$$A = \frac{Q_{PWWF}}{v_{s,MLSS}} = \frac{1{,}563}{4.77} = 328 \qquad \text{m}^2 \qquad (12.58)$$

4. Calculate the overflow for the state point diagram

The slope of the overflow line, *i.e.* the applied overflow rate or hydraulic loading, is calculated for PDWF as:

$$q_{I,PDWF} = \frac{Q_{PDWF}}{A} = \frac{938}{328} = 2.86 \qquad \text{m/h} \qquad (12.59)$$

For PWWF conditions, the overflow rate equals the design settling velocity, as prescribed by SHC II (see Eq. 12.58):

$$q_{I,PWWF} = v_{s,MLSS} = 4.77 \qquad \text{m/h} \qquad (12.60)$$

The overflow flux at the design MLSS (*i.e.*, the feed concentration) corresponds to the flux at the state point and is calculated as follows, for PDWF and PWWF, respectively:

$$J_{I,PDWF,F} = \frac{Q_{PDWF}}{A} \cdot X_F = q_{I,PDWF} \cdot X_F = 2.86 \cdot 3.5 = 10.0 \qquad \text{kg/m}^2.\text{h} \qquad (12.61)$$

$$J_{I,PWWF,F} = \frac{Q_{PWWF}}{A} \cdot X_F = q_{I,PWWF} \cdot X_F = 4.76 \cdot 3.5 = 16.7 \qquad \text{kg/m}^2.\text{h} \qquad (12.62)$$

In the state point diagram, SHC II is fulfilled if the state point lies below the gravity flux curve, which is ensured by the calculation of the settler surface area by Eq. 12.58.

5. Determination of minimum recycle ratio using the state point diagrams (SHC I)

The minimum recycle ratio is determined as the recycle ratio for which the underflow line is tangential to the descending limb of the gravity flux curve. This is done following an iterative procedure, applying different values of R, each time recalculating the position of the underflow line (Eq. 12.11) using updated values for the underflow flux (Eq. 12.9) and the total applied flux (Eq. 12.10).

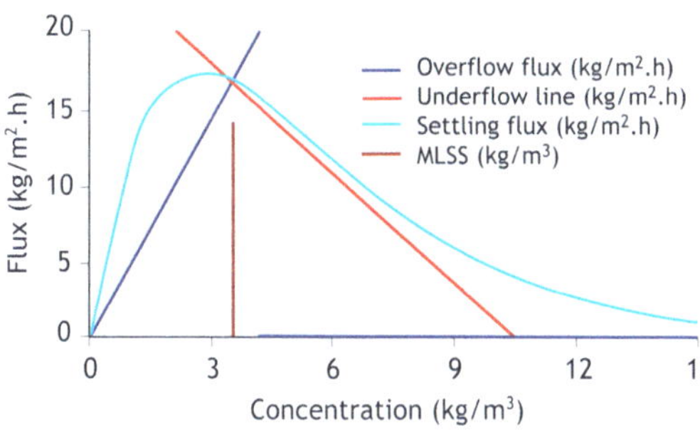

Figure 12.20 SPD at PWWF at R = 0.5.

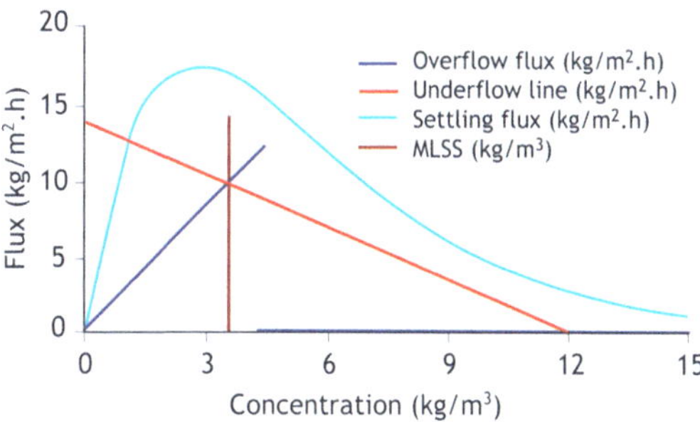

Figure 12.21 SPD at PDWF at R = 0.4.

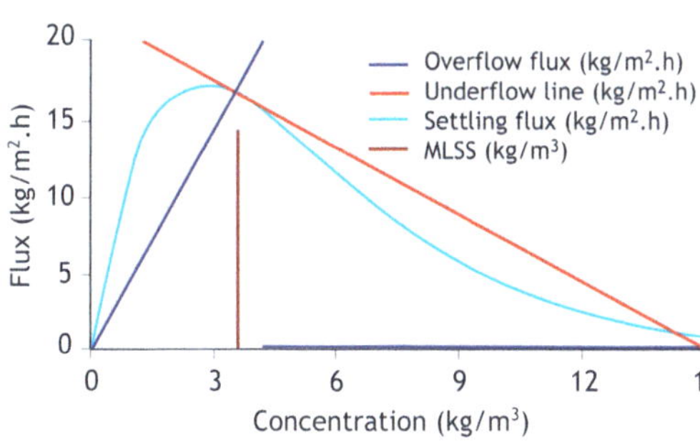

Figure 12.22 SPD at PWWF at R = 0.3.

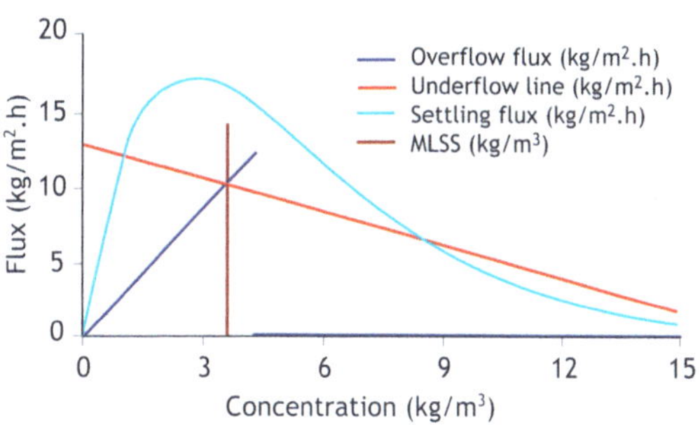

Figure 12.23 SPD at PDWF at R = 0.25.

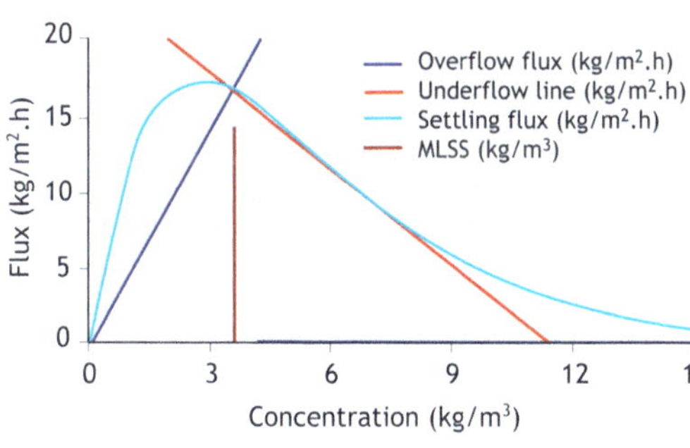

Figure 12.24 SPD at PWWF at $R_{min,PWWF}$ = 0.44.

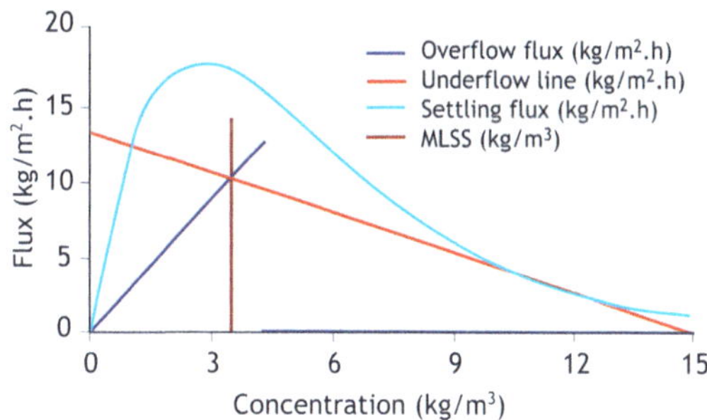

Figure 12.25 SPD at PDWF at $R_{min,PDWF}$ = 0.30.

Applying an initial value for the recycle ratio of R = 0.5 for PWWF conditions and R = 0.4 for PDWF conditions results in the state point diagrams of Figure 12.20 and Figure 12.21, respectively. In both cases, the underflow line is under the gravity flux curve above the feed concentration, which means that the recycle ratio is too high. Applying a recycle ratio of R = 0.3 for PWWF conditions and R = 0.25 for PDWF conditions, the underflow lines intersect with the gravity flux curve above the feed concentration, which means that the recycle ratio is too low (the state point diagrams in Figure 12.22 and Figure 12.23). After some trial and error, the minimum recycle flow at PWWF and PDWF conditions is found as $R_{min,PWWF}$ = 0.44 (Figure 12.24) and $R_{min,PDWF}$ = 0.30 (Figure 12.25), respectively.

6. Calculate recycle flow rates and recycle concentrations
Taking into account $R_{min,PWWF}$ = 0.44 and $R_{min,PDWF}$ = 0.30, the minimum recycle flow rates required to fulfil SHC I for PDWF and PWWF are calculated as:

$$Q_{R,PDWF} = Q_{PDWF} \cdot R_{min,PDWF} = 938 \cdot 0.30 = 281 \qquad \text{m}^3/\text{h} \qquad (12.63)$$

$$Q_{R,PWWF} = Q_{PWWF} \cdot R_{min,FWWF} = 1{,}563 \cdot 0.435 = 688 \qquad \text{m}^3/\text{h} \qquad (12.64)$$

The corresponding hydraulic underflow rates under PDWF and PWWF conditions are obtained (see Eq. 12.9) as:

$$q_{R,PDWF} = \frac{Q_{R,PDWF}}{A} = \frac{281}{328} = 0.86 \qquad \text{m/h} \qquad (12.65)$$

$$q_{R,PWWF} = \frac{Q_{R,PWWF}}{A} = \frac{688}{328} = 2.10 \qquad \text{m/h} \qquad (12.66)$$

and the underflow fluxes at design MLSS (*i.e.*, the feed concentration) are calculated from Eq. 12.9 as:

$$J_{R,PDWF,F} = 0.87 \cdot 3.5 = 3.05 \qquad \text{kg/m}^2.\text{h} \qquad (12.67)$$

$$J_{R,PWWF,F} = 2.07 \cdot 3.5 = 7.25 \qquad \text{kg/m}^2.\text{h} \qquad (12.68)$$

The total applied flux at design MLSS under PDWF and PWWF conditions follow from eqs. 12.10, 12.61, 12.62, 12.67 and 12.68 as:

$$J_{AP,PDWF} = J_{I,PDWF,F} + J_{R,PDWF,F} = 10.0 + 3.05 = 13.05 \qquad \text{kg/m}^2.\text{h} \qquad (12.69)$$

$$J_{AP,PWWF} = J_{I,PWWF,F} + J_{R,PWWF,F} = 16.7 + 7.25 = 23.95 \qquad \text{kg/m}^2.\text{h} \qquad (12.70)$$

Assuming no solids in the clarifier effluent, the sludge concentration in the recycle stream can be calculated from Eq. 12.21 which, applied to the given PDWF and PWW conditions, results in:

$$X_{R,PDWF} = \frac{J_{AP,PDWF} \cdot A}{Q_{R,ADWF}} = \frac{13.0 \cdot 328}{938} = 15.0 \qquad \text{kg/m}^3 \qquad (12.71)$$

$$X_{R,PWWF} = \frac{J_{AP,PWWF} \cdot A}{Q_{R,PWWF}} = \frac{23.9 \cdot 328}{1,563} = 11.6 \qquad\qquad kg/m^3 \qquad (12.72)$$

7. Iterate R_{min} values to calculate variables for the Ekama D&O chart
As an alternative to using the state point diagram, the minimum recycle ratio can be determined from the Ekama Design & Operation (D&O) chart.

The SHC II line is easily drawn as a straight horizontal line representing the settling velocity at the design MLSS (= feed) concentration, $v_{s,MLSS} = 4.77$ (from Step 2., Eq 12.57). The latter value corresponds to the overflow rate at PWWF conditions: $v_{s,MLSS} = Q_{PWWF} / A$, see Eq. 12.58).

The SHC I line is determined by equations 12.9 and 12.10 in the main textbook for the given Vesilind parameters and the given feed concentration, respectively:

$$\frac{Q_I}{A} = \frac{v_0}{R} \cdot \frac{1+\alpha}{1-\alpha} e^{\frac{-p_{hin}(1+R)\cdot X_F \cdot (1+\alpha)}{2R}} = \frac{16.8}{R} \cdot \frac{1+\alpha}{1-\alpha} e^{\frac{-0.36(1+R)\cdot 3.5 \cdot (1+\alpha)}{2R}} \qquad m/h \qquad (12.73)$$

in which:

$$\alpha = \sqrt{1 - \frac{4R}{0.36 \cdot (1+R) \cdot 3.5}} \qquad\qquad (12.74)$$

The value of Eq. 12.73 and Eq. 12.74 are calculated for a range of increasing recycle ratio values R, until the SHC I line intersects with the SHC II line, *i.e.*, until $Q_I / A = v_{s,MLSS}$ (Figure 12.26). The minimum recycle ratio for peak wet weather conditions is thus found as $R_{min,PWWF} = 0.44$. In Excel, this is done by using the built-in 'goal seek' function (see the spreadsheet 'Chapter 12 Design example 2.xlsx').

The safe operating zone in the Ekama D&O chart is the zone below SHC II and to the right of SHC I (the green zone in Figure 12.24). For the clarifier not to be overloaded, the overflow rate Q_I / A should be below the SHC I line in the case $R < R_{min,PWWF}$; the overflow rate Q_I/A should be lower than $v_{s,MLSS}$ in the case $R > R_{min,PWWF}$. From the SHC I line, the minimum recycle rate for PDWF conditions (*i.e.*, corresponding with $Q_{PDWF} / A = 938 / 328 = 2.86$) is found as $R_{min,PDWF} = 0.3$. The values for the minimum recycle ratios at PWWF and PDWF conditions found in the Ekama D&O chart correspond with the values found in the state point diagram (Figure 12.22 and Figure 12.23, respectively).

To complete the Ekama D&O chart, the criterion boundary line is drawn based on Equation 12.11 in the textbook:

$$\frac{Q_I}{A} = \frac{v_0}{e^2 \cdot R} = \frac{16.8}{e^2 \cdot R} \qquad\qquad (12.75)$$

The criterion boundary line gives the boundary between the lower recycle ratios where a critical flux can be found (SHC I applies) and higher recycle ratios where a critical flux cannot be found (only SHC II applies).

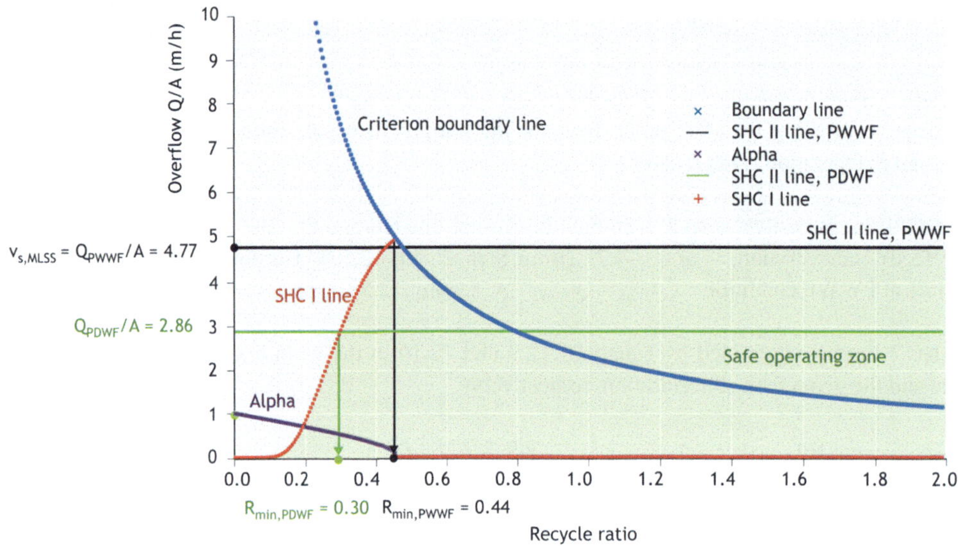

Figure 12.26 Ekama D&O chart. The green zone indicates the safe operating zone.

8. Calculate surface area and recycle pump capacity with safety threshold
Taking into account the safety factor F_{corr} (Table 12.4), the settler surface area (Eq. 12.58) is recalculated as:

$$A_{design,flux} = A \cdot F_{corr} = 328 \cdot 1.25 = 410 \qquad\qquad m^2 \qquad (12.76)$$

Due to practical considerations (standard design), the actual area chosen could be somewhat larger than the theoretically required (Eq. 12.26), for instance:

$$A_{design,flux} = 450 \qquad\qquad m^2 \qquad (12.77)$$

The recycle pump capacity is calculated according to the minimum recycle flow rates required at PDWF (and PWWF) conditions. Given that $Q_{R,PDWF} = 281$ m³/h (Eq. 12.63) and $Q_{R,PWWF} = 688$ m³/h (Eq. 12.64), say $Q_{R,PWWF} = 700$ m³/h, the decision is made to install two pumps of 350 m³/h each. During dry weather conditions, only one pump will be operational, while during rainy weather conditions both pumps will be in use.

Taking into account the settler area including the safety factor (Eq. 12.77), some of the settler specifications need to be recalculated. A summary of the final settler design specifications is given in Table 12.5.

<u>Empirical design</u>
In this example, the recycle ratio for empirical design is set at $R_{PDWF} = 0.6$ for PDWF conditions and at $R_{PWWF} = 0.4$ during PWWF conditions (from Table 12.4).

The recycle pump capacities at PDWF and PWWF are thus calculated as:

$$Q_{R,PDWF} = R_{PDWF} \cdot Q_{PDWF} = 0.6 \cdot 938 = 563 \qquad\qquad m^3/h \qquad (12.78)$$

$$Q_{R,PWWF} = R_{PWWF} \cdot Q_{PWWF} = 0.4 \cdot 1{,}563 = 625 \qquad\qquad m^3/h \qquad (12.79)$$

The corresponding recycle pump capacity amounts to:

$$Q_{R,Design} = Max(Q_{R,PDWF}; Q_{R,PWWF}) = 625 \qquad\qquad m^3/h \qquad (12.80)$$

The selection of the required settler surface area is based on the envisaged maximum surface overflow rates (hydraulic loading) for average dry weather flow and peak wet weather flow conditions, which are envisaged as $SOR_{ADWF} = 1$ m/h and $SOR_{PWWF} = 2.5$ m/h, respectively. In addition, the clarifier should not be loaded higher than 6 kg/m²/h during average dry weather and 15 kg/m²/h during wet weather conditions (*i.e.*, maximum solids loading specification). Each of the above four specifications will lead to a different required clarifier area, the largest of which will be selected, in this case 693 m² (Eq. 12.36). The calculations are provided in the design sheet in the spreadsheet 'Chapter 12 Design example 2.xlsx' (Figure 12.27). A number of additional settler specifications can be calculated and are summarized in Table 12.5.

The settler surface area based on maximum overflow rates at ADWF and PWWF are calculated as:

$$A_{design,SOR,ADWF} = \frac{Q_{ADWF}}{SOR_{ADWF}} = \frac{625}{1} = 625 \qquad\qquad m^2 \qquad (12.81)$$

$$A_{design,SOR,PWWF} = \frac{Q_{PWWF}}{SOR_{PWWF}} = \frac{1{,}563}{2.5} = 625 \qquad\qquad m^2 \qquad (12.82)$$

while the settler surface area based on the maximum solids loading at ADWF and PWWF is found as:

$$A_{design,SLR,ADWF} = X_F \cdot \frac{Q_{ADWF} + Q_{R,PDWF}}{SLR_{ADWF}} = 3.5 \cdot \frac{625 + 563}{6} = 693 \qquad\qquad m^2 \qquad (12.83)$$

$$A_{design,SLR,PWWF} = X_F \cdot \frac{Q_{PWWF} + Q_{R,PWWF}}{SLR_{PWWF}} = 3.5 \cdot \frac{1{,}563 + 625}{15} = 510 \qquad\qquad m^2 \qquad (12.84)$$

The settler surface area based on empirical design, and thus becomes:

$$A_{design,emp} = Max\left(A_{design,SOR,ADWF}, A_{design,SOR,PWWF}, A_{design,SLR,ADWF}, A_{design,SLR,PWWF}\right) = 693 \ m^2 \qquad (12.85)$$

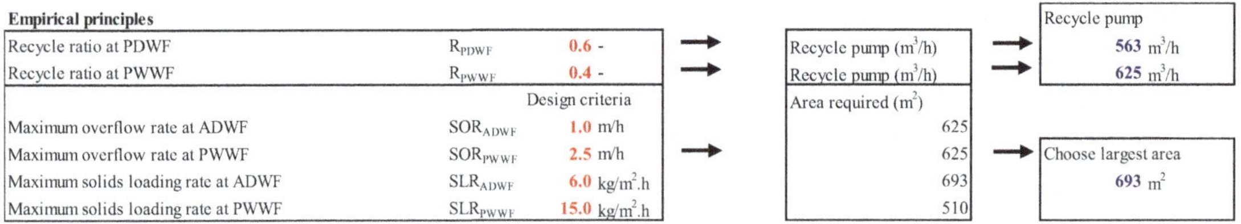

Figure 12.27 Empirical clarifier design results from the design sheet of the 'Final Settling Design Spreadsheet'.

WRC design

The critical hydraulic underflow rate corresponding with the given $SSVI_{3.5} = 48$ mg/l (Table 12.4) is calculated according to Eq. 12.12 in the main textbook, as:

$$
\begin{aligned}
q_{R,crit} &= 1.612 - 0.00793 \cdot SSVI_{3.5} - 0.0015 \cdot \left(MAX\left(0, SSVI_{3.5} - 125\right)\right)^{1.115} \\
&= 1.612 - 0.00793 \cdot 48 - 0.0015 \cdot \left(MAX\left(0, 48 - 125\right)\right)^{1.115} = 1.23
\end{aligned}
\qquad \text{m/h} \qquad (12.86)
$$

and the required surface area is obtained from:

$$
\begin{aligned}
A_{design,PWWF} &= X_F \cdot \frac{Q_{PWWF}}{306.86 \cdot SSVI_{3.5}^{-0.77} \cdot q_{R,crit}^{0.68} - X_F \cdot Q_{R,crit}} \\
&= 3.5 \cdot \frac{1{,}563}{306.86 \cdot 60^{-0.77} \cdot 1.23^{0.68} - 3.5 \cdot 1.23} = 401
\end{aligned}
\qquad \text{m}^2 \qquad (12.87)
$$

Taking into account a safety factor of 25 % ($F_{corr} = 1.25$), the surface area becomes:

$$
A_{design,WRC} = A_{design,PWWF} \cdot F_{corr} = 401 \cdot 1.25 = 501
\qquad \text{m}^2 \qquad (12.88)
$$

The PWWF overflow rate is then:

$$
q_{PWWF} = \frac{Q_{PWWF}}{A_{design,WRC}} = \frac{1{,}563}{501} = 3.1
\qquad \text{m/h} \qquad (12.89)
$$

The recycle pump capacity corresponding with the obtained values for the critical hydraulic underflow rate and the settler surface area is then calculated and, for practical purposes, rounded off as 650 m^3/h:

$$
Q_R = A_{design,WRC} \cdot q_{R,crit} = 501 \cdot 1.23 = 618 \, (650)
\qquad \text{m}^3/\text{h} \qquad (12.90)
$$

The calculations are provided in the design sheet of the spreadsheet 'Chapter 12 Design example 2.xlsx' (Figure 12.28). Additional settler specifications are summarized in Table 12.5.

WRC principles

Calculate Critical Recycle Rate	$q_{R,crit}$	1.23 m/h
Calculate Area based on $SSVI_{3.5}$ for PWWF		401 m²
Calculate Area based on $SSVI_{3.5}$ for PDWF		241 m²

$\longrightarrow$

Choose larger area	401 m²
Area with safety	**501** m²
PWWF recycle flow	618 m³/h
Choose Pump	**650** m³/h
PWWF overflow rate	3.1 m/h

Figure 12.28 WRC principle-based clarifier design in this example; a screenshot of the design sheet of the spreadsheet 'Chapter 12 Design example 2.xlsx'.

ATV design

The settled volume of the MLSS under the test conditions (DSV_{30}), Eq. 12.13 in the main textbook, which is calculated for this example as:

$$DSV_{30} = X_{F,DWF} \cdot DSVI = 3.5 \cdot 60 = 210 \qquad\qquad \text{ml/l} \qquad (12.91)$$

The design steps below are followed:

1) The permissible overflow rate, q_O, is calculated as Eq. 12.15 in the textbook:

$$q_O = 2{,}400 \cdot DSV_{30}^{-1.34} = 2{,}400 \cdot 210^{-1.34} = 1.86 \qquad\qquad \text{m/h} \qquad (12.92)$$

2) q_O must be smaller than $q_O\text{,max} = 1.6$ m/h, which is not the case, so in what follows the latter value is used.

3) The settler surface area follows from the permissible overflow rate and the influent PWWF flow rate:

$$A_{design,ATV} = \frac{Q_{PWWF}}{Min\left(q_O, q_{O,max}\right)} = \frac{1{,}563}{1.6} = 977 \qquad\qquad \text{m}^2 \qquad (12.93)$$

4) For practical considerations, the latter value is rounded off as:

$$A_{design,ATV} = 1{,}000 \qquad\qquad \text{m}^2 \qquad (12.94)$$

5) The maximum recycle solids concentration that can be reached under ADWF and PWWF conditions is calculated according to Eq. 12.16 and Eq. 12.17 in the main textbook, respectively:

$$X_{R,ADWF} = \frac{1{,}200}{DSVI} = \frac{1{,}200}{60} = 20 \qquad\qquad \text{g/l} \qquad (12.95)$$

$$X_{R,PWWF} = X_{RAS,ADWF} + 2 = 20 + 2 = 22 \qquad\qquad \text{g/l} \qquad (12.96)$$

6) The necessary recycle flow is calculated from the solids mass balance over the settler (Eq. 12.18 in the main textbook), both for ADWF and PWWF, as:

$$Q_{R,ADWF} = \frac{X_F}{X_{RAS,ADWF} - X_F} \cdot Q_{ADWF} = \frac{3.5}{20 - 3.5} \cdot 625 = 133 \qquad \text{m}^3/\text{h} \qquad (12.97)$$

$$Q_{R,PWWF} = \frac{X_F}{X_{RAS,PWWF} - X_F} \cdot Q_{PWWF} = \frac{3.5}{22 - 3.5} \cdot 1{,}563 = 296 \qquad \text{m}^3/\text{h} \qquad (12.98)$$

The recycle pump capacity is rounded up, to a practical value of:

$$Q_R = 300 \qquad \text{m}^3/\text{h} \qquad (12.99)$$

for all weather conditions.

The calculations are provided in the design sheet of the spreadsheet 'Chapter 12 Design example 2.xlsx' (Figure 12.29). Additional settler specifications are summarized in Table 12.5.

ATV principles

Permissible overflow rate (depends on DSV30)	q_O	1.86 m/h		Choose area	**1,000** m²
Maximum q_O	$q_{O,max}$	**1.60** m/h		ADWF recycle flow	133 m³/h
area based on smaller of the two above rates	A	977 m²	➡	PWWF recycle flow	296 m³/h
Maximum RAS, ADWF	$X_{RAS,ADW}$	20 g/l	➡	Choose pump	**300** m³/h
Maximum RAS, PWWF	$X_{RAS,PWW}$	22 g/l	➡		

Figure 12.29 ATV principle-based clarifier design for this example; a screenshot from the design sheet of the spreadsheet 'Chapter 12 Design example 2.xlsx'.

STOWA design

1) The permissible overflow rate is calculated based on the DSV_{30} (Eq. 12.19 in the main textbook), as follows:

$$q_O = \frac{1}{3} + \frac{200}{DSV_{30}} = \frac{1}{3} + \frac{200}{210} = 1.86 \qquad \text{m/h} \qquad (12.100)$$

2) The sludge volume loading rate is calculated according to Eq. 12.101, taking into account eqs. 12.91 and 12.100:

$$q_{sv} = q_O \cdot DSV_{30} = 1.86 \cdot 210 = 270 \qquad \text{l/m}^2.\text{h} \qquad (12.101)$$

3) The sludge volume loading rate q_{sv} must be between 300 and 400 l/m².h, which is not the case (Eq. 12.101). As a result, its value is set to $q_{sv} = 300$ l/m².h. The permissible overflow rate is recalculated by rearranging Eq. 12.101, as follows:

$$q_O = \frac{q_{sv}}{DSV_{30}} = \frac{300}{210} = 1.43 \qquad\qquad \text{m/h} \qquad (12.102)$$

4) The settler surface area follows from the permissible overflow rate and the influent ADWF flow rate according to Eq. 12.103:

$$A_{design,ADWF} = \frac{Q_{ADWF}}{q_O} = \frac{625}{1.43} = 438 \qquad\qquad \text{m}^2 \qquad (12.103)$$

5) The settler surface area based on PWWF conditions, taking into account a maximum reduction on the MLSS concentration of 30 % (corresponding with a reduction factor $f_{red} = 0.7$), results from Eq. 12.104:

$$A_{design,PWWF} = f_{red} \cdot \frac{Q_{PWWF}}{q_O} = 0.7 \cdot \frac{1,563}{1.43} = 766 \qquad\qquad \text{m}^2 \qquad (12.104)$$

6) The settler surface area is taken as the maximum of Eq. 12.103 and Eq. 12.104, and rounded up:

$$A_{design,STOWA} = above\,Max(A_{design,ADW}, A_{design,PWWF}) = 800 \text{ m}^2.$$

7) The recycle pump design is identical to the ATV method; the selected recycle pump has a capacity of 300 m³/h.

The calculations are provided in the design sheet of the spreadsheet 'Chapter 12 Design example 2.xlsx' (Figure 12.30). Additional settler specifications are summarized in Table 12.5.

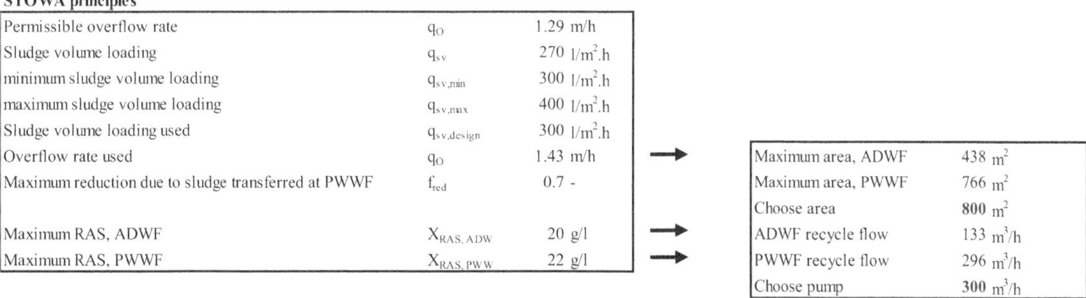

Figure 12.30 STOWA principle-based clarifier design; a screenshot from the design sheet of the spreadsheet 'Chapter 12 Design example 2.xlsx'.

Summary and discussion

An overview of the design specifications obtained with the different methods is given in Table 12.5. The design input indicates a sludge with good settling properties as DSVI is 60 ml/g. In addition, the settler load is in the middle range (MLSS is 3.5 kg/m³) and the wet weather peak is not extreme. The combination of these factors results in a relatively small settler area which is indicated in the design procedures. The ATV design stands out with the conservative approach for the maximum overflow rate of 1.6 m/h which results in a higher settler area.

Table 12.5 Summary of the final settler design detailed calculations. The calculations are provided in the summary sheet of the spreadsheet 'Chapter 12 Design example 2.xlsx'.

Parameter	Unit	Flux theory	Empirical	WRC	ATV	STOWA
Area, A	m^2	**450**	**693**	**501**	**1,000**	**800**
At ADWF						
Overflow rate, Q_{ADWF}	m^3/h			**625**		
Surface overflow rate, SOR_{ADWF}	m/h	1.39	0.90	1.25	0.63	0.78
Recycle flow rate, $Q_{R,ADWF}$	m^3/h	350	563	**650**	**300**	**300**
Hydraulic underflow rate, $q_{R,ADWF}$		0.78	0.81	**1.23**	0.30	0.38
Recycle ratio, R	-	0.56	0.90	1.04	0.48	0.48
RAS concentration, $X_{R,ADWF}$[1]	kg/m^3	9.75	7.39	6.87	10.79	10.79
Solids loading rate, SLR_{ADWF}[2]	$kg/m^2.h$	7.58	**6**	8.90	3.24	4.05
At PDWF						
Overflow rate, Q_{PDWF}	m^3/h			**938**		
Surface overflow rate, SOR_{PDWF}	m/h	2.08	1.35	1.87	0.94	1.17
Recycle flow rate, $Q_{R,PDWF}$	m3/h	**350**	**563**	**650**	**300**	**300**
Hydraulic underflow rate, $q_{R,PDWF}$	m/h	0.78	0.81	**1.23**	0.30	0.38
Recycle ratio, $R_{min,PDWF}$	-	**0.30**	**0.6**	0.69	0.32	0.32
RAS concentration, $X_{R,PDWF}$[1]	kg/m^3	12.88	9.33	8.55	14.44	14.44
Solids loading rate, SLR_{PDWF}[2]	$kg/m^2.h$	10.01	7.58	11.08	4.33	5.41
At PWWF						
Overflow rate, Q_{PWWF}	m^3/h			**1,563**		
Surface overflow rate, SOR_{PWWF}	m/h	3.47	2.26	**3.1**	1.56	1.95
Recycle flow rate, $Q_{R,PWWF}$	m^3/h	**700**	**625**	**650**	**300**	**300**
Hydraulic underflow rate, $q_{R,PWWF}$	m/h	1.56	0.90	**1.23**	0.30	0.38
Recycle ratio, $R_{min,PWWF}$	-	**0.44**	**0.4**	0.42	0.19	0.19
RAS concentration, $X_{R,PWWF}$[1]	kg/m^3	11.31	12.25	11.91	21.73	21.73
Solids loading rate, SLR_{PWWF}[2]	$kg/m^2.h$	17.60	11.05	15.44	6.52	8.15

[1] See Eq. 12.55

[2] See Eq. 12.56

12.4 EXERCISES

Final clarifier design principles (exercises 12.4.1-12.4.4)

Exercise 12.4.1

List the functions of a secondary settler tank.

Exercise 12.4.2

List the three main types of (gravitational) final settlers. What are the typical add-ons to improve their performance?

Exercise 12.4.3

Which considerations determine the settler type selection?

Exercise 12.4.4

What kind of operational problems can occur with inadequate settler implementation (design and/or operation)?

Sludge settleability and design measurements (exercises 12.4.5-12.4.8)

Exercise 12.4.5

What is the purpose of SVI tests? What are typical values? What are the main advantage and the main drawback of this method? How can the latter be overcome?

Exercise 12.4.6

What is the Zone Settling Velocity (ZSV)? Draw three ZSV test charts for an ideal settling sludge for low, medium, and high sludge concentrations.

Exercise 12.4.7

Identify the phases and the resulting zones of the ZSV test.

Exercise 12.4.8

Describe the Vesilind function.

Flux theory (exercises 12.4.9-12.4.10)

Exercise 12.4.9

Draw a typical total flux curve and discuss its constituents. Show the trend of bulk flux and gravity flux as well.

Exercise 12.4.10

Explain what the critical recycle flow rate and the lowest limiting concentration are.

State point analysis (exercises 12.4.11-12.4.13)

Exercise 12.4.11

What is state point analysis (SPA) and what are the associated considerations? List the variables used to draw the state point diagram, which forms the basis for SPA.

Exercise 12.4.12

Define Solids Handling Criterion (SHC) I and II types and their 3 conditions.

Exercise 12.4.13

What is the design principle for final settlers in terms of SHC I and II?

Ekama D&O chart (exercises 12.4.14-12.4.16)

Exercise 12.4.14

What is the Ekama D&O chart and how does it evaluate the Sludge Handling Criteria SHC I and SHC II?

Exercise 12.4.15

Explain the meaning of the criterion boundary line.

Exercise 12.4.16

How is the minimum recycle ratio determined based on the Ekama D&O chart?

Extreme manifestation of sludge bulking in a secondary settler (photo: D. Brdjanovic).

ANNEX 1: SOLUTIONS TO EXCERSISES
Final clarifier design principles (solutions 12.4.1-12.4.4)
Solution 12.4.1
Secondary settling tanks have three distinct functions:

- clarification, providing a clear effluent with suspended solids removal up to 99.9 %,
- thickening of sludge, in the blanket, to provide a concentrated recycle stream and thus lower the RAS flow,
- sludge storage, usually on a temporary basis, to cope with overload events.

Solution 12.4.2
The three main different types of (gravitational) final settlers are:

- circular clarifiers with a radial flow pattern,
- rectangular clarifier with a horizontal flow pattern,
- deep clarifiers with a vertical flow pattern.

Their operation can be improved by:

- a flocculation well, reducing the effluent solids concentration by promoting flocculation,
- scum baffles or other scum removal mechanisms,
- baffles for flow diversion and energy dissipation,
- lamellae, *i.e.*, slanted tubular or plate structures, to reduce the settling time.

Solution 12.4.3
The settler type selection is determined by the available space, cost limitation, plant loading, and performance.

Solution 12.4.4
The following are typical operational problems with final settlers:

- Shallow tanks (less than 2.5-3 m side water depth) are susceptible to solids overload, leading to scouring (loss of solids) from the sludge blanket or gross clarifier failure due to the sludge blanket reaching the effluent weirs. Maintaining a low sludge blanket level via higher recycle rates (RAS) is therefore important in shallow clarifiers.
- Uneven flow distribution between parallel clarifiers can cause overload in some final settlers.
- Incorrectly levelled weirs result in uneven weir loading, thus creating local upwelling from the blanket.
- Wind exposure can affect the flow circulation pattern and result in uneven weir loading.
- Sudden temperature change may cause an unexpected layer inversion, resulting in flotation of the sludge blanket. Freezing in cold climates may cause ice build-up.
- Recycle problems caused by suction pipe design.
- Algae on weirs can result in uneven weir load.
- Anaerobic clumps, which are floating due to methane gas generation, resulting from misaligned sludge removal.
- Birds can add extra nutrient load to the effluent.
- Bulking sludge, due to bad sludge settling properties (filaments).
- Rising sludge, due to nitrogen gas bubbles produced by denitrification in the clarifier.

Sludge settleability and design measurements (solutions 12.4.5-12.4.8)

Solution 12.4.5

The Sludge Volume Index (SVI) is a common measure for the settleability of sludge. It describes the volume of 1 g sludge (MLSS) after 30 minutes of settling. It is measured by taking 1 litre of sludge sample in a graduated cylinder, and after initial mixing, leaving it to settle for 30 minutes, after which the sludge volume is read. The sludge concentration is measured from a MLSS test. Well-settling sludge has a typical SVI value of 50 ml/g; an SVI value of 150 ml/g or higher indicates potential filamentous bulking.

The main advantage of the SVI test is that it is easy to perform (simplicity). The main drawback is that the results may be influenced by the type of sludge and by the MLSS concentration: sludges which settle fast or tests done at lower MLSS concentrations are mostly finished settling in 30 minutes, so the SVI is then an indicator of sludge compactability rather than settleability.

The results of the SVI test are made more standard by:

- diluting the sludge sample so that the settled volume after 30 minutes is 150-250 ml, *i.e.*, diluted SVI test = DSVI test.
- diluting the MLSS to a concentration of 3.5 gMLSS/litre and slowly stirring the settling vessel, *i.e.*, stirred SVI test performed at 3.5 gMLSS/litre = $SSVI_{3.5}$.

Solution 12.4.6

The Zone Settling Velocity (ZSV) is the velocity with which the sludge settles, more specifically the subsidence rate of the sludge blanket (solid/liquid interface). In a ZSV test, the ZSV is recorded in time at a certain MLSS concentration.

Three ZSV test charts are given below, for low, medium, and high sludge concentrations (1,000, 4,000 and 8,000 gMLSS/m^3, respectively). At low MLSS the sludge quickly settles, and it starts limiting the settling velocity, while at high MLSS the sludge settles slowly and steadily.

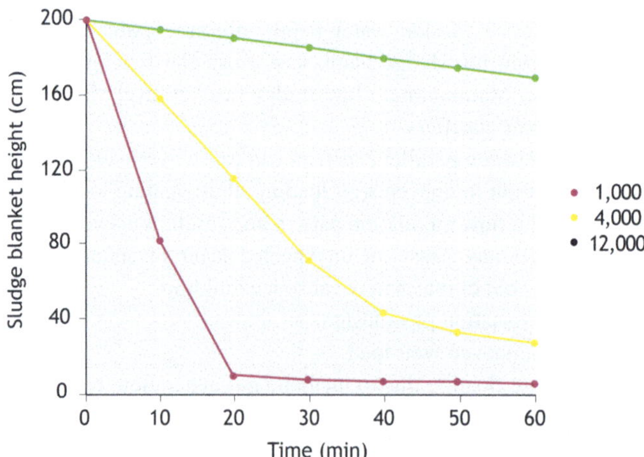

Figure 12.31 ZSV test results for varying MLSS concentrations (1,000, 4,000 and 8,000 gMLSS/m^3).

Solution 12.4.7

The ZSV test consists of the following three phases, which are reflected in three distinct zones in the obtained ZSV test curves (see Figure 12.31):

(1) the lag phase (1-2 min),
(2) the linear settling phase (3 to 30 min); a slope gives the ZSV of the sludge at the concentration with which the column was filled,
(3) an onset compression phase from the bottom, characterized by a gradual decrease in settling velocity.

Four zones can be identified in the vessel during the test: clear supernatant (discrete settling), MLSS (zone settling), a transitional zone, and sludge (a compression zone).

Solution 12.4.8

The Vesilind function describes the relation between the zone settling velocity (ZSV) and the MLSS concentration in the tank, reflecting that a higher MLSS results in a lower zone settling velocity $v_s = v_0 \cdot e^{-p_{hin} \cdot X}$ here v_s is the zone settling velocity (m/h), X denotes the MLSS concentration (g/l or kg/m^3), v_0 is the initial settling velocity (m/h), and p_{hin} is the hindered settling parameter of the Vesilind function (m^3/kg).

Flux theory (solutions 12.4.9-12.4.10)
Solution 12.4.9

The total flux (kg/m^2.h) consists of the sum of the gravity flux (also termed settling flux) and the bulk flux. The gravity flux (J_S, in kg/m^2.h) is the mass of solids transported under the influence of gravity-induced settling, $J_S = v_s \cdot X$, with v_s the zone settling velocity (m/h) at a given MLSS concentration (X, in kg/m^3). The gravity flux curve expresses the gravity flux as a function of the MLSS concentration. It first increases with increasing MLSS concentration, because of higher X, and then decreases, because of reduced settling velocity v_s; the gravity flux has a maximum usually at X = 2-3 kg/m^3 (Figure 12.32, left side).

The bulk flux (J_B, in kg/m^2.h) is the flux associated with the downward flow of the solids caused by the sludge recycle, $J_B = (Q_R / A) \cdot X$, for a given recycle flow rate Q_R (m^3/h) and settler area A (m^2). The bulk flux linearly increases with increasing MLSS concentration (Figure 12.32, right side, red curve).

The total flux curve is obtained as the sum of the gravity flux curve and the bulk flux curve and is displayed in Figure 12.32 (right side, blue curve).

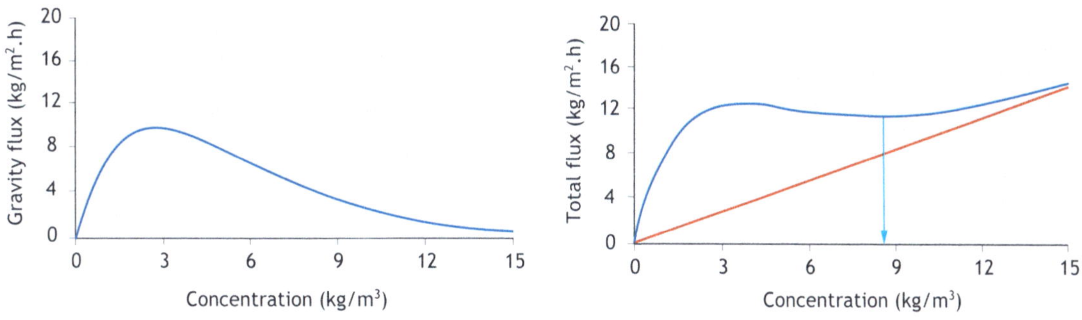

Figure 12.32 Gravity (settling) flux curve (left side), bulk flux curve (right side, red) and total flux curve (right side, blue).

Solution 12.4.10

The total flux curve displayed in Figure 12.32 has a local minimum around 8.5 kg/m³ MLSS, which is termed the limiting concentration of the clarifier at the given underflow rate. By increasing the underflow rate, the bulk flux and total flux increase for all the MLSS concentrations (the curves rotating counter clockwise). The underflow rate where the local minimum shifts into an inflexion point is called the critical recycle flow rate; the corresponding limiting MLSS concentration is termed the lowest limiting concentration.

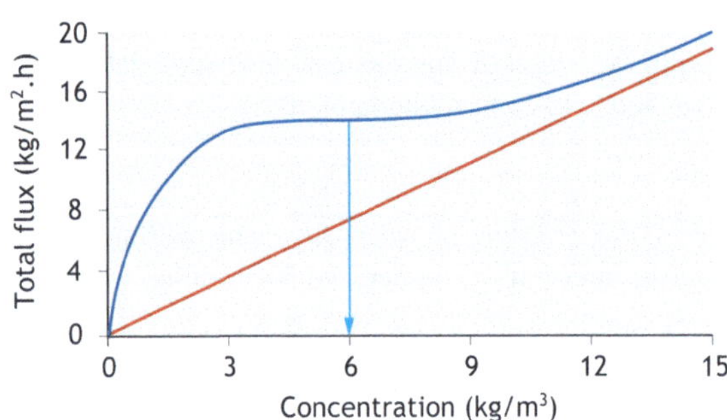

Figure 12.33 Total flux curve at critical recycle flow; determination of lowest limiting concentration.

State point analysis (solutions 12.4.11-12.4.13)

Solution 12.4.11

State point analysis is a graphical way of determining the operating conditions of the clarifier, based on its solids mass balances. The SPA method assumes steady-state conditions, it considers only the vertical dimension, it does not take into account compression, and it neglects effluent solids. The state point diagram represents various fluxes in a clarifier (the Y axis) as a function of solids concentration (the X axis). It is based on the gravity (settling) flux, on which the overflow line, the underflow line, and the feed (concentration) line are superimposed.

Solution 12.4.12

SHC I evaluates whether the applied flux to the clarifier (*i.e.*, the mass of solids applied per unit settler area) is lower than the minimum total flux (the total flux being the sum of the gravity and bulk fluxes), which is required for the clarifier not to be overloaded. In the state point diagram, SHC I is evaluated by the position of the underflow line relative to the descending limb of the gravity curve:

- SHC I is not met if the underflow line crosses the descending limb of the gravity flux curve (Figure 12.6 and Figure 12.22).
- SHC I is critical if the underflow line is tangential to the descending limb of the gravity flux curve (Figure 12.10 and Figure 12.24).
- SHC I is satisfied if the underflow line is below the descending limb of the gravity flux curve (Figure 12.8 and Figure 12.20).

SHC II requires the applied flux to be lower than the gravity flux at feed concentration. In other words, the applied overflow rate must be less than the zone settling velocity of sludge at the feed concentration. It is determined by the position of the state point relative to the gravity curve: SHC II is not met if the state point is above the gravity flux curve, SHC II is critical if the state point is on the gravity flux curve, while SHC II is satisfied if the state point is below the gravity curve.

All the nine potential combinations with SHC I and SHC II that are fulfilled, critically fulfilled, or not fulfilled, are depicted in Figure 12.26 of the main textbook.

Solution 12.4.13

The design principle is to define the clarifier surface area (A) and the recycle ratio (Q_R) to meet the SHC I and SHC II under any expected MLSS concentration and peak wet weather flow (PWWF) combination. It can be noted that SHC II is only determined by A, while SHC I is influenced by both A and Q_R. Consequently, SHC II is used to determine the minimum setter area, after which SHC I is applied to determine the minimum recycle ratio.

Ekama D&O chart (solutions 12.4.14-12.4.16)

Solution 12.4.14

The Ekama D&O chart reorganizes the information available in the flux theory and the state point diagram. It plots the overflow rate (the Y axis) against the recycle ratio (the X axis).

SHC II is judged based on a straight horizontal line representing the settling velocity at the feed concentration. Above this line, SHC II is not satisfied; the clarifier is overloaded. Below this line, *i.e.* under under-loaded conditions, SHC I remains to be evaluated.

The SHC I line describes the maximum allowable solids flux in terms of the recycle ratio. It defines a series of critical loading conditions related to the minimum flux, which are equivalent to the recycle line being tangential to the gravity flux in the state point diagram (during state point analysis).

Solution 12.4.15

Above a certain recycle ratio (R) it is not possible to determine a critical concentration and minimum solids flux. The criterion boundary line represents the boundary between lower recycle ratios where a critical flux can be found and higher recycle ratios where it cannot be found.

Solution 12.4.16

The minimum recycle ratio can be obtained from the Ekama D&O chart by reading the x value of the chart at the intersection of the SHC II and SHC I lines. For recycle ratios lower than this value, the settler is overloaded because SHC II or both SHC I and SHC II have not been met.

REFERENCES

Chen G.H., Van Loosdrecht M.C.M., Ekama G.A. and Brdjanovic D. (eds.) (2020). Biological Wastewater Treatment: Principles, Modelling and Design. ISBN: 9781789060355. IWA publishing, London, UK.

Ekama G.A., Barnard J.L., Günthert F.W., Krebs P., McCorquodale J.A., Parker D.S. and Wahlberg E.J. (1997). Secondary Settling Tanks: Theory, Modeling, Design and Operation. *IAWQ Scientific and Technical Reports* #6, IAWQ London.

NOMENCLATURE

Symbol	Description	Unit
A	Area of the settler	m^2
a	SHC I line determining parameter	-
DSV_{30}	Settled volume of MLSS under test conditions	ml/l
F_{corr}	Safety factor on area	-
f_{red}	Maximum reduction due to sludge transferred at PWWF	-
G	Velocity gradient	1/s
J_{AP}	Total applied flux	$kg/m^2.h$
J_B	Bulk flux	$kg/m^2.h$
J_I	Overflow rate flux	$kg/m^2.h$
$J_{I,F}$	Limiting flux, corresponding to X_L	$kg/m^2.h$
J_S	Gravity flux	$kg/m^2.h$
J_R	Underflow rate flux	$kg/m^2.h$
$J_{R,F}$	Underflow rate flux at feed MLSS	$kg/m^2.h$
PF_{DW}	Diurnal peaking factor	-
PF_{WW}	Storm peak factor	-
p_{hin}	Hindered settling parameter	l/g or m^3/kg
Q_{ADWF}	Average dry weather flow rate	m^3/h
q_I	Hydraulic loading or overflow rate	m/h
Q_I	Influent flow	m^3/h
q_o	Permissible overflow rate	m/h
$Q_{o,max}$	Maximum overflow rate	m/h
Q_{PDWF}	Peak dry weather flow rate	m^3/h
Q_{PWWF}	Peak wet weather flow rate	m^3/h
Q_R	Recycle flow	m^3/h
q_R	Hydraulic underflow rate	m/h
$q_{R,crit}$	Critical underflow	m/h
q_{sv}	Sludge volume loading rate	$l/m^2.h$
$q_{sv,min}$	Minimum sludge volume loading	$l/m^2.h$
$q_{sv,max}$	Maximum sludge volume loading	$l/m^2.h$
R	Recycle ratio (Q_R/Q_I)	-
R_{init}	Initial recycle ratio	-
R_{min}	Minimum recycle rate at PDWF	
SOR	Surface overflow rate	m/h
SLR	Solids loading rate	$kg/m^2.h$
v_o	Initial settling velocity	m/h
v_s	Settling velocity	m/h
x	MLSS concentration in the various ZSV tests	g/l or kg/m^3
X	Solids concentration	kg/m^3
X_F	Feed concentration	kg/m^3
X_L	Limiting concentration	kg/m^3
X_R	Recycle concentration	kg/m^3

Abbreviation	Description
ADWF	Average dry weather flow
DSVI	Diluted sludge volume index
DWA	Deutsche Vereinigung für Wasserwirtschaft, Abwasser und Abfall (German organization for waste management, sewage and waste)
DWF	Dry weather flow
MLSS	Mixed liquor suspended solids
PFDW	Peak dry weather flow
PWWF	Peak wet weather flow
RAS	Return activated solids
SHC	Solids Handling Criteria
SPA	State Point Analysis
SPD	State Point Diagram
$SSVI_{3.5}$	Stirred sludge volume index test performed at 3.5 g/l MLSS concentration
STOWA	Stichting Toegepast Onderzoek Waterbeheer (Dutch Foundation for Applied Water Research)
SVI	Sludge volume index
SZSV	Stirred zone settling velocity
ZSV	Zone settling velocity

Wastewater treatment infrastructure is best visible during (re)construction of a plant - settling tanks at WWTP Harnaschpolder, Den Hoorn, The Netherlands (photo: Water Board Delfland).

doi: 10.2166/9781789062304_0439

13

Membrane bioreactors

Xia Huang, Fangang Meng, Kang Xiao, Shuai Liang and Jiao Zhang

13.1 INTRODUCTION

Chapter 13 on Membrane Bioreactors (MBRs) in the book *Biological Wastewater Treatment: Principles, Modelling and Design* (Chen *et al.*, 2020) introduces the principles, configurations, treatment performance and mechanisms as well as membrane fouling that severely impact MBR processes. Comprehensive strategies for membrane fouling control, which are of great significance for a consistent operation, are summarized. All of these lay a solid foundation for the readers to understand and master the design, operation and maintenance of MBR-based plants. This chapter aims to further guide the readers through the principles, treatment mechanisms, and operational key points to design and assess MBR wastewater treatment systems.

13.2 LEARNING OBJECTIVES

After the successful completion of this chapter, the reader will be able to:

- Describe the membrane processes and the corresponding principles and mechanisms.
- Describe the different configurations of MBRs and list the principal features of each configuration.
- Discuss the advantages of MBRs for municipal wastewater treatment and explain the mechanisms behind their higher capability compared to other systems to remove pollutants including organic matter, nutrients, and emerging pollutants.
- Analyse and characterize the foulants and fouling layers in MBRs, and propose applicable control strategies regarding the fouling types.
- Formulate an MBR-based treatment process with regard to a specific influent quality and discharge standard, with designs for pre-treatment, biological treatment, and membrane processes.
- Recognize and match the knowledge learned with diverse practical applications of MBRs treating municipal wastewater, industrial wastewater, and leachate.

13.3 EXAMPLES

Example 13.3.1

Assessment of the contribution ratios for different treatment units and effects on the removal of typical pollutants including chemical oxygen demand (COD), biochemical oxygen demand (BOD), NH_4^+-N, total nitrogen (TN), total phosphorus (TP), and trace organic pollutants (TrOPs) in an MBR process.

An AAO-MBR municipal wastewater treatment plant with an average treatment capacity of 50,000 m³/d has been in operation for one year. The main process comprises an anaerobic tank (A1), an anoxic tank (A2), an aerobic tank (O), and a membrane tank (M) (illustrated in Figure 13.1). Activated sludge in the M tank is recirculated to the O tank at a ratio of 400 %. Meanwhile, there are two internal recirculations in the system: (*i*) from the end of the O tank to the front of the A2 tank at a ratio of 500 %, and (*ii*) from the end of the A2 tank to the beginning of the A1 tank at a ratio of 200 %. The M tank is equipped with hollow-fibre polyvinylidene fluoride ultrafiltration membrane modules (nominal pore size: 0.04 mm). Other operational information is listed in Table 13.1.

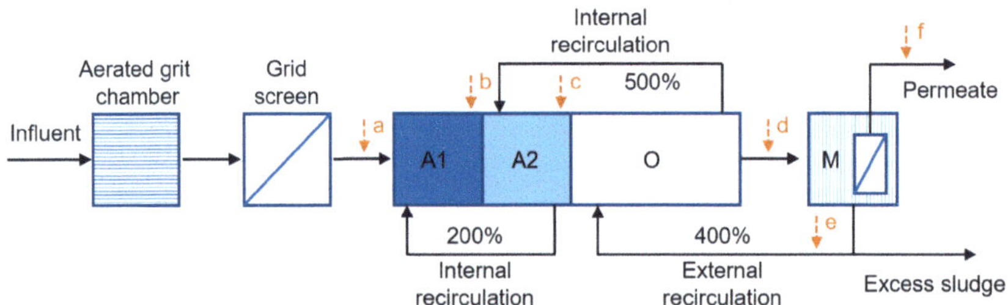

Figure 13.1 Flow diagram of the full-scale AAO-MBR process (A1: anaerobic tank; A2: anoxic tank; O: aerobic tank; M: membrane tank). The arrows denote the sampling locations.

Table 13.1 Operational information of the AAO-MBR process.

Process	HRT (h)	MLSS (g/l)	DO (mg/l)
A1	2	4.5	0.1
A2	5	8.0	0.4
O	7	9.5	3.6
M	0.5	11.5	4.9

HRT: hydraulic retention time; MLSS: mixed liquor suspended solids; DO: dissolved oxygen.

To assess the treatment contribution ratios, the effluents of different treatment units were periodically sampled (see the sampling locations in Figure 13.1) for water quality measurements. Bisphenol A (BPA) was selected as a representative compound of TrOPs for assessment. The averaged data of the water characteristics in the different stages are presented in Table 13.2.

Taking BPA as a representative pollutant, the contribution ratios of the different effects (*e.g.*, sludge adsorption, biological degradation, membrane retention, etc.) to pollutant removal can be further analysed. For this purpose, the contents of BPA in the mixed liquor and sludge flows along the different system stages were determined separately. The measured data are listed in Table 13.3. It is widely accepted that the removal of TrOPs can be due to several possible reasons such as: (*i*) adsorption by activated sludge, (*ii*) biological degradation, and (*iii*) membrane rejection. The role of adsorption can be assessed via analyses of the sludge-phase concentration and sludge-water partition coefficient (k_p). Since the adsorption of TrOPs to sludge usually follows a linear pattern (*i.e.*, an approximation of the Langmuirian isotherm at low concentrations), the concentration effect may be excluded by dividing the sludge-phase concentration by the aqueous phase concentration. The yielded k_p, which is analogous to the adsorption equilibrium constant, essentially reveals the true adsorption propensity. The k_p can be calculated and adopted to describe the partition characteristics of BPA between the sludge and the aqueous phases in the full AAO-MBR process. The k_p (l/gMLSS) is defined as:

$$k_p = \frac{C_s}{C_w}$$

where C_S is the concentration in the sludge phase (ng/gMLSS) and C_W is the concentration in the aqueous supernatant phase (ng/l).

Table 13.2 Effluent water qualities of different treatment units.

Unit	Sampling location	Symbol	Concentration					
			COD (mg/l)	BOD (mg/l)	NH$_4^+$-N (mg/l)	TN (mg/l)	TP (mg/l)	BPA (ng/l)
A1	a	C_0	400	300	40	60	8	300
	b	C_1	210	145	17	30	4	120
A2	c	C_2	130	90	7	18	2.1	80
O	d	C_3	90	60	2	13	1.1	60
M	e	C_4	60	20	0.1	9	0.1	40
	f	C_5	40	5	0.1	9	0.1	30

Table 13.3 Measured BPA contents in the supernatant and sludge at different sampling locations (sampling locations a, b, c, d, e and f indicated in Figure 13.1) at the MBR plant.

BPA content	Unit	Influent (a)	A1 (b)	A2 (c)	O (d)	M (e)	Effluent (f)
In sludge, C_s	ng/gMLSS	-	1,600	1,550	1,500	1,560	-
In supernatant, aqueous concentration, C_w	ng/l	300	120	80	60	40	30

Based on the previous information, calculate:

a. The contribution ratios of the different treatment stages (*i.e.*, A1, A2, O, and M) to the removal of the different pollutants.
b. The contribution ratios of different effects (*e.g.*, sludge adsorption, biological degradation, membrane retention, etc.) to the BPA removal (as an example of pollutants) in the treatment process.

Solution

a. Assuming that the flow rate of the excess sludge is very low (and can be ignored compared to the influent flow rate), the overall removal rate of a specific pollutant can be calculated as:

$$R_t = \frac{C_0 - C_5}{C_0} \cdot 100 \ \%$$

For the A1 tank, the mass balance equation for a specific pollutant can be expressed as:

$$C_0 Q + 2C_2 Q = 3C_1 Q + 3\Delta C_{A1} Q$$

where Q is the influent flow rate and ΔC_{A1} is the concentration change after the A1 tank treatment. Therefore, the contribution ratio of A1 (denoted as T_{A1}) to the overall removal of a specific pollutant can be calculated as:

$$T_{A1} = \frac{3\Delta C_{A1}}{C_0 - C_5} \cdot 100 \ \% = \frac{C_0 + 2C_2 - 3C_1}{C_0 - C_5} \cdot 100 \ \%$$

For the A2 tank, the mass balance equation for a specific pollutant can be expressed as:

$$3C_1 + 5C_3 = 8C_2 + 8\Delta C_{A2}$$

where ΔC_{A2} is the concentration change after the A2 tank treatment. Therefore, the contribution ratio of A2 (denoted as T_{A2}) to the overall removal of a specific pollutant can be calculated as:

$$T_{A2} = \frac{8\Delta C_{A2}}{C_0 - C_5} \cdot 100 \ \% = \frac{3C_1 + 5C_3 - 8C_2}{C_0 - C_5} \cdot 100 \ \%$$

For the O tank, the mass balance equation for a specific pollutant can be expressed as:

$$6C_2 + 4C_4 = 10C_3 + 10\Delta C_O$$

where ΔC_O is the concentration change after the O tank treatment. Therefore, the contribution ratio of O (denoted as T_O) to the overall removal of a specific pollutant can be calculated as:

$$T_O = \frac{10\Delta C_O}{C_0 - C_5} \cdot 100 \ \% = \frac{6C_2 + 4C_4 - 10C_3}{C_0 - C_5} \cdot 100 \ \%$$

For the M tank, the mass balance equation for a specific pollutant can be expressed as:

$$5C_3 = C_5 + 4C_4 + 5\Delta C_M$$

where ΔC_M is the concentration change after the M tank treatment. Therefore, the contribution ratio of M (denoted as T_M) to the overall removal of a specific pollutant can be calculated as:

$$T_M = \frac{5\Delta C_M}{C_0 - C_5} \cdot 100 \ \% = \frac{5C_3 - C_5 - 4C_4}{C_0 - C_5} \cdot 100 \ \%$$

Based on the above equations, the contribution ratios of the different treatment units can be calculated. The obtained results are listed in Table 13.4.

Table 13.4 Calculated contribution rates for the different treatment units to the removal of typical pollutants.

Unit	Contribution rate, %					
	COD	BOD	NH_4^+-N	TN	TP	BPA
T_{A1}	8.33	15.25	7.52	11.76	2.53	37.04
T_{A2}	11.11	5.08	12.53	21.57	8.86	7.41
T_O	33.33	6.78	56.14	27.45	25.32	14.81
T_M	47.22	72.88	23.81	39.22	63.29	40.74

b. As for the contribution of different effects on the removal of BPA in the treatment process, it is apparent (according to the calculated results in Table 13.4) that the adsorption of BPA by activated sludge may be sufficiently rapid relative to the hydraulic retention time in the anaerobic tank (2 h). The decreases in the BPA concentration in the A1, A2, O, and M tanks can be roughly attributed to the sludge adsorption and biological degradation processes. In the M tank, the decrease ratio r_{rej} owing to physical rejection (by the membrane and/or its fouling layer) can be calculated as:

$$r_{rej} = \frac{C_4 - C_5}{5\Delta C_M} = \frac{C_4 - C_5}{5C_3 - C_5 - 4C_4} = 9.1 \ \%$$

The adsorption of BPA by activated sludge could be an important removal pathway. The k_p can be calculated and the results are listed in Table 13.5.

Table 13.5 Sludge-water partition coefficient (k_p) of BPA in different sludge samples taken during the process.

Unit	A1	A2	O	M
BPA, k_p	13.3	19.38	25	39

The k_p value increased along with the process flow, demonstrating that the adsorption ability of activated sludge to BPA increased with the change in operating conditions.

Example 13.3.2

How can the transmembrane pressure (TMP) be determined?

TMP denotes the pressure difference across a membrane. It is a means to assess membrane fouling in MBRs. The TMP can be calculated as TMP = feed pressure – permeate pressure.

The following example is a lab-scale MBR (Figure 13.2). The relevant parameters are given as: the height between the membrane module and the gauge h_1 = 30 cm, the height between the liquid level and the gauge h_2 = 20 cm, and the height between the membrane module and the liquid level h_3 = 10 cm. These parameters are typical for lab-scale MBRs with little friction or local head loss. The readings of the vacuum gauge are −5 kPa, −10 kPa, and −25 kPa, corresponding to the stages 1, 2 and 3, respectively (Figure 13.3). Calculate the corresponding TMP_1, TMP_2, and TMP_3.

Solution

According to the equation $TMP = P_1 - P_2$, where P_1 and P_2 are denoted as external and internal pressure of the membrane module, respectively.

$$P_1 = \rho \cdot g \cdot h_3$$

$$\rho = 1{,}004 \qquad\qquad kg/m^3$$

$$g = 9.8 \qquad\qquad N/kg$$

$$P_2 = P + \rho g(h_1 + h_\lambda + h_\zeta)$$

h_λ and h_ζ are denoted as friction and local head loss, respectively. Both these two parameters are assumed to be negligible in this case because of the simple pipeline connection in the lab-scale MBR. In large-scale MBR plants, these losses should be taken into consideration due to their complicated pipeline systems. Then,

$$TMP = P_1 - P_2 = -P - \rho \cdot g\left(h_1 - h_3\right) = -P - \rho \cdot g \cdot h_2$$

$$TMP_1 = 5 - 1{,}004 \cdot 9.8 \cdot 0.2 / 1{,}000 = 3.03 \qquad\qquad kPa$$

$$TMP_2 = 10 - 1{,}004 \cdot 9.8 \cdot 0.2 / 1{,}000 = 8.03 \qquad\qquad kPa$$

$$TMP_3 = 25 - 1{,}004 \cdot 9.8 \cdot 0.2 / 1{,}000 = 23.03 \qquad\qquad kPa$$

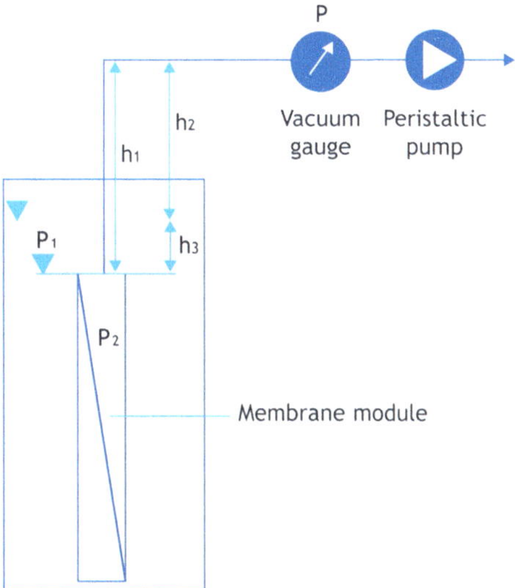

Figure 13.2 The lab-scale MBR setup showing the height between the membrane and gauge, the height between the gauge and liquid level, and the height between the membrane and liquid level (h_1, h_2 and h_3, respectively).

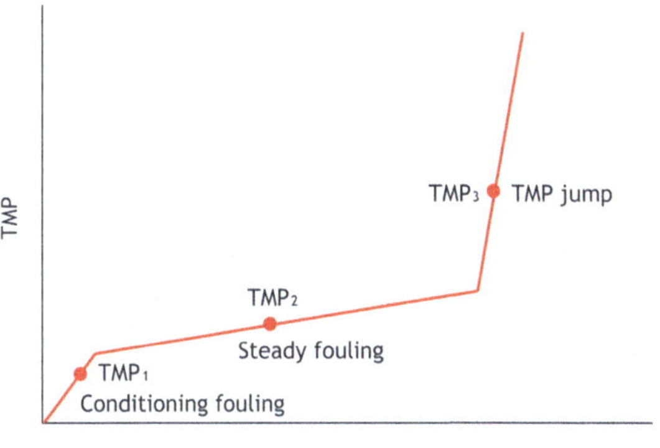

Figure 13.3 The typical fouling process.

Example 13.3.3

What is the contribution of the different resistances to the total resistance to membrane filtration and how can they be determined?

The total resistance of the fouled membrane (R_{tot}) can generally be quantified as the sum of the intrinsic resistance of the pristine membrane (R_m), the resistance of the cake layer (bio-cake deposition, R_C), and the resistance of the gel layer and pore fouling (R_g):

$$R_{tot} = R_m + R_C + R_g$$

A resistance-in-series model is usually applied to calculate the fouling resistance. According to Darcy's law,

$$J = \frac{\Delta P}{\mu \cdot (R_m + R_C + R_g)}$$

Where J ($m^3/m^2.s$) represents the permeate flux; ΔP (Pa) is the TMP; and μ (Pa.s) is the viscosity of the permeate solution, which is similar to water at a given working temperature. Then, each kind of resistance can be calculated as follows.

First, the permeate flux of the fouled membrane (J_1, $m^3/m^2.s$) is determined with pure water under a given TMP (ΔP_1, Pa). Then, the total resistance to filtration ($R_{tot,1}$) can be evaluated according to the following equation,

$$R_{tot,1} = R_m + R_C + R_g = \frac{\Delta P_1}{\mu \cdot J_1}$$

Afterwards, the membrane is physically cleaned to remove the bio-cake (e.g., the membrane surface is scrubbed with a sponge or flushed with tap water). Then, the permeate flux of the cleaned membrane (J_2, $m^3/m^2.s$) is determined with pure water under a given TMP (ΔP_2, Pa). Based on these values, the total resistance to filtration ($R_{tot,2}$) can be evaluated according to the following equation,

$$R_{tot,2} = R_m + R_g = \frac{\Delta P_2}{\mu \cdot J_2}$$

Subsequently, the membrane is chemically cleaned to remove the gel layer and pore fouling (e.g., NaOCl and citric acid cleaning). Based on these values, the total resistance to filtration ($R_{tot,3}$) can be evaluated according to the following equation,

$$R_{tot,3} = R_m = \frac{\Delta P_3}{\mu \cdot J_3}$$

where J_3 ($m^3/m^2.s$) is the permeate flux of the chemically cleaned membrane; ΔP_3 (Pa) is the TMP under the same conditions.

Therefore,

$$R_C = R_{tot,1} - R_{tot,2}$$
$$R_g = R_{tot,2} - R_{tot,3}$$

Example: a lab-scale MBR is operated for 20 d and the fouled membrane is used to calculate the fouling resistance. The parameters are as follows: $\Delta P_1 = 1.2 \cdot 10^4$ Pa, $J_1 = 7.5$ $l/m^2.h$, $\Delta P_2 = 1.0 \cdot 10^4$ Pa, $J_2 = 18.5$ $l/m^2.h$, $\Delta P_3 = 4.27 \cdot 10^3$ Pa, $J_3 = 21.55$ $l/m^2.h$, and $\mu = 1.0 \cdot 10^{-3}$ Pa.s.

Solution

$$R_{tot,1} = 1.2 \cdot 10^4 \cdot 3{,}600 / (1.0 \cdot 10^{-3} \cdot 7.5 \cdot 10^{-3}) = 5.76 \cdot 10^{12} \qquad 1/m$$

$$R_{tot,2} = 1.0 \cdot 10^4 \cdot 3{,}600 / (1.0 \cdot 10^{-3} \cdot 18.5 \cdot 10^{-3}) = 1.95 \cdot 10^{12} \qquad 1/m$$

$$R_{tot,3} = 4.27 \cdot 10^3 \cdot 3{,}600 / (1.0 \cdot 10^{-3} \cdot 21.55 \cdot 10^{-3}) = 7.13 \cdot 10^{11} \qquad 1/m$$

$$R_C = R_{tot,1} - R_{tot,2} = 3.81 \cdot 10^{12} \qquad 1/m$$

$$R_g = R_{tot,2} - R_{tot,3} = 1.24 \cdot 10^{12} \qquad 1/m$$

$$R_m = R_{tot,3} = 7.13 \cdot 10^{11} \qquad 1/m$$

Example 13.3.4

Process design for a typical AAO-MBR plant for municipal wastewater treatment.

The example is a typical AAO-MBR plant whose daily treatment capacity (Q) is 150,000 m³/d. The target water quality of the influent and effluent is given in Table 13.6. Design the biological treatment units (tank volumes, recirculation flow rates, excess sludge production, and oxygen demand) and the membrane filtration system (membrane area and aeration demand). Typical parameters for the design can be taken from Table 13.2 in Chen *et al.*, 2020.

Table 13.6 Wastewater characteristics of the influent and effluent.

	COD	BOD$_5$	TN	NH$_4^+$-N	TP	SS
Influent (mg/l)	400	180	45	30	5	300
Effluent (mg/l)	50	10	15	5	0.5	10

Solution

Step 1: Selection of process flow

The whole process flow of the example is shown in Figure 13.4. The biological treatment section of the process is also depicted in Figure 13.19A in Chen *et al.*, 2020.

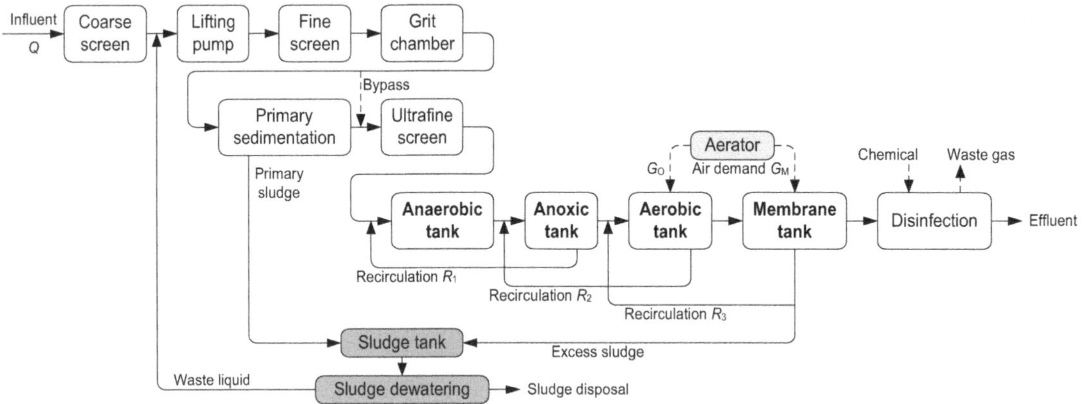

Figure 13.4 Flow chart showing the designed AAO-MBR process.

Step 2: Design of pretreatment and primary treatment
For details refer to Section 13.5.2 in Chen *et al.*, 2020.

Step 3: Design of biological treatment units
a.   Selection of design parameters
The parameter values for the design of biological treatment units in the AAO-MBR process are selected according to Table 13.2 in Chen *et al.*, 2020, and they are given here in Table 13.7.

Table 13.7 Selected parameters for the design of biological treatment units.

Description	Symbol	Unit	Selected value
MLSS concentration in the membrane tank	X_M	gMLSS/l	10
Proportion of volatile MLSS in the total MLSS	y	kgMLVSS/kgMLSS	0.6
Maximum specific growth rate of nitrifying bacteria	μ_{nm}	1/d	0.66
Half-velocity constant for ammonia utilization	K_n	$mgNH_4^+$-N/l	0.75
Observed yield of solids	Y_t	$kgMLSS/kgBOD_5$	0.45
Maximum specific ammonia utilization rate	v_{nm}	$kgNH_4^+$-N/kgMLSS.d	0.04
Specific denitrification rate	K_{dn}	$kgNO_3^-$-N/kgMLSS.d	0.06
Recirculation ratio (from anoxic to anaerobic tank)	$R_{A2 \rightarrow A1}$	-	1.5
Recirculation ratio (from aerobic to anoxic tank)	$R_{O \rightarrow A2}$	-	4
Recirculation ratio (from membrane to aerobic tank)	$R_{M \rightarrow O}$	-	5
Hydraulic retention time (HRT) of the anaerobic tank	t_{A1}	h	1.2

b. Calculation of tank volumes

Aerobic zone

The volume of the aerobic zone (V_{OM1}) includes that of the aerobic tank (V_O) and that of the membrane tank excluding the membrane cassette volume (V_{M1}). The ammonia concentration in the aerobic zone (N_{OM1}) is supposed to be equal to that in the effluent, i.e., 5 mg/l from Table 13.6. The temperature selected is 20 °C. The equations quoted are all from Chen *et al.*, 2020.

Option 1:

To calculate V_{OM1} using Equation 13.1 from Chen *et al.*, 2020, it is first necessary to estimate the sludge age and MLSS concentration in the aerobic zone.

According to Equation 13.3 from Chen *et al.*, 2020, the specific growth rate of nitrifying bacteria is estimated to be $\mu_n = 0.574$ 1/d. The minimum sludge age in the aerobic zone is then estimated, according to Equation 13.2, to be $\theta_{OM1} = 5.23$ d, given a safety factor of 3.

According to the mass balance, the MLSS concentration of the aerobic tank (X_O) can be calculated from that of the membrane tank (X_M), i.e., $X_O = X_M R_{M \rightarrow O} / (1 + R_{M \rightarrow O}) = 8.33$ g/l. Then the tank volume-weighted average MLSS concentration in the aerobic zone (X_{OM1}) is somewhere between 8.33 and 10 g/l; for simplicity of calculation here 8.89 g/l is used as an example (corresponding to a tentative $V_O : V_{M1}$ ratio of 2:1; this ratio can be rechecked later and V_{M1} is known from the design of the membrane tank). Therefore, according to Equation 13.1, $V_{OM1} = 6,750$ m^3.

Option 2:

This option is to calculate V_{OM1} from the ammonia utilization rate using Equation 13.1 from Chen *et al.*, 2020. The specific ammonia utilization rate is estimated to be $v_n = 0.0348$ kgNH$_4^+$-N/kgMLSS.d, according to Equation 13.6. The rate of biomass discharge from the system is calculated to be $\Delta X_v = 6,885$ kgMLVSS/d according to Equation 13.5. Then according to Equation 13.4, $V_{OM1} = 9,362$ m^3.

Anoxic zone

The volume of the anoxic zone (V_{A2}) is calculated from the denitrification rate and N mass balance according to Equation 13.7 from Chen *et al.*, 2020. The denitrification rate K_{dn} is set to be 0.06 kgNO$_3^-$-N/kgMLSS.d as shown in Table 13.7. According to the mass balance, the MLSS concentration in the anoxic tank is calculated to be $X_{A2} = X_O R_{O \rightarrow A2} / (1 + R_{O \rightarrow A2}) = 6.66$ g/l. Then according to Equation 13.7, $V_{A2} = 9,125$ m^3.

Anaerobic zone

The volume of the anaerobic zone (V_{A1}) is calculated from the HRT of the anaerobic tank (set to be 1.2 h, as shown in Table 13.7) according to Equation 13.8 from Chen *et al.*, 2020: $V_{A1} = Qt_{A1} / 24 = 7,500$ m^3.

Tank dimensions

The biological tanks are designed to be rectangular, with a total depth of 7.25 m and an effective water depth of 6 m.

c. Calculation of recirculation flow rates

According to Equations 13.9-13.11 from Chen *et al.*, 2020, the recirculation flow rates of the mixed liquor between the different tanks are calculated as: $Q_{M \rightarrow O} = QR_{M \rightarrow O} = 750,000$ m^3/d, $Q_{O \rightarrow A2} = QR_{O \rightarrow A2} = 600,000$ m^3/d, and $Q_{A2 \rightarrow A1} = QR_{A2 \rightarrow A1} = 225,000$ m^3/d, respectively.

d. Calculation of excess sludge production

The excess sludge production (ΔX) is estimated from the observed yield of biomass (Y_t) which is set to be 0.45 kgMLSS/kgBOD as given in Table 13.7. According to Option 2 for Equation 13.12 from Chen *et al.*, 2020, the calculated result is 11,475 kg/d. Given that the excess sludge is discharged from the membrane tank with $X_M = 10$ gMLSS/l, the average flow rate of excess sludge is then 1,147.5 m^3/d.

e. Calculation of aeration demand for biological reactions

The oxygen demand for biological reactions in the aerobic zone (O) can be calculated from the total demand for oxidizing organic carbon (O_s), the demand for nitrification (O_n), the offset for denitrification (O_{dn}), and the offset due to recirculation of oxygen-rich mixed liquor from the membrane tank (O_m). According to Equations 13.15-13.18 from Chen *et al.*, 2020, $O_s = 27,708$ kgO$_2$/d, $O_n = 13,236$ kgO$_2$/d, $O_{dn} = 10,428$ kgO$_2$/d, and $O_m = 4,500$ kgO$_2$/d (assuming C_{omd}, the dissolved oxygen concentration in the recirculated flow from the membrane tank, to be 6 mg/l). Then the net oxygen demand is $O = 15,844$ kgO$_2$/d, according to Equation 13.14.

To convert the oxygen demand into the standard-state amount (at 20 °C and 1 bar in clean water), the α factor for oxygen transfer in the sludge mixed liquor is estimated, from the MLSS concentration in the aerobic tank, to be 0.873 according to Equation 13.20. Then, letting the parameters $\beta = 0.95$, $P = 101.325$ kPa, $C_o = 1.5$ mg/l, $C_{os(20)} = 9.08$ mg/l, and $h = 6$ m, the standard-state oxygen demand is calculated to be $O_{std} = 17,097$ kgO$_2$/d according to Equation 13.19 from Chen *et al.*, 2020.

The air supply of the blower is then calculated to be $G_O = 6.36 \cdot 10^5$ Nm^3/h according to Equation 13.21, if it is tentatively assumed that the blower has an oxygen transfer efficiency η_A of 40 %.

Step 4: Design of the membrane filtration system

a. Flux

The average flux (J_{avg}) is designed in the range of 15-25 l/m^2.h. For example, it can be 20 l/m^2.h.

b. Membrane area

The total membrane area (A_M) is calculated from the total flow rate, the average flux and a safety factor ($F_M = 1.1$), according to Equation 13.24 from Chen *et al.*, 2020, to be 343,750 m^2. If each membrane cassette has a membrane area of 1,250 m^2, then 275 cassettes are required.

c. Tank dimensions

Assuming the membrane cassettes have a packing density of 100 m^2/m^3 (Judd and Judd, 2011) the volume occupied by the cassettes is 3,437.5 m^3. The total volume of the membrane tank should consist of the volumes of the cassettes, the inter-cassette space, and the space reserved for future use. These volumes are related to the layout requirement of the membrane cassettes. The membrane tank is designed to be rectangular, and sufficiently large to ensure the hydraulics of the tank. For instance, it is recommended to position the cassette a minimum distance of 0.5 m from the water surface, the tank wall, and any other cassette.

d. Aeration demand

The specific aeration demand per membrane area is set to be 0.3 Nm^3/m^2.h for scouring of hollow-fibre membranes, as suggested in Section 13.5.4.3 in Chen *et al.*, 2020. The air supply is then designed to be $G_M = SAD_m \cdot A_M = 103,125$ Nm^3/h, according to Equation 13.27 from Chen *et al.*, 2020.

Summary

A summary of the design procedure and results is given here in Table 13.8.

Table 13.8 Summary of the procedure and results for the process design example.

Item	Equation and symbol	Result	Table or equation (Chen *et al.*, 2020).
1. Selection of process flow (AAO-MBR process)			
2. Design of pretreatment and primary treatment			
3. Design of biological treatment units			
3.1. Selection of design parameters			
MLSS in the membrane tank	X_M	10 gMLSS/l	Table 13.2
MLVSS/MLSS	y	0.6 kgMLVSS/kgMLSS	Table 13.2
Maximum specific growth rate of nitrifying bacteria	μ_{nm}	0.66 1/d	Table 13.2
Half-velocity constant for ammonia utilization	K_n	0.75 $mgNH_4^+$-N/l	Table 13.2
Observed yield of solids	Y_t	0.45 $kgMLSS/kgBOD_5$	Table 13.2
Maximum specific ammonia utilization rate	v_{nm}	0.04 $kgNH_4^+$-N/kgMLSS.d	Table 13.2
Specific denitrification rate	K_{dn}	0.06 $kgNO_3^-$-N/kgMLSS.d	Table 13.2
Recirculation ratio ($A_2{\rightarrow}A_1$)	$R_{A2{\rightarrow}A1}$	1.5	Table 13.2
Recirculation ratio ($O{\rightarrow}A_2$)	$R_{O{\rightarrow}A2}$	4	Table 13.2
Recirculation ratio ($M{\rightarrow}O$)	$R_{M{\rightarrow}O}$	5	Table 13.2
HRT of the anaerobic tank	t_{A1}	1.2 h	Table 13.2
3.2. Calculation of tank volumes			
3.2.1. Aerobic zone – Option 1			
Specific growth rate of nitrifying bacteria	$\mu_n = \dfrac{\mu_{nm} \cdot N_{OM1} \cdot 1.07^{(T-20)}}{K_n \cdot 1.053^{(T-20)} + N_{OM1}}$	0.574 1/d	Equation 13.3
Minimum sludge age in the aerobic zone	$\theta_{OM1} = F \cdot \dfrac{1}{\mu_n}$	5.23 d	Equation 13.2, putting F = 3
Minimum volume of the aerobic zone	$V_{OM1} = \dfrac{Q \cdot (S_0 - S_e) \cdot \theta_{OM1} \cdot Y_t}{1,000 \cdot X_{OM1}}$	6,750 m^3	Equation 13.1, X_{OM1} obtained from mass balance

3.2.2. Aerobic zone – Option 2

Specific ammonia utilization rate	$v_n = \dfrac{v_{nm} \cdot N_{OM1} \cdot 1.07^{(T-20)}}{K_n \cdot 1.053^{(T-20)} + N_{OM1}}$	0.0348 kgNH$_4^+$-N/kgMLSS.d	Equation 13.6
Rate of biomass discharge from the system	$\Delta X_v = y \cdot Y_t \dfrac{Q \cdot (S_0 - S_e)}{1,000}$	6,885 kgMLVSS/d	Equation 13.5
Volume of the aerobic zone	$V_{OM1} = \dfrac{Q \cdot (N_{k0} - N_{ke}) - 124 \cdot \Delta X_v}{1,000 \cdot X_{OM1} \cdot v_n}$	9,362 m^3	Equation 13.4, X$_{OM1}$ obtained from mass balance

3.2.3. Anoxic zone

Volume of the anoxic zone	$V_{A2} = \dfrac{Q \cdot (N_{t0} - N_{te}) - 124 \cdot \Delta X_v}{1,000 \cdot X_{A2} \cdot K_{dn} \cdot 1.026^{(T-20)}}$	9,125 m^3	Equation 13.7, X$_{A2}$ obtained from mass balance

3.2.4. Anaerobic zone

Volume of the anaerobic zone	$V_{A1} = \dfrac{Q \cdot t_{A1}}{24}$	7,500 m^3	Equation 13.8

3.3. Calculation of recirculation flow rates

Recirculation flow (M→O)	$Q_{M \to O} = Q \cdot R_{M \to O}$	750,000 m^3/d	Equation 13.9
Recirculation flow (O→A$_2$)	$Q_{O \to A2} = Q \cdot R_{O \to A2}$	600,000 m^3/d	Equation 13.10
Recirculation flow (A$_2$→A$_1$)	$Q_{A2 \to A1} = Q \cdot R_{A2 \to A1}$	225,000 m^3/d	Equation 13.11

3.4. Calculation of excess sludge production

Excess sludge production rate	$\Delta X = \dfrac{Y_t \cdot Q \cdot (S_0 - S_e)}{1,000}$	11,475 kg/d	Equation 13.12
Average flow rate of excess sludge	$Q_w = \dfrac{\Delta X}{X_M}$	1,147.5 m^3/d	Equation 13.13

3.5. Calculation of aeration demand for biological reactions

3.5.1. Oxygen demand

Oxygen demand for organic carbon oxidation	$O_s = \dfrac{1.47}{1,000} \cdot Q \cdot (S_0 - S_e) - 1.42 \cdot \Delta X_v$	27,708 kgO$_2$/d	Equation 13.15
Oxygen demand for nitrification	$O_n = 4.57 \cdot [\dfrac{Q \cdot (N_{k0} - N_{ke})}{1,000} - 0.124 \cdot \Delta X_v]$	13,236 kgO$_2$/d	Equation 13.16
Oxygen demand for denitrification offset	$O_{dn} = 2.86 \cdot [\dfrac{Q \cdot (N_{t0} - N_{te})}{1,000} - 0.124 \cdot \Delta X_v]$	10,428 kgO$_2$/d	Equation 13.17
Oxygen demand for sludge recirculation offset	$O_m = \dfrac{1}{1,000} \cdot Q_{M \to O} \cdot C_{omd}$	4,500 kgO$_2$/d	Equation 13.18, assuming $C_{omd} = 6$ mg/l

Net oxygen demand	$O = (O_s + O_n - O_{dn}) \cdot \dfrac{V_O}{V_{OM1}} - O_m$	15,844 kgO$_2$/d	Equation 13.14
3.5.2. Air supply			
Alpha factor	$\alpha = k_1 \exp(-k_2 \cdot X_O)$	0.873	Equation 13.20, $k_1 = 1.7$, $k_2 = 0.08$, X_O obtained from mass balance
Standard-state oxygen demand	$O_{std} = \dfrac{O \cdot C_{os(20)}}{1.024^{(T-20)} \cdot \alpha \cdot [\beta \cdot C_{os(T)}(1 + \dfrac{\rho \cdot g \cdot h}{2 \cdot P}) - C_o]}$	17,097 kgO$_2$/d	Equation 13.19, T = 20 °C, $\beta = 0.95$, h = 6 m, $C_o = 1.5$ mg/l
Standard-state air supply	$G_O = \dfrac{O_{std}}{0.28 \cdot \eta_A} \cdot \dfrac{100}{24}$	$6.36 \cdot 10^5$ Nm^3/h	Equation 13.21, assuming $\eta_A = 40$ %
4. Design of the membrane filtration system			
Average flux	J_{avg}	20 l/m^2.h	
Total membrane area	$A_M = \dfrac{Q}{0.024 \cdot J_{avg}} \cdot F_M$	343,750 m^2	Equation 13.24, assuming $F_M = 1.1$
Number of membrane cassettes	$n_M = \dfrac{A_M}{A_{M0}}$	275	Equation 13.25, given $A_{M0} = 1{,}250$ m^2
Membrane aeration demand	$G_M = SAD_m \cdot A_M$	103,125 Nm^3/h	Equation 13.27, assuming $SAD_m = 0.3$ Nm^3/m^2.h

13.4 EXERCISES

Membrane separation principles (exercises 13.4.1-13.4.3)

Exercise 13.4.1

What are the typical membrane processes driven by pressure? What are the corresponding minimum sizes of matter that these membrane processes can reject?

Exercise 13.4.2

What is the mechanism of size sieving in membrane separation?

Exercise 13.4.3

Given that a feed liquor contains macro-molecular protein, virus, *E. coli*, and colloids, can these constituents be removed in a microfiltration (MF) process?

Introduction to membrane bioreactors (exercises 13.4.4-13.4.6)

Exercise 13.4.4

What are the principal features of MBR technology and which is the most essential?

Exercise 13.4.5

What are the major configurations of MBR? What are their merits and shortcomings?

Exercise 13.4.6

What do you think is the most efficient membrane module? Describe its characteristics and the reason(s) why you believe it is efficient.

Wastewater treatment performance and effluent quality (exercises 13.4.7-13.4.11)

Exercise 13.4.7

Describe and explain the advantages of the MBR process in removing ordinary pollutants compared with other biological treatment processes. Which mechanisms do you think are behind its advantageous performance?

Exercise 13.4.8

Describe and explain the advantages of the MBR process in removing emerging pollutants compared to other biological treatment processes. Which mechanisms do you think are behind its advantageous performance?

Exercise 13.4.9

Describe and explain the advantages of the MBR process in removing pathogens compared with other biological treatment processes. Which mechanisms do you think are behind its advantageous performance?

Exercise 13.4.10

In an MBR process, what are the effects of the solids retention time (SRT) on mixed liquor properties and its subsequent influence on phosphorus removal? If we extend the SRT indefinitely until no sludge is discharged, how will the inorganic composition ratio of the sludge change?

Exercise 13.4.11

Given the effluent quality data of an MBR process (Figure 13.5) at different stages (Table 13.9), calculate and analyse the contribution ratios of biological degradation, membrane retention, and fouling layers (*e.g.*, gel layer and cake layer) retention to the removal of typical pollutants (*i.e.*, SS, COD, N, and P), pathogens, and TrOPs.

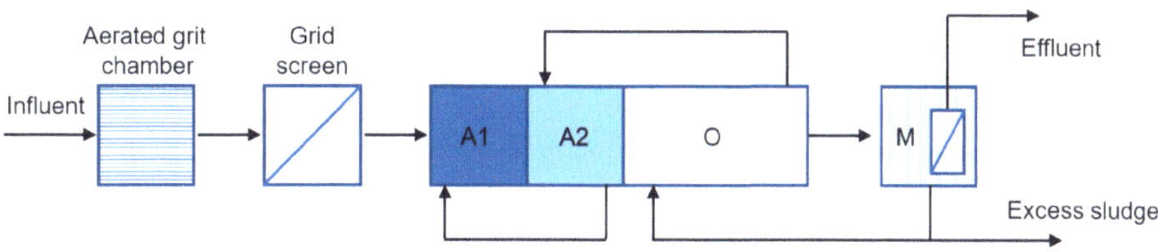

Figure 13.5 Schematic of a full-scale AAO-MBR process.

Table 13.9 Measured water quality data at different locations in the AAO-MBR process.

	Influent	A1	A2	O	M	Effluent	Membrane effluent[1]	Membrane effluent[2]
COD (mg/l)	350	300	200	100	50	30	35	35
TN (mg/l)	70	50	40	30	20	10	10	10
TP (mg/l)	5	4	3.5	1.5	1	0.5	0.5	0.5
E. coli (log pfu/100 ml)	6	5	4.5	4	3.5	0.1	0.1	0.1
TrOP (ng/l)	260	100	80	60	40	20	25	23

[1]After physical cleaning
[2]After chemical cleaning

Membrane fouling and control (exercises 13.4.12-13.4.16)
Exercise 13.4.12
What are the main substances that cause fouling in an MBR-based process and how can they be characterized? Describe their general roles in reversible and irreversible fouling.

Exercise 13.4.13
List the methods widely used for assessing the filterability of mixed liquor. What are the compositions of the total resistance to membrane filtration and how can they be calculated?

Exercise 13.4.14
How can fouling stages be differentiated? What are the main biopolymers responsible for fouling at different stages? Discuss their importance for the implementation of fouling control strategies.

Exercise 13.4.15
The fouling control approaches are mainly categorized as physical, chemical, and biological approaches. Give three typical examples of these approaches and compare their advantages and disadvantages.

Exercise 13.4.16
Discuss the relationships between aeration scouring, sludge properties and membrane fouling, and provide some feasible aeration modes for membrane fouling control that consume less energy.

MBR plant design, operation and maintenance (exercises 13.4.17-13.4.22)

Exercise 13.4.17

What are the differences in the pretreatment facilities between MBR and conventional activated sludge (CAS) processes?

Exercise 13.4.18

Discuss the possible routes of mixed liquor recirculation in an MBR-based process, evaluate acceptable ranges of the recirculation ratios, and explain.

Exercise 13.4.19

What are the differences between MBR and CAS processes in the design parameters for the biological section, and why?

Exercise 13.4.20

Discuss the factors that could influence oxygen transfer and the alpha factor in an MBR system, and their implications for MBR design.

Exercise 13.4.21

Discuss the factors that could render the excess sludge dewaterability of an MBR system different from that of a CAS system.

Exercise 13.4.22

In a process design for a typical AAO-MBR plant for municipal wastewater treatment, design the biological treatment units (tank volumes, recirculation flow rates, excess sludge production and oxygen demand) and the membrane filtration system (membrane area and aeration demand). The treatment capacity is 10,000 m^3/d. The quality of the influent and effluent is shown in Table 13.10. This plant has a primary sedimentation unit. The ratio of water volumes in the aerobic tank and the membrane tank is 2:1. The water depth of the aerobic tank is 5.8 m. The operating temperature is 15 °C and the atmospheric pressure is 1 bar. The clean-water saturated DO concentrations (at 1 bar) at 20 and 15 °C are 9.08 and 10.07 mg/l, respectively. The ratio of saturated DO concentration in the sludge to that in the clean water is 0.95. The oxygen transfer efficiency of the blower is 30%. The membrane is hollow-fibre with a packing density of 75 m^2/m^3 of the membrane cassette volume. The running flux is 25 $l/(m^2\ h)$.

Table 13.10 Influent and effluent quality for the design exercise.

	COD	BOD$_5$	TN	NH$_4^+$-N	TP	SS
Influent (mg/l)	400	250	50	40	10	200
Effluent (mg/l)	50	10	15	5	0.5	10

Practical application (exercises 13.4.23-13.4.24)

Exercise 13.4.23

An MBR wastewater treatment plant (capacity 10,000 m^3/d) was built with a project investment of 7 million USD 10 years ago, with membrane modules accounting for 15 % of the project investment. The lifespan of the membrane modules was 5 years, making the membrane depreciation 10 % of the operating expenditure. The current price of the same membrane module has reduced by 50 % compared with 10 years ago, and the lifespan has been extended to 6 years. Thus, what is the approximate project investment to build a new MBR plant now if the other costs are the same as 10 years ago? What is the approximate proportion of membrane depreciation cost in the total operating expenditure currently?

Exercise 13.4.24

Discuss the application potential of MBRs in your city/country, considering both technological and economic aspects, including the effluent standards, industry/wastewater features, the local environmental conditions, the project investment, footprint, operation expenditure, economic features, etc. (open discussion).

ANNEX 1: SOLUTIONS TO EXERCISES

Membrane separation principles (solutions 13.4.1-13.4.3)

Solution 13.4.1

The membrane processes driven by pressure include microfiltration, ultrafiltration, nanofiltration, and reverse osmosis. The corresponding minimum sizes of rejected matter are as follows: 0.1 μm, 0.002–0.1 μm, ~0.001 μm (100–1,000 Da), and < 0.001 μm (solutes), respectively.

Solution 13.4.2

The size-sieving mechanism is a physical process to retain the components that are larger than the membrane pore size, and the smaller components can penetrate through the membrane pores.

Solution 13.4.3

According to size sieving, MF can remove *E. coli*, while it cannot remove macro-molecular proteins. The removal efficiencies of viruses and colloids depend on their specific sizes. Viruses and colloids larger than 0.1 μm can be removed.

Introduction to membrane bioreactors (solutions 13.4.4-13.4.6)

Solution 13.4.4

MBR technology has a small footprint, good effluent quality, independent control of SRT and HRT, and small sludge production. The most essential feature is the separation of SRT and HRT, which allows the accumulation of activated sludge and furtherly enhanced biological degradation, a reduced footprint, and reduced sludge production. The separation of SRT and HRT relies on the membrane process to retain the activated sludge flocs in the reactor.

Solution 13.4.5

MBRs mainly have two configurations: side-stream configuration and submerged configuration. With side-stream MBRs it is easy to conduct frequent and/or *in situ* membrane cleaning and replacement, but they usually consume a large amount of energy. Submerged MBRs are less energy-intensive than side-stream MBRs because the aeration can control membrane fouling by its scouring effects. Both configurations can easily be integrated with other activated sludge processes.

Solution 13.4.6

This is an open question. The membrane modules and their characteristics are described in sections 13.2.4 and 13.2.5 in Chen *et al.*, 2020. Additional information can be found from the corresponding official websites and other literature.

Wastewater treatment performance and effluent quality (solutions 13.4.7-13.4.11)

Solution 13.4.7

There are several technical advantages of MBR that make it excellent for ordinary pollutant removal. The membranes can directly reject most sludge flocs, microorganism cells, and suspended solids. This ability to reject is usually enhanced when the membrane pores are narrowed due to adsorption/blocking of foulants or the membrane surface is covered by a dynamic foulant layer, enabling partial rejection of biopolymer clusters and debris, colloidal COD/N/P, and even soluble substances.

MF or UF membrane filtration, as a physical process, is not likely to straightforwardly alter the nature of microorganisms. However, it can enhance the biological process due to the high rate of retention of the microorganisms and partial rejection of the pollutants. The complete separation of HRT and SRT enables MBR to maintain a high sludge concentration and a low food-to-microorganism (F/M) ratio (*i.e.*, a lower sludge loading), and to keep the microorganisms in the endogenous respiratory period for more thorough degradation of the pollutants. Also, due to membrane rejection, nitrifying bacteria are normally easier to enrich in the system and more resilient at low temperatures, which is critical for the stable removal of NH_4^+-N and beneficial for overall TN removal.

However, for biological P removal, MBR is a double-edged sword. On the one hand, MBR can have a high sludge concentration and a high P content per unit mass of sludge (*i.e.*, an overall high P concentration in the sludge phase), and the membrane can reject some of the colloidal P as well; on the other hand, the typically longer SRT (15-30 d) of MBR means a slower discharge of the excess sludge from the system.

Solution 13.4.8

For TrOPs which are readily degradable, the removal efficiencies by MBR and CAS are normally comparable, whereas for refractory TrOPs, the MBR is potentially more effective in biological degradation. For TrOPs with strong hydrophobicity that are readily adsorbable by the sludge, the higher sludge concentrations, as well as probably higher extracellular polymeric substances (EPS) contents in the MBR, are conducive to the TrOPs' adsorption. For TrOPs that are less hydrophobic but still adsorbable by the particles and colloids in the mixed liquor, the MBR is potentially advantageous in rejecting these chemicals and improving the effluent quality. Overall, the removal efficiencies are jointly influenced by the mixed liquor properties (*e.g.*, MLSS and EPS) and the operational conditions (such as SRT).

TrOPs in the MBR system can undergo adsorption, degradation, rejection, and evaporation. They can be adsorbed by the bacterial cells, EPS, suspended solids, and colloids in the mixed liquor. After being adsorbed into the sludge phase, the pollutants may then undergo biodegradation. The MF or UF membrane has a marginal effect on the direct retention of TrOPs. However, the adsorption and size exclusion by the gel/cake layer are not negligible. The TrOPs can be readily rejected when they are adsorbed in the sludge phase or bound to colloids.

Solution 13.4.9

Membrane rejection is usually superior to gravitational settlement in bacteria removal. Compared with chemical disinfection technologies (such as chlorination, UV radiation, and advanced oxidation), membrane-based rejection of the pathogens is a gentle physical process that can bypass the risks of disinfection by-products, germ resistance, and involuntary mutations. MF or UF membranes can effectively reject protozoa, helminths, fungal spores, and bacteria, because their sizes are usually larger than the membrane pores.

Solution 13.4.10

MBR system, the long SRT is normally considered as an advantage, since a longer SRT results in higher biomass concentration, thus giving rise to a higher treatment efficiency. Nevertheless, a longer SRT also leads to the accumulation of inactive biomass and inorganic substances, thereby adversely affecting sludge activity. Generally, the sludge concentrations in the MBRs will increase with a longer SRT, whereas the ratio of volatile suspended solid to the total solid will decrease.

However, for biological P removal, MBR acts like a double-edged sword. On the one hand, MBR can possess a high sludge concentration and a high P content per unit mass of sludge and the membrane can reject some of the colloidal P as well; on the other hand, the typically longer SRT (15-30 d) of MBR means a slower discharge of the excess sludge from the system.

Solution 13.4.11

For a treatment plant during steady operation, the decrease in the foulant content in the A1, A2, O and M tanks can be attributed to biological degradation. The further decrease after the mixed liquid passes through the membrane can be attributed to the rejection by the membrane, which includes rejection by the intrinsic membrane matrix and the formed cake layer or gel layer on the fouled membrane. It is commonly accepted that the physical cleaning procedure can remove the cake layer while the subsequent chemical cleaning procedure can further remove the gel layer. Therefore, by calculating the difference in membrane rejection before and after the physical and chemical cleaning, the contribution by the cake layer and gel layer to foulant retention can be analyzed. The calculated results are listed as follows:

	Contribution ratios (%)							
	Biological degradation					Membrane retention	Cake layer retention	Gel layer retention
	Influent	A1	A2	O	M	Effluent	Effluent[1]	Effluent[2]
COD (mg/l)		93.75				3.75	1.25	1.25
TN (mg/l)		100				0	0	0
TP (mg/l)		88.9				8.9		2.2
E. coli (log pfu/100 ml)		42.4				57.6	0	0
TrOP (ng/l)		91.7				4.98	2.075	1.245

[1]After physical cleaning
[2]After chemical cleaning

Membrane fouling and control (solutions 13.4.12-13.4.15)

Solution 13.4.12

The main fouling-causing substances include polysaccharides, proteins, humic substances, and microorganisms and their cell debris.

These fouling-causing substances can be characterised using a series of advanced techniques. For example, the concentrations and functional groups of fouling-causing substances can be characterised using photometric methods, including UV-vis spectroscopy, excitation-emission matrix (EEM) fluorescence spectroscopy, and Fourier transform infrared (FTIR) spectroscopy. Fluorescence *in situ* hybridization coupled with confocal laser scanning microscopy (FISH–CLSM) and high-throughput sequencing technology can be used to characterise the microbial community of fouling-related microorganisms.

Hydrophobic humic substances tend to adsorb to membranes, altering the surface properties of the membranes and narrowing membrane pores. As a result, the modified hydrophobic membrane surface facilitates the deposition of hydrophilic biomolecules, such as polysaccharides. This enhancement can be attributed to stronger interactions between polysaccharides and hydrophobic humic substances than those with hydrophilic membranes. In addition, the non-covalent network formed by proteins and polysaccharides has high membrane-fouling potential. Proteins and humic substances can also contribute to membrane fouling development via direct deposition in the bio-cake (reversible fouling) or the inter-molecule interactions. In particular, the presence of multi-valent cations in wastewater streams can considerably increase the gel layer formation (irreversible fouling) on membranes.

Solution 13.4.13

A number of methods have been widely used for the assessment of filterability of mixed liquor, such as the critical flux determination by the flux-step method, Delft Filtration Characterization Method (DFCM) (De la Torre *et al.*, 2009), MBR-VITO fouling measurement (Huyskens *et al.*, 2012), Berlin Filtration Method (BFM) (De la Torre *et al.*, 2009), and Sludge Filtration Index (SFI) (Thiemig, 2012).

The total resistance of the fouled membrane can be generally quantified as the sum of the intrinsic resistance of the clean membrane (R_m), the resistance of the superficial deposition (bio-cake deposition, R_c), and the resistance of the gel layer and pore fouling (R_g).

$$R_{tot} = R_m + R_C + R_g$$

The resistance-in-series model is usually applied for fouling resistance calculation. According to Darcy's law,

$$J = \frac{\Delta P}{\mu \cdot (R_m + R_C + R_g)}$$

Where J ($m^3/m^2.s$) represents the permeate flux; ΔP (Pa) is the TMP; and μ (Pa.s) is the viscosity of the permeate solution, which is similar to water at a given working temperature. Then, each kind of resistance can be calculated as follows.

Firstly, the permeate flux of the fouled membrane (J_1, $m^3/m^2.s$) is determined with pure water under a given TMP ΔP_1 (Pa). Then, the total resistance to filtration ($R_{tot,1}$) can be calculated, according to:

$$R_{tot,1} = R_m + R_C + R_g = \frac{\Delta P_1}{\mu \cdot J_1}$$

Afterwards, the membrane is physically cleaned to remove the bio-cake (*e.g.*, the membrane surface is scrubbed with a sponge or flushed with tap water). Then, the permeate flux of the cleaned membrane (J_2, $m^3/m^2.s$) is determined with pure water under a given TMP (ΔP_2, Pa). Based on these values, the total resistance to filtration ($R_{tot,2}$) can be calculated according to the following equation,

$$R_{tot,2} = R_m + R_g = \frac{\Delta P_2}{\mu \cdot J_2}$$

Subsequently, the membrane is chemically cleaned to remove the pore fouling (e.g., NaOCl and citric acid cleaning). Based on these values, the total resistance to filtration ($R_{tot,3}$) can be calculated, according to:

$$R_{tot,3} = R_m = \frac{\Delta P_3}{\mu \cdot J_3}$$

where J_3 (m^3/m^2.s) is the permeate flux of the chemically cleaned membrane; ΔP_3 (Pa) is the TMP in the same conditions; and μ (Pa.s) is the permeate viscosity.

Therefore,

$$R_C = R_{tot,1} - R_{tot,2}$$
$$R_g = R_{tot,2} - R_{tot,3}$$

Solution 13.4.14

In general, three-stage fouling development is typically encountered in an MBR-based process. Stage 1 is an initial short-term rapid rise in TMP, which is caused by initial adsorption of solutes and pore blocking. Stage 2 is a long-term slow rise in TMP due to further blocking and gel layer formation. In Stage 3, the so-called 'TMP jump' occurs (Figure 13.3). The sudden jump in TMP is because of cake layer formation and is closely related to the sudden increase in the concentration of EPS at the bottom of the bio-cake. Therefore, the EPS produced by the deposited microbial cells becomes a major foulant after the TMP jump, whereas the soluble microbial products (SMP) compounds play an important role in the initial fouling. For fouling control, the operating period of Stage 1 and Stage 2 should be extended, and there should be no operation in Stage 3. Therefore, a sub-critical flux operation is usually applied in full-scale MBR plants.

Solution 13.4.15

Some typical examples of physical, chemical, and biological approaches and their advantages and disadvantages are summarised as following:

	Typical examples	Advantages	Disadvantages
Physical approaches	Intermittent filtration, mechanically-assisted aeration scouring, ultrasonication, vibration	No chemical reagents required, easy to be applied for practitioners, less membrane damage	Less effective, only remove reversible fouling
Chemical approaches	Cleaning in place (CIP), chemically-enhanced backflush (CEB) and recovery cleaning	Efficient cleaning and convenient operation	Potential membrane damage, adverse effects on microbial processes, produce toxic by-products
Biological approaches	Quorum quenching, protozoans and metazoans predation	Efficient to clean the biofilm fouling comprised of living microorganisms, less membrane deterioration, no formation of by-products	Expensive, less efficient to clean the membranes clogged by inorganic compounds, dead cells

Solution 13.4.16

Aeration is required in aerobic MBRs for providing DO to microbes, which puts the sludge flocs in a suspended state and mitigates membrane fouling. MBRs have a wide range of aeration rate, *i.e.*, the specific aeration demand per membrane area in previous works ranged from less than 100 to over 1,000 l/h.m^2 (Meng *et al.*, 2017). Generally, an increase in aeration intensity mitigates membrane fouling because it facilitates physical cleaning due to the increase in cross-flow velocity and shear stress. However, excessively high aeration increases both total and irreversible fouling rates. This is mainly attributed to the fact that the particle size distribution of the activated sludge shifts to lower size classes, and thus the concentrations of submicron particles, soluble microbial products, and biopolymers increase during high aeration. The shift of particle size to smaller flocs and fragments probably increases the total fouling resistance through the formation of a less permeable cake layer, while high concentrations of submicron particles probably increase irreversible fouling through pore blocking. Thus, aeration optimization, including aeration rates, bubble size and aeration modes, is crucial in membrane fouling control. For instance, intermittent and pulsed aeration are efficient methods to control membrane fouling with less energy consumption. Also, high aeration during backwash and low aeration during filtration can greatly reduce the aeration demand compared with conventional MBRs. In addition, automatic aeration control is also of great interest for optimization of aeration.

MBR plant design, operation and maintenance (solutions 13.4.17-13.4.22)

Solution 13.4.17

The biggest difference in the preliminary treatment between MBR and CAS lies in the design of the screen. The screen is used to remove large suspended and floating objects, fibrous material and particulate matter to protect subsequent facilities in the whole process. CAS employs a coarse screen (with a slot width or mesh size of 16-25 mm) and a fine screen (with a slot width or mesh size of 1.5-10 mm). MBR employs not only coarse and fine screens, but also ultrafine screen (with a slot width or mesh size of 0.2-1 mm) to remove hair and fibrous material more thoroughly to protect the membrane fibres from entanglement and the module/aerator from clogging. An example of the flow chart of the pretreatment facilities of MBR is shown in Figure 13.4.

Solution 13.4.18

The mixed liquor recirculation routes are designed to fulfil the requirements of N and P removal and sludge distribution among different units. The primitive version of an MBR process, *i.e.*, the aerobic-MBR (O-MBR) process comprised of an aerobic zone and the membranes, removes SS, BOD and NH$_4^+$-N. If the membranes are located in a separate tank to the aerobic tank (*i.e.*, a membrane tank), the mixed liquor in the membrane tank needs to be recirculated to the aerobic tank to avoid over-concentration of solids in the membrane tank. The recirculation ratio is recommended to be 4-6, such that the MLSS concentrations in the membrane tank and aerobic tank are at a ratio of approximately 1.2:1 according to mass balance.

To enable biological denitrification to remove TN, an anoxic zone should be incorporated, which forms the anoxic/aerobic-MBR (AO-MBR) process. This requires the mixed liquor in the aerobic zone to be recirculated to the anoxic zone. The recirculation ratio should be high enough to ensure a sufficient proportion of TN to be denitrified and a sufficient MLSS concentration in the anoxic zone, but should not be too high to disturb the anoxic environment for denitrification. This recirculation ratio is recommended to be 3-5.

To enable biological TP removal, an anaerobic zone is needed, which is usually seen in an anaerobic/anoxic/aerobic-MBR (AAO-MBR) process. The mixed liquor in the anoxic zone is normally recirculated to the anaerobic zone at a ratio of 1-2. An excessively high recirculation ratio will shorten the real

HRT for anaerobic TP release and also disturb the anaerobic environment because the anoxic mixed liquor contains electron acceptors (such as NO_3^--N).

The mixed liquor recirculation routes can be altered to form a series of variants of the AAO-MBR process, as listed in Table 13.10 (Xiao *et al.*, 2014).

Table 13.10 Examples of several variants of the AAO-MBR process (adopted from Xiao *et al.*, 2014).

Process flow[1]	Feature
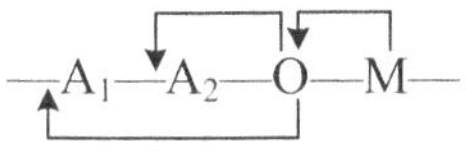 —A₁—A₂—O—M—	Simultaneous N and P removal, but with a potential adverse impact of aerobic recirculation on the anaerobic zone
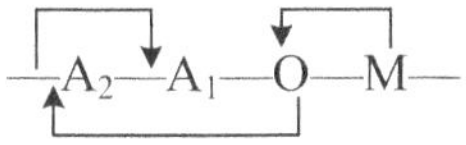 —A₂—A₁—O—M—	Bypassing reverse AAO-MBR to save carbon source for denitrification and diminish the impact of aerobic recirculation on the anaerobic environment
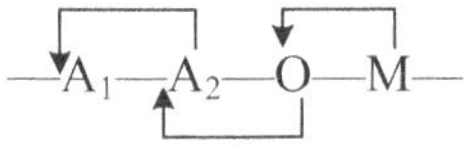 —A₁—A₂—O—M—	The default version of AAO-MBR in this textbook, *i.e.*, a University of Cape Town-type MBR (UCT-MBR)
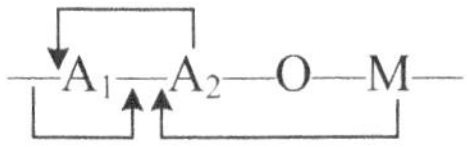 —A₁—A₂—O—M—	Bypassing UCT-MBR to save carbon source for denitrification
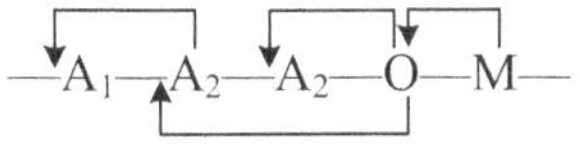 —A₁—A₂—A₂—O—M—	Modified UCT-MBR with one more anoxic zone to enhance endogenous denitrification
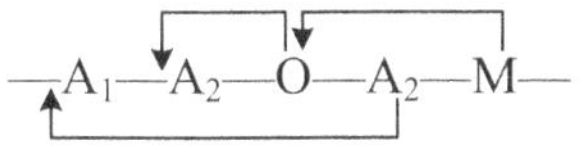 —A₁—A₂—O—A₂—M—	AAOA-MBR with a second anoxic zone after the aerobic zone to enhance endogenous denitrification
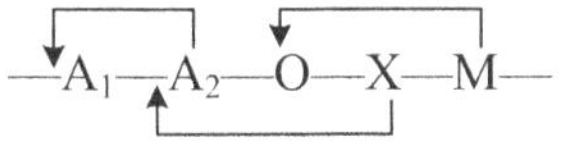 —A₁—A₂—O—X—M—	An additional zone (X) after the aerobic zone, switchable between anoxic and aerobic, to increase the flexibility of process adjustment for denitrification

[1] A_1 = anaerobic zone; A_2 = anoxic zone; O = aerobic zone; X = switchable zone between anoxic and aerobic; M = membrane tank.

Solution 13.4.19

The typical parameters for the design of the biological treatment units are given in Section 13.5.3.1 in Chen *et al.*, 2020. The major differences between the MBR and CAS processes are given below with explanations.

The total SRT of MBR (15-30 d) is typically longer than that of CAS (10-20 d for a conventional AAO process). This is because the membrane can completely reject suspended solids, which enables complete decoupling of SRT and HRT. A longer SRT facilitates the accumulation of nitrifying bacteria and increases TP content of the sludge. However, an excessively high SRT will restrict the excess sludge discharge, which is unfavourable for TP removal and will also lead to over-concentration of MLSS (especially inorganic MLSS), which impairs the operation of the system.

The MLSS concentration of MBR (6-15 g/l in the hollow-fibre membrane tank and 10-20 g/l in the flat-sheet membrane tank) is much higher than that of CAS (2.5-4.5 g/l in the aerobic zone). This is due to the complete rejection of MLSS by the membrane and the long SRT of MBR.

The sludge loading rate of MBR (0.03-0.1 kgBOD$_5$/kgMLSS.d) is typically lower than that of CAS (0.05-0.15 kgBOD$_5$/kgMLSS.d), due to the higher MLSS concentration of MBR.

The MLVSS/MLSS proportion of MBR (0.4-0.7) is normally lower than CAS fed with the same wastewater, because of the better degradation of the organics and the more complete membrane retention of inorganic MLSS in the MBR.

The synthesis yield of biomass of the MBR is similar to that of CAS (0.3-0.6 kgMLSS/kgBOD$_5$), but the endogenous decay coefficient of MBR (0.05-0.2 1/d) is considered to be larger than that of CAS (0.04-0.075 1/d). The observed yield of solids of MBR (0.5-0.9 kgMLSS/kgBOD$_5$, without primary sedimentation for instance) is thus designed to be slightly lower than that of CAS (0.6-1 kgMLSS/kgBOD$_5$), considering the longer SRT and smaller sludge loading rate that facilitate endogenous decay.

The maximum specific growth rate of nitrifying bacteria of MBR (~0.66 1/d) appears comparable with that of CAS, but the growth of nitrifying bacteria is facilitated by the longer SRT and can be more resilient at low temperatures in MBR than in CAS, which can influence the safety factor of the growth rate under these conditions.

Solution 13.4.20

The factors that could influence the oxygen mass transfer coefficient (k_La) include aeration intensity, bubble size, diffuser submergence depth, biomass concentration and extracellular polymeric substances (EPS) content (Xu *et al.*, 2017). The alpha factor is influenced by mixed liquor properties and aeration conditions. For coarse-bubble aeration, the alpha factor is jointly affected by MLSS concentration, EPS content and aeration intensity. For fine-bubble aeration, the alpha factor is heavily influenced by MLSS concentration (Xu *et al.*, 2017).

For the design of the biological section of an MBR process, the fine-bubble mode is usually adopted for biological aeration. Thus, the alpha factor can be estimated from the MLSS concentration in the aerobic zone (X_O), using Equation 13.20 from Chen *et al.*, 2020, *i.e.*, $\alpha = 1.7\exp(-0.08X_O)$. Note that the parameters in this equation are very different from those for CAS cases because MBR is different from CAS in MLSS concentration and mixed liquor properties.

Solution 13.4.21

The excess sludge dewaterability can be influenced by multiple factors, especially the EPS content and physicochemical properties. Since MBR has longer SRT, higher MLSS concentration and a lower sludge loading rate, the microbial status and hence the EPS properties can be different from that of CAS. In addition, sludge bulking is tolerable for MBR operation due to complete rejection of the bacteria. The sludge dewaterability deteriorates if there is sludge bulking.

Solution 13.4.22

a. Process selection

The same AAO-MBR process flow may be selected as that shown in Figure 13.4. See the example for the design of the pre-treatment and primary treatment units.

b. Design of the biological treatment units

The typical ranges of the design parameters are given in Table 13.2 in Chen *et al.*, 2020. The exact values selected are shown here in Table 13.11.

Table 13.11 Selected parameters for the design exercise.

Description	Symbols	Units	Values
MLSS concentration in the membrane tank	X_M	gMLSS/l	11.25
Proportion of MLVSS/MLSS	y	kgMLVSS/kgMLSS	0.5
Observed yield of solids	Y_t	kgMLSS/kgBOD$_5$	0.4
Synthesis yield of biomass	Y	kgMLSS/kgBOD$_5$	0.5
Endogenous decay coefficient	k_d	1/d	0.13
Maximum specific growth rate of nitrifying bacteria	μ_{nm}	1/d	0.66
Maximum specific ammonia utilization rate	v_{nm}	kgNH$_4^+$-N/kgMLSS.d	0.04
Half-velocity constant for ammonia utilization	K_n	mgNH$_4^+$-N/l	0.75
Specific denitrification rate	K_{dn}	kgNO$_3^-$-N/kgMLSS.d	0.05
Safety factor for the growth rate of nitrifying bacteria	F	-	3
Recirculation ratio (from the anoxic to the anaerobic tank)	$R_{A2\to A1}$	-	1.5
Recirculation ratio (from the aerobic to the anoxic tank)	$R_{O\to A2}$	-	4
Recirculation ratio (from the membrane to the aerobic tank)	$R_{M\to O}$	-	5
HRT of the anaerobic tank	t_{A1}	h	1.5
Conversion rate of sludge from feed SS	f	gMLSS/gSS	0.6
Average DO concentration in the aerobic tank	C_o	mg/l	1.5
DO concentration in the recirculation flow from the membrane tank	C_{omd}	mg/l	6
Density of the mixed liquor	ρ	g/cm^3	1.004
Safety factor for the membrane area	F_M	-	1.2
Running period of membrane filtration	τ_1	min	8
Relaxation period of membrane filtration	τ_0	min	2
Specific aeration demand with respect to the membrane area	SAD_m	Nm^3/m^2.h	0.3
Specific aeration demand with respect to permeate flow	SAD_p	Nm^3/m^3-permeate	15

According to mass balance,

$$X_O = X_M \cdot \frac{R_{M \to O}}{1 + R_{M \to O}} = 9.38 \qquad \text{g/l}$$

$$X_{A2} = X_O \cdot \frac{R_{O \to A2}}{1 + R_{O \to A2}} = 7.5 \qquad \text{g/l}$$

$$X_{A1} = X_{A2} \cdot \frac{R_{A2 \to A1}}{1 + R_{A2 \to A1}} = 4.5 \qquad \text{g/l}$$

Given the water volumes in the aerobic tank and the membrane tank (excluding membrane cassettes) is at the ratio of 2:1 (*i.e.*, $V_O = 2V_{M1}$),

$$X_{OM1} = \frac{X_O \cdot V_O + X_M \cdot V_{M1}}{V_{OM1}} = \frac{2 \cdot X_O + X_M}{2 + 1} = 10 \qquad \text{g/l}$$

Given $N_{OM1} = 5$ mg/L and temperature $T = 15\ °C$, the specific growth rate of nitrifying bacteria is estimated:

$$\mu_n = \frac{\mu_{nm} \cdot N_{OM1} \cdot 1.07^{(T-20)}}{K_n \cdot 1.053^{(T-20)} + N_{OM1}} = \frac{0.66 \cdot 5 \cdot 1.07^{(15-20)}}{0.75 \cdot 1.053^{(15-20)} + 5} = 0.42 \qquad \text{1/d}$$

$$\theta_{OM1} = F \cdot \frac{1}{\mu_n} = 3 \cdot \frac{1}{0.42} = 7.11 \qquad \text{d}$$

$$V_{OM1} = \frac{Q \cdot (S_0 - S_e) \cdot \theta_{OM1} \cdot Y_t}{1,000 \cdot X_{OM1}} = \frac{10,000 \cdot (250 - 10) \cdot 7.11 \cdot 0.4}{1,000 \cdot 10} = 683 \qquad \text{m}^3$$

This is considered to be the minimum volume of the aerobic zone required to fulfil the requirement of nitrification. Another option is to calculate V_{OM1} from the ammonia utilization rate and N mass balance:

$$v_n = \frac{v_{nm} \cdot N_{OM1} \cdot 1.07^{(T-20)}}{K_n \cdot 1.053^{(T-20)} + N_{OM1}} = \frac{0.04 \cdot 5 \cdot 1.07^{(15-20)}}{0.75 \cdot 1.053^{(15-20)} + 5} = 0.0256 \qquad \text{kgNH}_4^+\text{-N/kgMLSS.d}$$

$$\Delta X_v = y \cdot Y_t \cdot \frac{Q \cdot (S_0 - S_e)}{1,000} = 0.5 \cdot 0.4 \cdot \frac{10,000 \cdot (250 - 10)}{1,000} = 480 \qquad \text{kgMLSS/d}$$

$$V_{OM1} = \frac{Q \cdot (N_{k0} - N_{ke}) - 124 \cdot \Delta X_v}{1,000 \cdot X_{OM1} \cdot v_n} = \frac{10,000 \cdot (40 - 5) - 124 \cdot 480}{1,000 \cdot 10 \cdot 0.0256} = 1,137 \qquad \text{m}^3$$

This volume is larger than 683 m^3 and is considered to be safer for nitrification. Thus, we adopt 1,137 m^3 as V_{OM1}. The volumes of the aerobic tank and the membrane tank (excluding membrane cassettes) are then estimated to be $V_O = 758$ m^3 and $V_{M1} = 379$ m^3, respectively.

The volumes of the anoxic zone (V_{A2}) and the anaerobic zone (V_{A1}) are calculated as:

$$V_{A2} = \frac{Q \cdot (N_{t0} - N_{te}) - 124 \cdot \Delta X_v}{1,000 \cdot X_{A2} \cdot K_{dn} \cdot 1.026^{(T-20)}} = \frac{10,000 \cdot (50-15) - 124 \cdot 480}{1,000 \cdot 7.5 \cdot 0.05 \cdot 1.026^{(15-20)}} = 881 \qquad \text{m}^3$$

$$V_{A1} = \frac{Q \cdot t_{A1}}{24} = \frac{10,000 \cdot 1.5}{24} = 625 \qquad \text{m}^3$$

Then the overall weighted-average sludge concentration (X) is:

$$X = \frac{X_{A1} V_{A1} + X_{A2} V_{A2} + X_{OM1} V_{OM1}}{V_{A1} + V_{A2} + V_{OM1}} = 7.87 \qquad \text{g/l}$$

The excess sludge production rate (ΔX) can be calculated accurately as:

$$\Delta X = \frac{Y \cdot Q \cdot (S_0 - S_e)}{1,000} - k_d \cdot V_t \cdot Xy + \frac{f \cdot Q \cdot (SS_0 - SS_e)}{1,000} = \frac{0.5 \cdot 10,000 \cdot (250-10)}{1,000}$$

$$-0.13 \cdot (1,137 + 881 + 625) \cdot 7.87 \cdot 0.5 + \frac{0.6 \cdot 10,000 \cdot (200-10)}{1,000} = 989 \qquad \text{kgMLSS/d}$$

or estimated roughly as:

$$\Delta X = \frac{Y_t \cdot Q \cdot (S_0 - S_e)}{1,000} = \frac{0.4 \cdot 10,000 \cdot (250-10)}{1,000} = 960 \qquad \text{kgMLSS/d}$$

The average flow rate of excess sludge discharge (Q_w) is then:

$$Q_w = \frac{\Delta X}{X_M} = \frac{989}{11.25} = 87.9 \qquad \text{m}^3/\text{d}$$

The recirculation flow rates are:

$$Q_{M \to O} = Q \cdot R_{M \to O} = 50,000 \qquad \text{m}^3/\text{d}$$

$$Q_{O \to A2} = Q \cdot R_{O \to A2} = 40,000 \qquad \text{m}^3/\text{d}$$

$$Q_{A2 \to A1} = Q \cdot R_{A2 \to A1} = 15,000 \qquad \text{m}^3/\text{d}$$

The total oxygen demands for organic carbon oxidation (O_s), for nitrification (O_n), for denitrification offset (O_{dn}), and for sludge recirculation offset (O_m) are calculated as:

$$O_s = \frac{1.47}{1,000} Q \cdot (S_0 - S_e) - 1.42 \cdot \Delta X_v = \frac{1.47}{1,000} \cdot 10,000 \cdot (250 - 10) - 1.42 \cdot 480 = 2,846 \qquad \text{kgO}_2/\text{d}$$

$$O_n = 4.57 \cdot [\frac{Q \cdot (N_{k0} - N_{ke})}{1,000} - 0.124 \cdot \Delta X_v] = 4.57 \cdot [\frac{10,000 \cdot (40 - 5)}{1,000} - 0.124 \cdot 480] = 1,327 \quad \text{kgO}_2/\text{d}$$

$$O_{dn} = 2.86 \cdot [\frac{Q \cdot (N_{t0} - N_{te})}{1,000} - 0.124 \cdot \Delta X_v] = 2.86 \cdot [\frac{10,000 \cdot (50 - 15)}{1,000} - 0.124 \cdot 480] = 831 \quad \text{kgO}_2/\text{d}$$

$$O_m = \frac{1}{1,000} \cdot Q_{M \to O} \cdot C_{omd} = \frac{1}{1,000} \cdot 50,000 \cdot 6 = 300 \qquad \text{kgO}_2/\text{d}$$

Then the net oxygen demand for biological reactions in the aerobic zone (O) is:

$$O = (O_s + O_n - O_{dn}) \cdot \frac{V_O}{V_{OM1}} - O_m = (2,846 + 1,327 - 831) \cdot \frac{2}{3} - 300 = 1,929 \qquad \text{kgO}_2/\text{d}$$

The alpha-factor for oxygen transfer is:

$$\alpha = k_1 \cdot \exp(-k_2 \cdot X_O) = 1.7 \cdot \exp(-0.08 \cdot 9.38) = 0.803$$

The standard-state oxygen demand (at 20 °C and 1 bar in clean water) is calculated as:

$$O_{std} = \frac{O \cdot C_{os(20)}}{1.024^{(T-20)} \cdot \alpha \cdot [\beta \cdot C_{os(T)} \cdot (1 + \frac{\rho \cdot g \cdot h}{2 \cdot P}) - C_o]}$$

$$= \frac{1,929 \cdot 9.08}{1.024^{(15-20)} \cdot 0.803 \cdot [0.95 \cdot 10.07 \cdot (1 + \frac{1.004 \cdot 9.8 \cdot 5.8}{2 \cdot 101.325}) - 1.5]} = 2,282 \qquad \text{kgO}_2/\text{d}$$

The standard-state air supply is then:

$$G_O = \frac{O_{std}}{0.28 \cdot \eta_A} \cdot \frac{100}{24} = \frac{2,282}{0.28 \cdot 30\,\%} \cdot \frac{100}{24} = 1.13 \cdot 10^5 \qquad \text{Nm}^3/\text{h}$$

c. Design of the membrane filtration system

The average flux is calculated from the running flux:

$$J_{avg} = J_1 \cdot \frac{\tau_1}{\tau_1 + \tau_0} = 25 \cdot \frac{8}{8+2} = 20 \qquad \qquad \text{l/m}^2.\text{h}$$

The total membrane area is calculated as:

$$A_M = \frac{Q}{0.024 J_{avg}} \cdot F_M = \frac{10,000}{0.024 \cdot 20} \cdot 1.2 = 25,000 \qquad \qquad \text{m}^2$$

From the membrane cassette packing density of 75 m²/m³, the volume occupied by membrane cassettes is 25,000 / 75 = 333 m³. Thus, the total volume of the membrane tank (occupied by water and membrane cassettes) is:

$$V_M = 379 + 333 = 712 \qquad \qquad \text{m}^3$$

The air supply for membrane aeration is estimated via any of the below:

$$G_M = SAD_m \cdot A_M = 0.3 \cdot 25,000 = 7,500 \qquad \qquad \text{Nm}^3/\text{h}$$

$$G_M = SAD_p \cdot \frac{Q}{24} = 15 \cdot \frac{10,000}{24} = 6,250 \qquad \qquad \text{Nm}^3/\text{h}$$

The larger one, 7,500 Nm³/h, is selected as the designed air supply for membrane aeration.

Practical application (solutions 13.4.23-13.4.24)
Solution 13.4.23
a. Current total investment = 7 · (1–15 %) + 7 · 15 % · 50 % = 6.48   million USD
b. Current membrane depreciation = 7 · 15 % · 50 % / 6 = 0.09   million USD/year
    Previous membrane depreciation expenditure = 7 · 15 % / 5 = 0.21   million USD/year
    Other operating expenditure = 0.21 / 10 % · (1 − 10 %) = 1.89   million USD/year.
    Current operating expenditure = 0.09 + 1.89 = 1.98   million USD.

Thus, current membrane depreciation proportion = 0.09 / 1.98 = 4.5 %.

Solution 13.4.24
This is an open question. Necessary information about MBR plants includes the establishment time, the influent and effluent quality, the treatment capacity, and information about the installed membrane modules (for example, brand, pore size, and area), etc. The reason why MBR was selected as the treatment process? What was the cost, and what was the environmental effect? What is local people's attitude to building an MBR? If it is an upgrading project, a comparison between MBR and the previous treatment process is welcomed in order to highlight the features of these processes.

REFERENCES

Chen G., Ekama G.A., van Loosdrecht M.C.M. and Brdjanovic D. (2020). *Biological Wastewater Treatment Principles, Modeling and Design*, 2nd Edition. IWA Publishing, London, UK.

Judd S. and Judd C. (2011). The MBR Book: Principles and Applications of Membrane Bioreactors for Water and Wastewater Treatment, 2nd ed. Butterworth-Heinemann, Oxford.

De la Torre, T., Iversen, V., Moreau, A. and Stüber, J. (2009) Filtration charaterization methods in MBR systems: A practical comparison. Desalination and Water Treatment 9(1-3), 15-21.

Huyskens, C., De Wever, H., Fovet, Y., Wegmann, U., Diels, L. and Lenaerts, S. (2012) Screening of novel MBR fouling reducers: Benchmarking with known fouling reducers and evaluation of their mechanism of action. Separation and Purification Technology 95, 49-57.

Meng, F., Zhang, S., Oh, Y., Zhou, Z., Shin, H.S. and Chae, S.R. (2017) Fouling in membrane bioreactors: An updated review. Water Research 114, 151-180.

Thiemig, C. (2012) The importance of measuring the sludge filterability at an MBR - Introduction of a new method. Water Science and Technology 66(1), 9-14.

Xiao K., Xu Y., Liang S., Lei T., Sun J., Wen X., Zhang H., Chen C. and Huang X. (2014). Engineering application of membrane bioreactor for wastewater treatment in China: Current state and future prospect. *Frontiers of Environmental Science and Engineering*, 8, 805-819.

Xu Y., Zhu N., Sun J., Liang P., Xiao K. and Huang X. (2017). Evaluating oxygen mass transfer parameters for large-scale engineering application of membrane bioreactors. *Process Biochemistry*, 60, 13-18

NOMENCLATURE

Symbol	Description	Unit
A_M	Total membrane area	m^2
A_{M0}	Membrane area per unit membrane cassette	m^2
C_o	Average DO concentration in the aerobic tank	mg/l
C_{omd}	DO concentration in the recirculated flow from membrane to aerobic tank	mg/l
$C_{os(20)}$	Clean-water saturated DO concentration at 1 bar at 20 °C	mg/l
$C_{os(T)}$	Clean-water saturated DO concentration at 1 bar at T °C	mg/l
C_S	Pollutant concentration in the sludge phase	ng/gMLSS
C_W	Pollutant concentration in the aqueous phase	ng/l
C_0	Measured concentration at the location a (Figure 13.1)	mg/l or ng/l
C_1	Measured concentration at the location b (Figure 13.1)	mg/l or ng/l
C_2	Measured concentration at the location c (Figure 13.1)	mg/l or ng/l
C_3	Measured concentration at the location d (Figure 13.1)	mg/l or ng/l
C_4	Measured concentration at the location e (Figure 13.1)	mg/l or ng/l
C_5	Measured concentration at the location f (Figure 13.1)	mg/l or ng/l
F	Safety factor for estimation of the minimum sludge age in the aerobic zone	-
F_M	Safety factor for membrane area calculation	-
f	Conversion rate of sludge from SS	gMLSS/gSS
G_M	Standard-state air supply of blower aeration to the membrane tank	Nm^3/h
G_O	Standard-state air supply of blower aeration for biological reactions	Nm^3/h
g	Gravitational acceleration	m/s^2
h	Water depth of the aerobic tank	m
J_1	Running flux of a filtration cycle	l/m^2.h
J_{avg}	Average flux of a filtration cycle	l/m^2.h

Symbol	Description	Units
K_{dn}	Specific denitrification rate	$kgNO_3^- \text{-} N/kgMLSS.d$
K_n	Half-velocity constant for ammonia utilization	$mgNH_4^+ \text{-} N/l$
k_p	Sludge-water partition coefficient	-
k_1	The first empirical parameter for estimation of the α factor	-
k_2	The second empirical parameter for estimation of the α factor	-
k_d	Endogenous decay coefficient	d^{-1} or $1/d$
k_p	Sludge-water partition coefficient	$l/gMLSS$
$k_L a$	Oxygen mass transfer coefficient	$1/h$
N_{k0}	Kjeldahl N concentration of the aerobic tank influent	mg/l
N_{ke}	Kjeldahl N concentration of the membrane effluent	mg/l
N_{OM1}	Ammonia concentration in the aerobic zone	mg/l
N_{t0}	TN concentration in the influent of the biological system	mg/l
N_{te}	TN concentration in the effluent of the biological system	mg/l
n_M	Number of membrane cassettes	-
O	Oxygen demand for biological reactions in the aerobic zone	kgO_2/d
O_{dn}	Oxygen demand offset by denitrification	kgO_2/d
O_m	Oxygen amount brought by recirculated liquid from the membrane tank to the aerobic tank	kgO_2/d
O_n	Oxygen demand for nitrification	kgO_2/d
O_s	Oxygen demand for oxidizing organic carbon	kgO_2/d
O_{std}	Standard-state oxygen demand at 20 °C and 1 bar in clean water	kgO_2/d
P	Actual atmospheric pressure	kPa
Q	Designed flow rate	m^3/d
$Q_{A2 \to A1}$	Recirculation flow rate from anoxic to anaerobic tank	m^3/d
$Q_{M \to O}$	Recirculation flow rate from membrane to aerobic tank	m^3/d
$Q_{O \to A2}$	Recirculation flow rate from aerobic to anoxic tank	m^3/d
Q_w	Average flow rate of excess sludge discharge	m^3/d
$R_{A2 \to A1}$	Recirculation ratio from anoxic to anaerobic tank	-
$R_{M \to O}$	Recirculation ratio from membrane to aerobic tank	-
$R_{O \to A2}$	Recirculation ratio from aerobic to anoxic tank	-
R_t	Overall removal rate of a specific pollutant	-
R_{tot}	Total resistance of fouled membrane	$1/m$
R_m	Intrinsic resistance of pristine membrane	$1/m$
R_C	Resistance of cake layer	$1/m$
R_g	Resistance of gel layer and pore fouling	$1/m$
S_0	Feed BOD_5 concentration	mg/l
S_e	Effluent BOD_5 concentration	mg/l
SAD_m	Specific aeration demand with respect to the membrane area	$Nm^3/m^2.h$
SAD_p	Specific aeration demand with respect to the permeate flow	$Nm^3/m^3\text{-permeate}$
SS_0	Feed SS concentration of the biological system	mg/l
SS_e	Effluent SS concentration of the biological system	mg/l
T	Temperature of the mixed liquor	°C
T_{A1}	Contribution ratio of A1 tank (Figure 13.1) to the overall removal of a specific pollutant	-

Symbol	Explanation	Unit
T_{A2}	Contribution ratio of A2 tank (Figure 13.1) to the overall removal of a specific pollutant	-
T_O	Contribution ratio of O tank (Figure 13.1) to the overall removal of a specific pollutant	-
T_M	Contribution ratio of M tank (Figure 13.1) to the overall removal of a specific pollutant	-
t_{A1}	HRT of the anaerobic tank	h
V_{A1}	Volume of the anaerobic tank	m^3
V_{A2}	Volume of the anoxic tank	m^3
V_M	Volume of the membrane tank	m^3
V_{M1}	Volume of the membrane tank excluding membrane cassette	m^3
V_O	Volume of the aerobic tank	m^3
V_{OM1}	Volume of the aerobic zone, i.e. sum of V_O and V_{M1}	m^3
V_t	Total volume of biological tanks	m^3
v_n	Specific ammonia utilization rate	$kgNH_4^+\text{-}N/kgMLSS.d$
v_{nm}	Maximum specific ammonia utilization rate	$kgNH_4^+\text{-}N/kgMLSS.d$
X	Weighted average of biological tank MLSS concentrations	g/l
X_{A2}	MLSS concentration in the anoxic tank	g/l
X_M	MLSS concentration in the membrane tank	g/l
X_O	MLSS concentration in the aerobic tank	g/l
X_{OM1}	Weighted average of the aerobic and membrane tank MLSS concentrations	g/l
Y	Synthesis yield of biomass	$kgMLVSS/kgBOD_5$
Y_t	Observed yield of solids	$kgMLSS/kgBOD_5$
y	Proportion of volatile MLSS in the total MLSS	kgMLVSS/kgMLSS
ΔX	Excess sludge production rate	kgMLSS/d
ΔX_v	Biomass discharge rate	kgMLVSS/d
ΔC_{A1}	Concentration change after the A1 tank treatment (Figure 13.1)	mg/l or ng/l
ΔC_{A2}	Concentration change after the A2 tank treatment (Figure 13.1)	mg/l or ng/l
ΔC_O	Concentration change after the O tank treatment (Figure 13.1)	mg/l or ng/l
ΔC_M	Concentration change after the M tank treatment (Figure 13.1)	mg/l or ng/l

Greek symbols	Explanation	Unit
α	Ratio of the oxygen transfer coefficient in the sludge to that in clean water	
β	Ratio of saturated DO concentration in the sludge to that in clean water	
η_A	Oxygen transfer efficiency of the blower	%
θ_{OM1}	Minimum sludge age in the aerobic zone	d
μ_n	Specific growth rate of nitrifying bacteria	d^{-1}
μ_{nm}	Maximum specific growth rate of nitrifying bacteria	d^{-1}
ρ	Density of the mixed liquor	g/cm^3
τ_0	Relaxation period of a filtration cycle	minute
τ_1	Running period of a filtration cycle	minute
φ_M	Apparent membrane packing density in the membrane tank	m^2/m^3

Abbreviation	Description
A1	Anaerobic tank
A2	Anoxic tank
AAO-MBR	Anaerobic/anoxic/aerobic-MBR
AAOA-MBR	AAO-MBR with a post-anoxic tank
AO-MBR	Anoxic/aerobic-MBR
BFM	Berlin filtration method
BOD	Biochemical oxygen demand
BPA	Bisphenol A
CAS	Conventional activated sludge
CEB	Chemically enhanced backflush
CIP	Cleaning in place
COD	Chemical oxygen demand
DFCM	Delft filtration characterization method
DO	Dissolved oxygen
EEM	Excitation-emission matrix
EPS	Extracellular polymeric substance
F/M	Food-to-microorganism ratio
FISH-CLSM	Fluorescence *in situ* hybridization coupled with confocal laser scanning microscopy
FTIR	Fourier transform infrared
HRT	Hydraulic retention time
M	Membrane tank
MBR	Membrane bioreactor
MF	Microfiltration
MLSS	Mixed liquor suspended solids
MLVSS	Volatile MLSS
NH_4^+-N	Ammonium nitrogen
NO_3^--N	Nitrate nitrogen
O	Aerobic tank
O-MBR	Aerobic MBR
SFI	Sludge filtration index
SRT	Solids retention time
SS	Suspended solids
TMP	Trans-membrane pressure
TN	Total nitrogen
TP	Total phosphorus
TrOP	Trace organic pollutant
UCT-MBR	University of Cape Town-type MBR
UF	Ultrafiltration
UV-vis	Ultraviolet-visible
X	Switchable zone between anoxic and aerobic

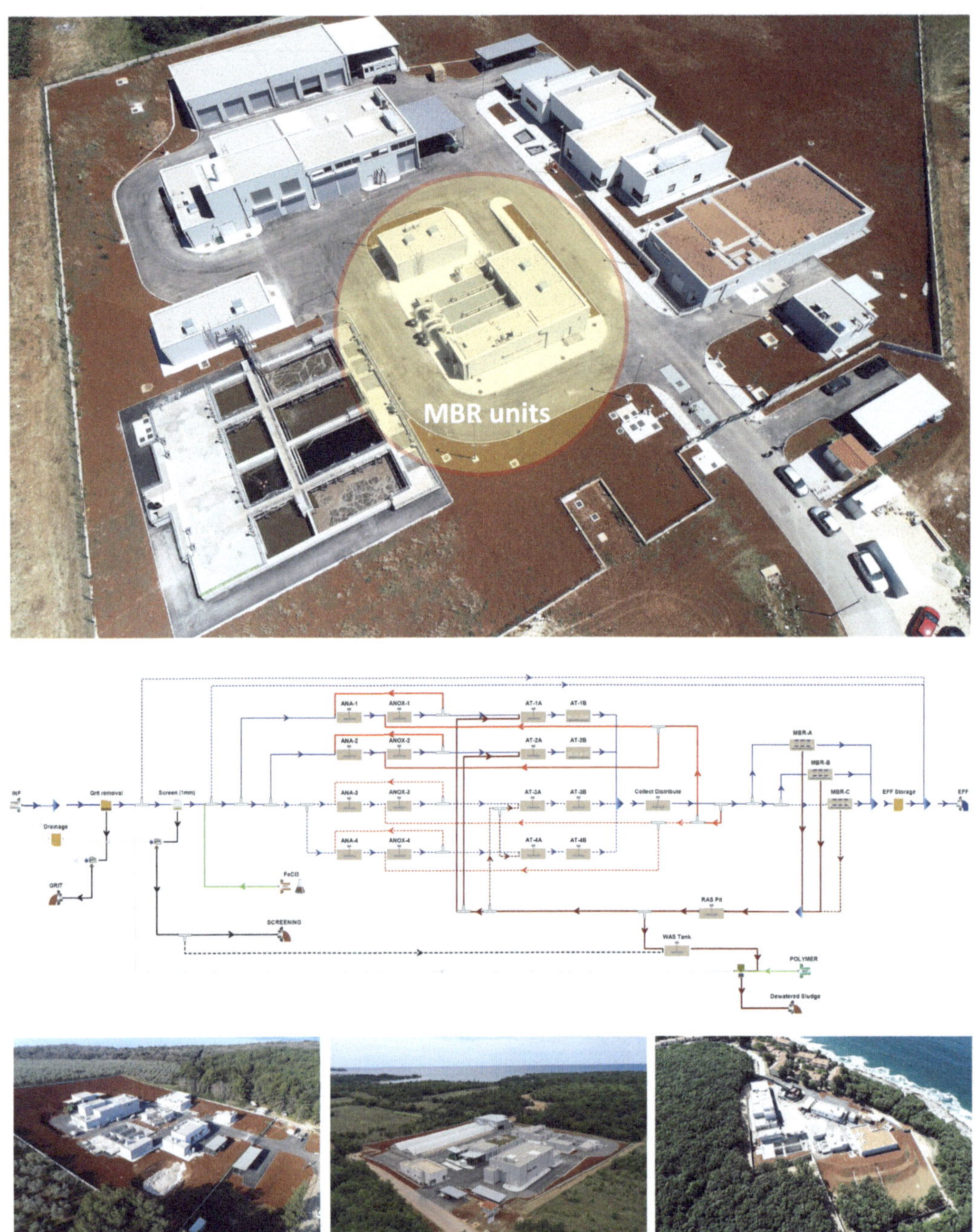

A textbook example of modelling of activated sludge systems (Chapter 14) with membrane bioreactors (Chapter 13) in the coastal city of Poreč in Croatia where four MBR plants were built as a part of a large environmental project for the protection of coastal waters and beaches, and reuse of effluent. Top: WWTP Poreč Jug (48,000 P.E.), middle: model scheme of the plant, bottom: WWTP Poreč Sjever (37,000 P.E.), Lanterna (30,000 P.E.) and Vrsar (22,500 P.E.). This project included study on integrated modelling of sewerage systems, treatment plant and bathing water quality. Photos: Odvodnja Poreč d.o.o.

doi: 10.2166/9781789062304_0475

14

Introduction to modelling activated sludge processes

Damir Brdjanovic, Carlos M. Lopez-Vazquez and Christine M. Hooijmans

14.1 INTRODUCTION

Chapter 14 on the modelling of activated sludge processes in the textbook *Biological Wastewater Treatment: Principles, Modelling and Design* (Chen *et al.*, 2020) presents the main concepts about how the biological processes that occur in these treatment plants can be mathematically represented. These processes mostly describe the relevant microbial conversions involved in the removal of specific compounds or elements. As such, a model aims to provide a purposeful representation of the most important processes occurring in the treatment units. An adequate and reliable description of the process is very useful to assess the performance of the treatment systems and explore potential upgrades and optimisation strategies. These strategies can be evaluated or used to improve the operation of the plants, increase their treatment capacity and decrease the operational costs, making the systems more cost-effective. This chapter aims to guide the reader through the contents of the chapter in the textbook in order to highlight and emphasise the most important principles involved when modelling activated sludge wastewater treatment plants. In addition, for further reading, Chapter 2 in the textbook presents the development of different modelling matrices to describe particular microbial conversions.

14.2 LEARNING OBJECTIVES

After the successful completion of this chapter, the reader will be able to:

- Describe the basic principles of activated sludge modelling, model components, matrix notation and stepwise development of biokinetic models.
- Explain the different activated sludge models and simulators and their applicability within the wastewater treatment context.

- Discuss the challenges of activated sludge modelling and the future trends.
- Explain the role of activated sludge modelling within the plant-wide and city-wide modelling approach.
- Discuss the development of sewage collection and wastewater treatment systems.
- Define the main characteristics and advantages and disadvantages of existing wastewater treatment technologies.
- Explain the main factors that have supported and led to the development of nutrient removal systems, instrumentation, control and automation, disinfection and micropollutant removal.
- Distinguish different resources that have been or could be recovered from wastewater.

14.3 EXERCISES

Exercise 14.3.1

Models are in essence a simplified representation of reality. How accurate should the models be? Would a simpler but less accurate model be preferable to a complex and more accurate model? Explain your answer.

Exercise 14.3.2

What is the difference between a black-box model and a glass-box model?

Exercise 14.3.3

A model is as accurate as the analytical methods used to determine its parameters. Is this true? Explain your answer and give a couple of supporting examples.

Exercise 14.3.4

Table 14.1 gives a model matrix of a simple wastewater treatment process. However, the stoichiometry matrix does not give two stoichiometric coefficients ($V_{1,O}$ and $V_{7,TSS}$). Calculate these values using the composition matrix.

Table 14.1 Example of stoichiometric matrix for activated sludge modelling (adapted from Gujer and Larsen, 1995)

Component	Oxygen	Inert	Substrate	Ammonia	Alkalinity	Biomass	Inert	Substrate	TSS	Rate
Symbol	S_O	S_I	S_S	S_{NH}	S_{HCO}	X_H	X_I	X_S	X_{TSS}	
Unit	gO_2	gCOD	gCOD	gN	mole	gCOD	gCOD	gCOD	gTSS	
Process	STOICHIOMETRY MATRIX									
Hydrolysis			1					−1	−0.75	r_1
Aerobic growth	−0.5		−1.5	−0.08	−0.005714	1			0.9	r_2
Lysis				0.07	0.005	−1	0.2	0.8	−0.12	r_3
Conservatives	COMPOSITION MATRIX									
ThOD-COD	−1	1	1	0		1	1	1		
N		0.02		1		0.08	0.05	0		
Charge				0.071429	−1					
Observables										
TSS						0.9	0.9	0.75		

Exercise 14.3.5

The following scheme shows the degradation path of COD in ASM3. Give the names of the conversions marked 1, 2, 3 and 4.

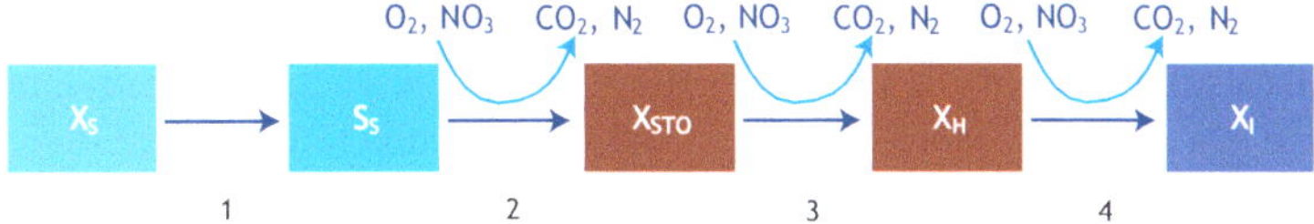

Exercise 14.3.6

For ASM1, the biodegradable COD in influent consists of two fractions. Give the name and abbreviation of these two fractions. What does i_{NXs} stand for?

Exercise 14.3.7

In ASM1, the stoichiometric coefficients of the process matrix for aerobic and anoxic growth of ASM1 can be calculated by setting up mass balances. Which three compounds can be balanced? Give three aspects that affect the conversions in the activated sludge but are not described by the activated sludge model. Which is the most important compound to be measured to evaluate the model?

Exercise 14.3.8

The following matrix does not show the stoichiometric coefficients and rate equations.

Component i List of processes j	1: S_O	2: S_S	3: X_H	Process rate equation ρ_j
1: Growth				
2: Lysis				

a) What are the units of the components?
b) How many and which mass balances are needed per process in order to determine the stoichiometric coefficients?
c) Write down the stoichiometric coefficients and rate equations. Include the assumption you made.

Exercise 14.3.9

Why is nitrite left out of the ASM1 model?

Exercise 14.3.10

When representing the hydraulics of the aeration tank in a full-scale WWTP, which compounds need to be measured across the tank? Under what conditions can the tank in the model be considered as fully mixed?

Exercise 14.3.11

Explain the role of the switching function in the model. Give an example of a switching expression used in the ASM1 model and explain how it works.

Exercise 14.3.12

Why are modern models COD-based (and not BOD-based)? Why is the oxygen concentration in ASM models expressed as a negative value of COD?

Exercise 14.3.13

During the model development carried out by Ekama and Marais (Chapter 14 in the textbook) a set of oxygen uptake rates (OUR) data was collected (Figure 14.1). At approximately 14 h, a sharp drop in OUR is observed. Explain the reason for this drop. Are there any other parameters related to this drop? If so, draw the possible graph of the concentrations of these parameters in the course of this experiment and explain the correlation with the OUR graph below.

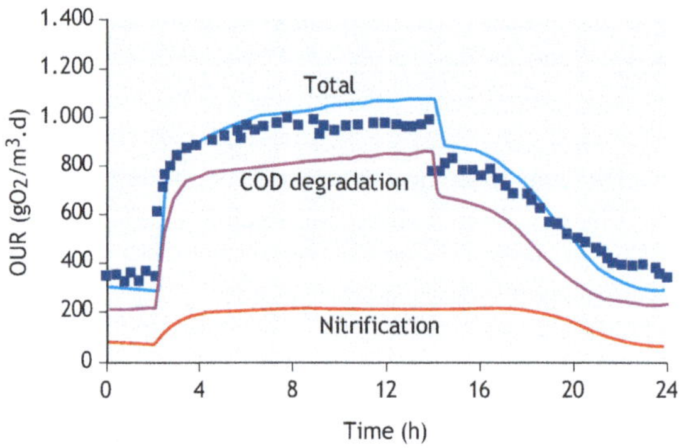

Figure 14.1 Comparison of experimentally observed values (data points) with theoretically predicted total oxygen uptake rate (light blue line). Oxygen uptake for COD degradation and nitrification are separated out (adapted from Ekama and Marais, 1978; Gujer and Henze, 1991, Figure 14.8 in Chen *et al.*, 2020).

Exercise 14.3.14

List the seven main steps to build a model for a wastewater treatment plant.

Exercise 14.3.15

It is unlikely that the first simulation will provide the user with the satisfactory results. This means that the user needs to calibrate the model. Which parameters of interest are usually calibrated in a biological nutrient removal (BNR) model, in which order?

Exercise 14.3.16

If your treatment plant is not designed for biological phosphate removal, would you still apply the activated sludge model that includes a bio-P removal on a non-bio-P plant? Explain your answer.

Exercise 14.3.17

Wastewater treatment plants have a lifecycle as depicted by Figure 14.2. At which stage of the plant's lifecycle is the modelling most cost-effective? Explain your answer. Relate it to the complexity of the models applied along the plant lifecycle.

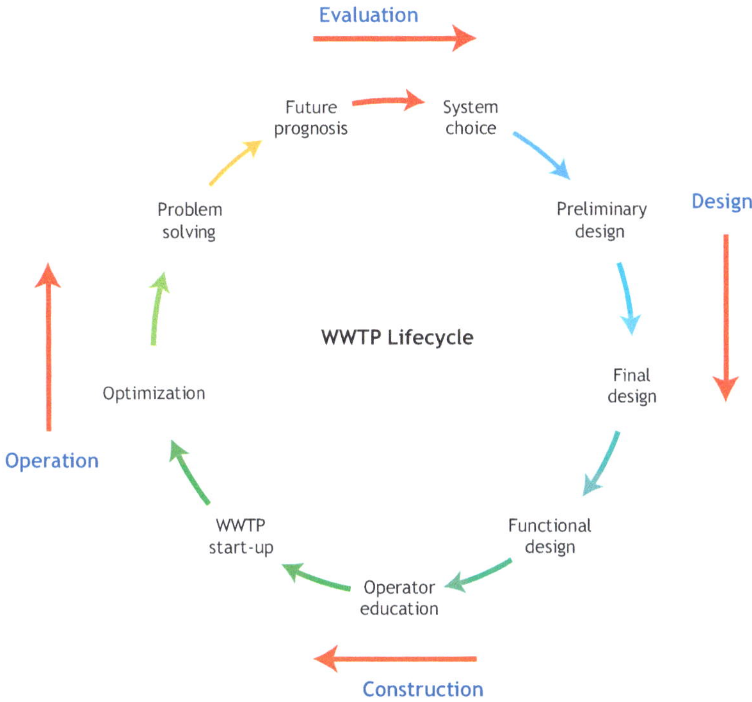

Figure 14.2 Lifecycle of wastewater treatment plants (source: Brdjanovic *et al.*, 2015).

Exercise 14.3.18

In some locations treated effluents are not disinfected, and therefore not free of pathogens, due to safe effluent discharges and restricted access to discharge points and receiving water bodies. However, in other regions, there is a need for pathogen removal from wastewater. ASM models do not include pathogen removal. Which pathogens should be included as indicators of faecal contamination? How would pathogen removal be modelled? Which factors could influence and therefore be included for model pathogen removal from wastewater?

Exercise 14.3.19

There are more than three decades of development and application of activated sludge wastewater treatment models. However, globally sewered sanitation serves only a fraction of the urban population. At the moment, modelling non-sewered sanitation systems is still in an early stage. How could the sewage modelling experience be extrapolated to non-sewered sanitation? In your answer choose one of the following onsite sanitation systems: containerized toilet, pit latrine, or septic tank and explain the main modelling principles.

ANNEX 1: SOLUTIONS TO EXERCISES

Solution 14.3.1

A model never exactly reflects reality. It merely needs to be sufficiently complex and accurate to satisfactorily describe the main and most relevant conversion processes in order to achieve the purpose and objectives of the modelling study.

Solution 14.3.2

A black-box model focuses on the plant influent and effluent with extremely limited (or no) involvement in or consideration of the processes happening inside the wastewater treatment units. Meanwhile, glass-box models have been developed to describe the metabolic routes inside the microorganisms. The application of either a black-box or a glass-box model depends very much on the purpose of the modelling study. For instance, black-box models (such as the ones based on the food to microorganisms (F/M) ratio) are good enough for the design of wastewater treatment systems. On the other hand, glass-box models (*e.g.,* metabolic models) have proven to increase the understanding of certain conversion processes and, consequently, used to improve the design and performance of the systems (such as in the enhanced biological removal process).

Solution 14.3.3

The required accuracy of the model depends on its purpose and application. Thus, the accuracy of the analytical methods influences the accuracy level of a model. For instance, to assess the effects of certain operating conditions on the sludge volume generated of a plant (often in tons of dry solids) or on the N_2O emissions (in kgN/day), the main trends are sufficient and an exact fitting of the model outcomes is not needed in view of the expected accuracy of the analytical determination methods. However, if the purpose is to model, for example, the ammonium effluent concentrations to meet an effluent discharge standard of less than 1 $mgNH_4\text{-}N/L$, then the analytical determination of ammonium needs to be more precise and the model accuracy (and likely also its complexity) be high enough to describe such ammonium concentrations.

Solution 14.3.4

$V_{1,O} = -0.5$, $V_{7,TSS} = 0.12$

Solution 14.3.5

1: Hydrolysis
2: Aerobic and anoxic storage
3: Aerobic and anoxic growth
4: Aerobic and anoxic endogenous respiration.

Solution 14.3.6

Ss (soluble biodegradable COD) and Xs (particulate biodegradable COD). $i_{N,Xs}$ is the nitrogen concentration related to or present in the particulate organic substrate Xs.

Solution 14.3.7

The three compounds that can be balanced are COD, N and charge. Three parameters or factors that affect the conversions but that are not included in the model are pH, toxicity and diffusion. Oxygen is the most important parameter that needs to be measured to evaluate the compartment model.

Solution 14.3.8

a) The units are mg COD/l, b) 1 mass balance, and c) if oxygen is in excess, the solution could be:

Component i List of processes j	1: S_O	2: S_S	3: X_H	Process rate equation ρ_j
1: Growth	$-\dfrac{1}{Y_H}+1$	$-\dfrac{1}{Y_H}$	$+1.0$	$\mu_H^{max} \cdot \dfrac{S_S}{K_S+S_S} \cdot X_H$
2: Lysis		$+1.0$	-1.0	$b_H \cdot X_H$

Solution 14.3.9

Most plants run at lower temperatures (< 20 °C), nitrite will remain very low, so from a mass balance perspective there is no need to take it into account. In the case of higher temperature or toxic events, nitrite might accumulate.

Solution 14.3.10

Oxygen, ammonium, nitrate and phosphate. When there are no concentration gradients measured.

Solution 14.3.11

A switching function is a saturation term (of the type $S / (K + S)$) whose main purpose is to stop a process when a relevant compound is no longer present.

As an example, if an ammonia switching function of the type $S_{NH} / (K_{NH,OHO} + S_{NH})$ is used in an expression that describes the growth process of ordinary heterotrophic organisms (with a relatively low value for the half-saturation concentration $K_{NH,OHO}$ of, for instance, less than 0.1 mgN/l), it will tend to a value of 1.0 at the usual concentrations observed in activated sludge systems (usually higher than 4-5 mg/l in the effluent of the plant): $S_{NH} / (K_{NH,OHO} + S_{NH})) = 4 / (0.1 + 4) \approx 1.0$. This will not limit or hinder the growth of ordinary heterotrophic organisms (as far as other relevant compounds are still present). However, if ammonia is fully removed and not present anymore, the result of the expression will be 0 ($0 / (0.1 + 0) = 0$), stopping the growth process.

Often, switching functions are mixed up with relevant saturation expressions and the differentiation between them is rather vague. Nevertheless, when the values of the half-saturation concentrations K are real parameters (calculated or estimated based on actual data to describe a conversion process in particular when the compound of interest tends to be limiting) the expression is a saturation term. On the other hand, when K is empirically assigned without looking into its influence on the conversion but only to stop the process, the expression is a switching function.

Solution 14.3.12

Modern models are based on the chemical oxygen demand (COD) and not on the biochemical oxygen demand (BOD) because COD makes it possible to perform a mass balance of the organic compounds, including the unbiodegradable organic compounds, present in the wastewater influent or generated during or from the biological conversion processes. This is also because the COD by definition measures the electron-donating capacity transferred to oxygen in order to oxidize the biodegradable organics. This makes it possible to precisely calculate the oxygen required for the aerobic processes to take place. On the other hand, the BOD technique is lengthy (it lasts at least 5 days, while the COD can be performed in 2 h) and rather sensitive to the influence of different factors (such as temperature, biodegradability, presence of toxic compounds) that can make it impossible to perform and track the conversion processes.

Furthermore, the COD mass balance can be applied to revise the correctness and accuracy of the organic matter removal models and calculate unknown stoichiometric parameters and values. This makes it a robust and reliable tool for modelling purposes.

Solution 14.3.13

Based on the OUR profiles shown in Fig. 14.1, the main trends (not plotted to scale) of the conversion profiles of COD and nitrogen compounds could be drawn as shown in Figure 14.3. As displayed in Figure 14.3a, due to the addition of the influent wastewater at around 2 h, the concentrations of soluble and particulate biodegradable organics increase (Ss and Xs, respectively). After this time, the influent Ss concentration is almost fully removed after 12 h and that of Xs around 20 h. These can cause the drops in DO observed around 2 h and 20 h, respectively. However, the OUR profile (Figure 14.1) does not reach zero after 20 h but instead approaches the OUR values observed before the addition of the influent wastewater. As described in the corresponding Chapter 14 in the textbook (Chen *et al.*, 2020), the OUR is caused by the endogenous respiration of the sludge, which occurs continuously and the model describes the lysis of the biomass (X_H) that leads to certain production of Xs and afterwards of Xs (known as the dead-regeneration process, see Ekama (2020) and Table 14.7 in the textbook).

Regarding the nitrogen compounds (Figure 14.3b), it is expected that there was certain nitrate (NO_3-N) observed before the addition of the influent wastewater. Then at around 2 h, the NH_4^+-N concentrations increase which are aerobically oxidized to NO_3^--N up to around 20 h. This could explain the drop in the OUR profile at that time.

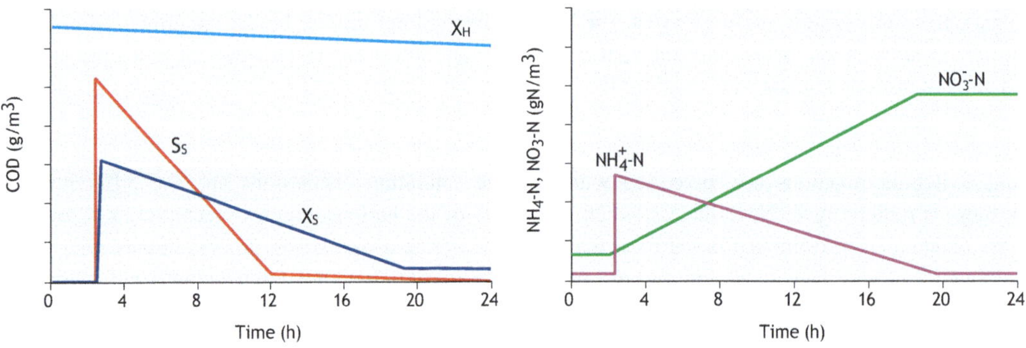

Figure 14.3 Main conversion trends potentially taking place in the aerobic experiments conducted by Ekama and Marais (Chen *et al.*, 2020) regarding the COD and nitrogen concentrations.

Solution 14.3.14

1. Definition of the model purpose or objectives of the simulation study.
2. Model selection: choice of models needed to describe the different plant units to be considered in the simulation, *e.g.,* selection of the activated sludge model, the sedimentation model, etc.
3. Hydraulics, *i.e.,* determination of the hydraulic models for the plant or plant components.
4. Wastewater and biomass characterization, including biomass sedimentation characteristics.

5. Calibration of the sludge model parameters.
6. Model verification.
7. Scenario evaluations.

Solution 14.3.15
Given that the influent characteristics, fractionation and hydraulics are correctly determined:

1. The aerated and non-aerated (anaerobic and anoxic) sections need to be well defined.
2. The solids retention time (SRT) needs to be adjusted or calibrated to describe the actual mixed liquor suspended solids concentration in the system and the sludge production; for this purpose, the phosphorus balance can be used.
3. The COD balance, including the effluent concentrations, needs to be adjusted.
4. The nitrogen (TKN) content present in the sludge waste needs to be adjusted to account for the actual nitrogen requirements for biomass synthesis.
5. The actual DO concentrations in the tanks need to be adjusted and assessed to describe the ammonia and nitrate concentrations in the treatment units and internal recirculation flows.
6. If the nitrogen profiles are not yet satisfactorily described, the maximum aerobic and anoxic kinetic rates of nitrifiers and denitrifiers and the anoxic kinetics of polyphosphate accumulating organisms (PAO), also their half-saturation concentrations (mostly for oxygen and key substrates) can be adjusted to fit the nitrogen balance (N load and denitrification).
7. The aerobic kinetic rates of PAO need to be adjusted to match the aerobic phosphorus profiles and the phosphorus content present in the sludge waste (also to assess whether the plant may have chemical P removal).

For further details, refer to Meijer and Brdjanovic, 2012.

Solution 14.3.16
Strictly speaking, if the plant is not designed to perform the enhanced biological phosphorus removal (EBPR) process, it is not necessary to use a model that considers the EBPR conversion processes because this will make the model more complex and modelling more demanding. However, inclusion of a bio-P component of the model is recommended because the model P balance gives a check on the data accuracy (since P is a conserved element), and SRT and sludge production.

Solution 14.3.17
Different models are used at different stages of the plant's lifecycle. At the early stages of the lifecycle, modelling is considered in general to be more cost-effective (for example, during the evaluation and design phases when the system choice, plant configuration, and the equipment (*e.g.,* aeration system), are selected, Figure 14.4) in comparison with the later stages. The saying 'If you are in the wrong train, all the stations are wrong' neatly describes the possible magnitude of the consequences of the errors that may have occurred in the early stages of the project. A simple model can often provide sufficient guidance to engineers less experienced in technology, units and equipment selection. From this perspective, simple models may save more money than, for example, a complex and demanding model used in optimization studies towards the end of the project cycle.

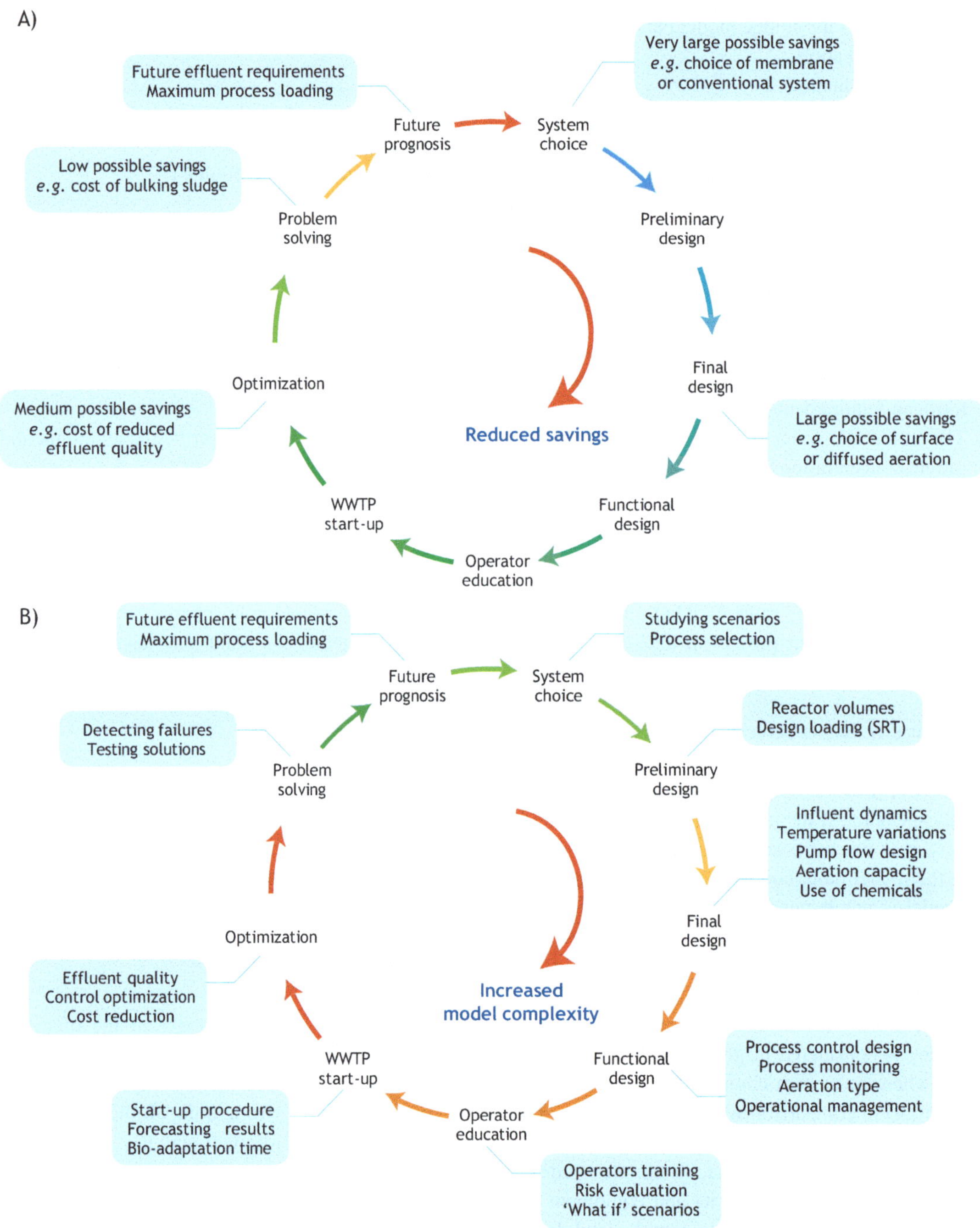

Figure 14.4 A) Possibilities for cost savings and B) model complexity during the lifecycle of wastewater treatment plants (source: Brdjanovic *et al.*, 2015).

Solution 14.3.18

Current ASM models do not include the expressions required to assess the inactivation of pathogens because conventional wastewater treatment systems hardly contribute to the removal of pathogens (besides certain natural systems, such as maturation ponds). For pathogen inactivation, ASM models need to be expanded to describe the disinfection processes that can take place in post-treatment units. Most of these systems are designed and operated as plug-flow reactors exposed to different types of disinfectant agents. The most common disinfection processes are chlorine (not recommended due to potential associated health issues), UV and ozone (O_3) (see Chapter 8 in the textbook, for further details). The disinfection processes driven by these disinfectants could be incorporated in the ASM models to describe the disinfection of selected key pathogens. For example, regarding faecal contamination, the inactivation of faecal coliforms could be incorporated. For other purposes and depending on the potential (re)use of the treated and disinfected effluents, other indicators could be added (*e.g.,* helminth eggs for irrigation purposes). Thus, the factors that influence the disinfection processes will need to be incorporated as well to provide an adequate description. These include, among others, the influence of suspended solids, pH and temperature. Also, certain processes will be process-dependent such as the hindrance caused by the suspended solids for UV disinfection and the gas-transfer expressions required to model the ozone supply. To incorporate pathogen removal, decay or inactivation, one or more mathematical equations will need to be incorporated in the model. More on modelling pathogen removal from wastewater and on faecal sludge can be found in Chapter 6 in Velkushanova *et al.,* 2021.

Solution 14.3.19

Velkushanova *et al.* (2021) present a few conceptual examples of modelling a containerized toilet, pit latrine, and septic tank in Chapter 6. Consult this chapter for further details.

The first model structure suggested is for a portable toilet (Figure 14.5). As discussed in Velkushanova *et al.* (2021)), this is usually a closed system with a short retention time (of maximum a few weeks). It is composed of three zones or phases: zone 1 where the sludge retains its physical properties, zone 2 where it is distributed and contains dissolved oxygen that drives certain aerobic conversions, and zone 3 where the conditions become anaerobic and anaerobic conversions take place. In this suggested model structure, it is assumed that the relatively short retention time (of a few weeks) does not allow the complete conversion and degradation of the organics. Consequently, only a partial degradation or conversion of the degradable matter is reached. There is no gas generation (since the conversions are not complete) and zone 4 is absent. When present, the function of zone 4 is to retain and accumulate the inert and non-degradable matter present in the influent or produced from the degradation processes. The fluxes of soluble (S) and suspended (X) compounds are indicated (Q_{1_2} and Q_{2_3}, for their transport from zone 1 to zone 2, $S_{FS,1\_2}$ and $X_{FS,1\_2}$, and from zone 2 to zone 3, $S_{FS,2\_3}$ and $X_{FS,2\_3}$, respectively) including the presence and transport of pathogens between zones ($X_{pathogens,inf}$, $X_{pathogens,1\_2}$, $X_{pathogens,2\_3}$). The system is fully closed and the only input is the discharge of faecal sludge, urine and water and the only output is the periodic emptying rate ($Q_{FS,emptying}$), resembling a fill-and-draw system. This is the simplest model structure for a faecal sludge system.

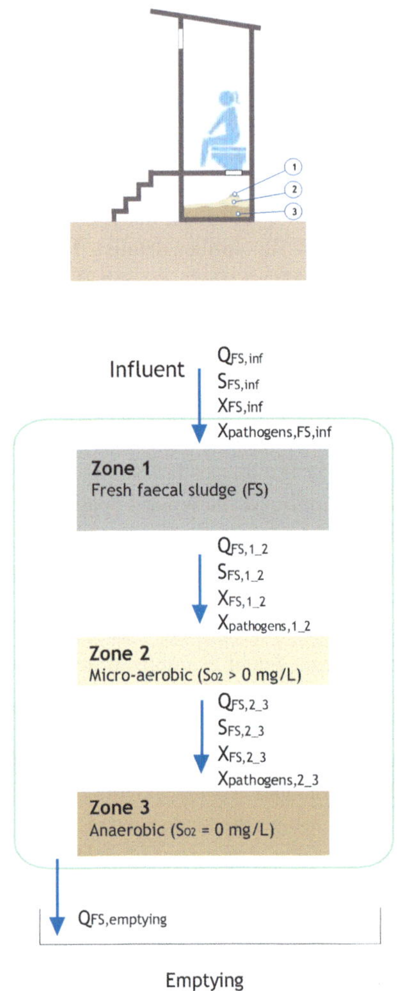

Figure 14.5 Portable toilet: suggested model structure (source: Velkushanova *et al.*, 2021).

Pit latrines are more complex than portable toilets (Figure 14.6). Although they are subject to some similar conditions, they have longer retention times (of several months and even years) that result in the full completion of the conversion processes (mostly the anaerobic ones). This implies that the kinetics will probably not play a major role and that stoichiometric relationships can be used to describe the conversion processes. This has already been observed in studies by Brouckaert *et al.* (2013) and Todman *et al.* (2015) who were able to model the filling rates of pit latrines using basic kinetic expressions. Moreover, pit latrines are prone to infiltration and percolation. Thus, besides the effects of the transport phenomena of the sludge matrix and associated processes between zones (*e.g.*, $Q_{FS,1\_2}$ that transports the soluble, $S_{FS,1\_2}$, and particulate concentrations, $X_{FS,1\_2}$ and $X_{pathogen,1\_2}$, from zone 1 to zone 2), pit latrines may also dilute their concentrations due to the infiltration of groundwater (*e.g.*, $Q_{infiltr,2}$ for the infiltration in zone 2) and/or concentrate the particulate compounds because of the percolation rates (*e.g.*, $Q_{exfiltr,2}$ to describe the exfiltration of compounds

$S_{FS,exfiltr,2}$, $X_{FS,exfiltr,2}$ and $X_{pathogens,exfiltr,2}$ from zone 2). Gases and inert and non-degradable matter ($S_{FS,U}$ and $X_{FS,U}$) are usually generated, since the anaerobic conversion processes are completed. On the one hand, this leads to the transport and diffusion of gases between zones (*e.g.,* $Q_{gas,2\_1}$ and $Q_{gas,3\_2}$ for the gas emissions from zone 2 to the atmosphere and from those of zone 3 to zone 2, respectively).

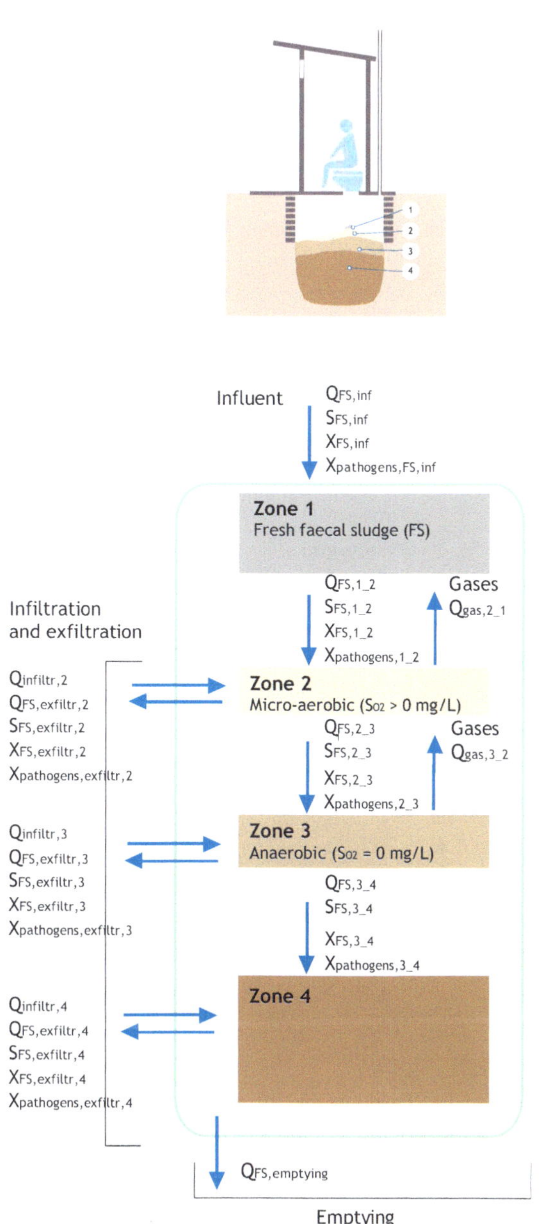

Figure 14.6 Pit latrine: suggested model structure (source: Velkushanova *et al.*, 2021).

On the other hand, due to inert and non-degradable products from the anaerobic processes remaining in zone 3, this also leads to their transport from zone 3 to zone 4 ($S_{FS,U,3\_4}$, $X_{FS,U,3\_4}$) and accumulation at the bottom of the system leading to the creation of an inert zone (zone 4). Similar to the portable toilets, the model structure of the pit latrine has one major input (the sludge feed, $Q_{FS,inf}$) and one major output (the emptying rate, $Q_{FS,emptying}$), but in addition the infiltration ($Q_{infilt,2}$, $Q_{infilt,3}$, $Q_{infilt,4}$) and exfiltration rates ($Q_{FS,exfiltr,2}$, $Q_{FS,exfiltr,3}$, $Q_{FS,exfiltr,4}$) that can affect each zone to different degrees. These also affect the soil and groundwater quality (due to the exfiltration of the soluble and particulate compounds (*e.g.,* the compounds, $X_{FS,exfiltr,4}$ and $X_{pathogen,exfiltr,4}$ flow from zone 4 into the ground).

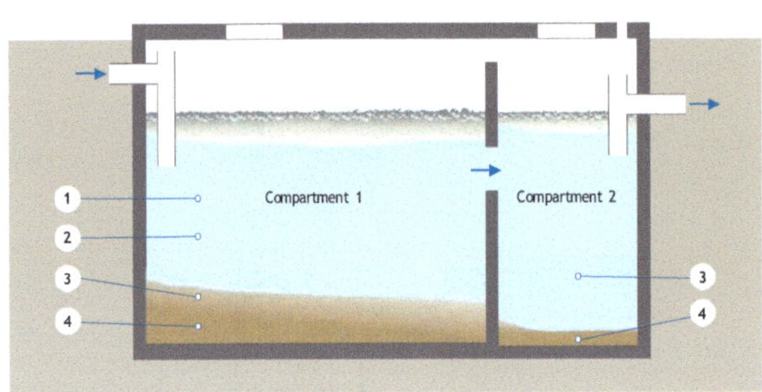

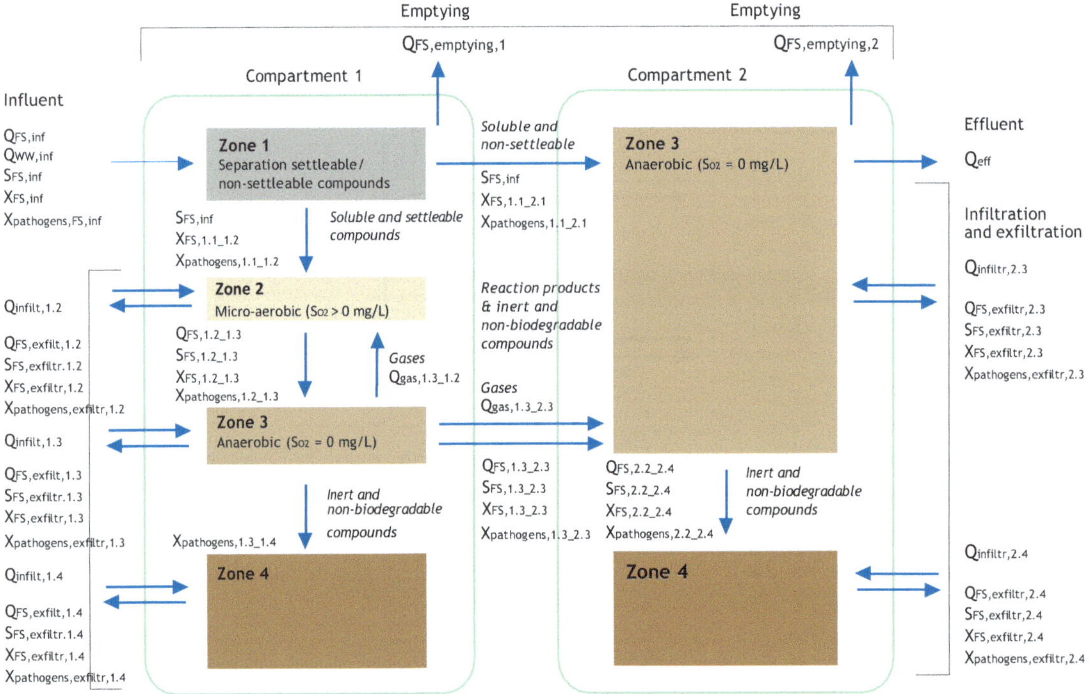

Figure 14.7 Septic tank: suggested model structure (source: Velkushanova *et al.,* 2021).

In contrast to pit latrines, septic tanks usually receive a combination of faecal sludge and water (domestic wastewater) and are usually divided into two compartments (Figure 14.7). They work in a continuous mode and have long retention times (of years) that, similar to pit latrines, will result in full completion of the conversion processes (mostly the anaerobic ones). This implies that stoichiometric conversion ratios can be sufficient to provide a satisfactory description of the processes that take place in these units. Septic tanks are also prone to infiltration and percolation issues. Therefore, they have well defined inputs ($Q_{FS,inf}$, $Q_{WW,inf}$) and output (Q_{eff}) but are prone to infiltration and percolation flows. Almost all the settleable solids present in the input tend to be retained in the 1st compartment while non-settleable solids flow to the 2nd compartment ($S_{FS,inf}$, $X_{FS,1.1\_2.1}$ and $X_{pathogens,1.1\_2.1}$).

The settleable solids need to be measured to split the flows between the two compartments. The expected low oxygen diffusion in the 2nd compartment and the split in the flow lead to the existence of four zones in the 1st compartment (similar to those proposed for pit latrines) but only two in the 2nd compartment. In the 1st compartment, most of the processes take place in the settleable solids and soluble components and, in the 2nd compartment, in the non-settleable solids and soluble components. In addition, the 2nd compartment receives the reaction products from zone 3 of the first compartment. Consequently, a higher accumulation of solids can be expected in the 1st compartment ($S_{FS,U,1.3\_1.4}$ and $X_{FS,U,1.3\_1.4}$) than in the 2nd compartment ($S_{FS,U,2.3\_2.4}$ and $X_{FS,U,2.3\_2.4}$).

REFERENCES

Brouckaert C.J., Foxon K.M. and Wood K. (2013). Modelling the filling rate of pit latrines. *Water SA*, 39(4), 555-562.

Brdjanovic D., Meijer S.C., Lopez Vazquez C.M., Hooijmans C.M. and van Loosdrecht M.C. (2015). Applications of activated sludge models. ISBN (e-book) 9781780404660. Published by IWA Publishing, London, UK.

Chen G., van Loosdrecht M.C.M., Ekama G.A. and Brdjanovic D. (eds) (2020). Biological wastewater treatment: principles, modelling and design. 2nd Edition. ISBN 9781789060355. Published by IWA Publishing, London, UK.

Ekama G.A. and Marais G.v.R. (1978). *The dynamic behaviour of the activated sludge process*. Research Report W27, Dept. of Civil Eng., University of Cape Town, Rondebosch, 7701, RSA.

Gujer W. and Henze M. (1991). Activated sludge modelling and simulation. *Wat. Sci. Tech.* **23**, 1011-1023.

Gujer W. and Larsen T.A. (1995). The implementation of biokinetics and conversion principles in ASIM. *Wat. Sci. Tech.* **31(2)**, 257-266.

Meijer S.C.F. and Brdjanovic D. (2012). *A practical guide to activated sludge modeling*. ISBN: 9789073445260. Published by UNESCO-IHE Institute for Water Education, Delft, The Netherlands.

Todman L.C., Van Eekert M.H.A., Templeton M.R., Hardy (née Kelly) M., Gibson W.T., Torondel B., Abdelahi F. and Ensink J.H.J. (2015). Modelling the fill rate of pit latrines in Ifakara, Tanzania. *Journal of Water, Sanitation and Hygiene for Development*, 5(1), 100–106.

Van Loosdrecht M.C.M., Nielsen P.H, Lopez-Vazquez C.M., Brdjanovic D. Ed. (2016) Experimental Methods in Wastewater Treatment. *IWA Publising*, pg. 350, ISBN 9781780404745 (Hardback) 9781780404752 (eBook)

Velkushanova K., Strande L., Ronteltap M., Koottatep T., Brdjanovic D. and Buckley C. (2021). Methods for faecal sludge analysis. ISBN(e-book): 9781780409122. Published by IWA Publishing, London, UK.

NOMENCLATURE

Symbol	Description	Unit
b_H	Specific rate of endogenous mass loss of ordinary heterotrophic organisms (OHOs)	1/d
F/M	Food to microorganism ratio or load factor (LF)	gCOD/gVSS.d
FSA	Free and saline ammonia	mgN/l
$i_{i,j}$	Fraction of compound i present in j	mg/mg
$i_{N,Xs}$	Fraction of nitrogen present in X_s	mg/mg
K	Half saturation constant	-
k_H	Maximum specific hydrolysis rate of SBCOD by OHOs under aerobic conditions	mgCOD/mgCOD.d
$K_{NH,OHO}$	Half saturation constant for growth of OHOs with nitrogen (FSA)	mgN/l
K_O	Half saturation constant for dissolved oxygen	mgO_2/l
$K_{O,A}$	Half saturation constant for nitrifiers for dissolved oxygen	mgO_2/l
$K_{O,H}$	Half saturation constant for OHOs for dissolved oxygen	mgO_2/l
K_S	Half saturation concentration for soluble organics utilization	mgCOD/l
K_x	Half saturation concentration for utilization SBCOD by OHOs	mgCOD/mgCOD.d
N	Nitrogen	mgN/l
q	Specific conversion rate	l/h
Q_{inf}	Influent flow rate	m^3/h
Q_{eff}	Effluent flow rate	m^3/h
$Q_{FS,emptying}$	Flow rate to de-sludge or empty the sanitation unit	m^3/h
Q_{i_j}	Flow rate from zone or compartment i to j	m^3/h
$Q_{FS,inf}$	Influent flow rate of faecal sludge into sanitation unit	m^3/h
$Q_{WW,inf}$	Influent flow rate of faecal sludge into sanitation unit	m^3/h
$Q_{infiltr\_i}$	Infiltration flow rate into compartment i	m^3/h
$Q_{exfiltr\_i}$	Exfiltration flow rate from compartment i	m^3/h
$Q_{gas\_i\_j}$	Gas flow rate from section or compartment i to j	m^3/h
r_i	Observed transformation rate for process i	$ML^{-3}T^{-1}$
S	Soluble concentration in bulk liquid	mgCOD/l
S_{HCO}	Bicarbonate concentration	mg/l
S_I	Soluble unbiodegradable COD concentration	mgCOD/l
S_{NH}	Free and saline ammonia concentration	mgFSA-N/l
S_O	Dissolved oxygen concentration	mgO_2/l
S_{FS,i_j}	Concentration of soluble compounds present in faecal sludge flowing from section or compartment i to j.	mgCOD/l
$S_{FS,inf}$	Concentration of soluble compounds present in faecal sludge in the influent of the containment unit	mgCOD/l
$X_{FS,inf}$	Concentration of particulate compounds present in faecal sludge in the influent of the containment unit	mgCOD/l
$X_{pathogens,FS,inf}$	Concentration of pathogens present in faecal sludge in the influent of the containment unit	ML^{-1}
S_S	Soluble readily biodegradable (RB)COD concentration	mgCOD/l

T	Time	h
V	Reactor volume	m^3
M	Mass	kg, g, mg
$v_{j,i}$	General stoichiometry term in model matrix for component i in process j	
X	Biomass concentration	$gCOD/m^3$
X_A	Nitrifier biomass concentration	$mgCOD/l$
X_H	Ordinary heterotrophic (OHO) biomass concentration	$mgCOD/l$
X_I	Unbiodegradable particulate organics from influent wastewater	$mgCOD/l$
$X_{pathogens,i,j}$	Concentration of pathogens flowing from section or compartment i to j.	MT^{-1}
X_S	Slowly biodegradable (SB)COD concentration	$mgCOD/l$
X_{STO}	Intra-cellularly stored organic concentration	$mgCOD/l$
X_{TSS}	TSS concentration in reactor	$mgTSS/l$
Y_H	Yield of OHOs	mgCOD/mgCOD

Abbreviation	Description
ADM	Anaerobic digestion model
ASM	Activated sludge model
ASM1	Activated sludge model No. 1
ASM3	Activated sludge model No. 3
BOD	Biological oxygen demand
BNR	Biological nutrient removal
COD	Chemical oxygen demand
CSTR	Complete stirred tank reactor
DO	Dissolved oxygen
EBPR	Enhanced biological phosphorus removal
IWA	International Water Association
N_2O	Nitrous oxide gas
OUR	Oxygen utilization rate
OHO	Ordinary heterotrophic organisms
PAO	Polyphosphate accumulating organisms
RBCOD	Readily biodegradable COD
SBCOD	Slowly biodegradable COD
SRT	Sludge retention time
ThOD-COD	Theoretical dissolved oxygen to COD ratio
TKN	Total Kjeldahl nitrogen
TSS	Total suspended solids
UV	Ultraviolet light
WWTP	Wastewater treatment plant

Greek symbol	Description	Unit
α	Symbol representing a stoichiometric formula	
μ	Specific growth rate of organisms	1/d
μ_A^{max}	Maximum specific growth rate of nitrifiers	1/d
μ_H	Specific growth rate of OHOs	1/d
μ_H^{max}	Maximum specific growth rate of OHOs	1/d
μ^{max}	Maximum specific growth rate of organisms	1/d
ρ_j	Kinetic rate of process j	$ML^{-3}T^{-1}$

Faecal and septic sludge characterisation is essential for development of citywide inclusive sanitation (CWIS) modelling. For more information on methods for faecal sludge analysis and experimentation, and on experimental methods in wastewater treatment, the reader is referred to Velkushanova *et. al.* (2021) and Van Loosdrecht *et al.* (2016), respectively (photo: K. Velkushanova).

doi: 10.2166/9781789062304_0493

15

Process control

Gustaf Olsson, Pernille Ingildsen and Bengt Carlsson

15.1 INTRODUCTION

The technical development of new processes, sensor and instrumentation technology, computer performance, communication technology including the Internet of Things, detection methods, control theory and artificial intelligence has made ICA (instrumentation, control and automation) widespread in all kinds of water operations. ICA will contribute to more efficient wastewater treatment, including many aspects of early warning systems, plant monitoring, and operator guidance. However, we emphasize that ICA cannot compensate for poor design or inflexible control handles. Therefore, the coupling of design and operation, known as control-integrated design, should be improved.

In addition to Chapter 15 in Chen *et al.* (2020) we also recommend Ingildsen and Olsson (2016) for a deeper understanding of control and operation of water resource recovery. The book is available open access athttps://www.iwapublishing.com/books/9781780407579/smart-water-utilities-complexity-made-simple.

15.2 LEARNING OBJECTIVES

Having studied Chapter 15, the reader should be able to:

- Identify the main driving forces for control and automation of wastewater treatment systems.
- Explain how external and internal disturbances appear, their typical timescales and their impact on thetreatment processes.
- Describe the ideas of feedback and feedforward control and their significance for a robust and resource-efficient operation.
- Describe basic requirements for successful control, such as sensors, actuators, signal communication,and testing.

- Explain the role of control and automation and understand basic control concepts.
- Identify how basic measurements couple to monitoring, early warning systems, and supervision.
- Describe the importance of dynamics (compared to static considerations) for measurements, for detection, and for control actions.
- Describe the structure of control systems.
- Know how to estimate the relationship between operating costs and control.
- Recognize motives for integrated operation and control, within the plant as well as between the sewersystem and the wastewater treatment system.
- Communicate with a control engineering specialist about the needs for control in your plant.

15.3 EXAMPLES

Example 15.3.1

Consider a nitrogen removal activated sludge process. The reactor consists of a pre-denitrification zone followed by an aerator for nitrification. Assume that the volumes are designed so that the effluent criteria are satisfied at 'normal' load.

a. Describe in words (and sketches) the consequences of:
   i. too low or too high dissolved oxygen (DO) content in the aerator,
   ii. low and high influent organic load (COD or BOD), respectively,
   iii. low and high influent ammonia nitrogen.
b. Describe some control actions to compensate for these conditions.
c. How can low organic load in the anoxic zone be compensated?
d. What happens if the nitrate recirculation rate becomes:
   i. too high?
   ii. too low?

Solution

a. *i: Too high DO concentration*: this is a waste of electrical energy. There is also a risk of DO recirculating back from the aerator to the anoxic zone in a pre-denitrifying plant. This would deteriorate the denitrification capacity.

Too low DO concentration: poor nitrification due to low nitrification rates; risk of nitrous oxide formation; and sludge settling properties can be at risk (see also Exercise 15.4.17).

ii: Too high organic influent load: the air supply capacity can be insufficient to handle extreme BOD/COD loads (this has happened in plants receiving wastewater from breweries or wine production). The influent nitrogen is key to determining the demand for aeration capacity. Typically, if the oxygen requirement for nitrification is satisfied, then it is also sufficient for BOD removal.

Too low influent organic load: there can be insufficient carbon source for the denitrification. Supplying an external carbon source will counter the insufficient carbon for the denitrification.

iii: High influent ammonia concentration: most of the influent nitrogen load is in the form of ammonia. Some of it is bound in organic matter. A higher load means a requirement for a larger treatment capacity, *i.e.* depending on how the plant responds it will require more aeration. The denitrification capacity can be too low (if there is insufficient organic carbon) and then the result will be increased total nitrogen in the outlet. A response could be to add an external carbon source. This will increase the denitrification rate and decrease the nitrate nitrogen in the effluent. There will. However, be an increasedcost for the carbon source.

Low influent ammonia concentration: this is not detrimental to the operation. However, a very low influent ammonia concentration could cause a risk to nitrifiers of wash-out. Also, the microorganisms need nutrients for their metabolism.

b. *Control actions:* DO control (see Section 15.8 in Chen *et al.,* 2020) is usually applied to ensure the aeration is increased when there is an increasing load. A cascading control can be applied withammonia sensors to ensure adjustment of the DO set-point. Control of the external carbon source canbe applied to ensure sufficient denitrification in spite of variation in the influent C/N ratio.

c. The addition of an external carbon source. The dosage can be controlled based on keeping a low concentration of nitrate in the end of the anoxic zone. If the nitrate level in the last anoxic zone is too high, then it does not help to increase the internal recirculation rate.

d. *Nitrate recirculation*: assuming a pre-denitrification plant, a too *low* nitrate recirculation rate means that the anoxic zone is not used to its full capacity. With no nitrate left, the anoxic zone can turn anaerobic toward the end of the reactor, causing phosphorous release. Too *high* a recirculation rate will make the denitrification capacity insufficient. Nitrate will exit the anoxic zone and re-enter the aerator. This is a waste of pumping energy. Generally speaking, a nitrate concentration of 0.1-0.3 mg/l at the end of the anoxic zone (or in the last anoxic zone) will be appropriate. With only one anoxic zone higher values of nitrate can be accepted. Note: this discussion is only valid for pre-denitrification plants; post-denitrification or alternating state plants will require other control methods.

Example 15.3.2

List examples of disturbances in a wastewater treatment plant. Specifically, list possible disturbances in the:

a. aerobic zone,
b. anoxic zone,
c. settler.

Solution

a. *Aerobic zone*: higher loads of organics or ammonia nitrogen will cause the oxygen demand to increase, requiring the airflow to increase. We might wish to increase the sludge concentration, but this will take longer. It can be done by decreasing the waste flow, which in turn increases the sludge retention time (but longer retention time increases the risk that the settler capacity is exceeded).

b. *Anoxic zone*: recirculated water entering the reactor should not have too high oxygen concentration. There are two ways to prevent recirculating water from the outlet of the aerator to the anoxic zone from becoming oxygen-rich: (1) keep the DO set-point near the aerator outlet low, and/or (2) let the recirculated flow enter a deoxygenation basin before entering the anoxic reactor. Denitrification can be insufficient due to too short residence time. The reason can be a too-high nitrate recirculation flow rate or a lack of easily available organic carbon source. The amount of BOD/COD available can be sufficient, but if it is not easily available it might be necessary to apply hydrolysis to make the carbon source easier to degrade biologically.

c. *Settler*: the sludge level can be too high, causing sludge escape. It can also be too low, so that the return sludge flow and waste flow become diluted.

Example 15.3.3

Consider important aspects of automation of a plant operation.

Solution

- What can be done manually and what needs to be automated? Fast changes (on a minute-to-minute timescale) ought to be automated. Very slow timescales can be handled manually. However, automatic supervision can be important for slow changes (on a daily and weekly timescale). A human being might not observe very small gradual changes until it is too late.
- Are the operators involved in the discussions? Are there plans for educating operators and maintenance personnel?
- Why do we need to control at all? Identify disturbances sources.
- How are various disturbances affecting the process?
- What can be measured online? How can we use this knowledge? Is there a maintenance plan for thesensors and actuators?
- How can the process be manipulated? Do we have sufficient control handles? How are valves, motors, compressors and pumps operated? Can flow rates be changed smoothly?
- Which criteria have to be satisfied? How is 'good' operation defined in the plant?
- Can we organize control actions on different timescales? Can we operate fast disturbances separatelyfrom slow and long-term disturbances?
- Identify different control structures and control strategies.

Example 15.3.4

For a deeper understanding of instrumentation, see Chapter 3 in Ingildsen and Olsson (2016).

Detections are made by different kinds of measurements and observations. Give some examples of:

a. binary information (such as on/off, yes/no),
b. analogue measurements,
c. human observations.

Solution

a. Some pumps or compressors are operated on/off. The control system should monitor if they are operating or not. Level sensors are often on/off, indicating a too-high or too-low level.
b. DO sensors or flow rate sensors indicate continuous measurements. Their sensor value can be digitalized and transferred to a SCADA system.
c. A human being can smell, detect patterns (such as air bubbles in a settler, indicating rising sludge), detect foam or other kinds of sludge formations, and notice strange noise from equipment. However, most importantly, a human being can investigate and troubleshoot in the event of new and recurring surprise events for which the plant is not prepared.

Example 15.3.5

Sensor dynamics are important in wastewater treatment systems. Consider the consequences of a slow sensor used for control.

Solution

If the response time of the sensor is not significantly shorter than the time constant of the process itself then the information to the controller is delayed too much and the control action can cause instability. Hence, controller tuning needs to take the sensor response time into account – this is generally done by reducing the sensitivity of the controller by reducing the gain, increasing the integral time, and omitting the differential part in a PID controller. In any case, a too slow sensor dynamics will detoriate the control performance.

Example 15.3.6
Give some reasons (economic and process control-related) to install variable speed pumping.

Solution
Smooth pumping is recommended so that no unnecessary turbulence of the water, which will influence the clarifier performance, is created. On/off pumping will create too sudden changes in the flow rate. Furthermore, if the pumped flow rate is either too high or too low, this will cause the pump to turn on and off more often. Also, using appropriate pumping can save energy. Additionally, reducing the on-off operation will reduce wear and tear of the pump as well as risk of water hammer. See exercises 15.4.1, 15.4.3 and 15.4.7.

Example 15.3.7
A simple proportional (P) controller calculates a control signal proportional to the difference between the reference value (set-point) and the measured value (this difference is usually called the control error). The goal is to make the error small or zero. However, a P controller might need a large gain to achieve this goal, which will create risks of instability. Even with a large gain in the P controller, the realized value of the process will always be different to the set-point, and the size of the difference depends on the gain.

a. The control signal can be complemented with a term that is proportional to the time integral of the control error. Then the controller becomes a PI controller. What is the advantage of a PI controller compared toa P controller? Discuss how the I part of the PI controller will influence the control. Are there special problems related to the I part of the controller?
b. A derivative (D) term can also be added, *i.e.* part of the control signal is proportional to the derivative (or rate of change) of the error. Discuss the advantages and potential problems with the D term in the PID controller.

Solution
a. When the error is integrated with time, even a small error adds up due to the integration. The controller reacts to the increasing integrated error and will correct until the error reaches zero. Hence, a PI controller in steady state will guarantee that the control error is zero, *i.e.* that the measured value will be equal to the set-point. Too low an integral time can cause the controller to become unstable (the controller is simply too ambitious). Too long an integral time on the other hand will make thecontroller too slow. Also, be careful to avoid integrator windup due to actuator saturation (reaching its upper or lower limits); any PI controller has to include an anti-windup feature. Most control tasks where a process is to be maintained at a certain set-point can be achieved successfully with a PI controller.
b. Knowing the derivative or rate of change of the signal can be helpful in order to counter fast variations. However, it is rarely used in wastewater treatment plants and if so the influence should be considered carefully to avoid the controller reacting to noise in the sensor signal. Due to this problem it is rare to see the D term activated for process control in water applications. For more details, see Chapter 5 in Ingildsen and Olsson (2016).

Example 15.3.8
What kind of conditions are favourable for the growth of filamentous organisms?

Solution
If the DO concentration is too low or there are starving conditions for microorganisms due to the shortage of substrate, then filamentous organisms can outcompete floc-forming organisms.

14.4 EXERCISES
This section contains further exercises for self-study. Solutions and comments are provided in the Annex1.

Exercise 15.4.1
An activated sludge process is provided with several recirculation flows. Nitrate recirculation was discussed in Example 15.3.1. For the recirculating flows below, discuss the consequences of either too low or too high flowrates. Note that the consequences can be both hydraulic and influence concentrations. Consider:

a) return sludge,
b) backwashing flow from effluent polishing filters,
c) reject flow from anaerobic sludge treatment back to the influent of the activated sludge process.

Exercise 15.4.2
Consider a pre-denitrification plant. Assume that the anoxic part and the aerator can be described as plug flow systems (or described as a series of reactors).

a) Describe with a simple sketch how the different carbon and nitrogen components vary along the aerator. Make two separate figures, one for carbon and one for nitrogen (ammonia and nitrate).
b) Consider the nitrate recirculation. Assume that we have 400 % recirculation, *i.e.* the recirculation flow rate is 4 times the influent flow rate. The return sludge flow rate is assumed to be 50 % of the influent flow rate. In steady state, how big a fraction of the nitrate will not be reduced to nitrogen gas in the plant? Explain this by making a simple mass balance of the nitrate. Assume ideal conditions with 100 % denitrification in the anoxic zone and 100 % nitrification in the aerobic zone
c) How would you control the nitrate recirculation? Which sensor information would you use?
d) How can we know that the denitrification rate is not limited by a lack of certain substances? What would you look for?
e) Compare the pros and cons of pre-denitrification versus post-denitrification?
f) What are the important limiting factors that must be checked if complete denitrification needs to occur (at least three factors)?

Exercise 15.4.3
It is important to handle hydraulic disturbances correctly. Suppose that a relatively large increase appears inthe influent water flow rate.

a) Explain what would happen if there is no control action taken (in the aerator or in the settler units).
b) Suppose that you know a few hours *ahead* that this flow increase will happen. How would you prepare the plant for the disturbance?

Exercise 15.4.4
List possible differences in disturbance patterns in different plant types, such as:

a) small municipal plants with mostly domestic waste,
b) small municipal plants with a major industry connected,
c) large municipal plants with mostly domestic waste,
d) large municipal plants connected to industries.

Consider their sources, their likely frequency, and magnitude.

Exercise 15.4.5
For each of the principal biological nutrient removal processes in the activated sludge process (carbon removal, nitrification, denitrification, and bio-P), list possible consequences of too much or too little of the following nutrients: VFAs, organic carbon, dissolved oxygen, ammonia, and phosphates.

Exercise 15.4.6
Give two examples of changes in process conditions that can affect the clarification and thickening processes.

Exercise 15.4.7
Assume that the influent flow rate is suddenly increased by 20 % due to a pump starting. Discuss how this flowrate change will manifest in the wastewater treatment system, and how various unit operations can be influenced. The timescale is important.

Exercise 15.4.8
We talk about 'data rich' and 'information poor'. Suppose that you see the record of a variable which is a straight line. Is that data information-rich?

Exercise 15.4.9
Explain in words the difference between a low-pass and a high-pass filter. List some possible applications of:

a. a low-pass filter,
b. a high-pass filter.

Exercise 15.4.10
Noise is defined as the measurement component that does not contain any information. However there are cases when noise contains information that helps to detect faults. Explain!

Exercise 15.4.11
Where is the best location for:

a. an ammonia sensor for control purposes?
b. a nitrate sensor?
c. a phosphate sensor (for chemical precipitation)?

Exercise 15.4.12

Why is it often favourable to combine feedforward with feedback? Explain what is needed to obtain a successful feedforward.

Exercise 15.4.13

Measurement signals (for example, DO concentrations) in a wastewater treatment plant are mostly noisy, and the noise can be related both to the sensor itself and to random process variations. Discuss ways to reduce the noise level by suitable signal treatment.

Exercise 15.4.15

Consider some nonlinear phenomena.

a. Give examples of nonlinear effects/terms in the DO dynamics.
b. What could be the effect of these nonlinearities if the DO is automatically controlled?
c. Is there a way to reduce the effect of the nonlinearities on the control system?
d. What type of nonlinearity occurs typically in a control valve?

Exercise 15.4.15

Suppose that a too-low DO level is detected in an aerator with closed loop DO control. What kind of causes should you look for?

Exercise 15.4.16

A toxic load enters an aerator that has automatic DO control. How can the DO control system detect the toxic load? Can the DO control system distinguish between a toxic disturbance and other load changes?

Exercise 15.4.17

In Example 15.3.1, we described DO set-point control based on ammonium measurements in the last aeration zone. In this way, we could save energy compared to the case of constant DO set-point control. It is important that the ammonium controller set-point is given a maximum and a minimum value. In other words: the DO set-points in the aerated zones are given maximum and minimum values. Give some reasons for this.

Exercise 15.4.18

Suppose that the DO level rises to about 9-10 mg/l from the normal operating range of 1-3 mg/l within a minute. What could be the reason?

Exercise 15.4.19

In a nitrogen removal plant the effluent standard often defines a limit on the total nitrogen out of the plant, *i.e.* the sum of ammonia N and nitrate N. In the operation of an N removal plant there is always a compromise on how to obtain minimum effluent total N. Discuss how to control the plant so that we will satisfy the total N condition and still save as many resources as possible.

ANNEX 1: SOLUTIONS TO EXERCISES
Solution 15.4.1

a. *Return sludge:* a high-return sludge can influence the settler performance, in particular when the return sludge flow rate is very high. The sludge level will decrease or disappear. A too-low return sludge flow rate can cause the sludge level to rise too much. Therefore, it is favourable to measure the sludge blanket level and control accordingly.

b. *Backwashing:* this often causes a sudden high hydraulic flow rate change in the plant inlet. It is important to make sure that the backwash flow can be stored temporarily and not returned to the inlet when there are high loads. Feedforward control from the filtering process to the plant influent could be considered.

c. *Anaerobic sludge:* this usually provides a high ammonia peak load. Therefore, the sludge load should be provided at low plant influent loads. See Section 10 in Chapter 15 in Chen *et al.*, 2020.

Solution 15.4.2

a. *Carbon and nitrogen profiles:* we assume that the DO concentration is sufficiently high. The heterotrophic organisms grow much faster than the nitrifiers. The carbon (BOD/COD) concentration decreases continuously and faster than the ammonia concentration. The nitrate concentration increases along the aerator, like a mirror of the decreasing ammonia.

b. *Nitrate recirculation:* assume ideal conditions with 100 % denitrification in the anoxic zone and 100 % nitrification in the aerobic zone. Then all the influent ammonia S_{NHin} is equal to the nitrate concentration S_{no} in the aerator effluent. Consider the flow rates out of the aerator. The nitrate recirculation Q_i is returned and all this nitrate will be denitrified. The nitrate returned via the return sludge flow Q_r will also be fully denitrified. Therefore, only the nitrate flow Q_{in} will exit the plant. The fraction is $Q_{in} / (Q_{in} + Q_r + Q_i)$. Assuming $Q_i = 4Q_{in}$ and $Q_r = 0.5Q_{in}$ the fraction becomes $1/5.5 = 0.18$. Hence $S_{no} = 0.18\ S_{NHin}$.

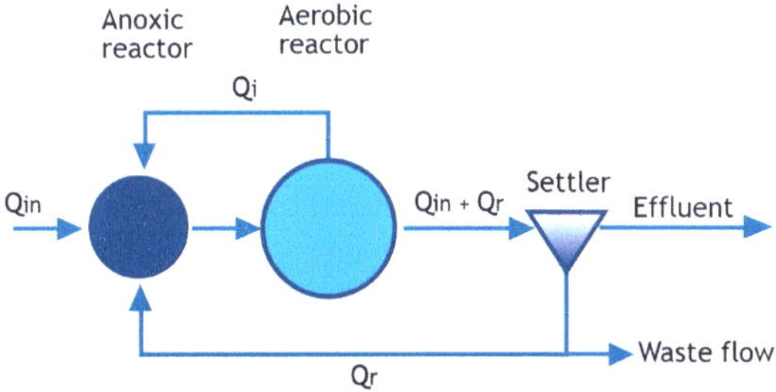

c. *Controlling nitrate recirculation:* measure the nitrate towards the outlet of the anoxic zone. If the nitrate concentration is very low then the recirculation flow rate Q_i can be increased. If the nitrate is higher than some predefined value, then the recirculation should be decreased (see also Example 15.3.1).

d. The denitrification rate could be limited due to lack of carbon source. Again, measure the nitrate in the last anoxic zone (see also Example 15.3.1).

e. *Pre- vs post-denitrification:* in pre-denitrification the carbon in the influent can be used as carbon source for denitrification. The obvious drawback is that nitrate has to be recirculated, which costs pumping energy. However, the pumps do not need to lift the water high, so the cost is defendable. Since some easily biodegradable carbon source is used in the anoxic zone, aeration energy is saved, because this part of the carbon load has been removed before the aerator. See also answer (b) for Example 15.3.2. As discussed in b., part of the nitrate cannot be denitrified in the plant. During post-denitrification all the nitrate from the aerator flows into the denitrification zone. Since organic carbon will be removed in the aerator an external carbon source needs to be added (at a cost) to achieve denitrification (compare with Example 15.3.1).

f. *Complete denitrification:* the reduction of nitrate to nitrogen requires an anoxic reactor. That is, no dissolved oxygen should be present in the reactor. There has to be sufficient biodegradable carbon available for the denitrification. The volume of the reactor has to be sufficiently large to allow time for full denitrification.

Solution 15.4.3

a. A large increase in the influent water flow will have two major impacts. Firstly, the residence time in the reactor will be shorter, which will lead to the risk of incomplete reaction time. Secondly, in extreme cases, it can result in washout. The flow rate of change will have a hydraulic impact on both the settler and clarifier (the two separate operations in the same settler unit) and create a turbulent flow, which can cause sludge washout.

b. By being able to predict an increasing flow rate it might be possible to offset the flow rate impact onthe plant by using an equalization basin. The main sewer can serve as such a basin to store the influent flow temporarily.

Solution 15.4.4

Usually, a large plant connects to a wider sewer network, compared to a small plant. Therefore, load variations in a small plant can be larger than those in a large plant. If industries are connected to the plant, their impact depends on the type of industry. Sometimes there will be a major industry connected to a small plant. For example, a large brewery or slaughterhouse can create large organic disturbances. On the other hand, if they are connected to a large plant, their impact is typically lower.

Solution 15.4.5

a. VFA and other organic carbon: there has to be a sufficient amount of VFA for P release in an anaerobic reactor and for denitrification in an anoxic reactor.

b. DO: in the anoxic zone the denitrification is hindered by too much DO. In the aerator it is the other way around: high DO favours a high nitrification rate.

c. Ammonia and phosphorous (see also Example 15.3.1): *too little* ammonia or phosphorous means that the activity of the microorganisms will decrease and hence over time the concentration of the bacteria can also decrease due to the low availability of 'food'. This can appear in industrial effluents, for example in the paper and pulp industry. Then nutrients have to be added. In municipal treatment, this rarely happens. If *too much* ammonia enters the plant, the aeration capacity might be too small. Similarly, the nitrification capacity can be reduced if influent water temperature is low and/or it contains toxic substances. Phosphorous removal using chemical precipitation can be supplemented by biological phosphorous removal if readily degradable organic matter is available. Anaerobic hydrolysis canconvert slowly degradable organic matter into more easily degradable organic matter.

Solution 15.4.6

a. The microbial sludge composition can influence both settling and clarification properties. If there are filamentous or free-swimming organisms, then the floc properties are unfavourable for settling and clarification. Typically, a too-low DO concentration in the aerator can inhibit the floc formation.

   b. On-off influent pump operation instead of smooth operation can disturb clarification by causing unwanted turbulence in the settler.

Solution 15.4.7

There are two different kinds of response. Firstly, as the influent flow rate suddenly increases, a 'hydraulic shock wave' will propagate along the plant. An increase in flow rate in the plant inlet will propagate some 20-40 minutes later (depending on the size of the plant) to the plant outlet. This in turn can have an impact on the clarifier operation. The second response relates to the decreasing retention time in the reactors. This will influence how the reactions manage to reach their completion. If sludge recirculation and internal recirculation are controlled proportionally to the influent flow rate, this can further amplify the impact of a flow-rate step change throughout the plant.

Solution 15.4.8

This depends on the circumstances. Generally, a straight line contains very little information. The sensor might have got stuck on a value. Usually large or small variations in a value contain more information. However, if the line is due to good control of *e.g.* DO, this is evidence of very good control. In this case the airflow rate is a good indicator of the reactor load. A competent observer can convert data to rich information just by looking at the graphics.

Solution 15.4.9

   a. Low-pass signal filtering will remove higher signal frequencies, such as noise, while retaining the essential signal information – it smooths the signal and makes it easier to see the general trend of the data. However, be aware that in the case of online filters, more smoothing results in more delay of the signal content.

   b. High-pass filters remove low signal frequencies and can be used to detect changes in a signal (like a derivative). In this way, sudden or fast changes can be detected. *See* Chapter 4 in Ingildsen-Olsson (2016).

Solution 15.4.10

Any sensor can present a noisy signal. For example, if a membrane becomes fouled, then the appearance of the noise can look different than if the membrane is clean. Automatic analysis of the noise character can reveal if it is time to clean the sensor. Noise is generally defined by being random, but occasionally, what at first glance appears random can contain some patterns, which means that some pattern can be recognized. Understanding this process can be valuable.

Solution 15.4.11

   a) Ammonia sensor: mostly the last aeration zone is used. The sensor can also be placed in the effluent, which has a more favourable environment. Then the response time is longer but also more representative of the effluent water quality (for legislative purposes). Ammonia sensors can also be placed close to the inlet to create a feedforward signal, but this is not common.

   b) Nitrate sensor: this should be located close to the outlet of the anoxic zone (see Exercise 15.4.2).

   c) A phosphate sensor for chemical precipitation ought to be located in the effluent of the flocculation chamber. The retention time in the flocculation chamber is short, typically around one hour, which is much shorter than the timeframe of phosphate variations in the chemical step. Chemical dosage can be based on feedback from this sensor.

Solution 15.4.12

The idea of feedforward is to measure a disturbance *before* it reaches the plant, for example, an increasing influent flow rate or an ammonia increase. If we know *exactly* how this disturbance is going to influence the plant, then the influence of the disturbance can be cancelled by the right actions. This is the idea of feedforward. However, we hardly ever have exact knowledge about how a disturbance influences the plant, since our model is not perfect. Therefore, we also need to measure the outcome of the disturbance, for example the effluent ammonia concentration, which is a feedback operation. Feedback then acts as a correction of a non-perfect feedforward. So, feedforward creates a quick but non-perfect correction to deal with a disturbance, while feedback will make an accurate but slower final correction. Controller tuning is critical in terms of both feedforward and feedback. A well-tuned feedforward control makes it possible to use a feedback controller which has been tuned to operate slowly, which reduces the risk of instabilities.

Solution 15.4.13

a. By measuring more often, the noise level can be suppressed by calculating the average value of the measurements. If we remove extreme measurements, then the result will be even more accurate. (Compare how athlete performances are judged in ski jumping or in figure skating by combining individual judges. Sometimes extreme judgements are removed).
b. Another possibility is to use a low-pass filter, for example exponential filtering, to suppress the noise level. *See* also Exercise 15.4.9. Refer to Chapter 4 in Ingildsen-Olsson (2016) for more details.

Solution 15.4.15

a. The oxygen transfer rate k_La is not proportional to the airflow rate. Therefore, it is nonlinear. At a low load, the k_La will increase more for an airflow increase compared to the case at a high load. Another nonlinearity is the factor DO_{sat}-DO that appears in the oxygen transfer term.
b. A DO controller will typically react more sensitively at a low load than at a high load. Additionally, Monod kinetics make the process rates non-linearly dependant on DO concentrations – this is the case for nitrification, denitrification as well as conversion of organic matter. A consequence of the different process gains at low and at high loads can cause oscillatory behaviour in some cases.
c. Another nonlinearity usually appears in valves. The flow rate may not be proportional to the valve opening. For example, the flow rate change is different for a valve opening change when the valve is almost fully closed compared to when it is almost fully open.
d. A remedy for these non-linearities is called *gain scheduling*. This is a way to automatically change the controller gain according to the size of the manipulative variable (control signal).

Solution 15.4.15

- Is the DO sensor calibrated?
- Are the blowers operating with full capacity?
- Is the airflow valve operating, or is it stuck in a position?

Solution 15.4.16

In principle, a decreasing load and a toxic disturbance will have a similar impact on the oxygen demand, so theDO controller cannot distinguish between the two. However, sometimes the timescale is different. If the oxygen demand decrease appears to be much quicker than a typical load decrease, this can be a probable reason for a toxic alarm.

Solution 15.4.17

Consider Figure 15.13 in Chapter 15 in Chen *et al.*, 2020. Assume that the ammonia load is very high. Then, more ammonia could not be oxidized even if the airflow increased. The aeration capacity is insufficient for the load. If the load is low, it is advisable to keep some aeration going, so that the reactor does not turn anoxic. There can also be a risk of nitrous oxide emission and deteriorating sludge settling properties.

Solution 15.4.18

This happens if the operator takes the DO sensor into the open air for a calibration check or cleaning. Subsequently, the DO control system should include a test; if the DO concentration increases at a certain speed, then the control system should keep the airflow constant until more normal DO concentrations are observed.

Solution 15.4.19

Bear in mind that nitrification has to occur before denitrification. This is the case both in pre-denitrification and in post-denitrification plants. So, we have to balance between oxidizing ammonia to nitrate and reducing nitrate N to nitrogen gas via denitrification. To oxidize ammonia N to nitrate N using nitrification is usually more expensive, since it requires energy for aeration. Normally it is cheaper to remove nitrate N using denitrification. In a pre-denitrification plant, there must be a sufficient carbon source, which in most cases is supplied by the BOD/COD in the influent. In a post-denitrification plant, external carbon has to be added because the BOD/COD has been biologically degraded by oxidation. The cost of external carbon should be compared with the cost of aeration. It is usually cheaper to reduce as much nitrate as possible and to minimize oxidation of ammonia.

REFERENCES

Chen GH., van Loosdrecht M.C.M., Ekama G.A. and Brdjanovic D. Ed. (2020) Biological Wastewater Treatment: Principles, Design and Modelling. 2nd edition. IWA Publishing, pg. 850. 9781789060355.

Ingildsen, Pernille, and Gustaf Olsson (2016) Smart water utilities: complexity made simple. ISBN13: 9781780407579. IWA Publishing, London, UK. Open access: https://iwaponline.com/ebooks/book/11/Smart-Water-Utilities-Complexity-Made-Simple

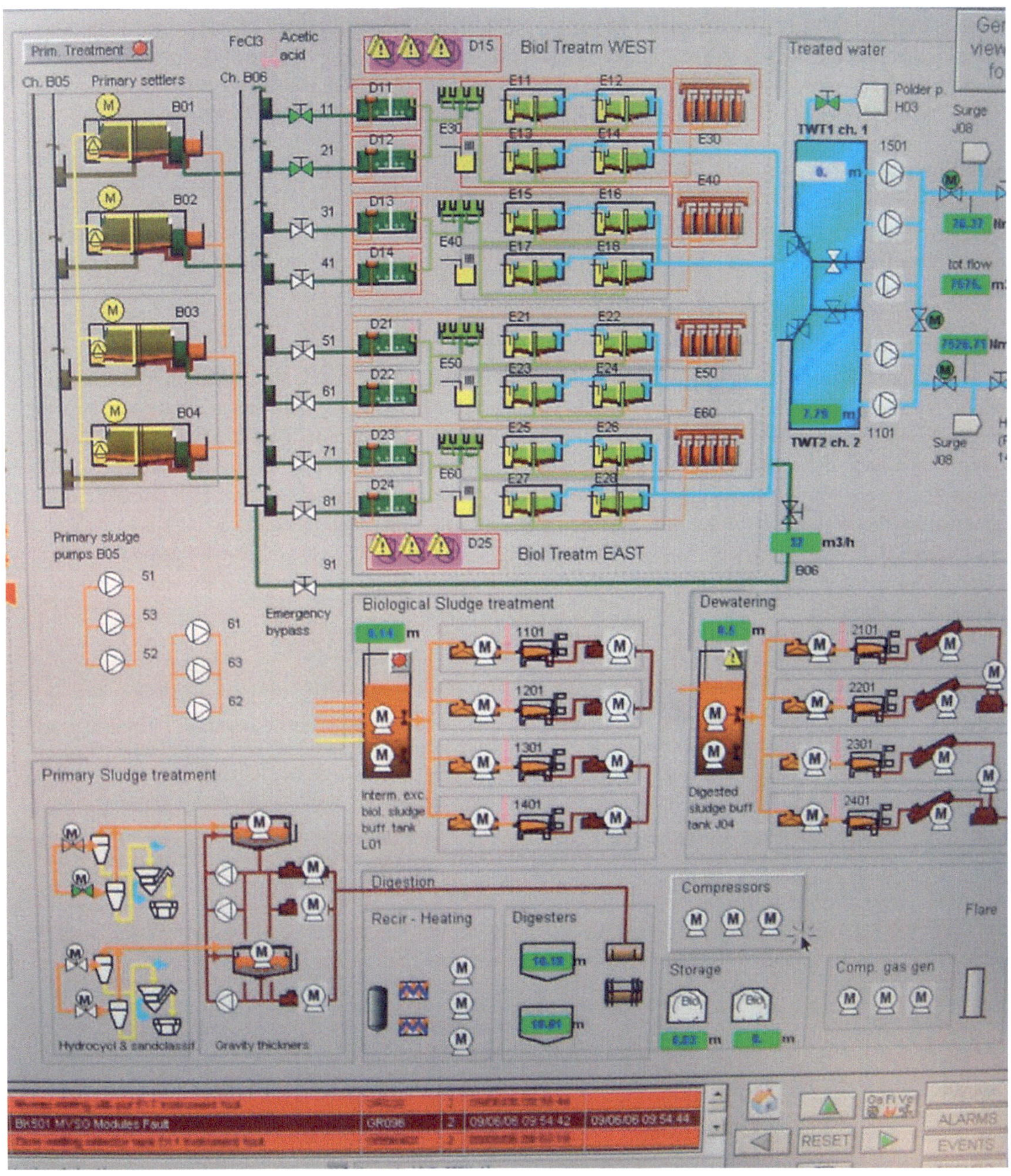

An example of a Supervisory Control And Data Acquisition (SCADA) interface used as a standard tool to operate and execute the process control in wastewater treatment practice (photo: D. Brdjanovic).

doi: 10.2166/9781789062304_0507

16

Anaerobic wastewater treatment

Jules B. van Lier, Nidal Mahmoud and Victor S. Garcia Rea

16.1 INTRODUCTION

Chapter 16 on anaerobic wastewater treatment in the textbook *Biological Wastewater Treatment: Principles, Modelling and Design* explains the principles of the anaerobic digestion (AD) process and its application for the treatment of industrial and municipal wastewater. The first section discusses the advantages of AD over aerobic treatment methods *e.g.,* energy savings, biochemical energy recovery, and reduced excess sludge production. After that, the basics of the microbiology, biochemistry, and thermodynamics of the process are explained, as well as the two metrics, chemical oxygen demand (COD) and total organic carbon (TOC). In AD, the incoming COD equals the outgoing COD. Therefore, it is considered a fundamental process control variable and is used to establish a COD balance over the anaerobic reactor. By means of a COD balance, an estimate can be made of the biogas and/or sludge production, depending on the available data on reactor performance and influent and effluent characteristics. In addition, by combining COD with TOC measurements the biogas composition can be predicted. The COD balance can also be used to determine the effects of alternative electron acceptors such as nitrate (NO_3^-) or sulphate (SO_4^{2-}) on *e.g.,* biogas production. However, readers should realize that not all organic compounds and/or reduced reaction products can be measured in COD analysis; examples are quaternary ammonium salts (such as betaine) and gases such as N_2 which will be formed upon reduction of NO_3^- and NO_2^- in anaerobic reactors. In the final part of the introductory sections, the theory behind biomass immobilization and its impact on anaerobic reactor technology is provided. Anaerobic sludge granulation, the phenomenon that has led to the worldwide breakthrough of high-rate anaerobic wastewater treatment, is discussed in detail. In the second part of the chapter, the different anaerobic reactor systems, their advantages and disadvantages, and the current state of their application are addressed. Special attention is given to the development of high-rate anaerobic reactors (HRARs) and their different configurations, such as the upflow anaerobic sludge blanket (UASB) and the expanded granular sludge bed (EGSB) reactor. After that, the design base and the main functional units of a UASB reactor are explained. The chapter mainly discusses the HRAR's applications in the industrial sector and closes with the experiences so far on municipal and domestic sewage treatment in warm climate countries, as well as the potentials for application in cold climate countries, and the possibilities for AD in novel sanitation concepts.

16.2 LEARNING OBJECTIVES

After the successful completion of this chapter, the reader will be able to:

- Explain the main biochemical, microbiological, and thermodynamic concepts of the anaerobic digestion process.
- Compare and contrast the advantages and disadvantages of applying anaerobic digestion in relation to aerobic methods for the treatment of industrial and municipal wastewater.
- Predict and calculate the theoretical methane production during the anaerobic digestion of organic matter.
- Calculate the energy recovery potential (in MJ/d or kW) when applying anaerobic treatment.
- Explain the effects of alternative (non-carbon) electron acceptors on the anaerobic digestion process and their impact on the microbiology related to them.
- Perform COD balances and apply them as a control parameter for anaerobic systems.
- Explain the concept of biomass retention and schematize the steps in the anaerobic granulation process.
- Compare and contrast the different anaerobic reactor systems.
- Explain the main characteristics of high-rate anaerobic reactors and their application for wastewater treatment.
- Explain when anaerobic treatment of domestic sewage can become advantageous, and what the current constraints are.
- Decide on the use of a particular anaerobic reactor configuration for the treatment of specific types of wastewater.
- Make a basic design of a UASB reactor.

16.3 EXAMPLES

Example 16.3.1

EGSB reactor for the treatment of wastewater from a food-processing industry

A food-processing industry is treating its wastewater with an expanded granular sludge bed (EGSB) reactor (Fig. 16.1). The wastewater and anaerobic process have the characteristics summarized in Table 16.1 and the stoichiometric parameters listed in Table 16.2. The food-processing wastewater is completely soluble and so without any suspended solids.

Table 16.1 Summary of the food-processing wastewater and effluent characteristics.

Description	Symbol	Value	Unit
Flow rate	Q_i	2,500	m^3/d
Total COD in the influent	COD_i	6.5	$kgCOD/m^3$
Substrate biodegradability		95	%
Concentration of excess granular sludge	$X_{t,w}$	9.0	%(w/w)
Total suspended solids in the effluent	$X_{t,e}$	400	$gTSS/m^3$
Total COD in the effluent	COD_e	1.0	$kgCOD/m^3$
SO_4^{2-} concentration in the influent	$S_{SO4,i}$	1.5	gSO_4^{2-}/m^3

Table 16.2 Stoichiometric parameters for the EGSB treatment process.

Parameter	Symbol	Value	Unit
Anaerobic sludge yield	$Y_{COD,An}$	0.07	$gCOD_{VSS}/gCOD$
Sludge VSS/TSS ratio	f_{VT}	0.7	$gVSS/gTSS$
CH4/COD ratio at standard temperature and pressure	$f_{CH4,COD}$	0.35	$l/gCOD$
COD/VSS ratio of the sludge	f_{CV}	1.42	$gCOD/gVSS$
$COD_{required}/SO_4^{2-}$ ratio	$f_{COD,SO4}$	0.67	$gCOD_{req}/g\ SO_4^{2-}$

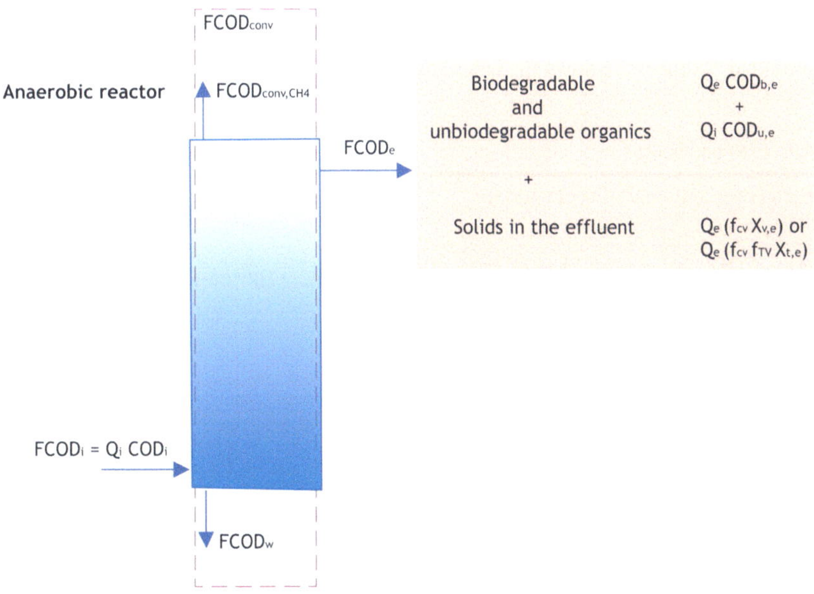

Figure 16.1 Schematic presentation of the (COD) mass balance in the EGSB reactor used for the treatment of the food-processing wastewater. Note the COD fractionation of the effluent. $FCOD_i$ = flux or mass flow rate of total COD into the reactor, Q_i = influent flow rate, COD_i = influent total COD, $FCOD_{conv}$ = flux or mass flow rate of COD converted, $FCOD_{conv,CH4}$ = flux or mass flow rate of COD converted to methane, $FCOD_w$ = COD flux of the excess of sludge (discharge of volatile suspended solids as COD), $FCOD_e$ = flux or mass flow rate of total COD in the effluent, Q_e = effluent flow rate, $COD_{b,e}$ = residual biodegradable COD present in the effluent, $COD_{u,e}$ = effluent unbiodegradable COD, f_{cv} = COD to VSS ratio of the sludge, f_{VT} = VSS to TSS ratio of the sludge, $X_{v,e}$ = volatile suspended solids in the effluent and $X_{t,e}$ = total suspended solids in the effluent

This example addresses the following points:

a) Calculate the COD removal efficiency $(E(\%)_{COD})$.
b) Calculate the amount of residual biodegradable COD in the effluent $(COD_{b,e}$ in mg/l).
c) Calculate the methane production $(FS_{CH4,prod})$ in m^3CH_4/d calculated at standard temperature and pressure (STP: 0 °C and 1 atm).

d) Calculate the reactor volume (V_r) required to achieve a minimum hydraulic retention time of 6 h and admissible organic loading rate (OLR) of 18 kgCOD/m³d.

e) Is the reactor volume limited by the hydraulic loading rate (HLR) or by the organic loading rate (OLR)?

f) Calculate the up-flow velocity (v_{UP}) considering a reactor height of 20 m.

g) Calculate the amount of excess wet granular sludge (Q_w) in m³/d that needs to be discharged from the reactor in order to maintain an equal sludge concentration (X_t) in the reactor.

h) Calculate the net yield of granular sludge ($Y_{TSS,COD}$) in % (kgTSS/kgCOD).

i) Following a change in the production process, the wastewater also contains 1,500 mg/l of SO_4^{2-}. Calculate the daily methane production ($FS_{CH4,prod}$) in the new situation assuming that all the sulphate is converted.

Solution

Using the data provided, Table 16.3 shows the calculation required to solve the questions.

Table 16.3 Answers to Example 16.3.1.

a) Treatment efficiency based on COD ($E(\%)_{COD}$)

$$E(\%)_{COD} = \frac{COD_i - COD_e}{COD_i} \cdot 100\% = \frac{6.5\,kgCOD/m^3 - 1.0\,kgCOD/m^3}{6.5\,kgCOD/m^3} \cdot 100\% = 84.6\%$$

b) Biodegradable COD in the effluent

$$COD_{b,e} = COD_e - \left(COD_{u,i} + COD_{vss,e}\right)$$

Where $COD_{u,i}$ is the unbiodegradable concentration of the COD_i and $COD_{vss,e}$ is the COD concentration corresponding to the VSS in the effluent. Thus:

$$COD_{u,i} = COD_i \cdot \left(1 - \frac{Substrate\ biodegradabilty}{100}\right) = 6{,}500\ mgCOD/l \cdot \left(1 - \frac{95}{100}\right) = 325\ mgCOD/l$$

$$COD_{VSS,e} = X_{t,e} \cdot f_{VT} \cdot f_{CV} = 400\ mgTSS/l \cdot 0.7\ mgVSS/mgTSS \cdot 1.42\ mgCOD/mgVSS = 398\ mgCOD/l$$

Therefore:

$$COD_{b,e} = COD_e - \left(COD_{u,i} + COD_{vss,e}\right)$$
$$COD_{b,e} = 1{,}000\ mgCOD/l - \left(325\ mgCOD/l + 398\ mgCOD/l\right) = 277\ mgCOD/l$$

c) Methane production under STP ($FS_{CH4,prod,STP}$ in m³/d)

For this calculation, consider the mass balance proposed in Figure 16.1:

$$FS_{CH4,prod,STP} = Q_i \cdot \left(COD_i - COD_{u,e} - COD_{b,e}\right) \cdot \left(1 - Y_{COD,An}\right) \cdot f_{CH_4,COD}$$

$$FS_{CH4,prod,STP} = 2{,}500 \ m^3/d \cdot \left(6{,}500 \ gCOD/m^3 - 325 \ gCOD/m^3 - 277 \ gCOD/m^3\right) \cdot$$

$$\left(1 - 0.07 \ gCOD_{VSS}/gCOD\right) \cdot \left(0.35 \ l_{CH_4}/gCOD\right) \cdot \left(1m^3 / 1{,}000 \ l\right) = 4{,}800 \ m^3 CH_4/d$$

d) Required reactor volume (V_r) with a minimum HRT of 6 h and an admissible OLR of 18 kgCOD/m^3.d

To determine the reactor volume, two equations are used, one based on HRT (Eq. 16.49) and the other one based on OLR (Eq. 16.59). Based on HRT we have:

$$V_r = HRT \cdot Q_i = 6 \ h \cdot 2{,}500 \ m^3/d = 625 \ m^3 \ .$$

However, with this volume, the OLR will be higher than the admissible one:

$$OLR = \frac{Q_i \cdot COD_i}{V_r} = \frac{2{,}500 \ m^3/d \cdot 6.5 \ kgCOD/m^3}{625 \ m^3} = 26 \ kgCOD/m^3.d$$

Therefore, we will determine the volume with a proposed OLR of 18 kgCOD/m^3.d:

$$V_r = \frac{Q_i \cdot COD_i}{OLR} = \frac{2{,}500 \ m^3/d \cdot 6.5 \ kgCOD/m^3}{18 \ kgCOD/m^3.d} = 903 \ m^3$$

As the second volume satisfies both the minimum HRT and the OLR, the volume dependent on the OLR will be used for the design.

e) Is the reactor volume limited by the hydraulic loading capacity or the organic loading capacity?

In question d, we determined that the organic loading rate was decisive for calculating the required volume. Therefore, we conclude that the reactor volume is organically limited. As a consequence, a reactor design based on HRT (a hydraulic constraint) will be organically overloaded.

f) Upflow velocity applying to a reactor height (h_r) of 20 m.

For calculating the up-flow velocity (Eq. 16.46), we first need to determine the area of the reactor:

$$A_r = \frac{V_r}{h_r} = \frac{903 \ m^3}{20 \ m} = 45.1 \ m^2$$

Then, by using Equation 16.46, we can determine the upflow velocity in the reactor as:

$$v_{up} = \frac{Q_i}{A_r} = \left(\frac{2{,}500 \ m^3/d}{45.1 \ m^2}\right) \cdot \left(\frac{1 \ d}{24 \ h}\right) = 2.3 \ m/h$$

g) Excess wet granular sludge (m³/d) that needs to be discharged from the reactor

First, we will use the mass balance (Fig. 16.1) again to determine the TSS produced ($FX_{t,prod}$):

$$FX_{t,prod} = Q_i \cdot \left(COD_i - COD_{u,i} - COD_{b,e}\right) \cdot \left(Y_{COD,An}\right) \cdot \frac{1}{f_{CV}} \cdot \frac{1}{f_{VT}} =$$

$$2{,}500 \text{ m}^3/\text{d} \cdot \left(6.5 \text{ kgCOD/m}^3 - 0.325 \text{ kgCOD/m}^3 - 0.277 \text{ kgCOD/m}^3\right) \cdot$$

$$\left(0.07 \text{ kgCOD}_{VSS}/\text{kgCOD}\right) \cdot \frac{1 \text{ kgVSS}}{1.42 \text{ kgCOD}_{VSS}} \cdot \frac{1 \text{ kgTSS}}{0.7 \text{ kgVSS}} = 1{,}038.4 \text{ kgTSS/d}$$

Once the daily production rate of TSS has been calculated we need to determine the rate of TSS that is washed out from the reactor ($FX_{t,e}$). Note that the influent solids concentration is assumed to be negligible.

$$FX_{t,e} = Q_i \cdot X_{t,e} = 2{,}500 \text{ m}^3/\text{d} \cdot 0.4 \text{ kgTSS/m}^3 = 1{,}000 \text{ kgTSS/d}$$

After calculating the amount of TSS that will be washed out from the reactor, it is possible to determine the amount of excess wet granular sludge that must be discharged ($FX_{t,w}$) by subtracting these two rates:

$$FX_{t,w} = FX_{t,prod} - FX_{t,e} = 1{,}038.4 \text{ kgTSS/d} - 1{,}000 \text{ kgTSS/d} = 38.4 \text{ kgTSS/d}$$

h) Net yield of granular sludge production (WW) (TSS/COD)

Using the mass balance in Fig. 16.1 again, we need to determine the amount of substrate that was converted ($FCOD_{conv}$):

$$FCOD_{conv} = Q_i \cdot \left(COD_i - COD_{u,e} - COD_{b,e}\right) =$$

$$2{,}500 \text{ m}^3/\text{d} \cdot \left(6.5 \text{ kgCOD/m}^3 - 0.325 \text{ kgCOD/m}^3 - 0.277 \text{ kgCOD/m}^3\right) = 14{,}745 \text{ kgCOD/d}$$

In the previous section g), it was determined that the net amount of excess sludge produced ($FX_{t,w}$) was 38.4 kgTSS/d. Therefore, the net yield of granular sludge production per COD removed (w/w) in expressed as the ratio kgTSS/kgCOD, can now be calculated:

$$Y_{TSS,COD} = \frac{FX_{t,w}}{FCOD_{conv}} \cdot 100\ \% = \frac{38.4 \text{ kgTSS/d}}{14{,}745 \text{ kgCOD/d}} \cdot 100\ \% = 0.26\ \% \ .$$

i) Methane production when 1,500 mgSO$_4^{2-}$/l is present in the influent

Once more, we will use the mass balance proposed in section c). First, we need to determine the organic mass, expressed in COD, that remains available for being used either for CH_4 production or SO_4^{2-} reduction ($FCOD_{available}$):

$$FCOD_{available} = Q_i \cdot \left(COD_i - COD_{u,e} - COD_{b,e}\right) \cdot \left(1 - Y_{COD}\right) =$$

$$2{,}500 \text{ m}^3/\text{d} \cdot \left(6.5 \text{ kgCOD/m}^3 - 0.325 \text{ kgCOD/m}^3 - 0.277 \text{ kgCOD/m}^3\right) \cdot$$

$$\left(1 - 0.07 \text{ kgCOD/kgCOD}\right) = 13{,}713 \text{ kgCOD/d}$$

Note that in this calculation, the produced effluent COD related to solubilised H_2S and HS^- resulting from sulphate reduction is ignored.

Now, the amount of COD available that will be used for the sulphate reduction ($FCOD_{SO4,red}$) can be estimated as:

$$FCOD_{SO_4,red} = Q_i \cdot S_{SO4,i} \cdot f_{COD,SO_4} = 2{,}500 \text{ m}^3/\text{d} \cdot 1.5 \text{ kgSO}_4^{2-}/\text{m}^3 \cdot 0.67 \text{ kgCOD}_{req}/\text{kgSO}_4^{2-} = 2{,}513 \text{ kgCOD/d}$$

Therefore, the methane production will be defined by:

$$FS_{CH4,prod} = \left(FCOD_{available} - FCOD_{SO_4,red}\right) \cdot f_{CH_4,COD} =$$

$$\left(13{,}713 \text{ kgCOD/d} - 2{,}513 \text{ kgCOD/d}\right) \cdot 0.35 \text{ m}^3_{CH_4}/\text{kgCOD} = 3{,}920 \text{ m}^3CH_4/\text{d}$$

Note that the COD removal efficiency will decrease because of the presence of inorganic COD in the form of HS^-/H_2S in the effluent.

A summary of the results obtained for the EGSB reactor treatment of the food-processing wastewater is presented in Table 16.4.

Table 16.4 Summary of EGSB reactor design results.

Description	Parameter	Unit	Value
1. Treatment efficiency			
	$E(\%)_{COD}$	%	84.6
2. Effluent COD			
Unbiodegradable COD in the effluent	$COD_{u,e}$	mgCOD/l	325
Effluent COD composed of VSS	$COD_{VSS,e}$	mgCOD/l	398
Residual biodegradable COD in the effluent	$COD_{b,e}$	mgCOD/l	277
3. Methane production under standard temperature and pressure			
Methane production rate	$FS_{CH4,prod}$	m³/d	4,800
4. Required reactor volume (minimum HRT = 6 h, admissible OLR = 18 kgCOD/m³.d)			
Reactor volume	V_r	m³	903
5. Reactor volume limited organically or hydraulically			
Organically limited			
6. Upflow velocity applying a reactor height of 20 m			
Area of the reactor	A_r	m²	45.1
Upflow velocity	v_{up}	m/h	2.3

7. Amount of excess wet granular sludge that needs to be discharged

Total suspended solids production rate	$FX_{t,prod}$	kgTSS/d	1,038.4
Total suspended solids mass flow in the effluent	$FX_{t,e}$	kgTSS/d	1,000.0
Total suspended solids to be discharged	$FX_{t,w}$	kgTSS/d	38.4

8. Net yield of granular sludge production (w/w)

COD mass flow converted	$FCOD_{conv}$	kgCOD/d	14,745
Net yield of granular sludge production per COD removed	$Y_{TSS,COD}$	%	0.26

9. Methane production when 1,500 $mgSO_4^{2-}$/l is present in the influent

COD mass flow available for CH_4 production or SO_4^{2-} reduction	$FCOD_{available}$	kgCOD/d	13,713
COD mass flow used for SO_4^{2-} reduction	$FCOD_{SO4,red}$	kgCOD/d	2,513
Methane production rate	$FS_{CH4,prod}$	m^3/d	3,920

Example 16.3.2

Replacement of aerobic with anaerobic wastewater treatment

A brewery is considering whether to implement anaerobic high-rate treatment instead of activated sludge treatment for its wastewater. The produced CH_4 will be converted into electricity using a combined heat-power (CHP) generator. The characteristics of the brewing wastewater, anaerobic high-rate process, and energy consumption/recovery are given in tables 16.5 and 16.6. Assume the same COD removal efficiency for the activated sludge plant with a reasonable energy usage (kWh/kgCOD) (Table 16.6). Calculate the annual energy benefit in € per year if an anaerobic reactor is installed instead of the activated sludge treatment.

Table 16.5 Summary of the brewing wastewater and AD treatment process characteristics.

Description	Symbol	Value	Unit
Influent flow rate	Q_i	5,000	m^3/d
Total COD concentration in the influent	COD_i	4.5	$kgCOD/m^3$
COD removal efficiency of the anaerobic reactor	$E(\%)_{COD}$	85	%

Table 16.6 Stoichiometric parameters and ratios for the AD treatment process of the brewery wastewater.

Parameter	Symbol	Value	Units
Energy generation efficiency of the CHP generator	$E(\%)_{CHP}$	40	%
Energy generation/CH_4 ratio at high heating value (HHV)	$f^{HHV}_{MJ,CH4}$	39.4	MJ/m^3CH_4
Energy generation/CH_4 ratio at low heating value (LHV)	$f^{LHV}_{MJ,CH4}$	35.5	MJ/m^3CH_4
Megajoule per kWh ratio	$f_{MJ,kWh}$	3.6	MJ/kWh
Anaerobic sludge yield	$Y_{COD,An}$	0.06	$gCOD_{VSS}$/gCOD
CH_4/COD ratio at STP	$f_{CH4,COD,STP}$	0.35	l/gCOD
Energy usage to COD removal ratio under aerobic conditions	$f_{kWh,COD,aer}$	0.75	kWh/kgCOD
Electricity price per kWh	$f_{€,kWh}$	0.25	€/kWh
CO_2 production/energy consumption ratio	$f_{CO2,kWh}$	0.8	$kgCO_2$/kWh
Price of carbon credits[1]	$€_{CO2}$	20	€/ton CO_2

[1] The price of carbon credits has continued to increase considerably over the years in view of the effects of climate change and the need to develop and implement more renewable energy sources.

The task of the reader is to:

a)   Explain how the brewery can obtain energy benefits and give three examples.
b)   Calculate the annual benefit (in €). Consider STP conditions and an average price of carbon credits ($€_{CO2}$) of €20/tonCO$_2$.

Solution

Using the data provided, Table 16.7 shows the calculations required to solve the tasks.

Table 16.7 Answers to Example 16.3.2.

---

a)   How energy benefit can be obtained.

1.   Energy recovery via CH$_4$ production.
2.   Avoidance of energy use for aeration.
3.   Via carbon credits.

b)   Calculate the annual energy benefit. Consider STP conditions.

Let us consider the next mass balance:

$$FCOD_i = Q_i \cdot COD_i = 5{,}000 \ m^3/d \cdot 4.5 \ kgCOD/m^3 = 22{,}500 \ kgCOD/d$$

$$FCOD_e = FCOD_i \cdot \left(1 - (E(\%)_{COD}/100)\right) = 22{,}500 \ kgCOD/d \cdot \left(1 - (85/100)\right) = 3{,}375 \ kgCOD/d$$

$$FCOD_{conv,VSS} = \left(FCOD_i - FOD_e\right) \cdot Y_{COD,An} =$$
$$\left(22{,}500 \ kgCOD/d - 3{,}375 \ kgCOD/d\right) \cdot 0.06 \ kgCOD/kgCOD = 1{,}148 \ kgCOD/d$$

$$FS_{CH4,prod} = \left(FCOD_i - \left(FCOD_e + FCOD_{VSS,e}\right)\right) \cdot f_{CH_4,COD} =$$
$$\left(22{,}500 \ kgCOD/d - \left(3{,}375 \ kgCOD/d + 1{,}148 \ kgCOD/d\right)^{a)}\right) \cdot 0.35 \ m^3CH_4/kgCOD = 6{,}292 \ m^3CH_4/d$$

Once the methane produced is known, we can calculate the electric energy production considering low heating value conditions ($E_{kWh,LHV}$) by:

$$E_{kWh,LHV} = FS_{CH4,prod} \cdot f_{MJ,CH4}^{LHV} \cdot f_{kWh,MJ} \cdot E(\%)_{CHP} =$$
$$6{,}292 \ m^3CH_4/d \cdot 35.5 \ MJ/m^3CH_4 \cdot \frac{1 \ kWh}{3.6 \ MJ} \cdot \frac{40}{100} = 24{,}819 \ kWh/d$$

To calculate the total annual savings, first, we need to determine the savings due to the electric energy production, avoidance of aeration, and carbon credits:

$$€_{E_{LHV,savings}} = E_{kWh,LHV} \cdot f_{€,kWh} = 24{,}819 \ kWh/d \cdot 0.25 \ €/kWh \cdot 365 \ d/yr = 2{,}264{,}729 \ €/yr$$

$$\text{€}_{aer,savings} = \left(Q_i \cdot COD_i \cdot E(\%)_{COD}\right) \cdot f_{kWh,COD,aer} \cdot f_{\text{€},kWh} =$$

$$\left(5{,}000 \text{ m}^3/\text{d} \cdot 4.5 \text{ kgCOD/m}^3 \cdot 0.85\right) \cdot 0.75 \text{ kWh/kgCOD}^{b)} \cdot 0.25 \text{ €/kWh} \cdot 365 \text{ d/yr} = 1{,}308{,}868 \text{ €/yr}$$

$$\text{€}_{CO_2 savings} = \left(Q_i \cdot COD_i \cdot E(\%)_{COD}\right) \cdot f_{kWh,COD,aer} \cdot f_{CO_2,kWh} \cdot \text{€}_{CO2}$$

$$\text{€}_{CO_2,savings} = \left(5{,}000 \text{ m}^3/\text{d} \cdot 4.5 \text{ kgCOD/m}^3 \cdot 0.85\right) \cdot$$

$$0.75 \text{ kWh/kgCOD} \cdot 0.8 \text{ kgCO}_2 / \text{kWh} \cdot (1 \text{ tonCO}_2 / 1{,}000 \text{kgCO}_2) \cdot 20 \text{ €/tonCO}_2 \cdot 365 \text{ d/yr} = 83{,}770 \text{ €/yr}$$

The total annual savings will correspond to:

$$\text{€}_{tot,savings} = \text{€}_{E_{LHV,savings}} + \text{€}_{aer,savings} + \text{€}_{CO_2,savings} =$$

$$2{,}264{,}729 \text{ €/yr} + 1{,}308{,}868 \text{ €/yr}^{b)} + 83{,}770 \text{ €/yr} = 3{,}657{,}367 \text{ €/yr}$$

[a] Note that even when the units refer to COD in kgCOD/d they express the COD flux caused by the presence of volatile suspended solids in the effluent; therefore, the VSS mass flow can be summed with the other kgCOD/d values.

[b] Any value between 0.5 and 1.0 kWh/kgCOD removed is considered reasonable, so the annual energy saving costs may range between €872,579 and €1,745,158.

The summary of the results obtained for the implementation of the anaerobic process is presented in Table 16.8.

Table 16.8 Summary of the results of the implementation of the anaerobic wastewater treatment process.

Description	Parameter	Unit	Value
1. Energy benefit examples			
Energy recovery via CH_4 production			
Avoidance of energy use for aeration			
Carbon credits			
2. Calculate the annual energy benefit. Consider STP and a carbon credit price of €20/tonCO$_2$			
Influent COD flux or mass flow rate into the reactor	$FCOD_i$	kgCOD/d	22,500
COD mass flow rate out of the reactor	$FCOD_e$	kgCOD/d	3,375
COD mass flow converted to VSS	$FCOD_{conv,VSS}$	kgCOD/d	1,148
Methane production rate	$FS_{CH4,prod}$	m^3/d	6,292
Energy production (lower heating value)	$E_{kWh,LHV}$	kWh/d	24,804
Annual savings – electric energy production	$\text{€}_{E,LHV,savings}$	€/year	2,264,729
Annual savings – avoidance of aeration	$\text{€}_{aer,savings}$	€/year	1,308,868[a]
Annual savings – carbon credits	$\text{€}_{CO2,savings}$	€/year	83,770
Total annual savings	$\text{€}_{tot,savings}$	€/year	3,657,367

[a] Energy saving costs may range between €872,579 and €1,745,158.

Note: sludge comprises both microbial growth yield and the fraction of non-digested primary sludge that leaves the reactor with the sludge discharge.

Example 16.3.3

UASB reactor design for municipal sewage treatment

A facility using rectangular UASB reactor units will be built for the treatment of municipal sewage from approximately 450,000 inhabitants. The city is served with a separate sewerage system that excludes stormwater from entering. Note that the size of each UASB reactor unit is limited to 2,000-2,500 m^3. The length and width of 1 UASB unit has consequences for the shape and construction of the gas-liquid-solids-separator (GLSS) devices to be mounted at the top part of the reactor. The width of the UASB unit agrees with the length of the GLSS devices. Each reactor unit will be equipped with 'n' devices. As a rule of thumb, the length of a reactor unit is n x 3 meters, while the length of a GLSS is < 10m. The wastewater characteristics are summarized in Table 16.9. The average temperature registered during the characterization was 25 °C and the atmospheric pressure 0.9 atm. The methane produced will be recovered and converted into electricity using a combined heat and power (CHP) generator with an efficiency of 40 % ($E(\%)_{CHP}$). Consider an energy price of €0.095 per kWh.

Using the data provided, propose a design for the UASB reactor (Fig. 16.2), based on the assumptions given in Table 16.10. Focus only on the reactor design and avoid preliminary treatment steps such as screens, grit removal chambers, or oil and grease traps. Assume there is no need for pre-acidification or an equalization tank. Consider the design parameters proposed in Table 16.10 and an expected COD removal efficiency of 75 %. Consider a methane production yield of 0.24 Nm^3CH_4 per kg COD removed under field conditions, and a CH_4 solubility of 29.6 ml/l at 1 bar and 25 °C. Finally, calculate the required sludge drying bed area for handling the excess sludge produced; assume a dry weight content of excess sludge of 9 %, a design (maximum) height of 0.2 m for the wet sludge in the drying beds ($h_{Xt,max}$), and a drying cycle of 7 days ($t_{dry,cycle}$).

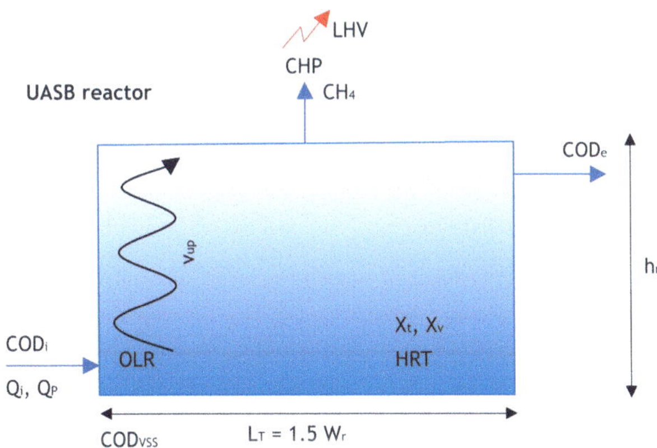

Figure 16.2 Scheme of a UASB reactor unit.

Table 16.9 Summary of the municipal wastewater characterization.

Description	Symbol	Value	Units
Influent flow rate	Q_i	50,000	m^3/d
Total COD in the influent	COD_i	0.9	$kgCOD/m^3$
Total suspended solids concentration in the influent	$X_{t,i}$	0.3	$kgTSS/m^3$
pH	pH	7.0	-
Average sewage temperature	T	25	°C
Micro and macro nutrients		Not limiting	
Peak[a] factor	p_f	1.2	-
Peak[a] duration	$p_{f,t}$	2	h/d

[a] Peak refers to the maximum flow of the diurnal flow pattern; note that rain/storm water is collected separately.

Table 16.10 Design parameters and assumptions for the UASB reactor and energy generation.

Description	Symbol	Value	Unit
Reactor height	h_r	5.5	m
Maximum volume per reactor unit	$V_{r,unit}$	2,000-2,500	m^3
Minimum HRT (if hydraulically limited)	HRT	8	h
Maximum OLR (if organically limited)	V_{OLR}	3	$kgCOD/m^3 \cdot d$
Reactor COD removal efficiency	$E(\%)_{COD}$	75	%
Sludge VSS/TSS ratio	f_{VT}	0.7	kgVSS/kgTSS
COD/VSS ratio of the sludge	f_{CV}	1.42	gCOD/gVSS
CH$_4$/COD ratio at STP (stoichiometric value)	$f_{CH4,COD,STP}$	0.35	$Nm^3CH_4/kgCOD$
CH$_4$ produced per COD removed under field conditions	$f_{CH4,COD,rem}$	0.24	$Nm^3CH_4/kgCOD$
CH$_4$ content in biogas	$f_{CH4,biogas}$	0.88	$Nm^3CH_4/m^3biogas$
Dry weight content of excess sludge	$X_{t,w,dry}$	9	%
Maximum height wet sludge in the drying bed	$h_{Xt,max}$	0.2	m
Drying cycle	$t_{dry,cycle}$	7	d
Energy generation efficiency of the CHP generator	$E(\%)_{CHP}$	40	%
LHV energy/CH$_4$ ratio	$f^{LHV}_{MJ,CH4}$	35.5	MJ/m^3CH_4
Ratio MJ/kWh	$f_{MJ,kWh}$	3.6^{-1}	MJ/kWh
Electricity price	$€_{kWh}$	0.25	€/kWh

Based on the information and data given in Example 16.33, determine the following:

a) Daily flow (Q_d)
b) UASB reactor volume
    b.1. Total UASB volume required.
    b.2. Whether the reactor is organically or hydraulically limited in its design
c) Number of UASB reactor units required
d) UASB reactor area
    d.1. Total UASB reactor area required
    d.2. Area per UASB reactor unit required

   e) UASB reactor length and width per unit
   f) Upflow velocity
   g) HRT for the specified design
   h) OLR for the specified design
   i) Methane and sludge production rates
      i.1. Methane dissolved in the effluent
      i.2. Methane recovered under field conditions
      i.3. Methane recovered at STP
      i.4. VSS production
      i.5. TSS production
   j) Area of the required sludge drying bed
   k) Energy production and financial benefit (€/year).

Solution

Using the data provided, Table 16.11 shows the calculation required to solve these questions.

Table 16.11 Answers to the Example 13.3.3.

a) Daily flow

$$Q_d = \left(Q_i \cdot p_f \cdot p_{f,t}\right) + Q_i - \left(Q_i \cdot p_{f,t}\right) =$$

$$Q_i \cdot \left(1 + p_{f,t} \cdot \left(p_f - 1\right)\right) = 50{,}000 \text{ m}^3/\text{d} \cdot \left(1 + (2 \text{ h}/24 \text{ h}) \cdot (1.2 - 1)\right) = 50{,}833 \text{ m}^3/\text{d}$$

b) UASB reactor volume [a)]

If hydraulically limited:

$$V_r = Q_i \cdot HRT = 50{,}000 \text{ m}^3/\text{d} \cdot 0.33 \text{ d} = 16{,}667 \text{ m}^3$$

If organically limited:

$$V_r = \frac{Q_i \cdot COD_i}{v_{OLR}} = \frac{50{,}000 \text{ m}^3/\text{d} \cdot 0.9 \text{ kgCOD/m}^3}{3 \text{ kgCOD/m}^3.\text{d}} = 15{,}000 \text{ m}^3$$

Since the second volume (15,000 m³) would not fulfil the minimum design HRT (7.2 h < 8.0 h), the bigger volume of 16,667m³ is chosen, which is rounded up to 16,700 m³. Therefore, we can conclude that the reactor is hydraulically limited.

c) Number of reactor units required

Considering a maximum volume per unit of 2,000-2,500 m³:

$$n_{r,unit} = \frac{V_r}{V_{r,max}} = \frac{16{,}700 \text{ m}^3}{2{,}000 \text{ m}^3} = 8.4 \quad \text{and} \quad n_{r,unit} = \frac{V_r}{V_{max}} = \frac{16{,}700 \text{ m}^3}{2{,}500 \text{ m}^3} = 6.7$$

Therefore, the number of required reactor units is between 6.7 and 8.4 units. Since construction works are preferably built with an even number of units, a total of 8 UASB reactor units is chosen.

d) UASB reactor area

According to Table 16.10:

$$A_{r,unit} = \frac{V_r}{h_r} = \frac{16{,}700 \text{ m}^3}{5.5 \text{ m}} = 3{,}036 \text{ m}^2, \text{ which can be rounded up to } 3050 \text{ m}^2.$$

Considering 8 UASB reactor units, the area per reactor unit will be:

$$A_{r,unit} = \frac{A_r}{n_{r,unit}} = \frac{3{,}050 \text{ m}^2}{8} = 381 \text{ m}^2$$

e) UASB reactor length and width per unit

For construction purposes, *i.e.,* the mounting of gas-liquid-solids-separator (GLSS) devices, UASB reactor units are rectangular with a width of about 10 m, agreeing with the length of the GLSS devices. The length of 1 UASB unit is n x 3, in which n is the number of GLSS device to be mounted.

Thus, per unit:

$A_{r,unit} = W_r \cdot L_r$, and $A_{r,unit} = 10 \cdot 3n$, where n = number of GLS devices.

So:

$381 = 30n \rightarrow n = 381 / 30 = 13$ (rounded value)

Therefore, the design size of the reactor unit area can be determined as:

$$A_{r,unit} = 10 \text{ m} \cdot 39 \text{ m} = 390 \text{ m}^2$$

With the resulting total UASB reactor area:

$$A_r = 8 \cdot 390 \text{ m}^2 = 3{,}120 \text{ m}^2$$

Note that the exact dimensions can be adjusted to ease the design and construction.

f) Upflow velocity [a]

$$v_{up} = \frac{Q_i}{A_r} = \frac{50{,}000 \text{ m}^3/\text{d}}{3{,}120 \text{ m}^2} \cdot \frac{1 \text{ d}}{24 \text{ h}} = 0.67 \text{ m/h}$$

g) HRT [a]

$$HRT = \frac{V_r}{Q_i} = \frac{n_{r,unit} \cdot h_r \cdot A_{r,unit}}{Q_i} = \frac{8 \cdot 5.5 \cdot 390}{Q_i} = \frac{17{,}160 \text{ m}^3}{50{,}000 \text{ m}^3/\text{d}} = 0.34 \text{ d} = 8.2 \text{ h}$$

h) OLR [a]

$$OLR = \frac{Q_i \cdot COD_i}{V_r} = \frac{50,000 \text{ m}^3/\text{d} \cdot 0.9 \text{ kgCOD/m}^3}{17,160 \text{ m}^3} = 2.6 \text{ kgCOD/m}^3.\text{d}^{[a]}$$

i) Methane and sludge production rates [a]

First, we will determine the COD mass flow rate ($FCOD_i$) flowing into the reactor:

$$FCOD_i = Q_i \cdot COD_i = 50,000 \text{ m}^3/\text{d} \cdot 0.9 \text{ kgCOD/m}^3 = 45,000 \text{ kgCOD/d}$$

Next, we will determine the COD mass flow rate converted into methane and withdrawn via excess sludge. The methanised fraction of COD can be calculated based on the observed m³CH4/kgCOD removed in comparison with the theoretical stoichiometric value.

$$\text{Observed} = f_{CH_4,COD,rem} = 0.24 \text{ m}^3CH_4/\text{kgCOD}$$

Note that this value includes the total amount of produced methane, both extracted as gas and dissolved in the effluent of the reactor.

$$\text{Stoichiometric} = f_{CH_4,COD} = 0.35 \text{ m}^3CH_4/\text{kgCOD}$$

The methanised amount of removed COD ($f_{CH4,COD,actual}$) = 0.24 / 0.35 = 68.6 %, whereas 31.4 % of the removed COD leaves the reactor with the excess sludge. It should be noted that the excess sludge consists of both newly grown sludge with an approximate yield of 5-10 % and (partly) stabilised primary sludge that entered the UASB reactor with the raw influent.

Calculating the COD converted to methane:

$$FCOD_{conv,CH4} = FCOD_i \cdot E(\%)_{COD} \cdot f_{CH4,COD,actual} = 45,000 \text{ kgCOD/d} \cdot 0.75 \cdot 0.686 = 23,153 \text{ kgCOD/d}$$

Therefore, the volumetric biogas production at STP:

$$FS_{CH4,prod,STP} = 23,153 \text{ kgCOD/d} \cdot 0.35 \text{ Nm}^3CH_4/\text{kgCOD} = 8,103 \text{ Nm}^3CH_4/\text{d}$$

And, by applying the combined gas law we can determine the methane production under field conditions (25 °C and 0.9 atm):

$$FS_{CH4,prod} = \frac{P_1 \cdot V_1}{T_1} \cdot \frac{T_2}{P_2} \cdot \left(\frac{1 \text{ atm} \cdot 8,103 \text{ m}^3CH_4/\text{d}}{273 \text{ K}} \right) \cdot \left(\frac{(273+25)\text{K}}{0.9 \text{ atm}} \right) = 9,828 \text{ m}^3CH_4/\text{d}$$

To answer Question i.i.1 (methane dissolved in the effluent), the CH_4 that escapes dissolved in the effluent (field conditions: 0.9 atm = 0.91 bar, and 25 °C) can be estimated as:

$$FS_{CH4,e} = Q_i \cdot S_{CH_{4,sol,std}} \cdot f_{CH_4,Biogas} \cdot P_{atm} =$$

$$50,000 \ m^3/d \cdot \frac{0.0296 \ m^3 CH_4}{1 \ m^3 \cdot 1 \ bar} \cdot 0.88 \cdot 0.91 \ bar = 1,185 \ m^3 CH_4/d$$

Being equal to a COD mass flow rate of:

$$FCOD_{CH_{4,e}} = S_{CH4,e} \cdot \frac{P_1}{T_1} \cdot \frac{T_2}{P_2} \cdot f_{CH_4,COD} =$$

$$1,185 \ m^3 CH_4/d \cdot \left(\frac{0.9 \ atm}{(273+25)K} \right) \cdot \left(\frac{273 \ K}{1 \ atm} \right) \cdot \left(\frac{kgCOD}{0.35 \ mN^3 CH_4} \right) = 2,792 \ kgCOD/d$$

Therefore, the recovered CH_4 via biogas exhaust is (solution to point i.i.2):

$$FS_{CH4,recov} = FS_{CH4,prod} - FS_{CH4,e} = 9,828 \ m^3 CH_4/d - 1,185 \ m^3 CH_4/d = 8,643 \ m^3 CH_4/d$$

Under STP conditions (answer to Question i.i.3):

$$FS_{CH4,recov,STP} = FS_{CH4,recov} \cdot \frac{P_1}{T_1} \cdot \frac{T_2}{P_2} = 8,643 \ m^3 CH_4/d \cdot \left(\frac{0.9 \ atm}{(273+25)K} \right) \cdot \left(\frac{273 \ K}{1 \ atm} \right) = 7,126 \ m^3 CH_4/d$$

And the COD removed with excess sludge:

$$FCOD_w = FCOD_i \cdot E(\%)_{COD} \cdot \left(1 - f_{CH4,COD,actual} \right) = 45,000 \ kgCOD/d \cdot 0.75 \cdot \left(1 - 0.686 \right) = 10,598 \ kgCOD/d$$

Corresponding to:

$$FX_{v,w} = FCOD_w \cdot \frac{1}{f_{CV}} = 10,598 \ kgCOD/d \cdot \frac{kgVSS}{1.42 \ kgCOD} = 7,463 \ kgVSS/d$$

$$FX_{t,w} = FX_{v,w} \cdot \frac{1}{f_{VT}} = 7,463 \ kgCOD/d \cdot \frac{kgTSS}{0.7 \ kgCOD} = 10,661 \ kgTSS/d$$

This TSS includes the newly grown sludge as well as settled and partly digested primary sludge.

j) Area of the sludge drying bed [a]

According to Table 16.10, the dry weight content of the excess sludge is 9%, and therefore the volumetric solids production rate will correspond to:

$$Q_w = FX_{t,w} \cdot X_{t,w,dry} = \left(10{,}661 \ \text{kgTSS/d}\right) \cdot \left(\frac{m^3}{90 \ \text{kgTSS}}\right) = 118.5 \ m^3/d$$

Considering a maximum height of the sludge bed of 0.2 m and 7 days for each drying cycle, the sludge drying bed area will correspond to:

$$A_{sdb} = Q_w \cdot \frac{1}{h_{Xt,max}} \cdot t_{dry,cycle} = \left(118.5 \ m^3/d\right) \cdot \left(\frac{1}{0.2 \ m}\right) \cdot 7 \ d = 4{,}148 \ m^2$$

k) Energy production and financial benefit [a]

The energy benefit due to the electricity production from the recovered methane will be calculated as showed:

$$E_{kWh,LHV} = FS_{CH4,recov,STP} \cdot f_{MJ,kWh}^{LHV} \cdot f_{kWh,CH_4} \cdot E(\%)_{CHP} =$$

$$7{,}126 \ m^3 CH_4/d \cdot 35.5 \ MJ/m^3 CH \cdot \frac{1 \ kWh}{3.6 \ MJ} \cdot 0.4 = 28{,}108 \ kwh/d$$

And the annual financial benefit is:

$$\text{€}_{E_{LHV,savings}} = E_{kWh,LHV} \cdot \text{€}_{kWh} = 28{,}108 \ kWh/d \cdot 0.25 \ \text{€/kWh} \cdot 365 \ d/yr = 2{,}564{,}855 \ \text{€/yr}$$

[a] These values were calculated using the non-peak flow. Therefore, the values will vary during the peak flow..

The summary of the calculation results for the UASB reactor design is shown in Table 16.12.

Table 16.12 Summary of the calculation results for the UASB reactor design.

Description	Parameter	Units	Value
1. Daily flow			
Average influent flow	Q_i	m³/d	50,000
Peak flow	Q_p	m³/d	50,833
2. UASB reactor volume			
Hydraulically limited	V_r	m³	16,700
Organically limited	V_r	m³	15,000
3. UASB reactor area			
Total area	A_r	m²	3,120
Area per UASB reactor unit	$A_{r,unit}$	m²	390
4. Number of required reactor units			
Number of required reactor units	$n_{r,unit}$		8
5. UASB reactor length and width			
Width of the reactor	W_r	m	10
Length of the reactor	L_r	m	39
6. Upflow velocity			
Upflow velocity	v_{up}	m/h	0.67
7. Hydraulic retention time			
Hydraulic retention time	HRT	h	8
8. Organic loading rate			
Volumetric organic loading rate	$_{VOLR}$	kgCOD/m³.d	2.6
9. Sludge production			
Mass flow rate into the reactor	$FCOD_i$	kgCOD/d	45,000
Methanised amount of COD removed	$meth_{COD}$	%	68.6
Mass flow rate converted into methane	$FCOD_{conv,CH4}$	kgCOD/d	23,153
Methane production at STP	$FS_{CH4,prod,STP}$	kgTSS/d	8,103
Methane production under field conditions	$FS_{CH4,prod}$	m³/d	9,828
Methane dissolved in the effluent	$FS_{CH4,e}$	m³/d	1,185
Methane recovered via biogas exhaust under field conditions	$FS_{CH4,recov}$	m³/d	8,643
Methane recovered via biogas exhaust under STP conditions	$FS_{CH4,recov,STP}$	m³/d	7,126
COD removed with excess sludge	$FCOD_w$	kgCOD/d	10,598
VSS production rate	$FX_{v,w}$	kgVSS/d	7,463
TSS production rate	$FX_{t,w}$	kgTSS/d	10,661
10. Area of the sludge drying bed			
Volumetric solids production rate	Q_w	m³/d	118.5
Area of the sludge drying bed	A_{sdb}	m²	4,148
11. Energy production and benefit			
Low heat value energy	$E_{kWh,LHV}$	kWh/d	28,108
Financial benefit	$€_{E,LHV,savings}$	€/year	2,564,855

16.4 EXERCISES

Sustainability and environmental benefits of the anaerobic process (exercises 16.4.1-16.4.3)

Exercise 16.4.1

Explain what anaerobic digestion is and what are the redox potentials associated to it.

Exercise 16.4.2

List six advantages of anaerobic over aerobic wastewater treatment methods.

Exercise 16.4.3

Calculate the annual benefit ($€_{tot,savings}$) due to energy recovery as CH_4 when a brewery installs anaerobic treatment instead of activated sludge for wastewater treatment. Assume a flow of 5,000 m^3/d with a COD concentration of 4,500 mg/l. The efficiency of the anaerobic reactor is 80 %. The energy content of 1 kgCOD equals 13.5 MJ giving a theoretical electric potential of 3.8 kWh. The price of electricity is 0.095 €/kWh.

Choose from the following:

a) Approximately €253,300.
b) Approximately €475,000.
c) Approximately €791,670.
d) Approximately €2,253,000.

Microbiology of anaerobic conversions (exercises 16.4.4-16.4.6)

Exercise 16.4.4

Give an example of an acetogenic conversion reaction and explain why this is the most difficult step in the entire digestion chain.

Exercise 16.4.5

Describe three differences between archaea and bacteria.

Exercise 16.4.6

Make a diagram of the different steps in the AD process and give the main microbial groups involved in each step.

Predicting the CH_4 content (exercises 16.4.7-16.4.9)

Exercise 16.4.7

Explain the differences between COD and BOD.

Exercise 16.4.8

An anaerobic reactor treats a flow of 50 m^3/h with a COD concentration of 3.0 g/l and BOD of 2.6 g/l. Calculate the daily methane production assuming a reactor efficiency of 90 %; ignore biomass yield.

Exercise 16.4.9

Determine the theoretical CH_4 and CO_2 content of the biogas produced by the degradation of glycine.

Impacts of alternative electron acceptors (exercises 16.4.10-16.4.12)

Exercise 16.4.10

List three of the most common electron acceptors that can be found in wastewater (apart from oxygen) and explain how and why they affect the methanogenic process.

Exercise 16.4.11

Using the example given in Exercise 16.4.8, recalculate the methane production if a) the influent contains SO_4^{2-} in a concentration of 2 g/l; and b) the influent contains NO_3^- in a concentration of 2 g/l.

Exercise 16.4.12

List four of the main problems found in anaerobic reactors when the influent to be treated has a high concentration of sulphate.

Working with the COD balance (exercises 16.4.13-16.4.15)

Exercise 16.4.13

Why is COD generally used to make a balance over an anaerobic reactor? Choose the correct answer(s).

a) The COD entering and exiting the reactor is the same, so a (mass) balance can be performed.
b) COD is easier to measure than BOD.
c) The COD of all the compounds in the wastewater is known, so it is possible to identify which compounds are being degraded and which not.
d) Most industries have test kits to assess the COD.

Exercise 16.4.14

A batch test was done to determine the SMA activity of an anaerobic sludge at high acetate concentrations (in terms of COD). To perform the experiment a 250 ml reactor with a working volume of 200 ml was used and 4.25 g of sodium acetate trihydrate ($CH_3COONa \cdot 3H_2O$) was added. After three days of incubation at 35 °C, a total biogas volume of 455 Nml was found. Determine the COD concentration in the batch reactor and perform a COD balance of the system to conclude if all the acetate was converted into methane.

Exercise 16.4.15

A food-processing industry produces a fully soluble effluent with a flow of 1,000 m³/d, a COD concentration of 7.5 kgCOD/m³, and a biodegradability of 90 %. The effluent has a COD concentration of 1.0 kgCOD/m³ and the excess sludge a concentration of 10 % (w/v). Considering a sludge yield of 10%, a COD/VSS ratio of 1.42, and a VSS/TSS ratio of 0.6, calculate the following:

1) Treatment efficiency based on the COD removal.
2) Methane production and expected biogas yield.
3) Sludge production in m³/d.

Immobilization and sludge granulation (exercises 16.4.16-16.4.18)
Exercise 16.4.16
Anaerobic wastewater treatment has developed over a period of 30-40 years from large-scale completely mixed tank systems to high-rate sludge bed systems. What was the key for this development?

a) Reactors were heated so reaction rates and bioconversions increased.
b) Hydraulic retention time was uncoupled from solids retention time by effective sludge retention.
c) Toxicity and inhibition problems could be avoided by increased knowledge of microbial conversion processes.
d) Dutch industries worked closely together with Dutch universities making new reactor designs.

Exercise 16.4.17
List the two most important conditions required to achieve biomass granulation in a reactor.

Exercise 16.4.18
Explain and schematize the granulation process mechanism.

Anaerobic reactors (exercises 16.4.19-16.4.21)
Exercise 16.4.19
List the main features of a reactor that is considered to be an HRAR system.

Exercise 16.4.20
For what type of wastewater is an anaerobic contact process a good choice?

a) Mainly soluble wastewaters.
b) Wastewaters with temperatures < 20 °C.
c) Food-processing wastewaters with COD concentrations up to 10 g/l.
d) Wastewaters with a high content of suspended solids.

Exercise 16.4.21
An industry needs to treat chemical wastewater containing formaldehyde which is a biodegradable but very toxic compound. The industry managers ask you to recommend the proper technology for the treatment of this type of wastewater. What type of reactor would you choose?

a) An EGSB reactor with an external recirculation pump.
b) A UASB reactor with a multi-layer gas withdrawal system.
c) An anaerobic filter reactor.
d) An EGSB reactor without an external pump but with a conical inlet device.

Upflow anaerobic sludge blanket reactor (exercises 16.4.22-16.4.24)
Exercise 16.4.22
Under what circumstances might sludge bed systems be less successful? More than one answer is possible.

    a) When pulp & paper wastewater is treated.
    b) When wastewaters with high salt concentrations are treated.
    c) When the wastewater has a high content of fat (oily compounds) and suspended solids.
    d) When the anaerobic sludge does not form granules.

Exercise 16.4.23
An anaerobic sewage treatment plant (UASB system) treats a flow of 80,000 m^3/d of raw sewage. It has a biodegradable COD concentration of 550 mg/l. The UASB reactor is characterised by a COD removal efficiency of 70% and a methane production of 0.18 m^3/kgCOD removed. Calculate the daily methane production in m^3/d without considering biomass growth. Because of industrial discharges to the sewerage network, the influent sulphate concentration of the UASB reactor increases to 250 mg/l whereas the influent COD concentration and the flow remain the same. Calculate the daily methane production in m^3/day for the new situation.

Exercise 16.4.24
A UASB reactor with a conventionally designed (reverse funnel) gas-liquid-solids (GLS) separator is treating a wastewater flow of 500 m^3/h and a COD concentration of 8 kg/m^3. The circular reactor has a diameter of 30 m and a height of 6 m and treats the wastewater at 20 °C with a CH_4-COD recovery efficiency of 90 %. Determine whether the GLS is adequate for the withdrawal of the produced biogas, which has a CH_4 content of 70 %.

Anaerobic process kinetics (exercises 16.4.25-16.4.27)
Exercise 16.4.25
Why might overloading with non-acidified substrate result in a low pH inside an anaerobic reactor? More than one answer is possible.

    a) Because acidifiers have high growth rates even at low pH.
    b) Because the production of VFAs by acidifiers exceeds the consumption of VFAs by acetogens and methanogens.
    c) Because methanogens are inhibited by accumulating acids.
    d) Because excessive CO_2 production during *e.g.*, sugar fermentation lowers the pH.

Exercise 16.4.26
What will happen to the pH when an overload of sodium acetate (NaAc) is fed to the reactor?

    a) pH will increase.
    b) pH will drop.
    c) pH will stay exactly the same.
    d) pH will firstly drop and then increase.

Exercise 16.4.27

Which of the following statements are true according to the ADM1 kinetic model?

a) The ADM1 kinetic model provides an analysis of stoichiometric conversions: COD balance, charge balance, and element balance.
b) The ADM1 kinetic model includes kinetic rates mostly based on the Monod kinetic equation.
c) Each conversion step in the ADM1 kinetic model is expressed by a specific set of parameters.
d) The ADM1 kinetic model contains information regarding the biochemistry of each microorganism involved in the AD process.

Anaerobic treatment of domestic and municipal sludge (exercises 16.4.28-16.4.30)

Exercise 16.4.28

Give five advantages and five constraints of applying AD for the treatment of municipal wastewater.

Exercise 16.4.29

Calculate the solids retention time (SRT) of a 1,000 m^3 UASB reactor with an average sludge concentration of 35 $kgTSS/m^3$, having a weekly sludge discharge of 20 m^3 with a TSS concentration of 10 % (m/v). The reactor treats completely soluble wastewater and is operated at an HRT of 12 h producing an effluent of 200 mgTSS/l.

Exercise 16.4.30

Explain the effect of temperature on the SRT of the anaerobic reactors.

ANNEX 1: SOLUTIONS TO EXERCISES

Sustainability and environmental benefits of anaerobic process (solutions 16.4.1-16.4.3)

Solution 16.4.1

Anaerobic digestion (AD) is a mineralization process in which carbon acts both as electron donor and electron acceptor (redox reaction) and organic matter is eventually converted into the most oxidized and most reduced form of carbon, *i.e.*, a mixture of CO_2 and CH_4, also called biogas. This process occurs under reducing conditions and, therefore, negative potentials reaching -300 to -500 mV are required for the reactions to proceed.

Solution 16.4.2

1) No energy is required for the main reactions to occur, *e.g.*, avoidance of aeration and therefore aeration equipment.
2) Energy can be recovered in the form of biogas.
3) Minimal excess sludge production (reduction of up to 90%).
4) Sludge produced is already stabilized.
5) Compact reactor system owing to high applicable COD loading rates reaching 20-35 $kgCOD/m^3$.
6) Nutrients such as ammonium (NH_4^+) and phosphate (PO_4^{-3}) remain in the treated wastewater for possible recovery, *e.g.*, in ferti-irrigation.

Solution 16.4.3

d) Approximately €2,253,000.

Microbiology of anaerobic conversions (solutions 16.4.4-16.4.6)

Solution 16.4.4

$$CH_3CH_2CO_2^- + 3H_2O \rightarrow CH_3COO^- + HCO_3^- + H^+ + 3H_2$$

For instance, the propionate acetogenic reaction has a positive Gibb's free energy change of +76 kJ/mol and thus cannot proceed spontaneously. This is the most difficult step in the reaction because it only proceeds when H_2 is continuously removed from the solution, improving the reaction's thermodynamics, by lowering the Gibb's free energy change to negative values.

Solution 16.4.5

1) Some archaea, such as acetoclastic and hydrogenotrophic methanogens, produce methane, but bacteria cannot.
2) Bacteria contain peptidoglycan in the cell wall whereas archaea do not (they mainly contain pseudopeptidoglycan).
3) Bacterial cell membrane mainly consists of fatty acids with ester bonds whereas archaeal cell membrane contains phytanyls linked through ether bonds.

Solution 16.4.6

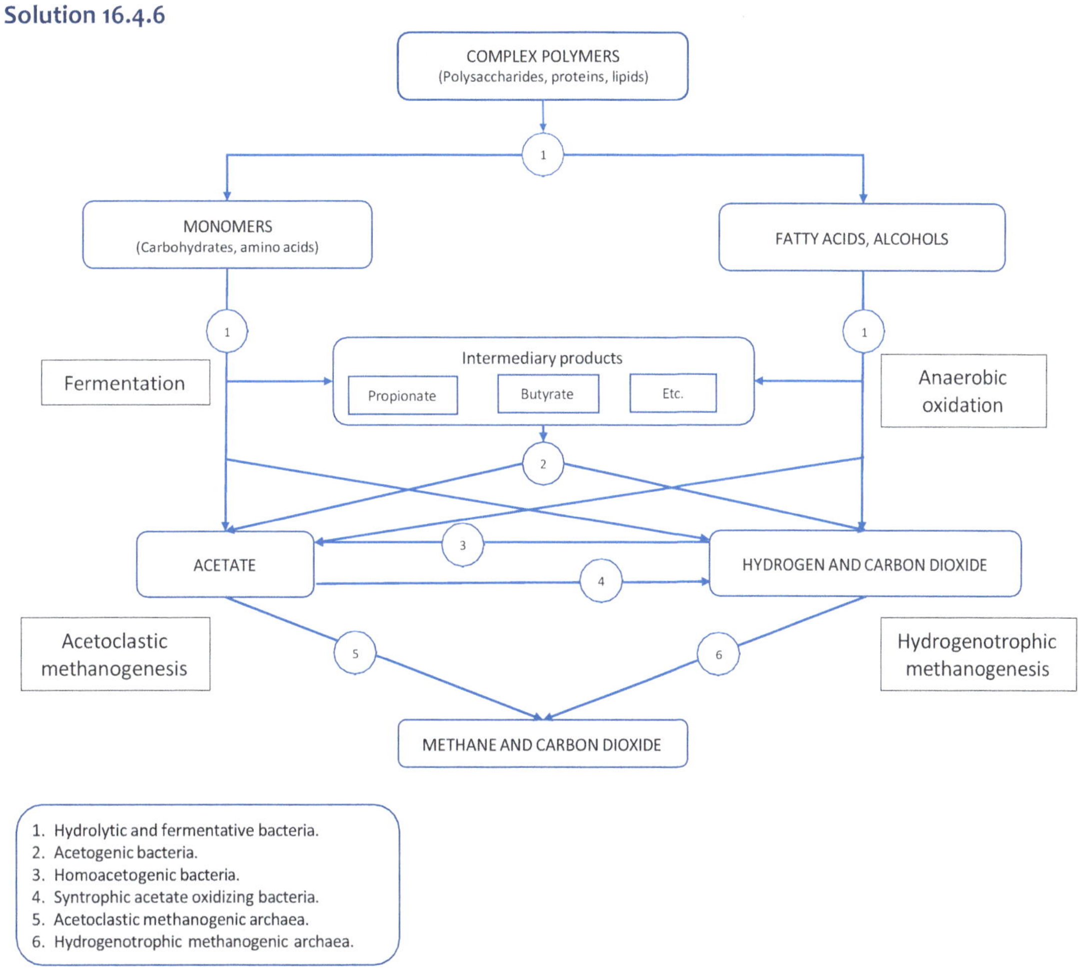

Figure 16.3 Different steps in the AD process and (physiological) microbial groups involved.

Predicting the CH₄ content (solutions 16.4.7-16.4.9)

Solution 16.4.7

COD stands for chemical oxygen demand. During the analysis all substances are measured that can be oxidized using a strong oxidant (dichromate). Both biodegradable and unbiodegradable matter is oxidized. Although more substances can be oxidized, the COD is mostly related to the organic compounds in a sample. It is important to remember that NH_4^+ is not oxidized and nor are quaternary ammonium salts, such as betaine. Experimentally, it is a quick test with an average duration of 3-4 h.

BOD stands for biochemical oxygen demand. It is a measurement of the quantity of substances, mainly organic matter, which can be oxidized by bacteria under aerobic conditions. This measurement considers only biodegradable matter, while NH_4^+ can also be oxidized. BOD is a lengthy test that usually takes 5, 7 or 10 days.

Solution 16.4.8

The daily methane production is 1,134 m^3/d.

$$CH_4 = 3,600 \ kgCOD/d \cdot 090 \cdot 0.35 \ m^3CH_4/kgCOD = 1,134 \ m^3CH_4/d$$

Solution 16.4.9

$CH_4 = 37.5 \ \%, CO_2 = 62.5 \ \%$

$$C_2H_5NO_2 + \left(2\frac{5}{4} - \frac{2}{2} + \frac{3 \cdot 1}{4}\right)H_2O \rightarrow \left(\frac{2}{2} + \frac{5}{8} - \frac{2}{4} - \frac{3 \cdot 1}{8}\right)CH_4 + \left(\frac{2}{2} - \frac{5}{8} + \frac{2}{4} + \frac{3 \cdot 1}{8}\right)CO_2 + NH_3$$

$$C_2H_5NO_2 + 0.5 \ H_2O \rightarrow 0.75 \ CH_4 + 1.25 \ CO_2 + NH_3$$

Impacts of alternative electron acceptors (solutions 16.4.10-16.4.12)

Solution 16.4.10

The ions sulphate (SO_4^{2-}), nitrate (NO_3^-), and oxidized iron (Fe^{3+}). The presence of these ions will decrease the methane production. This is because, thermodynamically, these molecules are better electron acceptors than carbon (C), meaning that *e.g.,* sulphate-reducing bacteria will outcompete methanogens in an anaerobic reactor and nitrate reduction can even completely halt methanogenesis.

Solution 16.4.11

With SO_4^{2-} present, the new daily methane production is 571 m^3/d.

$$FCOD_{SO4,red} = 1,200 \ m^3/d \cdot 2 \ kgSO_4^{2-}/m^3 \cdot 0.67 \ kgCOD/kgSO_4^{2-} = 1,608 \ kgCOD/d$$

$$FS_{CH4,prod} = \left(3,240 \ kgCOD/d - 1,608 \ kgCOD/d\right) \cdot 0.35 \ m^3CH_4/kgCOD = 571 \ m^3CH_4/d$$

With nitrate present, the new daily production will be 588 m^3/d:

$$FCOD_{NO3,rem} = 1,200 \ m^3/d \cdot 2 \ kgNO_3^-/m^3 \cdot 0.65 \ kgCOD/kgNO_3^- = 1,560 \ kgCOD/d$$

$$FS_{CH4,prod} = \left(3,240 \ kgCOD/d - 1,560 \ kgCOD/d\right) \cdot 0.35 \ m^3CH_4/kgCOD = 588 \ m^3CH_4/d$$

Solution 16.4.12

1) Methanogens can become toxified by high hydrogen sulfide (H_2S) concentrations.
2) The metallic parts of the reactors and pipelines corrode due to H_2S.
3) There is a decrease in the biogas quality due to the presence of H_2S.
4) There are bad odours in the vicinity of the reactor, due to the presence of hydrogen sulphide (H_2S).

Working with the COD balance (solutions 16.4.13-16.4.15)

Solution 16.4.13

a)   The COD entering and exiting the reactor is the same, so a (mass) balance can be performed.

Solution 16.4.14

COD concentration in the reactor 10 g COD/l (sodium acetate trihydrate COD = 0.47 g/g). Biogas produced 683 Nml, biogas expected 699 Nml (N = normal 0 °C and 1 atm). The CH_4 gap is 16 ml which agrees with a growth yield of 2.3 % for the methanogens, which is in the expected range. Therefore, we can consider that all the $COD_{acetate}$ was converted into CH_4.

Solution 16.4.15

Treatment efficiency based on COD removal = 86.7 %.

$$E(\%)_{COD} = \frac{6{,}500 \ \text{kgCOD/d}}{7{,}500 \ \text{kgCOD/d}} \cdot 100 = 86.7 \ \%$$

Methane production and expected biogas yield = 2,047.5 m^3CH_4/d and 2,925 m^3biogas/d.

$$FS_{CH4,prod} = 6{,}500 \ \text{kgCOD/d} \cdot (1 - 0.1) \cdot 0.35 \ \text{Nm}^3CH_4/\text{kgCOD} = 2{,}047.5 \ \text{Nm}^3CH_4/\text{d}$$

$$Q_{biogas} = 2{,}047.5 \ \text{Nm}^3CH_4/\text{d} \cdot \frac{1}{0.7} = 2{,}925 \ \text{Nm}^3\text{biogas/d}$$

Sludge production:

$$Q_W = 6{,}500 \ \text{kgCOD/d} \cdot 0.1 \ \frac{\text{kgCOD}}{\text{kgCOD}} \cdot \frac{1 \ \text{kgVSS}}{1.42 \ \text{kgCOD}} \cdot \frac{1 \ \text{kgTSS}}{0.6 \ \text{kgVSS}} \cdot \frac{m^3}{100 \ \text{kgTSS}} = 7.6 \ m^3\text{sludge/d}$$

Immobilization and sludge granulation (solutions 16.4.16-16.4.18)

Solution 16.4.16

b)   Hydraulic retention time was uncoupled from solids retention time by effective sludge retention.

Solution 16.4.17

1)   Dosage/feeding of the reactor with soluble substrates. 2) Reactor operation in an upflow mode with short hydraulic retention times.

Solution 16.4.18

1)   Nuclei growth, usually on inert organic or inorganic carriers. It is possible that the microorganisms aggregate in other already existing bacterial conglomerates.
2)   Film or aggregate formation and wash out of dispersed matter including microorganisms that were not retained in the granules.
3)   Formation of the 'first generation of granules'. These aggregates are mainly filamentous granules, voluminous and somewhat flocculent.
4)   Growth of the secondary nuclei, both in size and bacterial density.
5)   Aging or 'maturation' of the granules. In this step, compact and denser granules will displace the filamentous granules found in the initial stages.

Anaerobic reactors (solutions 16.4.19-16.4.21)

Solution 16.4.19

1) High retention of viable sludge in the reactor under operational conditions.
2) Sufficient contact between viable bacterial biomass and wastewater.
3) High reaction rates and absence of serious transport limitations.
4) Sufficient adaptation by the biomass to the prevailing conditions in the reactor.
5) Prevalence of favourable environmental conditions for all the required microorganisms in the reactor.

Solution 16.4.20

d) Wastewaters with a high content of suspended solids.

Solution 16.4.21

a) An EGSB reactor with external recirculation pump.

Up-flow anaerobic sludge blanket reactor (solutions 16.4.22-16.4.24)

Solution 16.4.22

b) When wastewaters with high salt concentrations are treated.
c) When the wastewater has a high content of fat (oily compounds) and suspended solids.
d) When the anaerobic sludge does not form granules.

Solution 16.4.23

Methane production, $FQ_{CH4} = 5{,}544$ m³CH₄/d.

Methane production after sulphate presence, $FQ_{CH4} = 3{,}132$ m³CH₄/d.

$$Q_{CH4} = 80{,}000 \ \text{m}^3/\text{d} \cdot 0.55 \ \text{kgCOD/m}^3 \cdot 0.7 \cdot 0.18 \ \text{Nm}^3\text{CH}_4/\text{kgCOD} = 5{,}544 \ \text{Nm}^3\text{CH}_4/\text{d}$$

$$FCOD_{SO4,red} = 80{,}000 \ \text{m}^3/\text{d} \cdot 0.25 \ \text{kgSO}_4^{2-}/\text{m}^3 \cdot 0.67 \ \text{kgCOD/kgSO}_4^{2-} -- = 13{,}400 \ \text{kgCOD/d}$$

After sulphate presence:

$$Q_{CH4} = \big((44{,}000 \cdot 0.7) \ \text{kgCOD/d} - 13{,}400 \ \text{kgCOD/d}\big) \cdot 0.18 \ \text{Nm}^3\text{CH}_4/\text{kgCOD} = 3{,}132 \ \text{Nm}^3\text{CH}_4/\text{d}$$

Solution 16.4.24

The maximum allowable superficial gas load is 2-3 m/h and calculated v_{up} of the gas is 2.73; therefore, the GLS is adequate for this reactor.

$$v_{up} = 8 \cdot \frac{90}{100} \cdot \frac{0.35}{0.70} \cdot \frac{(273 + 20)}{273} \cdot 0.71 = 2.73 \ \text{m/h}$$

Anaerobic process kinetics (solutions 16.4.25-16.4.27)

Solution 16.4.25

a) Because acidifiers have high growth rates even at low pH.

b) Because the production of VFAs by acidifiers exceeds the consumption of VFAs by acetogens and methanogens.

c) Because methanogens are inhibited by accumulating acids.

d) Because excessive CO_2 production lowers the pH.

Solution 16.4.26

a) pH will increase.

Solution 16.4.27

a) The ADM1 kinetic model provides an analysis of stoichiometric conversions: COD balance, charge balance, and element balance.

b) The ADM1 kinetic model includes kinetic rates mostly based on the Monod kinetic equation.

c) Each conversion step in the ADM1 kinetic model is expressed by a specific set of parameters.

Anaerobic treatment of domestic and municipal sludge (solutions 16.4.28-16.4.30)

Solution 16.4.28

Advantages:

1) Savings of up to 90 % in operational cost, no aeration needed.
2) 40 to 60 % reduction in investment costs as a simpler treatment train is needed.
3) Possibility of energy recovery as methane.
4) Reactor configurations that allow decentralized schemes.
5) A well-designed reactor, such as a UASB reactor, can partly filter helminth eggs.

Constraints:

1) Requirement of a polishing post-treatment.
2) Dissolved CH_4 in the effluent that will eventually escape into the atmosphere.
3) The CH_4 is not usually recovered or flared.
4) Little experience with full-scale application at moderate temperatures.
5) Other reduced gases such as H_2S may escape, causing odour problems in the vicinity of the treatment plant.

Solution 16.4.29

SRT = 51 d.

$$SRT = \frac{35 \text{ kgTSS/m}^3 \cdot 1{,}000 \text{ m}^3}{\left(286 \text{ kgTSS/d} + 400 \text{ kgTSS/d}\right)} = 51 \text{ d}$$

536

Solution 16.4.30

There is an inverse relationship between the operating temperature and the required SRT. This effect is because, as a rule of thumb, the SRT should be at least 3 times the doubling time of the biomass responsible for the rate-limiting step, which can be either the methanogens or the hydrolytic bacteria. When the temperature in the reactor is increased, the reaction and growth rates become higher, therefore decreasing the required time for sludge stabilisation or mineralisation. However, there is a maximum limit for the operating temperature, which is linked to the maximum temperature span of the respective organisms. Raising the temperature beyond the maximum temperature results in net decay.

NOMENCLATURE

Abbreviation	Description
HHV	High heating value energy
HLR	Hydraulic loading rate
HRT	Hydraulic retention time
LHV	Low heating value energy
OLR	Organic loading rate

Symbol	Description	Unit
€$_{aer,savings}$	Cost savings due to avoidance of aeration	€
€$_{CO2,savings}$	Cost savings due to carbon credits	€
€$_{E,LHV,savings}$	Cost savings due to electrical energy production at low heating value	€
€$_{kWh}$	Electricity price per kWh	€
€$_{tot,savings}$	Total cost savings	€
A_r	Area of the reactor	m^2
$A_{r,unit}$	Area per reactor unit	m^2
A_{sdb}	Area of the sludge drying bed	m^2
$COD_{available}$	COD available for CH_4 production or SO_4^{2-} reduction	kgCOD/d
COD_b	Biodegradable COD	mgCOD/l
$COD_{b,e}$	Biodegradable COD in the effluent (residual)	mgCOD/l
$COD_{b,i}$	Biodegradable COD in the influent	mgCOD/l
COD_{conv}	COD converted	kgCOD/d
COD_e	Total COD concentration in the effluent	mgCOD/l
COD_i	Total COD concentration in the influent	mgCOD/l
$COD_{SO4,red}$	COD concentration used for SO_4^{2-} reduction	kgCOD/d
$COD_{VSS,e}$	COD concentration in the effluent composed of volatile suspended solids	mgCOD/l
$E(\%)_{CHP}$	Energy generation efficiency of the CHP generator	%
$E(\%)_{COD}$	COD removal efficiency	%
$E_{kWh,LHV}$	Energy production considering low heating value conditions	kWh/d
$f_{CH4,biogas}$	CH_4 content in biogas	Nm^3CH_4/m^3biogas

$f_{CH4,COD}$	Stoichiometric CH_4 generated per COD removed	$l/gCOD$
$f_{CH4,COD,actual}$	Methanised amount of COD removed	$\%$
$f_{CH4,COD,rem}$	CH_4 production per COD removed in field conditions	$Nm^3CH_4/kgCOD$
$f_{CH4,COD,STP}$	CH_4/COD ratio at STP	$Nm^3CH_4/kgCOD$
$f_{CO2,kWh}$	CO_2 production/energy consumption ratio	$kgCO_2/kWh$
$f_{COD,SO4}$	COD/SO_4^{2-} ratio	$gCOD_{req}/g\ SO_4^2$
$FCOD_{available}$	COD mass flow available for CH_4 production or SO_4^{2-} reduction	$kgCOD/d$
$FCOD_{CH4,e}$	Flux or mass flow rate of methane lost through the effluent as COD	$kgCOD/d$
$FCOD_{conv}$	Flux or mass flow rate of COD converted	$kgCOD/d$
$FCOD_{conv,CH4}$	Flux or mass flow rate of COD converted to methane	$kgCOD/d$
$FCOD_{conv,VSS}$	Flux or mass flow rate of COD converted to volatile suspended solids	$kgCOD/d$
$FCOD_e$	COD flux or mass flow rate in the effluent of the reactor	$kgCOD/d$
$FCOD_i$	COD flux or mass flow rate into the reactor	$kgCOD/d$
$FCOD_i$	Flux or mass flow rate of total COD into the reactor	$kgCOD/d$
$FCOD_{NO3,rem}$	COD mass flow used for NO_3^- removal (denitrification)	$kgCOD/d$
$FCOD_{SO4,red}$	COD mass flow used for SO_4^{2-} reduction	$kgCOD/d$
$FCOD_w$	COD removed with excess sludge	$kgCOD/d$
f_{CV}	COD/VSS ratio of the sludge	$gCOD/gVSS$
$f^{HHV}_{MJ,CH4}$	Energy generation/CH_4 ratio at a high heating value (HHV)	MJ/m^3CH_4
$f_{kWh,COD,aer}$	Energy usage to COD removal ratio under aerobic conditions	$kWh/kgCOD$
$f^{LHV}_{MJ,CH4}$	Energy generation/CH_4 ratio at a low heating value (LHV)	MJ/m^3CH_4
$f_{MJ,kWh}$	Megajoule per kWh ratio	MJ/kWh
FQ_{CH4}	Methane mass flow rate	m^3/d
$FS_{CH4,e}$	Methane flux or mass flow dissolved in the effluent	m^3/d
$FS_{CH4,prod}$	Methane production mass flow rate under field conditions	m^3/d
$FS_{CH4,prod,STP}$	Methane production mass flow rate under STP conditions	m^3/d
$FS_{CH4,recov}$	Methane mass flow recovered via biogas exhaust under field conditions	m^3/d
$FS_{CH4,recov,STP}$	Methane mass flow recovered via biogas exhaust under STP conditions	m^3/d
f_{VT}	Sludge VSS/TSS ratio	$kgVSS/kgTSS$
h_r	Reactor height	m
$h_{Xt,max}$	Maximum height wet sludge in the drying bed	m
L_r	Length of the reactor	m
p_f	Peak factor	$-$
$p_{f,t}$	Peak duration	h/d
Q_{biogas}	Biogas mass flow rate	Nm^3/h
Q_e	Effluent flow rate	m^3/d
Q_i	Influent flow rate (daily average)	m^3/d
Q_p	Peak flow	m^3/d
Q_w	Flow rate of the excess of solids	m^3/d

$S_{CH4,e}$	Methane concentration dissolved in the effluent	m^3/d
S_{H2S}	Dissolved sulphide concentration	mgH_2S/l
$S_{SO4,i}$	Sulphate (SO_4^{2-}) concentration in the influent	gSO_4^{2-}/m^3
$t_{dry,cycle}$	Drying cycle	h
V_r	Reactor volume	m^3
$V_{r,max}$	Maximum volume of the reactor	m^3
$V_{r,unit}$	Maximum volume per reactor unit	m^3
$X_{t,w,dry}$	Dry weight content of excess sludge	%
$Y_{COD,An}$	Anaerobic sludge yield	$gCOD_{VSS}/gCOD$
$Y_{TSS,COD}$	Net yield of solids per COD removed	%

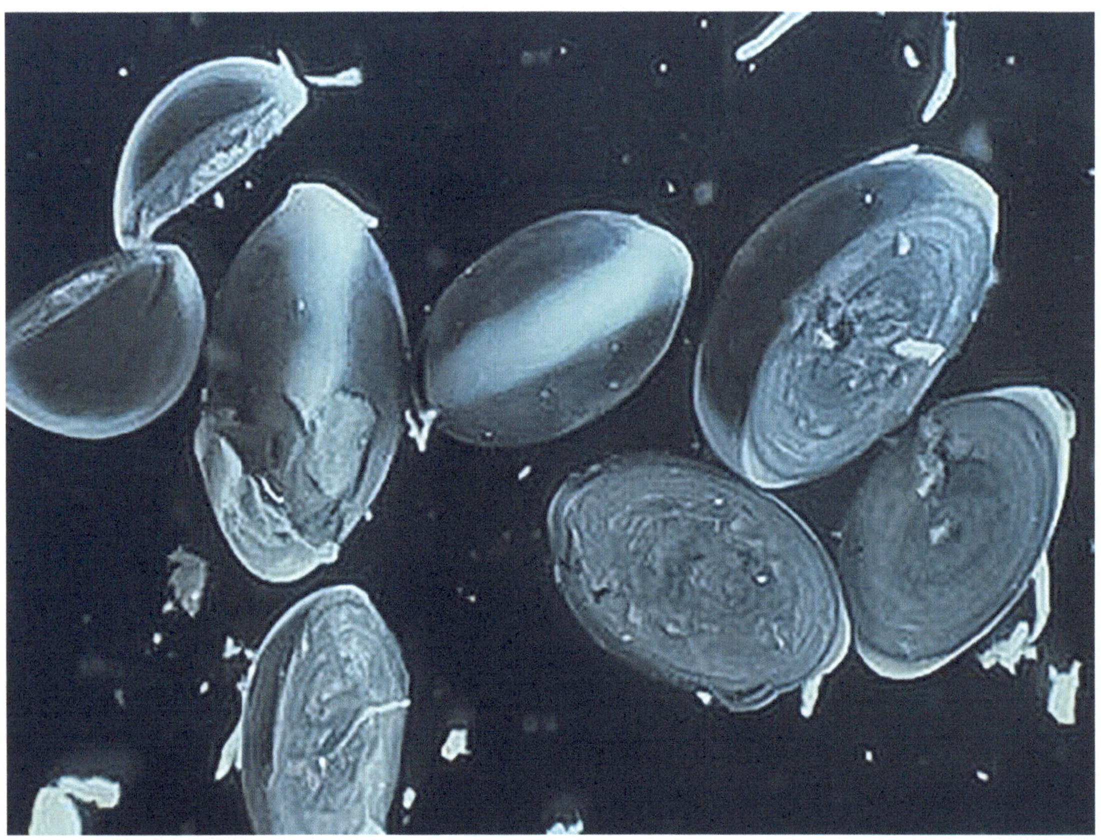

Anaerobic sludge (photo: J.B. van Lier).

doi: 10.2166/9781789062304_0539

17

Biofilm modelling and biofilm reactors

Eberhard Morgenroth and Kim H. Sørensen

17.1 INTRODUCTION

Chapters 17 and 18 in *Biological Wastewater Treatment: Principles, Modelling and Design* (*Chen et al.*, 2020) provide the background for analysing process kinetics and for biofilm reactors, respectively. The basic principles using activated sludge systems for organic carbon removal, nitrification, denitrification, and biological phosphorus removal are introduced in chapters 4-6. The same process objectives can be achieved in biofilm systems where the retention of active biomass does not require a settler, but active biomass grows in biofilms. This exercise chapter provides selected problems that are focused on detailed biofilm kinetics or on overall biofilm systems. Some of the exercises require the application of a simulator[1] to evaluate multi-component diffusion and competition between different groups of organisms.

17.2 LEARNING OBJECTIVES

After reading chapters 17 and 18 and completing these exercises, the reader should be able to do the following:

- Apply simple biofilm kinetics (first or zero-order rates inside the biofilm) to calculate substrate concentrations inside a biofilm, substrate flux, and overall biofilm reactor performance.
- List the information needed for the design, operation, and evaluation of biofilm reactors.
- Derive mass balance equations to design a biofilm reactor based on influent wastewater characteristics, treatment target, and biofilm parameters.
- Calculate the extent of substrate penetration based on bulk-phase substrate concentrations and biofilm parameters.

---

[1] Most commercial simulators include biofilm compartments that can be applied for these problems. Another approach is to use AQUASIM that is freely available at https://www.eawag.ch/de/abteilung/siam/software/.

- Determine whether an electron donor or electron acceptor are limiting the performance of a biofilm reactor and calculate the flux of both.
- Discuss the mechanisms of competition between different groups of organisms in a biofilm and how reactor design and operation can be used to influence this competition.
- Explain their reasons for choosing different types of reactors.
- Explain the mechanisms of retention and control of biofilms.
- Calculate the influence of small changes in (*i*) biofilm surface area, (*ii*) overall biofilm thickness, and (*iii*) boundary layer thickness on the biofilm reactor performance.
- Use a numerical model to evaluate the performance of a moving bed biofilm reactor and how it can be improved through changes in the reactor operation.
- Design a biofilm reactor based on recommended surface loading rates and then verify the design by mathematical modelling.

17.3 EXAMPLES

The reader should review examples 17.1 – 17.11 in Chapter 17 and the four levels of design in Section 18.3.1 in Chen *et al.*, 2020.

17.4 EXERCISES

Exercise 17.4.1

Simple biofilm kinetics: substrate fluxes into thick and thin biofilms assuming first-order kinetics.

Bulk-phase organic biodegradable substrate concentrations are 5 mgCOD/l. Calculate the substrate fluxes for a thick (500 µm) and a thin (50 µm) biofilm assuming first-order kinetics if $k_{1,F} = 2.4$ m^3/gCOD.d, $X_F = 10{,}000$ mgCOD/l, and $D_F = 8 \cdot 10^{-5}$ m^2/d.

Exercise 17.4.2

Simple biofilm kinetics: substrate penetration into thick and thin biofilms assuming first-order kinetics.

Calculate the substrate concentration at the base of the thick and the thin biofilms from Exercise 17.0. Then, calculate the distance from the surface of the biofilm where the substrate concentration is exactly 1 mgCOD/l.

Exercise 17.4.3

Biofilm kinetics: multi-component diffusion.

Calculate the organic substrate and electron acceptor (oxygen or nitrate) fluxes into a 500 µm thick biofilm assuming bulk-phase organic biodegradable substrate concentrations of 20 mg COD/l, zero-order kinetics, $k_{0,F} = 9.52$ 1/d, $X_F = 10{,}000$ mgCOD/l, and $Y_{OHO} = 0.63$ gCOD/gCOD. Diffusion coefficients are $D_{F,COD} = 8 \cdot 10^{-5}$ m^2/d, $D_{F,O2} = 2.1 \cdot 10^{-4}$ m^2/d, $D_{F,NO3} = 1.6 \cdot 10^{-4}$ m^2/d. Calculate the flux of the organic substrate and oxygen for an aerobic reactor (bulk-phase oxygen concentrations of 5 mg O$_2$/l). Compare your results with an anoxic reactor (bulk-phase nitrate concentrations of 5 mg NO$_3^-$-N/l).

Exercise 17.4.4

Biofilm kinetics: predict biofilm thickness.

The bulk-phase organic biodegradable substrate concentrations are 5 mgCOD/l and degradation is COD-limited. Calculate the thickness of the biofilm if the rate of detachment linearly depends on the biofilm thickness ($u_{d,S} = k_d L_F$), $k_d = 0.8$ 1/d, $k_{0,F} = 9.52$ 1/d, the yield is 0.63 gCOD/gCOD, the rate of inactivation is 0.5 1/d, $X_F = 10{,}000$ mgCOD/l, and $D_F = 8 \cdot 10^{-5}$ m^2/d. How will the biofilm thickness change if the bulk-phase

COD concentration, biofilm density, or the detachment rate constant are increased by a factor of two (evaluate one factor at a time)?

Exercise 17.4.5
Biofilm kinetics: influence of the biofilm thickness.

Evaluate the influence of the biofilm thickness on substrate removal in a biofilm. You can assume the bulk-phase concentration of the organic biodegradable substrate concentration to be 10 mgCOD/l and kinetic parameters are as used in Figure 17.8 in Chen *et al.*, 2020. You can assume that removal is limited by the organic substrate and not by oxygen. However, you need to determine what type of kinetics applies: first-order kinetics, zero-order partial penetration, or zero-order full penetration. You can ignore the influence of an external mass transport limitation (*i.e.* $R_L = 0$). Estimate the substrate flux for four different steady-state biofilm thicknesses of 10, 50, 100, and 800 μm. To what extent is this substrate flux into the biofilm reduced due to mass transport limitations compared to a substrate flux assuming that all of the biomass would be exposed to bulk-phase concentrations? Compare your limiting substrate fluxes.

Exercise 17.4.6
Simple reactor design: design of an MBBR based on flux.

Design an MBBR for nitrification assuming an ammonia flux of 1 gN/m^2.d. The influent flow rate is 2,000 m^3/d and the influent TKN concentration is 30 mgN/l. The supplier provides a specific surface area for the carrier of 500 m^2/m^3 and a maximum fill ratio of 60 %. Determine the necessary volume of the MBBR to reach an effluent ammonia concentration of 1 mgN/l.

Exercise 17.4.7
Predict reactor performance: an MBBR for carbon oxidation.

You are operating an MBBR that can be assumed to be completely mixed. The concentration of the soluble biodegradable organic substrate in the MBBR is 80 mg COD/l and the concentration of oxygen is 6 mg/l. Both the organic substrate and the oxygen diffuse into the biofilm where the organic matter is mineralized while consuming oxygen. You can assume zero-order kinetics and a deep biofilm in the MBBR. The zero-order removal rate for oxygen in this biofilm is equal to $k_{0,F} \cdot X_F$ where $k_{0,F} = 9$ gO$_2$/gCOD.d and $X_F = 14,000$ gCOD/m^3. You can assume that the diffusion coefficient of the substrate is equal to the diffusion coefficient of acetate and that the net biomass yield coefficient is equal to 0.4 gCOD/gCOD. What is the limiting substrate in the MBBR; is it the electron donor or the electron acceptor? How far does the limiting substrate penetrate into the biofilm (in μm)? Calculate the flux of the limiting substrate into the biofilm. Then calculate the flux of the non-limiting substrate into the biofilm.

Now assume that the influent organic substrate concentration is 120 mgCOD/l. Determine the required volume of your reactor, assuming a specific surface area (a$_F$) of 300 m^2/m^3 and an inflow of 10,000 m^3/d.

Exercise 17.4.8
Predict reactor performance: influence of bulk-phase oxygen concentrations on an MBBR.

Repeat Exercise 17.0 with bulk-phase concentrations of 2, 4, 8, and 10 mgO$_2$/l and compare the results with your solution, assuming a bulk-phase oxygen concentration of 6 mgO$_2$/l.

Exercise 17.4.9

Predict reactor performance: multiple reactors in series for carbon oxidation.

Evaluate the influence of mixing conditions on the design of a biofilm reactor operated for carbon oxidation. Your influent flow rate is 8,000 m^3/d containing 400 mgCOD/l that is assumed to be biodegradable and soluble. You are aiming for effluent COD concentrations of 10 mgCOD/l. Bulk-phase oxygen concentrations are controlled to 2 mg O_2/l. Assume a zero-order partially penetrated biofilm with $k_{0,F}$ = 9.52 1/d, X_F = 30 gVSS/l ($\approx$ 43 gCOD/l), and a yield of 0.4 gCOD/gCOD. You can assume that the diffusion coefficient of the substrate is equal to the diffusion coefficient of acetate. The supplier of biofilm carriers provides a specific surface area for the carrier of 500 m^2/m^3 and a maximum fill ratio of 60 %. Calculate the required volume for an MBBR assuming a single mixed reactor. Repeat your calculations now assuming three equally sized CSTR in series. How does the overall reactor volume compare for a single reactor or three reactors in series? How would the calculations change if the required effluent concentration were 1 mgCOD/l instead of 10 mgCOD/l?

Exercise 17.4.10

Reactor design and performance: design of a biofilm reactor (MBBR).

You are designing a biofilm reactor to treat a waste stream of 20,000 m^3/d containing soluble degradable organic substrate at a concentration of 800 mgCOD/l. Your target effluent organic substrate concentration is 2 mgCOD/l. Determine the volume of the MBBR if you can assume first-order kinetics and parameters as in Exercise 17.0 and a biofilm thickness of 300 μm, a yield of 0.4 gCOD/gCOD, and a specific surface of your biofilm support in the reactor of 300 m^2/m^3. Assume for all your calculations that removal is limited by the organic substrate and not by oxygen. Calculate the required hydraulic retention time and the oxygen demand in terms of oxygen flux into the biofilm (in gO_2/m^2.d) and in terms of overall oxygen demand (in gO_2/d). How would the size of the reactor change if the target effluent concentration were 4 mgCOD/l? How would the overall volume of the MBBR change if you built two equal-size MBBRs in series instead of one large completely mixed MBBR? Discuss the validity of the assumptions that were made in your different calculations.

Exercise 17.4.11

Reactor design and model-based evaluation: design and verification of an MBBR for carbon oxidation and nitrification.

Design an MBBR for an influent waste stream of 15,000 m^3/d containing 400 mgCOD/l and 30 mgTKN/l. Effluent requirements are 30 mgCOD/l and 0.5 mgNH$_4^+$-N/l.

a. What additional information do you need for your design calculations?
b. Estimate the volume of a single-tank MBBR using the design surface loadings from Table 18.3 in Chen *et al.*, 2020
c. Implement your MBBR in simulation software and evaluate your design from b. for steady-state conditions. Note: your results will depend on the specific model implementation of your software and the default parameters that you are using.
d. Evaluate the influence of variable influent loading on effluent COD and ammonia concentrations.
e. Compare the performance of the single-tank MBBR with a three-tank MBBR with the same overall volume.
f. How would you combine the MBBR with solids removal and why?

Exercise 17.4.12

Reactor design and model-based evaluation: design and verification of an MBBR for carbon oxidation and nitrification/denitrification.

Repeat Exercise 17.0 now with the goal of achieving < 10 mg total nitrogen in the effluent by introducing denitrification.

Exercise 17.4.13

Qualitative evaluation: systematic comparison of biofilm reactor types.

The biofilm reactors described in Chapter 18 have different characteristics. Complete Table 17.1 below with brief statements.

Table 17.1

		Trickling filter	RBC	Biofilters	MBBR	IFAS	AGS	MABR
a.	Are biofilm carriers fixed within the reactor or suspended?							
b.	Does this type of reactor require backwashing to remove excess biofilm? (Y/N)							
c.	Does the suspended biomass contribute significantly to overall removal? (Y/N)							
d.	If there is aerobic growth, how is oxygen supplied?							
e.	Typical specific surface area in m^2/m^3							
f.	Suitability of the system for nitrogen removal (*i.e.* nitrification and denitrification)? (Y/N)							
g.	Suitability of the system for enhanced biological phosphorus removal? (Y/N)							
h.	Ease of operation (very simple/complex)							

Exercise 17.4.14

Systems analysis: variation of parameters.

You are operating an MBBR for the aerobic oxidation of organic carbon. You can assume that removal is limited by the organic substrate and you have a thick biofilm (*i.e.* organic substrate concentrations at the base of the biofilm are very low). Complete Table 17.2 below to indicate what effect a small increase in a variable in the first column has on the system performance (columns two, three, and four).

Table 17.2

Variable	Effluent organic substrate concentration	Substrate flux into the biofilm	Removal efficiency
Influent Q			
Influent organic substrate concentration			
D_F			
X_F			
Biofilm thickness			

ANNEX 1: SOLUTIONS TO EXERCISES
Solution 17.4.1
Simple biofilm kinetics: substrate fluxes into thick and thin biofilms assuming first-order kinetics.

With equations 17.14 and 17.15:

$L_{crit} = 57.7$ μm
$J_{LF} = 6.93$ g/m^2.d (for $L_F = 500$ μm) or 4.85 g/m^2.d (for $L_F = 50$ μm)

Solution 17.4.2
Simple biofilm kinetics: substrate penetration into thick and thin biofilms assuming first-order kinetics).
With Equation 17.13:

$S_F(x = L_F) = 0.00173$ mgCOD/l (for $L_F = 500$ μm) or 3.57 mgCOD/l (for $L_F = 50$ μm)

By trial and error (or using a mathematical solver) using Equation 17.13 find

$S_F(x = 92.9$ μm$) = 1$ mgCOD/l (for $L_F = 500$ μm)

For the 50 μm thick biofilm the substrate concentration is higher than 1 mgCOD/l throughout the biofilm.

Solution 17.4.3
Biofilm kinetics: multi component diffusion.
Determine the limiting substrate for aerobic oxidation of organic substrate with Equation 17.64

$\gamma_{e.d.,e.a.} = 0.75 \rightarrow$ electron donor is potentially limiting

Calculate COD flux using Equation 17.28

$J_{LF,COD} = 17.45$ gCOD/m^2.d

Confirm partial penetration of COD with Equation 17.25

$\beta = 0.37 < 1 \rightarrow$ partial penetration

Calculate oxygen flux with Equation 17.65

$J_{LF,O2} = 6.46$ g O_2/m^2.d

Now evaluate anoxic oxidation of organic substrate with Equation 17.64

$\gamma_{e.d.,c.a.} = 0.86 \rightarrow$ electron donor is potentially limiting

As is the case with aerobic oxidation, COD is the limiting substrate. Thus, the same COD flux as for the aerobic case. Calculate nitrate flux using Equation 17.68 ($J_{LF,NO3} = J_{LF,COD} (1 - Y_{OHO})/(2.86$ gCOD/gNO$_3$-N)

$J_{LF,NO3} = 2.26$ gNO$_3$-N/m^2.d

Solution 17.4.4
Biofilm kinetics: predict biofilm thickness.

With Equation 17.76

Table 17.3

Parameter	Base case	$S_B = 10$ mgCOD/l	$X_F = 20{,}000$ mgCOD/l	$k_d = 1.6$ 1/d
L_F	423 µm	598 µm	299 µm	262 µm
Change compared to base case	100 %	+41 %	−29 %	−38 %

Solution 17.4.5
Biofilm kinetics: influence of biofilm thickness.

Calculate substrate fluxes with equations 17.14 and 17.15 (first-order), 17.28 (zero-order partially penetrated), and 17.31 (zero-order fully penetrated) and choose the smallest flux (highlighted in red in Table 17.4).

Table 17.4

L_F	10 µm	50 µm	100 µm	800 µm
$J_{LF,1}$	2.35 gCOD/m^3.d	9.62 gCOD/m^3.d	12.9 gCOD/m^3.d	13.8 gCOD/m^3.d
$J_{LF,0,p}$	12.3 gCOD/m^3.d	12.3 gCOD/m^3.d	12.3 gCOD/m^3.d	12.3 gCOD/m^3.d
$J_{LF,0,f}$	0.952 gCOD/m^3.d	4.76 gCOD/m^3.d	9.52 gCOD/m^3.d	76.2 gCOD/m^3.d
$min(J_{LF})$	0.952 gCOD/m^3.d	4.76 gCOD/m^3.d	9.52 gCOD/m^3.d	12.3 gCOD/m^3.d
β	13.0	2.59	1.30	0.16
Type of kinetics	Zero-order fully penetrated	Zero-order fully penetrated	Zero-order fully penetrated	Zero-order partially penetrated
Extent of mass transport limitations	No influence, fully penetrated biofilm	No influence, fully penetrated biofilm	No influence, fully penetrated biofilm	Only 16 % of the conversion compared to where all biomass is exposed to bulk-phase concentrations

Solution 17.4.6

Simple reactor design: design of an MBBR based on flux.
With the ammonia flux given in the problem statement calculate the necessary surface area and volume with equations 18.2 and 18.3.

$A_F = 58,000 \ \text{m}^2$
$V_R = 193.3 \ \text{m}^3$

Solution 17.4.7

Predict reactor performance: an MBBR for carbon oxidation.
With Equation 17.25

$\gamma_{\text{e.d.,e.a.}} = 1.9 \rightarrow$ process is potentially limited by oxygen (electron acceptor)

With Equation 17.26

Penetration depth for oxygen $= \beta_{O2} \cdot L_F = 126 \ \mu\text{m}$

With Equation 17.28

$J_{LF,O2} = 15.9 \ \text{gO}_2/\text{m}^2.\text{d}$

With Equation 17.65

$J_{LF,COD} = 26.5 \ \text{gCOD}/\text{m}^2.\text{d}$

With equations 18.2 and 18.3

$A_F = 15,091 \ \text{m}^2$
$V_R = 50.3 \ \text{m}^3$

Solution 17.4.8

Predict reactor performance: influence of bulk-phase oxygen concentrations on MBBR.

Table 17.5

Bulk-phase O_2 concentration	2 mgO$_2$/l	4 mgO$_2$/l	6 mgO$_2$/l	8 mgO$_2$/l	10 mgO$_2$/l
$\gamma_{\text{e.d.,e.a.}}$	3.29	2.32	1.90	1.64	1.47
$\beta_{O2} \cdot L_F$	72.9 μm	103.1 μm	126.2 μm	145.7 μm	162.9 μm
$J_{LF,O2}$	9.2 gO$_2$/m^2.d	13.0 gO$_2$/m^2.d	15.9 gO$_2$/m^2.d	18.4 gO$_2$/m^2.d	20.5 gO$_2$/m^2.d
$J_{LF,COD}$	15.3 gCOD/m^2.d	21.6 gCOD/m^2.d	26.5 gCOD/m^2.d	30.6 gCOD/m^2.d	34.2 gCOD/m^2.d
A_F	26,138 m^2	18,482 m^2	15,091 m^2	13,069 m^2	11,689 m^2
V_R	87.1 m^3	61.6 m^3	50.3 m^3	43.6 m^3	39.0 m^3

Solution 17.4.9

Predict reactor performance: multiple reactors in series for carbon oxidation.

With Equation 17.64

$\gamma_{e.d.,e.a.} = 1.35 \rightarrow$ oxygen is potentially limiting

With Equation 17.28

$J_{LF,O2} = 12.8 \ gO_2/m^2.d$

With Equation 17.65

$J_{LF,COD} = 21.4 \ gCOD/m^2.d$

With equations 18.2 and 18.3

$A_F = 146,024 \ m^2$
$V_R = 486.7 \ m^3$

When there are three CSTR in series, the overall volume will not change for an effluent concentration of 10 mg COD/l as in each of the three reactors the conversion will be oxygen-limited, resulting in the same flux for oxygen and substrate in each of the three reactors.

For effluent COD concentrations of 1 mgCOD/l the system will be COD-limited and no longer oxygen-limited.

With Equation 17.64

$\gamma_{e.d.,e.a.} = 0.135 \rightarrow$ oxygen is potentially limiting

With Equation 17.28

$J_{LF,COD} = 7.85 \ gCOD/m^2.d$

With Equation 17.65

$J_{LF,O2} = 4.71 \ gO_2/m^2.d$

With equations 18.2 and 18.3

$A_F = 406,805 \ m^2$
$V_R = 1,356 \ m^3$

Solution 17.4.10

Reactor design and performance: design of a biofilm reactor (MBBR).

Table 17.6

Case	$S_B = 2$ mgCOD/l (one tank)	$S_B = 4$ mgCOD/l (one tank)	$S_B = 2$ mgCOD/l (two tanks)	Equation
L_{crit}	57.7 µm	57.7 µm	57.7 µm	17.14
$J_{LF,COD}$	2.77 gCOD/m².d	5.54 gCOD/m².d	55.4 gCOD/m².d (first tank) 2.77 gCOD/m².d (second tank)	17.15
A_F	5,759,422 m²	2,872,494 m²	274,258 m² (first tank) 274,258 m² (second tank) 548,516 m² (both tanks)	18.2
V_R	19,198 m³	9,575 m³	914 m³ (first reactor) 914 m³ (second reactor) 1,828 m³ (both reactors)	18.3
HRT	0.96 d	0.48 d	0.09 d	
$J_{LF,O2}$	1.66 gO₂/m².d	3.33 gO₂/m².d	33.3 gO₂/m².d (first tank) 1.66 gO₂/m².d (second tank)	17.65
Oxygen demand	9,576 kgO₂/d	9,552 kgO₂/d	9,576 kgO₂/d	
Comments			WARNING: Two of the underlying assumptions are violated in the first tank: (1) oxygen rather than COD will be limiting in the first tank and (2) at these high concentrations the removal can no longer be described using first-order kinetics. This means that the results must NOT be used. See discussion below.	

For bulk-phase concentrations of 2 or 4 mgCOD/l it can be reasonable to assume that COD removal can be described using first-order kinetics and that COD removal is COD-limited rather than oxygen-limited. Bulk-phase oxygen concentrations were not provided in the problem statement and the limiting substrate could not be calculated explicitly. Compare the bulk-phase COD concentrations with Figure 17.8 and Table 17.4 in Chen *et al.*, 2020.

When separating the MBBR into two tanks in series, the COD concentration in the first tank will be much higher than the 2 or 3 mgCOD/l. In the solution presented above, the effluent COD concentration from the first tank was adjusted to obtain two equal-sized tanks. This effluent COD concentration from the first tank is 40 mg COD/l. However, in the first tank both assumptions – COD being the limiting substrate and COD removal can be described using first-order kinetics – are no longer valid. Values in the table presented above are numerically correct but should not be used. This example demonstrates that hand calculations assuming specific kinetics or limiting substrates can be tricky when conditions change. This makes numerical simulations of biofilms that do not require these assumptions very attractive (see the following exercises using a simulator).

Solution 17.4.11

Reactor design and model-based evaluation: design and verification of an MBBR.

a. The following additional information must be considered when answering this problem:

Influent dynamics: Treatment plant performance is usually limited during dynamic loading. Thus, the design calculations and a model-based evaluation must consider the dynamics of influent flow rate and composition. Influent dynamics are either measured for a given system or have to be assumed (Section 3.12 in Chen *et al.*, 2020; Figure 3-3 in Metcalf & Eddy Inc. *et al.*, 2013; or Section 5.5.2.1 in Rieger *et al.*, 2013). Simulations can be performed for repeating diurnal simulations (24 h) for a typical dry weather day, for weekly, monthly, or seasonal variations, or for specific events, such as a sudden load increase (Rieger *et al.*, 2013). In the current exercise diurnal variations of flow (small and large), total COD, and TKN in the influent are being considered (Figure 17.1).

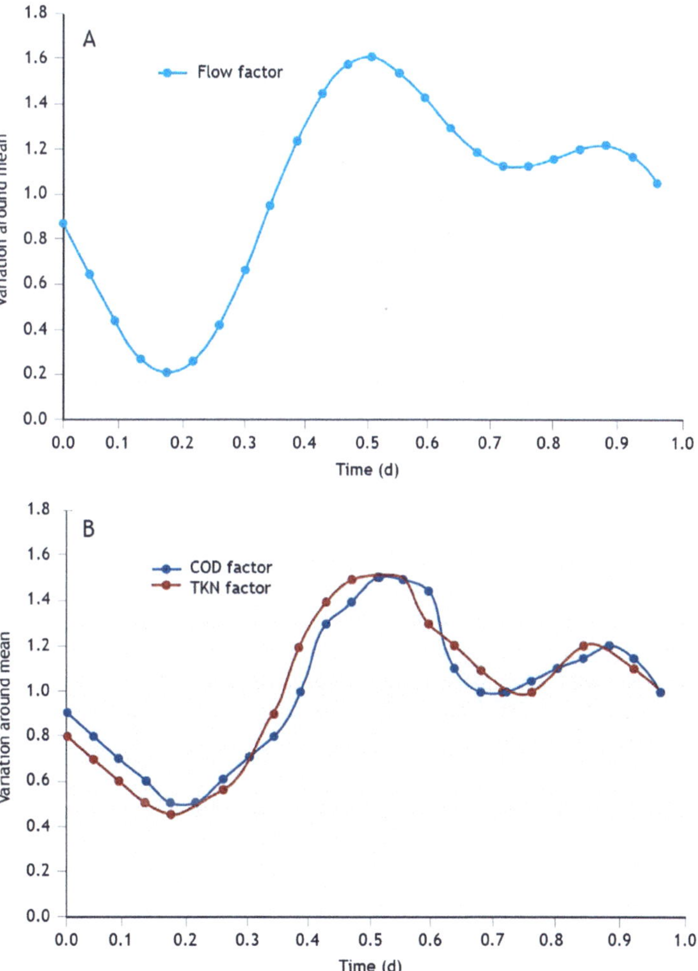

Figure 17.1 Variation of influent flow rates (A) and variation of COD and TKN concentration in the influent (B) assumed for dynamic simulations.

<u>Type of sampling:</u> To what extent a treatment plant exceeds effluent requirements during dynamic loading depends on the influent dynamics and the type of sampling. Some countries evaluate treatment plant performance based on 24-h average samples while other countries consider grab samples. Target effluent concentrations must always be considered together with the type of sampling as effluent concentrations in grab samples can be much higher than 24-h averages. In the current exercise both 24-h averages and maximum effluent concentrations will be discussed.

<u>Wastewater characterization:</u> Mathematical modelling requires that measured wastewater characteristics (in the current exercise, total COD and TKN) are converted into state variables (*e.g.* readily biodegradable COD, slowly biodegradable COD, soluble non-biodegradable COD, particulate non-biodegradable COD, ammonia, organically-bound nitrogen, and so on (Section 3.2 in Chen *et al.*, 2020). Most commercial simulators automatically translate measured wastewater characteristics into state variables. Note that it is the responsibility of the engineer to understand the underlying assumptions made in this conversion. In the current exercise the conversion is done by the software SUMO assuming that the MBBR is treating raw wastewater. SUMO translates the influent COD of 400 mgCOD/l into 143 mgCOD/l of readily biodegradable COD, 146 mgCOD/l of slowly biodegradable COD, 19 mgCOD/l of soluble unbiodegradable COD, 72 mgCOD/l of particulate unbiodegradable COD, and 20 mgCOD/l of heterotrophic bacteria. The influent TKN of 30 mgN/l is translated into 21 mgN/l of ammonia, 3.8 mgN/l of soluble biodegradable organic N, and 0.8 mgN/l of particulate biodegradable organic N. Detailed information is available in the supplementary information (https://doi.org/10.25678/00055H).

<u>Model:</u> In the current exercise the model Mini Sumo (http://www.dynamita.com/) is being used that is very similar to the activated sludge model No. 1 (Henze *et al.*, 1987).

<u>Specific surface area of the MBBR media:</u> In the current exercise a specific surface area $a_F = 300$ m^2/m^3 is assumed. This is a typical value for media with a specific surface area of 500 m^2/m^3 and a fill ratio of 60 % (Table 18.1).

<u>Bulk-phase oxygen concentrations:</u> The performance of biofilm reactors is usually directly related to bulk-phase oxygen concentrations (Figure 18.12 in Chen *et al.*, 2020). However, there is a trade-off between reducing the required reactor volume and increasing the energy requirement to achieve high bulk-phase oxygen demands. Temporarily increasing bulk-phase oxygen concentrations is one way for an operator to increase the treatment capacity during peak loading. In the current exercise we will evaluate treatment plant performance with bulk-phase oxygen concentrations of 3 mgO_2/l and compare with three-tank MBBR with carbon-oxidizing MBBR aerated to 2 mgO_2/l and nitrification MBBRs aerated to 4 and 5 mgO_2/l.

<u>Temperature:</u> Temperature influences bacterial growth rates and oxygen transfer. In the current exercise we will evaluate treatment plant performance at 12 °C.

<u>Biofilm parameters:</u> Parameters describing the detailed implementation of a biofilm are often hidden as default parameters in a simulator. However, reactor performance is significantly influenced by parameters describing the biofilm, and the engineer using the simulator is responsible for understanding the choices that were made when implementing the biofilm. In the current exercise we will assume a biofilm thickness $L_F = 400$ μm, a boundary layer thickness $L_L = 50$ μm, and a maximum biofilm density $X_F = 25,000$ gVSS/m^3. The biofilm is discretized with 4 layers.

b. Estimate the volume of the MBBR based on BOD design surface loading $B_{A,BOD}$ = 4 gBOD/m^2.d and ammonia surface loading $B_{A,N}$ = 0.8 gN/m^2.d for combined organic substrate and ammonia oxidation (Table 18.3 in Chen *et al.*, 2020). Assume COD/BOD = 2. With equations 18.4 and 18.15

$A_{F,BOD}$ = 15,000 m^3/d · (200 gBOD/m^3) / (4 gBOD/m^2.d) = 750,000 m^2
$A_{F,N}$ = 15,000 m^3/d · (30 gN/m^3) / (0.8 gN/m^2.d) = 562,500 m^2
$A_{F,tot}$ = 1,312,500 m^2

V_R = 4,375 m^3

c. A treatment plant consisting of a single-compartment MBBR was implemented and evaluated using a simulator. Under steady-state conditions, effluent ammonia concentrations below 0.5 mg N/l are achieved with a single MBBR at the design volume (Table 17.7). In fact, the volume of the MBBR can be reduced by 20 % and effluent ammonia concentration will still meet the requirements.

Table 17.7 Effluent of a single-tank MBBR with steady-state influent.

V_R (single tank)	Percent of design volume	Effluent readily biodegradable COD	Effluent ammonia	Effluent nitrate
(m^3)	(%)	(mgCOD/l)	(mgN/l)	(mgN/l)
4,375	100	2.6	0.45	13.4
3,500	80	3.1	0.53	12.9
2,625	60	4.0	0.72	11.2
1,750	40	6.0	4.37	4.4
875	20	13.3	16.07	0

d. The influence of variable readily biodegradable COD and ammonia concentrations on effluent concentrations are shown in Figure 17.2 and Table 17.8. While the single-tank MBBR achieved the effluent ammonia target of 0.5 mgN/l with steady-state influent, dynamic influent resulted in the plant violating effluent requirements for all types of sampling – average effluent ammonia concentrations are 1.07 mgN/l (24-h average).

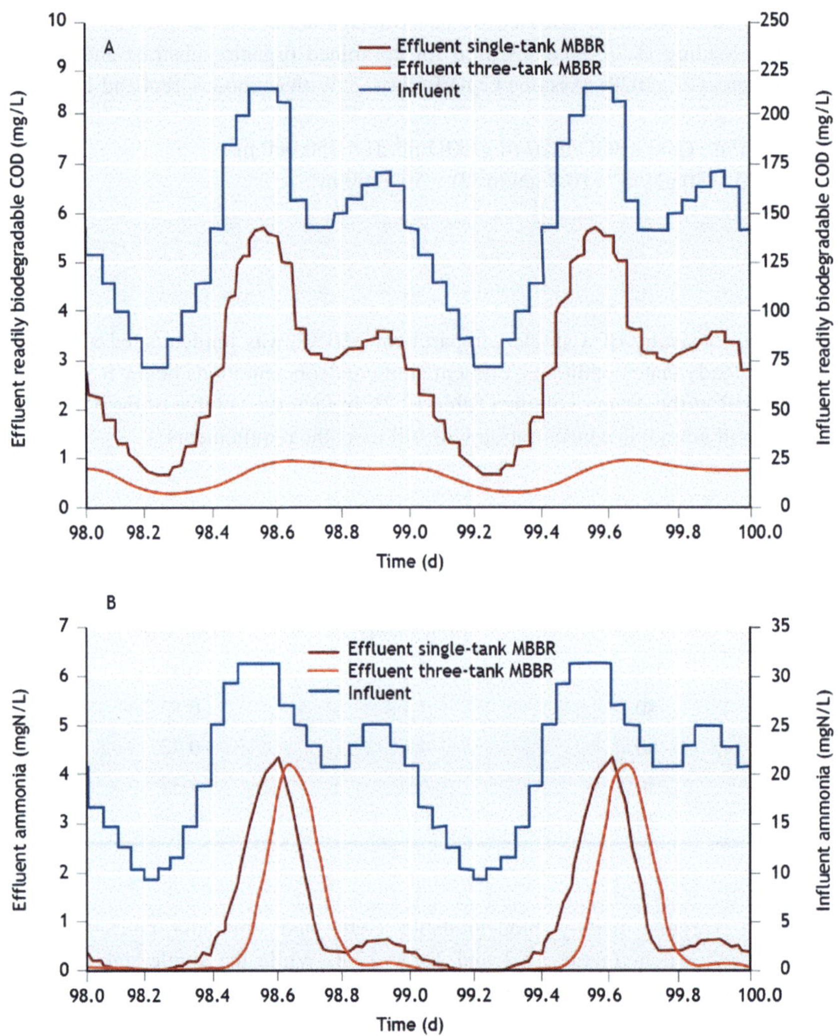

Figure 17.2 Influent and effluent dynamics for readily biodegradable COD (A) and ammonia (B) for single-tank and three-tank MBBRs.

Table 17.8 Average and maximum values for effluent readily biodegradable COD and ammonia for different reactor configurations (single-tank MBBR, three-tank MBBR with all reactors operated at 3 mgO₂/l or with a carbon-oxidizing reactor operated at 2 mgO₂/l and nitrification reactors operated at 4 mgO₂/l (N1) and 5 mgO₂/l (N2). The three tanks are referred to as C (carbon-oxidizing MBBR), N1 (first nitrifying MBBR), and N2 (second nitrifying MBBR).

	Effluent readily biodegradable COD (mgCOD/l)			Effluent ammonia (mgN/l)		
	Single-tank MBBR	Three-tank MBBR	Three-tank MBBR with adjusted DO	Single-tank MBBR	Three-tank MBBR	Three-tank MBBR with adjusted DO
Steady-state	2.6	4.4 (C)	4.4 (C)	0.45	3.49 (C)	8.78 (C)
		0.9 (N1)	0.9 (N1)		0.31 (N1)	0.65 (N1)
		0.6 (N2)	0.6 (N2)		0.14 (N2)	0.18 (N2)
24-h average	2.9	5.3 (C)	5.9 (C)	1.07	7.70 (C)	10.93 (C)
		1.0 (N1)	1.0 (N1)		2.01 (N1)	2.17 (N1)
		0.7 (N2)	0.7 (N2)		0.83 (N2)	0.73 (N2)
Maximum 2-h average	5.6	11.6 (C)	15.8 (C)	4.12	16.38 (C)	19.47 (C)
		1.6 (N1)	1.8 (N1)		7.49 (N1)	7.67 (N1)
		1.0 (N2)	1.0 (N2)		4.00 (N2)	3.43 (N2)
Maximum	5.7	11.9 (C)	16.6 (C)	4.43	16.97 (C)	20.03 (C)
		1.6 (N1)	1.9 (N1)		7.74 (N1)	7.93 (N1)
		1.0 (N2)	1.0 (N2)		4.27 (N2)	3.70 (N2)

e. In practice, the overall volume of an MBBR system is separated into separate tanks. A three-tank MBBR system is compared with the single-tank MBBR where the total volume of both systems is identical. In the three-tank system the first tank contains $A_{F,BOD}$, the second tank 2/3 of $A_{F,N}$ and the third tank 1/3 of $A_{F,N}$.

$V_{R,C} = 2,500$ m³
$V_{R,N1} = 1,250$ m³
$V_{R,N2} = 625$ m³
$V_{R,total} = 4,375$ m³

The three-tank MBBR system was implemented in the simulator and the results are given in Figure 17.2 and Table 17.8. These show that the three-tank MBBR can achieve effluent ammonia concentrations below 0.5 mgN/l for 24-h average sampling. Two approaches for operating the three-tank MBBR are shown. The first approach is operating all the reactors at the same target oxygen concentration of 3 mgO₂/l. The second option is operating the nitrification tanks at 4 and 5 mgO₂/l for the first and the second nitrification tank, respectively. It can be seen that increased aeration provides a benefit for 2-h average samples and maximum effluent ammonia concentrations.

From this part of the exercise, it can be seen that there is benefit in separating the carbon oxidation and nitrification into separate tanks. The first carbon-oxidizing tank will have some limited nitrification where nitrifiers in the biofilm will be oxygen-limited. Nitrification will mostly take place in the two nitrification tanks. This separation of carbon oxidation and nitrification has further consequences. Biofilms in the first carbon-oxidizing tank will be fluffier and the nitrifying biofilms denser with implication for the external mass transfer resistance. A next step in modelling could be to consider different biofilm thicknesses, densities, and external boundary layer thicknesses for the first carbon-oxidizing tank and the second and third nitrifying tanks, respectively (Rittmann *et al.*, 2018).

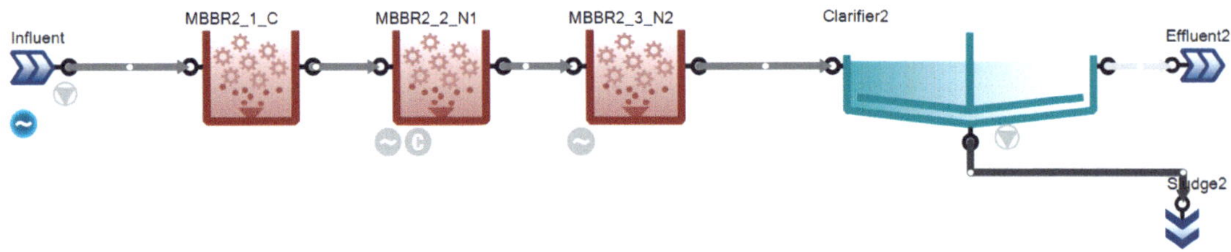

Figure 17.3 Implementation of the three-tank MBBR system followed by a clarifier in SUMO www.dynamita.com/). Software implementation and complete results are available at https://doi.org/10.25678/00055H.

f. Effluent from the three-tank MBBR before and after sedimentation is compared in Table 17.9 . While a clarifier is not needed in an MBBR system to retain active biomass in the reactors, solids removal is required to maintain effluent water quality. Effluent suspended solids contain significant amounts of organic carbon and organic nitrogen. Effluent COD is dominated by soluble non-biodegradable COD (19 mgCOD/l) from the influent. Effluent soluble biodegradable COD is negligible. That means that influent wastewater characteristics in terms of non-biodegradable soluble COD will determine the required solids removal to meet effluent requirements.

Some closing remarks:

The purpose of this exercise was to demonstrate the benefit for the design engineer of combining calculations done by hand and design based on simplified surface loading with mathematical modelling using a commercial simulator. Today's simulators are very powerful and allow a range of biofilm reactor types to be modelled. However, in the end the person using a simulator must understanding the background and the relevance of the model parameters. Ultimately, the biofilm model will never be smarter than the biofilm modeller. Rittmann *et al.* (2018) provide some general principle on using and calibrating biofilm reactor models.

The current exercise provided a first iteration on evaluating system performance. The reactor volume was calculated based on design surfaces fluxes and modelling was used to evaluate the influence of reactor configuration (single MBBRs or three MBBRs) and bulk-phase oxygen concentrations. In practice, modelling results would be used further to adjust the overall reactor volume. Control of aeration would take into account the ammonia concentrations in reactors and air flow would not only be influenced by oxygen demand but to provide sufficient mixing in the MBBR.

Table 17.9 Effluent of the third tank in the three-tank MBBR configuration with adjusted DO (2, 4, and 5 mgO₂/l in the three MBBRs, respectively) and of the subsequent secondary clarifier (steady-state simulations).

Parameter	Unit	Effluent of three-tank MBBR	Effluent of secondary clarifier
Readily biodegradable substrate	mgCOD/l	0.6	0.6
Slowly biodegradable substrate	mgCOD/l	12.6	0.3
Soluble unbiodegradable organics	mgCOD/l	19.0	19.0
Particulate unbiodegradable organics	mgCOD/l	72.0	1.7
Endogenous decay products	mgCOD/l	11.7	0.3
Ordinary heterotrophic organisms	mgCOD/l	145.4	3.5
Aerobic nitrifying organisms	mgCOD/l	0.60	0.01
Total chemical oxygen demand	mgCOD/l	262.1	25.5
Volatile suspended solids	mgVSS/l	173.5	4.2
Total suspended solids	mgTSS/l	208.5	5.0
Ammonia	mgN/l	0.18	0.18
Soluble Kjeldahl nitrogen (SKN)	mgN/l	1.01	1.01
TKN	mgN/l	15.3	1.5

Solution 17.4.12

Reactor design and model-based evaluation: design and verification of an MBBR for carbon oxidation and nitrification/denitrification.

As a first iteration, the available tanks in the three-tank MBBR from Exercise 17.10 are reconfigured to achieve nitrogen removal. The first tank is not aerated and an internal recirculation from the last to the first tank is introduced (recycle flow rate is 300 % of the influent flow rate). The overall volume of 4,375 m³ is maintained but distributed differently (anoxic MBBR = 1,000 m³, first aerobic MBBR = 1,800 m³, second aerobic MBBR = 1,575 m³). Bulk-phase oxygen concentrations are 4 and 5 mgO₂/l in the first and second aerated MBBRs, respectively. Steady-state results are summarized in Table 17.10. It can be seen that nitrogen removal to below 6 mgN/l can be achieved corresponding to 80 % nitrogen removal. A next step would be to further optimize the tank volumes, oxygen setpoints, and the internal recycle.

In 17.11, oxygen transfer requirements are shown for the different MBBR configurations. It can be seen that denitrification not only improves effluent water quality but also reduces the overall aeration demand as nitrate produced in the system can be used as the electron acceptor instead of oxygen to oxidize the organic carbon.

Table 17.10 Steady-state performance of a three-tank MBBR operated for nitrogen removal with 4 and 5 mgO₂/l in the second and third tanks, respectively.

Parameter	Unit	Influent	First tank (not aerated)	Second tank (aerated)	Third tank (aerated)	Effluent of secondary clarifier
Readily biodegradable COD	mgCOD/l	143	19.2	2.9	1.0	1.0
Slowly biodegradable COD	mgCOD/l	146	47.4	33.3	23.8	0.6
Total COD	mgCOD/l	400	285.8	270.6	263.3	25.9
Ammonia	mgN/l	21	4.9	0.8	0.2	0.2
Soluble Kjeldahl nitrogen	mgN/l	25	6.1	1.8	1.1	1.1
TKN	mgN/l	30	18.6	15.2	14.8	1.6
Nitrate	mgN/l	0	0.6	3.7	4.1	4.1

Table 17.11 Oxygen demand for aeration for single-tank MBBR and three-tank MBBR operated for carbon oxidation and nitrification (Exercise 17.4.11) or for carbon oxidation and nitrogen removal (Exercise 17.4.12). Bulk-phase oxygen concentrations of 2, 4, and 5 mgO₂/l in the first, second, and third MBBRs, respectively (except in the first anoxic tank when operated for nitrogen removal)

Configuration	Oxygen transfer requirement (kgO₂/d)			
	First tank	Second tank	Third tank	Total
Single-tank MBBR	3,023	-	-	3,023
Three-tank MBBR operated for carbon oxidation and nitrification	1,833	868	167	2,869
Three-tank MBBR operated for carbon oxidation and nitrification/denitrification	17 [a]	2,029	608	2,654

[a] No active aeration. However, the simulation considers oxygen transfer through the reactor surface.

Solution 17.4.13

Qualitative evaluation: systematic comparison of biofilm reactor types.

Table 17.12

		Trickling filter	RBC	Biofilters	MBBR	IFAS	AGS	MABR
a.	Are biofilm carriers fixed within the reactor or suspended?	Fixed	Fixed	Fixed	Suspended	Suspended	Suspended	Fixed
b.	Does this type of reactor require dedicated backwashing events to remove excess biofilm? (Y/N)	N	N	Y	N	N	N	Y [a)]
c.	Does the suspended biomass contribute significantly to overall removal? (Y/N)	N	N	N	N	Y	Y	N
d.	In the case of aerobic growth: how is oxygen supplied?	Typically: ventilation using natural convection brings in oxygen, in some cases forced ventilation could be provided.	Rotation of disks provided aeration- when the biofilm is out of water	Oxygen is provided by diffusion introducing air in at the bottom of the filter. As the bubbles rise to the top, oxygen transfer occurs.	Oxygen is provided by diffused-air aeration.	Oxygen is provided by diffused-air aeration.	Oxygen is provided by diffused-air aeration.	If the contaminant to be removed is 'reduced', then oxygen is provided through the base of the biofilm through the membrane.
e.	Typical specific surface area in m^2/m^3	50-100 (Table 18.1 in Chen *et al.*, 2020)	100-200 (Table 18.1)	1,000-3,600 (depending on material used for filter) (Table 18.1)	300 (Table 18.1)	300 (Table 18.1)	2,000-3,000 (Table 18.1)	150 - 500
f.	Suitability of the system for nitrogen removal (*i.e.* nitrification and denitrification)? (Y/N)	N	N	Y	Y	Y	Y	Y
g.	Suitability of the system for enhanced biological phosphorus removal? (Y/N)	N	N	N	N	Y	Y	Y
h.	Ease of operation (very simple/complex)	Very simple	Very simple	Requires specialized engineering implementation and suitable backwash intensity and frequency.	Simple	Simple	Simple but requires good understanding of granulation.	Simple

[a)] In some systems dedicated backwashing events and in other systems continuous shearing through coarse bubble aeration.

Solution 17.4.14

Systems analysis: variation of parameters.

Table 17.13

Variable	Effluent organic substrate concentration	Substrate flux into the biofilm	Removal efficiency
Influent Q	Increase	Increase	Decrease
Influent organic substrate concentration	Increase	Increase	Decrease
D_F	Decrease	Increase	Increase
X_F	Decrease	Increase	Increase
Biofilm thickness	No change	No change	No change

REFERENCES

Chen G., Ekama G.A., van Loosdrecht M.C.M. and Brdjanovic D. (2020). *Biological Wastewater Treatment Principles, Modeling and Design*, IWA Publishing.

Henze M., Grady C.P.L., Gujer W., Marais G.v.R. and Matsuo T. (1987). A general model for single-sludge waste-water treatment systems. *Water Research* 21(5), 505-515.

Metcalf & Eddy Inc., AECOM (Firm), Tchobanoglous G., Stensel H.D., Tsuchihashi R. and Burton F.L. (2013). *Wastewater engineering: treatment and resource recovery*, 5th Ed., McGraw-Hill Higher Education, New York.

Rieger L., Gillot S., Langergraber G., Ohtsuki T., Shaw A., Takacs I. and Winkler S. (2013). *Guidelines for Using Activated Sludge Models*, IWA Publishing.

Rittmann B.E., Boltz J.P., Brockmann D., Daigger G.T., Morgenroth E., Sorensen K.H., Takacs I., van Loosdrecht M.C.M. and Vanrolleghem P.A. (2018). A framework for good biofilm reactor modeling practice (GBRMP). *Water Science & Technology* 77(4), 1149-1164.

NOMENCLATURE

Refer to nomenclature in chapters 17 and 18 in Chen *et al.*, 2020.

IWA Publishing's authorised EU representative for General Product
Safety Regulations is Diane D'Arras, 15 rue Duret, 75116 Paris,
France, e-mail: safety@iwap.co.uk.

Printed and bound by CPI Group (UK) Ltd, Croydon, CR0 4YY

14/05/2026

02110308-0001